Experimental Hydrodynamics for Flow Around Bodies

Experimental Hydrodynamics for Flow Around Bodies

Viktor V. Babenko
Department of Information Systems in
Hydro-Aeromechanics and Ecology,
Institute of Hydromechanics of the
National Academy of Sciences of Ukraine,
Kyiv, Ukraine

Academic Press is an imprint of Elsevier
125 London Wall, London EC2Y 5AS, United Kingdom
525 B Street, Suite 1650, San Diego, CA 92101, United States
50 Hampshire Street, 5th Floor, Cambridge, MA 02139, United States
The Boulevard, Langford Lane, Kidlington, Oxford OX5 1GB, United Kingdom

Copyright © 2021 Elsevier Inc. All rights reserved.

No part of this publication may be reproduced or transmitted in any form or by any means, electronic or mechanical, including photocopying, recording, or any information storage and retrieval system, without permission in writing from the publisher. Details on how to seek permission, further information about the Publisher's permissions policies and our arrangements with organizations such as the Copyright Clearance Center and the Copyright Licensing Agency, can be found at our website: www.elsevier.com/permissions.

This book and the individual contributions contained in it are protected under copyright by the Publisher (other than as may be noted herein).

Notices
Knowledge and best practice in this field are constantly changing. As new research and experience broaden our understanding, changes in research methods, professional practices, or medical treatment may become necessary.

Practitioners and researchers must always rely on their own experience and knowledge in evaluating and using any information, methods, compounds, or experiments described herein. In using such information or methods they should be mindful of their own safety and the safety of others, including parties for whom they have a professional responsibility.

To the fullest extent of the law, neither the Publisher nor the authors, contributors, or editors, assume any liability for any injury and/or damage to persons or property as a matter of products liability, negligence or otherwise, or from any use or operation of any methods, products, instructions, or ideas contained in the material herein.

British Library Cataloguing-in-Publication Data
A catalogue record for this book is available from the British Library

Library of Congress Cataloging-in-Publication Data
A catalog record for this book is available from the Library of Congress

ISBN: 978-0-12-823389-4

For Information on all Academic Press publications
visit our website at https://www.elsevier.com/books-and-journals

Publisher: Matthew Deans
Acquisitions Editor: Brian Guerin
Editorial Project Manager: Emily Thomson
Production Project Manager: Nirmala Arumugam
Cover Designer: Matthew Limbert

Typeset by MPS Limited, Chennai, India

Contents

About the author		ix
Preface		xi
Annotation		xiii
List of symbols		xv

1 Nonstationary motion — 1
- 1.1 Drag of a disk and a sphere at the accelerated movement from a state of rest — 1
- 1.2 Influence of nonstationary on the mechanism of formation of hydrodynamic drag — 12
- 1.3 Drag coefficient of fluid bodies in acceleration and deceleration modes — 20
- 1.4 Energy aspects of the movement of bodies in the modes of acceleration and deceleration — 25
- 1.5 Features of the accelerated movement in water of bodies of various shapes — 31

2 Modeling of a waving fin mover — 41
- 2.1 Experimental investigation of waving wing — 41
- 2.2 Approximate method for calculating of the hydrodynamic characteristics of an oscillating wing — 44
- 2.3 Experimental investigation of the oscillating wing thrust — 48
- 2.4 Experimental investigation of hydrodynamic thrust generated by an oscillating wing — 52
- 2.5 Hydrodynamic characteristics of a rectangular rigid oscillating wing — 56
- 2.6 Experimental investigation of nonstationary vortex structures behind oscillating bodies — 65
- 2.7 Visualization of the vortex structures in the wake of the oscillating profiles — 75

3 Experimental investigations of the characteristics of the boundary layer on a smooth rigid plate — 93
- 3.1 Hydrodynamic stand of low turbulence — 93
- 3.2 Experimental methods for investigation the flow structures of a transition boundary layer of a rigid plate — 100

3.3	Experimental investigation of coherent vortex structures of the transition boundary layer of a rigid plate	103
3.4	Neutral curves of linear stability of a laminar boundary layer of a rigid plate	117
3.5	Distribution of disturbing motion over the thickness of a laminar boundary layer of a rigid plate	123
3.6	The physical process of laminar-turbulent transition of the boundary layer of a rigid plate	130
3.7	Laser Doppler velocity meter measurements of the structure of the boundary layer of a rigid plate in the beginning of turbulence	150
3.8	Boundary layer as a heterogeneous asymmetrical wave-guide	158
3.9	The influence of small and large disturbance on the characteristics of the boundary layer of a rigid plate	171

4 Experimental investigations of the characteristics of the boundary layer on a smooth rigid curved plate — **187**

4.1	Experimental equipment and methodology for investigation of Görtler stability	187
4.2	Experimental investigations of Görtler's vortex systems	191
4.3	Experimental investigation of Görtler stability on rigid curvilinear plates	200
4.4	LDV measurements of the profiles of the longitudinal velocity of the BL on rigid concave plates	206
4.5	Experimental construction of the Görtler neutral curve in the flow around rigid curvilinear plates	209
4.6	Model of the vortex structures in the vortex chamber	213
4.7	Görtler's vortices on a concave surface behind the vortex chamber nozzle	232

5 Experimental investigations of the characteristics of the boundary layer on the analogs of the skin covers of hydrobionts — **237**

5.1	Structures of elastic surfaces	237
5.2	Experimental investigation of coherent vortex structures in the transition boundary layer of elastic plates	259
5.3	Neutral curves of linear stability of a laminar boundary layer of elastic plates	272
5.4	Distribution of disturbing movements across the thickness of the laminar boundary layer of elastic plates	281
5.5	Laser Doppler velocity measurements of the structure of the boundary layer of elastic plates in the process of laminar-turbulent transition	295
5.6	Experimental construction of the Görtler neutral curve in the flow around elastic curvilinear plates	311
5.7	Determination of the mechanical characteristics of elastomers by the method of wave and pulse propagation	320

5.8	The investigation of oscillations on the surface of elastic plates during flow around a water stream	334
5.9	Physical substantiation of the interaction mechanism of the flow with an elastic surface	345
5.10	Boundary layer when flow around a controlled elastic plate	350

6 Experimental investigations of friction drag — 359

6.1	Control methods for coherent vortex structures of the boundary layer	359
6.2	Review of experimental investigations of integral characteristics of elastic plates	361
6.3	Experimental complex for investigations of friction drag	368
6.4	Experimental investigations of friction drag on elastic plates	376
6.5	Drag of longitudinally streamlined cylinders	389
6.6	Friction drags of elastic longitudinally streamlined cylinders	404
6.7	Influence of polymer additives on the friction drag of elastic cylinders	420
6.8	Engineering method for the selection of elastic plates to drag reduction	428

7 Combined methods of drag reduction — 437

7.1	Combined methods of drag reduction	437
7.2	Experimental investigations of bodies with xiphoid tips	441
7.3	Theoretical investigations of bodies with xiphoid tips	451
7.4	Brief review of the problem of injection of polymer solutions into the boundary layer	459
7.5	Experimental equipment and investigation methodology for the combined method of drag reduction	472
7.6	Drag of the model with the ogival tip and at an injection of polymer solutions into the boundary layer	480
7.7	Drag of the model with xiphoid tips and injection of polymer solutions into the boundary layer	490
7.8	Physical mechanism of the influence of xiphoid tip on drag reduction	498
7.9	Experimental investigations of the interaction of the boundary layer with the injected semilimited jet	506
7.10	Hydrodynamic peculiarities of the skin structure and body of the swordfish	515
7.11	Investigation of the influences of microbubbles injection into the boundary layer on the drag reduction (short review)	531
7.12	Experimental investigation of the influences of bubbles injection on the drag reduction	545
7.13	Modeling of disturbances development in the flow behind the ledge	560

8	**Conclusion**		**577**
	8.1	The problem of drag reduction, types of coherent vortex structures and combined methods of their control	578
	8.2	Body form optimization	579
	8.3	Principle of additivity of aquatic and underwater bodies	579
	8.4	Vortex chambers and control methods for coherent vortex structures in vortex chambers	580
	8.5	Environmental control concept of the coastal area	581
	8.6	Reduction friction drag	582
	8.7	Devices moving on and under water	586
	8.8	Technical proposals for two-medium devices	593
	8.9	Vortex devices	596
	8.10	CVS and thermal conductivity control	597
	8.11	Devices for environmental monitoring of the coastal area	599
	8.12	Adaptation of the form of the body	602
	8.13	Self-propelled buoys	602
	8.14	Underwater glider	602
	8.15	Nontraditional movers	602
	8.16	Changing the shape of the wing profile and its configuration	603
	8.17	Change wing span	603
	8.18	Changing the curvature of the wing profile	603
	8.19	Changing wing profile chords	603
	8.20	Maneuverability control when moving the device	604
	8.21	Vortex devices	604
	8.22	Combined drag reduction methods	604
	8.23	CVS control methods	605
	8.24	Heat and mass transfer	605
	8.25	Ecology	605

Bibliography   **607**
Index   **641**

About the author

Professor Viktor V. Babenko (born April 14, 1938) received a BDc in mechanical engineering from Aviation Institute, Moscow, in 1963, and a PhD in fluid and gas mechanics from the Institute of Hydromechanics of National Ukrainian Academy of Sciences in 1970 (Master of Technical Sciences) on the theme "The experimental research of hydro dynamical stability on rigid and elastic-damping surfaces," a Doctor of Technical Sciences in 1986 on the theme "Interaction of disturbing movements in a boundary layer," and a professor since 1990.

From 1963 to 1965 he has worked at the Kiev's aviation Antonov designer bureau.

Since 1965 he has worked at the Institute of Hydromechanics, Kiev, in the field of boundary layer research. From 1988 to 2000 he has worked a Department Head of the Institute of Hydromechanics. He has managed a variety of hydrodynamic and bionic research projects. Professor Viktor V. Babenko has developed several new measurement techniques for laminar and turbulent flows and designed some original equipment, devices, and apparatuses for hydrodynamic measurements. He has developed original methodologies in bionics, at research of the receptivity of boundary layer to 2D and 3D disturbances, at interaction between flow and compliant coating, at flow near different cavities, at flow in the vortical chamber, and at movement of high-speed surface devices. He has developed control methods of the coherent vortical structures arising at various types of flows.

Areas of competence include experimental fluid mechanics; stability and transition; turbulence; near-wall jets; hydrodynamics of the vortex chamber; flow in the cavities; flow control; coherent vortices structures; unsteady hydrodynamics; and bionics.

Viktor V. Babenko is a Fellow of National Committee of Ukraine on Theoretical and Applied Mechanics.

From 1989 to 2006 he was member of Scientific Council of the Institute of Hydromechanics NASU, Kiev, Ukraine. Since 2000 he has been a member of the Scientific Councils of the University of Civil Aviation, Ukraine, Kiev, and since 2007, the National Technical University of Ukraine "Kiev polytechnic institute."

Academic honor and professional recognition: Two bronzes medal on exhibition of national economy achievements; DINNIK and ANTONOV Prize Winner of NAS of Ukraine; and Honorable letter of Presidium of National Ukrainian Academy of Sciences.

Undergraduate courses taught: Mathematics; Hydraulics; Hydrosystems; Pneumatic Systems; and Pneumatic Automatics.

Short courses taught: Animal Mechanics and Bionics.

Viktor V. Babenko has prepared seven doctors of sciences (PhD).

Professor Viktor V. Babenko has published 268 publications in the basic and applied research areas of boundary layer flows, compliant coatings, unsteady hydrodynamics, bionics, and non-Newtonian fluids, flows control, and the characterization of vortical chamber. Published works include one-discovery, five books, 197 articles, and 71 inventions.

He participated in 70 All-Union and the international conferences.

In 1993–94 Viktor V. Babenko's research was financed with the International Science Foundation J. SOROS (Grand Number of Principals Investigators UAW000 and UAW200).

Professor Viktor V. Babenko in 1995–97 executed research pursuant to the contract made between Institute of a hydromechanics NANU, Kiev, Ukraine, and President "Cortana Corporation" K.J. Moore (Task 11 and 12 under DARPA Delivery Order 0011 of Contract MDA972-92-D-0011). By an outcome of this cooperation were not only scientific reports. Professor Viktor V. Babenko has received 10 American patents, one European Patent, and one International Patent.

In 1998–99 Professor Viktor V. Babenko has executed researches on hydrobionics in Berlin in Wissenschaftskolleg zu Berlin, Institute for Advanced Study, Berlin.

In 1999 for participation in activity of conference in Newport, Viktor V. Babenko has received financial support (Grant Number N00014-98-1-4040 Office of Naval Research International File Office—Europe Special Programs Assistant Code 240 223 Old Maryellen Road of Visiting Support Program Support for provides financial Support to visit to Naval Undersea Warfare Center).

In 2005–2006 Professor Viktor V. Babenko worked as visiting professor in Advanced SHIP Engineering Research Center (ASERC, Busan, Korea) at Pusan National University. He has executed research on physical features of a flow of elastic plates and methods of coherent vortical structure control of a boundary layer. Results of the executed work have been generalized in the monograph written at this time in English (in common with Professors Ho Hwan Chun and Inwon Lee).

Professor Viktor V. Babenko in 2008 and 2009 received financial support and visited the Islamic Republic of Iran to review the results of his research of gidrobionic.

Professor Viktor V. Babenko in 2008 received financial support and visiting the Ship Research Center (Wuxi, Chinese).

Professor Viktor V. Babenko in 2010 received financial support and visited the SHIP Research Center (Wuxi, Chinese) and at the same time, also participated in the 9th International Conference on Hydrodynamics (ICHD-2010), October 11–15, 2010, Shanghai, and Chinese.

In the encyclopedia of National Ukrainian Academy of Sciences, the short curriculum vitae of Professor Viktor V. Babenko is printed.

In the book MARQUIS "Who's Who: Who's Who in Science and Engineering," USA, 2008–2009, a short curriculum vitae of Professor Viktor V. Babenko is published.

Preface

Professor L.F. Kozlov, Head of the Department of "Hydrobionics and Boundary Layer Control," was the founder of investigations on new scientific directions of hydrobionics and the flow of elastic surfaces. He constantly developed numerous scientific and practical problems over the years. Since 1965, Ukraine and the Russian Federation began to conduct systematic investigations on bionics in various organizations, including various institutions of the National Academy of Sciences of Ukraine (NASU). Professor L.F. Kozlov (1927−87) organized bionic investigations in Ukraine and the department "Hydrobionics and Boundary Layer Control" at the Institute of Hydromechanics of the NASU. From 1967 to 1998, the journal *Bionika* was published, the editor-in-chief of which was Professor L.F. Kozlov. For several years L.F. Kozlov also coordinated hydrobionic investigations in Ukraine, published numerous articles on various issues of bionics and a number of monographs in particular, in 1983 he published a monograph "Theoretical Biohydrodynamics." Professor L.F. Kozlov organized a modern experimental basis for carrying out new scientific investigations. At the Institute of Hydromechanics of NASU, Kyiv, under his leadership, a large number of experimental apparatus, devices, and equipment were designed and manufactured, the DISA Elektronic hot-wire anemometer, Brul and Cer multichannel tape recorder were acquired. He instructed Viktor V. Babenko to master a new method of measuring the characteristics of the boundary layer based on the P.X. Wortmann tellurium method, and V.P. Ivanov to develop a laser anemometer, with the help of which numerous experimental investigations of the characteristics of the boundary layer were carried out.

L.F. Kozlov, V.M. Shakalo, and Ye.V. Romanenko found a corresponding decrease in the pulsation characteristics of speed and pressure in the boundary layer of dolphins on different phases of nonstationary movement when performing independent experimental investigations on dolphins moving in the channel.

Part I of this monograph presents the results of experimental investigations in hydrobionics obtained by the author and other researchers.

The J. Soros International Science Foundation (Grand Priest Number of UAW000 and UAW200) in 1993−94 sponsored the investigations of Professor Viktor V. Babenko.

In 1995−97 Professor Viktor V. Babenko conducted the investigations under a contract between the Institute of Hydromechanics of NASU, Kyiv, Ukraine, and Cortana Corporation, President K.J. Moore, USA (Task 11 and 12, DARPA Delivery Order 0011 of Contract MDA972-92-D-0011). The result of this collaboration were not only scientific reports. Professor Viktor V. Babenko received 11 US patents. In natural conditions on a high-speed vessel, a study was conducted of the effectiveness of some patents on combined methods of drag reducing. In 1995, Cortana

Corporation provided financial support for participation in the conference, which was held in Johns Hopkins University, Baltimore, USA, and the President of Cortana Corporation K.J. Moore provided great support in discussing and preparing three reports at the conference. These three reports systematized the main scientific results of experimental investigations, which formed the basis for the formulation of Part I of this monograph.

In 1998, 1999 Viktor V. Babenko conducted hydrobionic investigations and their systematization at the Wissenschaftskollegs zu Berlin (Institute for Applied Research, Berlin, Germany).

In 1999, Viktor V. Babenko received financial support for participation in a conference in Newport, (Grant No. 00014-98-1-4040, Office of the Marine International Research Unit—European special programs, Assistant Code 240 223 Old Maryellend Road, support program visits to the Naval Undersea Warfare Center).

In 2005–06 Professor Viktor V. Babenko worked as a visiting professor at the Engineering Center for Applied Research of the Ship (ASERC, Busan, Korea) at the National University of Busan (Director Professor H.H. Chun). The result of this work was a book published in English in 2012 by Viktor V. Babenko, H.H. Chun, Inwon Lee, *Boundary Layer Flow over Elastic Surfaces: Compliant Surfaces and Combined Methods for Marine Vessel Drag Reduction*. Amsterdam, Boston, Heidelberg, London, and others. Elsevier publishers. Butterworth-Heinemann: 2012, 631 p.

All this made it possible to continue investigations on the problems of hydrobionics in different directions. As a result, it was possible to formulate the main directions of investigations and systematize the results obtained in this book.

We bring sincere thanks to the specific individuals and organizations that provided financial support for the research. We thank K.J. Moore, President of Cortana Corporation for very important creative collaboration and creative discussions.

Part II of this monograph presents the results of verification of some of the obtained results of experimental studies given in Part I.

Annotation

Viktor V. Babenko sent the manuscript of the book to the Elsevier publishing house. The volume of the manuscript was large, so the Elsevier publishing house offered the author of the book to divide the manuscript submitted for publication into three books (three parts).

In the first book (Part I), a hydrobionic approach is developed for solving hydromechanical problems. The structural features of the body systems and their interaction during the movement of fast-swimming hydrobionts (dolphins, sharks, swordfish, and penguins) aimed at drag reduction and reducing energy consumption have been studied and systematized. The influence of the variable shape and protruding parts of the body during movement, as well as the nonlinearity and nonstationary motion of hydrobionts on the drag to movement is investigated. The principles of the movement of hydrobionts are formulated. Using the developed special equipment, devices, and techniques, experimental investigations on living dolphins were carried out. All this made it possible to detect previously unknown specific features of hydrobionts to reduce movement drag and minimize energy expenditure in the process of movement. This book is Part I (Experimental Hydrodynamics of Fast-Floating Aquatic Animals) published on June 12, 2020 (https://www.elsevier.com/books/experimental-hydrodynamics-of-fast-floating-aquatic-animals/vitaliiovych/978-0-12-821025-3). Anna Valutkevich, Acquisitions Editor | Animal Science, Dairy Science, Organismal & Evolutionary Biology, Mobile: +1 (857) 209-1578. Liz Heijkoop, Editorial Project Manager, Tel: 617-397-2887, Email: l.heijkoop@elsevier.com.

This is the second book, which is designated Part II for systematization. In Part II (Experimental Investigations Flow Around Bodies) of this monograph, hydrobionic modeling of individual systems and the detected features of the morphological structure of fast hydrobions was performed. Special devices, experimental installations, and methods of conducting experiments on hydrobionts analogs have been developed. Numerous physical and integral measurements have been performed, and new theoretical generalizations have been developed. In the second book, references are made to the material in Part I along the way.

Viktor V. Babenko also prepared the third book (Part III), which will be presented for the Elsevier publishing house after the publication of the second present book (Part II). The results of the experimental studies carried out in the published first book (Part I) and this book (Part II) made it possible to discover new phenomena and patterns, which are formalized in the form of discoveries and inventions and will be partially presented in the third book (Part III). The third book (Part III) will provide examples of new technical solutions and technologies developed, which are partially implemented in practice. Numerous examples of the implementation of the obtained results of experimental investigations are given.

List of symbols

x, y, z longitudinal, normal, and transversal axes of a Cartesian coordinate system
x_1 valid coordinate of a place of measurement
$\bar{y} = y/\delta$ dimensionless vertical coordinate
b width; transversal distance between elastomers layer in composite
h width of the hydrodynamic channel; height of a disposition of the lower edge of an ejector above a streamlined surface; thickness of the water layer which are taking place under a membrane; thickness of a flexible surface
l, L length; l—designated length besides a symbol l in various parts designate length of a way of mixing scales of turbulent vortices in the viscous sublayer
S wetted surface; area of a crack
F wetted area
s thickness of a slot of an ejector
d, D diameter
R radius; resistance force—$R_{turb} = 0.62 \times \rho \times (U^2/2) \times d \times l$
δ thickness of a boundary layer
δ^* theoretical value of displacement thickness
δ_m^* measured value of displacement thickness
δ^{**}, δ_2 momentum thickness
$H = \delta^*/\delta^{**}$ form parameter of a velocity profile
$\bar{\delta}_{lam} = 4.64 \left(\frac{\nu x_0}{u_0}\right)^{1/2}$, $\bar{\delta}_{lam} = \alpha \frac{\nu}{\sqrt{\tau_w/\rho}}$ viscous sublayer thickness
$\alpha = \frac{\bar{\delta}_{lam} u_l}{\nu}$ empirical constant
k, K_s height of element of a roughness; k—spring constant; k—kinetic energy of turbulent fluctuation motion; $k = 2\pi f/u = 2\pi/\lambda$ wave number
$K_{lim} = K_s U_\infty/\nu$ parameter of a roughness
c_a, C_L lift coefficient
C_f, c_f, c_w, C_{xi} coefficient of resistance
$\lambda, C_f = 2\tau_w/\rho U_\infty^2$ local skin friction coefficient
$C_F = \frac{2P}{S\rho U_\infty^2}$, $C_F = \frac{1}{x}\int_0^x C_f dx' = 2\frac{\Theta}{x}$ complete friction coefficient or the mean friction drag coefficient
Δy and Δz cross and transversal amplitudes of disturbing oscillation
A_{vibr}, A_v amplitude of oscillation of a vibrator tape
A_{do} amplitude of disturbing oscillation
$\varepsilon = \frac{\sqrt{\frac{1}{3}(\overline{u'^2} + \overline{v'^2} + \overline{w'^2})}}{U_\infty}$; T_u degree of turbulence
$\bar{k} = E/\varepsilon$ wave-guide amplification coefficient
$E = \sqrt{\overline{u'^2}}/U_\infty$ maximal value longitudinal pulsating velocity
t time

List of symbols

u, v, and w longitudinal, normal, and transversal component of a velocity of the resultant flow

U, V, and W longitudinal, normal, and transversal component of a velocity of the basic flow

u′, v′, and w′ longitudinal, normal, and transversal component of a disturbing flow

$\sqrt{\overline{u'^2}}, \sqrt{\overline{v'^2}}, \sqrt{\overline{w'^2}}, \overline{u'v'}, \ldots$ averaged over time components of a turbulent pulsation of velocity

$\dfrac{2\overline{u'v'}}{U_\infty^2}$ Reynolds stress

τ_0 tangent pressure on a wall

u_H value of a longitudinal component velocity on the exterior boundary of a boundary layer

ψ flow function

$q = \rho V^2/2$ dynamic pressure

p pressure of the basic flow, $\sqrt{\overline{p'^2}} = 0.5 \rho U_\infty^2 \, \text{Re}^{-0.3}$

P resulting pressure

p' pressure of the disturbing flow

$\dfrac{2\sqrt{\overline{(p')^2}}}{\rho U_\infty^2} = \alpha_\tau \lambda -$ energy of pulsations of pressure p'

α_τ coefficient of proportionality

$C_p, \overline{P}_i = \dfrac{P - P_\infty}{\frac{1}{2}\rho U_\infty^2}, \overline{p} = \dfrac{p}{\rho u_H^2}$ dimensionless pressure; coefficient surface-pressure distributions

$\alpha = \dfrac{2\pi}{\lambda}$ wave number of disturbing oscillation; $k = 2\pi f/u = 2\pi/\lambda$ wavenumber

f and λ frequency and length of the wave harmonic

u mean velocity in the point of measure

λ length of the oscillation wave

λ_x disturbing wave length of longitudinal vortices in direction of x-axis

$\tilde{\lambda}_x = \lambda_x/\delta_*$ dimensionless length of the oscillation wave

λ_z three-dimensional disturbing wave length in direction of z-axis

$\beta = \beta_r + i\beta_i$ complex frequency of disturbing oscillation

$\beta_r = 2\pi n$ circular frequency

$n = 0.159 \dfrac{\beta_r \nu}{U_\infty^2} \dfrac{U^2}{\nu}$ frequency of disturbing oscillation

Δn frequency span of disturbing oscillation

$T = \dfrac{1}{n} -$ cycle of vibration; viscous sublayer regeneration period; integral limit scale; elastomer tension

β_i increase coefficient

$c = \dfrac{\beta}{\alpha} = c_r + ic_i$ complex velocity of disturbing movement

c_r distribution velocity of waves of disturbing movement

c_i increase coefficient

$\alpha\delta$ dimensionless wave number of disturbing oscillation

$\dfrac{\beta_r \nu}{U_\infty^2}; \omega_r$ dimensionless frequency of disturbing oscillation

$\dfrac{c_r}{U_\infty} -$ dimensionless velocity of distribution of disturbing movement

μ dynamic viscosity coefficient

List of symbols

$\nu = \mu/\rho$ kinematics viscosity coefficient
$\rho = \gamma/g$ density of a liquid (mass of unit of volume)
γ specific gravity (weight of unit of volume)
g gravitational acceleration
ρ_m density of a material of a flexible surface
Re Reynolds number calculated on the x
Re$_\delta$ Reynolds number calculated on thickness of a boundary layer
Re* Reynolds number calculated on the displacement thickness
Re** Reynolds number calculated on the momentum thickness
$Re_1 = \dfrac{U_\infty}{\upsilon} -$ single Reynolds number
Nu Nusselt number
χ empirical constant of turbulent flow; $l = \chi y$
$\varepsilon_\tau = |\overline{u'v'}|/(|\partial U/\partial y|)$ turbulent viscosity coefficient, dissipation velocity of turbulence energy for unit mass
$k^+ = k\, u^+/\nu$ dimensionless kinetic energy of turbulent fluctuation motion
$P^+ = \dfrac{-\upsilon\overline{u'v'}}{u_\tau^4} \cdot \dfrac{dU}{dy}$ velocity of generated of the turbulent energy in a turbulent boundary layer
$D^+ = \dfrac{\upsilon^2}{u_\tau^4} \cdot \left(\dfrac{dU}{dy}\right)^2$ velocity of dissipation of turbulent energy of the averaged motion of a turbulent boundary layer
$y^+, y^* = y\, u^*/\nu$ dimensionless distance from a wall
u/u_* dimensionless velocity
$u_\tau, v_* = \sqrt{\tau_w/\rho}$ dynamic velocity; friction velocity; skim friction stress; dimensionless thickness of viscous sublayer
λ_z wave length of the viscous sublayer longitudinal vortices
$\lambda_z^+ = \lambda_z u_\tau/\nu$ dimensionless wave length of the viscous sublayer longitudinal vortices
$T^+ = T u_\tau^2/\nu, \ T^+ = \left(\dfrac{u_\tau \overline{\delta_l}}{4.64\nu}\right)^2$ period of bursting from a viscous sublayer of a turbulent boundary layer
$f^+ = (T^+)^{-1}$ frequency of bursting from a viscous sublayer of a turbulent boundary layer
$K_S^+ = K_S u_\tau/\nu$ dimensionless factor of a roughness
$\overline{\tau} = \dfrac{\tau}{\rho u_H^2}$ dimensionless of shear stresses
τ shear stresses (force on unit of the area)
τ_w shear stresses on a wall, $\tau_w = \mu\left(\dfrac{\partial u}{\partial y}\right)_0$
Π Coel's parameter
ΔB additive constant
$R_{u'u'(0,r,0)} = \dfrac{\overline{u_1' u_2'}}{\sqrt{\overline{u_1'^2}}\sqrt{\overline{u_2'^2}}}$, $R_{u'u'(0,r,0)}$, $R_{u'u'(r,0,0)}$, $R_{u'u'(0,0,r)}$ correlation factor, autocorrelation factor
$R_{u'u'(0,r,0,t)}$ space-time correlation
$R(r) = \displaystyle\int_0^\infty E(k)\cos kr\, dr -$ correlation function

$E(r) = \dfrac{2}{\pi} \int_0^\infty R(r)\cos kr\, dr\ -$ spectrum function

$\Phi(\omega);\ E(k) = \dfrac{F(\omega)u}{2\pi \Delta f}$ spectral functions

Δf infiltration band-pass filter width

$L = \int_0^\infty R_{(0,r,0)}dr,\ L = \int_0^1 R_{(0,r,\delta,0)}d(r/\delta)$ integral scale or macro scale of the turbulence

$R_i = \dfrac{u_1'(\xi)u_1'(\xi + x_i)}{\sqrt{u_1'^2(\xi)u_1'^2(\xi + x_i)}}\ -$ coefficients of transverse correlation between the longitudinal components of the velocity fluctuations

$R_i(\tau) = \dfrac{\overline{u_i'(t)u_i'(t+\tau)}}{\sqrt{\overline{u_i'^2}}}\ -$ autocorrelation coefficients of longitudinal fluctuation velocity

$Ca = \rho U_\infty/E$ Cauchi parameter

θ corner of shift of phases between a stress and deformation

σ stress of viscoelasticity material

$E = \sigma/\varepsilon$ module of elasticity of an elastic material

ε elastomer strain; relative cross deformation of polymeric materials, apparent kinematics viscosity of turbulent flow

$t_i,\ h$ thickness of a flexible surface

t_m thickness of a membrane

H thickness polyurethane sheet

$c = \sqrt{E/\rho}\ -$ group velocity of the elastic wave

$\lambda = c_\tau$ wave length

c_R phase velocity of Raleigh's wave

c_t phase velocity of a flat cross-section wave

$C_m = \sqrt{(T/M)}$ phase velocity of oscillation propagation, velocity of the forced fluctuations on a dolphin's skin surface

$\omega = 2\pi f$ circular frequency, limiting frequency of fluctuation of an elastic material

$\Omega_i(\omega)$ areas of spectral functions $\Phi(\omega)$ of fluctuation of the boundary layer

$\Psi\Delta_i(\omega)$ energy dissipated in elastic plate; areas of fluctuation energy of the boundary layer absorbed by the plate at different modes of a flow

K_w coefficient of oscillation frequency of elastic surface

$Z = \dfrac{\sqrt{\overline{p'^2}}}{\sqrt{\overline{a'^2}}}\ -$ complex coefficient of rigidity

$\sqrt{\overline{a'^2}}$ amplitude of the surface oscillations

$A(f)/\sqrt{\overline{a'^2}}$ oscillation spectrums of elastic surface

$p(f)/\sqrt{\overline{p'^2}}$ spectrums of pressure fluctuations on the elastic plates

$G^* = G' + iG'';\ E^* = E' + iE''$ complex modulus

$tg\delta = G^2/G\phi = \Delta/\pi(1 - \Delta^2/4\pi^2)$ dissipation factor, loss angle tangent

$\Delta = \ln(A_n/A_{n=1})$ logarithmic damping decrement

$\psi = \dfrac{\Delta W}{W} = \dfrac{2\pi}{Q} 2\pi tg\delta\ -$ loss coefficient

List of symbols

ΔW energy absorbed by a unit of the material volume
W energy saved by the unit of the material volume
Q parameter, characterizing the amplitude increasing with resonance
$K = 1 - h/h_0$ parameter of damping properties; damping coefficient
h_0 fall height
h recoil height
$K = -P/\varepsilon_\nu$ volumetric module
$\varepsilon_\nu = \Delta V/V$ three-dimensional strain
V volume of a elastomer specimen
$I(\tau) = 1/E(\tau); J(\tau) = 1/G(\tau)$ elastic compliance of an elastic material
η lateral of the wall; elastomer stringiness; viscosity of an elastic material; coefficient of absorption small ball beat, damping factor
$\tau = -\eta/E$ elastomer relaxation time
ξ longitudinal strain of the wall
T_F tension of an element of an elastic material
M, m oscillating mass of an element of an elastic material
$S, E' = E/t_i$ rigidity of an elastic material
P_i generalized force, size of compressing force
$\mu = \bar{\varepsilon}/\varepsilon$ Poisson coefficient
$\bar{\varepsilon}, \varepsilon$ relative cross strain and relative longitudinal strain
$\nu = 1/\mu$ cross (lateral) number
$\tau^* = G\gamma$ shear stress
G module of rigidity
$\gamma = \frac{\Delta l}{l} = tg\beta$ relative shear of upper side of the unit cube
$\lambda = l/l_0$ strain in the direction of stress line
l_0 length of the edge of unstrained cube
$\omega_{a.res.} = \sqrt{4k/m_1}$ antiresonance frequency
$f_P = \dfrac{U_\infty}{2\pi\delta}$ natural frequency of membrane vibration
ζ factor of a damping
$K_1 = E_{dynam}/E_{static.}; k_2 = E_{long}/E_{transv.}$ anisotropy factor
$\delta_i = \ln\left(\dfrac{A_i}{A_{i+1}}\right) -$ logarithmic decrement
$K_a = \dfrac{h_i - h_{i+1}}{h_i} 100 -$ absorbing factor
A_i and A_{i+1} amplitudes of the preceding and the following wave
h_i and h_{i+1} fall height and recoil height of indentor
D diffusivity of the polymeric component
C concentration of polymer solution
$c_Q; c_q; C_q = Q/US$ discharge coefficient of polymer solution
$Q = V/t$ discharge of polymer solution
U towing velocity
V volume of an injected liquid
$C_\mu = 2Qu_{sl}/\rho U^2 S = 2Cq\, \rho_c\cdot u_{sl}/\rho U$ factor of quantity of movement
$\xi(Re) = \left(C_{xrig.} - C_{xi}\right)/C_{xrig.}$ coefficient of relative changing in friction
u_{sl} velocity of injection of a liquid through a slot
$\Delta C_f, \% = 100 \cdot (C_f - C_{f\,inj})$. C_f efficiency of drag reduction as a percentage

Indices

Lower

t	parameter of turbulent motion
H	value of a parameter on the exterior boundary of a boundary layer
∞	in approach flow
$0, w$	value of a parameter on a on a wall
Σ	total value
n	nasal part
los. st.; s.l.	loss of stability
cr.	critical
m	middle
eff	effective value of a parameter
i	number of a node of a difference grid along a coordinate x
j	number of a node of a difference grid along a coordinate y
o	parameter in initial cross section of a boundary layer (at $x = x_o$)
max	maximum value of a parameter
sl	slot
ppm	parts per million

Upper

$+$	nondimensional parameter in the correspondence with the law of a wall
$\sim$	result of the first stage of the numerical method
$\approx$	result of the second stage of the numerical method

Abbreviations

BL	boundary layer
TBL	turbulent boundary layer
T.-S., T.-Sh., T.-Sch.	Tollmien-Schlichting wave
CVS	coherent vortices structures
LVS	longitudinal vortices systems
LDMV	laser Doppler measuring of velocity
CC	continuous coating
CCF	current-carrying fabric
LS	longitudinal strips
RF	rubber film
FP	foam polyurethane
FL	foam latex
PPU	penopoliuretan
PU	polyurethane
FE	foam elastic
SAA	same as above
DISA	thermo anemometer
PEO, POE	polyethylene oxide (WSR-301)
OT	ogive tip
SXT	short xiphoid tip
LXT	long xiphoid tip
MB	microbubbles

Nonstationary motion

1.1 Drag of a disk and a sphere at the accelerated movement from a state of rest

In Refs. [160,161], the coefficients of the nonstationary resistance of the disk and the sphere were experimentally determined during their accelerated motion from a state of rest. The resistance of a disk during accelerated motion in the direction of its normal was experimentally studied by many authors, for example, in Ref. [316], in which, based on systematic model tests, an approximate dependence of the nonstationary resistance coefficient of the disk on the dimensionless acceleration was established in a certain range of motion parameters. In Ref. [191], when experimentally determining the resistance force of a sphere during its accelerated motion in a liquid, there is a significant scatter of the measured values of this force, which sometimes differ several times during the transition from one experiment to another with other things being equal. In such cases, it is not possible to obtain a certain value of the coefficient of nonstationary resistance of the sphere for given specific motion parameters.

Towing investigations of models of these bodies (Fig. 1.1) were carried out in the high-speed hydrodynamic basin of the Institute of the National Academy of Sciences of Ukraine, Kyiv. The working part of the pool has a length of 140 m, width -4.0 m, depth -1.8 m. Disks and spheres had different diameters from 180 to 300 mm. Speed the towing carriage was changed within 1−25 m/second, the range of Reynolds numbers was $4 \times 10^3 - 4 \times 10^6$. The acceleration of the movement of the towing car was regulated in the range from 1 to 7.0 m/s².

One type of nonstationary movement was investigated—forward aperiodic accelerated motion of a body from a state of rest to a state of steady motion at a constant speed (Fig. 1.2). In this case, the acceleration of motion intensively varied from zero to a predetermined value, was kept for some time approximately constant, and then monotonously decreased to zero. During the tests, instantaneous values of the speed of movement, acceleration, and force of the hydrodynamic resistance of the body were continuously measured. In this work, it is assumed that the nonstationary force of the hydrodynamic resistance of the body $R_x(t)$ depends on the fluid density ρ, the kinematic viscosity ν, the body diameter D, the body velocity $v(t)$, and its acceleration $v'(t) = dv/dt$.

From analysis of dimensions, the following dependencies of dimensionless parameters can be obtained, which are similarity criteria when determining the

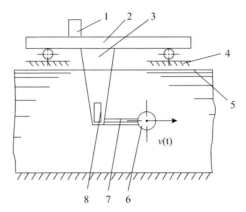

Figure 1.1 The scheme of the experimental setup: (1) equipment, (2) towing carriage, (3) vertical knife, (4) rails, (5) water level in the water channel, (6) model, (7) drag, and (8) strain gauge dynamometer [116,159].

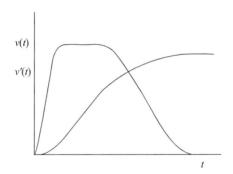

Figure 1.2 The characteristic form of the law of change in speed and acceleration during testing [116,159].

resistance of two geometrically and dynamically similar body motions with acceleration in a viscous fluid:

$$C_{xn} = f(Re, W) \tag{1.1}$$

$$C_{xn} = f(Re, Re^*) \tag{1.2}$$

where C_{xn}=coefficient of nonstationary resistance of the body; Re=Reynolds number; W=dimensionless acceleration; and Re^*=acceleration number, which can be considered as an analog of the Reynolds number.

$$C_{xn} = \frac{8R_x(t)}{\pi\rho v^2(t)D^2}, \quad W = \frac{Dv'(t)}{v^2(t)},$$

$$Re = Dv(t)/\nu, \quad Re^* = \frac{D^3 v'(t)}{\nu^2}.$$

An approximate relationship can also be considered [259,260,316]:

$$C_{xn} = f(W) \tag{1.3}$$

which does not take into account the influence of the Reynolds number.

The obtained experimental results are analyzed in the dimensional form, as well as in the form of Eqs. (1.1)–(1.3). Typical realizations of the nonstationary resistance curve for a disk with a diameter of 254 mm and for a sphere with a diameter of 245 mm, which moved from rest under the same law with an acceleration of about 2 m/s^2, are shown in Figs. 1.3 and 1.4. Curve (1) of the instantaneous values

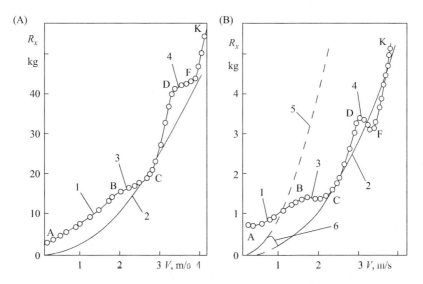

Figure 1.3 A typical view of the resistance curve of the disk (A) and the sphere (B), moving with acceleration [116,159].

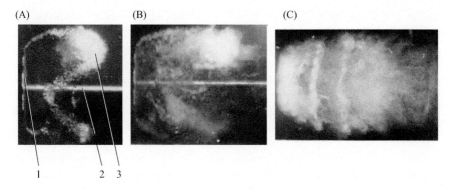

Figure 1.4 Photographs of the formation of the primary ring vortex (A, B) and the system of ring vortices (C) during the acceleration of the disk movement: (1) ring disk, (2) holder, and (3) form of ring vortex [116,159].

of nonstationary disk resistance versus instantaneous speed values is presented in Fig. 1.3 together with the stationary resistance curve (2) of the same disk with the corresponding towing speed values. Curve (2) is a monotonically increasing branch of the parabola $R_{xo}=av^2$ with a vertex at the origin with $a=\frac{1}{8}\pi C_{x0}\rho D^2$, which $C_{xo} \approx 1.12-1.16$ is the coefficient of stationary resistance of the disk, which does not depend on the Reynolds number in the considered speed range. Curve (1) is formed near the stationary resistance curve, but has a more complex form, which is a reflection of complex physical processes that accompany the nonstationary movement of the disk in a real fluid.

At the beginning of motion from the state of rest, the resistance of the disk is formed mainly by inertial forces determined by the attached mass of the disk and the acting acceleration. The viscous part of the resistance in this case only appears, increasing from zero, with increasing speed. This corresponds to the initial segment of the segment AB curve of nonstationary resistance. In general, in the section AB (Fig. 1.3A) behind the disk, the ring vortex first appears and then develops, increasing in size (Fig. 1.3A). This vortex with a continuously increasing mass, carried by the disk into accelerated motion, significantly increases the inertial part of the resistance of the disk. This is the reason that the nonstationary resistance of the disk in this area significantly exceeds the stationary. Having reached certain critical dimensions, at which it can no longer be involved by the disk in accelerated motion, the ring vortex is detached from the disk.

In this case, the fluid mass dragged into accelerated motion decreases sharply, which leads to a slower growth in the resistance of the disk with increasing speed. This corresponds to the section BC of the nonstationary resistance curve on which the first "hump of resistance" (3) is formed in this place. Instead of a detached vortex behind the disk, the second ring vortex arises and develops, increasing in size, which is also carried away in accelerated motion for some time. As a result, the nonstationary resistance starts to increase again with increasing speed more intensely than the stationary (part of the CD curve). Reaching a critical size and the second vortex is also detached from the disk. At the same time, the growth of nonstationary resistance slows down again (section DE) and the second "hump of resistance" (4) is formed. Following this, the third ring vortex develops (section EK). This process is repeated for some time and further (not shown on the graph), accompanied by the formation of new "humps of resistance." However, the development trend of this process is such that each subsequent vortex formed has a smaller scale than the previous one, and the time intervals between the breaks of the vortices are reduced (Fig. 1.4C). With a sufficiently long process of acceleration behind the disk, a cocurrent zone is formed, filled with small-scale vortices that move downstream, gradually dissipating in the surrounding liquid.

A similar picture can be observed in the graph of Fig. 1.3B for a sphere. However, it is necessary to note a number of features. The curve of stationary resistance of the sphere at the beginning of the motion coincides with the parabola of precrisis flow (5). Then, at a speed of about 0.7 m/second, there is a "crisis of resistance" section (6) in Fig. 1.3B and the curve of stationary resistance of the sphere goes to the parabola of super crisis flow (2). The nonstationary resistance of the sphere (1) behaves interestingly. The initial branch AB of this curve, which corresponds to the beginning of the dispersal of the sphere from a state of rest, is not formed on the basis of a parabola of precrisis flow around a sphere (5) at its steady motion, as soon as it follows the shape of the initial part of the parabola super crisis flow (2). In this case, the phenomenon of "crisis of resistance" does not arise. The visualization of the flow around the sphere in this case shows that the point of separation of the stream, originating at the beginning of the movement in the vicinity of the rear critical point of the sphere, moves upstream with increasing speed and, with a sufficiently long acceleration of the sphere, tends to the position of the separation point during its steady motion.

However, this movement of the separation point upstream is not uniform and strictly corresponding to the current values of speed and acceleration, but has the form of unordered oscillations around a certain center, which itself already tends to occupy the said position of the separation point with increasing speed during steady-state supercritical flow. The oscillations of the separation point are associated with the separation of large-scale vortex formations from the wake area. Each separation of the next vortex causes the separation point to move downstream, and the emergence and development of a new vortex formation causes the separation point to move upstream. Oscillations of the separation point lead to a narrowing or expansion of the wake zone behind the sphere, which affects the magnitude of the resistance force of the sphere. Moving the separation point downstream reduces the growth rate or even briefly decreases the resistance force, and moving the separation point upstream increases the growth rate of the resistance force. This is well illustrated by the form of (1) of the nonstationary resistance of a sphere in the graph of Fig. 1.3. In the section AB of this curve, the formation and development of a wake zone behind the sphere goes, the separation point is mixed up from the rear critical point of the sphere upstream, the resistance force is rapidly increasing. On the BC section, the first large-scale vortex breaks off, the separation point moves downstream, the resistance force briefly decreases, forming the first "hump of resistance" (3). On the CD section, the second vortex forms and develops, the separation point moves upstream; resistance force is increasing rapidly. Next, in the section DE, the second vortex breaks off; a second "hump of resistance" is formed (4). A third vortex develops on the EC section, and so on. With further accelerated movement of the sphere, the value of the resulting "humps of resistance" decreases, the time intervals between their formation decrease and the nonstationary resistance curve approaches closer to the stationary resistance curve during a supercritical flow.

A general view of the Eq. (1.3) of the nonstationary resistance coefficient on dimensionless acceleration, which is approximately the same for both the disk and the sphere and for other poorly streaming bodies, beyond the specific numerical values of the parameters, is shown in Fig. 1.5. This dependence is built for one cycle of body movement: acceleration–deceleration. When the body is accelerated from a state of rest to a state of motion with a constant speed, the dimensionless acceleration continuously changes from $+\infty$ to 0.

Point 0 on the x-axis corresponds to body motion with a constant speed. When the body brakes from a state of motion at a constant speed to a full stop, the

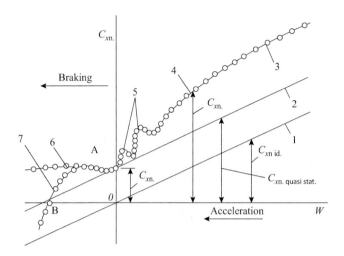

Figure 1.5 General view of the dependence of the nonstationary resistance coefficient of poorly streamlined bodies on dimensionless acceleration [116,159].

dimensionless acceleration varies from 0 to $-\infty$. On this graph, even without having the results of the experimental determination of a nonstationary body resistance force, you can display some information about the resistance coefficient of this force, based on the following considerations.

If the body moved rapidly in an ideal fluid, then the force of nonstationary resistance can be expressed by the equation:

$$R_{xn.\ id.} = K \times \rho \times V \times v', \tag{1.4}$$

where K is the coefficient of the attached body mass; and V is the characteristic volume of the body. For a disk and a sphere, $V = \frac{1}{6}\pi \times D^3$. Reducing the Eq. (1.4) to a dimensionless form, for the coefficient of nonstationary resistance of a disk and a sphere in an ideal fluid, we obtain the expression:

$$C_{xn\ id.} = 8R_{xn.\ id.}/\pi \times \rho \times v^2 \times D^2 = \frac{4}{3} K \times W \tag{1.5}$$

where K is 0.637 for a disk and 0.5 for a sphere.

Thus with a known value of the coefficient of attached body mass, the coefficient of nonstationary resistance of the body in an ideal fluid can be depicted as straight line (1) (Fig. 1.5) passing through the origin of coordinates with an angular coefficient determined by the value of the attached body mass. For a disk and a sphere, this angular coefficient is equal to $4/3K$. For any other body, it will be $2aK$, where a is the ratio of the characteristic volume of the body to the product of its characteristic area and characteristic linear size. Similarly, you can build a graph of the coefficient of nonstationary resistance of the body in a real fluid in a quasistationary representation. In accordance with Refs. [259,260], this coefficient can be represented as the sum of the body resistance coefficient determined according to stationary tests for instantaneous Reynolds number and the nonstationary body resistance coefficient in an ideal fluid:

$$C_{xn \text{ quasi stat.}} = C_{x0} + C_{xn \text{ id.}} = C_{x0} + \frac{4}{3} K \times W. \tag{1.6}$$

Here, C_{x0} is the coefficient of body resistance according to stationary tests. In general, for a disk in the zone of auto-similarity (with $Re \geq 10^3$), it can be assumed to be constant and equal to 1.12−1.16. For a sphere with $Re > 10^3$, this coefficient should be taken equal to approximately 0.09−0.15 that is the value that it takes in the supercritical flow of a stationary flow around the sphere. This follows from the fact that during the dispersal of a sphere from a state of rest, the phenomenon of a "flow crisis" does not occur, and the separation point, moving upstream from the rear critical point of the sphere, tends to the position of the separation point during supercrisis flow. Thus with known coefficients of stationary resistance and attached mass coefficient of nonstationary resistance of the disk and the sphere in a quasistationary representation can be represented in Fig. 7.5 straight line 2, cutting off on the ordinate axis a segment equal to C_{x0} and inclined to the axis a stasis's an angle equal to arctan $\frac{4}{3}K$.

If now in Fig. 1.5 we apply the experimentally obtained values of the coefficient of nonstationary resistance C_{xn} of a poorly flowing body according to the results of the implementation of one cycle of acceleration−deceleration, then we obtain the dependence presented in the generalized form of (3). At the beginning of motion from the state of rest for large values of W, this curve is located above the straight line (2) of the quasistationarity approximation and is approximately parallel to it. However, starting with some relatively small values of W, characteristic for each given body, (3) begins to approach straight line (2), forming a convex section (4), which, for example, for a disk and a sphere corresponds to the first "resistance hump" in Fig. 1.3. With further acceleration of the body, curve 3 more and more approaches the straight line (2) and at the same time can form "humps" (5), which are a reflection of the breakdown of large-scale eddies from the surface of the body. The number of "humps" depends on the duration of the process of acceleration of the body and the number of large-scale eddies, which during this time will have time to form and detach from the surface of the body. When the acceleration process stops and the body begin to move at a constant speed, (3) arrives at point A and the nonstationary resistance coefficient C_{xn} takes its stationary value C_{x0}. In the case of a short period of acceleration of the body, when large-scale vortices do not have time to form, and their separation from the surface of the body does not occur, (3) comes to point A smoothly, without forming "humps" (5).

During slow motion of a body from a state of steady motion with a constant speed to a state of rest, which corresponds to a change in W from 0 to $-\infty$, the curve of the nonstationarity resistance coefficient C_{xn} may behave differently depending on the intensity of the deceleration. With a relatively weak deceleration, it is in the region of positive values, forming branch (6). With a sufficiently intensive deceleration, this curve can cross the abscissa axis and move into the region of negative values, forming branch (7). In this case, the reverse flow around the body occurs when formed at previous time points the trail overtakes the slowing body and, as it were, falls on him from behind. It is interesting to note that if the movement is organized in such a way that the body moves for a while at a constant value $W=B$, then at that time it will move without hydrodynamic resistance. When moving not from a state of rest, the resistance coefficient of the body depends not only on the acceleration and deceleration mode, but also on the history of its movement, that is, from the driving mode that preceded the start of acceleration or deceleration.

It should also be noted that the dependence $C_{xn}=f(W)$, shown in Fig. 1.5 one curve (3), in fact, is not so unambiguous. Since different values of the Reynolds number may correspond to the same fixed values of dimensionless acceleration W, for different implementations of the acceleration cycle, the acceleration−deceleration $C_{xn}=f(W)$ dependence will be expressed by a series of curves similar to curve (3) and, generally speaking, not coinciding with together. This circumstance will be considered further on the specific experimental material for the disk and the sphere. In practice, when assessing the value of the coefficient of nonstationary resistance of a body, the question arises of whether the quasistationarity hypothesis is applicable for this purpose, the essence of which is expressed by Eq. (1.6). Some ideas about this can be made if we consider the function of the influence of nonstationary in the form of the ratio of the coefficient of nonstationary resistance of a body, determined experimentally, to the coefficient of nonstationary resistance in a quasistationarity approximation defined by Eq. (1.7), depending on the dimensionless acceleration:

$$\xi(W) = C_{xn}/C_{xn \text{ quasi stat}}.$$ (1.7)

The characteristic form of such a function for a poorly flowing body is shown in Fig. 1.6, from which it is clear that this function is close to unity and, therefore the hypothesis of quasistationarity is valid (you can take $\approx C_{xn \text{ quasi stat.}}$) only for relatively large and relatively small values of dimensionless acceleration W. However, there is some range of change W, where the function $\xi(W)$ differs significantly from

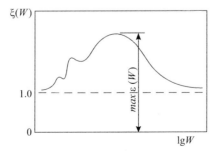

Figure 1.6 The characteristic form of the function of the influence of nonstationarity [116,159].

unity and, therefore in this range W, the quasi stationary hypothesis cannot be used when determining the value of the coefficient of nonstationary resistance of the body. The results of experiments with disks and spheres during their accelerated motion from a state of rest show that the range of variation of W, in which the quasistationarity hypothesis is not applicable, is approximately $0.01 \leq W \leq 100$ for a disk, and $0.02 \leq W \leq 50$ for a sphere. At that, the function $\xi(W)$ for a disk, it can reach values $\xi(W) \approx 2.5$ at $W \approx 2.0$ and, accordingly, $\xi(W) \approx 2.0$ at $W \approx 15.0$ for a sphere.

The results of experimental determination of the coefficients of nonstationarity resistance of a disk and a sphere in the form of dependences $C_{xn} = f(W)$ are shown in Fig. 1.7. They were obtained by implementing several dozen cases of accelerated movement of the disk and the sphere from the state of rest with various numerical variations of the law of motion, the general view of which is shown in Fig. 1.2. The same fixed values of the dimensionless acceleration W can correspond to different values of the Reynolds number. In Fig. 1.7, dashed straight lines (3) and (4) are plotted, which express the values of the nonstationary resistance coefficient determined by the Eq. (1.5) for the disk and the sphere, respectively, as they move in an ideal fluid. Dashed straight lines (5, 6) conventionally denote the values of the coefficients of the resistance of the disk and the sphere during their stationary motion in a real fluid. For a disk, this value is $C_{x0} = 1.12$, and for a sphere in this case, it is necessary to take the value of the drag coefficient for a supercritical flow around, which is equal on average to $C_{x0} = 0.1$. According to the general type of dependences $C_{xn} = f(W)$ for a disk and a sphere from the onset of accelerated motion from a state of rest to a mode of motion at a constant speed, which corresponds to a

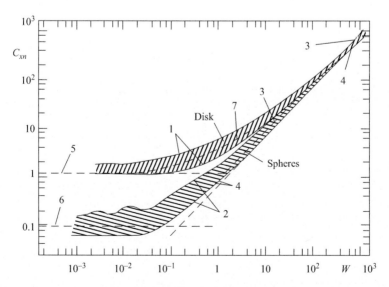

Figure 1.7 The coefficients of nonstationary resistance of the disk and the sphere during accelerated motion from a state of rest depending on the dimensionless acceleration [116,159].

change in W from $+\infty$ to 0, then it can be noted that everything reduces to transition of these dependencies on straight lines (3, 4) for large values of W to direct (5, 6) for small values of W, respectively. At the beginning of movement, for large W, the nonstationary resistance of the disk and the sphere are close in magnitude and their values almost coincide with direct lines (3, 4).

This is explained by the fact that for large W, inertial forces predominate in the composition of the body's nonstationary resistance, which is determined by the magnitude of the added masses. And the coefficients of the attached masses of the disk and the sphere are close in magnitude, which determines the direct proximity of (3, 4) to each other. But later, with the development of motion (with decreasing W), the nonstationary resistance of the body becomes increasingly important, and then the forces of viscous nature begin to prevail. In this case, the values of the nonstationary resistance coefficients of the disk and the sphere begin to move more and more away from straight lines (3, 4), simultaneously moving away from each other more and closer to their own values in the case of stationary motion, conventionally indicated by straight lines (5, 6).

Thus for large values of dimensionless acceleration W, the unsteady resistance of the body is close in magnitude to the unsteady resistance of this body in an ideal fluid and is easily determined with a known value of the attached mass of this body. At small values of dimensionless acceleration, the nonstationary resistance of the body approaches its value in steady-state motion. Between these extreme cases, there is a region of some average values of dimensionless acceleration, where nonstationary body resistance can be determined only on the basis of experimental data, since the hypothesis of quasistationarity is not applicable here, and the corresponding theoretical methods have not been developed [201,202]. The results of work [316] for determining the coefficient of nonstationary disk resistance are shown in Fig. 1.7 curve (7). On average, they agree well with the results of these studies, although they were obtained in a smaller range of changes in the value of dimensionless acceleration W. In order to avoid the marked ambiguity, which is inherent in the dependence $C_{xn}=f(W)$, an attempt was made to construct the dependence of the nonstationary resistance coefficient of the disk on the motion parameters in the form $C_{xn}=f(Re, W)$.

The main results of these studies are presented in Fig. 1.8 as Eq. (7.2). As the fixed values of the acceleration number Re^*, in these graphs, only the extreme bottom and top values Re^* (10^9 and 4×10^{10} for the disk; 10^9 and 6×10^{10} for the sphere) are allocated, within which systematic results were obtained during testing, as well as one intermediate value $Re^*=10^{10}$. For any other intermediate values Re^* within the indicated limits, the dependences $C_{xn}=f(Re, Re^*)$ flow equidistantly and can be approximately determined by interpolation. For comparison, Fig. 1.8 shows by dashed lines known dependences of the stationary resistance coefficient for a disk and a sphere [201–203], for which $Re^*=0$, in the range of Reynolds numbers $10^0 - 10^7$, are plotted with dashed lines. The presentation of the results obtained in the work against the background of these dependences determines their place in the space of motion parameters, which has physical and practical interest. So, it is clear that the region where the acceleration number Re^* changes from 10^9 down to zero

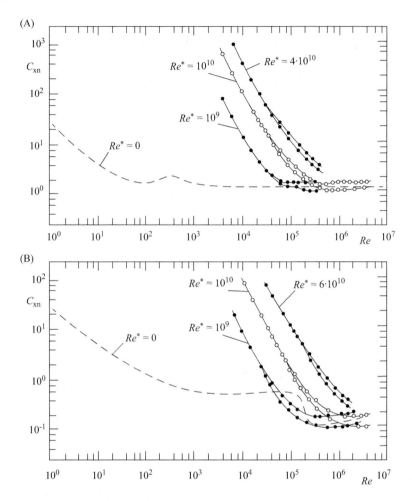

Figure 1.8 The coefficient of nonstationary resistance of the disk (A) and the sphere (B) during accelerated motion from a state of rest depending on the Reynolds number and the number of acceleration [116,159].

with $Re < 10^5$ remains unexplored. Data on the nonstationary resistance coefficient for $Re^* \geq 10^{11}$ with $Re > 10^5$ are also of practical interest.

An analysis of the results presented in Fig. 1.8 allows us to draw the following conclusions. The nonstationary resistance coefficient curves for fixed values of the acceleration number Re^* are arranged in the figures in ascending order of this number: the larger the acceleration number, the number. curves of the nonstationary resistance coefficient for any fixed number of acceleration Re^* with an increase in the Reynolds number initially decrease intensively and then smoothly reach one common level. This level for a disk approximately corresponds to the value of the coefficient of stationary resistance in the self-similarity zone.

For a sphere, this level in the surveyed range of variation of the Reynolds numbers approximately coincides with the value of the stationary resistance coefficient during a supercrisis flow. The curves of the nonstationary resistance coefficient do not reach the general level at the same time, but in order of increasing acceleration number Re^*. The larger the acceleration number, the higher the Reynolds numbers the corresponding nonstationary resistance coefficient curve goes to a common level. We note that for a sphere, the indicated general level is to some extent realized in the region of Reynolds numbers $Re < 2 \times 10^5$, in which, during steady motion, the sphere flows around in the precrisis mode.

This is evidenced by the fact that the curve of the coefficient of nonstationary resistance of a sphere with $Re^* = 10^9$ intersects the curve of the coefficient of stationary resistance of a sphere with a subcritical value of the Reynolds number (at $\approx 5 \times 10^4$) and reaches the level corresponding to the supercritical flow around the sphere. It follows that there are modes of nonstationary motion of the sphere, in which the coefficient of nonstationary resistance may be lower in magnitude than the corresponding coefficient of stationary resistance for the same Reynolds numbers. In general, the coefficient of nonstationary resistance of a poorly streamlined body substantially depends on both the Reynolds number Re and the acceleration number Re^*.

1.2 Influence of nonstationary on the mechanism of formation of hydrodynamic drag

In Ref. [585] the peculiarities of a nonstationary resistance of a vessel on the modes of acceleration–deceleration are considered. The need to assess the effect of a change in speed on the magnitude of the resistance to movement of the vessel arises when studying the period of its acceleration and deceleration. In the study of the motion of a vessel with variable speed, an equation of motion is compiled, in which, in addition to other external forces, hydrodynamic forces dependent on acceleration are included. In the simplest case of translational motion, this equation in the right-hand side contains hydrodynamic forces and the pull force Re of the tug or propulsion:

$$m \frac{dv}{dt} = - \lambda_{11} \frac{dv}{dt} - R_{xi} \pm P_e. \tag{1.8}$$

The first term on the right-hand side is the hydrodynamic force of inertial nature and is calculated taking into account the added mass λ_{11} [155,350,585]. The work to overcome this force is spent on changing the kinetic energy of the fluid surrounding the vessel during its motion with acceleration. Since the process of changing and the magnitude of kinetic energy, in addition to the shape of the body, significantly depend on the presence and properties of the fluid boundaries, the magnitude of the added mass λ_{11} will be different when the same body moves at a great depth or on a free surface, and also depends on the nature of the deformations free surface, that is, on the number of Fr. The number Re has some influence on the added mass; the viscosity of the fluid leads to a slight increase. However, in calculations, it is usually assumed that the effect of viscosity on λ_{11} can be neglected and the values of the added mass calculated theoretically when moving in an inviscid fluid are used. This technique is convenient to use to calculate the motion of bodies of the simplest forms in an unlimited fluid. For ships, sometimes λ_{11} values are used, which are known for ellipsoids. When moving near or on a free surface with speeds when observed wave formation, this method of approximate determination of the added masses becomes unsuitable.

The second term on the right side of Eq. (1.8) is the force of resistance to the movement of the vessel, calculated taking into account the effect on its magnitude of acceleration. In accordance with the general principles of the similarity theory, the resistance force when a body moves with variable speed can be calculated using the general formula in the form:

$$R_x = \xi \frac{\rho v^2}{2} \Omega, \tag{1.9}$$

where Ω is the wetted surface of the body in a static position, $\xi = f(Fr, Re, Sh)$, and v is the instantaneous values of speed. Moreover, the resistance coefficient, in addition to the similarity criteria Fr and Re, should depend on the parameter characterizing the acceleration of the body. If the characteristic values are taken as the instantaneous velocity v and the linear size of the body L, then the acceleration of the body in dimensionless form can be determined by a dimensionless parameter that represents a modification of the number Sh:

$$a_0 = \frac{L}{v^2} \frac{dv}{dt}, \tag{1.10}$$

As a result, we can assume that for a body of a given shape moving in a fluid with given boundaries, the drag coefficient in the general formula (1.9) is a function of the following form:

$$\xi = f(a_o, Fr, Re).$$

In reality, however, this dependence will be more complex. The instantaneous values of speed and acceleration, with the help of which the numbers Fr, Re and a_o are calculated do not sufficiently characterize the law of change in velocity at previous moments, that is, the history of the movement process. The influence of the prehistory of the process may be different for viscous and wave forces, but in the end we should expect some difference in the instantaneous values of resistance obtained, for example, as a result of body tests with positive (acceleration) or negative (deceleration) acceleration. Since the process of movement with variable speed is associated with the simultaneous action of a resistance force and inertial hydrodynamic force, direct dynamometric measurements do not allow in this case to isolate the resistance force. Therefore in the investigation of the movement of the vessel using the integration of Eq. (1.8), as well as to determine the resistance when moving with acceleration using approximate techniques. The first of these is to apply the stationarity hypothesis to calculate the force R_x. This hypothesis is that the instantaneous values of the force R_x when the body moves with acceleration are equal to its values when moving without acceleration with the corresponding speeds, that is, the coefficient ξ is assumed to be independent of the parameter a_o and the prehistory of the movement process. The values of the added mass are calculated in this case, as for movement in an inviscid fluid. This technique gives satisfactory results when moving elongated bodies, when there is no influence of the free surface on their hydrodynamic characteristics. More accurate results can be obtained if the resistance force at write with the acceleration R_{xn} in the form of the sum of two components:

$$R_{xn} = R_{xo} + \Delta R_n, \qquad (1.11)$$

where R_{xo} = resistance force calculated by the stationarity hypothesis; and ΔR_n = additional resistance caused by the influence of nonstationarity. The value of ΔR_n can be in general both positive and negative, and consists of an additional viscosity $\Delta R_{n \cdot \text{visc.}}$ and wave $\Delta R_{n.\text{wave}}$ resistances:

$$\Delta R_n = \Delta R_{n.\text{visc.}} + \Delta R_{n.\text{wave}} \qquad (1.12)$$

If Eq. (1.11) is substituted into Eq. (1.8) and the addend ΔR_n is combined with inertial force, then in the absence of thrust, the equation of motion can be written as:

$$m(1 + n_1)\frac{dv}{dt} = -R_{x0}. \qquad (1.13)$$

With the help of the coefficient n_1, the hydrodynamic forces caused by the nonstationarity of motion are taken into account. It is defined as

$$n_1 = \frac{\lambda_{11} + \Delta R_i / \frac{dv}{dt}}{m}, \qquad (1.14)$$

and depends on the shape of the vessel, the numbers Fr, Re, a_o, and the prehistory of the process and can be calculated relatively easily if there is a record of the change in the speed of the vessel or its model in time.

In order to clarify the effect of nonstationarity on water resistance to movement in the experimental basin of the Leningrad Shipbuilding Institute, special tests were carried out on the body of revolution with an elongation of about 1.15 (body length $L = 1.5$ m and maximum diameter $d = 0.21$ m) with three laws of nonstationarity [526]. Water resistance to nonstationary body motion was measured with specially designed strain gauge dynamometers. The speed of movement and the forces on the body were recorded using an oscilloscope. The error of force measurement was within 5%.

Nonstationary tests of the body of revolution under the free surface at immersion $H=0.7$ m ($H/d=3.33$) were carried out with three nonstationary laws: linear, exponential, and sinusoidal. When processing experimental results, it was thought that the resistance force water movement body of rotation R_n in the case of nonstationary mode of motion can be represented as:

$$R_n = R_{n.\text{visc.}} + R_{n.\text{wave}} + R_{n.\text{inert.}}, \qquad (1.15)$$

where $R_{n.\text{visc.}}$ = viscous resistance of the body in nonstationary motion; and $R_{n.\text{wave}}$, $R_{n.\text{inert.}}$ = respectively, the wave and inertial nonstationary resistance. Inertial resistance was determined by calculation:

$$R_{n.\text{inert.}} = (m + \lambda_{11})\frac{dv(t)}{dt},$$

where m = body mass; and λ_{11} = attached body mass. The Eq. (1.15) will be presented in the form of dimensionless resistance coefficients:

$$C_{xn} = C_{n.\text{visc.}} + C_{n.\text{wave}} + C_{n.\text{inert.}} \qquad (1.16)$$

where C_{xn} is the coefficient of total nonstationary resistance, which is a function of Re, Fr numbers, and dimensionless acceleration N:

$$N = \frac{\frac{dv(t)}{dt}L}{v^2(t)};$$

$C_{n.\text{visc.}} + C_{n.\text{wave}} + C_{n.\text{inert.}}$ = coefficients of nonstationary, respectively, viscous, wave, and inertial resistances.

When a body is immersed under the free surface at $H/d=3.5$, the wave resistance is insignificant and therefore is not considered further. Then Eq. (1.16) takes the form:

$$C_{xn} = C_{n.\text{visc.}} + C_{n.\text{inert.}} \qquad (1.17)$$

According to [269], the coefficient of viscosity is:

$$C_{n.\text{visc.}} = C_{\text{visc.}} + \Delta C_{n.\text{visc.}}(Re, Fr, N), \qquad (1.18)$$

where is $C_{\text{visc.}}$ = viscous drag coefficient, appropriate to the quasistationary mode of body motion; $\Delta C_{n.\text{visc.}}$ = coefficient of additional viscosity resistance due to nonstationarity of motion.

From the analysis of the obtained data, it can be seen that in the range of speed variation $0 \leq v \leq 3.0$ m/second, the linear and exponential laws give fairly close results. Fig. 1.9 shows the curves of additional viscosity resistance due to nonstationarity, as a function of the Re/N ratio of a body of revolution according to a linear law $v(t)=at$ with positive acceleration $a=0.1-0.5$ m/s^2 (curve (1)) and with negative acceleration $a = 0.1 - 0.3$ m/s^2 (curve 2) at $v(t) = v_o - at$.

Braking was carried out as follows: first, the small cart with the model of the body of rotation moved at a constant speed along the pool, and then its movement was slowed down according to a linear law. The tests of the model of a body of revolution were carried out in the range of Reynolds numbers $Re=(1.0-5.0) \times 10^6$. From Fig. 1.9, it can be seen that positive acceleration leads to an increase in viscosity resistance the greater, the smaller the ratio Re/N. Negative acceleration, on the contrary, reduces viscosity.

When carrying out towing tests of the body of rotation according to the sinusoidal law $v(t)=A \cdot \sin\omega t$ at the moment of time $t=0$, the cart with the body of rotation was at rest. At subsequent times, the small cart, along with the body, ran a path equal to half the length of the sinusoid, at the end of which its speed again became zero. Then this cycle was repeated again. The parameter ω (oscillation frequency) was chosen in such a way that one or several half-waves of a sine wave fit the length of the pool. Amplitude factor A allowed changing the speed of the body in its absolute value. The analysis of the oscillogram (Fig. 1.10) shows that when the body of rotation moves along a sinusoidal law, the maximum resistance force corresponds to the maximum acceleration, not the speed, as was supposed. With a further increase in speed to v_{max}, a decrease in the resistance force is observed. The maximum towing speed was 1.64 m/second.

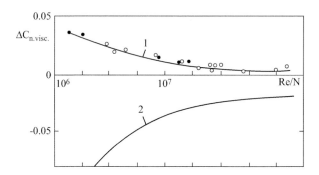

Figure 1.9 Curves of additional viscous resistance of a body of revolution with unsteady motion [52,455].

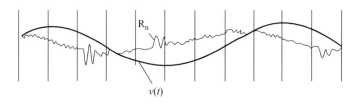

Figure 1.10 Oscillogram of nonstationary testing of a body of revolution according to the law $v(t) = A \cdot \sin\omega t$ (time scale = 1 s); transverse lines [52,455].

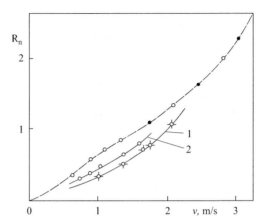

Figure 1.11 The curves of averaged resistance R_n, obtained when testing a body of revolution according to the "sinusoidal" law [52,455].

Fig. 1.11 shows the curves of averaged resistance, obtained by testing the body of rotation according to a sinusoidal law with half-periods $T = 2$ seconds (1) and $T = 4$ seconds (2). For comparison, the stationary resistance curve (dashed line). Averaged over half-period resistance was determined by the expression:

$$\overline{R}_i = \frac{2}{\dot{O}} \int_0^{\dot{O}/2} R_i(t) dt. \tag{1.18}$$

From Fig. 1.11, it can be seen that the averaged resistance curves are located below the corresponding stationary curve. This gives grounds to assume that when a body of revolution moves under a free surface according to a sinusoidal law, its energy consumption over a half-period, under certain conditions, may be less than with an appropriate stationary mode of motion. This can be explained by the fact that at negative accelerations an additional inertial force may act on the body, directed along the movement of the body and due to the presence of the boundary layer. In the works of Lighthill, it was pointed out that it was necessary to take into account the influence of viscosity on the added mass. The friction component of the added mass may turn out to be larger than the added body mass calculated on the assumption of the potentiality of motion. Obviously, this difference will be greater, the relatively thicker the boundary layer on the body.

Consider the problem of a nonstationary turbulent boundary layer on a flat plate. We write the integral ratio in the form:

$$\delta^{**} \frac{\partial H}{\partial t} + \frac{\partial \delta^{**}}{\partial t} + v \frac{\partial \delta^{**}}{\partial x} + \frac{\dot{v}}{v} H \delta^{**} = v \frac{\tau_\omega}{\rho v^2}, \tag{1.19}$$

where $H = \frac{\delta^*}{\delta^{**}}$; $\dot{v} = \frac{dv(t)}{dt}$.

When calculating a nonstationary boundary layer, it is usually assumed that the formulas for the velocity profile and the law of resistance are similar to the formulas for the stationary boundary layer [233]. Let the simple empirical Faulkner resistance law be true for a nonstationary boundary layer along a smooth plate:

$$\frac{\tau_w}{\rho v^2} = 0.00655 \text{Re}^{**-1/6}; \left(\text{Re}^{**} = \frac{v(x,t)\delta^{**}}{\nu} \right). \tag{1.20}$$

We introduce the form parameter f and the function ζ:

$$f(x,t) = \frac{\delta^{**}}{v} K(\text{Re}^{**})\left(v' + \frac{\dot{v}}{v}\right);$$

$$\zeta = \frac{\tau_w}{\rho v^2} K(\text{Re}^{**}), \tag{1.21}$$

where the factor K (Re**) is chosen so that the values of f and ζ are connected by universal dependencies:

$$\zeta = \zeta(f), H = H(f)$$

At present, the existence of the K (Re**) function with this property, as well as the universality of H, is unlikely, and the degree of approximation can be estimated by the results of experience, which, unfortunately, has not yet been established for a nonstationary boundary layer. In the following, it is assumed that H value slightly depends on t.

As a result of transformations, Eq. (1.19) applied to a flat plate can be reduced to the Cauchy task for the differential equation first order partial derivatives relative to the sought function f:

$$H_0 \frac{\partial f}{\partial t} + v(t)\frac{\partial f}{\partial x} + Bf - M = 0;$$

$$f(0,t) = 0; f(x,0) = 0, \tag{1.22}$$

where $B = 4.45\frac{\dot{v}}{v} - 1.33\frac{\dot{v}}{v}; M = 1.17\frac{\dot{v}}{v}; H_0 = 1.33$.

Fig. 1.12 shows the results of a numerical solution of Eq. (1.22), showing the effect of nonstationarity on the total friction resistance coefficient as a function of dimensionless acceleration. The coefficient $n = \frac{C_{F\text{nonst.}}}{C_{F\text{st.}}}$ is the ratio of the nonstationary coefficient of total friction resistance $C_{F\text{nonstationary}}$ to the corresponding stationary coefficient. In Fig. 1.12, it can be seen that a positive acceleration (solid line) leads to an increase in the magnitude of $C_{F\text{nonstationary}}$, and a negative acceleration (dashed line) to a decrease in this value compared to the corresponding stationary coefficient.

In Fig. 1.13, it can be seen that in some cases, the imposition of additional harmonic oscillations on the main translational motion can lead to a decrease in the friction resistance coefficient as compared with the stationary motion corresponding to the velocity v_0.

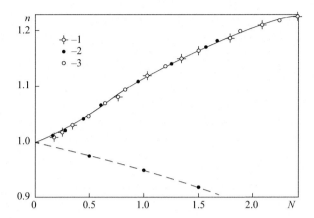

Figure 1.12 Curves of the values of n for a flat plate as a function $N = \frac{vL}{v^2}$ in the case of uniformly accelerated ($v(t)=at$; $a>0$) and uniformly slow ($v(t)=v_0 - at$) movements: (1) $a=0.1$ m/s^2, (2) $a=0.5$ m/s^2, (3) $a=1.0$ m/s^2 [52,455].

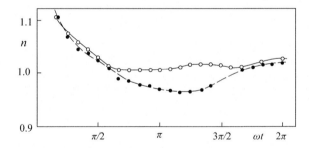

Figure 1.13 Curves n for a plate obtained for the law of motion $v(t) = v_0 (1 + \sigma \cdot \sin \omega t)$ [52,455].

In this Fig. 1.13, $\sigma=0.1$ (solid curve) and $\sigma=0.5$ (dashed curve); $\omega=0.5$ 1/second; $v_0 = 15$ m/second. The obtained experimental data and the calculation of the nonstationary boundary layer show that to reduce the resistance of water to motion, marine animals obviously use the corresponding nonstationary modes of motion. From the calculation of the coefficient n it can be seen that from the point of view of reducing friction resistance due to the use of a nonstationary mode of motion, the law is most appropriate when a short section of acceleration of a body with a positive acceleration follows, then replaced by a long section of body motion with negative acceleration.

Influence of the vortex generators installed on the leading edge of the wing on the unsteady aerodynamic characteristics of an oscillating wing with a flap was investigated in A.G. Shcherbon's Ph.D. thesis "Aerodynamic characteristics of a wing with vortex generators under unsteady flow conditions," performed at the National Aviation University, Kiev.

1.3 Drag coefficient of fluid bodies in acceleration and deceleration modes

Hydrobionts always swimming nonstationary (Part I, Section 2.7). To understand the laws of such a movement, numerous experimental studies of the nonstationarity of the motion of various canonical bodies were carried out in Refs. [159,455]. The resistance force R_n acting on the body of a hydrobiont in case of unsteady motion can be represented as two components: inertial $R_{\text{inert.}}$ and viscous $R_{\text{visc.}}$:

$$R_n = R_i + R_v. \tag{1.23}$$

The inertial component of the resistance R_i is proportional to the acceleration of motion and is determined by the formula $R_i = K \cdot \rho \cdot V \cdot v'_t$, where K is the coefficient of the attached body mass of the hydrobiont; V is the volume of the body of a hydrobiont; $v'_t = \frac{dv_t}{dt}$ — acceleration of motion; ρ is the mass density of water; t is time. The viscosity component of the resistance R_v is proportional to the square of the instantaneous velocity and is determined by the formula:

$$R_v = c_{xv} \frac{\rho v_t^2}{2} F,$$

where c_{xv} = viscous drag coefficient; v_t^2 = instantaneous speed; and F = area of the midsection of the body of a hydrobiont. The resistance force acting on the body during unsteady motion can also be determined by the formula:

$$R_n = c_{xn} \frac{\rho v_t^2}{2} F,$$

where c_{xn} = coefficient of nonstationary resistance of the body.

According to (1.1), we can write:

$$c_{xn} \frac{\rho v_t^2}{2} F = K \cdot \rho \cdot V \, v'_t + c_{xv} \frac{\rho v_t^2}{2} F, \text{ or}$$

$$c_{xn} = c_{xv} + 2K \frac{V}{F} \frac{v'_t}{v_t^2}$$

Assuming that $\frac{V}{F} = \frac{\delta LBH}{\beta BH} = \frac{\delta}{\beta} L$, you can write $c_{xn} = c_{xv} + 2K \frac{\delta}{\beta} \frac{Lv'_t}{v_t^2}$.

Here, L is the length of the body, B is the width of the body, H is the height of the body, δ is the coefficient of the overall fullness of the body, and β is the completeness of the mid-frame.

Denote $N = \frac{Lv'_t}{v_t^2}$ (Section 1.2), then we get:

$$c_{xn} = c_{xv} + 2K \frac{\delta}{\beta} N. \tag{1.24}$$

If in Eq. (1.24) we assume that the added mass coefficient is $K = K_\text{T}$, where K_T depends only on the shape of the body and is determined according to the scheme of potential flow of an ideal fluid around the body, and the viscous drag coefficient c_{xv} is equal to the drag coefficient of this body for a stationary flow around the instantaneous value speed c_{xV}^0, then we obtain the expression for the coefficient of nonstationary resistance in the quasistationary approximation [159]:

$$c_{xn}^{\text{quasist.}} = c_{xV}^0 + 2K_T \frac{\delta}{\beta} N. \qquad (1.25)$$

Thus the coefficient of nonstationary resistance of a body in the quasistationary approximation is a linear function of the dimensionless complex W, which is commonly called dimensionless acceleration.

To test the reliability of the dependence (1.25) and its applicability limits in determining the resistance force acting on a hydrobiont body during unsteady motion, experiments with a solid model, which can be considered as some analog of the hydrobiont body, were carried out at the Institute of Hydromechanics of the National Academy of Sciences of Ukraine (Kyiv). Experimental investigations were carried out in a high-speed hydrodynamic test bench, in which the speed of movement of the carriage in the speed range of 1−25 m/second is ensured. The tested model was a streamlined body, the front part of which was midshaped had the shape of a three-axial ellipsoid, and the rear part was a drop-shaped shape with a gathering of all forming at one point on the longitudinal axis of the body. The model is sealed and its mass in the process of the experiment was constant. The overall dimensions of the model were: length $L=0.780$ m, width $B=0.22$ m, height $H=0.4$ m, area of the midframe $F=0.069$ m^2, coefficient of overall fullness $\delta=0.4744$, coefficient of completeness of the midframe $=0.7854$, body volume $V=0.03$ m^3 (Section 1.2).

An experimental setup is shown in Fig. 1.1 (Section 1.1). The model of a streamlined body with its upper edge during the experiment sank to a depth h exceeding the length of the body L. In the course of the experiments, the model was accelerated from a state of rest and the model was braked from a state of motion at a steady rate. During the experiments, the acceleration of the movement of the towing carriage and body resistance force. The viscous drag coefficient c_{xV}^0 included in formula (1.25) generally depends on the Reynolds number, that is, $c_{xV}^0 = f(Re)$ (here $Re = V \times L/\nu$), where V is the model towing speed, L is the body length, and ν is the kinematic coefficient of viscosity of water. Fig. 1.14 shows the dependence of the viscosity coefficient of the model of the body on the Reynolds number obtained at $v_t' = 0$. For comparison, here are the results of two measurements of the coefficient of nonstationary force of resistance of the body c_{xn} during accelerated motion from a state of rest with approximately constant values of acceleration equal to respectively 0.8 and 7.0 m/c^2 depending on the instantaneous values of the Reynolds numbers. It is seen that the coefficient gradually decreases with increasing numbers Re, tending to a constant value when $Re > 4 \times 10^6$ (area of self-similarity). The crisis of resistance in the body of this form is not observed. The values of the coefficient c_{xn} differ significantly from the values of the coefficient c_{xV}^0 and the more, the greater the acceleration of motion. With an increase in the number of Re,

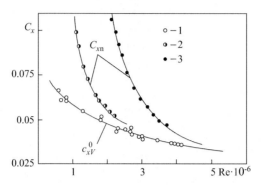

Figure 1.14 The dependence of the drag coefficient c_x on the Reynolds number for values v'_t of 0 (1), 0.8 (2) and 7.0 m/s² (3) [116,159].

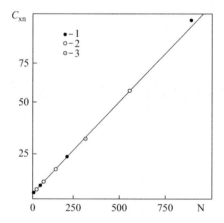

Figure 1.15 The dependence of the nonstationary drag coefficient c_{xn} at the beginning of the movement at $v'_t = 7$ m/second ($c_{xn} = (2\delta/\rho) \cdot K_\dot{O} \cdot N$: (1)–(3) are the mileage numbers [116,159].

the coefficient c_{xn} also decreases, but more intensively than the coefficient c^0_{xV}. For sufficiently large values of the number Re, the values of the coefficient c_{xn} more and more approach the values of the coefficient c^0_{xV}, and the later this happens, the greater the magnitude of the actual acceleration of motion.

The values of the coefficient c_{xn} differ significantly from the values of the coefficient c^0_{xV} and the more, the greater the acceleration of motion. With an increase in the number of Re, the coefficient c_{xn} also decreases, but more intensively than the coefficient c^0_{xV}. For sufficiently large values of the number Re, the values of the coefficient c_{xn} more and more approach the values of the coefficient c^0_{xV}, and the later this happens, the greater the magnitude of the actual acceleration of motion.

With increasing speed, the effect of acceleration on the drag force decreases. Fig. 1.15 shows the results of experimentally determining the coefficient c_{xn} in the form of the dependence $c_{xn} = f(N)$ at the beginning of the movement ($N \gg 1$). It can be seen that all experimental points are grouped around a common line, in contrast to the same results, which are presented in Fig. 1.14 as the dependence $c_{xn}=f(Re)$. Since at the beginning of the movement the body wrap is close to the potential one, then for the coefficient of attached body mass, $K=K_{\text{Ò}}'$ can be taken, and the coefficient of viscosity of the resistance c_{xv} can be neglected.

Then the coefficient of nonstationary resistance of the body will be determined by the expression:

$$c_{xn} \approx 2K_T \frac{\delta}{\beta} N. \tag{1.26}$$

Thus having an experimentally established dependence $c_{xn}=f(N)$ for the beginning of the movement, presented in this case in Fig. 1.15 straight line, and taking into account Eq. (1.26), it is possible to determine the coefficient of the attached body mass, depending on the shape of the body, from the expression $K_T \approx \frac{1}{2}\frac{\beta}{\delta}\frac{c_{xi}}{N}$, that for a given body will be $K_{\text{Ò}}' \approx 0.085$.

Fig. 1.16 shows the dependence $c_{xn}=f(N)$ for the case of developed accelerated body motion (small values of N) and for body motion in the braking mode ($n < 0$).

Here, as in the previous case (Fig. 1.15), it should be borne in mind that a full cycle of body movement (acceleration from rest, steady motion at constant speed and braking to a full stop) corresponds to a change in dimensionless acceleration N from $+\infty$ to $-\infty$. The values $N = \pm\infty$ correspond to the state of rest of the body before and after the completion of the movement. The range $0<N<\infty$ corresponds to the accelerated movement of the body, the value $N=0$ corresponds to steady-state movement of the body with a constant speed, and the range $-\infty<N<0$ corresponds to slow motion of the body. At the very beginning of the movement (Fig. 1.15) as $N \to +\infty$ also $c_{xn} \to \infty$. But then, with decreasing N, the coefficient c_{xn} acquires final values, and as $N > 0$ (Fig. 1.16), the values of c_{xn} continuously decrease. When $N=0$, which corresponds to the body moving at a constant speed, the curve c_{xn} comes to a point on the ordinate axis, corresponding to the resistance

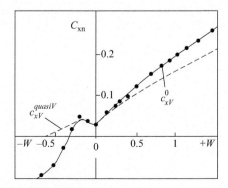

Figure 1.16 The dependence of the nonstationary resistance coefficient c_{xn} in the vicinity of the point $N = 0$ at $v_t' = 7.0$ m/s^2 [116,159].

coefficient of the body during steady motion. The position of this point on the ordinate axis depends on the number Re and can be determined for a given body by the dependence $c_{xV}^0 = f(Re)$ shown in Fig. 1.14.

In Fig. 1.16, besides the experimental curve $c_{xn} = f(N)$, the dashed curve of the quasistationary approximation is plotted, constructed in accordance with the Eq. (1.25), taking into account the $K_Ò \approx 0.085$ coefficient of the added mass determined for the beginning of movement and determined in accordance with Fig. 1.14 values of the viscous drag coefficient c_{xV}^0, which for small values of c_{xn} and N can no longer be neglected. It can be seen that the curves c_{xn} and in the region of small values of N do not coincide, although as $N \to 0$ they tend to a common value c_{xV}^0. This suggests that there is a range of N values in which the application of the quasistationary hypothesis when determining the nonstationary resistance coefficient can lead to significant errors.

To determine the order of these errors, the ratios $\xi(N) = c_{xn}/c_{xn}^{quasist.}$ were calculated between experimentally measured values of the nonstationary resistance coefficient of the body and its values in the quasistationary approximation determined by the Eq. (1.24) for different values of N. The graph of this relation is shown in Fig. 1.17, which shows that at $0.5 < \lg N < 1.5$, the use of the quasistationarity hypothesis in determining the nonstationary resistance coefficient of a given body leads to noticeable errors (more than 20% of the measured ranks).

Outside the specified range of N values, the use of the hypotheses of quasistationarity with respect to the body in question is permissible. The process of braking from the state of steady motion with constant speed is shown in Fig. 1.16. In this case, the dimensionless acceleration N goes into the region of negative values, and the inertial component of the drag coefficient changes sign. The curve $c_{xn} = f(N)$ intersects the axis of ordinates and then (with increasing negative values of N) slowly decreases in magnitude, remaining, however, in the region of positive values. The further character of the curve $c_{xn} = f(N)$ depends on the intensity of braking and, therefore on the balance between the viscous and inertial components of the drag coefficient.

If the whole process of braking to a full stop of the body proceeds under the condition that the inertial component of the drag coefficient is always less in absolute value than the viscosity component, then the curve $c_{xn} = f(N)$ will not cross the

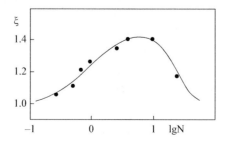

Figure 1.17 The graph of the function $\xi(N) = c_{xn}/c_{xn}^{quasist.}$ during the acceleration of the body from the state of rest [116,159].

abscissa axis, and as $N \to \infty$ the coefficient c_{xn} will tend to $+\infty$. But with more intensive braking, there can come a moment when the inertial and viscous components become equal in absolute value. If such a state could be fixed for a certain period of time, the body would move with zero hydrodynamic resistance. However, this mode of motion requires a complex law of change in acceleration over time and is difficult to achieve. Practically in such cases, the inertial component of the drag coefficient begins to exceed the viscous component in absolute value. The curve $c_{xn}=f(W)$ intersects the abscissa axis and goes into the region of negative values. At the same time, the body, instead of resistance from the fluid, begins to experience thrust. For this particular case of body motion (Fig. 1.16), this moment occurs at values of $N \approx -0.3$. If the above relation between the viscous and inertial components of the resistance coefficient is maintained until the body is completely stopped, then as $N \to -\infty$ also $c_{xn} \to -\infty$.

In Ref. [595], a justification of the experimental procedure was carried out in the study of periodic motions of bodies in a viscous incompressible fluid.

1.4 Energy aspects of the movement of bodies in the modes of acceleration and deceleration

The results of experimental investigations of the motion of bodies in the modes of acceleration and deceleration are given in Ref. [455]. Some crustaceans (Crustacea, Copepoda) have a substantially unsteady cycle of movement. In Ref. [456], it was determined that their instantaneous speed can reach 0.3 m/second at an average speed of 0.11 m/second, and the acceleration phase is approximately 30% of the time of the entire cycle of movement. Approximately the same results were obtained in experimental work in the study of swimming of swimming beetles [461]. It was also noted in Part I, Sections 1.2, 5.1 that the geometrical shape of hydrobionts changes in different phases of the cycle of movement. Such a change in shape, apparently, leads to a constant change in its hydrodynamic characteristics during the cycle of motion.

In Sections 1.2, 1.3 it is shown that fast acceleration alternating with slow deceleration may have advantages in terms of energy as compared with moving at a constant speed. Despite the fact that during acceleration, the inertial component of the hydrodynamic force increases, however, the displacement of the boundary layer on the curved surface of the body towards the stern reduces the width of the wake and thus reduces the viscous component of the hydrodynamic resistance [429, 456,481,485].

To determine the energy costs required for the organization of movement from alternating periods of acceleration and deceleration, experimental investigations were conducted in the high-speed hydrodynamic basin (Institute of Hydromechanics, National Academy of Sciences of Ukraine, Kyiv), towing tests on two models: the sphere and the streamlined body. The geometric dimensions of the models are given in Sections 1.1, 1.3. Photos of models are shown in Fig. 1.18. The arrow shows the direction of

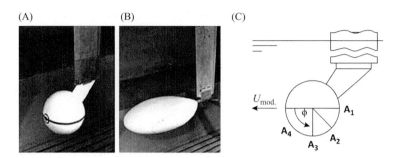

Figure 1.18 The appearance of the models of the sphere (A), the streamlined body (B), and the layout of the drainage holes on the model of the sphere (C) [116,159].

movement of the models. To visualize the flow around the models, a coloring solution was fed through the drainage holes to the boundary layer. Drainage holes were located in the longitudinal-vertical plane of the models. Drainage holes were located in the rear critical point of sphere A_1, and in the lower part of the sphere at points A_2, A_3, and A_4 at values of the central angle φ, respectively, 135°, 90°, and 45°. Drainage pipes were brought to these holes inside the sphere, which were brought out to the surface through a pylon where they were connected to a container for a solution of the coloring matter.

The nonstationary mode of movement of the test bodies was organized by imposing on the stationary translational movement of the towing carriage of the test basin of the longitudinal oscillatory motion. The oscillatory movement was organized using a special generator of longitudinal vibrations of large amplitude mounted on the towing carriage. Moreover, for each specific speed of movement of the towing cart, the mode of operation of the generator was chosen in such a way that at the initial moment of each oscillatory cycle the speed of the model relative to the basin was close to zero.

When determining the nonstationary hydrodynamic forces acting on the body, the quasistationary theory was most widely used (Sections 1.1–1.3). At the same time, the nonstationary coefficient of hydrodynamic resistance was obtained in the form (1.25), Section 1.3. For some bodies of canonical form, for example, for a sphere, one can obtain the drag coefficient in the quasistationary approximation in the form:

$$c_{xn}^{\text{quasist.}} = c_{xV}^0 + \frac{4}{3}KN, \tag{1.26}$$

where C_{xV}^0 is the viscous drag coefficient. The volume of the sphere is equal $V = \frac{1}{6}\pi D^3 = \delta D^3$, and the area of the midsection of the sphere is equal $F = \frac{1}{4}\pi D^3 = \beta D^3$ (here D=the diameter of the sphere). The test results will be analyzed in the future, taking into account these formulas.

As an integral criterion characterizing the effect of motion nonstationarity on the energy characteristics of the test bodies, a ratio was chosen between the work of A_N done by the body to overcome the nonstationary force of hydrodynamic resistance and the work of A_S performed by the same body, under the condition same distance for the same period of time, that is $K = A_N/A_S$.

At the same time, the work to overcome the unsteady force of hydrodynamic resistance A_N was calculated by the formula $A_N = \int_0^T R_N(t) \cdot x(t) dt$, where T is the period of motion, $R_N(t)$ is the instantaneous force of hydrodynamic resistance, $x(t)$ is the way traveled by the body during time t, and the work to overcome the stationary force of hydrodynamic resistance A_S was calculated using the formula $A_S = R_S S$ (here $S = \int_0^T x(t) dt =$ the way traveled by the body during T; $R_S = C_{xV}^0 (Re) \cdot \frac{\rho \cdot v_0^2}{2} F$ is the resistance force of the body as it moves at a stationary speed (here ρ is the mass density of water, v_0 is the stationary speed of movement, F is the cross-sectional area of the body).

Towing tests were carried out at four fixed values of the speed of movement of the towing car $v_0 = 0.433$; 0.714; 1.07; and 1.35 m/second, which made it possible to observe the patterns of behavior of the hydrodynamic resistance force of the sphere in the precrisis, crisis and super-crisis wrapping around. With a constant amplitude of $a = 0.375$ m, the period of longitudinal oscillations of the pylon in accordance with the speed of movement of the towing cart was $T = 5.88$; 3.52; 2.36; and 1.85 seconds. At the same time, the largest oscillation period corresponds to the lowest speed of the towing car. Fig. 1.19 shows a characteristic view of the kinematic parameters, hydrodynamic forces and the ratio of work when the sphere moves in acceleration and deceleration modes.

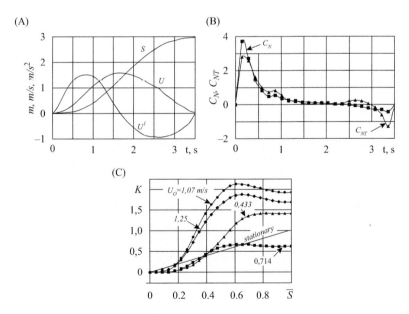

Figure 1.19 The characteristic form of kinematic parameters (A), hydrodynamic forces (B) and the relation of works (C) when the sphere moves in the mode of acceleration and deceleration [116,159].

At the beginning of the cycle, the velocity and acceleration of the sphere are close to zero (Fig. 1.19A). The dependence of the acceleration within the cycle is close to sinusoidal, and the speed of the sphere, obtained by imposing reciprocating motion on the uniform movement of the towing carriage, remains all the time positive. In this case, the way sphere traveled in one cycle is approximately ten times its diameter. Fig. 1.19B shows the characteristic behavior of the nonstationary hydrodynamic force coefficient during the oscillatory cycle:

$$C_N(t) = \frac{R_N(t)}{(\rho \cdot U^2(t))/2}. \tag{1.27}$$

It also shows the dependence of the nonstationary hydrodynamic force coefficient, calculated from the quasistationary theory $C_{NT}(t) = c_{xn}^{quasist.}$. It can be stated that the formula obtained in Ref. [455] fairly well reflects the dependence of the nonstationary hydrodynamic force coefficient in the greater part of the oscillatory cycle. The difference is observed only at the beginning and at the end of the cycle, when the nonstationarity of the N motion is greatest.

Dependencies of the ratio of work when the sphere moves in the acceleration and deceleration modes, shown in Fig. 1.19C, are of the greatest interest. It turned out that in the initial period of the oscillatory cycle, the motion of the sphere with acceleration in the energy plan has an advantage over the motion with a uniform speed (the mode of motion of the sphere with a uniform speed is plotted as a straight line on the graph). However, then this advantage is lost and only one of the nonstationary modes, in general, turned out to be advantageous in terms of energy in comparison with the regime of motion with a constant speed. This is the mode in which the towing car was moving at a speed of $U_0 = 0.714$ m/second, and the period of oscillatory movements was $T = 3.52$ seconds. A detailed analysis of the kinematic characteristics of this mode of motion of the sphere showed that this regime exactly corresponds to the "crisis of resistance" on the sphere by the Reynolds numbers. To clarify the reasons for this phenomenon, a flow was visualized around the sphere during its acceleration (Fig. 1.20A) and deceleration (Fig. 1.20B).

When the sphere is accelerated (the upper series of photographs, Fig. 1.20A), the separation point of the boundary layer moves to the stern. The wake formed behind the body narrows, breaks away and lags behind the body. This leads to a decrease in the viscous component of the hydrodynamic resistance of the body.

When the sphere is braked (the lower series of photos, Fig. 1.20B), an increase in the thickness of the boundary layer is observed. When the velocity of the sphere decreases, the vortex wake formed behind the body continues to move for some time at the previously reached speed. In the final phase of deceleration (photo B-3 in Fig. 1.20B), the vortex wake catches up with the body and "hits" on it. This can even lead to a change in the direction of the viscous component of the hydrodynamic resistance of the body.

The energy characteristics of a streamlined body with no "crisis of resistance" turned out to be significantly different. Fig. 1.21 shows the characteristic view of the kinematic parameters, hydrodynamic forces and the ratio of work during the

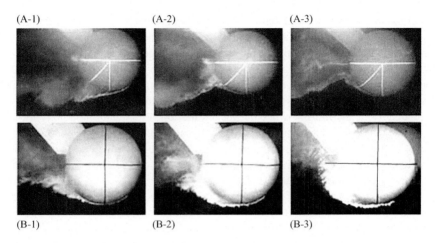

Figure 1.20 Displacement of the flow separation point during acceleration (A-i) and deceleration (B-i) of the sphere [116,159].

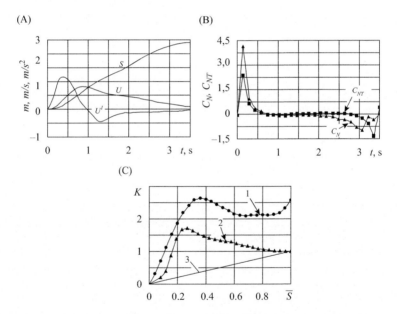

Figure 1.21 The characteristic form of the kinematic parameters (A), hydrodynamic forces (B) and the ratio of work (C) when the movement of the flow in acceleration and deceleration mode; (C): (1) slow acceleration; (2) fast acceleration; (3) stationary [116,159].

movement of a streamlined body in acceleration and deceleration modes. In the studied range of kinematic parameters, the stationary mode of motion in the energy plan was always more favorable than the mode of motion obtained by superimposing the reciprocating motion of the pylon to the stationary motion of the towing cart (Fig. 1.21C). By controlling the speed of movement of the towing car within the oscillatory cycle, the kinematic parameters of the movement of the streamlined body were organized, the characteristic view of which is shown in Fig. 1.21A. Distinctive feature of this driving mode is the fact that within the oscillatory cycle the acceleration period was significantly less than the braking period. Fig. 1.21B compares the dependence of the measured hydrodynamic resistance coefficient with the dependences obtained from the quasistationary theory. Fig. 1.21C shows the dependences of the ratio of work during the motion of a streamlined body moving in the mode of fast acceleration and slow deceleration. As can be seen from the graph, with such kinematic parameters of motion from an energetic point of view, the mode of motion with rapid acceleration and slow deceleration turned out to be the same with uniform motion.

From an energy point of view, a nonstationary mode of motion, consisting of alternating periods of acceleration and deceleration, may be preferable in comparison with the stationary mode of motion. The determining factor is the presence of a hydrodynamic resistance crisis in the body. The alternating periods of acceleration and deceleration are a powerful means of controlling the flow regime in the boundary layer and, accordingly, the crisis of hydrodynamic resistance. The effectiveness of this control is greatly enhanced by reducing the duration of acceleration and increasing the deceleration time (fast acceleration and slow deceleration).

V.V. Babenko proposed to check the possibility of controlling the vortex structures arising during the flow around a sphere with the help of a velocity head. For this purpose, through holes with a diameter of 0.07 m were made on the outer surface of the sphere along the longitudinal axis (Fig. 1.18A). V.V. Moroz performed visualization of the flow around vortex structures under various modes of motion (Fig. 1.22). In Fig. 1.22A, the visualization of the flow around a sphere during stationary motion was carried out in the same way as shown in Fig. 1.18C. In the precrisis mode (Fig. 1.22A, on the left two frames), the flow around a sphere is completely determined by external flow, and the flow through axial through holes on the sphere does not affect the structure and shape of the vortices.

In a supercritical flow around (Fig. 1.22A, on the right, the last frame), the jet flowing through the rear through hole on the sphere pushes the external flow back. In this case, the flow around becomes similar to the first frame of the precrisis flow (Fig. 1.22A on the left), but there are no vortex structures behind the sphere due to the sucking force of the axial vortex flowing through the axial hole. The energy of the axial vortex is proportional to the velocity of the incident flow. Such a flow pattern is similar to the flow around the aft part of the ship hull, when a suction force occurs during the operation of the screw, eliminating the flow separation on the hull in the vicinity of the screw.

To clarify the effect of through holes on the vortex structure that occurs when the sphere is flown around, the experiments, the results of which are shown in Fig. 1.22A, were repeated. In this case, the visualization was carried out only through one tube, installed along the axis of the central through hole in the back of the sphere. It can be seen that during the precrisis flow around the jet flowing through the axial hole in the sphere is completely absorbed and eroded by external flow around the sphere—the first frame on the left in Fig. 1.22B. With an increase in the velocity of

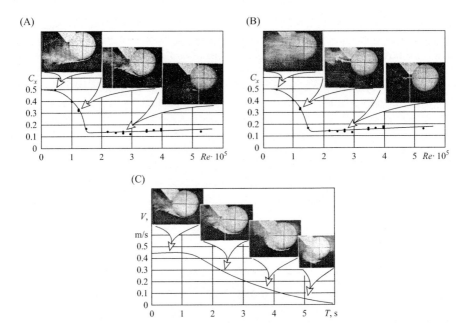

Figure 1.22 Control of vortex structures in the flow around a sphere in various modes of motion: (A) in the precrisis and super-crisis modes, (B) the same as (A), but visualization was carried out through an axial tube, (C) motion deceleration [116,159].

motion, the energy of the jet flowing through a pass-through hole in the sphere increases and stabilizes the vortex structures of the external flow of the sphere—the second frame from the left in Fig. 1.22B. In case of a supercritical flow (Fig. 1.22B, on the right, the last frame), the jet flowing through the rear a pass-through hole on the sphere fully stabilizes the vortex structures of the external flow around the sphere. Fig. 1.22C shows the structures of the vortices that occur when a sphere is flown around, depending on the flow velocity in the process of breaking the motion of the sphere (compare with Fig. 1.20B). It can be seen as the vortex structures in the wake of the sphere by inertia overtake the sphere and "lean on" it. The energy of the flow in the axial hole sharply decreases, with the partially vortex flow in the wake flowing outside the sphere in the opposite direction and increasing in thickness, and part of the vortex structure penetrates into the sphere through a pass-through hole and flows out through the front part of the hole against the flow.

1.5 Features of the accelerated movement in water of bodies of various shapes

In accordance with the idea and methodology of work [522], V.V. Babenko proposed to perform a similar experimental research, which A.G. Belousov performed at a modern technical level. In Ref. [164], the results of experimental investigations of rectilinear translationaerencel motion in water of bodies of various shapes that begin to move from a state of rest and reach, in some cases, a state of movement with a certain steady velocity V_{steady} are given. Investigations were carried out in a vertical hydrodynamic pipe with a height of 3.6 m and a diameter of 0.29 m (Fig. 1.23). The channel is made of viniplast pipe and has twenty transparent Plexiglas windows, arranged in pairs against each other and evenly along the entire length of the pipe.

The water level in the pipe remained constant. The volume of water in the pipe was 224 L. Inside the pipe, along its axis, a thin string with a diameter of 0.8×10^{-3} m made of noncorroding wire was stretched. The string with its lower end was fixedly fixed on the base of the pipe, and the upper—in a special device designed for its tension. This device was fixed on a platform mounted above the top edge of the pipe. The string served as a guide for the body moving in a pipe vertically downwards.

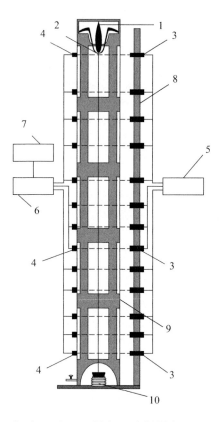

Figure 1.23 Hydrodynamic pipe scheme: (1) is model; (2) is temporary mobile support; (3) is illuminators; (4) is photodiode sensors; (5) is illuminators power supply; (6) is light-beam oscilloscope; (7) is source of standardized electromagnetic signals; (8) is rod for mounting illuminators; (9) is hydrodynamic pipe; and (10) is shock-absorber [52,455].

The tension of the string was chosen so as to ensure the movement of the body vertically. At the bottom of the pipe there was a trap, which quenched the energy of movement of the model and was part of the lifting device with which the model was removed from the pipe. The working section of the installation was considered the section of the pipe from the first sensor of the recording device, located 0.34 m below the water level, to the last, located at a height of 0.05 m above the trap.

Spatiotemporal characteristics of the motion of the models were recorded in 12 planes perpendicular to the pipe axis. In the registration planes, an illuminator emitting a flat beam of light (a light knife) was placed along the diameter line of the opposite transparent pipe walls and the sensor was a receiver. The optical system of the illuminator focused the beam on the photosensitive surface of the sensor. The sensor served as a semiconductor photodiode. The model moving in the pipe alternately blocked the light beams, and the sensors alternately recorded an almost instantaneous decrease in the intensity of the light flux. The sensors worked in the source mode of the Electro-Motive Force (EMF) and individually connected to the galvanometers of the light-beam oscilloscope. The overlap by the moving model of the light beam entailed a change in the deviation of the galvanometer frame by an angle proportional to the value of the EMF of the sensor. The magnitude of the deflection angle of the galvanometer frame was continuously recorded as a function of time on the oscilloscope. The time of overlapping of the beam of light was equal to the time during which the deflection angle of the galvanometer frame was minimal. The results of the subsequent decoding of the records determined the temporal characteristics of the movement of the models. The time constant of the photodiode is 10^{-5} seconds, and the EMF value of the sensor changed from its maximum value to its minimum value on average in 2×10^{-3} seconds. The diameter of the transparent window of the photodiode is 4×10^{-3} m.

The results recorded by the registration system were deciphered using calibration signals, which were continuously recorded on the oscilloscope with independent galvanometers. These signals were rectangular electromagnetic pulses with a frequency of 100 and 1000 Hz. The maximum absolute error of time measurement was 5×10^{-4} seconds, and the error of the results at speeds of models that reached 4 m/second at the end of the working section did not exceed 1.5%. The maximum error of the results of the obtained spatial-temporal characteristics of the movement, taking into account the final conditions, did not exceed 3%. As objects of research, flat rectangular plates and models were used, the shape of which is described by the equations of rotation bodies or their combination. Schematic illustrations of models representing the paraboloid of rotation $y^2 = a - bx$, the ellipsoid of rotation $\frac{x^2}{a^2} + \frac{y^2}{b^2} = 1$ and the combination of two paraboloid of rotation, one part of which is described by the equation $y^2 = \frac{R^2}{H_1}x$, and the second $y^2 = \frac{R^2}{H_2}x$, are shown in Fig. 1.24. The experiments also used models of a cone and a straight circular cylinder with flat ends. The nasal and tail parts of the cylinder and the end of the cone could be replaced with hemispheric extremities or bodies of rotation, since all models had a standardized connection unit. In turn, the models of rotation bodies could be articulated with any bow or tail tips, depending on the specific task.

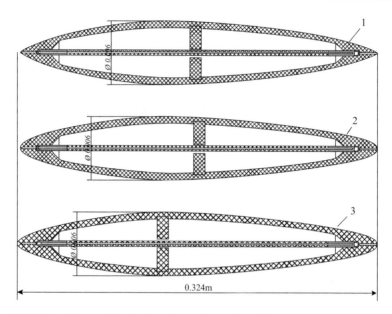

Figure 1.24 The scheme of the device of experimental models: (1) is paraboloid of rotation, (2) is ellipsoid of rotation; (3) is combination of two paraboloids of rotation [52,455].

Models were made of plexiglass. The length of the models was 0.324 m, and the diameter was 0.06 m. The weight of the models in the water was their driving force. Its invariance at the time of measurement was provided by careful sealing of the internal cavities of the models. The magnitude of the driving force of the models was determined by hydrostatic weighing and was monitored in the assembled and sealed form at the end of the measurements. Using hydrostatic weighing, the volume of models was determined. The change in the mass of the models was made by loading their inner cavity with lead shot. The model was cohesive with a metal tube inside the model. Threaded connection of the tube and the model allowed sealing the model for an arbitrarily long period of time. The holes drilled along the axis of the nose and tail extremities of the model, coincided with the axis of the inner tube-tie. Through these holes, and hence through the entire model, a wire string was passed along the longitudinal axis of the hydrodynamic channel. This provided a vertical axial movement of the model in the pipe. To obtain the reference characteristics of the movement, rectangular flat steel and aluminum plates measuring 0.32×0.10 and 1.5×10^{-3} m thick were used. The surface of the plates was either chromed to obtain a hydraulically smooth surface or painted with mineral synthetic paints, depending on the research tasks. The plates were attached to the string guide with special earrings that have long inner circular through holes. A string guide was passed through these holes to stabilize the movement of the plate along the axis of the hydrodynamic channel. The length and diameter of the holes in the earrings—mounts were chosen based on their role of sliding bearings. The water temperature in the pipe was measured before and after testing. To change the mode of water flow in the boundary layer of models and plates, thin wire turbulizers with a diameter of 0.5×10^{-3} m were installed on their nasal extremities. The turbulizer was installed in the transverse direction to its surface at a distance of 8×10^{-1} m from the leading edge.

The dependencies of the path traveled by the plates in accelerated motion are presented in Fig. 1.25A, and the dependences of the speed of movement of these plates, obtained by differentiating the laws of the path of their movement, in Fig. 1.25B. The graphs clearly show that in accelerated motion, the state of the boundary layer significantly affects the characteristics of the motion of objects. At the initial sites of the trajectory, differences in the character of the motion are not significant. This phenomenon may be due to the developing process of formation of the boundary layer, since in bodies that begin to move from a state of rest, it is completely formed on a section of the path comparable to their size.

Accelerated movements of the combined model, sphere and flat plates were also investigated (Fig. 1.26). The nose part of the combined model is made in the form of a hemisphere with a diameter of 60 mm; its middle part is made in the form of a segment of a straight circular cylinder of the same diameter, and the tail part in the form of a paraboloid of rotation.

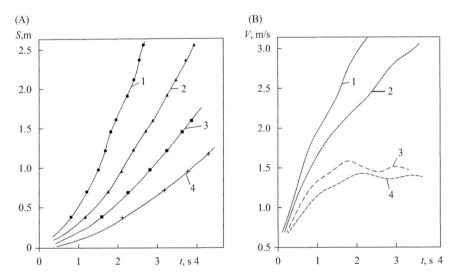

Figure 1.25 Patterns of ways (A) and speeds (B) of flat plates: (1, 2) is steel plate; (3, 4) is dural plate; (1, 3) is without turbulizer; and (2, 4) is with turbulizer [52, 455].

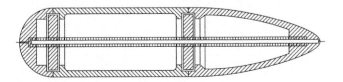

Figure 1.26 The scheme of the combined model [52,455].

The length of the model was 261 mm. The ball was 60 mm in diameter, and rectangular flat plates had dimensions of 320 × 100 mm and were cut from a metal (steel, duralumin, or aluminum) sheet in order to change their weight characteristics with the same geometry of the plates.

The weight characteristics of the combined models and the ball were changed by filling their internal waterproof cavities with lead shot. All objects of study began to move from a state of rest in the upper part of the pipe and moved under the action of a constant magnitude force equal to the difference between the weight of the object in the air and the Archimedean buoyant force in water. The magnitude of the driving force of objects was determined by hydrostatic weighing. The values of their weight characteristics are given in Table 1.1.

The driving force of the combined model and the ball was set so that its value takes the values $0.5 \times P_0$; P_0 and $1.5 \times P_0$ for the combined model and $0.5 \times P_0$; P_0; $1.5 \times P_0$ and $2 \times P_0$ for the sphere, where P_0 is their weight in the air with zero buoyancy. All study objects had a hydraulically smooth surface.

According to the results of processing the values of the space-time characteristics of the motion of each of the objects, graphs of their way, speed and acceleration were constructed as a function of time. Curves showing the nature of the change in their speed are presented in Fig. 1.27. Curves (1–3) illustrate the change in speed of the combined model for the values of the driving force $0.5 \cdot P_0$; P_0 and $1.5 \cdot P_0$, respectively. Curves (4)–(7) illustrate the change in the speed of the ball for values $0.5 \times P_0$; P_0; $1.5 \times P_0$ and $2 \times P_0$, respectively. Curves (8)–(10) are the changes in the velocities of flat plates, for which the values of masses and driving forces are given in Table 1.1 under the corresponding numbers. The greater the mass of the model (Table 1.1), the greater the speed of movement of the model (Fig. 1.27). It can be seen from the graph that the combined model does not move with the steady-state speed V_{steady} in any of the experiments performed. In this case, its maximum value of the velocity value each time at the end of the path near the bottom of the pipe. The Reynolds numbers, taken along the model length and the values of

Table 1.1 The weight of the investigated models.

Model	Weight in the air (H)	Weight in the water (H)
Combined	8.102	2.714
Combined	10.80	5.38
Combined	13.43	8.051
Sphere	1.673	0.568
Sphere	2.229	1.123
Sphere	2.769	1.664
Sphere	3.348	2.242
Flat plate	0.893	0.572
Flat plate	1.774	1.140
Flat plate	4.028	3.516

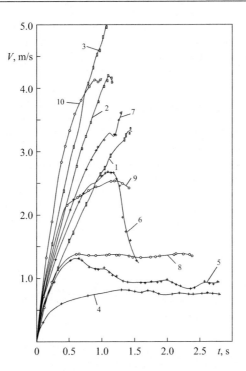

Figure 1.27 Dependencies of the speeds of the models on time: (1), (2), (3) is combined body of rotation; (4), (5), (6), (7) is spheres; (8), (9), (10) is flat plates [52,455].

the instantaneous velocity at which the monotonically varying acceleration acquires an oscillatory nature, take the values $Re_1 = 4.5 \times 10^5$, $Re_2 = 5.6 \times 10^5$ and $Re_3 = 4.8 \times 10^6$.

Fig. 1.28 presents the curves of the dependence of acceleration on time for the combined model (the notation is the same as in Fig. 1.27). It can be seen from the figure that as the magnitude of the driving force increases, the frequency and amplitude of oscillation of the acceleration values increase. Moreover, the nature of the occurrence of these oscillations is fundamentally does not change and in no case does the magnitude of the acceleration take on negative values. In Ref. [394], it was shown that for bodies of similar shape about 90% of the body surface has a continuous flow around.

For the sphere, the picture of the change in its speed is completely different. From the graph shown in Fig. 1.28, it can be seen that curves *4* and *5*, corresponding to the values of the driving force $0.5 \times P_0$ and P_0, respectively, demonstrate the state of the moving ball with the steady-state speed of motion V_{steady} with small fluctuations of its values. The Reynolds numbers at which monotonously varying acceleration acquires an oscillatory character are equal to $Re_4 = 4.5 \times 10^5$ and $Re_5 = 7.2 \times 10^5$, respectively, and correspond to the region of the crisis of resistance [521]. The nature of the movement of the ball changes dramatically with increasing driving force to values of $1.5 \times P_0$ and $2 \times P_0$: curves (6), (7), respectively. Its

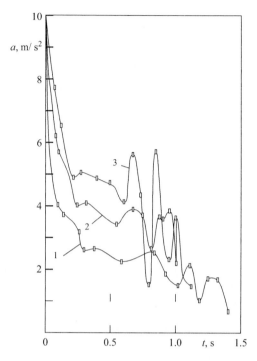

Figure 1.28 The dependence of the acceleration of the combined body of rotation on time [52,455]. Curves (1)–(3) are the same as in Fig. 7.27.

speed quickly increases to a certain maximum value, and then drops sharply, approaching the value of V_{steady}. The corresponding Reynolds numbers are $Re_6 = 1.2 \times 10^6$ and $Re_7 = 1.3 \times 10^6$, that is, in this case, there is a supercritical flow around. A feature of the acceleration graphs of the ball is the fact that in all cases not only positive acceleration values are observed, but also negative (Fig. 1.29). Curves (8)–(10) in Fig. 1.27 (plates) demonstrate the possibility of achieving a state with a steady speed V_{steady}, but with different features: unlike a ball, the plate at the initial stage of movement demonstrates a monotonic change in speed and acceleration to some characteristic only for this plates, values, and only then the oscillatory nature of changes in speed and acceleration begins to appear (Figs. 1.27, 1.30). Since the lower limit of the critical Reynolds number for an ideally smooth plate is $Re_{cr} = 3.2 \times 10^5$ [521], and for these plates the acceleration becomes oscillatory with the values $Re_8 = 5.4 \times 10^5$, $Re_9 = 8.8 \times 10^5$ and $Re_{10} = 1.2 \times 10^6$ accordingly, in these cases, the flow around the surface of the plates is mixed.

The results obtained indicate that for all models at a certain stage of the movement, a monotonic change in time of speed and acceleration acquired an oscillatory character, which is not reflected in [552]. There is not only an increase in the frequency of oscillations of the acceleration values, but also an increase in their amplitudes. It is clearly seen that the nature of the change in speed and acceleration, for

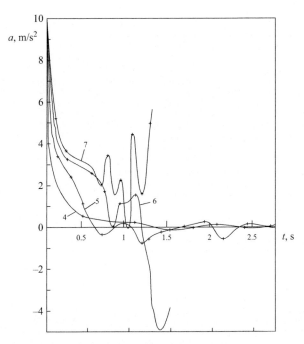

Figure 1.29 The dependence of the acceleration of the ball from time to time. Curves (4)−(7) are the same as in Fig. 1.27 [52,455].

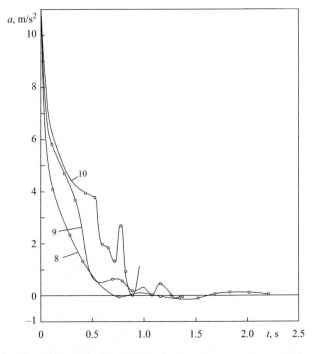

Figure 1.30 The dependence of the acceleration of a flat plate on time [52,455]. Curves (8)−(10) are the same as in Fig. 7.27.

example, a ball does not repeat the nature of the change in speed and acceleration of the combined body, while neither one nor the other repeats the character of the change in velocity and acceleration of a flat plate. If we assume that the oscillatory nature of changes in the characteristics of motion of models is associated with the process of vortex formation and the formation of a wake, then we can conclude that each moving body has its own, different from any other, mode of transition from one state with a steady speed $V_{\text{steady}1}$ of movement to another that has the speed $V_{\text{steady}2}$, and the body's drag coefficient C_x is determined not only by the Reynolds Re, Froude Fr, and dimensionless acceleration criteria N_w, but also by Strouhal Sh [259,526].

Detailed knowledge of such features will make it possible to predict the hydrodynamic properties of natural bodies and technical devices, as well as confirm that the prehistory of the flow at the stage of transition from the state from $V_{\text{steady}1}$ to the state from $V_{\text{steady}2}$ has a decisive influence on the flow around the body.

Experimental studies of nonstationary motion of the ball were performed at much lower speeds. In the preceding paragraphs, the characteristics and structure of the wake behind various bodies are studied for stationary flow and at relatively small Reynolds numbers. In these studies, the motion of various bodies was simulated under laminar and turbulent boundary layers. The different regime of the boundary layer determines the nature of the formation of a wake behind the bodies. Therefore specific characteristics of speed and acceleration were obtained during the unsteady motion of the bodies studied. In addition, for the first time, results were obtained for various well-streamlined bodies.

Modeling of a waving fin mover

2.1 Experimental investigation of waving wing

Theoretical investigations of waving fin mover were performed in Refs. [219–222,266–268,293,294,377,379,381,413,418,419,481,482,485,511,514,616]. The features of determining the effectiveness of the propulsion complex of hydrobionts were studied in Refs. [505,506].

Investigations by German scientists are briefly presented in Ref. [298]. It is shown that the body of anesthetized and dead fish when towed in a test basin has a resistance of almost five times more than a rigid model with the same Reynolds numbers. This is due to the self-oscillation mode with large amplitude of the elastic, but dead, long fish body, in the absence of a stabilizing movement of the active action of the fins in live fish. The laws of vortex formation behind the caudal fin of nonstationary swimming marine animals and for rigid models are investigated. Filming revealed that, depending on the modes of movement of marine animals, the following three systems of vortex formation in the wake is inherent. With the accelerated movement of the fish, the so-called active impact of the caudal fin, behind the body a propulsive vortex track is formed with two rows of vortices arranged in a staggered pattern. Between these rows there is a wave-like wake stream in which the liquid is thrown back. With a uniform movement of the fish, a "neutral" strike of the caudal fin, the vortex are located behind the fish's body not in a staggered manner, but alternate with a reciprocal rotation in the same longitudinal row, there is no pronounced wake. With the slow motion of the fish, the so-called passive strike of the caudal fin, a two-row vortex track and a backward wake are formed, in which the liquid flows in the direction of the fish movement. These studies were carried out at low Reynolds numbers; nevertheless, they clearly show the close relationship between thrust and resistance forces with various vortex formation systems. These three modes of vortex formation in the wake, characteristic for swimming of marine animals, were artificially reproduced on an experimental setup with a small rigid model of a waving fin with a profile length of 0.22 m and a thickness of 0.08 m with Reynolds numbers of the order of $Re \approx 10^4$.

The results of tests on the influence of the angle and speed of the fin rearrangement during torsional fluctuations on the forces acting on the fin are also given. If the angle of the transfer is increased slowly, then the lift coefficient C_y of the fin increases almost linearly to its critical maximum value, preceding the separation of the boundary layer (BL) ($\alpha_{кр} \approx 12$ degrees), and the flow can be considered as stationary. If the angle of change is changed quickly, then the continuous flow around the fin and the linear change of C_y can drag to a much larger critical angle ($\alpha_{cr} \approx 40$ degrees), while the nonstationary value of C_y increases, for example, by three to four times. This is due to the fact that due to the fast rerun of the fin, conditions for the separation of the BL are not created.

Experimental investigations of the hydrodynamic characteristics of waving wings were conducted at TsAGI named N.E. Zhukovsky [272]. It is shown that waving wings can simultaneously perform the functions of both a propulsor and a bearing surface with a sufficiently high efficiency—about $\eta = 0.6-0.7$ and higher, which is close to the efficiency of a conventional propeller. In the experiment, a waving wing rectangular in plan with a chord b moved with a constant horizontal velocity U_0, simultaneously performing harmonic translational vertical oscillations with amplitude α and torsional angular oscillations around the horizontal axis with amplitude β, with an angular oscillation frequency $\omega = 2\pi n$. The work presents graphs of the values of the propulsion coefficients C_T, C_N, and η as a function of the characteristic kinematic parameters—the relative velocity of the oscillating wing λ_p and the Strouhal number Sh_ω:

$$\lambda_p = U_0/\alpha\omega, \text{Sh}_\omega = \omega b/U_0$$

The relative velocity λ_p characterizes the maximum inclination of the trajectory of the axis of rotation of the wing to the direction of translational motion. This parameter is similar to the relative protrusion of the propeller and is useful in the case of a flapping wing with two degrees of freedom (with translational and angular oscillations) in that the experimental points of the propulsive coefficients depending on λ_p fall on one curve for different values of U_0, relative amplitude $\bar{a} = a/b$ and circular frequency of oscillations ω.

Tests of waving wings were carried out in a TsAGI towing tank on a specially manufactured oscillatory unit placed on a towing carriage. Using the unit, the wing was informed of vertical oscillations with an amplitude of $a = $ 15; 30; 50; 70 mm with an average angle of attack $\alpha_0 = $ 0–5 degrees and angular oscillations with amplitude $\beta \leq 20$ degrees with any given angle of phase δ and frequency $n \leq$ 3.5 Hz. Two models of significantly different types of wings were tested (with NACA-0015 and TsAGI-KV-I-7 profiles) with the same value of the relative span $\lambda = l/b = 4$, rectangular in plan, with rounded lateral edges. The stationary hydrodynamic characteristics of both wings are shown in Fig. 2.1 with the same relative immersion of the wings below the surface of the water $\bar{H} = H/b = 1.65$. The NACA-0015 profile with a chord $b = 0.175$ m, relative thickness $\bar{c} = 15\%$, optimum in aerodynamic quality, $K = C_y/C_x$ and, as relatively thick, is preferable for an oscillating wing, which operates in a wide range of angles of attack. The TsAGI-KV-1–7 profile with a chord $b = 0.15$ m relative thickness $\bar{c} = 7\%$ was chosen for reasons of maintaining a stable position when changing the angles of attack and corresponds to the small thickness of the wings usually used for hydrofoil vessels. The axis of rotation of the wing is located on a quarter of the chord from the leading edge of the wing (approximately in the center of pressure of the wing).

The best propulsive qualities of thrust and efficiency are achieved in the case of waving wing oscillations with two degrees of freedom, when the angular oscillations are ahead of the translational phase $\delta = \pi/2$. Then the horizontal projection of the

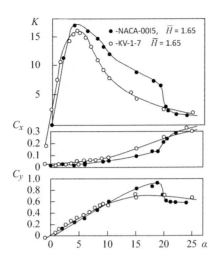

Figure 2.1 Stationary hydrodynamic characteristics of models of two rectangular in plan wings relative span $\lambda = 4$ with profiles NACA-0015 15% thick, with a chord $b = 0.175$ m and TsAGI-KV-I-7 with a thickness of 7%, with a chord $b = 0.15$ m, tested as waving wings in a towing tank with relative immersion $\overline{H} = H/b = 1.65$ [272].

instantaneous lifting force Y_i during the entire period of oscillations is directed toward the wing movement, that the resistance force. is, traction that most of the time exceeds the horizontal projection of the instantaneous force X_i: the resistance force.

Another component of the thrust force is due to the additional suction force arising on the nose of the waving wing, as shown by experiments with the waving wing horizontally moving at a constant speed, performing only one angular oscillation around a certain axis. The average angle of attack α_0 significantly affects the propulsive characteristics of the waving wing; with increasing α_0, the thrust force and efficiency decrease due to the increase in wing resistance. The increase in the bearing properties of this wing, that is, the average for the period of vertical force, leads to a decrease in its propulsive characteristics—thrust and efficiency. Of the two tested waving wings, the thicker wing with the NACA-0015 profile has significantly higher propulsive qualities compared to the thin wing of the TsAGI-KV-I-7. In cetaceans, waving fin movers are relatively thick (Part I, Section 2.4).

The main dimensionless kinematic parameters that determine the propulsive characteristics of a wing oscillating with two degrees of freedom are: relative speed of motion λ_p, amplitude of angular oscillations β, phase shift between angular and vertical oscillations δ. When changing the relative amplitude of the vertical translational oscillations in the range $\overline{a} = 0.1 - 0.47$, in the range of relative speed $\lambda_p = 2-7$, the action curves of the waving wing $C_T(\lambda_p)$, $C_N(\lambda_p)$, and $\eta(\lambda_p)$ are almost independent of the relative amplitude $\overline{a}$. Some test results of both models are shown in Fig. 2.2. One can see a significant difference in the characteristics of both tested wings.

With the aim of substantiating the method of approximate estimation of propulsive characteristics of a waving wing according to the stationarity hypothesis,

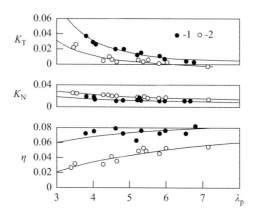

Figure 2.2 The propulsive characteristics of two waving wings with joint reciprocating vertical and torsional oscillations: (1) NACA-0015, $\bar{a} = 0.285$; (2) TsAGI-KV-1–7, $\bar{a} = 0.467$; in both cases, the same values: $\alpha_0 = 3.7$ degrees, $\beta = 3$ degrees, $\delta = \pi/2$, $\bar{H} = 1.65$ [272].

experiments were carried out to measure the instantaneous values of lift and resistance of the oscillating wing. The linear character of the dependence $C_{yi}(\alpha_i^0)$ is preserved in a much wider range of variation of the instantaneous angle of attack $\alpha_i = 0 - 20$ degrees than with a stationary flow around (Fig. 2.3). The instantaneous values of the resistance coefficient C_{xi}, corresponding to the maximum instantaneous angles of attack α_i, turn out to be close to stationary values only at small instantaneous angles of attack $\alpha_i < 8$ degrees. With an increase of $\alpha_i > 8$ degrees, the nonstationary resistance of the wing significantly increases in comparison with the stationary resistance. To assess the hydrodynamic interaction of two wings in the TsAGI experimental towing tank, experiments were carried out in which another stationary wing of similar dimensions was installed behind the waving wing KV-1-7 at different distances, with an angle of attack corresponding to a zero value of its lift. Experiments have shown that in this case, the second wing does not have a noticeable effect on the hydrodynamic characteristics of the first flapping wing, even at small distances between them.

The review article [508] carried out a thorough analysis of the main investigations of the formation of the vortex wake of flying and floating vertebrates. There are pictures of visualization of vortices in the wake and diagrams of the vortex track in the wake of living objects, obtained in the works of numerous authors.

2.2 Approximate method for calculating of the hydrodynamic characteristics of an oscillating wing

In Refs. [381,403,419,616] theoretical models of swimming of aquatic organisms with the help of a fin mover are given. In Ref. [514], a method for calculating the

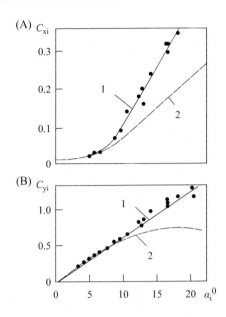

Figure 2.3 The results of experiments on the measurement of instantaneous nonstationary values (1) of the coefficients of resistance C_x (A) and the lifting force C_y (B) the waving wing with the TsAGI-KV-1-7 profile for given: $\bar{a} = 0.467$; $\alpha_0 = 3.7$ degrees, $\beta = 3$ degrees, $\delta = \pi/2$, $\bar{H} = 1.65$, in comparison with stationary tests (2) [272].

propulsive hydrodynamic characteristics of an oscillating wing of finite span was considered. The results of the investigations performed are compared with the data of TsAGI experiments given in Section 2.1. When describing the mechanism of movement of fast-swimming fish and dolphins, creating propulsive vehicles with flapping wings [266], the propulsive characteristics of the oscillating wings of finite span are of practical interest. Investigation of these characteristics in the framework of the ideal fluid model in a linear formulation is given in Refs. [162,163,266]. In Ref. [163], the problem of determining the hydrodynamic characteristics of a nonstationary flow of a viscous fluid around a wing profile was considered. A method for calculating the propulsive hydrodynamic characteristics of an oscillating wing of finite span based on the use of a viscous fluid model with a quasistationary approach, the Prandtl carrier line scheme, and flat section hypotheses are presented. The integral-differential Prandtl equation is solved by the method of spline functions [422]. A computer program was compiled, with the help of which a numerical experiment was carried out for specific wings of finite span.

Let the wing of finite span move with constant speed V_o and perform vertical oscillations according to the law $y = a \cdot \cos(\omega t)$ and angular oscillations around a certain selected axis according to the law $\beta_i = \beta \cdot \cos(\omega t + \delta)$, where t is time, β is the amplitude of angular fluctuations; a is the amplitude of vertical oscillations; δ is phase angle of advance by angular oscillations of vertical oscillations; ω is the

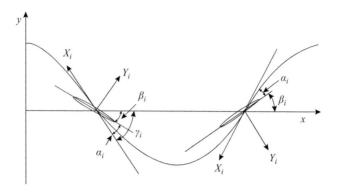

Figure 2.4 Movement of the oscillating wing [514].

angular velocity of oscillations (Fig. 2.4). Oscillations occur relative to the constant angle α_0. The instantaneous angle of attack is to be

$$\alpha_i = \alpha_0 + \beta \cos(\omega t + \delta) + \mathrm{arctg}\left(\frac{\alpha\omega}{V_0}\sin(\omega t)\right)$$

To be able to use the quasistationary approach, it is necessary that the step $\lambda_{\mathrm{st.}} \geq 4$ and the Strouhal number Sh <1. The average integral characteristics: thrust, power, efficiency was chosen in the same way as in Ref. [272].

The viscous fluid model allows us to obtain the value of the full drag of the wing and clarify the value of the lift force. The full drag of the wing is the sum of the inductive and profile resistance:

$$C_x = \frac{1}{l}\int_{-1/2}^{1/2}\left(C_{xp}(x) + C_{xi}(x)\right)dx,$$

where C_{xp} is the profile resistance, C_{xi} is the inductive resistance due to the last span of the wing; l is the length of the wing.

The influence of viscosity on the external flow can be taken into account using the method proposed in Ref. [420], whereby the displacement thickness is increased on the profile and the zero current line: for the new profile, the angle of attack decreases and the circulation on it will be less than the circulation on the main profile in the absence of a BL. Thus, the circulation can be written in the form:

$$G = k_1 k_2 G_0,$$

where G_0 is the circulation in an ideal fluid; k_1, k_2 are coefficients calculated using tabular functions [236]. Viscous fluid model allows determining the profile resistance

$$C_{xp} = 2\overline{V}_{(1)}^{3.2}\delta_{(1)}^{++},$$

where $\overline{V}_{(1)}$, $\delta_{(1)}^{++}$ is the external flow velocity and the momentum thickness at a given edge, calculated using tabulated functions [528,529]. When tabulating functions in Refs. [529,530], approximate formulas for the characteristics of the BL on the profile obtained in Ref. [529] are used, and for calculating the velocity of a potential flow around an arbitrary profile, an approximate formula for six-parameter profiles is used [503]. The quantities of interest to us depend on small parameters related to the geometry of the profile and the angle of attack:

$$\bar{\delta} = \gamma/\pi; \lambda_0 = \sqrt{\frac{\bar{r}_0}{2}};$$

$$\lambda_1 = \frac{1}{32}\frac{\bar{c}(7-10\bar{x}_c)}{\bar{x}_c^{3/2}(1-\bar{x}_c)^{5/2}} - \frac{\pi}{32}\bar{\delta}\frac{5-6\bar{x}_c}{\bar{x}_c^{1/2}(1-\bar{x}_c)^{3/2}} - \frac{3}{4}\frac{\lambda_0}{\bar{x}_c},$$

$$\lambda_2 = \frac{3}{128}\frac{\bar{c}(2\bar{x}_c-1)}{\bar{x}_c^{7/2}(1-\bar{x}_c)^{5/2}} - \frac{\pi}{128}\bar{\delta}\frac{1-2\bar{x}_c}{\bar{x}_c^{5/2}(1-\bar{x}_c)^{3/2}} + \frac{1}{16}\frac{\lambda_0}{\bar{x}_c^3},$$

$$\mu_0 = -\frac{1}{4}\frac{\bar{h}(1-6\bar{x}_h+6\bar{x}_h^2)}{\bar{x}^2{}_h(1-\bar{x}_h)^2},$$

$$\mu_1 = -\frac{1}{4}\frac{\bar{h}(1-\bar{x}_h)}{\bar{x}^2{}_h(1-\bar{x}_h)^2},$$

$$\nu = -\frac{(2-\bar{\delta})C_y}{4\pi}l^{-(\lambda_0+0.5\lambda_1+0.625\lambda_2)} \simeq \alpha + \mu_0 + \frac{1}{2}\mu_1,$$

where γ is the angle between the tangents in the trailing edge; $\bar{r}_0$ is relative radius of curvature at the leading edge; $\bar{c}$ is relative profile thickness; $\bar{h}$ is relative concavity of the profile; $\bar{x}_c, \bar{x}_h$ is abscissas corresponding to the maximum $\bar{c}, \bar{h}$; α is the angle of attack (in radians); and C_y is lift coefficient. These results in Ref. [529] were used to calculate the characteristics of the BL (laminar and turbulent), which are expressed by approximate formulas:

$$a = a_0 + \delta a_1 + \lambda_0 a_2 + \lambda_1 a_3 + \lambda_2 a_4 + \mu_0 a_5 + \mu_1 a_6 + \nu a_7.$$

Functions a_i are calculated and tabulated in Ref. [529]. In this work, we calculated the thrust coefficient, the power expended ratio, and the efficiency for rectangular wings with an elongation $\lambda = 4$. The axis of angular oscillations was located at a distance of one quarter of the chord from the front edge of the wing. The following profiles are used:

NACA $- 15, \lambda = 4, \alpha_0 = 3.7°, \beta = 3°, \lambda_{st.} = 3 \div 8, \bar{c} = 15\%, \bar{a} = 0.285,$

TsAGI-KV, $\lambda = 4, \alpha_0 = 3.7°, \beta = 3°, \lambda_{st.} = 3 \div 8, \bar{c} = 7\%, \bar{a} = 0.467.$

The results of these calculations (solid lines) are compared with experimental data [272] (Section 2.1, Fig. 2.2). As can be seen from Fig. 2.5, the proposed method gives good results when $\lambda_{st.} \geq 4$ and has, in addition, the following

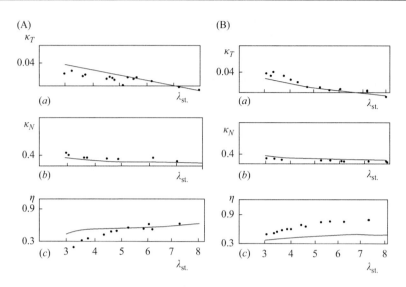

Figure 2.5 Coefficients K_T (a), power consumption K_N (b) and the efficiency η (c) of the wings TsAGI-KV (A) and NACA-15 (B). For TsAGI-KV $\bar{a} = 0.467 = 0.467$, $\beta = 3$ degrees, $\delta = 90$ degrees, $\alpha_0 = 3.7$ degrees, $\bar{c} = 7\%$. For NACA-15 $\bar{a} = 0.285$, $\beta = 3$ degrees, $\delta = 90$ degrees, $\alpha_0 = 3.7$ degrees, $\bar{c} = 15\%$ [514].

advantages: (1) wing profiles can have an arbitrary geometry; (2) the model of a viscous fluid makes it possible to obtain values of resistance close to reality and refined values of lifting force; and (3) the program compiled for this method allows you to quickly obtain comprehensive information about the characteristics of the wing. The main disadvantage of the discussed method is determined by the limited use quasistationary approach.

2.3 Experimental investigation of the oscillating wing thrust

Some results of an experimental investigation of the thrust generated by one wing and a system of two wings performing torsional oscillations are given in Ref. [267]. The main attention is paid to the investigation of the dependence of the thrust force on the amplitude and frequency of oscillations, as well as the phase shift between oscillations of neighboring wings. The law of torsional oscillations is selected taking into account the creation of thrust at the initial stage of the movement of the wing from a state of rest. The experimental installation was designed and manufactured in accordance with the conditions of the experiment, which was carried out in the hydrodynamic channel of the Novosibirsk Institute of Water Transport Engineers. Under the hull of a model boat with a flat bottom 1.5×0.6 m; two symmetrical wings with a chord

$b = 0.25$ m and a length $l = 0.57$ m were installed. An electric motor with a power of 0.6 kW was installed on the model case; reduction gear with interchangeable gears, making it possible to obtain the oscillation frequency of the wings $f = 3.1$; 5.5 and 16 Hz; two rods connecting the wings to the respective gears of the gearbox; two side washers, between which wings are placed, which can perform torsional oscillations relative to the axis of rotation, located at a distance of a quarter of the chord from the leading edge of the wing. Changing the amplitude of the oscillations of the wings and the phase shift between them is made by installing the fingers of the cranks on different radii of the gear wheels of the gearbox. The dependence of the thrust force on the amplitude and oscillation frequency of one wing in the mode of mooring the vessel when the flow velocity $V = 0$ was studied. The wing was installed under the bottom of the model so that the average angle of attack was zero. The distance of the wing in the middle position from the bottom is $h = 0.1$ m ($b/h = 2.5$). The oscillation mode of the wing without incident flow corresponds to the infinite value of the Strouhal number $k = \omega c/V$ ($\omega = 2\pi f$, c is the half chord of the profile). In this case, the linear theory of a wing in a flat nonstationary flow gives for the average for the period of oscillations the values of the thrust force $\overline{R}_T$ [266]:

$$\overline{R}_T = \frac{\pi}{8} \rho S (\alpha_0 \omega c)^2 \left(\frac{1}{2} - \frac{x_0}{c}\right)^2, \qquad (2.1)$$

where $S = b \cdot l$; ρ is the density of the liquid; α_0 is amplitude of torsional oscillations, radians; x_0 is the coordinate of the axis of rotation, measured from the middle of the profile in the positive direction to the rear edge (for the axis of rotation located at a quarter of a chord from the leading edge of the wing, $x_0/c = -1/2$).

We introduce the dimensionless coefficient of the average thrust force:

$$C_T = 2R_T / \rho S (\omega A)^2, \qquad (2.2)$$

where $A = \alpha_0 (c + |x_0|)$ is the maximum linear amplitude of the wing oscillations. From formulas (2.1), (2.2) it follows that as $k \to \infty$, the theory predicts a quadratic dependence of the thrust force on the amplitude and frequency of the wing torsional oscillations. At the same time, the coefficient C_T remains constant.

The results of the experiment, shown in Fig. 2.6, showed that in a wide range of frequencies and amplitudes of wing oscillations, a quadratic dependence $\overline{R}_T$ on α_0 and f does indeed take place. However, at high frequencies (in the experiment at $f = 16$ Hz), a significant violation of the quadratic dependence $\overline{R}_T$ on α_0 is observed. Moreover, when $\alpha_0 > 10$ degrees decrease occurs $\overline{R}_T$, which indicates a separation mode of flow past the wing. It should be noted that formula (2.1) was obtained for the case of wing oscillations in an infinite fluid, whereas in the experiment the wing oscillates near a flat bottom (screen). To estimate the effect of the parameter b/h, a special calculation was carried out on the basis of [335]. The results of this calculation (Fig. 2.7) showed a small effect of the parameter b/h on the magnitude of the thrust force in the considered range of its change.

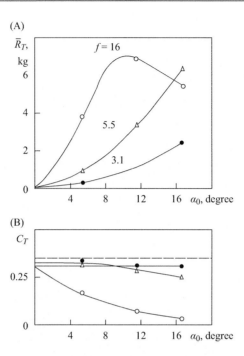

Figure 2.6 The influence of the oscillation frequency f on the dependence of the thrust force (A) and coefficient (B) of the oscillating wing on the amplitude of torsional oscillations α_0 at $V=0$, the dashed line is theoretical calculations [267].

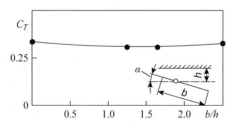

Figure 2.7 The theoretical dependence of the thrust force coefficient C_T on the parameter b/h at $\alpha_0 = 0.3$, $\omega c/V = 2\pi$, $x_0/c = -0.5$ [267].

The dependence of the thrust force on the speed of movement of the boat model in the mode of towing with torsional oscillations of one wing was also experimentally investigated. In the experiment, the average values of the resistance force of the model R_D and the difference between the forces of thrust and resistance were measured. The measurements were carried out on the control section of the model towing at a constant speed. The force R_D was also determined for the model body without a wing. The frequency of oscillation of the wing, taking into account the results of tests on the mode of mooring the vessel is 3.1 and 5.5 Hz, the amplitude of oscillation varied in the range of 11.5−18 degrees. The results of the experiment are shown in Fig. 2.8.

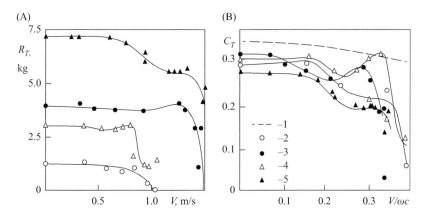

Figure 2.8 The dependence of the thrust force (A) and coefficient (B) of an oscillating wing on the flow velocity in the case of torsional oscillations with $f = 3.1$ and 5.5 Hz ($h = \infty$): (1) theoretical data; (2) $\alpha = 12$ degrees, $f = 3.1$ Hz; (3) $\alpha = 11.5$ degrees, $f = 5.5$ Hz; (4) $\alpha = 18$ degrees, $f = 3.1$ Hz; (5) $\alpha = 16.5$ degrees, $f = 5.5$ Hz [267].

They show a weak dependence of the average thrust force $\overline{R}_T$ on the speed V in a large interval of its change and a sharp drop $\overline{R}_T$ with a further increase in V. At the same time, with increasing wing oscillation frequency, an increase in V is observed, at which the thrust force $\overline{R}_T$ drops. Such a character of the change in thrust force from speed corresponds to the wing theory in Fig. 2.8 shows the theoretical dependence of the coefficient C_T on nonstationary flow, according to which the value $\overline{R}_T$ in the case of torsional oscillations decreases with a decrease in the Strouhal number $\omega c/V$. $V/\omega c$ in the case of profile oscillations in an $V/\omega c$ in the case of profile oscillations in an unlimited fluid ($h = \infty$).

This dependence is qualitatively consistent with the experimental data. However, the experimental values of C_T from $V/\omega c$ turn out to be different depending on the frequency and amplitude of the wing oscillations. The thrust force generated by a system of two oscillating wings was also investigated. It should be noted that such studies were carried out in Reference [272], which did not receive a noticeable mutual influence of the wings (Section 2.1). In our experiment, which was carried out at relatively large amplitudes of oscillations and close distances between the wings ($0.2b$), the effect of mutual influence turned out to be very significant. The wings oscillated according to the law $\alpha_1 = \alpha_{01}\cos(\omega t + \mu)$, $\alpha_2 = \alpha_{02}\cos(\omega t)$, where $\alpha_{02} = 18$ degrees, and the amplitude α_{01} and the phase shift μ changed.

Fig. 2.9 shows the results of the experiment on the mode of mooring the model. They show that the total thrust force of the wing system depends on the ratio of the amplitudes of oscillation and phase shift between them. At small values of the α_{01}/α_{02} ratio, there is little and hydrodynamic interaction between the wings, which is manifested in a weak dependence of the thrust force on the phase μ. In particular, when the second wing oscillates in the presence of a fixed first, the wing system develops practically the same thrust as with the vibrations of one wing.

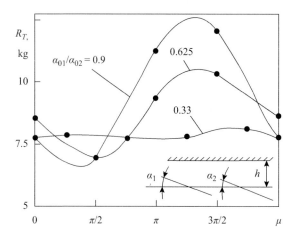

Figure 2.9 The dependence of the total the thrust force of two oscillating wings on the phase shift between them ($V = 0$) [267].

The hydrodynamic interaction of the wings increases with the ratio of the amplitudes of their oscillations to unity. So at $\alpha_{01}/\alpha_{02} = 0.906$, a change in the phase of oscillations of one wing with respect to the other leads to a change in thrust force twice. In this case, the maximum thrust is achieved at $\mu = 270$ degrees and almost coincides with the thrust developed by two isolated wings. Similar results were obtained when investigation of the oscillations of two wings on a towed boat model. The corresponding dependences of the total thrust force $\overline{R}_T = \overline{R}_{T1} + \overline{R}_{T2}$ on the flow velocity have the same character as for a single wing.

A good agreement was obtained between the experimental data and the available theoretical results obtained on the basis of the theory of a wing in a nonstationary flow.

2.4 Experimental investigation of hydrodynamic thrust generated by an oscillating wing

In Ref. [333], the magnitudes of the hydrodynamic emphasis created by a rigid rectangular wing with an elongation $\lambda = b/l = 3$, which undergoes transverse and angular oscillations in the incoming fluid flow, are experimentally investigated. The dependences of the hydrodynamic emphasis created by an oscillating wing on the frequency and amplitude of transverse and angular oscillations are obtained. To measure the forces acting on a wing oscillating in a fluid, a special experimental installation has been designed and manufactured. The experimental installation was based on the drive scheme [272], in which the wing was oscillated by two rods attached to the front and rear edges of the wing and reciprocating along its axis (Section 2.1). The harmonic law of oscillation of the wing was chosen, which is close to sinusoidal [332] (Part I, Section 2.7, Fig. 2.30). Investigations were conducted on an experimental sinus installation, consisting of a drive and two sinus mechanisms. The oscillatory

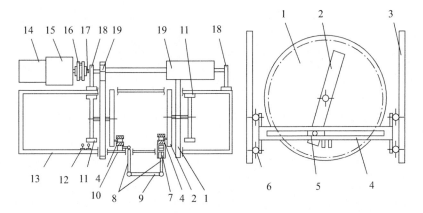

Figure 2.10 The scheme of the experimental installation, designations are given in the text [333].

mechanism of the experimental installation is made in the form of a rocker sinus mechanism (Fig. 2.10 on the right). On the same axis with the driven cylindrical gear wheel (1), the link (2) is seated, which is connected to the carriage (4) through the slider (5). When the link (2) rotates evenly, the carriage (4), supported by four rollers (6) on the guides (3), makes vertical oscillations according to the law $y_1 = A_0 \cos\omega t_i$, where y_1 is instantaneous coordinate of the vertical movement of the carriage; A_0 is the maximum linear amplitude of oscillations; $\omega = 2\pi f$ is the circular frequency; f is the oscillation frequency; and t_i is time. The diagram of the experimental sinus installation is shown in Fig. 2.10 left. The wing (9) through two hinges along the root chord with the help of two biases (8) and two strain gauge elements (7) and (10) was attached to the carriages (4) of sinus mechanisms. Strain gauge element (7) is made in the form of a two-component strain gauge beam, with which the vertical and horizontal components of hydrodynamic forces arising on the oscillating wing were measured, and strain gauge element (10) in the form of a single component strain gauge with a hinge, with the help of which the second vertical component of hydrodynamic forces was measured.

The wing (9) through two hinges along the root chord with the help of two biases (8) and two strain gauge elements (7) and (10) was attached to the carriages (4) of sinus mechanisms. For greater uniformity of rotation of the wings, sinus mechanisms have additional flywheels (11) on their axes of rotation. Rotation of wheels (1) of sinus mechanisms was carried out from two driving gear wheels (19) mounted on one drive shaft (17), which was mounted on frame (13) through two supports (18). One wheel (19) was completed large width, which allowed changing the location of a single sinus mechanism and exploring the wings with different size of the root chord. The drive of the unit was a direct current motor (14) with a reduction gear (15) and an elastic coupling (16), with which it was connected to the drive shaft (17). The armature winding and the serial motor excitation winding were supplied from a rectifier to which it came from a voltage regulator. On parallel winding of the electric motor, a direct current with a voltage of 24 V was supplied from the rectifier. The frequency of the wing oscillations was monitored by a frequency meter connected to the photosensor (12), which was mounted on the shaft of the sinus mechanism. The change in the amplitude of the angular oscillations β_0 was made by specifying a certain magnitude of the shift angle φ in phase of the oscillations of the carriage of one sinus mechanism with respect to another.

Investigations were carried out in a hydraulic channel with reversed motion at flow velocities equal to 0; 0.31; 0.52; and 0.67 m/s. The cross section of the working part of the hydraulic channel is $H \times B = 0.42 \times 0.62$ m. The amplitude of the transverse oscillations of the wing was 0.06 m and 0.1 m, the angular amplitude changed in the course of the experiments from 0 to 18 degrees with a step of 3.6 and 6 degrees. Stable uniform operation of the sinus installation was provided with wing oscillations with a frequency of 1−2.5 Hz. The wing with the NACA-0015 profile [392], the chord $b = 0.14$ m and the span $l = 0.42$ m was made of an aluminum alloy hollow so that its weight was equal to the weight of the water displaced by the wing. Depth of the wing in the neutral position was $h/b = 1.4$.

Fig. 2.11 shows the change in hydrodynamic forces on an oscillating wing during one cycle of oscillations, when the wing oscillated with an amplitude $A_0 = 0.06$ m,

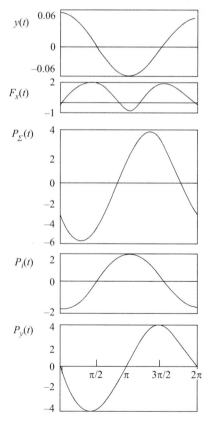

Figure 2.11 Patterns of measured values for the period of oscillation of the wing [333].

frequency $f = 2$ Hz and the phase angle of oscillations of the sinus mechanisms $\varphi = 36$ degrees, with the amplitude of angular oscillations $\beta_0 = 14.3$ degrees, the speed of the incident flow $V_x = 0.31$ m/s.

The trajectory of the point located on the wing chord at a distance of $0.5b$ from the profile nose is shown in Fig. 2.11 as a dependence of $y(t)$. The trajectories of the front and rear edges of the wing have the same shape, but are shifted in phase by the values of $+\varphi/2$ and $-\varphi/2$. Fig. 2.11 shows the horizontal and vertical components of the forces acting on an oscillating wing: $F_x(t)$ is the hydrodynamic thrust created by the wing; $P_\Sigma(t)$ is the sum of the vertical components of the hydrodynamic forces $P_y(t)$ and inertial forces $P_i(t)$. The total force was obtained from an experiment with wing oscillations in a fluid flow by analyzing oscillograms; inertial force—when the wing oscillation in the air. The vertical component of the hydrodynamic forces is obtained by subtracting the graph $P_i(t)$ from the graph $P_\Sigma(t)$. The phase shift present on the oscillogram (Fig. 2.7) with a change in the total vertical force with respect to the horizontal force $F_x(t)$ is determined only by the presence of inertial forces and will be the greater, the greater the inertial forces $P_i(t)$.

The oscillograms (Fig. 2.11) were used to determine the average value of the hydrodynamic stop F_x for the oscillation period, and a number of dependences $F_x = f(A_0, f, \beta_0, V_x)$ were constructed. By the known formulas, it is possible to determine the relative step λ_p, the thrust coefficient k_F and the efficiency η [272].

As a preliminary result, an example is given in the form of the dependences $F_x = f(A_0, f)$. Fig. 2.12A shows the dependence $F_x(f)$ for a wing that performs transverse oscillations without reruns, i.e. the amplitude of the angular oscillations $\beta_0 = 0$ degrees. Curves (1) and (3) characterize the oscillation of the wing with an amplitude of $A_0 = 0.06$ m at flow velocities $V_x = 0.67$ m/s (1) and $V_x = 0.31$ m/s (3). Correspondingly, curves (2) and (4) characterize the oscillation of the wing with an

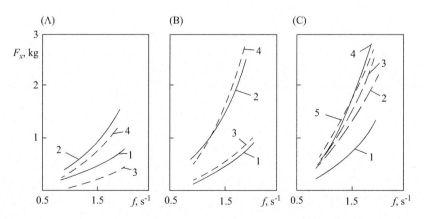

Figure 2.12 Dependence of the hydrodynamic thrust value F_x of the oscillating wing on the frequency f of the amplitude A_0 of transverse and angular oscillations without shift (A), with the shift and amplitude of angular oscillations (B) and the amplitude of angular oscillations (C). For (A) $\beta_0 = 0$ degrees; for (B) $\beta_0 = 180$ degrees; for (C): (1) $\beta_0 = 0$ degrees; (2) $\beta_0 = 6$ degrees; (3) $\beta_0 = 12$ degrees; (4) $\beta_0 = 18$ degrees; (5) $\beta_0 = 24$ degrees [333].

amplitude $A_0 = 0.1$ m at the same flow velocities. As the amplitude of A_0 oscillations increases 1.7 times, the hydrodynamic thrust F_x of an oscillating wing increases by about 2.8–3.2 times, which corresponds to the quadratic dependence of F_x on A_0.

Fig. 2.12B shows the dependences $F_x(f)$ for the same wing with the same values of the amplitude A_0 of the transverse oscillations and velocities of the incident flow, but there is already a wing overlay and an amplitude of angular oscillations $\beta_0 = 18$ degrees (and it should be noted that with amplitude $A_0 = 0.1$ m, the phase angle of oscillations of the sine mechanisms of the installation is $\varphi = 27$ degrees, and with the amplitude $A_0 = 0.06$ m, the angle $\varphi = 45$ degrees). The dependence of the hydrodynamic thrust F_x of an oscillating wing on the magnitude of the amplitude of angular oscillations β_0 ($A_0 = 0.1$ m = const and $V_x = 0.31$ m/s = const) is shown in Fig. 2.12C. The greatest influence of the magnitude of the amplitude of the angular oscillations of β_0 on the magnitude of F_x for small values of β_0. Thus, as β_0 increases from 0 to 6 degrees, the magnitude of the thrust F_x increases by approximately 70% (1) and (2), and as β_0 further increases twice and threefold (2)–(4), the F_x value increases only by 20% and 40%, respectively. With an increase in the amplitude of angular oscillations (5), the hydrodynamic thrust F_x does not actually increase and with a further increase in the value of β_0, the value of F_x decreases.

Thus, an increase in the amplitude of the angular oscillations at constant frequency f and the amplitude A_0 of the transverse oscillations of the wing increases the value of the hydrodynamic thrust only to certain limits. There is an optimal value of the amplitude β_0 for each mode of operation of the oscillating wing.

2.5 Hydrodynamic characteristics of a rectangular rigid oscillating wing

Experimental investigations of the hydrodynamic characteristics of a rigid rectangular-shaped wing of elongation $\lambda = 3$ with a symmetric profile NACA-0015 [392] (relative profile thickness $\overline{C} = 15\%$) were performed in Ref. [261] on an experimental installation developed by V.P. Kayan (Section 2.4). On the front there was a two-component strain gauge device for measuring horizontal and vertical forces acting on a wing oscillating in the liquid, and on the rear one—one component, which served to measure only the vertical force. The tests were carried out with the speeds of the incident flow $V_x = 0$; 0.31; 0.52; 0.67 m/s (Section 2.4). The vertical coordinates of the centers of the front and rear hinges on the edges of the wing y_1 and y_2 relative to the neutral axis of oscillation at each time point were determined by the formulas:

$$y_{1i} = A_0 \cos \omega t_i, \tag{2.3}$$

$$y_{2i} = A_0 \cos (\omega t_i - \varphi), \tag{2.4}$$

where A_0 is the specified maximum vertical amplitude of oscillations, $\omega = 2\pi f$ is the circular frequency, f is the frequency of the oscillations of the wing, t is the time, φ is the phase angle of the vertical oscillations of the front and rear edges of the wing.

The instantaneous vertical velocity of the hinge centers V_{y1i} and V_{y2i} was determined by the formulas:

$$V_{y1i} = -A_0\omega \sin \omega t_i, \tag{2.5}$$

$$V_{y2i} = -A_0 \sin(\omega t_i - \varphi), \tag{2.6}$$

and the instantaneous vertical velocity of the center *(c)* of pressure *(pr)* of the oscillating wing was determined as the resultant velocity V_{y1i} and V_{y2i}:

$$V_{yc.pr.i} = V_{y1i}\left(1 - \frac{z}{a}\right) + V_{y2i}\left(\frac{z}{a}\right), \tag{2.7}$$

where a is the distance between the centers of the hinges, equal in this case to 0.15 m; z is the distance from the center of the front hinge to the center of pressure of the wing profile. The integration over the oscillation period was determined by oscillograms the magnitude of the hydrodynamic thrust $\overline{F}$ created by an oscillating wing [332]. The coefficient of hydrodynamic thrust K_T oscillating wing obtained by the formula:

$$K_T = \frac{2\overline{F}}{\rho V^2 s}, \tag{2.8}$$

where ρ is the density of water; s is the area of the wing; $V = \sqrt{V_x^2 + (A_0\omega)^2}$ is the amplitude of the instantaneous values of the velocity of the flow around the wing. The use of the total velocity V makes it possible to avoid infinite values K_T when the wing is operating on mooring modes (that is, when $V_x = 0$).

By setting the shift angle of the oscillation phase φ of two sinus mechanisms on the installation, the amplitude β_0 of the wing angular oscillations was thus varied in a wide range of values ($\beta_0=0-18$ degrees). The values of the coefficients K_T obtained in this way, depending on the values of the relative arrival of the oscillating wing $\lambda_p = V_x/A_0\omega$ and the values of β_0, are presented in Fig. 2.13. In investigations, the relative immersion of the wing is $H = h/b = 1.4$, where h is the average for the period value of the immersion value of the trailing edge of the wing; b is wing chord ($b = 0.14$ m). The axis of rotation of the wing is located in the middle of the wing chord.

Curves (1) and (5) correspond to plane-parallel oscillations of the wing (angular amplitude $\beta_0=0$ degrees); curves (2)–(4) is oscillations with angular amplitudes $\beta_0=7.2$, 10.8, and 17.8 degrees, curves (6)–(8) is with angular amplitudes $\beta_0=6$, 12.1, and 18.2 degrees, respectively. Fig. 2.13 shows the influence of the magnitude of the amplitude β_0 of angular oscillations on the magnitude of the coefficient of hydrodynamic thrust, especially for small values of β_0, which is consistent with Ref. [272] (Section 2.1, Fig. 2.2). In the whole investigated range of values $\lambda_p = 0.2-1.7$, the coefficient K_T

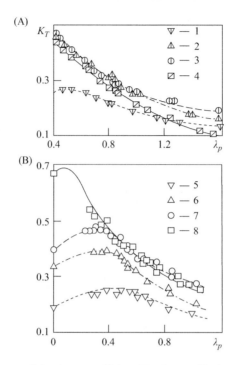

Figure 2.13 Dependences of the thrust coefficient K_T of an oscillating wing on the relative input λ_p at A_0 equal to 0.06 m (A) and 0.1 m (B) at different amplitudes β_0 of wing angular oscillations, degrees: (1); (5) 0 degrees; (2) 7.2 degrees; (3) 10.8 degrees; (4) 17.8 degrees; (6) 6 degrees; (7) 12.1 degrees; (8) 18.2 degrees [261].

reaches a maximum at $\beta_0 = 8-12$ degrees, and a further increase in the amplitude value β_0 does not lead to an increase in the coefficient K_T, and when a certain value β_0 begins to decrease, the value K_T starts to decrease (Fig. 2.13, (4)). Fig. 2.13B clearly shows the maxima of the coefficients K_T at different values of the angular amplitude β_0, and with its increase the values of the relative intake λ_p decrease, at which the maxima K_T are reached (at $\beta_0 = 0$ degrees, $\lambda_{opt} \approx 0.45$, with $\beta_0 = 12.1$ degrees $\lambda_{ort} \approx 0.3$ magnitude K_T increases almost twice).

Hydrodynamic efficiency η oscillating wing was determined by the formula:

$$\eta = \overline{N}_x / \overline{N}_y, \tag{2.9}$$

where $\overline{N}_x = \overline{F} V_x$ is useful work performed by an oscillating wing in one oscillation period; $\overline{N}_y = \frac{1}{T_0} \int_0^{T_0} |P_{yc.pr.i} V_{yc.pr.i}| dt$ - the average for one period of oscillations $T_0 = 1/f$ power spent on the vertical oscillations of the wing. The vertical component of the hydrodynamic forces $P_{yc.pr.i}$ at each moment of time was determined as the sum of the values of the components P_{y1i} and P_{y1i} on the front and rear rods of the experimental installation, that is

$$P_{yc.pr.i} = P_{y1i} + P_{y2i} \tag{2.10}$$

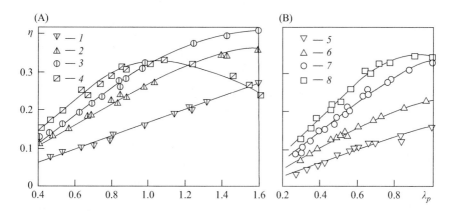

Figure 2.14 Dependences of the hydrodynamic efficiency η of an oscillating wing on the relative input λ_p at A_0 equal to 0.06 m (A) and 0.1 m (B), with different amplitudes β_0 of angular oscillations of the wing, degree: (1; 5) 0 degrees; (2) 7.2 degrees; (3) 10.8 degrees; (4) 17.8 degrees; (6) 6 degrees; (7) 12.1 degrees; (8) 18.2 degrees [261].

The vertical components of the hydrodynamic forces P_{y1i} and P_{y2i} were determined as the difference between the instantaneous vertical forces recorded using strain gauges during experiments in water and the inertial forces recorded during wing oscillations in air [333] (Section 2.4). The value of the vertical velocity V_{yi} depended on the phase position of the sinus mechanism, the vertical amplitude and frequency of oscillations and was determined by calculation.

The values of the efficiency η obtained by formula (2.9), depending on the values of λ_p of the oscillating wing, are presented in Fig. 2.14. In the entire investigated range of relative values of λ_p, an increase in the amplitude of angular oscillations β_0 leads to an increase in the efficiency η. Thus, an increase in the amplitude β_0 from 0 to 6−7 degrees leads to an increase in the efficiency η by more than 1.5 times. A further increase in the amplitude of β_0 leads to a slight increase in the values of the efficiency η to some maximum value, after which a further increase in the amplitude of β_0 leads to a decrease in the efficiency (Fig. 2.14 (4), (8)). At a certain magnitude of the angular oscillation amplitude β_0=const, the η (λ_p) curve has a maximum at some particular value of λ_{opt}, and with a decrease in the amplitude of β_0, the value of λ_{opt} increases.

Due to the fact that the water flow level in the hydraulic tray did not exceed 0.4 m ($H = 1.4$), with an increase in the oscillation amplitude A_0, a noticeable influence of the flow boundaries on the hydrodynamic characteristics of the oscillating wing was observed. Fig. 2.15 shows the dependences K_T (λ_p) (1) and (2) and η (λ_p) (3) and (4) with β_0=const. When the relative depth $\bar{h}' = h'/b \leq 1$ (h' is the distance to the extreme points of the wing oscillation range from the water flow boundary), its decrease significantly affects the values of the coefficients of the hydrodynamic thrust k_T and the hydrodynamic efficiency η. So, at small values of the relative flow of an oscillating wing (λ_p<1), a decrease in the value h' by 30% leads to an increase in the values of the coefficients K_T and η by 10%−20%.

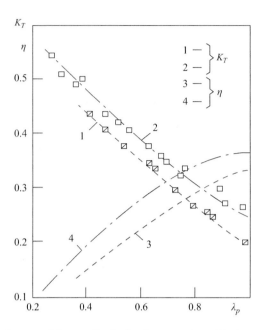

Figure 2.15 The influence of the amplitude A_o of transverse oscillations of an oscillating wing on the hydrodynamic characteristics K_T (1; 2) and η (3; 4) at $\beta_0 = 18 \pm 0.2$ degrees = const: (1; 3) $A_0 = 0.06$ m; (2; 4) $A_0 = 0.1$ m [261].

Fig. 2.16 shows for comparison the hydrodynamic characteristics of rigid rectangular oscillating wings, obtained in Ref. [261] (1) and in Ref. [272] (2) and (3) under similar experimental conditions (Table 2.1). Coefficient of power consumption K_N was determined by the formula:

$$K_N = \frac{2\overline{N}_y}{\rho V^2 S A_\omega}. \tag{2.11}$$

Fig. 2.17 presents a comparison of the experimental results obtained with data from [462], which presents the results of experimental studies of an oscillating wing in a hydraulic channel. The wing is rigid, rectangular with an elongation $\lambda = 2.8$, thickness $\bar{c} = 20\%$ (the shape of the profile is not specified). The flow velocity in the tray $V_x = 0.3$ m/s. The oscillations of the wing were in the horizontal plane, the drive is made in the form of a vertical shaft, lowered into the water, which made angular oscillations with amplitude of 29 degrees. Horizontal levers are rigidly attached to the shaft, to the ends of which the wing is hinged. The axis of rotation of the wing passes through its front edge. The transverse amplitude of oscillation of the front edge of the wing $A_0 = 0.12$ m. At the attachment of the wing and lever was an elastic connection with variable flexural rigidity EI, which, depending on the hydrodynamic head, allowed the wing to deflect at some angle β_0 relative to the lever. Hydrodynamic thrust was measured by the strain element on the drive shaft. The nature of curve (4) is explained by

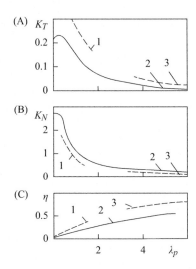

Figure 2.16 The dependences of the hydrodynamic characteristics K_T (A), K_N (B), and η (C) on the relative input λ_p for oscillating wings with different profiles (see Table 2.1) [261].

Table 2.1 Characteristics of the curves in Fig. 2.16.

Parameter	Curves 1	Curves 2	Curves 3
Profile	NACA-0015	KV-1-7	NACA-0015
Relative thickness $\bar{c}$ %	15	7	15
Extension λ	3	4	4
Amplitude A_o, m	0.06	0.07	0.07
Relative amplitude $\bar{A}$	0.428	0.467	0.467
Angle amplitude β, degree	3.6	3	3
Deepening $\bar{H}$	1.4	1.65	1.65

the fact that with increasing frequency of transverse oscillations of the wing, the hydrodynamic head and deflection angle of the wing with respect to the lever increase (angle β is maximum when the wing passes through the neutral axis of oscillation when the transverse velocity of the wing is maximum). With an increase in the stiffness of an elastic compound (5), the angles β decrease, and the efficiency η decreases (this also confirms the position of (1−3)).

Experiments conducted with a rigid relatively thick oscillating wing in the range of values of relative input $\lambda_p = 0.2-1.7$ showed the possibility of obtaining coefficients of a hydrodynamic thrust K_T, reaching values 0.5−0.6 at $\eta = 0.1-0.4$. With an increase in λ_p, the coefficient K_T decreases, and the efficiency η increases, reaching the values $\eta = 0.7-0.8$ at $\lambda_p = 4-6$ (Fig. 2.16). The hydrodynamic characteristics of an oscillating wing are very significantly influenced by the maximum angle of oscillation amplitude β_0, with each value of λ_p corresponding to the optimum amplitude β_0.

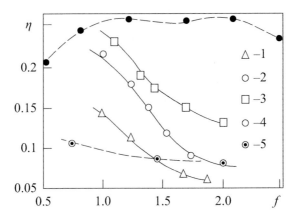

Figure 2.17 Dependences of the efficiency η of an oscillating wing with different amplitudes of angular oscillations β_0 (degrees) on the oscillation frequency f with A_0 equal to 0.1 m (1–3) and 0.12 m (4): β_0 for (1) 6 degrees; (2) 12.1 degrees; (3) 18.2 degrees; (4; 5) β_0=different from [462]; (4) EI=5.6 kg·cm, (5) EI = 103 kg·cm [261].

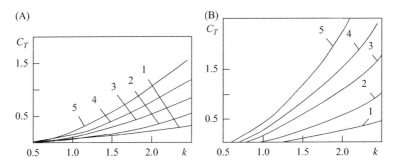

Figure 2.18 Dependences of the C_T coefficient of the oscillating wing on the Strouhal number k and the relative amplitude of the linear oscillations $\overline{A}$ at β_0 equal to 0 degrees (A) and 18 degrees (B) and the following values $\overline{A}$: 0.333 (1); 0.5 (2); 0.666 (3); 0.827 (4): 1.0 (5) [262].

In Ref. [262], the results of similar experimental studies with V_x = 0; 0.3; 0.55; 0.75 m/s. The relative immersion of the wing was H = 1.55. The magnitude of the linear amplitude of oscillations A_0 in the experiments was set equal to 0.04; 0.06; 0.08 and 0.1 m, the magnitude of the oscillation angular amplitude β_0, depending on the setting of the angle φ, was (0–21.4) degrees, with a step of about 3 degrees. The installation provided stable wing oscillations with a frequency of 0.5–2.5 Hz. Measurements of the F, P_{y1i}, and P_{y2i} values obtained on the oscillograms, and during the oscillation period were subjected to statistical processing, the integration was determined by the average $\overline{F}$ thrust force during the oscillation period.

The results are presented in Fig. 2.18 for two values of the angular amplitudes of the oscillations of the wing β_0=0 and 18 degrees in the form of the dependence of the thrust coefficient C_T on the Strouhal number k (Sections 2.1 and 2.3):

$$C_T = \frac{2\overline{F}}{\rho V_x^2 S}, \qquad (2.12)$$

$$k = \frac{\omega b}{V_x}, \qquad (2.13)$$

Curves (1)–(5) correspond to the dependencies $C_T(k)$ when $\overline{A} = A_0/b = 0.333$; 0.50; 0.666; 0.827; 1.0, respectively. Curve (5) is obtained by extrapolation according to the graph $C_T = f(\overline{A})$. The magnitude of the coefficient of thrust C_T is directly proportional to the values of k and $\overline{A}$, and the dependence of $C_T = f(\overline{A})$ is close to quadratic and can be represented by an equation:

$$C_T = C_1 \overline{A}^2 \qquad (2.14)$$

where the value of the coefficient C_1 is a function of the oscillation frequency of the wing and for the range of Strouhal numbers $k = 0.5-2.5$ can be represented by an equation:

$$C_1 = 0.3\, k^n \qquad (2.15)$$

where $2 < n < 3$, and when $\beta_0 = 0$ degrees, the exponent n is about 2, and with increasing magnitude of the angular amplitude of oscillations β_0 also increases. If we consider an oscillating wing as a propulsor, then the change in its propulsive characteristics (thrust coefficient and efficiency) will be more interesting to consider depending on the magnitude of the relative come on mover $\lambda_p = V_x/A_0\omega$, which most fully characterizes the kinematic parameters of the oscillating wing. In this case, the coefficient of thrust is more convenient to present in the form:

$$K_T = \frac{2\overline{F}}{\rho W^2 S}, \qquad (2.16)$$

where $W = \sqrt{V_x^2 + (A_0\omega)^2}$, that will allow to avoid large values of the coefficient of thrust C_T for very small values of λ_p. It should also be noted that λ_p and k are inversely proportional values, linked by the expression $\lambda_p = (\overline{A}K)^{-1}$.

Fig. 2.19 shows the dependences $K_T(\lambda_p)$ at $\overline{A} = 0.333$ (A) and $\overline{A} = 0.667$ (B) for different amplitudes of angular oscillations $\beta_0 = 0 - 21.4$ degrees with a step of 3 degrees (curves (1)–(8)).

The character of the experimental dependencies $K_T(\lambda_p)$ indicates a very significant effect of the magnitude of the angular amplitude of β_0 on the magnitude of the thrust coefficient k_T and is in good agreement with the data given in Figs. 2.13, 2.15, and 2.16 for a geometrically similar, but larger wing with a slightly different mounting principle to him of leverages.

All dependences $K_T(\lambda_p)$ have pronounced maxima (in the region of small values of λ_p), and the value $\lambda_{p,opt}$ at which $k_T = \max$ decreases with increasing β_0 and A_0.

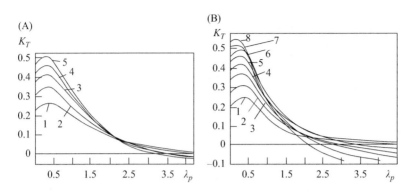

Figure 2.19 Dependences of the oscillating wing K_T thrust coefficient on the relative arrival of λ_p at values $\overline{A}$ equal to 0.333 (A) and 0.667 (B) and amplitudes β_0 wing angular oscillations, degrees: 0 degrees (1); 3 degrees (2); 6 degrees (3); 9 degrees (4); 12 degrees (5); 15.1 degrees (6); 18.2 degrees (7); 21.4 degrees (8) [262].

When A_0=const, almost all points $k_{T\max}$ lie on one line, the slope of which is the same on all graphs $K_T(\lambda_p)$. When β_0=const, the value K_T = max decreases with increasing A_0. With an increase in the relative input λ_p, that is, with an increase in V_x or a decrease in A_0 and f, the values K_T quickly decrease, eventually reaching values K_T = 0 (except for the curve $K_T(\lambda_p)$ for β_0=0 degrees), and the greater the angular amplitude of oscillations β_0, the lower the values of λ_p it happens. With an increase in the value of λ_p, there is also a change in the direct dependence of the K_T on the value of β_0 (at λ_p=0−0.6) on the opposite (at λ_p>1.5−3.0). The intersection points of two adjacent curves (λ_p) (for example, curves $K_T(\lambda_p)$ for angular amplitude β_0 equal to 3 and 6 degrees; 6 and 9 degrees; 9 and 12 degrees, etc.) are located along the abscissa axis of the graph K_T, λ_p in a certain sequence and the segments between these points of the curve $K_T(\lambda_p)$ are the line of maximum values K_T. For example, in the range of values λ_p=1.7−2.2 at A_0=0.04 m, the values K_T will be maximum at β_0=6 degrees, in the range λ_p=1.5−1.35, respectively, K_T = max at β_0=12 degrees.

The second main propulsion characteristic of a propulsion unit with a working body in the form of a rigid harmonically oscillating wing is the hydrodynamic coefficient of efficiency η, which is determined by the formula (2.9). The vertical component of the hydrodynamic forces $P_{yч.д.i}$ is determined by the formula (2.10). The dependences of the efficiency η of an oscillating wing obtained by formula (2.9) on the values of λ_p and β_0 for $\overline{A}$=0.333 (A) and 0.667 (B) are presented in Fig. 2.20. For small values of λ_p <1, a direct proportionality of the value of η with respect to the value of β_0 is observed. An increase in the amplitude of angular oscillations β_0 by half (from 3 to 6 degrees or from 6 to 12 degrees) increases the efficiency η by 1.5 times. All curves $\eta(\lambda_p)$ at β_0=const have pronounced maxima (except for the curve for β_0=0 degrees) and with increasing β_0 the value $\lambda_{p,opt}$ at which η = max decreases. All the graphs $\eta(\lambda_p)$ converge at the origin of the coordinates of the graph η, λ_p, since for λ_p=0 and V_x=0, respectively, and $\overline{N}_x$ also equal to zero.

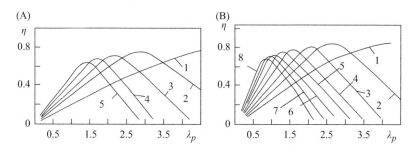

Figure 2.20 Dependences of the hydrodynamic efficiency η of an oscillating wing on the relative arrival of λ_p at values $\overline{A}$ equal to 0.333 (A) and 0.666 (B) and amplitudes β_0 of wing angular oscillations, degrees: 0 (1); 3 (2); 6 (3); 9 (4); 12 (5); 15.1 (6); 18.2 (7); 21.4 (8) [262].

After reaching the value $\eta =$ max, all curves $\eta(\lambda_p)$ change the direct dependence of η on β_0 and reverse, and with a further increase in λ_p, the efficiency values η decrease to zero, and the points of intersection of the curves $\eta(\lambda_p)$ with the abscissa of the η, λ_p graph coincide with the points of intersection of the curves $K_T(\lambda_p)$ with the same axis, and the greater the magnitude of the amplitude β_0, the smaller the value of λ_p this occurs.

Thus, an oscillating wing thruster can have a high efficiency value (about 80% at $\lambda_p > 1$), but at the same time, the values of the thrust coefficient have small values ($k_T < 0.1$). High values K_T ($K_T > 0.5$) can be obtained for small values of λ_p ($\lambda_p < 0.5$), but the efficiency values do not exceed 0.1–0.3.

Integrated hydrodynamic characteristics of an oscillating wing [366] were investigated in the hydrodynamic tube of V.Ye. Pyatetsky [330].

2.6 Experimental investigation of nonstationary vortex structures behind oscillating bodies

To understand the physical processes during nonstationary flow around bodies, three installations were developed, with the help of which one can visualize vortex structures behind oscillating bodies. In this case, the well-known property of fluid flows was taken into account, namely, when Reynolds numbers are $Re \leq 10^3$, viscous forces are small compared to inertial ones, and therefore inertial forces play a determining role in the formation of vortex structures [148,163]. If the vanishing points are identical, the macrostructure of the flow pattern formed at $Re = 10^5 - 10^8$ in the experiment with the numbers $Re \leq 10^3$ can be fairly accurately reproduced. The microstructure of such flows differs. Transformations of a laminar BL descended from a body into a turbulent flow may occur with the obvious presence of various forms of instability. However, often in the first approximation, information on the microstructure can be neglected and refined, if necessary, in subsequent experiments. This type of currents includes the continuous flow around the wings, the separation flow around bodies with sharp edges, the flow around profiles with corners, etc.

The form of oscillation of profiles can be of three types: pure transverse oscillation: "immersion" (Sections 2.1 and 2.2), pure rotation around the transverse axis: "swing" (Section 2.3), and joint movement of these types: "swing-immersion" (Sections 2.4 and 2.5).

In Refs. [293–297,595], various theoretical and methodological issues of an oscillating axisymmetric rigid wing are considered. In Part I, Section 2.4 the form, geometrical characteristics and other parameters of the fin mover of high-speed hydrobionts are given. Taking into account the "principles" of hydrobionics (Part I, Section 1.2) and other materials given in the previous sections, it becomes obvious that for a complete simulation of the mover complex of hydrobionts it is necessary to take into account the joint work of the stem and the fin mover, as well as the peculiarities of changing the shape of the caudal fin movements at different swimming speeds (Part I, Sections 4.2, 5.1, etc.). When modeling the fin mover of hydrobionts, it is also important to visualize the structure of the stream flowing around the fin and the wake behind it. To accomplish this task, studies have been performed during the flow around an oscillating wing in the "swing-immersion" mode [295]. In the study of the hydrodynamic characteristics of oscillating bodies, one has to deal with a large number of dimensionless defining parameters. In the study of a problem with two degrees of freedom (with linear body oscillations across the stream, along with angular oscillations), which is implemented on the installation below, numerous variations of the defining parameters of bodies are possible. The laws of linear and angular oscillations can be arbitrary. The condition of periodicity is only imposed on them. This is one of the reasons for using the cam drive in the design. We assign the law of linear and angular displacements of the body, respectively, in the form:

$$y = \frac{a_{0y}}{2} + \sum_{k=1}^{\infty}(a_{ky}\cos k\omega t + b_{ky}\sin k\omega t),$$

$$v = \frac{a_{0v}}{2} + \sum_{k=1}^{\infty}(a_{kv}\cos k\omega t + b_{kv}\sin k\omega t),$$

where a_{ky}, b_{ky}, a_{kv}, b_{kv}, etc., are the coefficients of the Fourier series. We take the speed of the incident flow V_0 as the scale of velocity, and the characteristic size of the body b as the scale of length. As a result, we obtain the following similarity criteria: Reynolds number $\mathrm{Re} = \frac{V_0 b^2}{\nu}$; Strouhal number: $\mathrm{Sh} = \frac{b\omega}{V_0}$; $\bar{a}_{0y}$, $\bar{a}_{ky}$, $\bar{b}_{ky}$, $\bar{a}_{0v}$, $\bar{a}_{kv}$, $\bar{b}_{kv}$, where $\bar{a}$ and $\bar{b}$ are dimensionless coefficients of the laws of living.

For practical purposes, it is enough to take into account six harmonics. The dimensionless geometric characteristics of bodies are not considered here. The question of the similarity criteria for this problem is discussed in more detail in Ref. [269]. The coefficients of the series for linear and angular displacements allow us to determine the area swept by the oscillating body, and the range of oscillations. This allows the oscillation parameters of the model to be matched with the

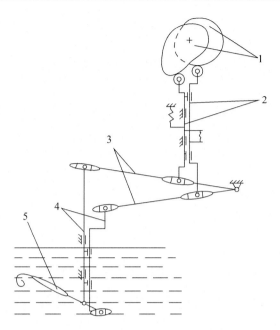

Figure 2.21 Kinematic scheme of the experimental installation: (1) cams; (2) roller pushers; (3) wings; (4) pushers of the model drive mechanism; and (5) the model under study [295].

conditions of the unperturbed flow of the working part of closed hydrodynamic tunnel. It should also be taken into account that at small flow velocities of 2−5 cm/s, the thickness of the BL rapidly increases. Conventional recommendations were used to eliminate the effect of flow boundaries [269].

The kinematic scheme of the installation is shown in Fig. 2.21. Two cams (1) form which are made in accordance with the given laws of motion of the test body, mounted coaxially on the same shaft. To prevent slippage of one relatively other and improve the reproduction of the laws of motion, the shaft with cams is driven during rotation by an electric motor through a reducer (not shown in the diagram). With cams two pushers (2) interact, which at the ends have rollers and move appropriate wings (3). Springs on pushers are necessary for guaranteed tracking the rollers of the laws of motion recorded on the cams.

At a distance from the point of attachment of the pushers to the wings there is a mechanism (4) for driving the model, which also consists of two pushers. It would be possible to build a model drive mechanism directly using followers after cams. However, this would lead to a significant increase in cams sizes. The application of the earlier-mentioned principle allowed reducing the dimensions of the experimental installation, which were: length=0.44 m, width=0.32 m, height with the maximum omitted model=0.7 m. Maximum amplitude of transverse oscillations= ±0.09 m, angular oscillations= ±35 degrees, vertical speed amounted to $V_y = 0.005-0.01$ m/s. The experiments used a rectangular wing with a span of 0.18 m, with an elongation $\lambda = 4$ and a relative profile thickness $\bar{c} = 6\%$.

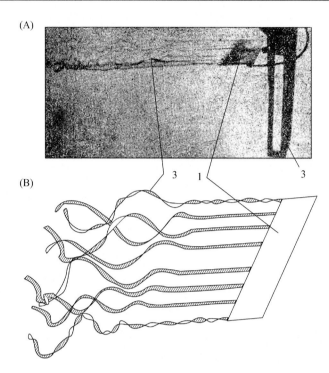

Figure 2.22 Photograph (A) and diagram (B) of the visualization of vortex structures behind the wing at an angle of attack $\alpha = 0$ degrees: (1) wing; (2) wing oscillation mechanism; (3) tinted streams for visualizing the wing track [295].

Experimental investigations were performed in Refs. [330,504]. Similar investigations performed in Ref. [558]. Fig. 2.22 shows the results of visualization of a flow around a wing with angle of attack $\alpha = 0$ degrees (A: photograph of vortex structures behind the wing, B: behavior diagram visualizing discrete streams), the flow velocity was 0.05 m/s. It is seen, that the jets of the colored fluid overflow into the stream in the trailing edge of the wing, indicate the instability of fluid layers, descending from the upper and lower wing surfaces. At first, waves rendered in trickles wear linear character, but then downstream become nonlinear.

Behind the rear edge of the wing, at a distance of two chords of the wing, the flat wake becomes unstable, plane waves arise, which very quickly become three-dimensional and quickly transform into longitudinal vortex structures. In the flow behind the side edges of the wing, characteristic end vortex bundles arise, forming the inductive resistance of the wing. As the flow rate increases, the pattern of flow behind the rear edge of the wing will change: the three-dimensional vortex bundles behind the rear edge of the wing will rush to the side edges of the wing and roll up into a pair of end vortices forming the inductive resistance of the wing. By increasing or decreasing the speed of the incident flow, it is possible to adjust the distance from the place where the disturbances occur to the trailing edge of the wing.

Modeling of a waving fin mover 69

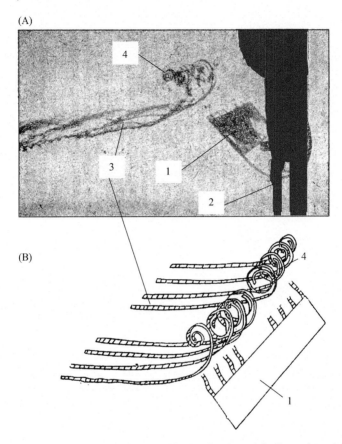

Figure 2.23 Photograph (A) and scheme (B) of visualization of the formation of the over clocking vortex behind the wing: (1) wing; (2) wing oscillation mechanism; (3) tinted streams for visualizing the wing trail; (4) over clocking (accelerating) vortex [295].

Fig. 2.23 shows the moment of formation of the overclocking vortex. The flow velocity and the speed of the vertical movement of the wing are about 0.02 m/s. The wing is inclined at an angle of 35 degrees to the horizon. The form of overclocking vortices varies along the span of the wing. In the area of the end chord, the overclocking vortices are smaller. This can be explained by the magnitude of the bevel flow, different in scope. At the trailing edge, formations are observed that are oscillatory in nature, which is not related to the loss of stability of the flow, but may be due, most likely, to small vibrations of the trailing edge of the wing as a result of the operation of the electric motor.

In Fig. 2.24, the vortex bundles descending from the lateral edges of the oscillating wing and the accelerating vortices at the places where the wing is shifted are clearly visible. Downstream, the "accelerating vortices" are transformed into a spherical formation (see the upper vortex shown in Fig. 2.24B on the left), the dimensions of which increase downstream. As already mentioned earlier, the

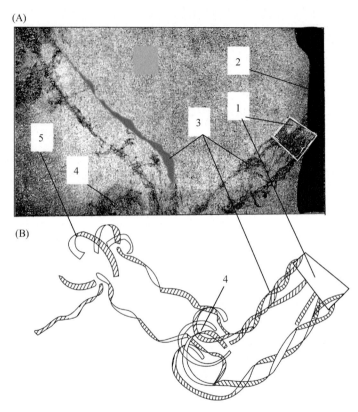

Figure 2.24 Photograph (A) and scheme (B) of visualization of the vortex pattern behind an oscillating wing: (1) wing; (2) wing oscillation mechanism; (3) tinted trickles for visualizing the wing trail; (4) overclocking (accelerating) vortex; (5) transverse vortex at the upper wing transfer [295].

macrostructure of such flows in general terms coincides with the results of investigations of oscillating wings carried out at higher Reynolds numbers than in this experiment. The experimental data shown in Fig. 2.24 show that the traditional calculated scheme of a swirling wake behind a flapping wing significantly simplifies the model of a wake behind an oscillating wing. The microstructure of the flow in this case is complex and needs additional investigations.

In Ref. [236], the formation of thrust by the caudal fin of bottlenose dolphins (*Tursiops truncatus*), which swam in a large external drive (Fig. 2.25), was experimentally investigated. The digital particle image velocimetry (DPIV) method was used, in which the visualization of the vortices in the wake of the dolphin's tail fin was made with the help of microbubbles, which were produced in a narrow layer from the finely porous hose and air source. The plane of the bubbles is oriented along the middle plane of symmetry of the animal. The movement of the bubbles was tracked by a high-speed video camera. Dolphins swam at speeds of 0.6−3.7 m/s, Reynolds numbers were $1.6 \cdot 10^6 - 7.7 \cdot 10^6$.

Figure 2.25 Visualization using microbubbles of the vortex wake behind the tail fin of a dolphin Tursiops truncates [236].

The amplitude of the tail fin oscillations was 0.30 ± 0.12 m and was located within $0.14-0.47$ m, which was $5.8\%-19.4\%$ of the body length. This was less than the typical amplitude of oscillation, equal to 20% of the body length of the animals. The frequency of oscillation of the fin was $0.5-2.9$ Hz at a low swimming speed. With dorsoventral oscillations, pairs of antirotating vortices are produced during each oscillation cycle. The calculated thrust varied within $50.8-700.4$ N for small octillation amplitudes and $334.0-1467.6$ N for large octillation amplitudes that the dolphin produced during acceleration after rest. In Fig. 2.25 it is seen that the shape of the dolphin's tail fin during the oscillation process bends under the influence of water resistance. The shape of the posterior edge of the fin is also clearly visible—the features of the cetacean fin mover are discussed in detail in Part I, Sections 2.4, 4.2, and 5.1. This feature of the caudal fin leads to the formation of vortex structures in the wake behind the fin, which rapidly dissipate. Experimental investigations of the vortex wake behind an oscillating rigid rectangular symmetric profile have shown that lateral vortex harnesses and accelerating vortices that are formed during wing overshoot are preserved in the wake. In the future, it is advisable to experimentally investigate a form of the wing profile that, with the maximum possible approximation, would model the tail movers of hydrobionts.

For an experimental investigation of the flow around oscillating bodies with their complex oscillations, the problem arose of modeling the flow and determining the hydrodynamic characteristics of bodies moving in an arbitrary manner. V.V. Babenko proposed an appropriate installation scheme (Fig. 2.26), which was implemented by V.N. Khatuntsev [297]. This installation allows you to simultaneously move the body in the plane along three degrees of freedom, using various laws of motion. Moreover, thanks to special electronic devices, these laws can be changed. The installation is a tank (1) filled with water and measuring $1.2 \times 0.82 \times 0.16$ m. Along it, on its sides, are guides (2), along which the main carriage 12 moves.

To move the trolley, a cable drive (13) and a stepper motor (14) are used. Perpendicular to the main movement, the trolley (11) moves along the main trolley.

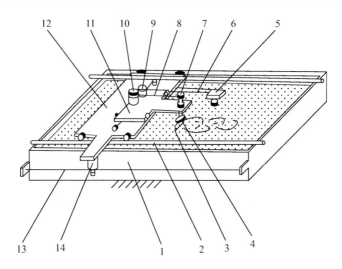

Figure 2.26 The scheme of the experimental installation: (1) tank with water; (2) guides; (3) test body; (4) strain gauges; (5) camera; (6) bracket; (7) pulley; (8) rubber mounts; (9), (10), (14) stepper motors; (11) trolley; (12) main trolley; (13) cable drive [297].

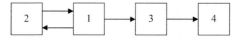

Figure 2.27 The block diagram of the control unit hydrodynamic installation: (1) control unit; (2) permanent storage device; (3) power amplifier; (4) motor [297].

To move it, a cable drive with a stepping motor (9) is also used. On the trolley there are one more stepping electric motor (10) rotational motions of the model of the body (3), which is attached to the axis installed in the bearings on the trolley (11). In the upper part of the axis there is a pulley (7), which with the help of a flexible coupling (8) is driven into the rotational-oscillatory motion by a stepping electric with the drive (10). On top of the pulley (7), the bracket (6) is fixed, on which, in turn, the camera (5) is fixed. The loads acting on the model of the body (3) are measured by strain gauges (4). All three electric motors are controlled by a special electronic device. Its block diagram for one motion component is shown in Fig. 2.27.

Control system works as follows. The digital code corresponding to a specific law of motion and recorded in the persistent storage device (2) is processed in control unit (1). During processing, the motion parameters are set: maximum and minimum speeds, acceleration, repetition period, number of cycles. In addition, in block (1), different laws of motion can be established: separately for the acceleration mode, separately for the braking mode. The output control signal in the form of a sequence of pulses through the power amplifier (3) is fed to the executive stepping motor (4). The design of the control unit allows you to quickly change the law of motion for each component.

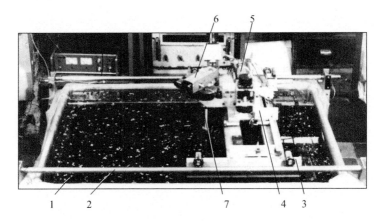

Figure 2.28 Photograph of the experimental installation: (1) tank with water; (2) guides for the main carriage; (3) main carriage; (4) carriage for transverse displacements; (5) stepper motor for moving the carriage; (6) stepper motor for turning the body under study; (7) cylinder for suspensions of the body under study [297].

Fig. 2.28 shows a photograph of the indicated experimental installation. This installation provides for the measurement of body model movements in all three degrees of freedom (the sensors in Figs. 2.26, 2.28 are not shown). Two methods for determining the kinematic characteristics were considered: inertial (using accelerometers) and potentiometric, which in this case turned out to be simpler and more accurate. Therefore, the measuring system records the displacements, and to obtain the corresponding values of the velocities and accelerations, they must be differentiated. In this case, the measurement of forces has specificity in that it is necessary to measure quantities whose maximum does not exceed 10 g. Therefore, it follows from the calculations that it is possible to use semiconductor tensors with a strain sensitivity coefficient $S = 80 - 100$.

Strain gauges are located above the surface of the water. Installation of strain gauges and their connection to the measuring system is carried out by elastic conductors. These measures are necessary in order to improve the elastic properties of the measuring elements of strain gauges. The rigidity of the mechanical part of the strain gauges is high enough so that the natural oscillations of the strain gauge more than an order of magnitude higher than the frequency of the load oscillations in the experiment. This makes it possible to reduce the calculated dynamic error in determining forces to a value (0.5%–1.0%) and not to take it into account when processing experimental data. For the same reason, the phase shift between the measured value and the signal at the output of the measuring system is quite small, and it can also be ignored during processing.

A two-dimensional flow pattern around a moving body was visualized using markers floating on the surface of the fluid. Markers are circles of wax paper with a diameter $(1-1.6) \cdot 10^{-3}$ m. The position of the markers was fixed with the help of several flash lamps, triggered by the action of the driver, which makes it possible to obtain several consecutive positions of the markers on a single frame of photographic film. The visualization methodology is based on the principles considered in Refs. [467,468], with the difference that in this case it is not the instantaneous velocity field that is determined, but the velocity field in the time interval between the first and last flash lamp in Lagrange variables.

Photographing is carried out either with a camera mounted on a tripod near the installation, or with a camera rigidly connected to the body under study. The latter option is more preferable, since in the connected coordinate system the body contours are clear, and the markers give tracks consisting of four points in accordance with the number of flash lamps.

As an example in Fig. 2.29 shows the visualization patterns of the flow around a flat plate $4 \cdot 10^{-2}$ m wide, located perpendicular to the direction of movement along reservoir (1): in Fig. 2.28—the main carriage moves from right to left. In Fig. 2.29A picture of the formation of vortices behind the transverse plate when moving from a state of rest with a constant velocity of ~ 0.1 m/s, and Fig. 2.29B when the plate moves from a state of rest with a constant acceleration of 0.1 m/s^2. Both pictures were photographed when the plate has passed a distance equal to two chords from the moment of start.

In the first case, the formed vortex structure tends to rapidly collapse. Fig. 2.29 shows the initial stage of this process. The form of the confusion zone behind the plate is not limited. There are secondary vortex formations, the growth of which turns the trail into

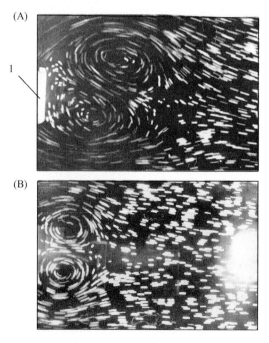

Figure 2.29 Examples of visualization of the flow around a transverse plate with different laws of motion: (A) plate movement from rest at a constant speed of 0. 1 m/s; (B) plate movement from rest at a constant acceleration of 0.1 m/s^2; (1) towed plate [297].

the Karman path. In the case of motion with constant acceleration (Fig. 2.29B), a clear picture of the wake is recorded. This is, as before, two symmetric elongated vortices formed at the time of the start of the plate. Resistance forces in the first and second cases will differ. When moving at a constant speed, the energy used to move the plate is spent on vortex formation in the stream. In the second case, there is no energy exchange between the wake zone and the main stream. Energy is only initially spent on the formation of this zone, and subsequently is spent mainly on overcoming the inertial force acting on the accelerated moving body. Moreover, the role of the additional "added mass" is played by the wake-up zone, therefore, energetically more advantageous from the point of view of reducing the resistance force will be such modes of movement, in which the energy spent on the accelerated movement of the body will be less than the vortex formation energy in the process of movement of the same body with constant by speed.

The experimental setup presented allows solving a wide range of scientific and technical problems. It can be used to obtain complex data, including the kinematic characteristics of the velocity field of a fluid over a finite period of time, with simultaneous recording of integral characteristics of loads and the laws of body motion in three degrees of freedom for bodies of different configurations. It is a universal tool for the instrumental study of a number of tasks in the field of nonstationary fluid mechanics.

2.7 Visualization of the vortex structures in the wake of the oscillating profiles

To understand the physical processes of flow oscillating profiles were carried out numerical and experimental studies of the wings, oscillating in the "swing," "immersion," and "swing-immersion" modes [65,72,121,264,265,301,318,437,442]. In Ref. [458], a viscous incompressible fluid flow around a oscillating surface was theoretically investigated. V.V. Babenko designed a hydrodynamic stand of small turbulence and a number of devices for carrying out various experimental investigations, mainly for BL tasks [18,139,378]. V.V. Babenko, together with M.V. Kanarsky, on this hydrodynamic stand, carried out investigations of the BL with the help of a DISA thermoanemometric apparatus, and together with V.P. Ivanov with the help of a laser anemometer developed and manufactured by V.P. Ivanov.

S.A. Dovgy performed extensive experimental investigations on the hydrodynamic stand with the help of the earlier-mentioned equipment on various types of wings and plates oscillating in the "immersion" mode. The hydrodynamic stand length was 7 m, and the working area was 3 m, the cross section of the working section was 0.09×0.25 m, the working speed range was 0.03–1.5 m/s. The main attention was paid to the visualization of vortex structures on the surface of oscillating bodies and in the wake of them at different oscillation frequencies. The vortex wake of oscillating bodies in the reverse motion was investigated in three cases: near the water surface, in the undisturbed flow and near the streamlined flat bottom, that is, in the BL that occurs at the bottom of the hydrodynamic stand. Other tasks

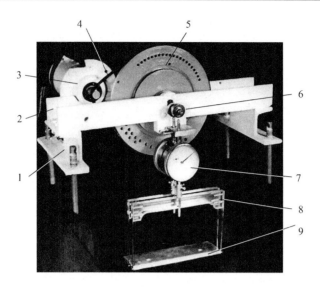

Figure 2.30 Device for transverse oscillations of various bodies. Designations are given in the text [66,139,378].

were investigated; in particular, oscillation of jointly fixed oscillating profiles and jointly fixed transversely streamlined cylinders, flow around the vortex structures of the track behind the profiles placed downstream, flow around the projections located at the bottom of the hydrodynamic stand, structures of flow around the end edge of the oscillating profile and others.

Fig. 2.30 shows a photograph of a device designed and manufactured by V.V. Babenko for the implementation of mechanical oscillations of thin plates and wing profiles [66,139,378], modified by S.A. Dovgy for the study of oscillating profiles. The device consists of two channel (1), which was installed on top of the side walls of the working section of the hydrodynamic stand. On top on the channels (1), (2) is fixed, on which a electric motor (3) is installed, whose shaft speed changes smoothly in a predetermined range. On the motor shaft, a sleeve was fixed with a groove in which a rubber belt (4) was installed, connected to disk (5). This disk was rigidly fixed on axis 6, mounted with two bearings on the side walls of the channel (2). On one side, axis (6) was fixed with bearings two cylinders having different defined eccentricities. From the bottom, on a channel (2), a bracket was mounted, on which two dial gauge (7) were fixed, the axes of which rested through springs into two cylinders fixed to axis 6. In the disk (5), through holes were made. A small-sized bulb with a focusing lens is fixed on the device so that the beam travels in the direction of the through holes, and a photo-diode is installed on the opposite side of the disk, allowing the angular rotation of the disk to be fixed. There are several cylinders with the same outer diameter, the axial holes of which are drilled with different eccentricities, which allows providing given amplitude of oscillations of the axis of the mass when the axis (6) rotates. At the bottom, on the axes of the dial gauge (7), the holders (8) of the body under study are installed. In Fig. 2.30, a flat plate (9) with a width of 2–3 mm, with a span of 150 mm or a NACA-0015 wing profile with chords $b = 2$ mm; 15 mm and 25 mm were mounted on the holder (8). The device made it possible to investigate the structure of the wake behind bodies oscillating in the mode of "immersion," "rocking," or complex motion.

Visualization in the experiments of S.A. Dovgy was carried out by the method of coloring jets. A comb was made, consisting of thin medical needles, the inner diameter of which was $(0.4-0.8) \cdot 10^{-3}$ m. A liquid of different colors was fed to each needle, the Reynolds number at the output was $Re < 40$, so that no comb would be introduced into the flow behind the comb disturbances. Each coloring fluid was connected to a separate balloon, which was located at a different height, so that a jet flowed out of the corresponding needle at a speed equal to the local flow velocity along the height of the cross section of the working section of the hydrodynamic test bench. This made it possible to obtain smooth colored jets, which did not introduce additional shear stresses to the flow. Various chemical solutions were used as dyes: fuchsine, fluorescing, methyl blue and violet indicators. The density of the dye solutions was controlled so that the density of the solutions coincided with the density of water.

In Section 2.6, Fig. 2.22 shows a photograph of the visualization of vortex structures in a span behind the wing at an angle of attack $\alpha = 0$ degrees. It is seen that the flow around the wing is plane-parallel. At the first moment of the beginning of the oscillations of the "swing-immersion" type, an accelerating plane-parallel vortex is formed (Fig. 2.23), behind which a plane-parallel wake is maintained, but with a steady oscillatory motion, the terminal vortex cords quickly form (Fig. 2.24). This pattern is typical for large oscillation amplitudes. In S.A. Dovgy experiments, an oscillation mode of the "immersion" type was investigated, while the oscillation amplitude was small due to the small depth of the flow in the working area of the hydrodynamic stand. Figs. 2.31–2.35 show the results of investigations performed with the arrangement of oscillating wings on the boundary of the BL thickness in the undisturbed flow. All investigations were conducted at the beginning of the work area in the same vertical plane located along the longitudinal axis of the work area with low speeds that in the flow during the experiments there were no disturbances that could affect research results.

Fig. 2.31 shows the results of visualization of oscillations of a small wing with a chord $b = 2$ mm and with an amplitude of oscillation $A = 2$ mm, located on the outer boundary of BL ($h = \delta = 15$ mm), with an incident flow velocity $U_\infty = 30$ mm/s. The number (1) indicates the bottom of the work area, the number (2) indicates the conditional position of the flow boundary, the numbers (3) indicate the vertical marks applied to the transparent side walls of the work area. The distance between marks (3) is 100 mm; however, it is necessary to introduce corrections for the size obtained in the photograph between these marks when calculating the parameters of the vortex structures. The number (4) indicates the location of the wing. The visualization was made by streams of two colors: on top of the wing the trickle was red (red color), and from below, under the wing—blue (blue color).

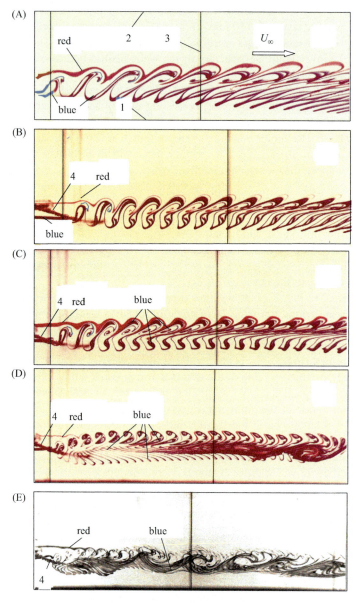

Figure 2.31 Visualization of the oscillation of a small wing with a chord $b = 2$ mm with an amplitude of oscillation $A = 2$ mm, located on the outer boundary of the boundary layer ($h = \delta = 15$ mm), free-stream velocity $U_\infty = 30$ mm/s, oscillation frequency $f = 1$ Hz (A); 1.5 Hz (B); 2 Hz (C); 3 Hz (D); 3.5 Hz (E): (1) bottom; (2) water surface; (3) vertical marks; (4) wing.

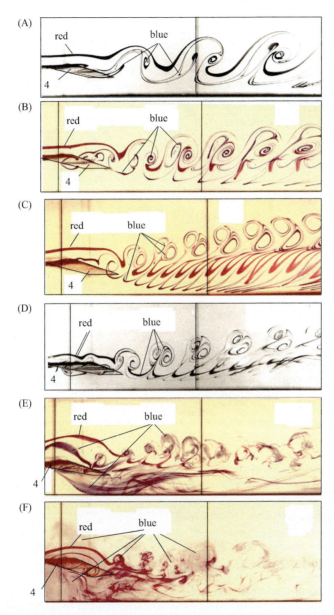

Figure 2.32 Visualization of the oscillation of the middle wing chord $b=15$ mm with the amplitude of oscillation $A=2$ mm, located on the outer boundary of the BL ($h=\delta=15$ mm), free-stream velocity $U_\infty=30$ mm/s, oscillation frequency $f=0.5$ Hz (A); 1 Hz (B); 1.6 Hz (C); 2.1 Hz (D); 2.5 Hz (E); 3 Hz (F). Designations are the same as in Fig. 2.31.

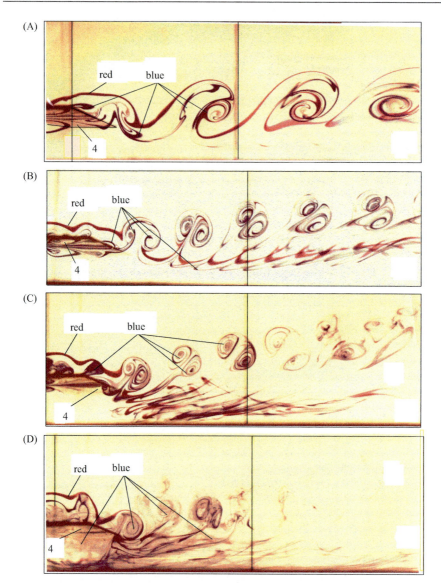

Figure 2.33 Visualization of the oscillation of a large wing with a chord $b = 25$ mm with an amplitude of oscillation $A = 1$ mm, located on the outer boundary of the boundary layer ($h = \delta = 15$ mm), free-stream velocity $U_\infty = 30$ mm/s, oscillation frequency $f = 0.5$ Hz (A); 1 Hz (B); 1.6 Hz (C); 2.0 Hz (D). Designations are the same as in Fig. 8.31.

Figure 2.34 Visualization of the oscillation of a small wing with a chord $b = 2$ mm with an amplitude of oscillation $A = 2$ mm, located in an unperturbed flow ($h = 35$ mm), free-stream velocity $U_\infty = 30$ mm/s, oscillation frequency $f = 0.5$ Hz (A); 1 Hz (B); 1.5 Hz (C); 2.0 Hz (D); 2.6 Hz (E); 3 Hz (F); 3.5 Hz (G). Designations are the same as in Figs. 2.31, (5) wing restraint.

Modeling of a waving fin mover

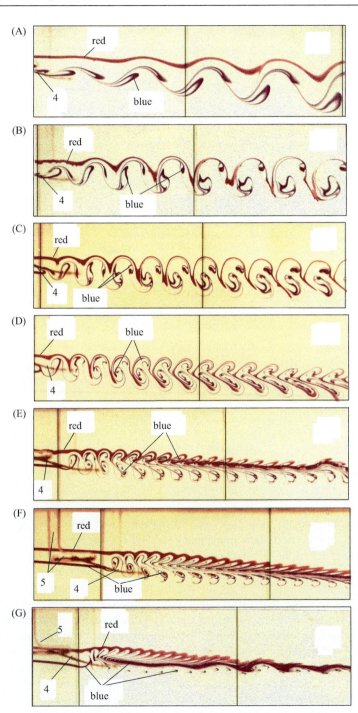

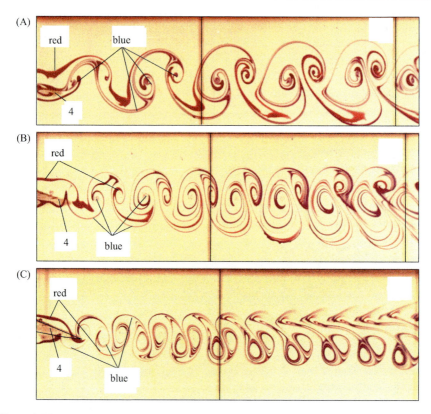

Figure 2.35 Visualization of the oscillation of the middle wing chord $b = 15$ mm with the amplitude of oscillation $A = 2$ mm, located in the undisturbed flow ($h = 35$ mm), free-stream velocity $U_\infty = 30$ mm/s, oscillation frequency $f = 0.5$ Hz (A); 1.1 Hz (B); 1.6 Hz (C). Designations are the same as in Fig. 2.31.

Despite the small amplitude of oscillation and minor wing chord, in Fig. 2.31A, at the oscillation frequency $f = 1$ Hz, a vortex wake of large amplitude of the order of the BL thickness (approximately seven times greater than A) is formed immediately behind the oscillating wing. The perturbations caused by the vortex wake behind the wing almost did not penetrate into the BL at a distance of 100 mm, but at the BL boundary the velocity of propagation of the perturbations is slowed down. Therefore, there is a distortion of the symmetry of the vortices behind an oscillating wing. In the upper region of the vortex wake, the velocity of the unperturbed while increases the speed of propagation of the vortex disturbances, which rise up, so that the thickness of the wake increases. Downstream, the velocity of propagation of the lower part of the wake is constantly slowed down in the BL, and in the upper part of the wake, the velocity of propagation of the vortex wake increases under the influence of the velocity of the main stream. This leads to the dissipation of the vortex. When the wing moves downward, a fluid located near the BL (blue color) rushes upwards in the form of a thin vortex layer, bending the upper contour (red). When the wing moves upward, a layer of red liquid bends to the top. Under the influence of the velocity of the main outflow from the upper part of the vortex, the layer of liquid painted in blue is blown off and twisted down. The upper layers of the flat vortex track, painted in red, do not penetrate and do not mix with the lower layers of the vortex track, painted in blue. Thus, an asymmetrical vortex wake behind an oscillating wing is formed. This and all subsequent photographs capture the development of a single vortex in time. Thus, in the process of development of the first vortex, all components of the vortex develop. The vortex pitch in the vortex track is approximately 25 mm.

At the oscillation frequency $f = 1.5$ Hz (Fig. 2.31B), the character of the vortex is similar to that shown in 2.31 A, but when the wing moves downward, the first vortex in the vortex path is formed by moving the liquid up (blue color) upward, so the first vortex is more vertical. During the subsequent movement of the wing up the liquid on top of the wing twists down (red color) under the influence of the incident flow. With further interaction of the vortex with BL and undisturbed flow in the lower part of the vortex track, a swirling area (blue color) is formed in the vortex, which rotates counterclockwise (against the flow). This leads to the fact that after approximately 50 mm in the vortex track, a gradual increase in the vortex slope along the flow is observed in the upper part, and in the lower part, the vortices under the braking effect of the BL remain in the vertical position and only 100 mm from the wing gradually increase the slope. In the vortex track, a region is formed with a pronounced slope in the upper part and an almost vertical position of the vortices in the lower region. At the same time, a fixed area is clearly visible in the longitudinal axial part of the vortex track located in the wake of the neutral position of the wing, and the thickness of the vortex track decreases compared to the wing oscillation frequency $f = 1$ Hz (Fig. 2.31A). The vortex pitch in the vortex track has decreased and is approximately 14 mm.

With further increase in the frequency of oscillation of the wing to $f = 2$ Hz (Fig. 2.31C), the aforementioned picture of the formation of the vortex track becomes more intense: the upper part of the vortices more intensively twists clockwise (downstream), and the lower part of the vortices more intensively twists counterclockwise (upstream). Therefore, the bending of the vortices is clearly formed and the inhibited area in the longitudinal axial part of the vortex track is formed faster, and the thickness of the vortex track becomes even smaller. The vortex pitch in the vortex track has decreased further and is about 10 mm.

As the wing oscillation frequency increases to $f = 3$ Hz (Fig. 2.31D), immediately behind the wing, the fluid located under the wing (blue color), when the wing moves down, is thrown into the undisturbed flow in the upper part of the vortex wake as a pair of vortices, rotating upstream and downstream, and in the lower part of the vortex wake, weak traces of inhibited vortices remain in accordance with Fig. 2.31C. Paired vortices rise up above the oscillating wing, while their constituent vortices, rotating against the flow are located above the vortices, rotating along the stream. The upper fluid layers above the wing (red color) are sucked in and placed between the upper rows of rotating pairs of vortices (blue color). Immediately after the oscillating wing, the vortex pairs vigorously rise up and after about 50 mm at a given flow rate U_∞, the pair vortices in the upper part of the vortex wake begin to deform under the action of the main flow: the first vortices rotating against the flow decrease their propagation velocity in the flow, and the below them, the second vortices, rotating downstream, increase their propagation velocity, overtake the upper vortices and stretch along the stream, similarly to the upper vortices when the wing oscillates with the previous frequency. In this case, the fluid layers tinted in red are drawn into the longitudinal axial space of the vortex layer. Approximately 60 mm from the oscillating wing in the axial longitudinal part of the vortex wake, a sinusoidal movement of the mixed layers of blue and red colors (upper and lower layers of the vortex wake) is formed. In this layer, periodically, at the maxima of the sinusoid, complex powerful vortices are formed, rotating along the flow (seen at the end of Fig. 2.31D). The vortex pitch in the vortex track has decreased further and is about 9 mm.

As the oscillation frequency increases to $f = 3.5$ Hz (Fig. 2.31E), paired eddies are formed immediately behind the wing, similar to those shown in Fig. 2.31D, and a complex sinusoidal motion is formed. In Fig. 2.31D, the first powerful transverse vortex was formed only 160 mm from the wing, and in Fig. 2.31E, the first powerful transverse vortex formed after 60 mm. The distance between the vortices is about 25 mm (4 times less than at $f = 3$ Hz). The pitch of the primary vortices in the vortex lane has further decreased and is about 6 mm, but after 9 primary vortices large transverse vortices have formed, the spacing between which is 33 mm.

Common to all of Fig. 2.31 is that the dominant in the formation of the wake behind the oscillating small wing is the fluid located under the wing (blue color). She is thrown up above the wing and forms a vortex lane. The fluid located above the wing (red color) is sucked between the formed vortices and does not significantly affect the energy of the formed vortices.

Fig. 2.32 shows the results of visualization of oscillations of an average wing with a chord $b = 15$ mm and with an amplitude of oscillation $A = 2$ mm, located on the outer boundary of the BL ($h = \delta = 15$ mm), the flow velocity $U_\infty = 30$ mm/s. With an increase in the wing chord 7.5 times as compared with Fig. 2.31 already at the oscillation frequency $f = 0.5$ Hz (Fig. 2.32A) a similar vortex wake is formed behind the oscillating wing, as at the oscillation frequency $f = 1$ Hz of the small wing (Fig. 2.31A). The difference is that the amplitude of the vortex track in the wing trail with a chord $b = 15$ mm is larger than that of the wing with $b = 2$ mm, and the vortex trail penetrates deep into the thickness of the BL. The spacing of the vortices in the longitudinal direction of the small wing is approximately 25 mm, and that of the middle wing with $b = 15$ mm is more than two times larger. The basic patterns of the formation of eddies behind the wing are the same as those of the small wing. But, as can be seen in Fig. 2.32A, already within the third main vortex of the vortex, the second vortex almost formed, which rotates upstream and is located above the core of the main vortex rotating along the stream—similar structures of the vortices are observed near the small wing in Fig. 2.31D. So the pair vortex began to form.

Fig. 2.32B shows the results of visualization of oscillations of the middle wing ($b = 15$ mm) with a frequency $f = 1$ Hz. Separate perturbations are formed above and below the wing so that immediately behind the trailing edge of the wing in the first vortex of the vortex track a pair vortex is immediately formed—the same vortex recorded in Fig. 2.31D at a frequency $f = 3$ Hz. The step between the double vortices is 9 mm, and in Fig. 2.31E the step between the large eddies is the same as in Fig. 2.32B.

When the middle wing oscillates at $f = 1$ Hz, the size of the double vortices more than doubles and the pitch alternation of double vortices increases almost four times compared to similar vortices that occur when the small wing oscillates with a frequency $f = 3$ Hz. In addition, the thickness of the vortex sheet behind the wing increases significantly, and the decelerated lower parts of the vortices descend in the laminar BL below the critical layer equal to the extrusion thickness [378]. Outer shell of the double vortex when moving along the stream rotates with the left part upstream, and with the right part along the stream, while inside this vortex shell a pair of internal vortices is formed, the one closest to the wing rotates inside the shell against the stream, and the farthest from the wing rotates along the stream. The disturbances generated by this wing are much larger than those of the small wing, therefore the fluid layers located above the wing (red color) are sucked in by the right outer rotating outer contour of the double vortex envelope so that, as the vortex moves, rotate around the inner second vortex, rotating along the stream. With the further development of the vortex track, as can be seen in Fig. 2.32B, the layers of liquid painted in red color are partially separated from the small second vortex, braked and lowered into the BL and dissipate there. The inhibited layers of the vortices in the BL fall below the critical layer. The spacing of the pair of vortices in the longitudinal direction is approximately 25 mm.

Fig. 2.32C shows the results of visualization of oscillations of the middle wing ($b = 15$ mm) with a frequency $f = 1.6$ Hz. The cited features of the development of vortices with the oscillation frequency $f = 1$ Hz (Fig. 2.32B) are preserved even with an increased oscillation frequency. However, at $f = 1.6$ Hz, the disturbances seem to repel from the rigid surface of the bottom of the working section. Therefore, the vortex path with paired vortices rises up above the wing. The oscillating wing imparts high energy to the pair of vortices, in which the swirling fluid layers, located above the wing (red color), are rotated by the outer shells of the pair of vortices down against the flow into the retarded region of the BL. Inside the pair of vortex in the right vortex, the fluid flowing from the area under the wing (blue color) rotates along the flow. Fluid from the outer region of this vortex is captured by a vortex located on the left and rotating upstream (see the first pair vortex behind the wing) and is ejected between the pair vortices down into the BL region. As a result, in the area of the outer boundary of the BL, two types of flow are formed: at the top, the vortex path of paired vortices will propagate along the stream, and below the upper border of the BL are the inhibited liquid region located above the wing (red color) and below this region the inhibited liquid region located under the wing (blue color). The closer to the streamlined wall, the slower the propagation speed of the inhibited fluid, while this fluid reaches the almost streamlined bottom of the working area. A similar picture is observed in Fig. 2.31B. The spacing of the vortices in the longitudinal direction decreases and is approximately 20 mm.

Fig. 2.32D shows the results of visualization of oscillations of the middle wing (b = 15 mm) with a frequency $f=2.1$ Hz. The regularities of the formation of the vortex track of paired vortices in the wake of the oscillating wing are kept similar for the frequency of the oscillation of the wing with $f=1.6$ Hz. The difference is that the vortex structures are formed above and below the surfaces of the wing, and not behind the wing. In addition, the lower hindered areas of the vortex track do not descend into the BL, but gradually rise above the thickness of the BL, form a sinusoidal wake like the one shown in Fig. 2.31C for oscillation of a small wing with a frequency $f=3$ Hz. The vortex spacing in the longitudinal direction has increased and is about 33 mm.

Fig. 2.32E shows the results of visualization of the average oscillations ($b=15$ mm) with a frequency $f=2.5$ Hz. The energy of the disturbing motion generated by the wing has increased. As a result, when moving upwards, the wing captures the fluid located under the wing and throws this fluid (blue color) and the fluid located above the wing (red color) into the undisturbed flow. Double vortex-like vortices are formed above the wing, which are strongly deformed in the wake. When the wing moves down, the fluid located under the wing (blue color) is intensively discarded. This liquid is discarded from the bottom of the working area and moves with dissipated layers in the BL.

Fig. 2.32F shows the results of visualization of oscillations of the middle wing ($b=15$ mm) with a frequency $f=3$ Hz. The energy of the disturbing motion generated by the wing increased even more. The result was a suction force under the wing. When moving up the wing captures the fluid located under the wing and twists it in the form of a single vortex located under the wing in front of the wing and behind the rear edge of the wing (blue color). The liquid located above the wing (red color) is sucked into the wing in contrast to Fig. 2.32E and forms a single vortex behind the wing, which collapses in the wake behind the wing. When the wing moves down, the fluid located under the wing (blue color) and above the wing (red color) is intensively sucked in. The liquid (red color) is thrown from the bottom of the working area, sucked under the wing and forms a double vortex near the front edge of the wing, and behind the wing is displaced by dissipated layers moving in the BL. The liquid (blue color) is ejected by a dissipated cloud against the flow; it surrounds the wing in height equal to the thickness of the BL, and moves in large swirling lumps downstream.

Fig. 2.33 shows the results of visualization of oscillations of a large wing with a chord $b=25$ mm and with an amplitude of oscillation $A=1$ mm, located on the outer boundary of the BL ($h=\delta=15$ mm), the free-stream velocity $U_\infty=30$ mm/s. The wing chord increased 12.5 times as compared with Fig. 2.31 and 1.87 times as compared with Fig. 2.32. With the oscillation frequency $f=0.5$ Hz (Fig. 2.33A), a similar vortex wake is formed behind the oscillating wing, as with the oscillation frequency $f = 0.5$ Hz of the middle wing (2) (Fig. 8.32A). The oscillation amplitude of the large wing was smaller. The vortex path in the wake of a large wing is formed differently than the middle wing (Fig. 8.32A). So immediately above the wing near the rear edge from below the wing, a vortex (blue color), formed near the rear edge, is ejected upwards from under the wing. This vortex rises up and forms a double vortex (blue color) above the profile and behind the rear edge of the wing, which sucks and twists the liquid above the upper surface of the wing above it. From the bottom surface of the wing, a layer of fluid (blue color) is sucked into the primary double vortex, which forms the outer surface of the second double vortex, on the outer surface of which the fluid layer moves from the outer surface of the wing (red color). This layer covers the outer surface of the double vortex and rotates along the inner bottom surface of the first double vortex, forming a small second vortex, in the core of which layers of blue and red colors rotate. Below from the system of double vortex two layers of liquid flow out: from the second small vortex a layer of red color flows out, which outside covers the third double vortex. From the inner surface of the first vortex of a pair of vortices, a layer of liquid is blue color, which forms the outer surface of the subsequent double vortex. When moving away from the wing at pair vortices, the inclination to the streamlined bottom of the working section decreases and the sizes of pair vortices increase. When a large wing oscillates, a paired vortex formed immediately behind the trailing edge, and when the middle wing oscillated (Fig. 2.32A), only the second vortex almost formed inside the third vortex, which rotates upstream and is located above the core of the main vortex rotating along the flow—similar structures eddies are observed near the small wing in Fig. 2.31.

Fig. 2.33B shows the results of visualization of oscillations of a large wing with a frequency $f = 1$ Hz. Already on the leading edge of the oscillating wing, small vortices were formed above and below, which distorted the rectilinear shape of the streams, visualizing the flow around the wing profile.

When flowing around the middle wing (Fig. 2.32B), a clear configuration of the double vortex formed in the second pair vortex behind the rear edge of the wing. When flowing around a large wing, a clear configuration of a double vortex formed behind the trailing edge of the wing, moreover, compared with the middle wing, the shape of the first vortex in the pair vortex is practically preserved, and the shape of the second and subsequent vortices in the pair vortex is deformed and increases. When the large wing oscillates, the pair vortex rises up, remains as a solid vortex, is carried away by an undisturbed flow along the work area, and bends to the bottom surface. The lower layers of paired vortices of the red color located above the blue color first, they descend down behind the wing inside the BL, and then rise to the outer boundary of the substation. When the middle wing oscillates, these layers descend to the bottom of the work area. The spacing of the pair of vortices in the longitudinal direction is approximately 30 mm.

Fig. 2.33C shows the results of visualization of oscillations of a large wing with a frequency $f = 1.6$ Hz. As in Fig. 2.33B, small vortices formed on the front and rear edges of the oscillating wing above and below, which distorted the straight-line shape of the streams, visualizing the flow past the wing profile. When flowing around the middle wing (Fig. 2.32C), two types of flow are formed on the outer boundary of the BL: the vortex path of paired vortices spreads along the stream and upwards, and upward, and the inhibited region of fluids (red and blue color) lies below the upper boundary of the BL. When flowing around a large wing, the paired vortex path rises intensively upward into the main flow and after the fourth pair vortex; the form of paired vortices dissipates. The second type of flow is absent—below the BL, all the structures (red and blue color) at a distance of 50 mm from the wing descend to the bottom of the working area, then repel slightly upward and become blurred. The spacing of the pair of vortices to their destruction in the longitudinal direction increases and is approximately 33 mm. The resulting flow pattern resembles rather a flow pattern around the middle wing with a wing oscillation frequency $f = 2.1$ Hz (Fig. 2.32D).

Fig. 2.33D shows the results of visualization of oscillations of a large wing with a frequency $f = 2$ Hz. Already above the wing a large vortex was formed, which above and behind the rear edge of the wing formed a double large vortex. The next double vortex is blurred and consists from blue color. Behind this second pair vortex is a deformed third pair vortex of the red color, the shell around which is blurred and consists of a blurred blue color. The next fourth double vortex (blue color) is almost blurred. All four vortices are located at a distance of about 60 mm, and the pitch between them is 20 mm. Further, all the vortices are destroyed, and only the blurred trace of the blue color is visible. The area located under the wing consists of a blurred blue color, which is caused by the suction force arising under the large oscillating wing. This area is located down to the bottom. Behind the trailing edge, layers of liquid from the outer shell of the first pair vortex are red color and fall vertically downwards and are arranged below the blue color in a thin layer. This layer is then washed up as the lower part of the subsequent pair vortices and, after their destruction, moves in the BL as a mixture of longitudinal layers of red color and mostly blue color. The given picture of the flow visualization around a large oscillating wing is similar to the picture of the flow around the middle wing, oscillating at a frequency $f = 3$ Hz (Fig. 2.32E).

Thus, the peculiarities of the formation of a vortex wake behind oscillating wings located on the outer boundary of the BL and having (conditionally) small, medium and large wing chord values at a low velocity of the main flow $U_\infty = 30$ mm/s are considered. Found a number of features:

- For all the examined wings, the vortex wake pattern is similar at low oscillation frequency $f = 0.5$ Hz. In small wings, pair vortices did not form due to the low energy of the disturbing motion in the wake behind the wing.
- Paired vortices behind the middle and large wings began to form already at $f = 0.5$ Hz and formed at $f = 1$ Hz, and in the small wing only at $f = 3$ Hz.
- For a small wing, the vortex wake is located in the area of the outer boundary of the BL, and for the large and medium wings, at all investigated oscillation frequencies, the vortex wake captures the entire area of the BL.
- The fluid located under the wings (blue color) constitutes the main area of the vortices. This is due to the fact that the visualization jet (red color), supplied above the wing, is located at a greater distance from the surface of the wing profile, while the jet, supplied

below the wing (blue color), and is very close to the surface of the wing. This is done in order to explore a larger cross section of the vortices in the wake.
- In the middle wing at $f=2.5$ and 3 Hz, and in the large wing at $f=1.6$ and 2 Hz, a suction force occurs under the oscillating wings.

The vortex wake behind the small and medium wings located in the undisturbed flow was also investigated. Fig. 2.34 shows the results of visualization of oscillations of a small wing with a chord $b=2$ mm and an amplitude of oscillation $A=2$ mm located in an unperturbed flow ($h=35$ mm), whose speed was $U_\infty=30$ mm/s. As in the previous cases, the main energy in the wake of the oscillating wing is formed by the flow flowing around the bottom profile (blue color). In Fig. 2.34A small wing oscillates at a frequency $f=0.5$ Hz. Footprint behind the wing has a sinusoidal shape. Only at the maxima of the jets of blue color nascent elongated vortices, blown away by the main flow, twist in and against the flow.

In Fig. 2.34B, the small wing oscillates with a frequency $f=1$ Hz. At this frequency, the sinusoidal shape of the wake behind the small wing changes: the vortices, blown off from the maxima of the oscillating wake, acquire a rounded shape; a circular vortex is gradually formed, streamlined outside and above (red color). This symmetrical structure differs from the shape of the vortices behind a small wing at the same oscillation frequency when the wing is located in the upper boundary of the BL (Fig. 2.31A). The vortex pitch is approximately 33 mm.

In Fig. 2.34C, a small wing oscillates at a frequency $f=1.5$ Hz. In each rounded vortex, the size and intensity of internal vortices (blue color) increase, and the vortices located below have a significantly larger size of the vortices located above. The movement in these vortices is complex. So a piece of liquid in each circular vortex from the bottom rises up to the bottom of the small vortex. Here the direction of movement is divided: on the right and on the left, the particles rise up, but on the right, the particle twists up against the flow, rises up and down, then moves against the flow, forming a small vortex. To the left of this vortex, particles of a liquid of blue color move upward along the contour of the rounded vortex in the flow, descend down the contour of the rounded vortex, then spin in the small internal vortex located inside the large vortex above the internal small vortex, and descend to form the lower internal vortex. Since the energy of the lower small vortex is greater than the energy of the upper small vortex, the lower part of the rounded vortex begins to deform—ahead of the upper part of the rounded vortex. The pitch of the eddies decreases and is approximately 20 mm. Such a pattern of formation of a vortex wake is completely different from a similar pattern at the same frequency of wing oscillation, but at its location on the outer boundary of the BL (Fig. 2.31B).

When the frequency of the oscillation of the small wing increases to $f=2$ Hz (Fig. 2.34D), two small closed vortexes form at the bottom and at the top inside the circular vortex. The upper small vortex rotates downstream, and the lower small vortex is upstream.

These two small vortexes increase the deformation of the rounded vortex, which becomes arrow-shaped, moreover, due to the higher energy of the lower small vortex, the lower part of the arrow-shaped common vortex is pulled out along the stream more than its upper part. In this case, the vortex pitch decreases even more and is approximately 15 mm. Such a pattern of formation of a vortex wake is completely different from a similar pattern at the same frequency of wing oscillation, but at its location on the outer boundary of the BL (Fig. 2.31C).

At the oscillation frequency of the small wing $f=2.6$ Hz (Fig. 2.34F), the vortex pitch significantly decreases and the deformation rate of the main vortex increases. Already the third vortex behind the wing begins to deform, and beyond the eighth vortex, an axial flow of both colors of liquids is formed, with remnants of swept vortices along the periphery. The same picture is observed in Fig. 2.34F with $f=3$ Hz.

With a small wing oscillation frequency $f=3.5$ Hz (Fig. 2.34G), an arrow-shaped vortex forms immediately behind the wing, along the longitudinal axis of symmetry of which the axial flow of both colors of liquids, which has remnants of arrow-shaped vortices peripherally, becomes similar to that shown in Fig. 2.34E and F. After 50 mm, this axial flow acquires a sinusoidal form, which downstream transforms into a nonlinear counterflow wave (compare with Fig. 2.31E).

Fig. 2.35 shows the results of visualization of oscillations of an average wing with a chord $b=15$ mm and with an amplitude of oscillation $A=2$ mm located in an unperturbed flow ($h=35$ mm), whose speed was $U_\infty=30$ mm/s. In Fig. 2.35A mean the wing oscillates with a frequency $f=0.5$ Hz. A laminar BL is formed on the upper and lower surfaces of the wing located in the flow. When the wing moves down, the liquid (blue color) located under the wing in its BL comes off the trailing edge of the wing and rises up and under the action of the incident flow, it curves along the stream smoothly along the arc, forming the outer shell of a large vortex (in Fig. 2.35A—this is the first vortex behind the wing).

At the same time, the liquid located in BL below the wing's farther from its surface is also directed upward behind the rear edge of the wing, but already at some distance from the rear edge of the wing. This liquid has a more contrasting color, since it is directed upwards inside the outer shell of a large vortex that is forming, as it were, in its inhibited region behind the vortex shell, which has a less contrasting color, since the vortex shell is washed away by the incoming flow. Under the "protection" of the shell, the second layer of fluid (blue color) has a greater contrast also because it is directed from the wing's BL in height farther from the wing surface; therefore, the local velocity of this layer is higher than that of the fluid forming the vortex shell. Due to this, below the rear edge of the wing a small vortex forms and twists against the flow inside the large vortex. When the wing moves downward, the fluid located above the wing (red color) accelerates and begins to flow around the outer shell of a large vortex. In this case, from the maximum of the outer shell of a large vortex, a part of its surface comes off, and a second small vortex is formed at the top, located outside the shell and rotating along the flow. The liquid, colored in red, flows around the second small vortex, distorts its shape and flows around the outer shell of the developing large vortex. These photos capture the development of a single vortex in time. In the process of development of the first vortex, all components of the vortex develop. So already on the second vortex you can see a fully formed twin vortex. At the same time, in the second and subsequent stages of development of a large vortex, the external liquid (red color) is sucked into the upper small vortex and rotates in it together with the blue color. The step between the pair of vortices was about 50 mm. All subsequent stages of development of the first vortex show that the outer shell of the large vortex below, being in the inhibited region under the action of the first small vortex rotating against the flow, is deformed, since in this region the local velocity of propagation of the wake behind the oscillating wing decreases, and the slope changes pair vortex. A similar form of a double vortex is fixed in Fig. 2.32B when the same wing oscillates at a frequency $f=1$ Hz at the outer boundary of the PS. But the slope of the pair vortex is directed in the opposite direction under the action of the velocity of the main flow.

In Fig. 2.35B, the oscillation frequency of the middle wing increased and amounted to $f=1.1$ Hz. With increasing frequency of wing oscillations, all the above-mentioned features of the formation of a pair vortex are preserved. The outer shell of a large vortex formed a liquid of red color. The size of the first small vortex, inside of which the red color rotates, has increased. The step between the pair of vortices decreased and amounted to about 30 mm.

In Fig. 2.35, the oscillation frequency of the middle wing increased further and amounted to $f=1.6$ Hz. The development of the pair vortex was fixed in the same way as in the previous cases, but starting from the fourth vortex, the pair vortices turned downwards, and the tails left above, blown off by the main stream. This can be explained by the fact that at such a frequency of oscillation the wing began to generate thrust, which increased the speed in the wake of the oscillating wing. The step between the pair of vortices decreased and amounted to about 25 mm before the coup of the pair of vortices, and after the coup, 20 mm.

In [437], for the oscillation mode of the "swing-immersion" wing with large oscillation amplitude, a qualitatively similar pattern of formation of pair vortices was recorded.

Experimental investigations of the characteristics of the boundary layer on a smooth rigid plate

3

3.1 Hydrodynamic stand of low turbulence

The results of the investigations [5,15,28,42,43,45,62,63,68,71,73,76,77,94,115,123,198] revealed a number of physical features of the swimming of aquatic organisms, which in technical investigations have not yet found wide understanding and development. Body systems of aquatic organisms interact with each other in the process of movement. In aquatic organisms, automatic and active control takes place (Part I: Experimental Hydrodynamics of Fast-Floating Aquatic Animals):

- Changes in body shape and fins.
- Changes in the geometric parameters of the skin.
- Regulation of the mechanical characteristics of the skin by tension of the skin with muscles, as well as with the help of the circulatory system and innervations.
- Regulation of skin and body temperature.
- Regulation of vibration characteristics of external covers, etc.

Hydrobionic investigations also allowed detecting (Part I):

- Features of nonstationary movement, allowing reducing the resistance and economically using the energy of hydrobionts.
- Methods of control and stabilization of some coherent vortex structures (CVS) of the boundary layer (BL).
- Features of the interaction in the boundary layer of various disturbances entering in the BL from its different boundaries.
- Combined drag reduction methods.
- Features of fluid flow through a single slot and a system of slots.
- Influence of the xiphoid tip and specific deepening on the body of swordfish.
- Influence of microbubbles located in the feather cover of penguins on their swimming speed, especially when they leave the water.
- Features of the adaptive shape of the body and its individual parts, etc.

Verification of some of these identified features of the functioning of hydrobionts was performed, in particular, in Chapter 1, Nonstationary motion, nonstationarity of the movement of hydrobionts, and Chapter 2, Modeling of a waving fin mover, features of the waving fin mover of hydrobionts. In Chapter 6 (Part I), Experimental investigations of friction drag, experimental investigations of some characteristics of the BL of hydrobionts were performed under their various swimming regimes, and the mechanical characteristics of the skin of hydrobionts were measured. Received results showed that the BL of hydrobionts differ significantly from the known analogous classical characteristics of the BL during the flow around modern technical objects. All this has substantiated the implementation of verification investigations of the characteristics of BL on the analogs of the skin of hydrobionts.

To perform the indicated verification investigations, it is necessary to investigate the characteristics of the laminar, transitional and turbulent BL in the flow past a rigid smooth plate, and then conduct similar investigations on analogs of the skin of hydrobionts. V.V. Babenko designed a hydrodynamic stand of low turbulence (Fig. 3.1) [18,139,378]. When designing the stand, the installation scheme given in Ref. [611] was adopted as a prototype. Calculation of hydraulic losses, dimensions and geometry of the stand units is made in accordance with the main provisions when designing aero-hydrodynamic pipes [378]. Fig. 3.1 shows a hydrodynamic stand with a length of 7 m. Water using a centrifugal pump (1) through nozzles (3) enters the damper (4). Then through the confuser (5), made of organic glass, water enters the working part (6) and then through the diffuser (8) into the water collector (9), from where the hoses (12) flow into the supply tank (11) and then enter the pump (1). The operating speed range is 0.05−1.5 m/s.

The device of devices for experimental investigations of hydrodynamic stability and their photographs are given in [18,523,612]. In order to eliminate the influence of vibrations arising from the pump operation, rubber hoses (12) were used to measure the stability of the laminar BL, and the pump was installed on a massive foundation, isolated from the foundation of the experimental stand. For the same reason, an unclosed hydrodynamic stand was selected, and the stand is located on massive

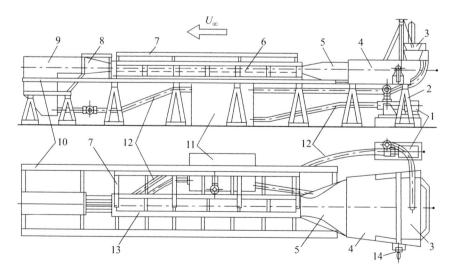

Figure 3.1 Hydrodynamic stand of small turbulence [18,139,378].

frame (2). In cross section, nozzles (3) have a truncated cone shape and are located along the width of damper (4). In the upper part of the nozzle, a distributor is installed for uniform water supply throughout the nozzle. Inside the nozzle, in its lower part, there are two stainless steel nets, between which there is a layer of aluminum chips (3) cm thick. Water flows from the pump through the distributor into the nozzles, penetrates through the layer of metal chips into the lower part, then goes through the labyrinth partition in the upper part of the nozzle and a uniform stream flows into a damper (4). With the help of a winch (14), the height of the nozzle relative to the damper is adjusted. The nozzles are fixed on the cables going to the winch and separated from the supporting frame and damper with rubber gaskets. The side walls of the damper in the area between the grids and the confuser are lined on the inside with sheets of Plexiglas, and the bottom of the damper is leveled with ceresin. A removable cap is installed on top of the damper. The damper smoothly passes into the confuser, the contraction coefficient of which is (10). The junction between the confuser (5) and the working part 6 is also aligned with ceresin. The diffuser (8) has an adjustable cap, which allows for smooth docking with the cap of the working part when its angle of inclination changes. The short diffuser is joined to the r-shaped water collector (9), in which the grids are installed to eliminate the effect of the reverse flow of water on the flow in the working section, which is caused by low speeds during the experiments. A removable cap is installed on the sump. The working part has a removable cap (13) with levers (7), and channel (2) is fixed on frame (2).

The main design features of the hydrodynamic stand are enclosed in the working part, the cross section of which is shown in Fig. 3.2. The working part (1) (length: 3 m, width: 0.25 m, height: 0.09 m) is structurally designed so that it allows to install the second upper bottom (7) to conduct research on various ways to control the BL. In necessary for addition, in the space between the first (8) and second (7) bottom you can place the devices carrying out various experiments. Double bottom allows you to perform a distributed angular suction (drain) of the angular BL at the bottom in the working part. The cap (4) designs also allows you to suck out (drain) the corner BL that forms in the upper corners, and drain it, like the lower corner BL, into the collector (9). This allows you to increase the effective width of the working section. Provided for draining the BL along the front and rear edges of the working area. The design allows you to horizontally install the second bottom by applying three types of adjustment screws.

Cap in the working part is made removable and hermetically attached to the working part. It is placed on the frame, which with the help of the adjusting screws mounted on the racks can change the inclination with respect to the bottom, located horizontally. By varying the inclination of the cover and the amount of suctioned angular BL, uniformity of speed and a small degree of turbulence along the entire working section at different speeds are achieved. The cap is lifted with the help of levers fixed on racks (2). The side walls of the working part are made of silicate glass with a thickness of 7 mm, and the cap are made of organic glass with a thickness of 10 mm. This allows the use of visual methods for measuring the velocity field and qualitative research methods. Due to the need to change the position of the support

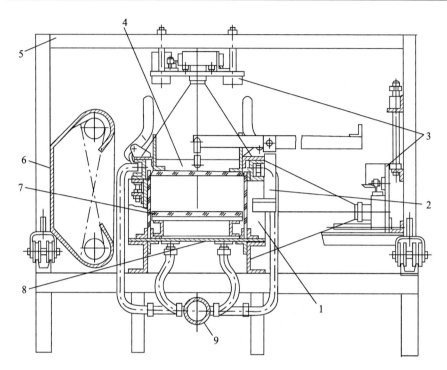

Figure 3.2 The cross section of the hydrodynamic stand [18,139,378].

frame, the side windows on the first floor and in the area of the racks are fixed with solid putty (70% chalk, 15% red lead or lead fiber, 15% natural drying oil), and around the support frame: with soft putty (70% beeswax, 20% rosin, 10% grease). To record the measurement results, a shooting cart (5) was designed, moving along the working part. It allows you to separately or simultaneously photograph the velocity field in a fixed place or along the working part, moving with the flow velocity. The cart rests on four rubber wheels needed to dampen vibrations caused by the actuation of the camera shutter. Spring stops with ball bearings allow the camera dolly to move smoothly and straightforwardly. You can move the side of the shooting panel (3) in two directions, and the top, in one. Screen (6) is installed on the film trolley, which is necessary for photographing the velocity field.

Fig. 3.3 shows a photograph of a device designed and manufactured by V.V. Babenko for the implementation of mechanical vibrations of thin plates and wing profiles [139,378]. The device consists of a frame (1), which was mounted on top of the side walls of the working section of the hydrodynamic stand. On top of the frame (1), a channel (2) is fixed, on which are installed two DC electric motors (3) MN-145 and (5) MU-24, the shaft speed of which changed smoothly in the specified range. Sleeves with grooves were fixed on the shafts of electric motors. Along the longitudinal axis of symmetry of the device on the side walls of channel (2), axis (9) is fixed with the help of bearings. In the middle on the axis

Experimental investigations of the characteristics of the boundary layer on a smooth rigid plate

(A) 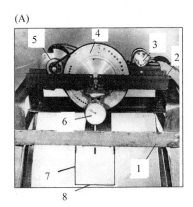 (B)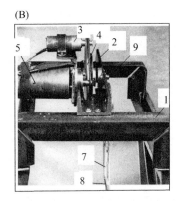

Figure 3.3 Device for transverse oscillations of different bodies. The designations are given in the text: (A) front view and (B) side view [139,378].

(9), the disk (4) is rigidly fixed, consisting of two disks: large and small diameter. On the end surfaces of both disks grooves are made. With the help of rubber wheels, installed in these grooves and in the grooves of bushings fixed on the shafts of electric motors, it is possible to alternately connect the MN-145 engine to the central disk (4) having a small diameter and the MU-24 engine to the central disk (4) having a large diameter. This allows you to adjust in a wide range the frequency of rotation of the disk (4), which has through holes. A small electric lamp with a focusing lens is fixed on the device so that the beam passes in the direction of the through holes, and a photodiode is installed on the opposite side of the disk, which makes it possible to fix the angular velocity of disk rotation. A channel is attached to the channel (2) at the bottom where mass (6) is fixed. The top of the mass axis rests on the cylinder fixed on axis (9). There are several cylinders with the same outer diameter, the axial bores of which are drilled with different eccentricities, which allows providing given amplitude of oscillation of the axis of the mass at axis rotation (9). At the bottom, on the axis of the micro displacement meter, the holder (7) of the body under study is mounted. In Fig. 3.3, various details were mounted on holder (7) for carrying out relevant experiments. A small electric lamp with a focusing lens is fixed on the device so that the beam passes in the direction of the through holes, and a photodiode is installed on the opposite side of the disk, which makes it possible to fix the angular velocity of disk rotation.

To П-shaped beam (2) is attached to the bottom of the sleeve, in which the micro displacement meter (6) is fixed. The top of the axis of the micro displacement meter rests on the cylinder fixed on the axis (9). There are several cylinders with the same outer diameter, the axial holes of which are drilled with different eccentricities, which allows providing the specified amplitude of the axis oscillations during the rotation of the axis (9). At the bottom of the micro displacement meter axis, the holder (7) of the body under study is mounted. In Fig. 3.3, various details were mounted on holder (7) for carrying out relevant experiments.

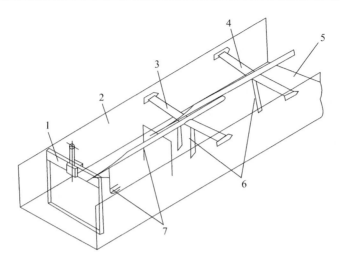

Figure 3.4 Scheme of the supports and the oscillator in the working part of the hydrodynamic stand: (1) vibrator, (2) side walls, (3) support for injection in tellurium streams in boundary layer, (4) support for fixing velocity profiles, (5) bottom of the working part, (6) cathodes, (7) anodes [18,378].

For placement in the flow of tellurium wires, various types of supports were designed and tested. As a result, the investigation of their properties and to reduce the random error when measuring on rigid and flexible plates, supports made of organic glass was used (Fig. 3.4). Tellurium wire was soldered to a needle fixed in the nose of the support (4) for measuring the velocity profile, or to a flat brass strip fixed in the nose of the support (3) for measuring the amplitudes of current-line oscillations. Insulated wires serving as anodes were soldered to the needle and the brass strip. The cathodes 6 are made of stainless steel strips placed downstream in the fairing, and its wire is mounted into the supports (3) and (4) [18]. To visualize the current lines between the cathode and the anode of the holder (3), a voltage of 10–20 V was applied, and to visualize the velocity profiles, a pulse voltage of 500–600 V was applied to the holder (4) in 0.5 s. In the working area (2), (5) at a distance of 10–15 cm from the oscillating tape, mounted on the support (1), a support (3) was installed with tellurium wires to obtain tellurium streams, even lower, at a distance of 15–20 cm from them, and a support (4) with tellurium wires was placed for visualizing speed profiles. The measurement technique is described in detail in Refs. [18,139,378].

In Fig. 3.5 photographs of unified microcoordinate and generators of spatial disturbations (vortex formers) are given. The microcoordinate consists of a unified bracket (1), which is mounted on the cart or on the frame of the working part when the cap is raised (Fig. 3.2). On the bracket (1) fixed micrometers (2) and two guide axes (4), along which the second micrometer moves. On both micrometers (2), axles (7) with hinged fastening are installed to organize their translational movements. On these axes, standardized supports (5), (6) are fixed for fastening tellurium wire and vortex

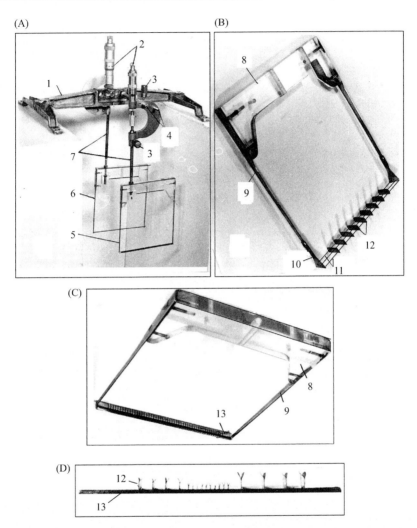

Figure 3.5 Microcoordinates (A) and supports with longitudinal vortex generators (B, C, D): (1) unified bracket, (2) micrometers, (3) clamp, (4) axis guides, (5) support for longitudinal vortex generators, (6) support for tellurium wire, (7) rods for installation of supports, (8) plate, (9) streamlined rods, (10) nozzles, (11) strings, (12) vortex generators, (13) guide plate for installation of vortex generators.

generators. The spatial disturbance generator consists of the same unified parts as the flat disturbance generator (Fig. 3.3). On streamlined rods (9) (Fig. 3.5B) fastened nozzles (10), through the holes in the lower part of which three strings with a diameter of 0.08 mm are stretched. The vortex generators (12) are mounted on these strings and can be placed with different pitch. In Fig. 3.5C a streamlined plate (13) with slots on the upper surface of the plate is attached instead of strings. The pitch of the longitudinal slits is 2 mm. In these slots are installed generators of longitudinal vortices of various sizes. In Fig. 3.5C, instead of strings, a streamlined plate (13) is attached with slots on the upper surface of the plate. The pitch of the longitudinal slits is 2 mm. In these slots are installed generators of longitudinal vortices of various sizes.

When conducting experimental investigations of the physical picture of the structure of disturbances in the BL, the high-quality surface of tellurium wires is extremely important. Various methods for manufacturing high-quality tellurium wires are presented in Ref. [378], which makes it possible to reduce the complexity of their manufacture. A special method was developed for fabricating high-quality tellurium wire by vacuum deposition [205]. V.V. Babenko designed and made a device consisting of two crosses fixed on the edges of two cylinders, one of which went inside the other and was spring-loaded like a rod in a ballpoint pen. A thin wire made of constantan was wound between the crosses, while the spring was squeezed. During the manufacture of tellurium wires, when heated, their length increased, the spring straightened, and the wires were always in a strained position. All parts of the device were made of quartz glass, except for the spring. The standard quartz ampoule is divided into two zones: the first, in which metallic tellurium was located, and the second, in which there was a device for winding tellurium wire. When pumping air from the ampoule (up to 10^{-4} mm Hg), the entire first zone of the ampoule is placed in the stove. The ampoule is heated to 200°C–300°C for 5–10 min. At this time, tellurium is degassing. After heating, the vial is advanced so as to be placed in the second zone of the furnace. The second zone is heated to a temperature of 400°C–450°C, and in the first zone the temperature rises sharply to the melting temperature of tellurium. At the same time, tellurium vapors are deposited on a constantan wire with a thin uniform layer. Using ampoules of different diameters, using this method, tellurium can be sprayed onto bodies of various configurations. A good metallic deposition quality is indicated by a silver metallic deposit on the finished product. Through the wire thus obtained, current can be passed up to 200 times.

Various aspects of hydrodynamic stability were investigated in Refs. [17,19–23,26,378]. Classical curves of neutral stability are obtained, new dependences of neutral curves are proposed, as well as limit neutral curves. The influence of turbulence of the main flow and amplitude parameter on hydrodynamic stability is investigated. The growth coefficients of the disturbing motion, as well as the distribution of instability parameters across the thickness of the BL, have been investigated. The hydrodynamic instability is investigated with a positive pressure gradient and a nonsinusoidal nature of the disturbing motion.

3.2 Experimental methods for investigation the flow structures of a transition boundary layer of a rigid plate

When studying the natural transition of the BL, research is usually performed along the longitudinal axis of the flow. This technique is justified only in the study of the initial stages of the transition, in which there is a flat perturbing motion. In the subsequent stages of the transition, the flow becomes periodic in the transverse direction.

3.2.1 The method of investigation the nonlinear stages of the natural transition

Initially, the BL was comprehensively investigated over the entire cross section along the working section using the tellurium method. Tellurium wires were placed at different distances from the second bottom and in various places in the transverse direction. The investigation was carried out at U_∞ = const along x, and due to a change in the number of Re in certain places in x, various stages of the transition were recorded. Based on the analysis of the $U(z)$ profiles, characteristic points for y and z were determined, at which the kinematic characteristics of the BL were investigated using a laser Doppler velocity meter (LDVM) and a DISA thermo-anemometric instrument. All these measurements were performed with a low degree of turbulence. Then, in the BL in the region of the strain gage insert in the second bottom of the working section, the peculiarities of the alternation of transition stages were investigated at x = const and U_∞ = var. Weight measurement cycle was repeated with an increased degree of turbulence of the main flow.

Variants of a curvilinear bottom were installed in the working section, a curvilinear cover was also installed with a small curvature of the bottom, and a flow in a curvilinear channel was investigated. The specified measurement cycle was repeated.

The same method was used to study the flow on various elastic surfaces.

3.2.2 Method for the study of hydrodynamic stability

The main methods of experimental construction of neutral curves were developed by Shubauer and Scramstad [523] based on the theoretical investigations of Tollmien and Schlichting and described in detail in Refs. [156,167,378,518−521,523,612]. In these investigations, the disturbing motion was formed using a thin oscillating tape. Shubauer and Scramsted experimentally investigated hydrodynamic stability using two sensors of the hot-wire anemometer. In our experiments with a fixed value of the x coordinate, the amplitudes of the oscillations of the disturbing motion of different frequencies were photographed simultaneously at different distances y from the bottom. According to the data obtained for different x, dependences of the amplitudes of oscillations or growth factors on the wave number were built, according to which for a fixed value of x a pair of wave number values with a zero growth factor was determined, and then a neutral curve was constructed. In the first method of constructing a neutral curve, in accordance with the graph of a neutral curve, it is as if moving parallel to the abscissa axis at some constant value of $\beta_r\, v/U_\infty^2$. In other words, they study the behavior of a fixed frequency oscillation when moving it along a plate. In the second method of constructing a neutral curve, in accordance with the graph of a neutral curve, on the contrary, it seems to move parallel to the ordinate axis for a fixed value of the number Re^*: investigate the behavior of disturbing oscillations of different frequencies for a fixed value of x. In this case, oscillations with frequencies located in the middle of the instability region along a neutral curve, where the magnitude of the increase is maximum, do not have time to decay as one moves toward the second branch of the neutral curve. This is due to the amplitude of the disturbances introduced into the BL. The observed oscillation amplitudes with frequencies corresponding to the second branch of the neutral curve are maximum. In Ref. [378], other methods for the experimental construction of neutral curves and features of the investigation of hydrodynamic stability were considered. The investigation of hydrodynamic stability with an increased degree of turbulence of the main flow was carried out with the aid of LDVM [41,314,315].

3.2.3 The method of the investigation of Benny & Lin vortices and Görtler stability

When investigating the nonlinear development of disturbances in the transitional BL of a flat plate, it is of interest to study longitudinal vortex formations of the Benny & Lin vortex type. The same technique also allows experimentally investigating Görtler stability. At the same time, the degree of turbulence of the main flow should also be minimal, so that the stages of the natural transition alternate as slowly as possible and do not interact with the introduced disturbances. To create Benny & Lin vortex-type vortex flows, four types of longitudinal vortex generators (LVG) were used. They were similar in shape, as they were made on the same stamp, but differed in length and height: LVG V1 had a height of $3 \cdot 10^{-3}$ m and a length of $12 \cdot 10^{-3}$ m, respectively, V2 had a height of $5 \cdot 10^{-3}$ m and length $15 \cdot 10^{-3}$ m, V3: $7 \cdot 10^{-3}$ and $18 \cdot 10^{-3}$ m, V4: $15 \cdot 10^{-3}$ and $18 \cdot 10^{-3}$ m.

When studying the patterns of development of Benny & Lin vortices in a flat BL, LVG were placed in various places along the working area, and downstream at a distance of $5 \cdot 10^{-2}$ m from them—tellurium wire to photograph the velocity profile in the transversal direction. The rest of the method was similar to the method of studying linear stability. Introduced into the BL at $x = $ const various LVG, varied the value of U_∞ and the length λ_z by changing the distance between neighboring LVG. The profiles $U(z)$ were photographed at different distances from the bottom. To clarify the details, photographs were taken in certain places along z lines of the current and the $U(y)$ profiles, which were also measured with the aid of the LDMV.

For the investigation of Görtler stability, three types of curvilinear second bottom were established. Görtler stability was investigated in the same way as Benny & Lin vortex stability: in various regions of curvilinear and straight sections with and without a cap, longitudinal vortex systems of various shapes and sizes were introduced, the patterns of development of which were studied depending on U_∞. Measuring the patterns of the development of vortex systems along x at $U_\infty = $ const, $R = $ const and $\lambda_z = $ const seem to be moving along the Görtler stability graph along the line $P = $ const. Varying alternately the values of R, U_∞, and λ_z, it is as if moving along the Görtler stability graph along various lines P. This determines the right branch of the neutral Görtler curve and the region of maximum amplifications of vortex systems. Along the line P = const, one can also "move forward" if at R = const, x = const and λ_z = const for different U_∞. If at R = const, U_∞ = const and x = const change λ_z, then according to the schedule of the Görtler curve it is as if to move along the line G = const. Varying alternately the values of R and U_∞, as if to move along the Görtler curve along different lines G. Thus we can determine the lower minimum value of G and the left branch of the neutral curve.

Comparing the measurement results at λ_z = const, U_∞ = const and x = const for different R, as if moving along the Görtler curve along the lines $\alpha\delta^{**}$ = const.

The proposed method makes it possible to experimentally determine points on the Görtler graph in the indicated ways.

3.2.4 The method for studying the sinusoidal development of systems of longitudinal vortices

The above are experimental research methods for the most characteristic stages of the transition. Similarly, one can study the development of the BL during the natural transition at all stages, including the phase of sinusoidal development of the systems of longitudinal vortices formed in the BL. When installing two generators of simple plane disturbances (Figs. 3.3 and 3.5A) located in the transversal direction from each other (along z) so that their tapes of vibrators are arranged at a certain angle to the longitudinal axis x, a pair of oblique waves is formed. At the same time, the stages of nonlinear deformation of a plane wave are investigated.

The generator of complex flat perturbations (Section 2.7, Fig. 2.30) makes it possible to place several tapes of the vibrator at different distances from each other. By matching the cams accordingly, a simulation of the superposition of several plane disturbances is provided.

The final stage of the transition—the sinusoidal development of the longitudinal vortex systems in the transverse direction—is investigated using a modified disturbations generator. The BL contains the generators of the vortices installed on the stand (Fig. 3.5B) and oscillating in the transverse direction. The rest of the method of experimental research is similar to the above methods.

3.3 Experimental investigation of coherent vortex structures of the transition boundary layer of a rigid plate

When moving, most fish and birds flow around the stream, the BL of which is laminar due to the small size of their bodies. Part of large animals at high speeds of movement flow around the stream with a transitional BL. In the case when the speeds and sizes of animals are such that there is a turbulent BL on rigid motionless bodies, as a result of evolution, nature has developed mechanisms that allow to maintain a quasi-transitional flow regime in the BL to certain limits [29]. Therefore in order to be able to analyze the structural features of the external integuments of animals associated with the force effect of the flow around them, it is necessary first to study the physical picture of the process of turbulence in the BL of a longitudinally streamlined rigid plate. For this purpose, a specially designed hydrodynamic stand of low turbulence (Section 3.1) was used to carry out experimental studies of hydrodynamic stability in a water flow. The tellurium method used to measure the velocity field made it possible to obtain a physical picture of the flow in the BL.

The most detailed process of the beginning of turbulence is described in the works of Schlichting G. [518–521], Prandtl L. [501], Lin C.C. [414,415], Shen S.F. [527], Stuart J.T. [540], Drazin P.C, Reid W.H. [224], Kozlov L.F., Babenko V.V. [378], Boyko A.V. and others [183], and the physical picture of the beginning of turbulence—Knapp C.F. and Roach P.I. [347,348]. The following classification of transition stages was proposed in Ref. [540]: region I is the instability of small wave perturbations; Region II: three-dimensional wave amplification; area III zig-zag development with a vortex system along the stream; Region IV: concentration of the vortex tension and development of a layer with a shift; region V: destruction of the vortex; Region VI: development of a turbulent spots.

The dependence of the resistance of a smooth flat longitudinally streamlined plate on the Reynolds number is shown in Fig. 3.6, where the theoretical curves are given for the laminar flow law in the BL (1), for the transient flow law (2) and for the turbulent law (3). When conducting the experimental data, the Reynolds number loss of stability $Re_{l.s.}$, calculated from x was $5.4 \cdot 10^4$ [22]. Point A in Fig. 3.6 characterizes the number of $Re_{l.s.}$ and corresponds to the beginning of the region (1) of the occurrence of the transition region. From point A to point B, transition areas I–V consistently arise and develop, and from point B to point C, transition area VI [540] develops. Thus the whole process of the onset of turbulence should be

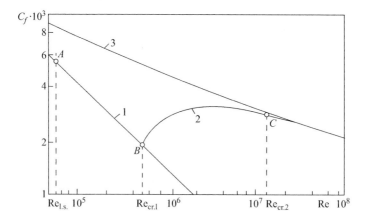

Figure 3.6 The laws of resistance of a smooth longitudinally streamlined flat plate with laminar (1), transition (2), and turbulent (3) flow regimes in the BL.

considered as a single process of flow development from point A to point B, characterized by three Reynolds numbers: $Re_{l.s.}$, $Re_{cr.1}$ and $Re_{cr.2}$.

Under the most favorable flow conditions, the transition of a laminar BL into a turbulent one will take place from point A to point B. However, such factors as initial flow turbulence, curvature, waviness, surface roughness and vibration, pressure distribution on the outer boundary of the BL, etc., can significantly accelerate the beginning of turbulence so that points B and C will begin to move to a practically fixed point A. And the distance between points A and B can, under unfavorable conditions, decrease to about a small value is the same as the distance between points B and C. In some cases (see, for example, Refs. [19,22]) point A, on the contrary, can move to the right simultaneously with points B and C and the distance between these points is substantially will increase. Below are the results of studying the physical picture of the flow in the BL I–III transition regions.

The velocity profiles in the undisturbed BL were investigated. Fig. 3.7 shows a photograph of the velocity profile obtained by visualizing of the flow by the tellurium method. The dimensionless velocity profile in Fig. 3.7A exactly coincided with Blasius' theoretical velocity profile [224,518,520,521]. This indicates that the tellurium method allows obtaining an instantaneous velocity profile in a laminar BL. The OX axis is located along the longitudinal axis of symmetry of the bottom of the work area in the direction of the main flow, the OY axis is directed vertically upwards and the OZ axis is directed to the right from the origin of coordinates in the transverse direction. It is clearly seen in the photograph of Blasius's profile that the velocity profile line is not blurred to approximately 0.7δ, and the velocity profile is greasy blurred in the region

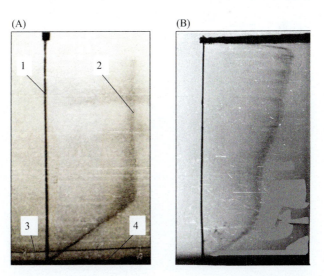

Figure 3.7 The velocity profile of the laminar boundary layer on a flat rigid plate with $x = 74$ cm from the beginning of the plate and $U_\infty = 10.5$ cm/s (A), with $x = 2430$ cm from the beginning of the plate and $U_\infty = 18$ cm/s (B): (1) tellurium wire; (2) velocity profile; (3) current line, visualized by tellurium; (4) plate.

from 0.7δ to $\sim 1.4\delta$. In the undisturbed flow, the velocity profile is blurred due to the increased velocity and diffusion of tellurium. It is not yet clear what caused this region of increased tellurium diffusion in the region of the outer boundary of the BL. Fig. 3.7B shows a profile of the velocity profile, measured at the end of the working section at a higher average velocity of the main stream. The velocity profile is fixed in excess, due to the formation of subsequent nonlinear transition stages in the BL.

Similar speed profiles were photographed when two tellurium wires were installed at different z coordinates. In all cases, the velocity profiles were identical, which confirms the low degree of flow turbulence and indicates a plane-parallel flow in the hydrodynamic bench. At the same time, when flat sinusoidal disturbances were introduced at $y = 0.3\delta$, in the region of maximum amplitudes of the disturbing motion, the velocity profiles had bends in accordance with the phases of the oscillations. Such speed profiles are fixed in Ref. [611] and will be partially present below when visualizing the flow by means of tellurium jets.

The speed profiles were photographed when the tellurium wires were parallel to the plate. No disturbances were introduced into the stream. Tellurium wire was placed along the OZ parallel to the flat bottom at a distance of 5.5 mm. The flow velocity $U_\infty = 13.8$ cm/s. Fig. 3.8 shows the velocity profiles $U(z)$, when the visualization picture was simultaneously photographed when viewed from the side and from above.

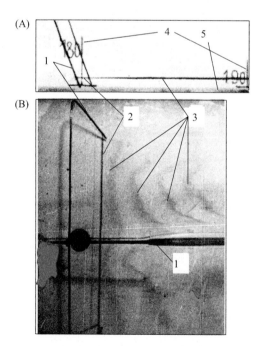

Figure 3.8 Simultaneous photographing of the velocity profile from the side (A) and from above (B): (1) holder for tellurium wires (Section 3.1, Fig. 3.3); (2) tellurium wires for visualizing the velocity profile; (3) velocity profiles; (4) distance markers from the beginning of the working section (cm); (5) flat bottom of the working area.

Visible deformation of the velocity profile in the XOZ plane when the tellurium wire was placed in the main flow, there was no deformation of the velocity profile. On the left in Fig. 3.8B, shadows are visible at the bottom of the work area from the stand and tellurium wire. The distance between the markers (4) is 10 cm.

Fig. 3.9 shows photographs of the velocity profile $U(z)$ at $Re \approx 0.8 \cdot 10^5$ and at different distances of tellurium wire from the bottom surface of the working section. The distance between the transverse marks on the bottom of the working part is 0.1 m; the longitudinal marks are made along the longitudinal axis of symmetry of the bottom. In Fig. 3.9, a generated specific system of longitudinal vortex structures is photographed in the working section of a hydrodynamic stand, the velocity profile $U(z)$ of which depends on the vertical coordinate. The maximum amplitude $U(z)$ is fixed at $y = 3 \cdot 10^{-3}$ and $5 \cdot 10^{-3}$ m. The BL thickness δ is about $12 \cdot 10^{-3}$ m at the measurement site. In the region of the critical layer (I), the velocity profile is $U(z)$ differs from the velocity profile at other vertical coordinates, in which a pair of longitudinal vortices is formed, characteristic of the working part of the hydrodynamic stand. These photographs show the dynamic structure of the longitudinal vortex structures, which near the outer boundary of the BL are smoothed so that a uniform profile $U(z)$ is formed on the outer boundary. The photographs show a tendency to twist the front of the velocity profile along the OZ axis (the appearance of vortex tubes)—there is a clear dark tellurium line behind which a diffusible tellurium cloud is formed.

Fig. 3.10 shows a diagram of the deformation of the velocity profiles $U(z)$ in the transverse direction during its development along the plate. It was found that the velocity profile in the transverse direction is more informative in order to illustrate the development of disturbances along x. According to the shape of the velocity profile in a flat XOY, it is almost impossible to measure such deformations as compared with the velocity profile in a flat XOZ.

Even with a low degree of flow turbulence and in the absence of forced perturbations, three-dimensional deformation of the plane natural perturbation in the BL is observed. This process is recorded by photographing the velocity profiles shown in Figs. 3.6 and 3.9. When the tellurium wire was placed at the coordinate of the critical layer (approximately the displacement thickness), the informatively of the velocity profile becomes especially significant. Depending on the coordinate, the development of such a deformation changes, this leads to the transformation of the initial small plane disturbances into small three-dimensional disturbances.

The development of disturbances along a flat plate during the natural development of the BL was also investigated using tellurium wires, from which tellurium jets flowed, propagating in the BL along the OX axis (Fig. 3.11). The support for tellurium wires was made in the form of a horizontal thin strip placed in the transverse direction, to which three short tellurium wires were soldered in the horizontal plane. At the ends of the horizontal strip was bent by 90 degrees. On each side, three short tellurium wires were soldered to these curved vertical end portions of the flat strip. This made it possible to produce tellurium jets simultaneously in the horizontal and vertical planes of the main flow, which made it possible to analyze the development of a disturbing motion in the flow volume. The flow visualization

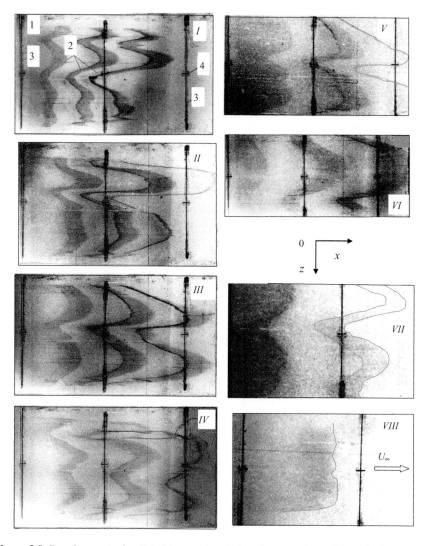

Figure 3.9 Development of a disturbing motion during the natural transition of a laminar boundary layer to a turbulent one on a rigid plate, depending on the coordinate of the tellurium wire $y_{t.w.}$ at $U_\infty = 6.7 \cdot 10^{-2}$ m/s and $x_{t.w.} = 1.1$ m: I-$y_{t.w.} = 3 \cdot 10^{-3}$ m; II-$y_{t.w.} = 4 \cdot 10^{-3}$ m; III-$y_{t.w.} = 5 \cdot 10^{-3}$ m; IV-$y_{t.w.} = 6 \cdot 10^{-3}$ m; V-$y_{t.w.} = 7 \cdot 10^{-3}$ m; VI-$y_{t.w.} = 8 \cdot 10^{-3}$ m; VII-$y_{t.w.} = 9 \cdot 10^{-3}$ m; VIII-$y_{t.w.} = 12 \cdot 10^{-3}$ m; (1) tellurium wire; (2) tellurium clouds; (3) transverse and (4) longitudinal marks.

showed that with a low degree of turbulence of the main flow, tellurium jets develop in a plane-parallel fashion without disturbances along the length. Fig. 3.11 shows simultaneous photographing from the side and from above both sinusoidal and nonsinusoidal oscillations of the disturbing motion. Visualization has shown

Figure 3.10 Scheme of deformation of a velocity profile $U(z)$ in a boundary layer: (1) the flat bottom of the working section; (2) tellurium wires; (3) velocity profile in the XOY plane; (4) velocity profiles in the XOZ plane in nondisturbed flow; (5) velocity profiles in the XOZ plane in a critical layer of the boundary layer.

that the disturbing motion is quickly transformed from flat to three-dimensional. When a wave is folded during nonsinusoidal oscillations, in contrast to sinusoidal oscillations, it is not a double thickness of the stream that is fixed, but an irregular solidified spatial loop. The development of the oscillation amplitude of the transverse velocity v' of the disturbing motion along the working part of the hydrodynamic test bench is shown in Fig. 3.12. In each photograph, the oscillation frequency was located near the frequency of the second neutral oscillation along a neutral curve [14]. It was found that at the beginning of the plate in the region of the point of loss of stability of the amplitude, the velocity v' oscillations are maximal and steepest, and the process of wave formation is more pronounced. Along with wrapping the wave, the comb was pulled forward and upward. The greater the x coordinates, the flatter the wave was.

In Fig. 3.12, it can be seen that the oscillations very quickly become nonlinear. At one wavelength, the crest of the wave is wrapped, and pulled forward. However, if this leads to the formation of only the outer part of the wave, the sinusoidal curve will practically follow further. After the second or third wave in the development of oscillations in the BL, the crest of the wave seems to acquire rigidity and propagates without deformation. It can be assumed that this form of wave deformation is determined by the reflection of the wave from the hard bottom. This reflection leads to a nonlinear development of a sinusoidal oscillation. As the Reynolds number increases, the wavelength increases. Wave development has a similar character, as in the zone of buckling.

Fig. 3.13 shows the process of the formation of waves when small sinusoidal disturbances are introduced into the BL and the propagation of a disturbing motion along the plate.

Three tellurium streams were photographed at the same time. The interval between frames was 0.5 s, the disturbance frequency was $n = 0.83$ Hz, the velocity of the unperturbed flow was $U_\infty = 10.5$ cm/s. The numbers in the photographs represent the geometrical coordinate x from the beginning of the work area. The figure shows how the crest of a wave is formed and folded, and the propagation of a disturbing movement across the BL occurs. Obviously, the wave characteristics (amplitude, length and shape) are different and depend on the distance from the bottom of the work area.

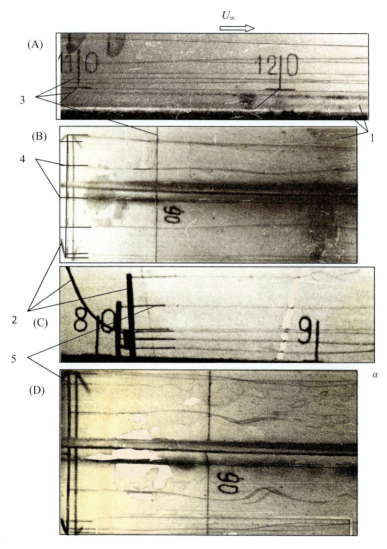

Figure 3.11 Simultaneous photographing of a disturbances motion from the side (A, C) and from above (B, D) with a sinusoidal (A, B) and nonsinusoidal (C, D) fluctuations at $U_\infty = 10$ cm/s: (A, B) $n = 0.3$ Hz; (C, D) $n = 0.4$ Hz; (1) bottom; (2) supports of tellurium wires; (3) distance markers (10 cm); (4) horizontal (along the bottom) tellurium wires; and (5) vertical (along side walls) tellurium wires.

Fig. 3.14 shows photographs of the visualization of the development of a disturbing motion in the BL as a function of the oscillation frequency. The characteristics of the disturbing motion essentially change at different oscillation frequencies. Evaluating the amplitude of oscillation and other parameters of the disturbing

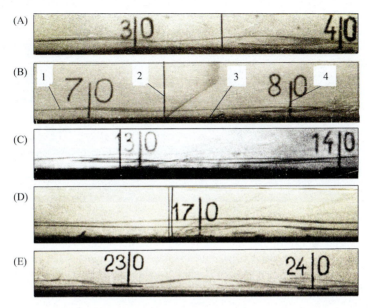

Figure 3.12 Photographing disturbing motion in various places along the plate: (A) $U_\infty = 10.5$ cm/s, $n = 0.55$ Hz; (B) $U_\infty = 10.5$ cm/s, $n = 0.5$ Hz; (C) $U_\infty = 10.5$ cm/s, $n = 0.4$ Hz; (D) $U_\infty = 16$ cm/s, $n = 0.37$ Hz; (E) $U_\infty = 16$ cm/s, $n = 0.3$ Hz: (1) tellurium jet; (2) tellurium wire; (3) bottom of the working section; and (4) distance marker from the beginning of the working section in cm.

motion, you can determine the frequency of neutral oscillations for a given measurement site.

Similar investigations were carried out with a positive pressure gradient on the inclined upper bottom of the working section (Fig. 3.15). The design of the work area was such that two bottoms were made: the bottom was stationary, and an autonomous bottom was installed above, which could change its position along the length of the working part (Fig. 3.2).

A positive pressure gradient during flow in the working area was created by lowering the right end of the upper bottom 3 cm until it stops at the bottom (the working part is 300 cm long). Therefore the work area was a long diffuser. All measurements were carried out on the upper bottom of the work area. On the side transparent walls of the work area, markers are applied that indicate the distance from the beginning of the work area. The BL was formed in the area of confuser. When measurements were taken at each coordinate of the measurement site, a velocity profile was photographed, from which the actual thickness of the BL was calculated, and the actual x coordinate was determined. Every 0.5 s, electrical contacts were closed, and a cloud of colloidal tellurium formed around the tellurium wire, which flowed downstream, visualizing the velocity profile. It was possible to obtain successive velocity profiles in order to investigate the dynamics of the deformation of the velocity profile under certain experimental conditions. In Fig. 3.15

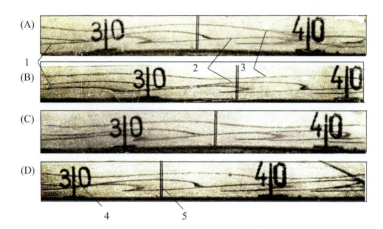

Figure 3.13 Photographs of the successive phases of the development of a disturbing motion: (A)–(D) consistent development in 0.5 s; (1)–(3) tellurium jets introduced at various distances y from the bottom; (4) distance markers from the beginning of the working section; (5) tellurium wires for measuring velocity profiles.

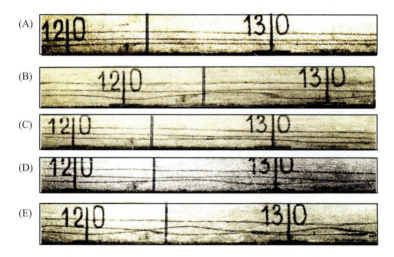

Figure 3.14 Photographs of the velocity amplitude v' as a function of the oscillation frequency at $U_\infty = 13$ cm/s: (A) $n = 0$ Hz; (B) $n = 0.68$ Hz; (C) $n = 0.79$ Hz; (D) $n = 0.87$ Hz; (E) $n = 1.3$ Hz. Designations as in Fig. 3.10.

two speed profiles and a tellurium jet are photographed. Vibrating tape was missing. Despite this, the tellurium jet registered oscillations, since natural disturbances began to develop in the BL with a positive pressure gradient. In Fig. 3.15 two speed profiles and a tellurium jet are photographed. Vibrating tape was missing. Despite

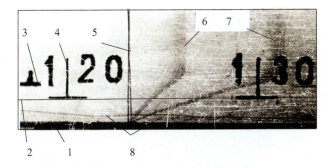

Figure 3.15 Photographs of the velocity profiles with a positive pressure gradient: (1) upper tilted bottom; (2) horizontal line of the upper bottom; (3) marker on the near wall of the working section; (4) marker on the far wall of the working section; (5) tellurium wires for measuring the velocity profile; (6) the first velocity profile; (7) development of the velocity profile in 0.5 s; (8) the shape of the current line, visualized by a tellurium stream.

this, the tellurium jet registered oscillations, since natural disturbances began to develop in the BL with a positive pressure gradient.

The velocity profile differs from that shown in Figs. 3.7 and 3.12B. With a positive pressure gradient, the velocity profile has a slight bend and becomes blurred above the thickness of the BL. Horizontal markers indicate the horizontal position of the second bottom. The markers on the far side wall of the work area are clear, and on the near side wall (on the left in the photo) are more blurred, as the camera focuses on the plane along the longitudinal axis of symmetry of the work area where the experimental investigations are performed. These markers allow the definition of distortion in photographs, since the photograph was taken with optical lenses that changed the focal length of the camera lens.

Fig. 3.16 shows the velocity profiles of $U(z)$ when photographing from above. At the beginning of the working part, where the pressure gradient influenced the flow slightly, the velocity profile was uniform. His deformation was regular. In the absence of a pressure gradient, the distortion of the velocity profiles was practically not observed in the same place.

Downstream, the three-dimensional deformation of the velocity profile has increased. Compared to similar measurements in a flow with a zero pressure gradient (Fig. 3.8), a positive pressure gradient essentially increased the three-dimensional deformation of the perturbations of the BL, at a lower velocity and coordinate. The results obtained are well consistent with the results of [611,613].

Fig. 3.17 shows the development of a sinusoidal disturbances motion in the BL with a positive pressure gradient. The characteristic of the disturbing motion has changed significantly compared with the flow at zero pressure gradients (Fig. 3.12). At the beginning of the working part, the perturbations became nonlinear (Fig. 3.17A). The shape of the crest of the wave changed in comparison with the nongradient flow: there was a slow movement of the wave on its reduced part, then

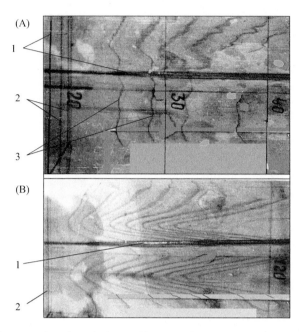

Figure 3.16 Photographing the velocity profiles $U(z)$ with a positive pressure gradient: (A) the beginning of the working part; (B) the middle of the working part; (1) tellurium wire support; (2) tellurium wires; (3) velocity profiles during their development along x.

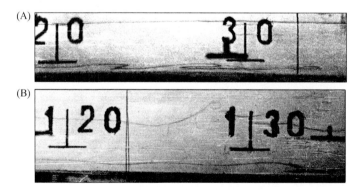

Figure 3.17 Photographs of the velocity amplitude v' in different places along the plate: (A) $U_\infty = 9.5$ cm/s, $n = 0.77$ Hz; (B) $U_\infty = 11.3$ cm/s, $n = 0.7$ Hz. Designations are the same as in Fig. 9.14.

the folding of the wave and the further propagation of tellurium jets in the form of a "frozen" loop.

The upper tellurium jets at the beginning of the working part moves without change in the region of the outer boundary of the BL (compare with Fig. 3.14). As the x coordinates increases, the disturbances propagate throughout the entire BL thickness (Fig. 3.17B).

In the hydrodynamic stand of small turbulence, the influence of the main flow turbulence, as well as the amplitude of the disturbing motion on the hydrodynamic stability, was experimentally investigated. Fig. 9.18A shows the results of measurements of the amplitudes of the disturbing motion as a function of the oscillation frequency and the measurement site at a speed of $U_\infty = 18$ cm/s and the amplitude of oscillation of the vibrator $A_v = 0.32$ mm. At a constant flow rate, the value of the oscillation frequency at $A = $ max decreases with increasing x. In each measurement site, the increase in the amplitudes of the oscillations to the maximum is more intense than their decrease.

Fig. 3.18B shows the experimental dependences of the amplitudes of the oscillations of the transverse velocities v' (solid curves) and w' (dashed curves) on the oscillation frequency of the disturbing motion, the measurement point and the velocity of the main flow with the amplitude of the vibrator $A_v = 0.24$ mm.

Measurements showed that in every place along the working part, with $U_\infty = $ const there is a narrow frequency range with a maximum amplitude of oscillation. Along the working section, the amplitude maximum moves to the lower frequency range. With an increase in the flow velocity, with the same x coordinate, the maximum of the oscillation amplitudes shifts in the opposite direction—toward higher frequencies. When the velocity values v' and w' were fixed, the maximum amplitudes in both cases coincided. Depending on the flow speed, at the specified distance between the source of the disturbances and the measurement station, the

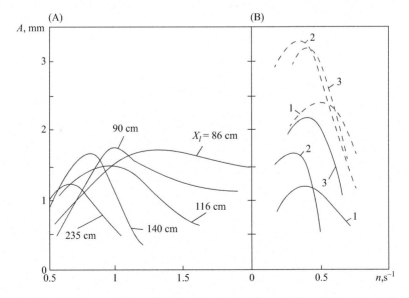

Figure 3.18 Dependence of the amplitudes of the transverse velocity v' (A) and speeds v' and w' (B) on the oscillation frequency and x coordinate: (1) $U_\infty = 9.5$ cm/s, $x = 98$ cm; (2) $U_\infty = 10$ cm/s, $x = 133$ cm; (3) $U_\infty = 13.5$ cm/s, $x = 280$ cm.

amplitude of the disturbances increased 5–10 times, and then, as a rule, their decrease was less intensively.

The influence of the location of the oscillating strip $\bar{y} = y/\delta$ over the BL thickness on the amplitude of fluctuation of transversal velocity v' of disturbances was also investigated. The dependences obtained are presented in Fig. 3.19A for $\frac{\beta_r \nu}{U_\infty^2} = 430 \cdot 10^{-6}$, Re* = 570 (solid lines) and for $\frac{\beta_r \nu}{U_\infty^2} = 255 \cdot 10^{-6}$, Re* = 810 (dashed lines). The dependences for speeds v' and w' have a similar character. The amplitudes of the velocities of the disturbing motion are maximal when the source of the disturbances lies within the limits (0.1–0.3) $\bar{y}$. Fig. 3.19B presents the results of investigations of the distribution of the amplitudes of the transverse velocities of the disturbing motion over the thickness of the BL at various values of A_v for $\frac{\beta_r \nu}{U_\infty^2} = 430 \cdot 10^{-6}$ and Re* = 570. The speed v' is directly proportional to the value of A_v. With $A_v > 0.7$ mm, the velocity v' became so significant (respectively, the velocities u' and w' increased) that the vortex formation occurred immediately behind the vibrator. Thus the maximum values of the component velocities of the disturbing motion are within (0.15–0.3) $\bar{y}$. The tellurium method allowed measurements to be carried out at velocities of the main flow, not exceeding 0.2 m/s. It has been established that when conducting investigations of hydrodynamic stability it is necessary to observe the following conditions. The degree of turbulence should not exceed 0.04% in water. The amplitude of the vibrator in water should be 0.2–0.5 mm at this speed range. The vibrating strip and the sensor must be placed at the same distance from the wall (0.1–0.3) $\bar{y}$. The transverse velocities v' created by the vibrator should not exceed 2% of the velocity of the main flow. Otherwise, the transition points substantially approach the point of loss of stability and, with a ratio of the indicated speeds of 2.5%, the transition occurs immediately behind the vibrator. The ratio of the degree of disturbing motion to the degree of turbulence should be about 100.

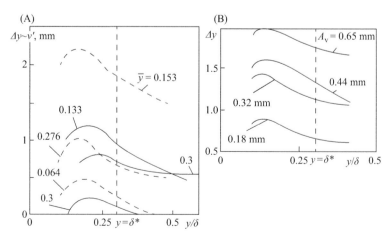

Figure 3.19 Distribution of the amplitudes of the velocity v' over the thickness of the boundary layer, depending on the values of $\bar{y}$ (A) and A_{vibr} (B).

The critical Reynolds numbers depend not only on the location of the source of the disturbance in the thickness of the BL, but also on the pressure gradient, roughness, Mach number, temperature, wall vibration, etc. The instability of the surface of a thin liquid film on a rotating disk was experimentally studied in [406].

3.4 Neutral curves of linear stability of a laminar boundary layer of a rigid plate

When solving real tasks of flow around bodies, there are various factors that affect the BL. First of all, investigations of the effect of pressure gradient on the transition and hydrodynamic stability were carried out [225,341–343,501,521,523]. Further, the factors affecting the BL from its lower boundary, on the side of the streamlined surface, were investigated. The influence of the dimensions of roughness elements on the transition, separation and heat transfer of streamlined profiles was investigated in Refs. [10,203,258,309,408,420,430,542,574]. The state of the streamlined surface has a significant impact on the characteristics of the transition BL. Therefore along with the roughness, the influence on the hydrodynamic stability of the waviness of the streamlined surface [226,309,319,408,545], its various oscillations [4,22,231,408,430,517,602], as well as the effects on the hydrodynamic stability changes in flow viscosity with changes in surface temperature [14,319,378,380]. In addition to the factors influencing the hydrodynamic stability from the lower boundary of the BL, in various practical problems of the BL it is also necessary to take into account factors acting "from the outside," that is, at the upper boundary of the BL. First of all, these include the degree of turbulence of the undisturbed flow [209,247,254,255,260,430,497], acoustic fields [260,319,380,408], flow compressibility [188,189,255] and its oscillations [349,378,472,473], as well as in the case of a flow past a stream of water, the influence of medium stratification, etc.

First of all, hydrodynamic stability was investigated under ideal flow conditions. In Section 3.2 and Ref. [378], the methods for the experimental construction of neutral curves using the tellurium method are considered in detail. Fig. 3.20 shows neutral curves in the coordinates of the dimensionless frequency with a longitudinal flow past a rigid plate. The obtained experimental data are consistent with the calculations of Lin and Shen, as well as with the experimental data of other authors.

Fig. 3.21 shows the neutral curves in the coordinates of the dimensionless wave number (A) and phase velocity (B). In Fig. 3.21B Schlichting's calculations [518,520] are indicated by a solid line, and the measurements of Shubauer and Scramsted [523] are indicated by a dashed line. A number of results of hydrodynamic stability studies are not shown in Figs. 3.20 and 3.21. During the measurements, it was found that in each place along the working part, oscillations were observed in a strictly defined frequency range, which in a dimensionless form is plotted on a graph of a neutral curve in the form of points (Fig. 3.21C).

Based on the limiting values of these points, a region was constructed outside which oscillations of any frequency in the BL were not observed. The curve

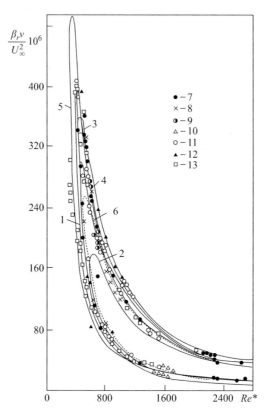

Figure 3.20 Neutral curves in the coordinates of the dimensionless frequency of disturbing oscillations in the flow past a flat plate. Calculated data: (1) Tollmien [554]; (2) Schlichting [516,521]; (3) Lin [414,415]; (4) Shen [527]; (5) Zaat [631]; (6) Barry and Ross in [378]. Experimental data: (7) Shubauer and Scramsted [524]; (8) Wortman [611]; (9) Hama [288]; (10) Burns in [378]; (11), (12) obtained by tellurium method [378]; (13) obtained using LDVM [41].

covering this area is called the limit neutral curve. Since the measurements were carried out mainly at the amplitude of the vibrator $A_v = 0.32$ mm and with the degree of turbulence $\varepsilon < 0.04\%$, then with an increase in these parameters, the limiting neutral curve probably acquires a certain value. Apparently, it limits the instability region associated with nonlinear effects and finite disturbance amplitudes. Limiting neutral curves were also constructed in the new coordinates of the neutral curve (Fig. 3.22 dash-dotted lines), which determine the dependence of the dimensionless wave numbers and phase velocities on the dimensionless frequency. The theoretical curves in this figure were constructed by the authors according to the data given in the works of Tollmin [554], Shen [527] and Schlichting [520,521].

The usual experimental neutral curves in these coordinates were not constructed. Compared to the data shown in Figs. 3.20 and 3.21, some features were found. It

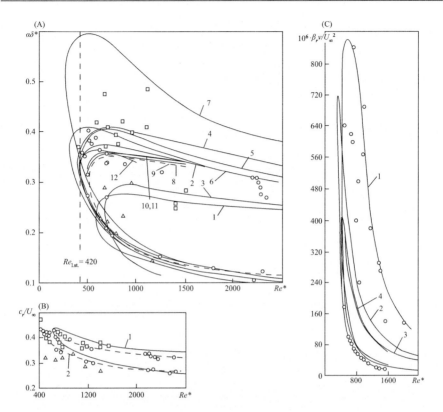

Figure 3.21 Neutral curves for the dimensionless wave number (A) and propagation velocity of disturbing oscillations (B): calculated data: (1) Schlichting [519]; (2) Tollmien [554]; (3) Pretsch [see in 378]; (4) Lin [414,415]; (5) Shen [527]; (6) Zaat [631]; (7) Korotkin [see in 378]; (8) Skripachev [see in 378]; (9) Kaplan [325]; (10) Brown [188]; (11) Kurtz [see in 378]; (12) Powers [see in 378]; experimental data; ○ Shubauer and Skramsted [523]; □ and △ Babenko [19–23]. Limit neutral curve (C) for the dimensionless frequency of disturbing oscillations (1) and ordinary neutral curves: (2) Korotkin [see in 378]; (3) Shubauer and Skramsted [523] and (4) Babenko [378].

turned out that the experimental points are placed at different speeds of the main flow U_∞ along certain lines (in Fig. 3.22 dashed lines). At different speeds, the coefficients of proportionality between the wavelength, the velocity of propagation of the disturbing motion and the frequency vary in a larger range than would be expected from theoretical data.

For high speeds, empirical dependencies are defined:

$$\alpha\delta^* = 1.4 \cdot 10^3 \frac{\beta_r \upsilon}{U_\infty^2} + 0.15 \tag{3.1}$$

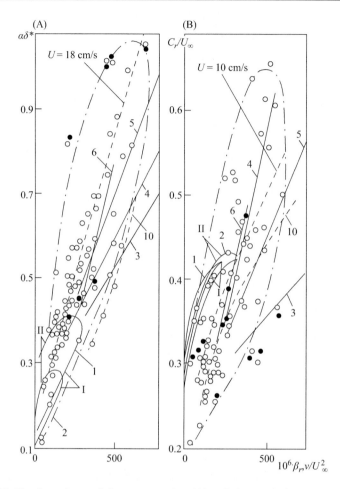

Figure 3.22 The dependence of the wave number (A) and phase velocity (B) on the frequency of disturbing oscillations. Calculated: (1) Tollmien [554] for dimensionless frequency and Shen [527] for phase velocity; (2) Schlichting [518,520,521]; points measurements [378]; I and II branches of the neutral curve; (3) $U_\infty = 9.5$ cm/s and (4) $U_\infty = 11.3$ cm/s (measurements with a positive pressure gradient); (5) $U_\infty = 9.1$ cm/s and (6) $U_\infty = 11.5$ cm/s (measurements with nonsinusoidal disturbances).

$$\frac{c_r}{U_\infty} = 0.7 \cdot 10^3 \frac{\beta_r \upsilon}{U_\infty^2} + 0.2. \tag{3.2}$$

Here $\alpha = 2\pi/\lambda$ is the wave number of disturbing oscillations; λ is the Tollmien–Schlichting (T–S) wavelength; δ^*-extrusion thickness, $\beta_r = 2\pi n$ is circular frequency; ν is coefficient of kinematic viscosity. The dark dots in Fig. 3.22 characterize the velocity w'. These points indicate that when a rigid plate is flown around, neutral curves can be constructed from the results of measurements of both amplitudes Δy and amplitudes Δz. The amplitudes Δx, measured with the aid of the LDVM [41], are in good agreement with the data of [378].

Table 3.1 shows the limiting values of characteristic values in comparison with similar data from other authors. The increase in ε ($\varepsilon > 1\%$) has led to the need to increase the amplitude of vibration of the vibrator by three times to conduct a study of hydrodynamic stability [41]. All the main characteristics of the disturbance turned out to be similar to those obtained by measurements in a stream with a small ε. The points of the neutral curve, found with the help of an LDVM, are in good agreement with the known theoretical and experimental data [520,523]. This confirms the conclusions obtained in studying the natural transition, in particular, on the unity of the laws of the development of disturbances in the BL under any conditions.

The growth ratios of the disturbing motion are experimentally determined. The method for determining the magnitudes of the disturbing motion was first developed in [521], where a formula was proposed (for a fixed value of the dimensionless frequency):

$$a(x_2) = e^{\int_{t_1}^{t_2} \beta_i dt} \qquad (3.3)$$

where $a(x_2) = A_2/A_1$ is the ratio of the amplitudes of the oscillations at the two compared points: x_1 and x_2. Given this, the expression (3.3) can be written in the form:

$$2.3 \lg \frac{A_2}{A_1} = \int_{t_1(x)}^{t_2(x)} \beta_i dt. \qquad (3.4)$$

Differentiating expression (3.4) with respect to time and taking into account that $dx/dt = c_r$, we obtain the formula for determining the growths:

Table 3.1 Limit values of neutral oscillations.

Parameter	Theoretical data: Schlichting Tollmin		Experimental data (approximated): Subhauer and Scramsted Babenko	
Re^*	575	420	450	404
$\frac{\beta_r \nu}{U_\infty^2}$	$178 \cdot 10^{-6}$	—	$400 \cdot 10^{-6}$	$405 \cdot 10^{-6}$
$\alpha \delta^*$	0.278	0.367	0.4	0.43
$\frac{c_r}{U_\infty}$	0.42	0.425	0.43	0.47

$$\beta_i = 2.3 c_r \frac{d\left(\lg \frac{A_2}{A_1}\right)}{dx}. \tag{3.5}$$

Taking into account the relations $c_r = \beta_r/\alpha$, $c_r = n \cdot \lambda$, $c_i = \beta_i/\alpha$ expression (3.5) takes the form:

$$c_i = 0.366 n \lambda^2 \frac{d\left(\lg \frac{A_k}{A_{k-1}}\right)}{dx}. \tag{3.6}$$

By (3.6) it is possible to determine the magnitude of the buildup experimentally.

The amplitudes of oscillations are determined at a fixed point x_o and at a point located at a variable distance x from it, not exceeding 20 cm. Then, graphical dependences of the logarithm of the ratio of these amplitudes on the distance $x-x_o$ are plotted. The average tangent of the angle of inclination of this graphical dependence determines in formulas (3.1) and (3.6) the derivative with respect to x. In Ref. [378], the results of measurements by Schubauer and Scramsted [523] of the magnitude of the growth of the disturbing motion, as well as the results obtained using the tellurium method, are presented.

The growth rate of unstable fluctuations of the laminar BL is also characterized by the growth factors c_i. Having performed the corresponding calculations using the formula (3.6), determine the values of the growth factors for fixed values of the Reynolds numbers (Fig. 3.23). The measured values of the growth factors were determined for oscillations of a given frequency, and the

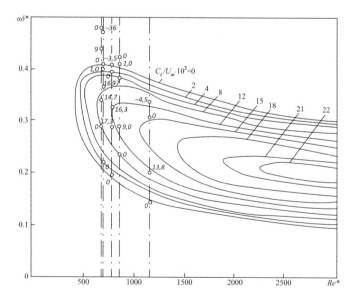

Figure 3.23 The growth curves of disturbances in the BL on a longitudinally streamlined plate according to Shen [527] ○ measurements [378].

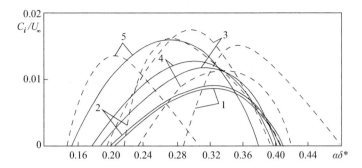

Figure 3.24 Dependence of growth factors on the wave number for numbers Re*: (1) 676, (2) 693, (3) 786, (4) 865, (5) 1160 [378].

wavelength of the oscillation was simultaneously recorded by a telluride method. This made it possible to determine the wave numbers of the oscillations and construct the dependences of the growth coefficients on the wave number (Fig. 3.24, dashed curves). Having determined the points Re* = const (Fig. 3.23, dash-dotted curves) with the growth curves of the values of dimensionless wave numbers, the calculated in Fig. 3.24 dependences of growth factors on wave numbers (solid curves). The data presented in Figs. 3.23 and 3.24 show that the experimentally obtained values of the coefficients of accretions best fit the calculations of Shen. It should be noted that the dependences shown in Fig. 3.24 are similar to the dependencies $A = f(n)$.

The experimental results of measurements of the distribution of the components of the longitudinal and transverse pulsation velocities and the kinetic energy of the disturbing motion over the thickness of the BL are given in [378]. Experimental investigations of hydrodynamic stability with a positive gradient and nonsinusoidal nature of the disturbing motion are performed.

3.5 Distribution of disturbing motion over the thickness of a laminar boundary layer of a rigid plate

When conducting experimental investigations of hydrodynamic stability, we studied the influence of the location of the vibrator tape over the thickness of the BL on the distribution of the characteristics of the disturbing motion over the thickness of the BL. Fig. 3.25 shows the results of measurements.

The regularities of the distribution of the quantities $\alpha\delta^*$ and c_r/U_∞ practically did not depend on the location of the source of disturbances along the thickness of the BL if small sinusoidal disturbances were introduced near the second neutral oscillation and in the range $\bar{y} = 0.1 - 0.3$. These data confirm the conclusion about the need for a controlled experiment to place the vibrator tape within $(0.1-0.3)\,\bar{y}$. With such $\bar{y}$, was investigated the regularities of the distribution of the quantities

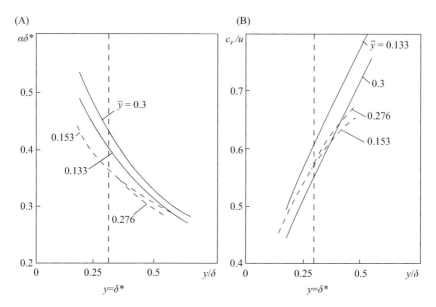

Figure 3.25 The distribution of the wave number, (A) and phase velocity (B) disturbing motion across the thickness of the boundary layer depending on the coordinate $\bar{y}$ of the vibrator tape at Re* = 570, $\frac{\beta_r \nu}{U_\infty^2} = 430 \cdot 10^{-6}$ (solid lines) and Re* = 810, $\frac{\beta_r \nu}{U_\infty^2} = 255 \cdot 10^{-6}$ (dashed lines).

$\alpha\delta^*$ and c_r/U_∞ with respect to δ for different values of Re and n (see Fig. 3.25B). In the area of $Re_{l.st.}$ (Re* = 404), these patterns differ from those measured with other Re* numbers, but for all cases they are determined by the oscillation frequency, and the values of $\alpha\delta^*$ differ from each other more than c_r/U_∞. Only in the area of loss of stability, the wave numbers change only slightly over the thickness of the BL, and as perturbations develop along x, the values of $\alpha\delta^*$ and c_r/U_∞ change dramatically and significantly depend on the oscillation frequency. Moreover, the values of c_r/U_∞ at Re* = 654 are maximal. If this is related to the stages of the natural transition, then with a given Reynolds number, the fifth stage should develop, at which the pulsation velocities are also maximum. With distance from the wall, the wavelength and its phase velocity increase.

Such a complex picture of the development of disturbances at the first stage of the transition allows us to conclude that under realistic conditions of flow around, disturbances entering the BL from below the wall or from the top of the undisturbed flow will cause oscillations that have different wavelengths and phase velocities across the BL, there will be a perturbation spectrum generated. Only in the critical layer in the region of $Re_{l.st.}$, oscillations will develop in accordance with the laws of the linear theory of stability. Consequently, in the natural transition, at the first stage, the disturbing motion will be complex, consisting of different modes, and only in the critical layer the oscillations having the frequency of the second neutral oscillation will play the main role in the development of disturbances. The emerging deformation of a plane wave due to its Görtler instability [36] leads to the fact that due to the data of Fig. 3.25 on the crest of a wave, the fluid particles in different places in z will move with different phase velocities, while the values of $\alpha\delta^*$ will change. This will lead to further development of the cascade transition process and an increase in the fullness of the spectrum of disturbing oscillations.

For a better understanding of the development of disturbances at the first stage, the dependences of $\alpha\delta^*$ and c_r/U_∞ on y/δ were measured at different amplitudes of vibrations of the vibrator tape. It turned out that varying the value of A within $0.1-0.5$ mm does not make significant changes in the data in Fig. 3.25. Only at $A_g \geq 0.65$ mm, these values, when moving away from the wall by more than $y/\delta = 0.5$, tend to limit values of 0.5 and 0.65, respectively [23]. Consequently, its parameters depend on the intensity of the disturbance. An increase in the magnitude of the disturbance leads to a nonlinear nature of its development. It is possible that at the sixth stage of the transition, the wavelength across the thickness of the BL will vary little: $\alpha\delta^* \approx 0.5$, $c_r/U_\infty \approx 0.65$. The magnitude of c_r/U_∞ is of the same order as in the measurements at the nonlinear stages of the Klebanov transition [343] and in Landal's calculations [397]. Fig. 3.26 shows the results of experimental studies of the influence of the oscillation amplitude of the optimally located δ vibrator ribbon on the distribution of the thickness of the BL of $\alpha\delta^*$ and c_r/U_∞ at the oscillation frequency of the disturbing motion $\frac{\beta_r \nu}{U_\infty^2} = 255 \cdot 10^{-6}$ and $\mathrm{Re}^* = 653$. The distribution of $\alpha\delta^*$ and c_r/U_∞ with

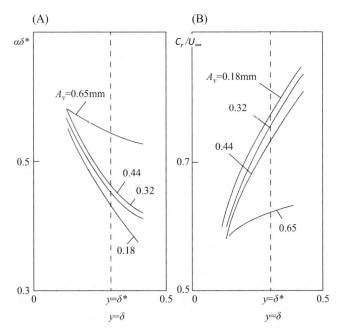

Figure 3.26 The influence of the amplitude of vibrations of the vibrator tape on the distribution of the wave number (A) and phase velocity (B) of the disturbing motion along the thickness of the BL.

respect to δ up to a certain value of the amplitude of oscillation of the vibrator practically does not change, but as the amplitude increases from $A = 0.44$ mm to $A = 0.65$ mm, $\alpha\delta^*$ and c_r/U_∞ tend to the limiting values throughout the entire thickness of the BL, respectively, 0.5 and 0.65 (for the given conditions of experiments).

Fig. 3.27 presents the results of measuring the distribution of the values of $\alpha\delta^*$ and c_r/U_∞ over the thickness of the BL. It has been found that with increasing the Reynolds number these values change over the thickness of the BL more dramatically. For each oscillation frequency, there is a definite character of the dependence $\alpha\delta^* = f(\bar{y})$, while the dependences $c_r/U_\infty = \varphi(\bar{y})$ for $Re^* = $ const do not depend much on frequency. As the value decreases, $\alpha\delta^*$ increases, that is, the wavelength of the perturbing motion decreases. With $\bar{y}$ decreasing, c_r/U_∞ also decreases. In Fig. 3.27B, it can be seen that only for some values of the oscillation frequency and Re^* numbers does the phase velocity in magnitude tend to U_∞ near the critical layer.

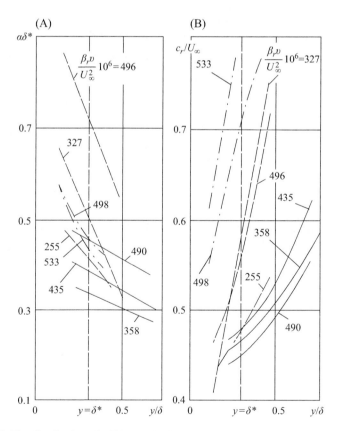

Figure 3.27 The distribution of $\alpha\delta^*$(A) and c_r/U_∞ (B) over the thickness of the BL depending on the oscillation frequency at $Re^* = 404$ (solid curves), $Re^* = 654$ (dash-dotted curves), $Re^* = 810$ (dashed curves).

The fundamental position of $c_r \approx U_\infty$ in the region of the critical layer at any frequencies of disturbing oscillations was not found, although measurements showed that there is a liquid layer in which the components of the velocity of the disturbing motion are maximal. Another fundamental property of hydrodynamic stability, namely, a change in the oscillation phase by 180 degree near the boundary of the BL [523,613], although it was found, but in the region of 30%−80% of the thickness of the BL depending on the oscillation frequency.

The results obtained made it possible to determine that errors in measuring stability may occur not only due to poor quality sensors or insufficient accuracy in determining the values of λ and c_r, but also due to inaccurate determination of the vertical coordinate of the disturbing motion.

Fig. 3.28 shows the results of measuring the distribution of c_r/u over δ (u is the local velocity at the measuring point, Table 3.2). The general nature of the laws is opposite to the laws shown in Fig. 3.25, which reflects the fact that the value of c_r increases with y/δ. The growth rate of the averaged velocity u at the measurement point with an increase in y/δ is ahead of the growth rate of velocity c_r, which determines the nature of the curves in Fig. 3.28. From the data presented, it follows that $c_r \approx u$ with $y = \delta^*$ only in the case when the oscillation frequency of the disturbing motion is equal to the frequency of the second neutral oscillation, and the disturbance is introduced and fixed in the region of the critical layer. In all other cases, when the frequency of the disturbing oscillation is more or less than the second neutral oscillation, $c_r \approx u$ only for $y/\delta = 0.4-0.6$, and the greater the ratio n/n_{II}, the higher the values of y/δ for the indicated speed equality and steeper curves. An increase in the amplitude of oscillation [Fig. 3.28 curve (8)] does not change this pattern; although $c_r \approx u$ with a smaller value of y/δ [Fig. 3.28, curves (6), (7)]. The general nature of the regularities is as follows: near the wall, c_r considerably exceeds u, and the larger, the larger x, and the higher $y/\delta = 0.4-0.6$ becomes less than u.

However, near the wall due to the viscosity vibrations subside. Despite large values of c_r near $\text{Re}_{l.st.}$, the curves do not vary so much with respect to y/δ than with large x.

Fig. 3.29 shows the dependences of c_r/u on the oscillation frequency. It turned out that if $y_v \approx y_{t.w.} \approx \delta^*$, then under all experimental conditions (x, U_∞) $c_r \approx u$ only for $n = n_{II}$, that is, at the frequency of the second neutral oscillation, and for $n \neq n_{II}$ $c_r/u > 1$. At a frequency of $n < n_{II}$, that is, in the region of unstable oscillations of the neutral curve, c_r is less different from u than when going through the neutral curve to the region of stable oscillations. For $n > n_{II}$, at the beginning, c_r increases sharply until it becomes greater than u by 1.2−1.5 times, and then with increasing n, the value of c_r gradually decreases. It is characteristic that as x decreases or as U_∞ increases, n increases, that is, $\beta_r/U_\infty \neq \omega_{rII}$, which is accompanied by alignment of curves [in Fig. 9.29A, curves (7), (12), and (13)], and in the $\text{Re}_{l.st.}$ region, almost all n, $c_r/u \approx 1$.

Thus depending on the frequency and place of the thickness of the BL introduced by disturbances, the speed of their propagation varies considerably. Only disturbances introduced in the region of the critical layer during $\text{Re}_{l.st.}$ propagate at a

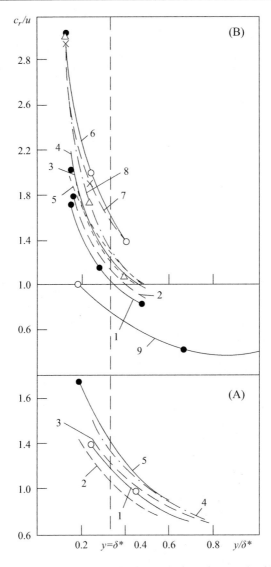

Figure 3.28 The distribution of the dimensionless velocity c_r/u over the thickness of the BL in the region of $Re_{l.st.}$ (A) and at $Re \gg Re_{l.st.}$ (B): (A) $x = 0.4$ m, $y_v/\delta = 0.248$, (1) $n/n_{II} \approx 1$, $n_{II} = \beta_r \cdot v/U_\infty^2 = 4.4 \cdot 10^{-4}$; (2) 1.02; (3) 0.82; (4) 1.12; (5) 1.29); (B) $x_1 = 1.3$ m, (1) $y_v/\delta = 0.256$ and $n/n_{II} \approx 1.05$; (2) 0.145 and 1.12; (3) 0.256 and 1.8; (4) 0.256 and 1.43; (5) 0.256 and 2.18; (6) $n/n_{II} = 2.34$ and $A_v = 0.32$ mm; (7) 2.34 and 0.18 mm; (8) 2.34 and 0.65 mm; (9) $n/n_{II} = 1$, membrane surface, experiments B37 [378].

speed of $c_r = u$ over a wide frequency range of oscillation. The disturbances generated, for example, by the wall (when $y/\delta < 0.33$), propagate faster, and those entering the BL from the outside ($y/\delta > 0.33$) are slower than u. In this case, it remains

Table 3.2 The conditions of the experiments are shown in Fig. 3.29.

№ curve	№ of experience	x_I, m	U_∞, m/s	u, m/s	$Y_{t,w} \cdot 10^3$, m	$y_v \cdot 10^3$, m	$\delta^* \cdot 10^3$, m	n_{II}	n_I	$\bar{y}$	y_v/δ	δ^*/δ
1	XXXI	0.86	0.18	0.059	3.5	2.0	3.79	1.25	0.53	0.31	0.176	0.3
2	XXXII	1.16	0.18	0.064	4.0	2.0	4.42	0.93	0.40	0.272	0.14	0.3
3	XXXIV	1.4	0.18	0.056	4.0	2.0	4.85	0.79	0.36	0.25	0.125	0.3
4	XXXV	0.9	0.18	0.050	3.0	2.0	3.88	0.98	0.51	0.23	0.16	0.3
5	XXXVIII	1.78	0.17	0.090	6.0	5.6	5.6	0.91	—	0.33	0.20	0.3
6	XLI	2.39	0.105	0.052	4.0	—	3.56	0.69	—	0.29	0.18	0.3
7	B4,B9	1.04	0.105	0.040	3.2	2.5	5.0	0.5	0.3	0.19	0.15	0.3
8	B10,B11	1.3	0.13	0.0572	3.2	2.9	6.28	0.5	—	0.155	0.145	0.3
9	membr.	—	—	—	5.9	5.15	—	—	—	0.285	0.256	—
10	membr.	—	—	—	1.0	—	—	—	—	0.482	—	—
11	B48	1.55	0.1	0.034	3.5	3.3	6.84	0.3	—	0.3	0.19	0.3
12	B59	0.51	0.117	0.042	3.5	2.9	4.96	1.0	—	0.29	0.22	0.3
13	B65	0.4	0.095	0.0333	3.5	3.0	4.22	0.7	—	0.31	0.17	0.3
14	b66	1.2	0.113	0.0396	4.0	—	5.7	0.45	—	0.25	0.2	0.3

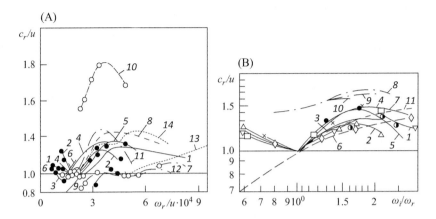

Figure 3.29 The dependence of the phase velocity on the oscillation frequency: (A) the designations of the curves are given in Table 3.2; (B) notation for curves (1), (2), (4) are given in Table 3.2, curves (3), (5), (6) and (7) correspond to the numbers of curves (5), (11), (13) and (8) in the Table 3.2; curves (8), (9), (10) correspond to the membrane surface, and (11) to the generalized dependence for a viscoelastic surface.

general that when these disturbances fall into the region of the critical layer, they propagate with the value of $c_r = u$ with $n = n_{II}$. The reverse is also valid: oscillations with frequencies greater than n_{II} propagate with a speed of $c_r > u$. This, which also follows from Fig. 3.29, is the basis for the beginning of additional harmonics, even if all the conditions are fulfilled and only the oscillation with the value $n = n_{II}$ is specified, as well as one of the reasons for the beginning and development of nonlinear stages of the transition.

Fig. 3.29B shows the data of Fig. 9.29A, dimensionless in frequency of the second neutral oscillation ω_{rII}. Taking into account the errors of the experiments, we obtain:

$$c_r/u = (n/n_{II} - 1)^2 \text{ at } n < n_{II}, \tag{3.7}$$

$$c_r/u = 1.6 \cdot \lg(n/n_{II}) + 1 \text{ at } n > n_{II}. \tag{3.8}$$

The distribution of the amplitudes of the velocities v' and u', the kinetic energy, and the local energy exchange of the main and disturbing motions along the thickness of the BL were also investigated.

3.6 The physical process of laminar-turbulent transition of the boundary layer of a rigid plate

The research results presented above provide an opportunity to answer questions that are extremely important for understanding the physical picture of fluid flow in the BL:

- What influences the occurrence of the initial stage of the transition?
- How the subsequent stages of transition occur and what explains the regularity of their alternation?
- What factors influence the transition process of the BL?
- What features the fluid flow has during the transition stages?
- What is the manifestation of the unity of the flow in the BL, etc?

The question also arises whether there can be such a flow regime at which a natural transition of the BL does not occur.

Even under the most ideal conditions, there are prerequisites for the emergence of turbulence inherent in the nature of the formation of the BL. The growth gradient of the BL thickness at the beginning of the streamlined plate is large, while the absolute value of the BL thickness in this place is small. Therefore the critical layer in which energy is exchanged between the main and perturbing motions is located quite close to the outer boundary of the BL. As can be seen in the photograph of the Blasius velocity profile (Section 3.2, Figs. 3.5, 3.10 and 3.13), the thickness of the BL and its outer boundary are clearly defined. Therefore at the beginning of the plate, the main flow is directed to the outer boundary of the BL at a large angle. Therefore additional stresses arise on the outer boundary, caused by the curvature of the stream lines of the unperturbed flow. With a slight irregularity of flow, due, for example, to its nonparallelism, these stresses will lead to the appearance of sinusoidal oscillations at the BL boundary. And since the critical layer is located in this place very close to the external boundary, these oscillations will easily pass into the critical layer, in which unstable oscillations may occur. As long as the flow past the plate is insignificant, the amplitudes of such random oscillations are small and they are quickly damped by the flow. But as the velocity increases, the energy of the main flow increases and the energy exchange in the critical layer increases, the amplitudes increase, which leads to the development of T–S waves. Therefore the stability of the flow at the beginning of the plate is small. This also explains the location points of buckling, its characteristic features and the shape of a neutral curve.

It is known that a body with elastic properties is characterized by a range of natural frequencies. Liquid, in particular water, also has a range of natural frequencies, due to the ratio of viscous and inertial forces. Moreover, if viscous forces in a liquid depend mainly on its temperature, then inertial forces depend on its flow velocity. Therefore at different flow rates, the ratio of viscous and inertial forces changes. Any disturbance introduced into the fluid will cause an oscillatory process in it in accordance with the parameters of the disturbing motion and the natural frequencies of the fluid. Therefore in all cases, in the initial stage of the interaction of the disturbing and basic movements, a sinusoidal process is inevitable. This is a physical substantiation of the shape of a neutral curve, which describes the region of natural frequencies of oscillations in a laminar BL.

Depending on various factors: the type of disturbing movement, its energy and frequency range, the location of the disturbing movement in x and y, the speed of the main flow and its degree of turbulence, sharpness and smoothness of the front edge of the plate, its roughness, etc.—a sinusoidal process will have a different frequency-amplitude nature and time of existence, that is, it will attenuate, increase or lead to turbulence.

Consider a few special cases of turbulence, using an experimentally obtained neutral curve (Section 3.4, Fig. 3.20). Let the initial flow turbulence is not a source of disturbing motion. Then, with a low degree of turbulence ($\varepsilon < 0.05\%$) and a small amplitude of disturbing motion ($v'/U_\infty \leq 1.5\%$), the spectrum of frequencies of disturbing motion with a range of order $\frac{\beta_r \nu}{U_\infty^2} \cdot 10^6 = 40 - 420$ (and higher), passing through the region of instability increase the amplitude of oscillation to a critical value, which will lead to the occurrence of turbulence. Such a disturbing movement will always stabilize. The low-frequency spectrum of perturbing motion with a range of $\frac{\beta_r \nu}{U_\infty^2} \cdot 10^6 = 0 - 40$ will pass the region of instability with a large length along the coordinate of the Reynolds numbers. This means that the development of turbulence here depends on the place along the x-axis of introducing disturbances into the BL. If the disturbance is brought to the point of loss of stability, then there is enough time to increase the amplitude of the disturbing movement above the critical value. In this case, according to the limiting neutral curves (Section 3.4, Fig. 3.22), even with a low degree of turbulence, a random disturbance with an amplitude less than 1.5% can cause an increasing process at any frequency of oscillation in the specified frequency range.

With an increase in the initial turbulence, the amplitude of the disturbing motion, or under other unfavorable conditions, the entire spectrum of unstable oscillations can lead to turbulence, since the increase in the amplitudes of the oscillations in the instability region of ordinary and limiting neutral curves occurs very quickly. The maximum amplitude of the disturbing motion according to the measurements made was achieved if the source of the disturbance was at a distance $(0.1-0.3)$ $\bar{y}$. Therefore the magnitude of the increase in the amplitudes of the oscillations also depends on where the disturbing motion was made and with what frequency.

Thus three cases can be noted:

- Only sinusoidal disturbances occur in the flow around the plate, which stabilize with time.
- Sinusoidal fluctuations increase, and the disturbance passes through all stages transition.
- Under certain conditions, the speed of increase is so great that the stages of the occurrence of turbulence follow each other almost instantly ("catastrophic" transition).

The second and third cases were studied in sufficient detail in Refs. [347,540].

Fig. 3.30 shows the schemes of various options behaviors of a disturbing motion, obtained on the basis of an analysis of the results given in Sections 3.2−3.4 [139,247,378]. Fig. 3.30A shows the behavior of a tellurium jet with an increase in the oscillation frequency at $x =$ const. With increasing frequency, the waveform and the crests change, the wave amplitude initially increases, and its length decreases. With further increase in frequency, the amplitude also decreases. With increasing frequency, the crest of the wave begins to deform, stretch forward and upstream, sharpens, acquires a saw-like shape, and then, when a certain frequency is reached,

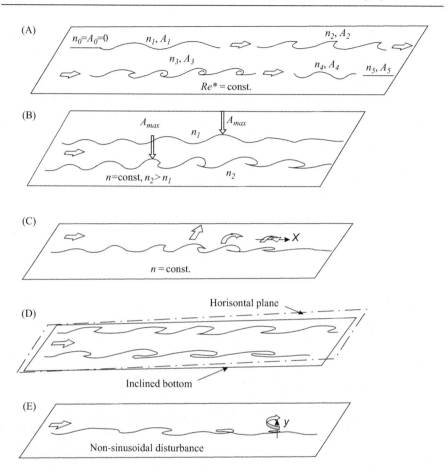

Figure 3.30 Schemes of disturbing motion in a longitudinal flow past a flat plate. The arrows indicate the direction of movement of the disturbing motion: (A) $Re^* = $ const; (B) $n = $ const, $n_2 > n_1$; (C) $n = $ const; (D) inclined bottom; (E) nonsinusoidal disturbance.

quickly turns and continues to move downstream in such a "folded" form without destruction and deformation (stabilization of disturbance). With further increase in the frequency, the inverse process of the crest of the wave occurs, as it were, but without the phase of formation of a saw-like shape. The amplitude of the wave becomes smaller, its length decreases even more, and the shape becomes smooth and purely sinusoidal. Finally, starting with a certain "limiting" frequency, with its further growth, the tellurium jet becomes even, as in the beginning of the experiments, before introducing disturbances into the BL. Such an evolution is similar to the picture of the development of a tellurium jet along x at $n = $ const, as well as the evolution of the flow of current lines in the subsequent stages of the transition. It was found that for each oscillation frequency there is a certain distance between the appearance of visible oscillations and the "folding" of the wave. This distance is called the zone or band of stabilization. With increasing frequency of oscillation, the stabilization zone initially decreases and then increases. Near the second neutral oscillation, it simultaneously moves against the flow to the strip of the vibrator.

The process of development of a disturbing motion presented in Fig. 3.30 corresponds to the advancement in the plane of the neutral curve parallel to the y axis along the straight line $Re^* = $ const. Fig. 3.30B shows a scheme of the behavior of a tellurium jet during its propagation along the plate with $n = $ const. When a tellurium jet that characterizes a disturbing motion propagates downstream, the oscillation amplitude gradually increases in accordance with the shape of a neutral curve. After passing the place on the plate corresponding to the second neutral oscillation for a given frequency, the amplitude begins to decrease. Observations on the propagation of disturbing oscillations of a constant but different frequency ($n_2 > n_1$) along the plate showed that with an increase in n, the amplitude of the tellurium je oscillates more sharply and decreases after the passage of the place corresponding to the second neutral oscillation is slower. In this case, depending on the measurement location and oscillation frequency, the waveform and the crest may become as shown in Fig. 3.25A, but with the value n_2 or n_3. Thus Fig. 3.30B shows the character of the development of current lines in the linear and nonlinear stages of the transition. The development scheme of the disturbing motion, shown in Fig. 3.30B, corresponds to the advancement in the plane of the neutral curve parallel to the x-axis along the straight line $\frac{\beta_r \nu}{U_\infty^2} = $ const.

With simultaneous photographing of tellurium jets from above and from the side, it was found that very quickly behind the vibrator tape a pure flat sinusoidal oscillation goes into a sinusoidal oscillation simultaneously in two mutually perpendicular XOY and XOZ planes (screw movement, subsequent nonlinear transition stages) (Fig. 3.30C). The wave crest after the phase of its transition from a pointed to rounded shape simultaneously twists along a helical line into a longitudinal vortex (along the x-axis). Moreover, if two tellurium jets move in parallel in the same XOZ plane, then they are "screwed" in opposite directions inwards toward each other. After turning the wave crest, the wave "collapses" and immediately stops oscillations in the XOY and XOZ planes, oscillations disappear and only a jet moving along the plate without shifting is observed (the oscillation is fully stabilized).

These experimental investigations made it possible to determine the physical characteristics of the point of stability loss (by the point of loss of stability we mean the point with x coordinate corresponding to the Reynolds number of stability loss at a given flow velocity). Given the results of these studies, it can also be argued that this point should be understood as a certain area that has a small length along the x-axis and is characterized by distinctive features in comparison with other areas along the plate:

- The amplitudes of the oscillations of the component velocities of the disturbing motion in this region are maximum.

- The wave of transverse oscillations is sharper, and the wave formation is more pronounced and the "folding" of the wave occurs faster.
- The frequencies of unstable oscillations are higher, as well as oscillation ranges with maximum amplitudes of unstable oscillations wider.
- The wavelengths of unstable oscillations are shorter.
- The phase velocity and especially the growth factors are large.
- The limiting frequencies are maximum.
- The all vibrations propagate at a rate of $c_r = u$.
- The fluctuations are concentrated in a narrow conical layer near $y = \delta^*$.

Fig. 3.30D shows a scheme of oscillations of tellurium jets flowing around an inclined plate, when a positive pressure gradient was formed on it. A specific wave crest has been detected. The crest was wrapped more slowly with increasing frequency, and the wave, without folding, moved in a flat loop located in the XOY plane, downstream as if in a "frozen" form. The waveform immediately acquired a nonlinear character.

Nonsinusoidal oscillations (Fig. 3.30E) resulted in a less pronounced sinusoidal wave, somewhat similar in shape to the wave shown in Fig. 3.30D. With increasing frequency, the tellurium jet twisted more sharply and longitudinal vortices formed faster. When the wave was folded, the longitudinal vortex rotated so that its axis became parallel to the y axis.

Flat disturbances develop in a conical layer, and the three-dimensional deformation arising on the crest under the action of the Görtler instability of the current lines of a plane wave, as it develops, increases this deformation and further alternates the transition stages.

In Ref. [342], the results of measurements in a wind tunnel on the thickness of a flat BL of transverse velocity components $W + w'$, which are formed when a plane wave disturbance is introduced into the BL with a frequency corresponding to the second branch of the neutral curve, are presented in a wind tunnel. The measurements were carried out in places along the z axis in the middle between the adjacent peaks and valleys (z_1 and z_2), and the oscillation phase of the specified wave was recorded in parallel. As a result of measurements, a three-dimensional deformation was detected on the crest of a plane wave. Similar results are given in Section 3.2, Figs. 3.6, 3.9, and 3.14.

The data cited in [547] indicate that the value of $W + w'$ is minimal on the concave portion of the streamlines and maximal on the convex part of the streamlines. In addition, it changes sign at $y/\delta = 0.2$, and profiles at z_1 and z_2 are mirrored. This also agrees with the data obtained by tellurium method [378], and suggests that on the crest of a plane wave a standing wave was formed in the z direction with a plane of symmetry at a height of $y/\delta = 0.2$ and at $\lambda_z/2$ equal to the distance between z_1 and z_2. Such a deformation of a plane wave represents the initial phase of the formation of longitudinal vortex formations. Such a deformation of a plane wave represents the initial phase of the formation of longitudinal vortex formations.

According to the Helmholtz theorem, the flow of the velocity vortex vector through an arbitrary cross section of the vortex tube is the same at a given moment of time along the entire tube, that is, $w_1\sigma_1 = w_2\sigma_2$. The vortex motion caused by the

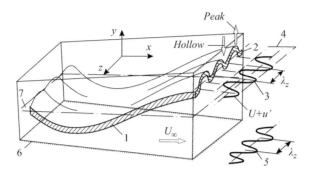

Figure 3.31 Evolution of nonlinear disturbances with a low degree of turbulence of the main flow: (1) Tollmien–Schlichting plane wave; (2) nonlinear secondary disturbances; (3) $U(z)$ velocity profile; (4) $U(z)$ velocity projection on plane XZ (top view); (5) projection on the XZ plane (top view); (6) conical layer of development of nonlinear perturbations; (7) the XZ plane of the critical layer.

Görtler instability of the current lines of a plane wave should be preserved at all phases of the wave. Near the crest, the vortex motion covers a small region along δ, σ_1 is small, and therefore w_1 is large. As a result, there is a clear manifestation of three-dimensional deformation. In the vicinity of concave streamlines, vortex motion can spread over a large thickness of the BL; σ_2 is large, w_2 is small. Note that the characteristic feature of an ordered vortex motion is compensation w' the value of u', where w' is small and u' increases.

Fig. 3.31 [36] shows a diagram of the formation of a three-dimensional deformation on the crest of a plane T–S wave. Confirmation of the occurrence of a vortex motion already at the initial stage of the transition are photos of the streamlines from the side and from above (Section 3.3, Figs. 3.16 and 3.17). When viewed from above, some asymmetry is noticeable: at the plane wave stage, a three-dimensional deformation arises, and at the stage of this deformation, prerequisites appear to oscillations of longitudinal vortex systems that are sinusoidal in the z direction. The laws governing the development of the T–S plane wave and Benny & Lin vortices along the length of the working section of the hydrodynamic stand were studied (Fig. 3.32). The obtained regularities on a logarithmic scale are presented in the form of straight lines. The scatter of points is due to the fact that the values of λ_x and λ_z were considered at different flow speeds and near the second neutral oscillation. If the wavelengths are considered only for the second neutral oscillation, then the error will significantly decrease. This allowed us to obtain the following empirical dependencies:

straight (1):

$$\lambda_x = 2.95 \cdot 10^{-4} \mathrm{Re}^{0.46} \text{ m (for a plane wave)}, \tag{3.9}$$

straight (2):

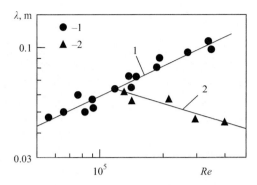

Figure 3.32 The change in the plane wavelength λ_x (1) and the wavelength of the longitudinal vortices λ_z (2) along a rigid horizontal plate.

$$\lambda_z = \text{Re}^{-0.24} \text{ m (for longitudinal vortices)}, \tag{3.10}$$

The dimensionless dependence of the wavelength of the longitudinal vortices has the form:

$$\lambda_z/\lambda_x = 3.4 \cdot 10^3 \text{Re}^{-0.7}. \tag{3.11}$$

Expressions (3.9) and (3.10) can also be obtained in a dimensionless form, using, for example, some characteristic linear scale of the BL.

As shown in Fig. 3.32, in the first stages of the transition, when a plane wave undergoes weak spatial deformations, its role is decisive. The small-scale wave recorded at the beginning of the working section along z has a length λ_z very close to λ_x. As it develops along the x-axis, the wavelength λ_z increases, as does λ_x, but, starting with $\text{Re} = (1.1-1.35) \ 10^5$, it decreases, while λ_x continues to increase. The flow patterns at the subsequent stage of the transition begin to be determined at the previous stage of the transition: even before the sixth stage of the transition occurs, a smaller-scale vortex structure begins to form at the fifth, and λ_z begins to decrease.

The presented results are well illustrated in Fig. 3.33, where the structure of the disturbing movement is shown at separate stages of the transition, a single process of which can be divided into the following stages (part I Section 2.3 Fig. 2.4 and Fig. 3.33):

1. Pure laminar flow in the BL to the Reynolds number loss of stability.
2. Amplification of plane waves of T–S.
3. Deformation of this wave in the XOZ plane from the point of view of an observer moving with the speed of wave propagation.
4. Deformation of this wave in the YOZ plane, the wave amplitude becomes a periodic function of z with "peaks" and "hollows" in this direction.
5. Amplification of the deformed wave, leading to the pulling of "peaks" forward and upward and the formation of longitudinally oriented "hairpin" (Λ-shaped) vortices.
6. Separation and "folding" of the head sections of these vortices and the merger of successive longitudinal parts of them—the formation of a vortex system with mutually opposite rotation of adjacent longitudinal vortices.

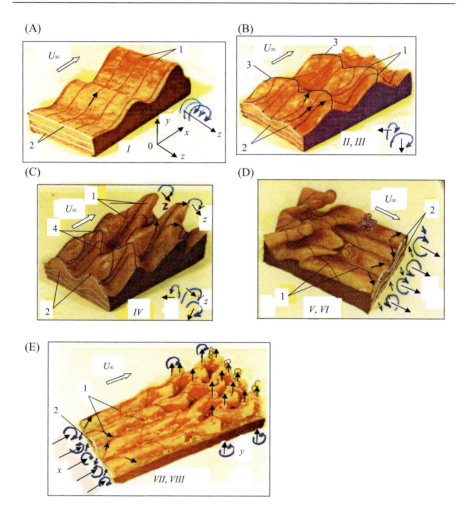

Figure 3.33 The photograph of the model of coherent vortex structures that are formed at the stages of the transition of the laminar boundary layer to the turbulent boundary layer in the flow past a rigid smooth plate: (A) stage I; (B) II and III stages; (C) stage IV; (D) V and VI stages; (E) VII and VIII stages of transition; (1) streamlines; (2) line of equal velocities; (3) sinusoidal wave in the horizontal XOZ plane; (4) sinusoidal wave in the XOZ plane, inclined at an angle to the horizontal plane of the streamlined plate.

7. Changing the shape and intensity of the vortices; the transition from straight to zigzag-like in the XOZ plane of the trajectory of their movement.
8. Separation of the peripheral parts of vortices curving along z. The formation of a turbulent boundary layer (TBL).

Depending on the external conditions, the alternation of these stages can occur at different rates. It is possible to merge individual stages or develop a turbulent BL immediately after the disruption of the heads of Λ-shaped vortices (after the fifth stage of the transition).

As the plane wave amplifies at stage I of the transition (Fig. 3.33A), the wave amplitude increases, increases the curvature of the streamlines. The fluid particles on the crest inhibited near the wall of the wave are carried to a higher velocity flow region and, due to its very small irregularity along z, experience different pressures on themselves. Figuratively, this can be imagined like the crest of a wave blown by the wind. This leads to the fact that on the crest the front of a plane wave begins to bend in the XOZ plane (the second stage of the transition).

Another reason for the deformation of the wave front is microscopic irregularities of the streamlined plate (roughness, waviness, scratches, etc.). Because stage *II* of the transition is at the beginning of the plate, the δ value is still small here. Therefore the front of a plane wave, when moving in a depression near a wall, experiences the influence of its irregularity and, falling on a crest, changes (heredity). In Fig. 3.33B it can be seen that the deformation of the wave front occurs both at the wall, in the trough, and on the crest of the wave.

Visualization has shown that the BL is very sensitive to the smallest irregularities at the bottom. This also explains the third reason for the deformation of the front of a plane wave: the microscopic irregularities of the leading edge with some delay (relaxation) are manifested in the modification of the wave shape.

At stage *II* of the transition, the Görtler instability of streamlines of a plane wave leads to a simultaneous deformation of the wave front not only in the XOZ plane, but also in the YOZ plane. From this it is clear why the spatial distortion of the disturbances was recorded in the experiments already at the first stage of the transition. At stage *II* of the transition, the streamlines have different curvatures depending on z (Fig. 3.33B). This leads to the dependence of the Görtler instability of a plane wave on z and accelerates the development of the spatial deformation of a plane wave.

In Refs. [299,300], it was shown how the primary vorticity will change in the BL. Since the streamlines are perpendicular to the wave front, this further increases the vorticity and the tendency to wrap up the lateral sides of the wave front deformed in the XOZ plane forward and down the helix between the extremes.

As soon as the disturbing motion becomes three-dimensional and the component w' appears, further development of the disturbances can be considered by analogy with the behavior of the vortex line: its curvature in any plane causes this plane to rotate in the direction opposite to the direction of rotation of the vortex line [350,420]. They lead to the fact that the regions in the region of the minimum of the front of the wave begin to descend, and in the region of the maximum they rise. The *III* stage of transition will come (already at the second stage, a standing wave can be seen in the vertical section of YOZ).

At the *VI* stage of transition, the process is intensified in accordance with the Helmholtz theorem. As can be seen from Fig. 3.33C, the "peaks" are pulled forward and up, and the streamlines, getting into more high-speed fluid layers, are accelerated and, heading at a greater angle to the wave front, increase its vorticity. This leads to the fact that at the tops of the "peaks" the vorticity increases sharply (the peripheral edges of the "hairpin-shaped" vortices are wrapped down the helix).

Under the influence of the velocity of the layers of the BL lying above, the upper part of the "hairpin-shaped" vortices stretches upward, and the apex leans toward the wall. As a result of a large twist of streamlines near the top (Fig. 9.33D), the head part of the vortex as if unscrews, and the incoming flow ("wind," higher-speed layers above the liquid) disrupts it and carries it away. The top of the vortex under the action of the vorticity of the peripheral parts and the "wind," freed from the base of the vortex, breaks into small vortices (explosion of a turbulent spot).

At stage V of the transition under adverse factors and high Reynolds stresses, the separation of the vortex head serves as a catalyst for the onset of a turbulent BL. If ε and U_∞ are small, then "decapitated" "hairpin-shaped" vortices catch up with each other and, due to the large angle of inclination to the wall and the strong vorticity of the lateral sides, they merge to form the classical Benny & Lin system of vortices. The features of a plane wave persist until now: if in Fig. 3.33A, B, C, for any z, the XOY plane is drawn, then a flat linear or nonlinear wave is fixed in it. After the onset of stage V, the plane wave completely disappears.

Due to the exchange of energy between the main and disturbing motions, the vorticity increases and the longitudinal vortices begin to grow in transversal cross section. Since they are constrained in the z direction, the growth first occurs along the y. As the Re number increases, the Reynolds stresses, the size and energy of the vortices increase. They begin to coarsen, merging with each other, which causes their nonsteady oscillations.

The unevenness of the wave front, which at the first stage of the transition caused the deformation of the front and the onset of the second stage, remains at all stages of the transition, in particular, it was fixed at the third stage as a zigzag-like oscillation in two adjacent layers of liquid (meandering). At the fourth stage, this will be manifested in the fact that the rows (Fig. 3.33C) with "hairpin-shaped" vortices during the development downstream will move with a shift, alternately overtaking each other. At the VII stage of the transition, such a motion is enhanced due to the growth of the energy of the vortices and leads to an increase in the zigzag-like development of the system of longitudinal vortices. And since the dimensions of the vortices and their energy are large (occupy the entire region δ), the process of oscillation of the longitudinal vortices in the z direction passes with a large increase.

After the zigzag development at the VII stage of the transition on the lateral surfaces of the longitudinal vortices due to their overtaking nature of development (and the appearance of shear stresses), vortices with a vertical axis of rotation are formed (Figs. 3.30 and 3.33E). The amplitude of oscillations along z of longitudinal vortices and the sizes of peripheral vortices increase until the peripheral vortices separate. To some extent, this process is reminiscent of the oscillation in the wind of a vertically arranged flag with a breakdown from the cloth alternately on both sides of the vortex sheet. After the peripheral vortex breaks down, a stage of the turbulent BL begins.

In Fig. 3.33D and E, from the front side of the photo, respectively, to the right and left, round arrows show the direction of rotation of the fluid in the longitudinal vortices in their cross sections. The vertical arrows located between round arrows show the direction of the flow of the fluid, determined by the rotation of the fluid in the longitudinal vortices. Accordingly, the arrows pointing up are called "peaks," and the arrows pointing down are called "hollows"—a similar scheme for the formation of longitudinal vortices is shown in Fig. 3.31 and in Part I, Section 5.2, Fig. 5.10.

Based on numerous experimental investigations, the results of which are systematized in Refs. [52,54,70,74,91,232,345,346], basic concepts were developed on the characteristic types of CVS in transition and TBL (Fig. 3.33). The main types of CVS in the transitional BL are analyzed in Refs. [60,139,247,345,378,540]. In Ref. [464], for the first time, the initial stages of development of disturbances in the transition BL with a flat surface flow were visualized. In Refs. [189,347], by visualizing smoke in a wind tunnel, various stages of the transition BL with flown around the bodies of revolution were photographed (Fig. 3.34).

In Ref. [457], images of CVS obtained with the help of smoke visualization are shown, in a transitional BL with flow around a stationary body they rotate. Similar visualization patterns were obtained on rotating cones [575]. Stuart was the first to formulate the features of the motion of disturbances at different stages of the development of a transitional BL [540]. On the basis of own results of experimental

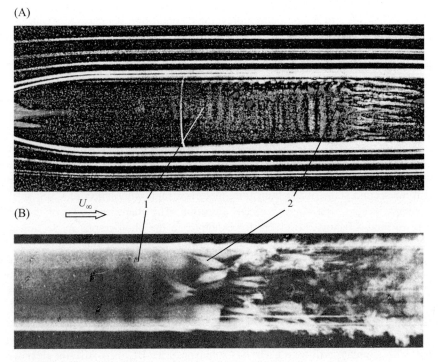

Figure 3.34 Smoke visualization of flow in the boundary layer of rotation bodies. Photograph by Brown [189] (A) and Knapp and Roache [347] (B): (1) Tollmien–Schlichting waves; (2) λ-shaped vortices.

studies and the results of other authors, Fig. 3.33 shows the developed and photographed model of the CVS in the transition BL. In all cases of two-dimensional or three-dimensional bodies, the law of evolution of the CVS in the BL is identical.

In Section 3.3, Fig. 3.6, the dependences of the resistance of a smooth flat longitudinally streamlined plate on the Reynolds number are given: point A characterizes the number $Re_{l.st.}$ and corresponds to the beginning of the first stage of the turbulence arise. From point A to point B successively appear and develop from the first to the fifth stages and from point B to point C—the sixth and seventh stages of the transition. When conducting these experiments, the Reynolds number of the loss of stability $Re_{l.st}$, calculated on the x-axis, was $5.4 \cdot 10^4$. The photograph of the model of CVS shown in Fig. 3.33 at the stages of transition from a laminar to a turbulent BL is also reflected in Part I Section 2.3, Fig. 2.4. Thus the whole process of the arise of turbulence should be considered as a single process of flow development from point A to point B, characterized by three Reynolds numbers: $Re_{l.st}$, $Re_{cr.1}$ and $Re_{cr.2}$. Under the most favorable flow conditions, the transition of a laminar BL into a turbulent one will take place from point A to point B (the first type of transition). However, factors such as the initial turbulence of the flow, curvature, waviness, roughness, surface vibration, pressure distribution on the outer boundary of the BL, and others, can significantly accelerate the occurrence of turbulence, so that point B and C will begin to move to the almost fixed point A. And the distance between chutes A and B can, under unfavorable conditions, decrease to a very small value, as well as the distance between points B and C (the second type of transition). When flow elastic surfaces, suction of the BL and in other cases, point A, on the contrary, can move to the right simultaneously with points B and C so that the distance between these points will increase significantly.

Numerous investigations by various authors have shown that the viscous sublayer of the turbulent BL has common features with a transition BL. We have proposed a hypothesis (part I Section 2.3 Fig. 2.4) [35,247], that CVS in the viscous sublayer of the TBL have a similar form, shown in Fig. 3.33, but with various degrading factors. Investigations have shown [170,247,378] that in this case, the first stage of the transition is practically absent in the transitional BL, and starting from the second stage of the transition, there is a rapid alternation of CVS, so that turbulence occurs after stage 5 of the transition. The shape and type of the CVS with worsening factors are given, for example, in the photographs in Section 3.2, Figs. 3.8, 3.13–3.15. Investigations have shown [170,247,378] that in this case, the first stage of the transition is practically absent in the transitional BL, and starting from the second stage of the transition, there is a rapid alternation of CVS, so that turbulence occurs after stage V of the transition. The shape and type of the CVS with worsening factors are given, for example, in the photographs in Section 3.2, Figs. 3.8, 3.13–3.15.

Experimental investigations of CVS in the transition BL perform in Refs. [247,378], simulate a flow in a viscous sublayer of a TBL. It is still not clear what the correlation is between the CVS of the outer region of the TBL and in its viscous sublayer. In our opinion, large vortices periodically separated from the outer boundary of the TBL under the influence of the main flow in a similar way, as shown in

Fig. 3.28D. Large vortices rush to the streamlined boundary and hit the outer boundary of the viscous region. At the same time they are destroyed by the area, as if flattened out and deform the viscous area. Such a picture is presented in the work of Shlanchauskas [528]. A sharp jump of the disturbance at the outer boundary of the viscous sublayer initiates periodic burstings from the viscous sublayer. After emissions in the viscous sublayer, a new viscous sublayer is formed in the release area. This pattern occurs periodically in the plane of the streamlined plate in different places. There is another point of view, according to which emissions from a viscous sublayer occur for the same reasons as described above, with the development of CVS transitional BL. Probably, both mechanisms of emission formation from a viscous sublayer take place: both under the influence of large external vortices and due to the development of CVS in a viscous sublayer.

This process of forming a new viscous sublayer was modeled experimentally. To do this, at the bottom of the working section (Section 3.1, Figs. 3.1 and 3.2) V.I. Korobov designed an insert in the working section of the hydrodynamic stand, containing a slot located across the flow, through which it was possible to "merge" the BL. The chamber for merge the BL was made autonomous and easily removable. The span of the slit was 0.2 m, and the slot width varied and was 0.5, 1.0, 1.5, 2.0 mm. From the chamber, the nozzle was brought out, which allowed with sufficient accuracy to determine the flow through the slot, which was regulated by changing the slot width and using a valve. Because the flow velocity was small, it was enough to regulate the flow through the slot to such an extent as to ensure at a given flow velocity the discharge of a volume of liquid equal to the thickness of the BL.

Fig. 3.35 shows photographs of velocity profiles without merge and with merge of the BL. In addition to consumption varied U_∞. Depending on the flow through the slot, the velocity profile became more filled up to the impact one. It should be noted that the velocity profile with suction has excesses. According to the obtained photos of the velocity profiles, the values δ^* and δ^{**} were determined graphically and the form parameter H. Without merge of the BL, the form parameter $H_o = 2.18$. Depending on the flow rate of the fluid through the slot, the value of H decreased ($H_1 = 1.58$, $H_2 = 1.45$) and for the impact velocity profile $H_3 = 1.26$. The obtained values of H agree well with the calculated data [376,380].

Fig. 3.36 shows the development of sinusoidal disturbances along the plate in the absence of a merge through the slot. On the left in the photographs you can see a comb of tellurium wires from which colloidal tellurium flows. To the right in the photographs are located the vibrator rods, on which the vibrator tape is fixed (Section 3.1, Fig. 3.3). In the photos of Fig. 3.36A, B, C, the oscillation frequency increases sequentially. At the same time, it is seen how the amplitude of oscillations increases.

In the photograph of Fig. 3.36B, the vibrator oscillated with the frequency of the second neutral oscillation. With further increase in the oscillation frequency (Fig. 3.36C), the oscillation amplitude decreases and the wave crest collapses, with the appearance of three-dimensional deformations and the appearance of nonlinear effects. It can be seen how the disturbing motion is distributed over the thickness of the BL (Section 3.2, Fig. 3.11). In Fig. 3.36D and E, the vibrator generates

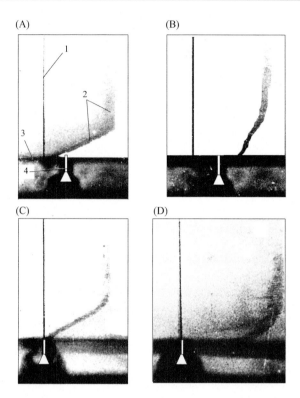

Figure 3.35 Photographs of the velocity profiles in front of the slit (A, B) and in the region of the slit (C, D) without merge of the boundary layer (A, C) and with merge of the boundary layer (B, D): (1) tellurium wire; (2) cloud of colloidal tellurium (profile longitudinal averaged velocity); (3) smooth horizontal bottom of the hydrodynamic stand; (4) slit for merge of the boundary layer.

disturbances with the frequency of the second neutral oscillation at different values of the flow rate of the fluid discharged through the slit.

Downstream, the amplitude decreased 4–5 times, and the wavelength increased 1.5–2 times, that is, the oscillation frequency was tuned. The stability of the BL has increased significantly. It can be seen that the amplitude of oscillation has decreased throughout the entire thickness of the BL. However, with a slight decrease in the flow rate Q (Fig. 3.36E); the amplitude of the oscillation increases and a high-frequency modulation of the plane wave appear. With an increase of the flow rate Q, it is noticeable that the BL (absolute stability and increased friction) is almost absent behind the slit. However, then there is a rapid increase in the thickness of the BL (large gradient δ). In some cases, a plane wave of large amplitude arose. In all three cases, the disturbance wavelength after the slip increased so much that for a given x, the oscillation was in the region of instability of the neutral curve. By adjusting the flow rate of the merged fluid through the slit, a partial or complete merge of the BL at the slot location is obtained. It can be seen that in all

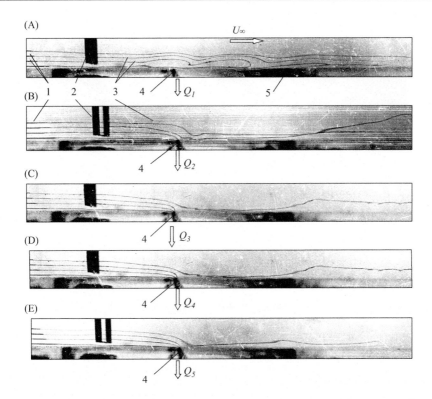

Figure 3.36 Modeling of a flow in a viscous sublayer of a turbulent boundary layer using a boundary layer suction through a slit: (1) tellurium wires; (2) vibrator; (3) tellurium current lines; (4) slit; (5) rigid plate; $Q_1 = 0$, $Q_2 < Q_3$, $Q_4 < Q_5$, Q_3 max.

cases the oscillations do not spread beyond the slot. However, the emerging new BL behind the slot under the action of large gradients of thickness δ becomes unstable. Significantly larger disturbances develop in it than in the BL to the slot.

Thus it can be argued that a new viscous sublayer with large gradients of external loads is formed in the viscous sublayer after ejection (bursting). These worsening factors in the new viscous sublayer will cause the formation of nonlinear phases of the CVSs, similar to the transition BL. As a result, the photos of the CVSs in the viscous sublayer of the TBL and in the transition BL have many common external features and characteristics.

Fig. 3.36 shows a picture of the formation of a new BL after it is merged through a slit, which models the flow in a viscous sublayer of the TBL and coincides well with the pattern of flow after ejection from a viscous sublayer [443–445].

Fig. 3.37 shows a photograph of the CVS of a TBL along the plate, obtained with the help of smoke visualization [575]. The TBL consists, as it were, of individual turbulence clouds moving along the plate. Between the clouds in dark color are the so-called pockets of the inhibited liquid (2), which are directed to the plate

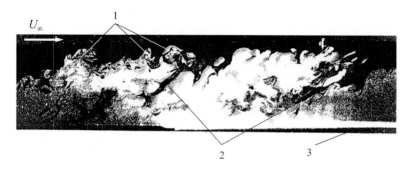

Figure 3.37 Photograph of the coherent vortex structures of a turbulent boundary layer along the plate: (1) large vortices on the outer boundary of the boundary layer; (2) pockets of braked liquid; (3) plate [575].

approximately at an angle of 45 degrees. Pocket pressure is greater than turbulence-filled clouds. Therefore the TBL does not merge into one structure, but moves in the form of alternating areas. The main flow "blows off" from the outer boundary of the BL large vortices (1), which rush to the streamlined plate. These vortices retain the velocity of movement of the order (0.8−0.9) U_∞, along with the resulting motion impulse toward the plate. Considering that the pressure in the pockets is higher than in the turbulent zone, these vortices do not cross the pockets, but, as it were, move along the pocket to the wall.

Fig. 3.38 shows a photograph of a TBL, obtained with the help of smoke visualization in transverse cross section relative to the longitudinal velocity [152]. It is seen that in the transversal direction there are pockets of inhibited fluid. The distance between the pockets remains constant at different Reynolds numbers. With increasing velocity, the thickness of the pockets decreases in both the longitudinal and transverse directions. Large vortex structures on the outer boundary of the TBL in accordance with Figs. 3.36 and 3.37 are called "hairpin-shaped vortex structures." Indeed, these vortex structures resemble the hairpin-like structures of the transition BL and Klein vortices in a viscous sublayer.

Fig. 3.39 shows a photograph of the CVSs in the viscous sublayer of the TBL, obtained using the method of hydrogen bubbles [345]. It can be seen that the Klein's vortices form after the development of zigzag previous vortex structures.

In 1980, we put forward a hypothesis [35]: in the viscous sublayer of the TBL, the development of CVS is about the same as in the transitional BL, but when exposed to deteriorating factors. In 1983, R.F. Blackwelder draws an analogy between flows in a transitional BL and TBL [170], and also, like other authors, argues that at the boundary of the fluid layers with large shear stresses, vortex structures develop. In particular, on the basis of his own experimental investigations of natural Görtler vortex systems, Blackwelder determined the wave number of the Görtler vortices (GV). The wavelength in the decay region of the ordered GV form it denotes Δ. Available data are presented in

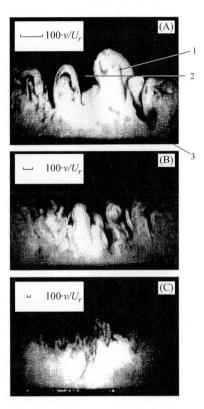

Figure 3.38 Photo of large coherent vortex structures of a turbulent boundary layer in the transverse direction with Reynolds numbers Re** = 600 (A), 1700 (B), 9400 (C): (1) vortices on the outer boundary of the boundary layer, (2) pockets of inhibited fluid, (3) plate [152].

dimensionless form $\Delta^+ = \Delta u_\tau/\nu$. Although the physical length Δ in experiments of various authors varies, Δ^+ has a constant value of about 10. Here Δ^+ denotes the variation of the used scale of viscosity ν and u_τ. Therefore for most amplifying disturbances, the wavelength $\lambda_x = 2\pi/\alpha_r$ should be $\lambda_x^+ \approx 150$. Some researchers determined the values of the dimensionless wavelength: [169] $\lambda_x^+ \approx 130$, [312] $\lambda_x^+ \approx 90$, [3] $90 < \lambda_x^+ < 230$. According to Blackwelder, $\lambda_x^+ \approx 175$.

Fig. 3.40 shows a scheme for the development of a CVS in a transitional and TBL [139]. The thickness of the transition BL varies in the transverse direction in accordance with the picture of the development of the CVS (Fig. 3.33). Such nonuniformity of thickness δ develops in the TBL as well, which is well seen from the visualization pictures given in Ref. [152] (Fig. 3.38). At high Reynolds numbers, such nonuniformity disappears due to the high mixing speed of the CVS at the outer boundary of the BL. When forming a scheme for the development of the CVS in the BL at different stages its development has been used numerous investigations of

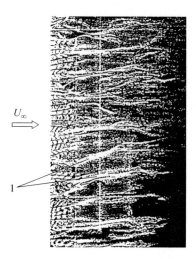

Figure 3.39 Photographs of the longitudinal coherent vortex structures (1) of a viscous sublayer of the turbulent boundary layer [345].

various authors, including their own results. To understand this scheme, it is necessary to take into account the photographs given in Refs. [113,139,443,445,524], and in Section 3.2, and also Figs. 3.30, 3.31, 3.33, 3.37–3.39.

In the laminar BL (2) of the flat plate (1) (Fig. 3.40), T–S plane waves (8) arise, which are transformed into three-dimensional waves (9). These waves have a sinusoidal shape in the *XOZ* and *YOZ* planes. In the process of its development, spike-like (hairpin) vorties (10) are formed (*V* stage of transition, Fig. 3.40). The head part of these vortices goes to the outer boundary of the BL, and hollow forms in the longitudinal direction in the vortex. Under the action of more high-speed shear layers of the BL, the head part (14) of the vortex (11) is separates, and the root part of the vortex (11) is folded into two longitudinal vortex (12) (*VI* transition step, Fig. 3.33). During the meandering of the longitudinal vortices (12), vertical vortices arise on their lateral surfaces (13). With the development of vortices (12), of the vortices (13) separation (*VII* stage of transition). Due to the destruction of the vortices (13), the destruction of the longitudinal vortices occurs and a TBL is formed. Longitudinal vortices can collapse without meandering in the process of increasing their size downstream. The development of a disturbing motion occurs in a narrow flat conical layer located in the BL with a plane of symmetry in the horizontal plane corresponding to the critical layer (Fig. 3.31).

Measurements of the kinematic characteristics made it possible to determine the existence of two regions in the BL in which an energy jump occurs. The first jump of energy occurs after the separation and destruction of the head parts of the hairpin vortices. After the destruction of the head parts, the energy of the BL sharply decreases. The second energy jump occurs after the destruction of the longitudinal vortices. At the same time, the energy of the BL also decreases and a viscous sublayer of the TBL is formed.

Experimental investigations of the characteristics of the boundary layer on a smooth rigid plate   149

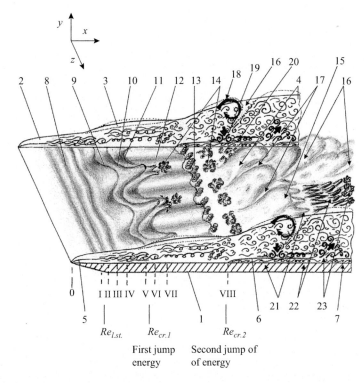

Figure 3.40 Coherent vortex structures development scheme in the boundary layer of a rigid smooth plate: (1) plate; (2) laminar; (3) transition; (4) turbulent boundary layer; (5) critical layer of laminar boundary layer; (6) viscous turbulent boundary layer sublayer; (7) renovated viscous sublayer; (8) flat Tollmien−Schlichting waves; (9) Tollmien−Schlichting three-dimensional waves; (10) hairpin-shaped vortices; (11) destruction of hairpin-shaped vortices; (12) longitudinal vortices; (13) vertical vortices; (14) separated vortices; (15) hairpin-like vortices of turbulent boundary layer; (16) transverse pockets of turbulent boundary layer; (17) longitudinal pockets of turbulent boundary layer; (18) "blown off" large vortices; (19) flattened this vortex; (20) formation of coherent vortex structures in the buffer area; (21) three-dimensional wave in the viscous sublayer; (22) Klein vortices; (23) emissions (bursting) from a viscous sublayer [139].

The TBL begins after the *VII* stage of transition, if ideal flow conditions are preserved. With various deteriorative factors, turbulence may occur after the *V* stage of transition. In both cases, the prehistory in the form of longitudinal vortex systems in the transitional BL continues to develop and in the formation of a TBL. In this case, the TBL is not a continuous array, but retains the typical features of the CVSs of the transition BL. There is a significant thickening of the TBL, which at a certain stage of development represents large hairpin-like structures (15), separated in the longitudinal direction by the pockets of the braking liquid (17), and in the transverse direction by the pockets (16). As in the transitional BL, the head parts of these hairpin-like structures are "blown away" now undisturbed flow. Large vortices (18) go downward into the lower velocity regions of the BL. When they hit the outer boundary of a viscous region, these vortices "flatten out" (19) and cause a deflection of the outer boundary and deformation of the flow in the viscous region. In the buffer region, with the destruction of these vortices, CVSs are formed in the form of longitudinal vortices (20). The induced deformations in the viscous region initiate bursting (23) from the viscous sublayer due to the acceleration of the CVSs development of the viscous sublayer.

As shown in Fig. 3.36, three-dimensional waves (21) are formed in a viscous sublayer (Fig. 3.40), similar to similar waves in a transition BL. Since the flow in the viscous sublayer undergoes intense disturbances from the TBL core, the perturbing motion in the viscous sublayer almost immediately from the linear wave stage transforms into three-dimensional waves, which later develop as Klein vortices (22) (Fig. 3.39). These Klein vortices develop in two ways: they increase in size and, like the harpin vortices of the transitional BL, are destroyed starting from the head. In the second case, this process is accelerated due to the action of large vortices (18), penetrating to the streamlined boundary. In both cases, the destruction of the Klein vortices leads to the destruction of the viscous sublayer and the bursting of the decelerated fluid from the viscous sublayer. Due to these two causes, constant statistically uncertain bursts from a viscous sublayer occur over the area of its external boundary. With the help of various averaging methods, it was possible to determine the probability characteristics of such bursts and the geometric dimensions of CVSs viscous sublayer.

Thus the thickness of the TBL has three areas containing characteristic CVSs:

- The region of the outer boundary of the TBL, in which large vortices are formed with a predominant axis of symmetry in the direction of the oz axis.
- The area of the buffer layer with the periodic occurrence of vortices with a predominant axis of symmetry in the direction of the axis ox.
- The viscous sublayer region with a Klein vortex system with a predominant axis of symmetry in the direction of the ox axis, and with periodic bursts upwards at an angle to the oy axis.

The model of CVS development in the transition BL (Fig. 3.40) is uniform. Similar to this model, the development of a CVS occurs at the initial stages of the development of a TBL throughout its thickness and in all flow regimes in a viscous sublayer of the TBL. With an increase in the Reynolds number, the sizes of the pockets (16), (17) decrease and can practically be ignored.

A model of a TBL in the form of large-scale motion is given in [528]: (18), (19) in Fig. 3.40.

3.7 Laser Doppler velocity meter measurements of the structure of the boundary layer of a rigid plate in the beginning of turbulence

The results of the visualization of the development of transition stages (Section 3.2) made it possible to determine the characteristic areas of the BL along the X, Y, Z

axes for further quantitative measurements using a LDVM developed and manufactured by V.P. Ivanov V.P. and V.A. Blokhin [39,313,315]. As with the visualization of the BL, measurements with using of the LDVM were carried out in a hydrodynamic stand of small turbulence (Section 3.1, Figs. 3.1 and 3.2). The placement of the equipment on the hydrodynamic stand (Section 3.1) is shown in Fig. 3.41.

On the V.I. Korobov designed trolley (19) with the help of a powerful coordinate device, a special mounting farm (10) is attached, on which the LDVM apparatus is located, developed by V.P. Ivanov [39,315]: total reflection prism (13), rotary prisms (14), (16), ray splitter (15), the focusing lens (18), the photodetector (21) and other equipment. Full reflection prism (13) is fixed motionless relative to the trolley (19). Laser (12) (LG-38) is mounted on a massive vibration-proof base so that its ray is parallel to the movement of the trolley along the Π-shaped beams (20) and hits the prism of total reflection (13) regardless of the location of the trolley. A photodetector (21), a focusing lens (18) and a ray splitter (15) are installed on special high precision devices (17). Such an arrangement of optical assemblies makes it possible to easily move the measuring volume $300 \times 30 \times 30$ μm inside the work area in three coordinates without additional adjustment of the LDVM.

The method of studying the nonlinear stages of the natural transition was as follows. First, using the tellurium method, the BL was thoroughly investigated over the entire cross section along the working section. Special attention was paid to

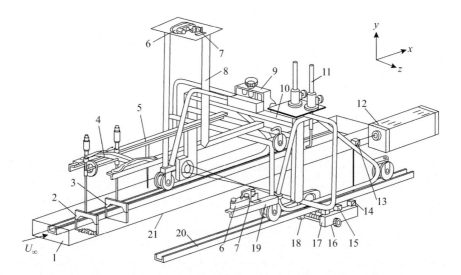

Figure 3.41 Placement on the trolley of equipment for the investigation of the boundary layer: (1) working section, (2) holder with vortex generators, (3) holder with tellurium wire, (4) microcoordinate, (5) anode, (6) camera, (7) autorun relay, (8) frame, (9) coordinate device, (10) mounting farm, (11) testers for positioning thermo anemometer sensors, (12) laser, (13) full reflection prism, (14), (16) rotating prisms, (15) ray splitter, (17) high precision device, (18) focusing lens, (19) trolley, (20) Π-shaped beams, (21) photodetector [247].

fixing the velocity field in the transversal direction, with the tellurium wire being placed at different distances from the second bottom. Such a study was carried out at U_∞ = const along the x-axis; at the same time, due to a change in the number of Re in certain places along the x-axis, various stages of the transition were recorded. Based on the analysis of the $U(z)$ profiles, characteristic points were determined along the y and z axes, at which the kinematic characteristics of the BL were studied using the LDVM and the thermo anemometer Disa Elektronik. All these measurements were performed with a low degree of turbulence. Then, on the strain gage insert, we studied the peculiarities of the alternation of transition stages at x = const and U_∞ = var. In accordance with this method, the measurements were carried out with an increased degree of turbulence and on a curvilinear bottom, which has different radii of curvature and is installed in the working section of the hydrodynamic stand.

The visualization was performed at the beginning of the working section, and the profiles of the averaged and pulsating longitudinal velocity of the BL were measured by V.P. Ivanov in ten sections along the working section (Fig. 3.42) located at a distance x equal to 0.15; 0.65; 1.15; 1.65; 1.87; 2.076; 2.23; 2.306; 2.43; 2.546 m from the junction of the bottom of the hydrodynamic stand with a confuser, with a flow velocity U_∞ = 0.6 m/s. The insert at the end of the work area was attached to a strain gage. In order not to introduce errors into the measurement results, a wedge was installed in the rear gap between the insert and the fixed part of the bottom of the working section, which pressed the insert tightly to the edge of the notch in the bottom and made the insert stationary. Profiles of longitudinal averaged and pulsation velocities are in good agreement with similar measurements by other authors [341–343,518,520,521]. The transitional BL can be investigated by varying the x coordinate or (for x = const) the value of U_∞. The second approach is preferable, but the analysis should take into account the critical values of Re numbers for proper understanding and modeling of the transition stages [378]. Therefore the remaining measurements were carried out in the ninth section (x = 2.43 m) on the inset for different U_∞. The analysis of the data obtained was carried out in accordance with the classification of the transition stages described in Section 3.6 (Fig. 3.33).

Fig. 3.43 shows a diagram of the velocity field in the formation of longitudinal vortex systems (Benny & Lin vortices) on a flat plate [612]. In accordance with the CVS formation model in the BL (Section 3.6, Fig. 3.33), starting from the V stage of the transition of the laminar BL to the turbulent one, systems of longitudinal vortices are formed in the BL. When flowing around a flat plate (5) in the later stages of the transition, pairs of longitudinal vortices (6) are formed in the BL, rotating in opposite directions. The distance between rotating coaxially vortices is indicated by the wavelength in the transverse direction λ_z. Depending on the direction of their rotation, fluid layers (1) form, with the direction of flow to the wall, conventionally denoted by the word "Hollow," and (2) with the direction of flow from the wall, conventionally denoted by the word "Peak." The shapes of the velocity profiles in various cross sections of the longitudinal vortices are denoted (3) $U(y)$ and (4) $U(z)$. In accordance with the scheme shown in Fig. 3.43 various variants of the distribution of velocity profiles in the BL are considered.

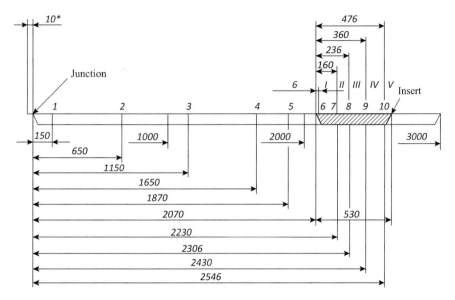

Figure 3.42 The layout of the cross sections along the working section of the hydrodynamic test bench for measuring the kinematic characteristics of the boundary layer.

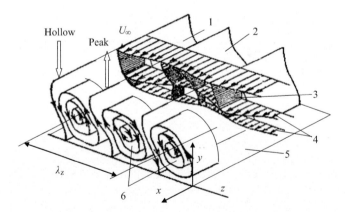

Figure 3.43 The structure of the longitudinal vortices (Benny & Lin vortices) formed in the boundary layer of the longitudinal plate at the later stages of the transition: (1) layer of fluid between the longitudinal vortices directed toward the wall ("Hollow"), (2) layer of fluid between the longitudinal vortices directed from the wall ("Peak"), (3) $U(y)$, (4) $U(z)$, (5) rigid plate, (6) pair of longitudinal vortices.

Fig. 3.44A shows the results of measuring the profiles of the longitudinal averaged speeds at various stages of the transition. Curves (1), (2) correspond to stage *I* of the transition according to Fig. 3.33A (Section 3.6), curve (3) *II*, *III* stages of transition (Fig. 3.33B), curve (4) stage *IV* of the transition (Fig. 3.33C), curves (5),

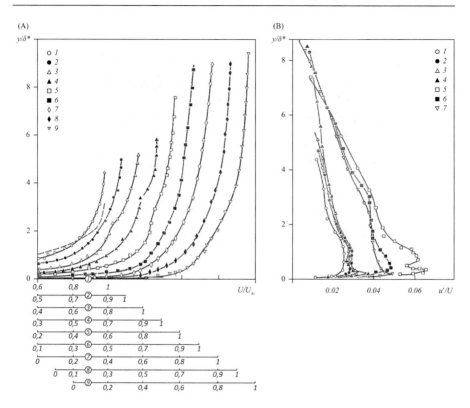

Figure 3.44 Profiles of averaged (A) and pulsation (B) velocities in the ninth section with: (1) $U_\infty = 0.09$ m/s; (2) 0.12; (3) 0.15; (4) 0.18; (5) 0.21; (6) 0.27; (7) 0.35; (8) 0.4; (9) 0.6 m/s [315].

(6) – stages V, VI of the transition (Fig. 3.33D), curves (7), (8) VII, VIII stages of the transition (Fig. 3.33E), curve (9) corresponds to the TBL.

Already at the first stage of the transition, the profile of the longitudinal averaged velocity, measured using LDVM, differed from the Blasius profile (dashed curve) due to the increased flow velocity (when visualizing the flow structure, $U_\infty = 0.1$ m/s). With the development of the BL and the alternation of transition stages, a characteristic deformation of the velocity profiles due to the structural features of the flow at each stage of the transition is found. The greatest deformation when measured along the axis of the working section turned out to be at $U_\infty = 0.21$ m/s ($\sim 6\% \ U_\infty$). In general, we can assume that the stages of the natural transition along the channel axis with $\varepsilon > 1\%$ are characterized by the deformation of the front of the $U(y)$ profile within the limits of $y/\delta^* = 1-4$. The instantaneous averaged velocity profile measured by a tellurium method (Fig. 3.5B) corresponds to velocity profile (3) (Fig. 3.44). The flow characteristics at individual stages of the transition are better traced by the pulsation profiles (Fig. 3.44B). For each stage of the transition, a characteristic profile shape was found. With increasing U_∞, the maximum of the pulsation velocity, characteristic for the first stage of the transition (stretched along y) first becomes "double-humped," then the vertical coordinates of the maxima approach each other, and their value increases. Then, in the area between these two maxima, one maximum is formed, decreasing in size, approaching the wall and sharpening until a characteristic turbulent form of the pulsation profile is formed. At the initial stages of transitions, the appearance of a double-humped shape of the pulsation profile is due to the flow structure given in References [347,348].

The upper maximum of the pulsating velocity is formed under the influence of deformation along the y in the z direction of the plane wave front on the crest caused by the Görtler instability of streamlines (Section 3.6, Fig. 3.31). It is located near the critical layer—with $y/\delta^* \approx 1$. Energy is expended on the formation and development of the indicated plane-wave deformation on the wave crest, as a result of which the flow in the region of the ridge is inhibited. Due to the laws of conservation and continuity, the flow under the ridge is accelerated, which leads to the formation of the lower maximum of the pulsation velocity. The visualization of the BL at further nonlinear stages of the transition showed that a floor-by-floor system of longitudinal vortices is formed in the BL, leading to an increase in the maxima of the two-humped front of the pulsating velocity profile. Modification of the flow structure at the subsequent stages of the transition determines the further development of the shape of the pulsation profile up to the turbulent profile. Although the alternation of transition stages is smooth, it can be assumed that curves (1)–(6) in Fig. 3.44 characterize the first six stages listed in Section 3.6 (Fig. 3.33), and curves (7)–(9), the seventh stage of transition. Moreover, for curves (1), (2), the number $Re_{\delta^*} = 290$; for (3) 450; (4) 520; (5) 800; (6) 1220 and (7) 2100.

Fig. 3.45 shows the dependence of the change in the maximum values of the velocity pulsations u' on the number Re. The values of u' are taken according to Fig. 3.44B and [315]. One can see an increase in the intensity of pulsations u' at the

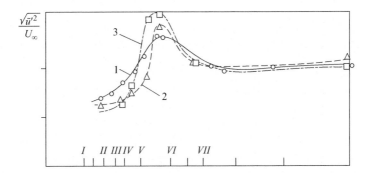

Figure 3.45 The value of the root-mean-square values of the longitudinal pulsation velocity at the *I-VI* stages of the transition: (1) measurements along the working section at $U_\infty = 0.6$ m/s; (2) measurements in the ninth section at U_∞ according to Fig. 3.32; (3) measurements of receptivity to flat disturbances.

nonlinear stages of the transition and a gradual decrease in the transition to the turbulent flow regime. The character of the curve is similar to that obtained in [378] by investigation a transition of a laminar BL to a turbulent one in a wind tunnel using a hot-wire anemometer. The data in Fig. 3.45 confirm the hypotheses [31,621] that the minimum coefficient of friction corresponds to the flow patterns in the BL at the *IV* and *VI* transition stages.

Fig. 3.46 shows the results of measurements using the LDVM averaged profiles of the longitudinal velocity of the BL in the ninth section in Fig. 3.41. At each stage of the transition, the corresponding measurements of the velocity profiles were performed for the values of z, the coordinates of which were determined in order to investigation the characteristic cross sections of the vortex system according to Fig. 3.33 (Section 3.6) and Fig. 3.43. Visualization has shown that as the BL develops, a stable axisymmetric system of longitudinal vortices with a wavelength $\lambda_z \approx 4.8 \cdot 10^{-2}$ m is formed in the working section of the hydrodynamic stand along its entire thickness. According to measurements of the velocity profiles $U(y)$ (Fig. 3.44A) shows that at the first two stages of the transition (Fig. 3.46A) at $U_\infty = 0.11$ m/s, the velocity profiles are filled, without kinks, and their difference between them depending on z is small and is due to the deformation of the front plane T–S waves in the plane *XZ*. At the third stage (at $U_\infty = 0.15$ m/s), the velocity profiles become more filled near the wall, the profile deformation increases and spreads to the greater thickness of the BL. The difference in the profiles with respect to z becomes more significant in accordance with the characteristics of the stage. At the fourth stage of the transition (at $U_\infty = 0.18$ m/s), the trends indicated for the third stage increase. The deformation of the velocity profile along the y increases, and several bends appear at the front of the profile.

The reverse kinks in the upper part of the profiles are apparently due to the heads of "hairpin-shaped" vortices, which can be considered as secondary longitudinal vortex systems of small sizes. In general, at these stages of the transition, with increasing velocity, the convexity of the profile near the wall increases, the thickness δ^*, δ and the degree of deformation of the front of the velocity profile along y. If at the *IV* stage the formation and growth of Λ-shaped vortices caused the profile deformation to occupy the largest area along the y ($y/\delta = 0.5-6$), then after the "folding" of the head sections of these vortices at the *V* stage of the transition, the profile deformation area decreased velocity along y ($y/\delta = 1-5$). At the same time, the strain value became maximum $(7\%-8\%)U_\infty$. The convexity of the profile near the wall has increased. In contrast to the previous stages of the transition, a noticeable regularity of the vortex structure appears: the velocity profiles (1)–(3) and (4)–(5), depending on z, become equidistant between each other, and the change in their deflections indicates the presence of a stable longitudinal vortex structure in the BL. At the *VII* stage (at U_∞, equal to 0.32 and 0.6 m/s), although the convexity of the profile near the wall increases slightly and δ continues to grow, the deformation of the velocity profile front decreases noticeably. The destruction of an ordered longitudinal vortex system and the formation of a TBL (at $U_\infty = 0.6$ m/s) leads to the complete absence of kinks in the velocity profile and its uniformity in z. If at the initial stages of transition (Fig. 3.46A) the deformation of the velocity profiles

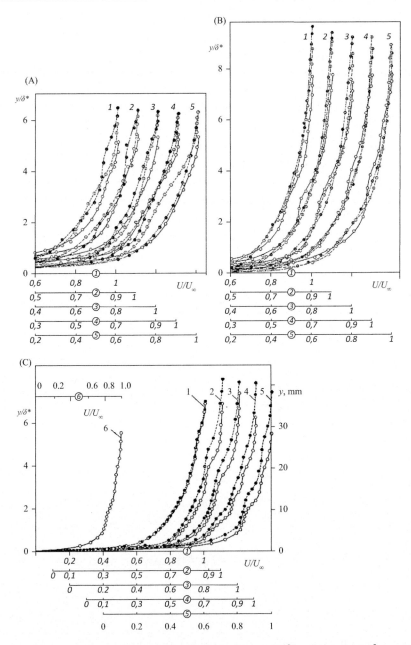

Figure 3.46 The velocity profiles measured at $z = 0$ (1), $z = 10^{-2}$ m (2), $z = 2 \cdot 10^{-2}$ m (3), $z = 2.5 \cdot 10^{-2}$ m (4), $z = 3 \cdot 10^{-2}$ m (5): (A) in the first 4 stages of the transition at $U_\infty = 0.11$ m/s (solid curves), 0.15 m/s (dashed) and 0.18 m/s (dash-dotted); (B) at the V and VII stages of the transition at $U_\infty = 0.21$ m/s (solid), 0.32 m/s (dashed) and 0.6 m/s (dash-dotted); (C) at the VI stage of the transition at $U_\infty = 0.26$ m/s, the solid curves are dimensionless, the dashed curves dimensional velocity profiles. Curve 6 is curve 5, shown on a different scale [32].

increases with the development of the flow, then at subsequent stages (Fig. 3.46B), starting from the *V* stage, this deformation decreases. We also note a sharp increase in δ at the *V* and *VII* stages of the transition.

Fig. 3.46C shows the development of flow along z at stage *VI*, when, after a breakdown of the heads of Λ-shaped vortices, a system of longitudinal vortices is formed in the boundary layer, which is characterized by a significant decrease in u' compared with stage *V*. At the same time, the main differences were in reducing the profile deformation and organizing a regular flow along z (all profiles became almost equidistant). At the *VI* stage, in addition, the vortex system increased in y and z. The above λ_z value of the vortex system, characteristic of a hydrodynamic test bench, was obtained from the visualization results of the BL and agrees well with these measurements. As can be seen from Fig. 3.46C, near the wall, the most convex profiles are (1), (3), and (4), the value of λ_z for which is $\sim 4.5 \cdot 10^{-2}$ m. It is seen that while the characteristic features of the profile remain well visible. Fig. 3.46C shows how the profile changes during the transition from absolute to dimensionless coordinates. It is seen that while the characteristic features of the profile remain well visible. The scale in which the construction is performed, sometimes significantly affects the shape of the curves [curves (5), (6)].

The results confirm the existence of certain CVSs in the BL when turbulence occurs and allow us to trace the gradual change in both the type of the average velocity profile and the dependence of the velocity pulsations on the transition stages, as well as to establish the correspondence of these profiles to certain stages. The existence of velocity profiles with kinks, which are characteristic of the *III-VI* stages of the transition, was discovered. The presence of a certain transverse structure of the BL in the transition region, associated with the appearance of a longitudinal vortex system, was also confirmed. According to the results of these measurements and visualization of the BL (Section 3.2), it is possible to construct the shape in sufficient detail and determine the size of the flow structures at certain stages of the transition.

An overview of the development of a flow in a pipe was made in [344].

3.8 Boundary layer as a heterogeneous asymmetrical wave-guide

The laws of development of disturbing motion along the thickness of the transition BL (Sections 3.3–3.7) revealed a number of features of the development of disturbances in the BL when a rigid plate flows around.

Analysis of the characteristics of the disturbing motion revealed that these characteristics are similar to the characteristics of fluctuations propagating in wave-guides. For the analysis of the revealed features, we note the basic formulations of the wave-guide [572], which will be applied to the BL when flowing around rigid and elastic surfaces.

A wave-guide is the site of environment limited in one or two directions and the employee for transfer of waves. A wave-guide can be a layer or a pipe filled by a liquid or gas, a core or a plate (solid wave-guides). Waves can propagate in a wave-guide in the form of a plane wave, the same as in infinite media (a layer or a tube with rigid walls), or as normal waves with a sufficient layer thickness. Such normal waves are formed at consecutive reflections from wall. In wave-guides probably simultaneous distribution of longitudinal and shift waves. In liquids and gases, only longitudinal waves can propagate, which are elastic waves whose propagation direction coincides with the direction of displacement and velocity of the particles of the environment. Basic parameters of a wave are the length of a wave, the period of fluctuation or frequency, a wave vector.

One of characteristics of a wave is polarization. The plane, in which there are fluctuations of a cross-section wave, is perpendicular to a direction of propagation. This feature of cross-section waves causes an opportunity of occurrence of the phenomenon of polarization of a wave, which consists in infringement of symmetry of propagation of indignations, for example, displacement and speeds, concerning a direction of propagation. Polarization can arise at distribution in the anisotropic environment, at refraction and reflection of a wave on border of two environments.

In Fig. 3.47, examples of polarizing waves are shown.

Wave dispersion is the dependence of the phase velocity of a monochromatic wave on frequency. Dispersion is caused by the physical properties of the environment, the presence of body boundaries, etc.

Nonlinear effects are a change of the form of a wave at its distribution. The account of nonlinear members of the equations of hydrodynamics and the equation of a condition is necessary for the description of nonlinear effects. A characteristic feature of nonlinear effects is their dependence on amplitude of a wave. From the spectral point of view, this process corresponds transfers of energy to the higher, more strongly absorbed harmonious making waves. In connection with that the form of a wave varies in process of its distribution, absorption of a wave also depends on distance.

Generation of harmonics occurs, when at collinear interaction and equality of frequencies $\omega_1 = \omega_2$ resultant wave is the harmonic $\omega_3 = 2\omega_1$. At absence of absorption for harmonics, linear growth of a harmonic in space is observed. Absorption of a sound limits linear growth of an acoustic harmonic in space. Therefore in the beginning the amplitude of the second harmonic grows linearly, then processes dissipation slow down its growth, stabilization of amplitude then there is its recession caused by attenuation of a harmonic is observed. The distance of stabilization is determined by a parity $l_{st} = \ln 2/2\alpha$, where α coefficient of absorption of the basic frequency. On distance $x > l_{st}$ processes dissipation lead to that the amplitude of the second harmonic exponentially falls with increase x.

Normal waves—the harmonious waves extending in a wave-guide without change of the form. In all wave-guides for each normal wave there is a so-called critical frequency below which it does not propagate and turns to fluctuation with amplitude varying along the wave-guide by exponentially law. Normal waves in plates and cores are harmonious elastic disturbances. Unlike elastic waves in

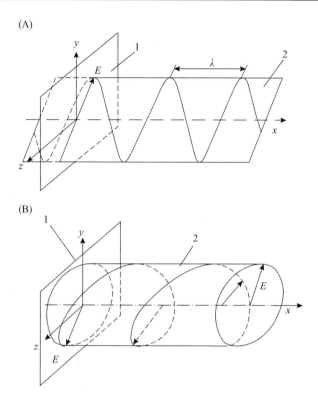

Figure 3.47 Types of waves polarization: (A) plane polarization wave, (B) circularly polarized wave; (1) plane of oscillations; (2) surface (plane) of polarization; x direction of a wave propagation; E vector representing the propagating disturbance [572].

infinity solid environments, normal waves in plates and bars satisfy not only to the equations of the theory of elasticity, but also boundary conditions on a surface of the plate or bar. In most cases, these conditions are reduced to absence of mechanical pressure on surfaces. Normal waves in plates and bars are the same elementary waves as longitudinal and shear mode in the unlimited environment, in the sense that any complex wave movement breaks up to the sum of normal waves and flux of elastic energy is equal to the sum of streams all normal waves. Normal waves in plates are subdivided into two classes: Lamb's waves, which have an oscillatory displacement of particles in a direction of wave propagation, which is parallel to a plane surface and perpendicular to the plane. The second class of a wave are the cross-section normal waves possessing only one component of displacement (which is absent in Lamb's waves), that is parallel to the plate plane and perpendicular to a direction of wave propagation (Fig. 3.48).

Thus deformation in a transverse normal wave is pure shear. By the character of deformation, transverse normal waves are divided into symmetric and antisymmetric ones. In symmetric waves movement occurs symmetrically relative to the median plane $z = 0$, and in different waves they move in each half-plane of a plate in antisymmetric waves. Normal waves, as well as Lamb's waves, are characterized by that fact at specified ω (circular frequency) and h (half of thickness of a plate) only a certain number of waves can extend. All properties of Lamb's waves and a normal wave are defined by parameters of elasticity and density of a material, frequency ω and the transversal sizes of a wave-guide.

At distribution of a normal wave in a bar, as well as in plates, only one normal wave low of each type can propagate at low frequencies. Thus a zero wave of longitudinal type has the phase and group speed equal $\sqrt{E/\rho}$, where E is Young's modulus and ρ is density of the material. Thus a zero waves are usual bending mode.

Superficial waves are the elastic waves propagating along a free surface of a solid body or on a border of a solid half-space with vacuum, gas, a liquid or another solid half-space. Elastic suface waves are a combination of nonuniform longitudinal and shift waves, the amplitudes of which exponentially decrease at moving away from a boundary. Surface waves are of two classes: (1) with vertical polarization, at which the vector of oscillatory displacement of particles of environment in the wave is located in a plane, perpendicular to a boundary surface (a vertical plane); (2) with horizontal polarization at which the vector of oscillatory displacement of particles of environment is parallel to a boundary surface and it is perpendicular to the direction of wave propagation.

The elementary and frequently observed waves with vertical polarization are Raleigh's waves, propagating along a boundary of a solid body with vacuum or rather rarefied gaseous environment. Raleigh's waves are the elastic waves propagating in a solid body along its free boundary and fading with depth. In flat Raleigh's wave, in homogeneous isotropic elastic half-space there are two components of displacement, one of which is directed along the direction of wave propagation (axis x), and the other component is perpendicular to the free border into the half-space.

Their energy is localized in the surface layer of thickness from λ to 2λ, where λ is wavelength. On depth λ density of energy in a wave is ≈ 0.05 the density near a

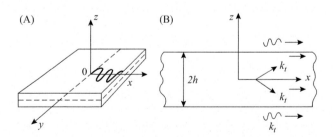

Figure 3.48 Transverse normal symmetric wave in a plate thickness $2h$ (A) and the same wave presented in the form of set of two shear modes, propagating at an angel to the direction of its propagation (B); k_t wave vectors of the shear mode forming a normal wave; x direction of wave propagation; y direction of oscillating displacement of particles [572].

surface. Particles in a wave move on ellipses, the greater axes of which are perpendicular to the boundary, and the small ones parallel to the direction of wave propagation. The phase speed of Raleigh's wave is $c_R \approx 0.9\, c_t$, where c_t is phase speed of a flat transverse wave (Fig. 3.49A).

In anisotropic environments, the structure and properties of Raleigh's wave depend on type of anisotropy and direction of wave propagation. These waves can propagate not only on flat, but also on a curvilinear free surface of a solid body. Thus their speed, distribution of displacement and stresses throughout depth, and a spectrum of allowable frequencies, which can become discrete instead of continuous, vary.

If a solid body borders with a liquid, a fading wave of Raleigh's type can propagate on the boundary. When propagating, this wave continuously radiates energy into a liquid, generating there a nonuniform wave moving away from the boundary (Fig. 3.49B). Phase speed of such wave is equal within percents to c_R, and coefficient of attenuation on wavelength is ~ 0.1. The means that on distance 10λ the wave fades approximately by e times. The distributions of displacement and stresses throughout depth in such a wave in a solid body are similar to distributions in Raleigh's wave.

An addition to a fading surface wave, a not-fading surface wave always exists on the boundary between a liquid and solid body, travels along the boundary border

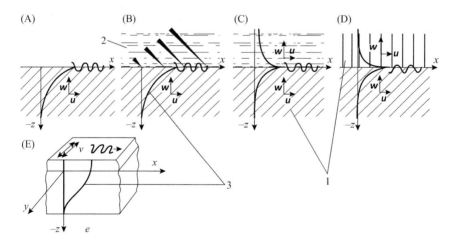

Figure 3.49 The scheme of superficial waves of various type: (1) hard environments; (2) liquid; (3) character of change of amplitude of displacement with removal from border of section of two environments; x direction of distribution of a wave; u, v, w components of displacement of particles in the given environment; (A) Raleigh's wave on free border of solid bodies; (B) fading wave of Raleigh's type on boundary of a solid body with a liquid (inclined lines in a liquid indicate wave fronts of a departing wave, their thickness it is proportional to amplitude of displacement); (C) not-fading surface wave on boundary of a solid body with a liquid; (D) Stoneley wave on the boundary between two solid bodies; (E) Lava wave on the border between solid half-space and solid layer [572].

with phase speed less than speed c_w of a wave in the liquid and speeds of longitudinal and transverse waves in the solid body. This surface wave is a wave with vertical polarization; it has a completely different structure and speed to a Raleigh's wave (Fig. 3.49C). It consists of a weak inhomogeneous wave in the liquid which amplitude slowly decreases moving away from the border, and two strongly inhomogeneous waves in the solid body (longitudinal and transverse). Owing to this, energy of the wave and movement of particles are localized in the liquid, instead of the solid body.

If solid environments have common boundary and their densities and modules of elasticity do not strongly differ, than Stoneley wave can propagate along the boundary (Fig. 3.49D). This wave consists as though of two Raleigh's waves—one in each environment. Vertical and horizontal components of displacement in each environment decrease moving away from the boundary so, that energy of the wave is concentrated in two BL of thickness $\sim \lambda$. Phase speed of Stonily waves is less than magnitudes of longitudinal and transverse speeds in both environments.

Waves with vertical polarization can extend on the border of solid half-space and liquid, or a solid layer, or even system of such layers. If thickness of layers is much less than wavelength, than movement in the half-space is similar to that in Raleigh's wave, and phase speed of the surface wave is close to c_R. Generally, movement can be such that energy of the wave will be redistributed between solid half-space and layers, and phase speed will depend on frequency and thickness of layers (look above about dispersion).

Apart from surface waves with vertical polarization (waves of Raleigh's type), there exist waves with horizontal polarization, Love waves. These waves can propagate on the border of solid half-space with solid layer (Fig. 3.49E). Such waves are transverse; they have only one component of displacement v, and elastic deformation in the wave represents pure shear. Displacement in a layer is distributed by cosine, and exponentially decreases with depth in half-space. Depth of penetration of the wave into half-space varies from shares of λ up to many λ depending on thickness of the layer h, frequency ω and parameters of the environment. The condition of existence of Love wave is defined by presence of a layer on half-space. At $h \to 0$, depth of penetration of the wave in half-space tends to infinity, and the wave becomes spatial one. Phase speed c of Love waves is limited by phase speeds of transverse waves in the layer and half-space $c_{t1} < c < c_{t2}$.

Acoustic wave-guides serve for propagation of waves of a zero mode (with uniform distribution of amplitude across cross-section) and other modes of fluctuations. Acoustic wave-guides on spatial waves are strips, tapes or wires in which certain normal waves are excited (Fig. 3.50):

$$c = c_t \left[1 + (a/l)^2 tg^2(kh) \right]^{-1/2} \tag{3.12}$$

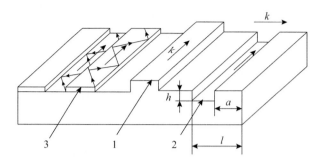

Figure 3.50 Types of acoustic wave-guides for surface acoustic waves: (1)—ledge; (2)—groove; (3)—metal film [572].

In Ref. [492] provides basic equations of distribution of fluctuations in a waveguide are written down. The intensity S of a sine wave changes in time and with distance by the law:

$$S = S_0 \cos(\omega t - kx + \phi) \quad (3.13)$$

where S_0 is amplitude of wave, ω circular frequency, $k = 2\pi/2$ wave vector, φ initial phase of the wave.

$$S = S_0 \cos\theta, \quad (3.14)$$

where $\theta = \omega t - kx + \varphi$, S = const at $\omega t - kx$ = const, i.e. $x = Vt$, $V = \omega/k$ is phase speed of the wave, frequency $f = V/\lambda$, where λ is wavelength, $\omega = 2\pi f$, period $T = 1/f$, $V = \lambda/T = \lambda f$.

The power of wave is equal to a square of intensity:

$$P = 2S^2 = 2S_0^2 \cos^2(wt - kr + w) = S_0^2[1 + \cos^2(wt - kr + w)] \quad (3.15)$$

Average power $P = 2S_0^2$.

By focusing attention on the environment in which the wave processes occur while studying a wave; it is possible to explore very complex interactions. However, the other approach exists. Behavior of many waves does not depend on specific nature and character of disturbances, related to the wave. It is possible to learn much about waves, treating them as certain kind, or mode, of disturbances of environment in which waves move. So the equation of a sine wave describes one particular type of waves or mode, one form among many others, which can propagate in an environment. (For example, two modes depending on the plane of polarization can be distributed on a string; yet two modes is a wave ahead and wave back).

When considering any disturbance of environment, it is necessary to take into account a probability of excitation of many modes at all possible frequencies. However, it is possible to investigate many interesting phenomena analyzing only one or two modes. The term "mode" for waves, propagating with constant speed in environments is used here. Partially it will concern to nonuniform environments.

Group speed is an inclination of a tangent to curve $\omega = f(k)$. The group speed is a speed of energy moving in direction x. If E is linear energy (energy per unit of length), power P is speed of a flow of energy:

$$P = V_{gr}E. \tag{3.16}$$

V_{gr} can be either greater or less than V_{middle}.

The electromagnetic waves can propagate in wave guides only in the case when their frequency is above so-called critical frequency, which we shall designate ω_{cr}. At power, equal to one ($P = 1$), the amplitude S_0 is great at small group speed and is small at large group speed:

$$P = V_{gr}E = AV_{gr}S_0^2. \tag{3.17}$$

If the properties of environment gradually change with distance (for example, waveguide, the diameter of which gradually decreases), the critical frequency ω_{cr1} of chosen mode grows with distance. If the diameter of a wave-guide changes rather slowly, the wave of given frequency is distributed almost in such way as if the properties of environment on its way do not change.

A wave of single power, which is distributed at some mode with frequency ω, exceeding the critical frequency, will now be considered. At the narrowing of a wave-guide ω_{cr1} is increased, and, hence, ω/ω_{cr1} decreases. Thus at distribution of a wave along narrowed wave-guide the group speed decreases and the amplitude S_0 is increased, that is, electrical field intensity is increased. At sufficient narrowing the frequency ω_{cr1} can be equal to ω, and then the chosen mode cannot distribute further on a pipe of a waveguide; it will be reflected so, that a wave with negative speed will run back along a waveguide. Based on formula (3.11) it is possible to conclude, that the amplitude S_0 should become infinite large in section of a waveguide, where $\omega = \omega_{cr1}$. Actually it is not so. All these ratios are not correct in case, when $V_{gr} \to 0$. However, in narrowed wave-guides, an electrical field of a wave really grows in accordance with reduction of section of a wave-guide and critical frequency grows.

The dependence k from ω can be presented by a power series:

$$k = k_0 + \left(\frac{\partial k}{\partial \omega}\right)_0 (\omega - \omega_0) + \frac{1}{2}\left(\frac{\partial^2 k}{\partial \omega^2}\right)_0 (\omega - \omega_0)^2 + \frac{1}{6}\left(\frac{\partial^3 k}{\partial \omega^3}\right)_0 (\omega - \omega_0)^3 + \cdots \tag{3.18}$$

Index "0" at individual derivatives shows that they depend on ω_0 and k_0.

Fig. 3.51 shows often used element of centimetric waves—such contour is known as directed branches of communication. A wave-guide is a metal pipe on which the electromagnetic wave can be distributed. Two identical wave-guides W_1 and W_2 have a common wall of some length L, in which apertures are done so that the wave of one wave-guide interacts or contacts with a wave of other wave-guide. In result some share of power P_1 in the first wave-guide is transferred to wave-guide W_2, and the part $(1-a) P_1$ remains in wave-guide W_1. The share of power, which is transmitted from one wave-guide to other depends on length of a way L on which the waves are connected, and on the force of connection, i.e. on the size of apertures, and on frequencies. At some value of connection $a = 1$, that is, all power passes from one wave-guide to the other.

In [492] methods of calculation of various wave-guides and characteristics of the fluctuations extending in wave- guides are described. In Fig. 3.52, the scheme of connection of two wave-guides is resulted. For such scheme, corresponding methods of calculations of the characteristic of a wave-guide are mentioned. Fig. 3.52 shows the scheme of wave parameters change in a narrowed wave-guide. By means of such schemes, it is possible to change parameters of waves propagating in wave-guides.

In the previous paragraphs, special attention was paid to the study of the development of the structure and properties of a perturbing motion over the thickness of the BL. In Section 3.5, the characteristics of a plane perturbing motion are presented in the form of dimensionless wave number and phase velocity, depending on the amplitude and frequency of oscillation, the thickness of the BL and the Reynolds number. In Section 3.3, Figs. 3.28 and 3.29, the dimensioning was performed not according to the velocity of the undisturbed flow, but according to the local velocity u at the point of measurement. The corresponding experimental dependences were constructed and analyzed. The main conclusions are that the propagation velocities of the introduced disturbances substantially depend on the frequency and place of their introduction through the thickness of the BL.

Only in the critical layer the given two-dimensional sinusoidal wave propagates with second neutral oscillation frequency according to the Orr-Sommerfeld equation as a T−S wave. Outside the critical layer and at other frequencies of disturbing oscillations in the BL, a packet of interacting harmonic oscillations develops. This can be presented, for example, as follows [185]:

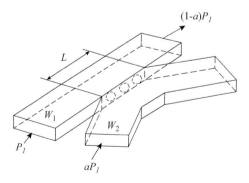

Figure 3.51 Scheme of two wave-guides with one common site of length L [572].

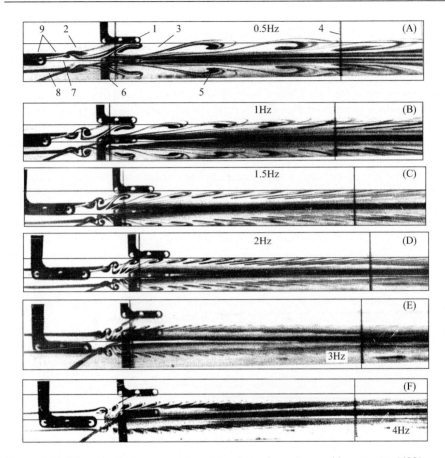

Figure 3.52 Diagram of wave parameters dependence from wave-guide geometry [492].

$$f(x,t) = F_0(x,t)\exp[i(k_0 x - \omega_0 t)] \qquad (3.19)$$

where $\omega_0 = \omega(k_0)$ is dependence of wave number upon wave frequency (dispersion); $F(x, t)$ is the envelope of harmonic oscillation set. The (3.19) equation specified a modulated harmonic wave. Having spread out function $\omega(k)$ in Taylor's line, receive:

$$f(x,t) = \int_{\infty}^{-\infty} \tilde{f}(k)\exp\left[i\left(k_0 x + \aleph x - \omega_0 t - \frac{d\omega}{dk}\left(\frac{\aleph t}{k_0} - \cdots\right)\right)\right]dx$$

where $f(k)$ is spatial Fourier spectrum of initial disturbance, $\aleph$ small parameter $k = k_o + \aleph$. Quantity $d\omega/dk = c_{gr}$ is wave group velocity. Thus Eq. (3.19) can be stated as follows:

$$f(x,t) = F_0(x - c_{gr}t)\exp[i(k_0 x - \omega_0 t)], \qquad (3.20)$$

This means that the envelope $F_0(x-c_{gr}t)$ is being propagated without changing its form with small t. These conclusions will be agreed with results of the present experiments. However, for a describe the actual picture of the development of wave disturbances in the BL, a more complex mathematical apparatus is required, since for large t and the envelope $F(x, t)$ it is characterized by dispersion, which leads to subsequent changes in the flow structure (alternation of transition stages).

The disturbing movement in transitional BL, except for dispersion, is characterized by one more property—nonlinear interaction. Measurements have shown, that even at very small ε, generation of two-dimensional disturbances in a BL is accompanied by nonlinear effects, characteristic for all stages of transition. As follows from the drawings given in the previous paragraphs, dispersion and $c_r(y/\delta)$ excite additional harmonics, which interact with each other in a nonlinear fashion. This is evident from the photographs of the visualization of the development of the disturbing movement given in Sections 3.3 and 3.6. Dependence of disturbing movement on amplitude is also an indication of nonlinearity. It is known, that if nonlinear effects predominate over dissipated ones [185], the sinus waves go over into saw-shaped while growing. It is registered in Sections 3.3 and 3.6.

It is known [185,572] that a nonharmonic wave, consisting of package of harmonic waves of different frequencies, during development changes the form because of dispersion. The relation between phases of the harmonic of a wave changes. The waveform modifies also owing to nonlinear properties of the medium.

With colinearity interaction and equality of frequencies $f_1 = f_2$ the second harmonic $f_3 = 2f_1$ is generated. The harmonic at first linearly rises and then stabilizes and attenuates. Length of the stabilization is determined by equation

$$l_{st} = \ln 2/2\alpha_i \qquad (3.21)$$

where α_i is basic frequency attenuation coefficient. For each frequency of disturbances introduced into the BL, a stabilization zone was recorded.

Taking into account data on the polarization of waves [139,572] and relying on the well-known electromechanical and electro acoustic analogy [310,532], it can be concluded that the transitional and turbulent (in its viscous sublayer) boundary layer can be considered as an asymmetric wave-guide [572].

Investigations of hydrodynamic stability [18,48] can be considered from the receptivity viewpoint (Section 1.11). Generated two-dimension wave, introduced into a BL, interacts with natural two-dimensional disturbances in the region where measurements are made. It is registered that the introduced wave, which has the frequency of II neutral oscillation in the measurement region, growths higher than others do. At other frequencies, interaction between introduced and natural disturbances occurs less intensively. For every introduced disturbance frequency, a stabilization region was registered (Sections 1.5, 1.7 and [18]).

The heterogeneous wave-guide is characterized by that, that its permittivity can change under the square law [537]:

$$k(r) = k_0 - k_r r^2 \qquad (3.22)$$

where k_0 is the dielectric constant of the outer surface; k_r is the dielectric constant in the direction of the wave-guide axis; r is the wave-guide radius. The dialectic permeability of a heterogeneous wave-guide can also be represented by another law, for example, an exponential law. An asymmetrical wave-guide, for example, flat, is characterized by the fact that the dielectric constant of its external surfaces is different (for example, metal, glass, air).

Parameters of a heterogeneous nonsymmetric wave-guide cannot be theoretically analyzed now. In Fig. 3.53, schemes of homogeneous and heterogeneous wave-guides are resulted. A symmetric heterogeneous wave-guide differs from a homogenous one with continuous changing of refraction factor across section. Because of this, a ray does not reflect from outer surfaces, but moves inside wave-guide following a sinus trajectory. Consequently, the heterogeneous wave-guide can be specific with considerably less losses compared to the homogeneous one.

In Refs. [58,60,378], it is shown that BL can be presented as a wave-guide with upper boundary moving asymptotic away from the lower one and penetrable for outer energy. There can be a doubt in legitimacy of the offered hypothesis to consider a BL as a nonlinear asymmetrical wave-guide. Standard wave-guides are characterized by the big throughput of waves moving inside of a wave-guide. According to offered hypotheses, a rigid surface from below and steady flow from above form such a wave-guide (BL), which is characterized by viscous dissipative properties, decelerating propagation of a wave in a BL. However, specific properties of a BL as wave-guide consist in that disturbing movement moves in such wave-guide with constant change of the form and polarization of fluctuation. The losses of energy due to dissipation are compensated by picking energy from the mean flow.

Disturbances in a free flow propagate almost without modification. However, when they are introduced in a BL and interact with it, they are amplified and lead to BL reaction, which clearly becomes known and can be registered. With this, certain structures of disturbance motion are created [378]. For example, with free stream turbulence intensity $\varepsilon < 0.05\%$ the maximum is $E = \sqrt{u'^2}/U_\infty$ [378]. In this case, the wave-guide amplification coefficient is:

$$\bar{k} = E/\varepsilon = 48 \qquad (3.23)$$

Although the disturbance motion in the wave-guide propagates in a narrow cone layer, the main part of its energy is concentrated in the critical layer of transitional BL or in the region of outer boundary of viscous sublayer of a TBL. Such features

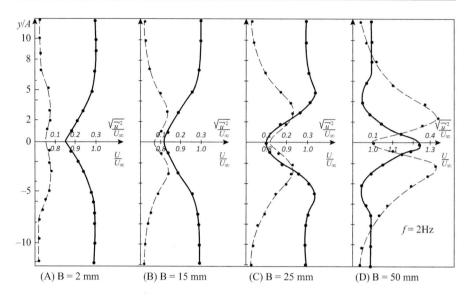

Figure 3.53 Schemes of homogeneous (A) and heterogeneous (B) types of wave-guides; k dielectric constant.

of a BL as wave-guide can be explained by two factors. First, physical characteristics of flow in a BL model the typical asymmetrical homogeneous wave-guide, described high throughput. Second, additional charging of energy from the basic movement compensates dissipative properties of a BL [378]. All this leads to that in such wave-guide disturbing movement extends in narrow cone layer, but its basic energy concentrates in area of a critical layer of transitional BL or in area of external border of the viscous sublayer of TBL.

One further property of this wave-guide should be noted—self-pumping of energy. As soon as disturbing movement begins dissipating (lose energy) due to dissipation and development throughout thickness of a BL, there comes first critical jump of energy (Fig. 1.47): heads of hairpin vortex collapse and disturbing movement again concentrates in narrower layer on y/δ, that is accompanied by reduction of pulsating velocities (Section 3.6, Fig. 3.40). At further "swelling" of disturbing movement there comes second critical jump of energy—TBL is formed. Transmission of energy in the wave-guide again concentrates in narrower layer on δ in the viscous sublayer. The same picture is observed also in the viscous sublayer. There happen, periodically and chaotically in the plane of the viscous sublayer, burstings from the viscous sublayer, after which new viscous sublayers are formed.

The amplification coefficient of the wave-guide (i.e., of a BL) depends on the transition stage and the outer disturbance magnitude. For example, at $\varepsilon < 1\%$ $\bar{k} = 7$. The amplification coefficient can be changed by changing the wall properties.

One more validation of the mentioned hypothesis is jumps of energy of disturbing movement during development of a BL by analogy to the condition of development in diverging or converging wave-guides [247,572].

3.9 The influence of small and large disturbance on the characteristics of the boundary layer of a rigid plate

In accordance with the ideas of Morkovina and Reshotko [454], the receptivity of the BL to external disturbances is understood as the reaction of the BL to various flat and spatial disturbances. Moreover, for each stage of the transition, the interaction in the BL of natural disturbances of the BL with artificially induced disturbances with a small degree of turbulence of the main flow is investigated [40,378].

In real conditions of the flow around bodies, there are always conditions leading to arise of various CVS, which can be divided into disturbances of groups I and II. Group I includes small-scale disturbances acting on the disturbing motion of the BL. For example, external disturbances, such as acting from the outer boundary of the BL is the turbulence of the external flow, its vortex and thermodynamic disturbances, acoustic and electrodynamics fields of the environment, etc. The internal disturbances of the first group act on the BL from its internal boundary, that is, from the streamlined surface. These include surface roughness, various ledges and grooves, surface vibration, its thermodynamic characteristics, etc.

Group II disturbances are large-scale disturbances acting on the entire BL. Such disturbances may occur, for example, in the following conditions:

- In the angular regions of the structures, for example, in the region of the body and wing.
- Junction two pairs of large longitudinal vortices (LLV) are formed. These vortices affect the entire thickness of the adjacent BL.
- When maneuvering on protruding parts of the body and on the side "leeward" (shaded) areas of the body, LLV are formed, affecting both the adjacent and downstream BL.
- Under real conditions of movement, the body performs oscillations of different amplitude and frequency. In the process of such oscillations on the "leeward" sides of the body and the elements of its construction, LLVs are formed, which due to the oscillatory motion of the body meander and nonstationary. The reason for the appearance of the LLV is the acceleration and deceleration modes and, as a consequence, the appearance of additional attached mass and retarded detachable "bubbles."

Currently, very little research has been done on the effect of such LLVs on the flow around the body. In cases where such investigations were carried out, only the structure and type of LLV arising from unsteady body movement were considered. Separate investigations of the effect of LLV on the characteristics of the BL were performed. The formulation and development of investigations of the influence of the LLVs on the characteristics of the BL and its CVS are relevant.

It is important to investigation the effect on the BL of external and internal disturbances of group I. The research of the problem of receptivity for the case of the interaction of natural CVSs of the BL at each stage of its development, including the turbulent flow regime, with disturbances of group II, has been completely insufficiently carried out. Finally, the third section of the problem of receptivity is practically not studied. These are the tasks of investigation of the features of CVSs formation, the interaction of disturbances of the I and II groups, and the interaction of the I and II groups with the CVSs of the BL.

Thus the current state of the problem of body flow and the calculation of the characteristics of the BL are idealized and do not take into account the real picture of the flow of fluid near the body. The results of existing calculations may be of a reference nature and need corrections caused by various disturbances acting on the BL, as well as on the nature of the flow around bodies, for example, when a LLV influences.

We have developed a methodology and experimental methods for investigation of the receptivity of the BL in a generalized representation. In connection with the above, the problem of receptivity of the BL consists of three directions:

- Receptivity by the BL of disturbances of group I (small disturbances of the CVS type).
- Receptivity by the BL of disturbances of group II (large disturbances of LLV type).
- Receptivity to the BL of disturbances of groups I and II in various combinations.

The development of disturbing motion at various stages of the natural transition depending on the degree of turbulence of the main flow is considered in Refs. [40,378]. The receptivity of the BL to flat and three-dimensional disturbances generated inside the BL was studied in Refs. [40,49,53,247,319,622].

The results of the receptivity investigation of small plane disturbances introduced in the unperturbed flow near the outer boundary of the BL are given in [319]. In this paper, we investigation the receptivity of small and finite plane disturbances generated near and on the outer boundary, as well as inside the BL. In all cases, the introduced disturbances in the linear instability diagram were characterized by a point located in a stable region

$$\mathrm{Re}_{\delta^*} \approx 200, \quad \frac{\beta_r \nu}{U_\infty^2} \approx 4 \cdot 10^3.$$

The equipment and methods for studying the receptivity of the BL are given in [40,53,247,378,622]. Disturbances were introduced into the BL or into the undisturbed flow using a device whose photo and construction are given in Section 2.7 in Fig. 2.30 and in Section 3.1 in Figs. 3.3–3.5.

Fig. 3.54 shows photographs of the visualization of a flat disturbing motion generated using an oscillating plate of width $B = 0.002$ m, located at a distance from the bottom of the working section $H = 0.045$ m. The oscillation frequency of the plate f was for (A) 0.5 Hz, (B) 1 Hz, (C) 1.5 Hz, (D) 2 Hz, (E) 3 Hz, (F) 4 Hz, oscillation amplitude $A = 0.001$ m, $U_\infty = 0.03$ m/s. The disturbing motion was visualized using a color jet flowing from a thin medical needle along the longitudinal axis of symmetry of the working section with a velocity equal to the local flow velocity. On the holder (1) a flat plate (2) is fixed (Section 2.7, Fig. 2.30), making

Experimental investigations of the characteristics of the boundary layer on a smooth rigid plate 173

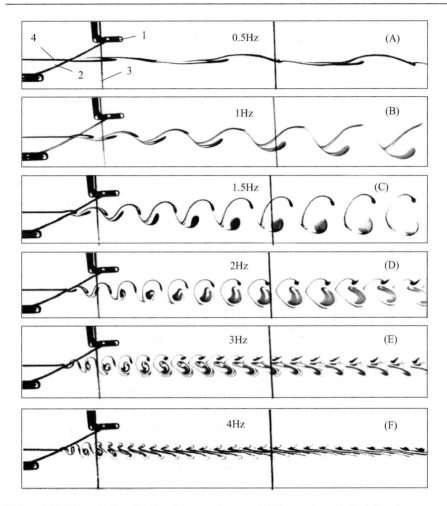

Figure 3.54 Photographs of a disturbing motion generated in an unperturbed flow by an oscillating plate of width $B = 0.002$ m, distance from the bottom of the working section $H = 0.045$ m, oscillation frequency for (A) $f = 0.5$ Hz; (B) 1 Hz; (C) 1.5 Hz; (D) 2 Hz; (E) 3 Hz; (F) 4 Hz, oscillation amplitude $A = 0.001$ m, $U_\infty = 0.03$ m/s: (1) holder; (2) oscillating plate; (3) distance marks on the front glass of the working section; (4) colored jet [50].

transverse small oscillations. Vertical marks (3) applied to the transparent side front glass of the work area. The distance between marks (3) is 0.1 m. The number (4) denotes a colored jet.

With a small oscillation frequency of the vibrator ribbon (Fig. 3.54A), the disturbance caused by the oscillating of the vibrator tape spreads in the form of a flat sine wave of constant length. The crests, which move at the same speed as the plane wave, are pulled down from the lower maxima of the wave.

When the oscillation frequency of the vibrator ribbon is 1 Hz (Fig. 3.54B), the disturbance retains the shape of a flat sine wave. In this case, the wavelength decreases approximately two times and during the first four waves the amplitude and wavelength increase, and then the wave amplitude continues to increase, and the wavelength decreases. With the development of a plane wave in the region of its upper and lower maxims, detachable secondary crests arise, and, above, these crests are poorly developed, and from below they are increasingly separated from the main wave. The visualization jet and the vibrator tape are located in the middle of the thickness of the water flow in the working section, so that the distance to the flow boundaries is 3δ ($\delta = 0.015$ m). The free surface of the flow in the working section BL is located at a distance of 2δ from the wave. The asymmetry of the vortex track in an undisturbed flow is associated with different boundary conditions above and below the propagating wave.

With vibrating tape frequency of 1.5 Hz (Fig. 3.54C), the disturbance saves the shape of a plane wave, which is very quickly behind the vibrator ribbon is deformed. Length the waves are significantly reduced, retaining the constancy of the wavelength from above and below, but from the bottom the wavelength is longer than that from above. From below, the wave crest rotates upstream and already after two wavelengths it is significantly separated from the main wave, which leads to the deformation of the waveform and the appearance of its nonlinearity.

At a frequency of oscillation of the vibrator ribbon of 2 Hz (Fig. 3.54D), the second crest from the bottom of the wave comes off and turns against the flow, so that the wave deforms and leads to the formation of a transversal cylindrical vortex rotating against the flow. On top of this vortex, an outer cylindrical vortex is formed, covering the top vortex from above. Thus the wave is very quickly transformed into cylindrical vortices, which increase the size downstream of the vertical. The shape of the vortex is deformed in cross section. When vortices propagate along the stream, their size grows very slowly. The deformation of the form of a perturbing motion is caused by the influence of various boundary conditions above and below the perturbations plane.

With oscillation frequency of the vibrator ribbon of 3 Hz (Fig. 3.54E), the pattern of deformation of the disturbing motion continues to change. Almost immediately behind the vibrator tape, a transversal cylindrical vortex is formed, rotating at the bottom against so far, which slows down the development of disturbances at the top and leads to an increase in the vortex at the bottom and significant deformation. With a given harmonic disturbances of the vibrator ribbon almost immediately after the ribbon a nonlinear vortex motion is formed. Gradually, both vortices dissipate and stop rotation, so that they move downstream as if in a frozen form.

With vibration frequency of a vibrator ribbon of 4 Hz (Fig. 3.54F), the pattern of development of a disturbance behind a vibrating vibrator ribbon is similar to that shown in Fig. 3.54E. The difference is that at $f = 4$ Hz, the process of deformation of the disturbing motion proceeds rapidly. After 2 and 3 oscillations of the vibrator ribbon, a dipole of a pair of vortices is formed in the wake, tilted against the flow. The upper small-scale transversal cylindrical vortex rotates downstream, and a similar vortex located below rotates upstream. This vortex slows down the upper vortex, which moves more slowly than the lower vortex, which leads to a strong deformation of the dipole and dissipation of the disturbing motion. Approximately 0.12 m from the vibrator strip, the lower vortex completely dissipates, and the upper vortex also gradually dissipates. This leads to degeneration and dissipation of disturbing motion.

Fig. 2.34 (Section 2.7) shows photographs under the same oscillation conditions as on Fig. 3.54, but at several other frequencies of oscillation of the vibrator ribbon. In addition, in Fig. 2.34, the visualization was performed in two color streams: at a certain distance from the vibrator ribbon, a red stream flowed from the bottom, and below it, a blue stream. This made it possible to consider the development of a disturbing motion in a thicker layer of liquid relative to the location of the vibrator ribbon. The visualization pictures obtained in Figs. 2.34 and 3.54 coincide and complement each other.

Fig. 3.55 shows photographs of the visualization of a flat disturbing motion generated by the same oscillating plate located on the upper boundary of the BL ($H = \delta = 0.01$ m). The oscillating frequency of the plate f was for (A) 0.5 Hz; (B) 1 Hz; (C) 2 Hz; D 3 Hz; $A = 0.001$ m, $U_\infty = 0.03$ m/s.

At a low frequency of vibration of the vibrator tape 0.5 Hz (Fig. 3.55A) in contrast to the Fig. 3.54A, the disturbance caused by the oscillation of the vibrator tape

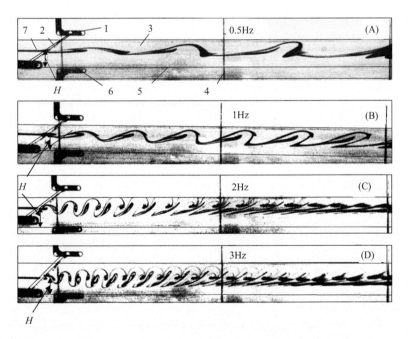

Figure 3.55 Photographs of disturbing motion caused by a plate oscillation of width $B = 0.002$ m and amplitude of oscillation $A = 1$ mm, located on the outer boundary of the boundary layer ($H = \delta = 0.01$ m), $U_\infty = 0.03$ m/s, $f = 0.5$ Hz (A), 1 Hz (B), 2 Hz (C): (1) holder; (2) oscillating plate; (3) bottom of the working section; (4) distance marks on the front glass of the working section; (5) longitudinal axis of symmetry of the bottom; (6) reflection on the bottom tool holders (1); (7) color jet; H distance of the plate from the bottom [50].

propagates at the upper boundary of the BL in the form of a flat sine wave only for a single wavelength, at the end of which the plane wave begins to deform. On the upper part, the plane wave rises above the upper boundary of the BL and enters the undisturbed flow. Therefore the velocity of propagation of the wave at the top becomes equal to the velocity of the main one for the time being. Neither the lower part of the wave propagates in the inhibited region of the BL nor does its propagation velocity become less than the velocity of the main flow. As a result, the upper crest of the wave is ahead of the lower crest, and the wave is deformed and becomes nonlinear.

With a vibration frequency of a vibrator ribbon of 1 Hz (Fig. 3.55B), this process of deformation of a plane sine wave is accelerated. In this case, a crest breaks down from the lower maximum of the wave and rises above the BL boundary and advancing with the velocity of the main flow. The wavelength decreases by approximately two times, as in Fig. 3.55A, but in both cases considered, the wavelengths are preserved when the disturbing motion propagates along the flow.

At a frequency of oscillation of a vibrator ribbon of 2 Hz (Fig. 3.55C), the disturbing motion is repulsed from the inhibited region of the BL and rises upward into the undisturbed flow. The upper region of the disturbing motion is increasingly stretched along the flow, and the lower part of the disturbance is strongly slowed down and flattened. As in Fig. 3.54D, a transversal cylindrical vortex rotating clockwise is formed almost immediately behind the vibrator ribbon. But due to the influence of the BL, this vortex is gradually deformed in the form of small-scale weak wave motion. The distance between the transversal vortices and the wavelength formed downstream are much smaller than in Fig. 3.55B. The shape of the vortices is deformed in the cross section, and the thickness of the perturbing motion decreases downstream.

With vibrating ribbon frequency of 3 Hz (Fig. 3.55D), the deformation pattern of the disturbing motion is almost the same as in Fig. 3.55C ($f = 2$ Hz), but the distance between the transversal vortices decreases even more. Compared with the development of a disturbing motion in an undisturbed flow (Fig. 3.54E), the influence of a solid wall (lower boundary of the BL) significantly distorts the disturbing motion.

Fig. 2.31 (Section 2.7) shows photographs under the same flow conditions as in Fig. 3.55, but with slightly different oscillation frequencies ribbon vibrator. In addition, in Fig. 2.31, the visualization was performed in two color jets: at the top, at some distance from the vibrator ribbon, a jet of red color was supplied, and under the ribbon—a jet of blue color. This made it possible to consider the development of a disturbing motion in a larger fluid layer relative to the location of the vibrator ribbon. Fig. 2.31A and 3.55B show the development of a disturbing motion at $f = 1$ Hz. The main patterns of disturbance development are identical in these photos, but when visualizing a disturbing movement in two streams, it is shown that the disturbing movement generated by the vibrator ribbon encompasses a much larger area of thickness of the BL and the adjacent area of the unperturbed so far. The same applies to comparing photos at other frequencies. In Fig. 2.31D, it is clear that in the process of development, the disturbing motion is transformed into a large-scale wave motion, which is then transformed into a large transversal cylindrical vortex. In Fig. 3.55D, only the beginning of the formation of wave motion was photographed.

Fig. 3.56 shows photographs of the visualization of a flat disturbing motion generated by the same oscillating plate located in the near-wall area of the BL in the region of the critical layer of the BL at $H = 0.003$ m ($\delta = 0.01$ m). The oscillating frequency of the plate f was for (A) 0.5 Hz; (B) 1 Hz; (C) 1.5 Hz; (D) 2 Hz; (E) 3 Hz; (F) 4 Hz; $A = 0.001$ m, $U_\infty = 0.03$ m/s. At the bottom of the photographs are mirrored visualizations of the disturbing motion, the holder and the vibrator ribbons in the bottom of the work area. In the investigation of hydrodynamic stability, it

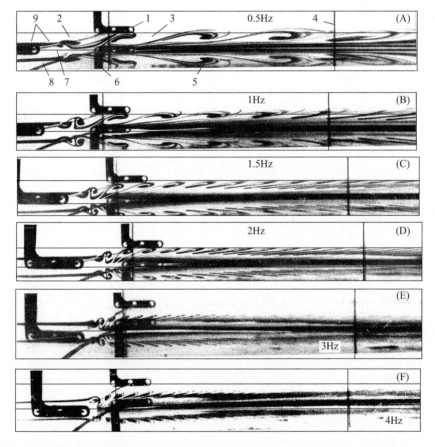

Figure 3.56 Photographs of disturbing motion caused by a plate oscillation of width $B = 0.002$ m with amplitude of oscillation $A = 1$ mm, located at $H = 0.003$ m ($\delta = 0.01$ m), $U_\infty = 0.03$ m/s, $f = 0.5$ Hz (A); 1 Hz (B); 1.5 Hz (C); 2 Hz (D); 3 Hz (E); 4 Hz (F): (1) – holder; (2) oscillating plate; (3) bottom of the working section; (4) distance tags on the front glass of the work area; (5) reflection at the bottom of the disturbing motion; (6) reflection at the bottom of the holder (1); (7) the longitudinal axis of symmetry of the bottom; (8) the reflection at the bottom of the oscillating plate (2); (9) a color jet [50].

was found that the maximum values of the constituent velocities of the disturbing motion are within (0.15–0.3) y/δ, and the critical layer, in which the components of the pulsating velocities are maximum, is located in the area of extrusion thickness ($y = \delta^*$).

With a small frequency of vibration of the vibrator ribbon 0.5 Hz (Fig. 3.56A), in contrast to Fig. 3.55A, a disturbance caused by the oscillation of the vibrator ribbon spreads above the critical layer immediately after the first oscillation of the vibrator ribbon in the form of a plane nonlinear wave. The disturbance seems to be repelled from the solid wall upwards, and the wave splits along the thickness of the BL. In the lower part of a nonlinear wave, the motion is inhibited; therefore the color dye is concentrated in this layer. Higher in y in a nonlinear wave, the dye content decreases as the local velocity increases over the thickness of the BL, and the dye diffuses accordingly. The disturbing motion propagates above δ^* and propagates in the layer (0.3–0.6) y/δ. Nonlinear wave in the process of propagation twists and pulls. The length of the nonlinear wave increases.

With an increase in the oscillation frequency of the vibrator ribbon (Fig. 3.56B) at $f = 1$ Hz, the nonlinear wavelength decreases by about 1.5 times and after 5 oscillations the nonlinear wave dissipates—its upper part is blurred by the increased velocity in the BL. The same process is observed when the oscillation frequency of the vibrator tape increases to 1.5 Hz (Fig. 3.56C). In this case, the length of the nonlinear wave becomes even smaller as it propagates along the stream and, as compared with the previous frequencies, is three times less as compared with the frequency of 0.5 Hz.

With the oscillation frequency of the vibrator ribbon 2 Hz (Fig. 3.56D), the length of the nonlinear wave is further reduced, and the dissipation rate increases. The disturbing motion propagates in the layer of ever smaller thickness, approaching the critical layer.

With a further increase in the oscillation frequency of the vibrator ribbon (Fig. 3.56E, F- 3 and 4 Hz), the process of dissipation of the disturbing motion is accelerated, and already after 8 lengths of nonlinear waves, the disturbing motion completely dissipates.

On the hydrodynamic bench (Section 3.1), in addition to visualizing the perturbing oscillations in the wake of the oscillating plate and thin profiles, with the help of LDVM (Section 3.7), measurements of the distribution of the averaged (solid curves) and pulsation (dashed) longitudinal velocity components located in an unperturbed flow, for different values of $B = 0.002$ m, 0.0I5 m, 0.025 m, and 0.05 m at $U_\infty = 0.04$ m/s (Fig. 3.57). The distance between the trailing edge of the oscillating plates and the measuring point is $l = 0.025$ m. The oscillating plates were located in the middle of the water layer thickness in the working section at $h = 0.045$ m. The oscillation amplitude is A = 0.001 m and the oscillation frequency is $f = 2$ Hz.

When a thin plate with a width of 2 mm oscillates at the measuring point, the pulsating velocities are almost uniformly distributed over the thickness: the decrease in the pulsation and averaged velocities is due to friction resistance on the surface of the oscillating plate. The damping of the pulsating velocities to the level

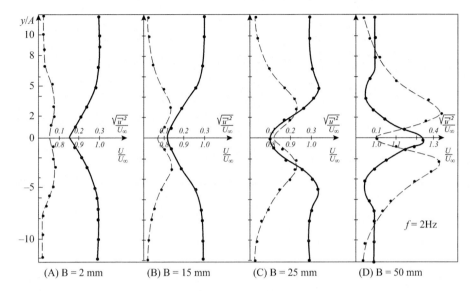

Figure 3.57 Distribution of averaged (solid curves) and pulsation (dashed) longitudinal velocity component in the wake behind oscillating plates located in the undisturbed flow, for different values of B: (A) 0.002 m; (B) 0.015 m; (C) 0.025 m; (D) 0.05 m. $U_\infty = 0.04$ m/s, the distance between the rear edge of the plate and the measuring point is $l = 0.025$ m, $h = 0.045$ m, $A = 0.001$ m, $f = 2$ Hz [50].

of the pulsating velocity in the undisturbed flow occurs approximately at a distance of 7 mm from the oscillating plate. The averaged longitudinal velocity in the wake behind the plate is 0.86 U_∞ and increases to the level of undisturbed velocity at the same distance from the oscillating plate. A similar distribution of velocities is observed in the wake of a thin profile oscillating in width $B = 15$ mm. At the same time, the damping of the pulsation velocity profile and the alignment of the averaged velocity profile to the magnitude of the main flow velocity occur at a greater distance from the oscillating profile — 15 mm. The minimum of the averaged speed decreased to 0.82 U_∞.

A similar distribution of velocities is observed in the wake of a thin profile oscillating in width $B = 15$ mm. At the same time, the damping of the pulsation velocity profile and the alignment of the averaged velocity profile to the magnitude of the main flow velocity occur at a greater distance from the oscillating profile — 15 mm. The pulsation velocity profile has a maximum at a distance of 2 mm and a minimum in the area of the oscillating profile. The minimum of the averaged speed decreased to 0.82 U_∞. This change is due to the fact that the area of the wetted surface of the oscillating profile is significantly larger than the corresponding area of the plate. Therefore the oscillating profile gives the flow of disturbing motion with greater energy.

With an increase in the width of the oscillating thin profile ($B = 25$ mm), the pulsating velocity profile degenerates at the same distance from the longitudinal axis of the oscillating profile, but the minimum and maximum of the profile of the pulsating velocity increase even more. The maximum of the pulsation profile is located approximately at the same distance as compared with the profile oscillations with $B = 15$ mm. The minimum of the averaged velocity profile decreases as compared with profile oscillations of width $B = 15$ mm, and the maximum of the averaged velocity profile appears at a distance of 5 mm from the longitudinal axis. This indicates that an oscillating profile with a width $B = 25$ mm pumps the increased energy of the disturbing motion into the flow in the layer, equal to the distance between the maxima of the pulsation and averaged profiles of the longitudinal velocity.

When a thin profile with a width of $B = 50$ mm oscillates, the shape of the pulsating and averaged profiles of the longitudinal velocity differs significantly from the previous cases of generation of a disturbing motion. The maxima of the pulsation profile of the velocity increase substantially and approach the longitudinal axis, and the minimum decreases slightly as compared with the previous case considered. The pulsation profile degenerates at a greater distance (14 mm) from the longitudinal axis. The profile of the averaged longitudinal velocity of the disturbing motion is completely modified. At a distance of 4 mm from the longitudinal axis, a minimum appears, and a maximum of averaged longitudinal velocity appears along the longitudinal axis of the oscillating profile. This indicates that the energy of the disturbing motion increases and is pumped into the undisturbed flow in the region of the longitudinal axis of the oscillating profile. This form of the longitudinal components of speed is characteristic when a thrust is formed by an oscillating profile on a moving body.

Fig. 3.58 shows the results of measurements of the averaged (solid curves) and pulsation (dashed) longitudinal velocity components obtained using the LDVM in the wake of the oscillating plate and thin profiles located on the outer boundary of the BL ($h = 0.015$ m). Values of B were 0.002 m, 0.015 m, 0.025 m, 0.05 m and 0 m, $U_\infty = 0.04$ m/s, $l = 0.025$ m, $A = 0.001$ m, $f = 2$ Hz. Experimental investigations with the help of LDVM were carried out at the beginning of the working section of the hydrodynamic stand (Section 3.7, Fig. 3.42) in order to exclude the effects of the CVS of the BL on the results obtained. The reference measurements are the profiles of averaged and pulsating longitudinal velocity in the absence of sources of disturbing motion in the flow—curves f in Fig. 3.58. The obtained measurement results correspond to the velocity profiles given in Fig. 3.44, curves (1), with $U_\infty = 0.09$ m/s and obtained with $x = 2.43$ m, $U_\infty = 0.1$ m/s. Small differences in the reference results from the data in Fig. 3.44 are related to the fact that the reference measurements were obtained for small values of x.

In Fig. 3.58, the pulsating velocity profiles when a plate oscillates of width $B = 2$ mm [curve (a)], unlike Fig. 3.57 (introducing disturbances in an unperturbed flow), are deformed in the region where the vibrator band is located, are significantly larger, so that maxims located in the region of the longitudinal axis of the oscillating ribbon.

There is a minimum in the averaged speed profile (Fig. 3.58) caused by the flow deceleration at the location of the ribbon in the same way as can be seen in

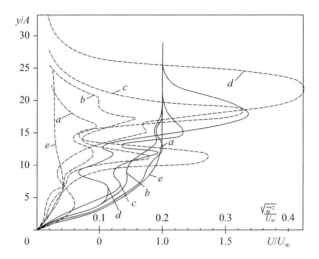

Figure 3.58 Distribution of averaged (solid curves) and pulsation (dashed) velocity components in the boundary layer behind oscillating plate and axisymmetric profiles located on the outer boundary of the boundary layer—at $h = 0.015$ m, for different values of B: $U_\infty = 0.04$ m/s, $l = 0.025$ m, $A = 0.001$ m, $f = 2$ Hz; (A)–(D) is the same as in Fig. 3.56; (E) is in the absence of oscillating plates [50].

Fig. 3.57B. When a thin profile with a width $B = 15$ mm [curve (b)] oscillates, the shape of the pulsating velocity profile is the same as that of curve (a). In accordance with the increased area of the oscillating profile, the maxima of the curve (b) increased significantly, with the lower maximum being greater than the upper maximum, possibly due to the influence of the days of the working section reflecting the pressure field caused by the oscillation of the profile. The shape of the averaged profile [curve (b)] is similar to the profile of the averaged velocity when the plate oscillates [curve (a)]. The profile deformation at the same time increased compared to the reference velocity profile.

In accordance with the increased wetted surface area, the oscillation of thin a profile with a width of $B = 25$ mm [curve (c)] led to a significant increase in both maxima of the pulsating velocity profile. The upper maximum is located significantly above the position of the oscillating profile, while, at least, the pulsating velocity continues to be, as in previous cases, in the region of the oscillating profile. The shape of the averaged profile [curve (c)] is similar to the profiles of the averaged velocity in the previous cases [curves (a), (b)]. The deformation of the averaged profile at the same time significantly increased compared with the reference profile. The averaged velocity profile in the region of the oscillating profile became larger than the reference profile. This, along with the upper maximum value of the pulsation profile, indicates that the energy of the perturbing motion is introduced into the main flow and, to a lesser extent, into the area located along the BL thickness $y/A = 10$. The oscillation of a thin profile with a width $B = 50$ mm [curves (d)] led to a further increase in the upper maximum of the pulsation and averaged velocity profiles. The absolute values of the maxima of these velocities have increased significantly. In accordance with the profiles of these velocities [curves (d)], it can be concluded that this oscillating profile contributes the energy of the disturbing motion to the undisturbed flow in the region $y/A = 22$ ($h = \delta = 0.015$) and to a much lesser extent in the region $y/A = 12$.

Fig. 3.59 shows photographs of the disturbing motion generated at the outer boundary of the BL ($H = \delta = 0.015$ m) by vibrations with a plate of width $B = 0.002$ m (*I*), the amplitude of vibration $A = 0.002$ m, and thin profiles with chords $B = 0.015$ m (*II*) and $B = 0.025$ m (*III*), oscillation amplitude $A = 0.001$ m: for *I* and *III*, the oscillation frequency is $f = 1$ Hz (A) and 2.5 Hz (B), for *II* – $f = 0.5$ Hz (A) and 1.6 Hz (B), flow velocity $U_\infty = 0.03$ m/s. The following designations were adopted: (1) the bottom of the working section, (2) the water surface, (3) vertical marks on the front glass, the distance between the marks is 0.1 m, 4 a plate or wing profile. A jet of red color is fed over the oscillating plate and profiles, and under them—a jet of blue color.

Multicolored jets make it possible to see (a series of measurements *I*) that in the process of development oscillations along the stream at $f = 1$ Hz and $B = 0.002$ m, the disturbing motion from below seems to be repelled by the retarded BL into an unperturbed flow (a blue jet) plates and profiles on top of them (red color), bends around a disturbing motion coming from below into an undisturbed flow. After about 0.05 m, blue jets continue to develop in an expanding conical layer: along the upper boundary of this layer, blue jets spread at the same height from the bottom equal to δ, and below they descend and after 0.2 m move along the upper border of the extrusion thickness of the BL. At the same time, the red jets do not fall below the thickness of the BL. At $f = 2.5$ Hz and $B = 0.002$ m, blue and red jets develop in the same way as with $f = 1$ Hz, only in a tapered layer narrowing along the length, the axis of which corresponds to the upper boundary δ. Through 0.07 m, the disturbing motion propagates in a "frozen" form in the form of a wave, the first length of which is $\lambda = 0.035$ m, and the second wavelength is $\lambda = 0.049$ m, with the crest of the second wave being wrapped. Comparison of this picture of the visualization of a disturbing motion at $f = 2.5$ Hz with the results of profile measurements velocities in the stream at $f = 2$ Hz (Fig. 3.57A) at a distance from the vibrator tape $l = 0.025$ m showed that measurements using the LDVM can reflect the averaged characteristic of the disturbing motion: in the profile of the averaged longitudinal velocity there are bends, corresponding to the visualization of the structure disturbing movement at the initial stage of its development.

In measurement series *II* (Fig. 3.59A) at $B = 0.015$ m and $f = 0.5$ Hz, the disturbing motion has a similar character as in series *I* at $B = 0.002$ m and $f = 1$ Hz. The difference is that the wetted surface area of a thin wing at $B = 0.015$ m is significantly larger than that of a thin strip at $B = 0.002$ m. Therefore the intensity and size of the disturbing movement is greater at $B = 0.015$ m. At $f = 1.6$ Hz and $B = 0.015$ m (*II* Fig. 3.59A), the intensity of the disturbing motion increased so much that when the wing profile oscillates downward, the blue and red jets rush down to the streamlined bottom of the working section. When the profile of the wing moves upward, a dipole consisting from a pair of transversal cylindrical

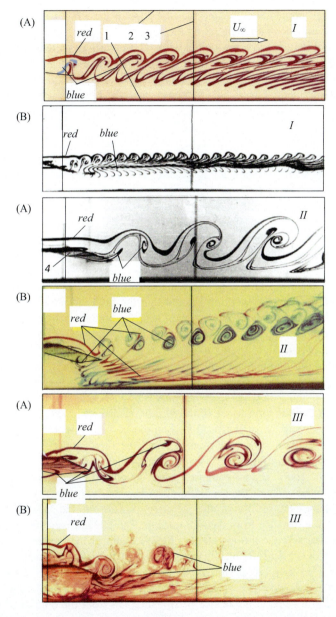

Figure 3.59 Photographs of the disturbing motion on the external boundary δ ($H = \delta = 0.015$ m), $B = 0.002$ m (*I*), $A = 0.002$ m; profiles with chords $B = 0.015$ m (*II*) and $B = 0.025$ m (*III*), $A = 0.001$ m: for *I* and *III* $f = 1$ Hz (A) and 2.5 Hz (B), for *II* f = 0.5 Hz (A) and 1.6 Hz (B), $U_\infty = 0.03$ m/s; (1) the bottom of the working area, (2) the water surface, (3) vertical marks on the front glass, (4) plate or wing profile.

vortices formed immediately behind the rear edge of the wing. Both vortices have a red jet on the outside and a blue jet on the inside. The upper vortex rotates counterclockwise, and the lower vortex—clockwise. These vortices increase in size as they move downstream and rise to the free surface. A veil of red jets and, to a lesser extent, of a blue jet, moves in the region of the critical layer.

The measurement results of velocity profiles in the stream at $f = 2$ Hz (Fig. 3.58B) at a distance from the vibrator tape $l = 0.025$ m correspond to the visualization picture near the trailing edge of the oscillating wing profile (Fig. 3.59B).

In measurement series *III* (Fig. 3.59A) at $B = 0.025$ m and $f = 1$ Hz, the disturbing motion has a similar character as in series *II* at $B = 0.015$ m and $f = 0.5$ Hz. The difference is that the collapse of the upper crest of the wave occurs more intensely and the red jet during oscillation is attracted to the surface of the profile, and then when the profile moves upwards it is thrown upwards, so that a specific transversal cylindrical vortex forms immediately on the rear edge of the wing. At $f = 2.5$ Hz and $B = 0.025$ m (*III* Fig. 3.59B), the intensity of the disturbing movement increased so much that when the wing profile oscillates from the front edge in the lower part of the wing profile as the wing moves down, the flow under the wing goes to the bottom of the working section that the blue jet moves down along the bottom surface and is directed downstream near the bottom surface. When the wing moves upward, the blue jet is sucked up by the wing and is carried down to the stream from the rear edge.

A jet of red during movement of the wing down is attracted to the surface of the profile already near the front edge of the wing, then the top is discarded when the wing moves up and then when the wing moves down it is again attracted to the profile surface so that the top surface of the profile reaches the rear edge places already above the upper surface of the profile, a dipole is formed, consisting of a pair of transversal cylindrical vortices rotating against the flow (upper vortex) and clockwise (lower vortex). In Fig. 3.59B, the second dipole diffuses as it travels along the stream, and the fourth dipole collapses, after which turbulent mixing of the entire flow occurs in the wake of the oscillating profile.

The measurement results of the velocity profiles in the stream at $f = 2$ Hz (Fig. 3.58C) at a distance from the vibrator tape $l = 0.025$ m correspond to the visualization picture near the trailing edge of the oscillating wing profile (*III* 3.59B).

A more detailed analysis of the visualization of the track behind the oscillating plate and thin profiles was performed in Section 2.7. So in Fig. 2.34 (Section 2.7) photographs are shown when visualizing the oscillation of a thin plate with a width of $B = 0.002$ m located at a distance of $H = 0.035$ m from the bottom of the working section of a hydrodynamic test bench. The oscillation amplitude is $A = 0.002$ m and the oscillation frequency is $f = 0.5$ Hz; 1 Hz; 1.5 Hz; 2.0 Hz; 2.6 Hz; 3 Hz; 3.5 Hz; $U_\infty = 0.03$ m/s. In Fig. 3.54 shows photographs when visualizing vibrations of the same plate at $H = 0.045$ m, $A = 0.001$ m, $U_\infty = 0.03$ m/s and $f = 0.5$ Hz; 1 Hz; 1.5 Hz; 2.0 Hz; 3 Hz; 4 Hz. Fig. 2.35 (Section 2.7) shows photographs of the visualization of the oscillation of the profile of a thin profile with a chord $B = 0.015$ m located at a distance of 0.035 m from the bottom of the working section. Fig. 2.31 (Section 2.7) shows photographs when visualizing the oscillation of a thin plate with a width of $B = 0.002$ m located at a distance $H = \delta = 0.015$ m from the bottom of the working section. The amplitude of the oscillation is $A = 0.002$ m and $f = 1$ Hz; 1.5 Hz; 2.0 Hz; 3 Hz; 3.5 Hz; $U_\infty = 0.03$ m/s. In Fig. 2.32 are photographs of the profile of a thin wing with a chord $B = 0.015$ m at $H = \delta = 0.015$ m, $A = 0.002$ m, $U_\infty = 0.03$ m/s and $f = 0.5$ Hz; 1 Hz; 1.6 Hz; 2.1 Hz; 2.5 Hz. In 2.33 are photographs of visualization of the oscillation of the profile of a thin wing with a chord $B = 0.025$ m at $H = \delta = 0.015$ m, $A = 0.001$ m, $U_\infty = 0.03$ m/s and $f = 0.5$ Hz; 1 Hz; 1.6 Hz; 2.0 Hz.

Thus one type of perturbing motion was investigated, the BL acting on the CVS, namely, disturbances arising behind harmonic transverse oscillations of a thin plate with width $B = 0.002$ m and thin symmetric profiles with $B = 0.015$, 0.025 and 0.05 m. The values of the upper limit δ and in the region of the displacement thickness, near the streamlined surface of the bottom of the work area. Depending on the magnitude of B, small or finite perturbations were generated in the form of a sine wave or various types of vortex structures. Depending on the location on the BL thickness, the disturbing movement interacted with various types of CVS BL as the disturbing movement moved along the flow.

The scale of the disturbing motion and its intensity varied. It was found that with harmonic oscillation of wings with a thin profile, various types of disturbing motion appeared: in the form of waves with varying wavelength, transversal cylindrical vortices of uniform or asymmetric diameter, dipoles consisting of a pair of transversal cylindrical vortices rotating in opposite directions, which advances were transformed into complex wave structures with variable wavelength.

Measurements carried out with the aid of the LDVM made it possible to detect the dependence of the kinematic characteristics of the flow on the place where the perturbing motion was introduced and its intensity. The occurrence of liquid layers along the thickness of the flow, whose velocity exceeded the undisturbed velocity of the main flow, was detected.

Thus an external disturbing movement under certain conditions changes the shape and structure and, depending on its development process, interacts with certain types of CVS BL.

Experimental investigations of the characteristics of the boundary layer on a smooth rigid curved plate

4.1 Experimental equipment and methodology for investigation of Görtler stability

The experiments were performed at the Institute of Hydromechanics of the National Academy of Sciences of Ukraine (Kiev) at the hydrodynamic stand of small turbulence, the device and features of which are given in Section 3.1, Chapter 3, Experimental Investigations of the Characteristics of the Boundary Layer on a Smooth Rigid Plate. Thermo-anemometer sensors and the laser anemometer optical system were installed on the trolley, the radiating tube of which was mounted at the side of the work area stationary (Section 3.7, Fig. 3.40). The optical system made it possible to measure at any place along the working area without adjusting it. The visualization of the flow field at any longitudinal coordinate x was carried out using the Wortmann tellurium method (Section 3.3). The hydrodynamic stand has a double bottom: the first bottom was solid metal (Section 3.1, Fig. 3.2). A flat replaceable plate was installed on this bottom, which had a metal frame on which a sheet of organic glass was fixed on top.

To investigation of the Görtler stability, V.V. Babenko designed three curvilinear plates and a curvilinear cover for a curvilinear plate with a curvature radius $R = 1$ m (Fig. 4.1). Curvilinear plates had the same design as the flat plate (second bottom): the frame of the plates had longitudinal and transverse power sets, to which curvilinear plates made of 0.01 m thick organic glass were attached to the outside. Adjustment screws were mounted in the frame below. The length and width of the three plates were equal to the same size of the interchangeable flat plate (second bottom) and, accordingly, the size of the work area. The initial and end parts of all curvilinear plates had flat sections (the initial part had a length of 0.5 m), which smoothly mated with curved sections. The radius of curvature of the plates R was 1; 4 and 12 m, and the curvature $k = 1/R$, respectively, 1; 0.25 and 0.083/m. The curvilinear section for a plate with $R = 1$ m ended at $x \approx 1.2$ m, at $R = 4$ m $x \approx 1.7$ m, at $R = 12$ m $x \approx 2.5$ m. In the investigation of Görtler stability, curvilinear plates were installed instead of a flat plate (second bottom) in the form of a second bottom. At the same time, the horizontal check of each curvilinear bottom was carried out in the same way as was done for the horizontal plate, including the horizontal check of the entire installation. After installing each curvilinear plate, the flow structure was visualized along the entire plate.

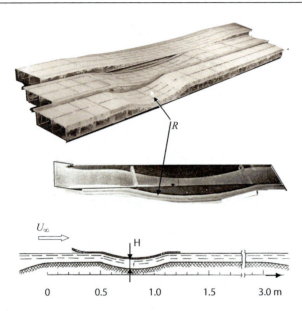

Figure 4.1 Curved plates (A), curved cover (B) and the relative position of the curved cover and the second bottom (C).

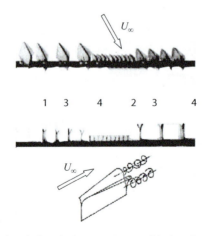

Figure 4.2 Three-dimensional disturbation generators: (1) plate for placement of vortex generators, (2) small VG1 vortex generators, (3) medium VG2, (4) large VG3.

Experimental investigation of the Görtler stability were carried out depending on the radius of curvature, the flow velocity and the wavelength λ_z in the transversal direction.

To determine the influence of the parameter $\alpha = 2\pi/\lambda_z$ on Görtler stability, V.V. Babenko developed and manufactured longitudinal vortex generators, the distance between which in the transverse direction determined the wavelength λ_z (Fig. 3.5, Section 3.1 and Fig. 4.2). Vortex generators were made of thin metal plates of various sizes. Each plate was bent in half in the longitudinal direction. After that, two obtained plates were unbent using a special device developed by V.V. Babenko, so that the distance between the two planes increased along the length of the plates. Both plates along the length were unfolded along the conical surface of the device so that on each plate in the upper part along its length a sector of the cone surface was formed. Thus, the wings were formed with a pair of petals deployed at the top of the conical surface that was the same for all the wings. The end sides of the wings at the top were rounded. The streamlined surface of the petals from the bottom was larger than the top, since the entire lateral surface of the wings flowed from below, and from above only a part of the lateral surface in the form of a sector of the conical surface.

In addition, from below, the area of the petals was in the region of the boundary layer (BL) thickness from zero to the height of the corresponding wing, and from above, the area of the petals was flowed only by the flow of the higher-speed upper BL layer located at the height of the wings. This is due to the fact that at the beginning of the wings their planes were compressed and further along the stream gradually opened along the conical surface. As a result, two fluid layers twisted to opposite sides met behind the rear edges of the petals—the higher-speed twisted upper layer merged with the lower-velocity layer of liquid twisted along the same surface and flowing around the plane of the petal from the lower surface. The upper layer of liquid was superimposed on the lower layer and even more twisted the lower layer of liquid. A pair of longitudinal vortices formed behind each wing. The size and intensity of these vortices was determined both by the size of the wings, and the distance between adjacent wings λ_z and the measurement site behind the wings (the degree of development of the longitudinal vortices).

The vortex generators had the dimensions: VG1 height $h = 3.0 \times 10^{-3}$ m and length $b = 9 \times 10^{-3}$ m; respectively VG2 5×10^{-3} m and 15×10^{-3} m; VG3 7×10^{-3} m and 18×10^{-3} m; VG4 15×10^{-3} m and 18×10^{-3} m. The opening angle of the wings, variable in height and in the extreme upper position, was $\beta = 10$ degrees. The dimensionless parameter b/h for VG1 and VG2 was 3.0, for VG3, 2.6, and for VG4, 1.2. Each wing is drilled in three places. Thin wires were pulled into the holes, which were tightened and fastened to the holder. This allowed in BL to introduce systems longitudinal vortices of various scales at different distances from the streamlined surface (Section 3.1, Fig. 3.5). In addition, a plate was made with a length of 0.12 m, a width of 7×10^{-3} m, and a thickness of 0.8×10^{-3} m (Fig. 4.2). In the cross-section, the plate from the bottom was flat, and the top was made as a circular segment with a diameter of 0.2 m. The plate was mounted on a streamlined surface across the stream (along the z axis). Along the flow on the plate, slots were made with a step of 2×10^{-3} m. The specified types of wings were installed on this plate in the slot, the distance between which varied (Section 3.1, Fig. 3.5).

The methodology for conducting an experimental investigation of the Görtler stability was the same as in the Wortmann study of the characteristics of the BL on a horizontal plate (Section 3.2). In accordance with Section 3.1, Fig. 3.3, holder with telluric wires were installed at various locations along curvilinear plates and velocity profiles and streamlines were photographed, fixing the characteristics of the BL along curvilinear plates in the absence of generators of longitudinal vortices. Then, various curvilinear vortex generators were installed on curvilinear plates (Fig. 4.2) and features of the development of introduced longitudinal vortex systems in various places along the plates were photographed: the interaction of these vortices with natural longitudinal vortices forming in the BL of curvilinear plates.

Fig. 4.3 shows a scheme for conducting experimental investigations of the Görtler stability. In the absence of longitudinal vortex generators, the structures of the Görtler vortices (GVs) on a curved plate were investigated. The characteristics of the geometric parameters of the formed GVs along a curvilinear plate at different velocities U_∞ are obtained. Layers of liquid were formed between the pairs of GVs, directed upward from the wall ("Peaks") or directed toward the wall ("Hollows"). If the size of the generators of the longitudinal vortices corresponded better to the size of the natural GVs and their mutual arrangement was such that λ_z of the natural GVs was close to λ_z of the introduced pairs of longitudinal vortices, in this case their resonant interaction occurred when photographing the velocities $U(z)$. The distance between the vortex generators λ_z corresponded to the wavelength of three-dimensional disturbances and varied within 0.004−0.032 m with a step of 0.002 m. The disturbances development was recorded by visualizing the flow by a tellurium method while controlling the parameters of the BL with a laser anemometer [47,313,314]. The investigation of the laws of three-dimensional, disturbing in the form of longitudinal vortices was carried out by placing the working section of the vortex generators in different places along the x axis. Downstream of them at a

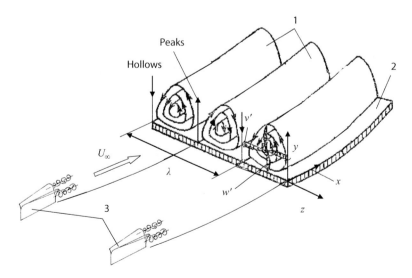

Figure 4.3 Scheme of interaction of Görtler vortices with longitudinal vortices introduced into the BL when flowing around a curvilinear plate: (1) Görtler vortices, (2) curvilinear plate, (3) generators of longitudinal vortices.

distance $x = 5 \times 10^{-2}$ m—tellurium wires were installed to photograph the visualized velocity profile in the transversal direction.

The disturbing development was recorded by visualizing the flow by a tellurium method while controlling the parameters of the BL with a laser anemometer [47,313,314]. The investigation of the laws of three-dimensional, disturbing in the form of longitudinal vortices was carried out by placing the working section of the vortex generators in different places along the x axis. Downstream of them at a distance $x = 5 \times 10^{-2}$ m—tellurium wires were installed to photograph the visualized velocity profile in the transversal direction. Vortex generators were installed on curvilinear plates in the BL at x=const, their sizes and wavelength λ_z were varied by changing the distance between them, as well as the value U_∞. The profiles $U(z)$ were photographed at various distances y from the wall. To clarify the details, the streamlines and profiles $U(y)$, which were also measured with the aid of the LDVM, were photographed in characteristic locations along z.

The parameters and patterns of development of the longitudinal vortex systems along U_∞, R, and λ_z=const determined, as it were, the movement of the control point according to the Görtler stability graph along the line $P = U_\infty \lambda_z^{1.5} \nu^{-1} R^{-0.5}$=const ($P$ is the wave parameter, ν is kinematic viscosity coefficient). Varying alternately the values of R, U_∞, and λ_z, as if moving along different lines P=const. In this case, the right branch of the neutral Görtler curve and the region of maximum amplifications of the longitudinal vortex systems were determined. The points of the neutral curve in experimental investigations are characterized by the formation in the BL of the most stable in space and time of the undulating profile $U(z)$.

If at U_∞, R, and x=const we change λ_z, then according to the Görtler curve we are moving along the line $G = U_\infty \delta_2^{1.5} \nu^{-1} R^{-0.5}$=const, where G is the Görtler parameter and δ_2 is the thickness of the loss of momentum. Varying alternately the values of R and U_∞, as if moving along different lines of G. Thus, it is possible to determine the lower, minimum value of G and the left branch of the neutral curve.

Comparing the measurement results for λ_z, U_∞, and x=const for different R, as if moving along the lines $\alpha_z \delta_2$=const, where $\alpha_z = 2\pi/\lambda_z$ is the wave number.

4.2 Experimental investigations of Görtler's vortex systems

Investigations of the flow structure along curved surfaces were carried out in Japan [3,312,547, etc.], in Europe [168,169,270,271, etc.], in America [151,168,169,515,516, etc.], in Canada [239–243, etc.], in Ukraine [12,13,34,44,126,247,283–287,591,620–629, etc.]. In Ref. [169], systems of GVs, formed on concave surfaces in the study in a water channel (Fig. 4.4), were obtained. Visualization was carried out using the method of hydrogen bubbles. The picture of the formation of coherent vortex structures (CVS) was recorded during the natural transition of the BL with flow around a curvilinear plate with a radius of curvature $R=0.5$ m (Fig. 4.4A) and $R=1.0$ m (Fig. 4.4B). The Görtler parameter $G=(U\delta_2/\nu)(\delta_2/R)^{0.5}$ was 9. Here, δ_2 is the thickness of the BL impulse loss, and ν is the kinematic viscosity coefficient.

Figure 4.4 Visualization of the transition on the concave wall in the water channel at $R = 0.5$ m, $G = 9$: (A) a natural transition; (B) artificial excitation of longitudinal vortices using regular local heating of the surface [169].

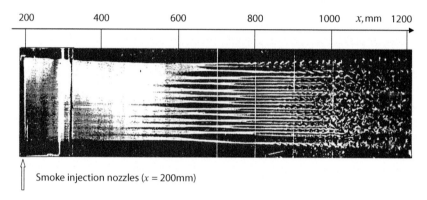

Figure 4.5 The visualization of the Görtler's instability in a wind tunnel at $U_\infty = 2.5$ m/s and $R = 1$ m [312].

The photograph shows the stages of the formation of Λ-shaped vortices, merging downstream into longitudinal vortices (Fig. 4.4A). When flowing around a curvilinear plate with $R = 1.0$ m and $G = 9$ (Fig. 40.4B), systems of longitudinal vortices were artificially stimulated with the help of regular local heating of the plate. In contrast to the natural transition, longitudinal vortices formed in the BL, which began to meander downstream, and secondary vertical vortices appeared on both sides in the periphery of the longitudinal vortices.

The regularities of the formation of CVS are similar for any cases of shear flows [74,139]. In Fig. 40.4, it can be seen that GVs are formed in a similar way, as shown on the CVS model of the transition BL (Section 3.6, Fig. 3.33). The horseshoe-shaped vortices that occur during the development of the Görtler's vortices are similar to lambda-like vortices in a transitional BL on a flat plate and Klein's vortices in a viscous sublayer of turbulent BL.

In Ref. [312], the GVs were visualized when flowing around a curvilinear plate in a wind tunnel ($R = 1$ m, $U_\infty = 2.5$ m/s). The photograph shows a section from $x = 0.2$ to $x = 1.2$ m (Fig. 4.5). The curvilinear plate width was 0.2 m. The visualization was carried out using smoke plumes blown through a system of nozzles located at a distance of 0.2 m from the beginning of the working section. Smoke jets were parallel on a flat surface and partially at the beginning of a curved surface (up to $x=0.5$ m).

In the region of $x=0.5-0.6$ m, the GVs began to form and the jets merged with each other in accordance with the λ_z of these vortices—the distance between the smoke jets increased. At $x=0.75$ m, the size of λ_z decreased again, and at $x=0.85$ m, vertical vortices began to form on the peripheral regions of the GVs. Unlike experiments [169], the GVs formed naturally, therefore they did not meander, and the vertical vortices that formed on the sides of the GVs were located symmetrically.

Fig. 4.6A shows a photograph of the development of the Görtler's vortices as seen from the butt-end of the working section, and Fig. 4.6B—photograph of the development of the Görtler's vortices at $x=0.8-1.1$ m [312]. This made it possible to trace the development of the transverse structure of the Görtler's vortices as they develop, up to the formation of a mushroom shape and their further destruction. White lines and areas in the photographs indicate areas of flow inhibition in which the speed of movement is less than in the intermediate black areas of the picture. For example, in that part of Figs. 4.4B and 4.5 ($x=0.8$ m), where straight light lines are fixed, pairs of longitudinal vortices are formed, rotating toward each other toward the wall. The current lines between these vortices are directed along a helix to the streamlined surface. This area of rotation to the wall is called the "hollows." Smoke fills this area between the whirlwinds. For neighboring pairs of vortices, the rotation is directed toward each other from the wall. This area between the vortices is called "peak." The flow in this area is directed upward along the y coordinate, where the velocities in the BL increase as compared with the location of the vortices, therefore, the smoke in these regions diffuses. The noted features correspond to Fig. 4.3 (Section 4.1).

Photographing with the butt-end (Fig. 4.6) made it possible to fix the formation of the Görtler's vortex system under the action of the centrifugal forces. The difference between the experimental forms of the longitudinal vortex structures and the theoretical calculations of Görtler was found.

Our experiments showed [34,247,626, etc.] that the Görtler's vortices that form on the concave surface during the transition from this surface to the horizontal plate continue to exist and affect the flow significantly further behind the concave section. In Ref. [534], these results were confirmed by experiments on profiles in a wind tunnel. Kohama performed experimental investigations in a wind tunnel with an open working section with a cross section of 0.4 × 0.4 m in DFVLR, Göttingen. The flow velocity was up to 12 m/s. Using the hot-wire anemometer and smoke imaging, the features of the flow around the NASA 998A supercritical (SC) wing profile model (LFC: laminar flow control) were investigated. The wing model imitated only one pressurized concave-convex side of the wing profile, and instead of the convex side of the profile there was a flat surface. Therefore, the pressure distribution along this wing layout did not match the specified wing type.

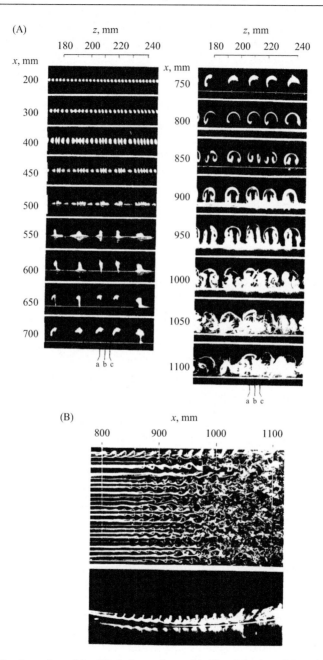

Figure 4.6 The formation of the Görtler's vortices with $U = 2.5$ m/s and $R = 1$ m: (A) top view, (B) side view [312].

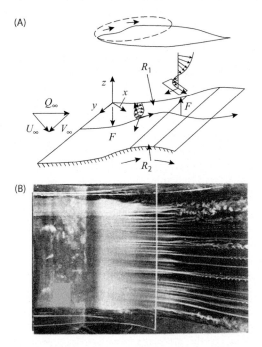

Figure 4.7 Stream-lining of the supercritical wing LFC NASA 998 A profile: (A) profile contours and velocity field in the area of wing concavity; (B) visualization of the flow field at $U_\infty = 8.9$ m/s [354].

Fig. 4.7 shows the profile form of a wing model [354]. The radii of curvature of the convexity and concavity of the wing profile model were the same and were 0.3 m. The nonbent section was 0.2 m before the beginning of the curvature. The characteristic Görtler's vortices in Fig. 4.7B are formed at the end of the concave section. On the convex side of the profile, secondary vortices are formed up to its middle, which downstream increase in the transverse direction.

The visualization of the Görtler's vortices showed that, at a speed of $U_\infty = 10.3$ m/s, the vortices become more stable, and the evolution of the development of the vortices shifts significantly downstream to the convex part of the wing profile model. Similar measurements were made in a water stream at a speed of 10 m/s (see Chomaz I.M., and H. Perrier in Ref. [412] p. 79). The radius of curvature was 0.15 m. The authors found that CV's begin to form on the concave surface and continue evolve on a convex surface.

In contrast to Ref. [354], fundamental research on the development of a class of laminarized S-shaped types of SC LFC profiles was made in Ref. [490] when they were flown around the air flow at a subsonic speed range. The free-stream number M_∞ ranged from 0.75 to 0.783. The wing profiles X63T18S, Y927S, X782S, PENIR2, X25 D5 were investigated. As in Ref. [354], the profile on the pressure side had a concavity in the nose and tail parts. The geometry of these profiles is given. In the analysis, the following parameters are varied: Mach number, lift coefficient, tailing ratio, angle of attack, and angle of deflection 0.115 of the tip edge of the flap chord, ratio of height and length of a supersonic bubble. The work analyzed the pressure distribution of these wing profiles for various options given parameters. The main task of the analysis was to determine the set of parameters at which the sizes of supersonic bubbles on both streamlined profile surfaces decreased. For each type of profile, a set of parameters was found with the minimum sizes of these bubbles. Defined by the author limit values of the coefficients of lift.

In Ref. [490] a detailed physical substantiation of the geometry of laminarized profiles is given. It is argued that the role of curvature in the nose of the profiles is to stabilize the disturbances of the BL of the Tollmien-Schlichting type. The formation of secondary transverse disturbances during the development of T-S waves is taken into account. It is briefly reviewed [243], in which Görtler stability is investigated in the presence of suction of the boundary layer. The advantage of S-shaped wing profiles in comparison with the BL suction method, which minimizes the disturbances in the BL and laminarization of the BL, is considered. As follows from the above results of the visualization of disturbances in the BL, the real physical effect in Ref. [490] is to stabilize the BL with the help of the GVs. The GVs formed in the front concave part stabilized the BL and shifted the region where the turbulent BL originated. In the aft part of the S-shaped profiles, concavity helps to reduce the area of separation of BL at zero and negative angles of attack, and at positive angles of attack Görtler's eddies prevent separation from the bottom edge of the profile and stabilize the BL at the top of the profile.

I.A. Skang (see Proc. Colloquium on Görtler's Vortex Flows. Euromech 261. Nantes, France, June 11–14, 1990; pp. 64, 65) investigated the peculiarities of the formation of Görtler's vortices on cones with a solution angle of 30–80 degrees. The measurements were carried out on a vertical installation, when an axisymmetric jet of water flowed from above. The flow varied from laminar to turbulent. A cone was placed along the jet axis, and the flow structure was fixed on its surface. Due to an abrupt change in the direction of the jet as it flows into the cone, stable Görtler's vortices formed on the surface of the cone. Photographs of such eddies are given.

Görtler is the founder of a new direction [270], which theoretically successfully developed in Refs. [12,13,157,239–243,247,283–287,534, etc.]. Tani [547,548], Wortmann [612], Bippes [168], Bottaro [182] experimentally confirmed the theory of Görtler. Further experiments were performed by other authors. The main method of conducting all the experiments was to determine the Görtler's parameter G and the wave number of the Görtler's vortices $\alpha\delta_2$, which are formed on a curvilinear surface. To calculate G and $\alpha\delta_2$, it is necessary to determine the pulse loss thickness δ_2. Since the velocity profile in the transverse direction essentially depends on the place of its determination relative to the Görtler's vortices, the value of δ_2 is determined with some degree of error. The velocity profiles were measured in various places along z with respect to the formed Gertler's eddies: in the "peaks" and "hollows," as well as in the intervals between them. For various forms of these longitudinal velocity profiles, the pulse loss δ_2 was calculated. The average value of these values was taken to calculate the specified parameters. The wave number $\alpha = 2\pi/\lambda_z$ determines the wavelength of the Görtler's vortices in the transverse direction. The calculation of the wavelength λ_z is determined either by the results of measurements of the velocity profiles in the transversal direction using a hot-wire anemometer, or on the basis of the visualization patterns obtained, for example, in Figs. 4.4–4.7. As can be seen in these figures, the distance between the Görtler's vortices in the transverse direction λ_z depends on the x and y coordinates of the measurement site. In the course of the development of the Görtler's vortices, the distance λ_z between the vortices and the value of δ_2 change. In addition, the thermo-anemometer sensor cannot be located close to the wall, since the measurement results are distorted near the wall. Thus, the parameter λ_z is determined with some degree of error. The calculated values were plotted on the Görtler's diagram as individual points, which were located mainly within the region of the Görtler's curve instability. Experimental points were located along straight lines $P = (U\lambda_z/\nu)(\lambda_z/R)^{0.5}$ or curves $\beta^* = \beta\delta_2/\nu$, characterizing the degree of enhancement of the development of Görtler's vortices. The curve corresponding to the lower branch of the neutral Görtler's curve has a coefficient $\beta^* = 0$.

P.R. Bandyopadhyay (see Refs. [150–152]) carried out thorough experimental studies of the turbulent boundary layer in a flow in an S-shaped tube with a cross-section of 25.4 cm (width) × 10.2 cm (height). The experiments were performed in an open air channel. The diffuser is made of fiberglass, inside which were installed honeycomb made of paper, and three grids to align and reduce the degree of turbulence of the main flow. The outlet section of the diffuser was 38.1 × 38.1 cm. A pipe made of four sections with a length of 50.4 cm adjoined the diffuser. The first section was straight; the second section was curvilinear, rising up. The third section was also curvilinear, but it was directed downwards. The fourth section was straight and ended with a confuser of 25.2 cm long. The radius of curvature was the same and was 50.4 cm. In the second section, the upper wall of the channel was concave, and the lower—convex, and in the third section—vice versa. The flow velocity was 9.0 m/s. The static pressure distribution was measured. The parameters of the BL were measured by a Pitot tube. Friction on the wall was measured with Preston tubes with a diameter of 0.71, 1.45, and 2.0 mm. A DISA 55P11 hot-wire anemometer instrument with a 1 mm wide wire gauge and a diameter of 5 microns was also used. Many new results were obtained in the paper. In particular, the interesting results of the velocity defect in the angular regions of the channel. All the results demonstrated the differences in the flow parameters of the flow around the convex and concave surfaces of the square tube. The distributions of wall friction and pressure were measured as a function of the coordinate along the length of the curved pipe and as a function of the z coordinate. According to the results of the distribution of wall friction in the transversal direction, the parameters λ_z of the Görtler's vortices were determined. The calculated corresponding parameters G and $\alpha\delta_2$ were plotted as corresponding points on the Görtler's curve calculated by Smith. In the same place, the measurement points in Ref. [547] are plotted for comparison. Comparison is made for turbulent and laminar flow patterns. A good agreement was obtained with Tanya's experiments: the points were in the region of instability of the Görtler's curve.

Fig. 4.8 shows the results of the study of Görtler stability with flow around rigid horizontal and inclined plates, as well as a horizontal membrane plate installed in the working section of the hydrodynamic stand [36]. The investigations were carried out using the Wortmann tellurium method [611] in the hydrodynamic stand of small turbulence (Section 3.1). To test this technique, experiments were performed with a flow around an inclined plate. In the working section of the hydrodynamic stand, an inclined flat plate was installed, the front edge of which was docked flush with the confuser, and the rear edge of the plate was lowered below the front edge of the diffuser by 0.03 m. At the end of the working section, this step allowed the curvature of the current lines and thereby simulates the flow on a concave wall, the radius of curvature of which was ~1.25 m. Depending on the flow velocity, this radius changed. The flow velocity varied in the range of 0.07–0.13 m/s. The curvature of the current lines was determined by photographs of the visualization of the current lines in the step area.

Photographing from above the velocity profiles in the transverse direction made it possible to determine λ_z (Section 3.3, Fig. 3.16), and photographing the velocity

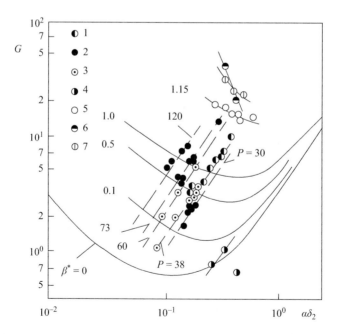

Figure 4.8 Stability of longitudinal vortex systems on concave and flat surfaces: Görtler's calculation—solid curves [270]; Tani and Aykhara experiments on concave surfaces [548]; dashed curves: (1) $R = 3$ m, $U_\infty = 7$ m/s ($P = 38$); (2) $R = 5$ m, $U_\infty = 3$ m/s ($P = 30$); 7 m/s ($P = 73$); 13 m/s ($P = 120$); (3) $R = 10$ m, $U_\infty = 11$ m/s ($P = 38$); 16 m/s ($P = 60$); measurements [36]: (4) at a flow around an inclined bottom; in the area of concave sections of current lines: (5) on a horizontal plate, (6) on a rigid plate, and (7) on a membrane plate.

profiles from the side allowed us to determine δ_2 (Section 3.3, Fig. 3.15). The obtained data allowed us to calculate the parameters G and $\alpha\delta_2$. At the beginning of the working section on an inclined rigid plate, a point was obtained [in Fig. 4.7, points (4)] located in a stable region of the Görtler's curve. This is due to the fact that the radius of curvature of the streamlines in this place is very large. GV systems have not yet been formed. At the end of the inclined bottom near the location of the step, the radius of curvature decreased, Görtler's vortices formed and points 4, connected by a straight line, are located near the neutral curve.

When plotting the linear stability diagram (Section 3.4) for fixed values of x and U_∞, the velocity profiles obtained by the tellurium method were taken, which were used to determine the δ_2 values. At the same time, the vibration frequencies corresponding to the values of the second branch of the neutral curve were recorded. Photographing the introduced small oscillations showed that in the frequency range corresponding to the second branch of the neutral curve, the amplitudes of the oscillations are maximum. These oscillations are located at small values of x. From the photographs of the oscillations with the maximum amplitudes in y, the radii of curvature of the current lines of the T-S wave were graphically determined. According to the results of photographing the velocity profiles in the transversal direction, the λ_z values of the distorted front of the velocity profiles were determined as a result of the development of the T-S wave. In accordance with the obtained values of the calculated parameters G and $\alpha\delta_2$ in Fig. 4.8 points 5 were plotted, which are located along the curve $\beta^* = 1.15$. The amplification coefficients of the Görtler's instability of secondary flows along the curvilinear surface of plane waves T-S when flowing around a flat rigid plate turned out to be significantly larger than the formed GV in the experiments of Tanya and Aihara, as well as around the inclined plate [point (4)]. This is due to the small radius of curvature of the T-S wave and large centrifugal forces. From this it follows that the curvature of the streamlines of a plane T-S wave leads to the formation of a substantial Görtler's instability and, as a result, to the formation of a deformation of the T-S wave in the transverse direction (Section 3.6, Fig. 3.31).

Similar measurements of the Görtler's instability of the T-S wave were performed at a flow around an inclined rigid plate [points (6)] and a membrane horizontal surface [points (7)], the device of which and the data of its mechanical characteristics are given in Ref. [378]. The Görtler's instability of T-S waves when flowing around these plates increases in comparison with a flat rigid plate. This is explained by the fact that when a sloped rigid plate is flown around at a greater length, there is a positive pressure gradient, and when flowing around a horizontal membrane plate, the T-S wavelength decreases and the T-S curvature radius decreases [378].

The parameters of the Görtler's vortices formed during the flow around spherical cavities [81,82], as well as in the flow on the inner surface of the circular tube of the vortex chamber during air inflow from the inlet nozzle [126] were also investigated. The stream entering the vortex chamber propagated along the internal curvilinear surface of the pipe in the form of a curvilinear parietal jet. The radius of the inner surface of the pipe was 0.05 m. The flow visualization given in Ref. [128] fixed the developing GV and made it possible to determine the value of λ_z. The δ_2 and U_∞ values were determined from the results of measuring the flow parameters formed on the inner surface of the pipe behind the inlet nozzle with the help of a DISA thermal anemometer. The calculated coordinates of the points are located on the diagram of the Görtler's curve in the region of instability. In Ref. [126], the results of experimental studies by other authors are presented in a diagram of the stability of GV.

4.3 Experimental investigation of Görtler stability on rigid curvilinear plates

In accordance with the methodology an experimental investigations of Görtler's stability, given in Sections 3.2, and 4.1, experimental investigations of the development of longitudinal vortices on concave plates, as well as in a cylindrical channel, were carried out (Section 4.1, Fig. 4.1). Experimental investigations of Görtler's stability were carried out by N.F. Yurchenko under the scientific guidance of V.V. Babenko. Investigations of vortex structures in the BL of curvilinear plates were carried out by tellurium-method F.X. Wortmann. Subsequently, vortex generators were installed on curved surfaces. The flow velocity and the x coordinate of the location of the vortex generators varied. The flow visualization thus allowed determining the nature of the interaction of natural longitudinal vortex structures with introduced longitudinal vortex structures in accordance with Fig. 4.3 (Section 4.1) [53,247]. The method of receptivity of BL of various disturbances made it possible to develop ways to control the kinematic characteristics of the Görtler's vortices and experimentally determine the neutral curve of the Görtler's stability.

Fig. 4.9 shows a photograph of the instantaneous velocity profile of the BL, obtained using of the tellurium-method Wortmann. After 0.5 s, the electrical contacts close and a voltage is applied to the tellurium wire, resulting in a colloidal

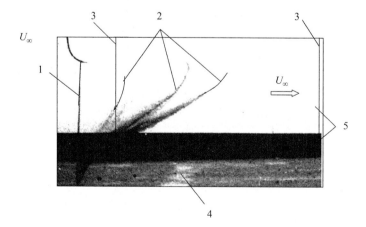

Figure 4.9 The velocity profile of the BL on a curvilinear rigid plate: (1) tellurium wire, (2) velocity profiles, (3) marks on the side glasses of the working section, (4) curvilinear bottom, (5) side walls of the working section.

tellurium around the tellurium wire, which is carried away by the flow, fixing the velocity profile (2) of the BL or current line, depending on the direction of the tellurium wire location.

Tellurium wire (1) was soldered to the holder (Section 3.1, Fig. 3.4). Marks (3) are marked on the side walls of the work area; the distance between the marks is 0.1 m. The side wall (5) of the work area is colored dark below the level of the horizontal surface of the work area. The velocity profiles in Fig. 4.9 near the curvilinear plate have a significant bend and characterize the peculiarity of the BL development along the plate length. Almost over the entire thickness of the BL in the inflection region, the velocity profile is blurred in accordance with the process of the formation of Görtler's longitudinal vortices, which in the process of natural development have an irregular complex nature and shape. It is interesting to compare the given velocity profiles with the shape of the velocity profile when flowing around a flat horizontal plate. In Fig. 3.7 (Section 3.3), the Blasius profile is fixed at the beginning of the plate, and the speed profile changes at the end of the horizontal plate: a bend appears in the middle part of the profile due to the development of subsequent stages of the transition.

The velocity profiles recorded during the flow around an inclined bottom (with a positive pressure gradient) have a slight bend and become blurred above the thickness of the BL (Section 3.3, Fig. 3.15).

In Section 3.3, Fig. 3.9 shows photographs of the velocity profile $U(z)$ at various distances of tellurium wire from the surface of the flat bottom of the working section of the $y_{T.w.}$ at $U_\infty = 6.7 \times 10^{-2}$ m/s and $x_{T.w.} = 1.1$ m. When flow curvilinear plates, the velocity profile $U(z)$ was also investigated at different coordinates $x_{T.w.}$, $y_{T.w.}$. Fig. 4.10 shows photographs of the velocity profiles $U(z)$ of the BL on curvilinear rigid plates with curvature radii of 1, 4, and 12 m. When analyzing the obtained results, it should be noted that in the initial diffuser section of curvilinear concave portions of the plates the positive pressure gradient affects the development of disturbances. At the confused part of the curvilinear part of the plates, the disturbances are stabilized under the influence of centrifugal forces and a negative pressure gradient. When the radius of curvature of the plate is $R=4$ m, before the beginning of the diffuser region, a strong curvature of the velocity profile in the transversal direction is observed in the horizontal part of the plate (Section 4.1, Fig. 4.1) at $x=0.3$ m (Fig. 4.10A) compared to the flow past a flat plate (Section 3.3, Fig. 3.9). The photographs of Figs. 4.10A−H, which record the flow around a curvilinear plate with a radius of curvature $R=4$ m, show the velocity profile $U(z)$. After 0.5 s, the electrical contacts were closed, therefore, for example, in Fig. 4.10, and at the second contact closure, the first profile develops in the flow and is located on the right of the photo. For the calculations, the first velocity profiles were taken into account.

Already at the entrance to the curvilinear part of the plate, a strong curvature of the $U(z)$ profile is observed (Fig. 4.10A). At the beginning of the diffuser part of the curvilinear plate at $x=0.6$ m (Fig. 4.9B), a steady in time deceleration of the flow along the channel axis occurs.

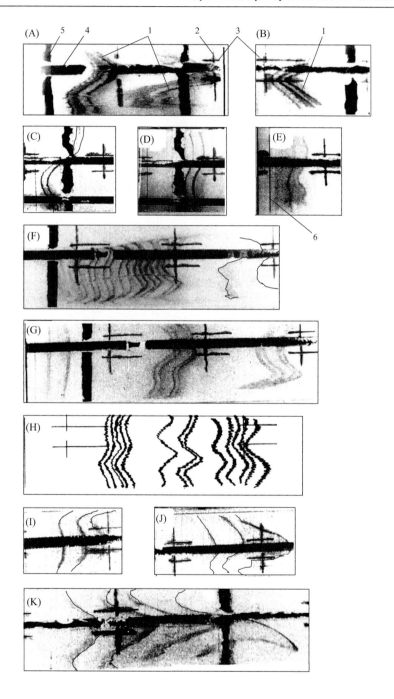

Measurements near the center of the curvilinear section (higher along the y axis—above the inhibited flow region) showed that the velocity profile $U(z)$ is slightly bent in the direction of the z axis: at $x=1.1$ m (Fig. 4.10C and D), the profile deformation is the same as with $x=1.52$ m (Fig. 4.10E). At the confused section at the end of the curvature at $x=1.8$ m (Fig. 4.10F and G) and beyond the section of curvature at $x=2.15$ m (Fig. 4.10H) the wavelength of the longitudinal vortex system and the amplitude of the deformation of the front of the $U(z)$ profile decrease—the flow stabilizes under the action of centrifugal forces. In this case, the nonstationarity of the distribution of $U(z)$ is manifested—much more intense than when a horizontal plate flows around. The change in the form of $U(z)$ occurs within 3–8 s, four to six times faster than when flowing around a horizontal plate, especially when $y>\delta_1$ (Fig. 4.10G and H). When $y<\delta_1$ (Fig. 4.10F), the nonstationarity manifests itself not so much in the change in the shape of $U(z)$, as in the oscillatory motion of disturbances in the transverse direction. This indicates a specific floor-by-stage development of disturbances. The zigzag behavior of the $U(z)$ profile starts at $y<\delta_1$ (Fig. 4.10F), and then spreads over the entire thickness of the BL. With an increase in the curvature radius of a curvilinear plate, the distribution of $U(z)$ over the entire length of the working section is constant: at $R=12$ m, $x=0.6$ m, $y=4\times 10^{-3}$ (Fig. 4.10I); $x=0.6$ m, $y=6\times 10^{-3}$ m (Fig. 4.10J). As the radius of curvature decreases ($R=1$ m), a more intense deformation of the velocity profile $U(z)$ occurs (Fig. 4.10K). Experiments have shown that with the natural development of the Görtler's vortices on the plates with $R=1$ m and 4 m on the horizontal part of the plate, behind the curvilinear section, longitudinal vortex systems are formed, the value of λ_z which was 0.28 m and 0.15 m, respectively. On the horizontal smooth plate with natural development at the same flow rates, λ_z is significantly larger (Section 3.2, Fig. 3.7). The detected differences in the vortex structures depend not so much on the radius of curvature, but rather characterize the difference in the flow type.

The flow was visualized also in the channel near the concave (Fig. 4.11) and near the convex (Fig. 4.12) walls. It has been found that in both cases the distribution of $U(z)$, although it remains nonstationary, is much more stable than in the open channel. Nonstationary flow consists mainly in the oscillation of the minimum velocity in the transversal direction; however, at higher speeds, $U(z)$ can be formed with

Figure 4.10 Photographs of the visualization of the $U(z)$ profiles in the flow around curvilinear plates depending on the x, y coordinates at $U_\infty = 0.04$ m/s: (1) tellurium cloud characterizing $U(z)$, (2) marks on the curvilinear bottom of the working section in the transversal direction, (3) marks at the bottom of the working section in the longitudinal direction, the distance between the longitudinal marks 0.1 m and between the transverse marks 0.5 m, (4) the longitudinal frame of the curvilinear plate located under the plate, (5) the transverse frame of the curvilinear plate, (6) tellurium wire. $R = 4$ m: (A) $x = 0.3$ m, $y = 6\times 10^{-3}$ m, $dP/dx = 0$; (B) $x = 0.6$ m, $y = 6\times 10^{-3}$ m, $dP/dx > 0$; (C) $x = 1.1$ m, $y = 1\times 10^{-2}$ m; (D) $x = 1.1$ m, $y = 2\times 10^{-2}$ m; (E) $x = 1.52$ m, $y = 1\times 10^{-2}$ m; (F) $x = 1.8$ m, $y = 4\times 10^{-3}$ m, $dP/dx < 0$, $y<\delta_1$; (G) $x = 1.8$ m, $y = 6\times 10^{-3}$ m, $dP/dx < 0$, $y<\delta_1$; (H) $x = 2.15$ m, $y = 6\times 10^{-3}$ m, $y<\delta_1$. $R = 12$ m: (I) $x = 0.6$ m, $y = y = 6\times 10^{-3}$ m; (J) $x = 2.22$ m, $y = 3\times 10^{-3}$ m. $R = 1$ m: (K) $x = 0.6$ m, $y = 4\times 10^{-3}$ m.

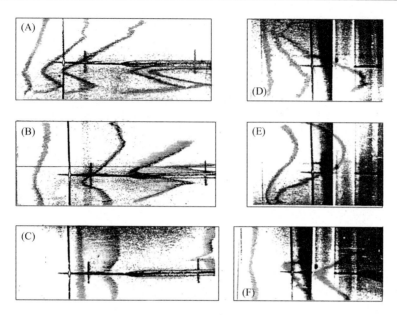

Figure 4.11 Copies of photographs of the development of a disturbing motion [evolution of the profiles $U(z)$] with a flow around a concave surface curved channel depending on x, y coordinates: $R = 1$ m, $U_\infty = 0.044$ m/s, $x = 0.8$ m: (A) $y = 2 \times 10^{-3}$ m; (B) $y = 6 \times 10^{-3}$ m; (C) $y = 1 \times 10^{-2}$ m; $U_\infty = 0.037$ m/s, $x = 0.93$ m; (D) $y = 5 \times 10^{-3}$ m; (E) $y = 1 \times 10^{-2}$ m; (F) $y = 1 \times 10^{-2}$ m.

approximately half as large λ_z. Compared to all other cases studied, the minimum speed on a concave wall is more pronounced. With distance from the wall and with increasing x, the character of the profiles $U(z)$ does not change (except for a natural decrease in the amplitude of a change in U with respect to z with increasing y). Only on a convex surface with $x > 1$ m, the profile becomes more unstable and tellurium lines blur faster. The thickness of the BL on the concave surface is about 1.5 times greater than on the convex one; in the core of the flow, the $U(z)$ profile is straightforward, that is, the flow is free from regular vortex disturbances here.

Investigations on a concave surface correspond to the type of the Görtler's flow, and the flow in the channel is more likely the type of the Taylor's flow.

The disturbing motion over surfaces with a concave portion is three-dimensional, starting with the smallest Reynolds numbers. This is proved by the visualization of both the $U(z)$ profiles and the streamlines. Photographing the visualized flow field at successive moments t_1 and t_2 shows that when viewed from above, the middle line hardly deviates from the axial line, along which, judging by the visualized $U(z)$ profiles, a "hollow" of the vortex system is formed (maximum velocity z): two the lines on both sides of the axial oscillate in antiphase. When viewed from the side, the oscillations of the three jets also differ; the extreme oscillates with small amplitude, almost merging among themselves in the projection on the xy plane, and the amplitude of the average is so large that the crest of the wave at the time t_2 is tilted. In addition, observation from above of a

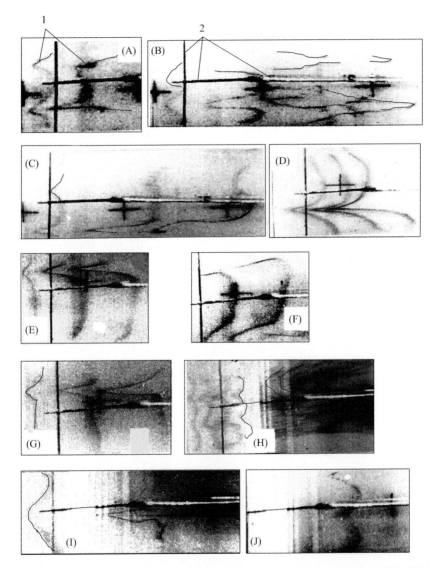

Figure 4.12 Photographs of the development of a disturbing motion [evolution of the U(z) profiles] with a flow along the convex surface of a curvilinear channel depending on the x, y coordinates: (1) tellurium clouds characterizing $U(z)$, (2) tellurium wire holder; $R = 1$ m, $U_\infty = 0.044$ m/s, $x = 0.8$ m: (A) $y = 2 \cdot 10^{-3}$ m; (B) $y = 6 \cdot 10^{-3}$ m; (C) $y = 1 \cdot 10^{-2}$ m; $U_\infty = 0.037$ m/s, $x = 0.93$ m: (D) $y = 5 \cdot 10^{-3}$ m; (E) $y = 1 \cdot 10^{-2}$ m; (F) $y = 1 \cdot 10^{-2}$ m; (G) $y = 1.5 \cdot 10^{-2}$ m; $U_\infty = 0.044$ m/s, $x = 1.05$ m: (H) $y = 1 \cdot 10^{-3}$ m; (I) $y = 5 \cdot 10^{-3}$ m; (J) $y = 1.5 \cdot 10^{-2}$ m.

vertical tellurium line [when visualizing profiles $U(y)$] showed that, propagating downstream, it rotates around the x axis so that its upper and lower ends move in opposite directions along z.

Fig. 4.12 shows photographs of the development of a disturbing motion in the form of $U(z)$ profiles when a curved surface of a curvilinear channel with a radius of curvature $R=1$ m flows past (Section 4.1, Fig. 4.1). Investigations [150] confirmed our results: on the concave and convex surfaces of the character flow is different. The main conclusions of the results obtained are as follows. Compared with a concave plate, a more ordered longitudinal vortex flow is formed in the channel, which has a difference depending on the measurement coordinate y—a floor-by-floor vortex structure is formed in the cross-section. In the middle of the channel length (Fig. 4.12H) a small-scale shape of the velocity profile $U(z)$ was formed at $y = 1 \times 10^{-3}$ m, which with increasing y coordinate (Fig. 4.12I and J) becomes large-scale, while in the longitudinal axial direction inhibition of speed $U(z)$, as in Fig. 4.12D, E, and G.

4.4 LDV measurements of the profiles of the longitudinal velocity of the BL on rigid concave plates

The fluid flow in a hydrodynamic test bench was experimentally investigated in Ref. [628] using an LDVM equipment [39,315]. Fig. 3.40 (Section 3.7) shows the scheme accommodation of the LDVM equipment on the hydrodynamic stand of small turbulence (Section 3.1). The experiments were performed on curvilinear plates (Section 4.1, Fig. 4.1).

Fig. 4.13 shows profiles of curved surfaces with a radius of curvature $R=1$ m (*I*), 4 m (*II*), 12 m (*III*), and also shows the change in velocity in the flow core along these surfaces. In addition, a cover was installed above the curvilinear plate with $R=1$ m to investigation the flow in a closed channel with a cylindrical section. At $U_\infty \approx 0.05$ m/s and $U_\infty \approx 0.1$ m/s, the maximum decrease in velocities in the concave part for surface *I* am 10% and 20%, respectively, for *II* 15% and 25%, for *III* 30% and 40%. Along the

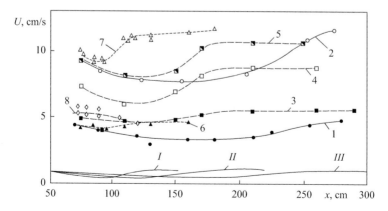

Figure 4.13 Profiles of concave surface areas with $R = 1$ m (*I*), 4 m (*II*), 12 m (*III*) and the distribution of the velocity of the unperturbed flow over curved plates: (1), (2) the distribution of U_∞ for *I*; (3), (4), (5) for *II*; (6), (7) for *III*; and (8) for the top cover of the cylindrical channel with $R = 1$ m [628].

axis of the cylindrical section, the speed increases by about 15% at $U_\infty = 0.05$ m/s and the distance between curved surfaces is $h = 0.07$ m. The kinematic picture of the flow was determined from the results of measurements of the averaged and pulsating profiles of the longitudinal velocity component using a laser anemometer and volumetric visualization of the BL by the tellurium method [622].

Along the longitudinal axes of the working section for different values of U_∞ were measured longitudinal averaged and pulsating BL velocities. In Fig. 4.14, such profiles in a half-size view are plotted along the length of the concave surface with $R = 12$ m. The visualization of the profiles $U(y)$ confirmed the measurement data and made it possible to identify one feature. It turned out that the type of visualized profiles changes over time. Therefore, the profile measured from point to point is obtained statistically averaged, and in appearance close to that which is observed most often in visualization.

The results of measurements of longitudinal averaged and pulsation velocities on a curved plate shown in Fig. 4.14 are compared with similar results for a flow around a horizontal flat plate (Section 3.7, Fig. 3.42) with the same values of $U_\infty(y)$. The difference in the results obtained is due to the diffuser and confused portions when flowing around a curvilinear plate, and, moreover, the results on a horizontal flat plate along the ordinate axis are at dimensioned to δ^*, while the curvilinear plates flow around the ordinate axis without dimensioning. Since H is small, the flow at the beginning of the curvature (Fig. 4.14 on the left) was practically not affected by the positive pressure gradient. Therefore, in both compared cases, the velocity profile shapes are similar. At the end of the curvilinear part, the influence of the negative pressure gradient was noticeable, which influenced the shape of the velocity profiles.

Similar profiles in dimensionless form for surfaces with different curvatures are shown in Fig. 4.15, while the dimensioning was performed using the extrusion thickness δ^*, as in the flow past a horizontal plate.

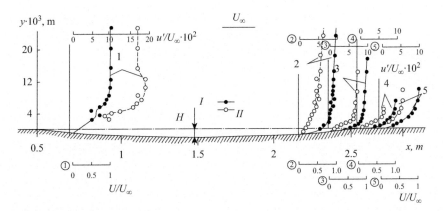

Figure 4.14 Averaged and pulsating longitudinal velocities of the BL on a curved plate with a radius of curvature $R = 12$ m: I profiles U/U_∞ (y); II profiles u'/U_∞ (y); H is the maximum deflection of the plate, cm; U_∞ = 0.082 m/s (1); 0.085 m/s (2); 0.097 m/s (3); 0.111 m/s (4); 0.117 m/s (5) [303].

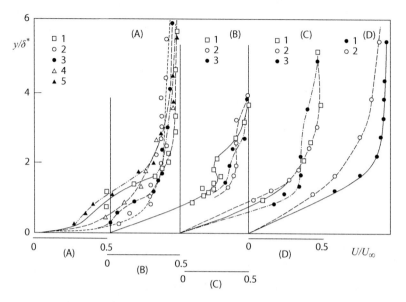

Figure 4.15 Dimensionless velocity profiles along a concave surface with $R = 12$ m (A), curve designations are the same as in Figure 10.13; $R = 4$ m (B): (1) $x = 0.73$ m, $U_\infty = 0.10$ m/s; (2) $x = 1.55$ m, $U_\infty = 0.10$ m/s; (3) $x = 1.8$ m, $U_\infty = 0.13$ m/s; $R = 1$ m (C): (1) $x = 1.15$ m, $U_\infty = 0.11$ m/s; (2) $x = 1.35$ m, $U_\infty = 0.08$ m/s; (3) $x = 1.35$ m, $U_\infty = 0.12$ m/s; for the concave wall of the cylindrical channel (D) in the core, $U_\infty = 0.1$ m/s and $x = 1.05$ m (1) and 2.3 m (2) [628].

The increase in speed (above the horizontal section $U_\infty > 0.05$ m/s) leads to sharp bends of the profiles in the diffuser part of the channel, separation of the BL with the formation of a stagnant region downstream and, finally, its attachment to the confused part. These effects also accompany a decrease in the radius of curvature of the surface with the constancy of other flow parameters. A characteristic distinguishing feature of the flow regime over the surfaces under consideration is an increase in the level of pulsations in the outer region to (0.0602) U_∞ and its monotonous decrease when approaching the wall. Only at the very beginning of the curvature, the maximum u'/U_∞ was fixed at a distance of $\sim \delta/3$ from the surface (Fig. 4.14, the profile of the pulsation velocity (1)), which can be explained by the preservation in this area of some features of the BL on the previous flat section.

The representation in the dimensionless form of the velocity profiles U/U_∞ (y/δ^*) revealed the presence of inflection points in them with $y/\delta^* = 0.6-1.2$. This characterizes the conditions for the occurrence of an inviscid instability and, as a result, creates the prerequisites for the formation of three-dimensional disturbances in the form of a system of longitudinal vortices. The visualization showed that under the experimental conditions a strong curvature of the $U(z)$ profiles takes place already at the entrance of the initial horizontal section. At the beginning of the diffuser part, a stable minimum of velocity in the BL along the channel axis is observed. In the confuser part of the curvilinear plate, the nonstationarity of the distribution of $U(z)$ manifests itself much more intensely than on a flat plate with the same degree of turbulence. Closer to the wall, nonstationarity is expressed not in a change in the shape of $U(z)$, but in the oscillatory motion of the BL along the z axis. Above the surface with a lower curvature of the plot, the distribution of $U(z)$ is rather stable. An increase U_∞ from 0.03 to 0.06 m/s in the confused section leads to a decrease in λ_z. Above the surface with a greater curvature, a more significant change in velocity occurs in the z direction, that is, an increase in the amplitude of the undulating profile $U(z)$.

The development of a three-dimensional structure of disturbances was also traced in a channel with a cylindrical section. Fig. 4.15 shows the profiles $U(y)$ for two values of Reynolds numbers $Re = 2.8 \times 10^5$ (1) and $Re = 1.5 \times 10^5$ (2).

4.5 Experimental construction of the Görtler neutral curve in the flow around rigid curvilinear plates

Figs. 4.16 and 4.17 show photographs of the visualization of the development of the $U(z)$ profiles along the x axis when flowing around curvilinear plates with a

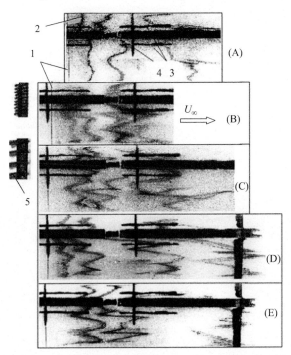

Figure 4.16 Photographs of the visualization of the $U(z)$ profiles in a flow around a curvilinear plate with $R = 4$ m, $U_\infty = 0.05$ m/s, $x = 2.1$ m, $y_{t.w.} = 0.002$ m: (A) without introducing disturbances, (B)–(E) upon introducing disturbances with $\lambda_z = 0.004$ m (B); 0.006 m (C); 0.016 m (D); 0.018 m (E). (1) tellurium wire, (2) tellurium "clouds," (3) longitudinal marks, (4) transverse marks on the bottom of the working section (distance between longitudinal marks 0.1 m and transverse marks 0.5 m), (5) vortex generators [140].

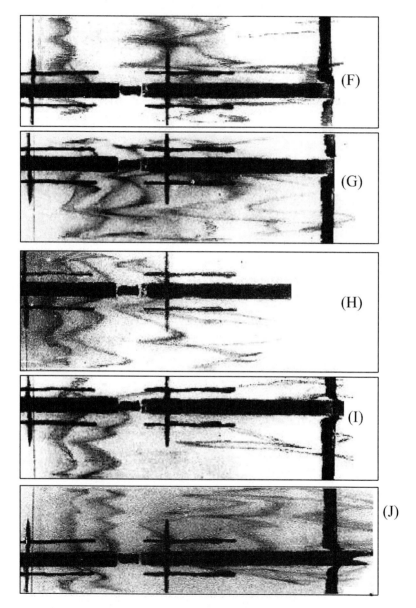

Figure 4.17 (E) Photographs of the visualization of the $U(z)$ profiles in a flow around a curvilinear plate with $R = 4$ m, $U_\infty = 0.05$ m/s, $x = 2.1$ m, $y_{\text{t.w.}} = 0.002$ m: (F)–(J) when introducing disturbances with $\lambda_z = 0.02$ m (F); 0.022 m (G); 0.024 m (H); 0.026 m (I); 0.028 m (J). The remaining data in Fig. 4.17 the same as in Fig. 4.16.

radius of curvature $R=4$ m when artificial longitudinal disturbances generated by the VG1 vortex generators (for Fig. 4.16B) and VG2 are introduced into the BL for the rest of the photos. The investigations were performed at $U_\infty=0.05$ m/s ($G = 1.45$) for different values of the distance λ_z between the vortex generators.

When analyzing the results given in Fig. 4.10 (Section 4.3) and Figs. 4.16 and 4.17, attention should be paid to the x coordinate of the measurement site. For all radii of curvature, the beginning of curvature is located at $x=0.5$ m, and the curvature ends at different x coordinates: for a plate with a radius of curvature $R=1$ m, $x \approx 1.3$ m, for $R=4$ m, $x \approx 1.9$ m, for $R=12$ m, $x \approx 2.5$ m. In Fig. 4.10H (Section 4.3) with $x=2.15$ m and $R=4$ m with natural development of disturbances recorded $\lambda_z=0.016$ m, and in 4.10 k at $x=2.22$ m and $R=1$ m $\lambda_z=0.028$ m. In Fig. 4.16A at $x=2.1$ m $\lambda_z=0.021$ m. The difference in wavelengths λ_z is due to the fact that during the experiments (Fig. 4.16) the flow velocity was greater, and the measurements were taken closer to the wall at $y_{t.w.}=0.002$ m. When analyzing the results obtained, one should take into account that the measurements were carried out on a flat surface of the plate approximately at a distance of 0.4 m behind the curvature. In this area, under the action of centrifugal forces, systems of longitudinal vortices were formed, on which systems of longitudinal vortices of various scales introduced into BL using superimposed vortex generators were superimposed. Tellurium wire (1) was placed in all the photographs approximately at a distance of 0.25 m from the location of the vortex generators. In Fig. 4.1B, the introduction of disturbances using VG1 vortex generators with $\lambda_z=0.004$ m does not lead to their attenuation, as the stability diagram requires, but to convert them into an enlarged structure with $\lambda_z=0.01$ m, with the development of which λ_z decreases to 0.008 m. Scale introduced disturbances doubled. Subsequently, the longitudinal vortex systems were introduced with larger vortex generators VG2. In Fig. 4.16 disturbances with $\lambda_z=0.006$ m were made in BL. A further increase in wavelength gives increasingly clearer wave patterns (Fig. 4.16D and E) with a tendency to form harnesses (Fig. 4.17F–H), indicating acceleration of the growth of the disturbing motion. At $\lambda_z=0.024$ m, this increase in growth stops: the amplitude $U(z)$ decreases in Fig. 4.17I and J as compared with Fig. 4.17H.

From here we can draw the following conclusions. The scale $\lambda_z=0.024$ m corresponds to disturbances with a maximum amplification. Formed longitudinal disturbances with $\lambda_z=0.012$ m are neutral. As λ_z increases from 0.012 to 0.024 m in the Görtler's diagram, the resulting picture reflects the displacement of a point along the abscissa to the left. At the same time, it intersects a series of equal gain curves $\beta\delta_2/\nu$ with increasing values of β until reaching maximum values determined by the maximum amplification curve. A similar trend was recorded on neutral curves of Tollmien-Schlichting waves. With a subsequent increase in λ_z, the characteristic point moves to curves with smaller values of β, which follows from a decrease in the amplitudes $U(z)$. Downstream propagation of a disturbance with a fixed value of λ_z is displayed on the Görtler's diagram by moving the point with coordinates $U_\infty\delta_2/\nu(\delta_2/R)$, $\alpha_z\delta_2$ upwards along the line $P=$const in accordance with the increase in δ_2. Then, for $G<1.5$, disturbances from a stable region with sufficient intensity can reach the instability region. For $G>1.5$, the $P=$const lines that begin in the stable region do not intersect the instability region; therefore, under such conditions, disturbances with stable parameters should attenuate as they propagate downstream.

In Refs. [40,47,53], the velocity profiles $U(y)$ were measured in characteristic places along z, which were determined according to the obtained visualization pictures (Figs. 4.16 and 4.17). Characteristic places are "peaks," "hollows," and the gap between them. The resulting velocity profiles allowed us to calculate the extrusion thickness at each measurement site. For each x, the corresponding three values of extrusion thickness δ_2 were obtained. The arithmetic mean of the extrusion thickness and the corresponding λ_z value calculated from photographs allowed us to determine the coordinates of a point on the Görtler's stability diagram (Fig. 4.18). Vertical line III corresponds to the value of $\alpha\delta_2$, calculated from the value of $\lambda_z = 0.045$ m, characteristic of the natural transition on a flat plate. In this area there should be a critical Görtler's number or an extremum of a neutral curve. Since the lower point was obtained for a flat plate under the condition $\delta_2/R = 10^{-4}$, that is, with a small curvature, the extrapolation made it possible to determine that $G_{cr} \approx 0.3$.

Table 4.1 shows the parameters of the experimental curves. Curves I and II are located to the right of the corresponding calculated curves. This is explained by the fact that the experiment was performed with a large value of the degree of turbulence of the main flow ($\varepsilon > 1\%$). In addition, the determination of the parameters of

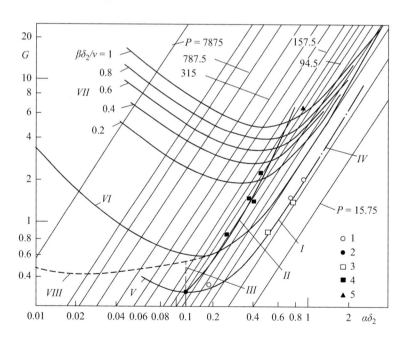

Figure 4.18 Görtler's stability diagram on a concave rigid surface [628]: I experimental neutral curve [(1), (3) flow past a surface with $R = 12$ m and 4 m]; II maximum amplification curve [(2), (4), (5) flow past a surface with $R = 12, 4$ and 1 m]; III the critical value line $\alpha_z\delta_2 = 0.12$ ($R \approx \infty$); IV experimental data [243]; VI, V extrapolation of curve I; VII neutral curve and equal amplification curves according to calculations [270]; VIII calculated neutral curve [534].

Table 4.1 Parameters of the Görtler's neutral curve.

R, m	U, m/s	x, m	$\lambda_z \times 10^2$, m	$\delta_2 \times 10^3$, m	$\alpha_z \delta_2$	G
12	0.03	0.6	2.4	2.8	0.73	1.5
12	0.035	0.73	2.4	3.5	0.92	2.1
12	0.03	2.2	4	1.2	0.18	0.36
12	0.08	2.2	2.4	1.45	0.38	1.48
4	0.03	1.8	1.6	1.3	0.5	0.84
4	0.03	1.8	3.2	1.3	0.25	0.84
4	0.05	2.1	1.2	1.5	0.78	1.45
4	0.05	2.1	2.4	1.5	0.39	1.45
4	0.128	1.8	1.6	1.1	0.43	2.4
1	0.087	1.35	1.2	1.75	0.92	6.3

the experimental points was carried out in the region of the horizontal sections behind the concave sections, since the shape of the Görtler's vortices manifested itself as a result of the susceptibility process downstream from the concave section.

Our experiments showed [34,247,627,629] that Görtler's vortices that form on a concave surface continue to exist on going to a horizontal plate and affect the flow significantly further behind the concave section.

4.6 Model of the vortex structures in the vortex chamber

Vortex chambers are widely used in energy and technological machines and apparatus for organizing their work processes and intensifying the transfer of mass, momentum and heat [281,283–287,395,396,598]. When using for the local swirling flow of twirlers of tangential-slit types, an unstabilized swirling flow of a rather complex structure takes place along the flow part [285,286,396]. In the near-wall and paraxial regions, there may be flows with longitudinal static pressure gradients of various signs. Along with radial pressure gradients, this greatly complicates the structure of flows and presents a problem when calculating internal swirling flows [541].

When formation of return flow zones, used, in particular, in the combustion chambers to prevent flame separation, conditions for intensification of turbulent mixing arise. At the same time, the radial movement of turbulent moles, caused by velocity pulsation and associated with performing work against centrifugal forces, can slow down transfer processes in the radial direction and contribute to stable stratification in density and temperature. This phenomenon is used to improve the characteristics of gas curtains in high-temperature installations and in plasma technology in order to protect the surface of the channels from high temperatures. The theoretical analysis and experimental data from Refs. [396,409,487] also indicate the possibility of a stabilizing and even blocking effect of swirling flow on turbulent transfer in a vortex chamber.

The decisive role in the formation of areas with an active and conservative nature of the effect of centrifugal mass forces on the flow structure has the laws of radial changes in the axial and transverse velocity components [286,287]. For flow parts of the vortex chamber, these laws are studied much more fully than for dead-end areas. The influence factor of the so-called "end effect" [263] is also noted, in particular, for vortex chambers with swirlers in the form of single nozzles [428,561−563] and belts of inlet windows of rectangular cross-section evenly distributed around the circumference [428,563]. In addition, investigations of the characteristics of vortex structures in vortex chambers are practically unknown. All this makes it difficult to further improve the vortex chambers as the most important elements of heat exchangers, chemical reactors, mixers, cyclone chambers and other energy and technological installations.

The purpose of the research was to identify CVS and their interaction with each other in the fluid flow in the vortex chamber. The visualization of the flow was made when photographing with a camera, high-speed movie camera and video camera. The detected characteristic sites were thoroughly investigated using thermoanemometric equipment, a laser beam and a mobile sensor for the direction of local currents. The resulting flow pattern in the vortex chamber was complemented by measurements using the Pitot micro tubes. The experimental setup and research methodology are given in Refs. [126,428].

The analysis of the obtained results allowed us to identify the following main areas of flow in the vortex chamber:

- Flow in the inner area of the vortex chamber.
- Flow in the area of the nozzle.
- Flow in the peripheral area of the vortex chamber.
- Flow in the butt area.
- Flow in the active area.

Below are the results of the investigation of the flow in the indicated parts of the vortex chamber, the interaction of the vortex structures in all parts of the vortex chamber is considered, and a model of the vortex flow structures in the vortex chamber is constructed [128].

V.N. Turick (Professor of the Department of Applied Hydromechanics and Mechatronics of NTUU Kiev Polytechnic Institute, Kiev) developed a vortex chamber [96,108,563], which has a circular cross-section with an internal diameter of 0.102 m, made of transparent Plexiglas and has a measuring section of 0.7 length m. Fig. 4.19 [570] shows a schema of an open-type wind tunnel modified by the proposal of V.V. Babenko and made by V.N. Turick by connecting the lemniscate entrance (2) and the entrance section (3) in the region of the end of the vortex chamber (photograph of the wind tunnel is shown in Fig. 7.45, Section 7.10). Experimental investigations were performed on a vortex chamber, the scheme of which is shown in Fig. 4.19 in the absence of sections (2), (3). Fig. 4.19B shows the longitudinal and cross-section of the vortex chamber (working area (1) in Fig. 4.19A). Fig. 4.19C shows a top view of the vortex chamber in the vicinity of the nozzle. V.V. Babenko suggested that the nozzle be executed as a turning point. V.N. Turick realized a rotary nozzle scheme to study various methods of controlling coherent vortex flow structures in a vortex chamber.

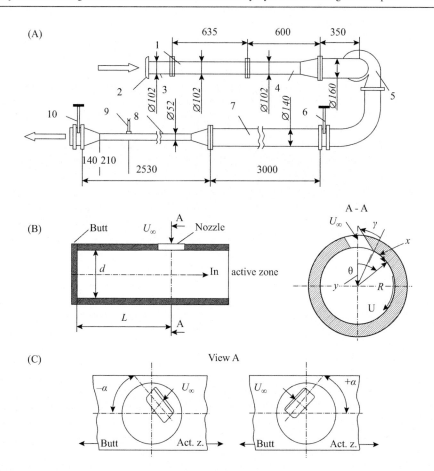

Figure 4.19 Scheme experimental setup (dimensions are given in mm). (A): (1) working area (vortex chamber); (2) lemniscate entrance; (3) entrance area; (4) soothing area and a diffuser, (5) centrifugal fan; (6), (10) gate; (7) pipeline, (8) flow meter section, (9) pneumometric tube; (B) vortex chamber and its cross-section; in top view of the rotary nozzle [96,108,563].

4.6.1 The flow in the inner area of the vortex chamber

The flow in the inner region of the vortex chamber was originally investigated using a Pitot tube and thermo-anemometric equipment, as well as a specially manufactured flow direction sensor [87,428,561,562]. Fig. 4.20 shows the main results of the investigations performed. The white color indicates the direction of flow at $x/d > 0$ (butt), and the gray color indicates the direction at $x/d < 0$ (flow toward the active region). Black color corresponds to a higher speed. The numbers indicate the isolines of equal velocities in each area of the flow. The figures correspond to the local flow rate, referred to the average flow rate in the active region of the vortex chamber. The highest speed and energy is recorded in the region of the vortices, the so-called "whiskers" (1). The

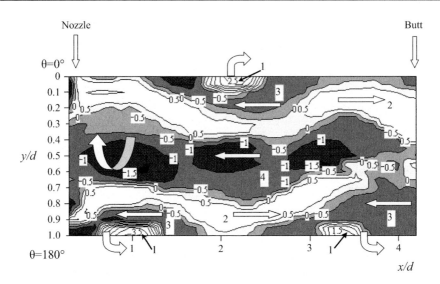

Figure 4.20 The distribution of the axial velocity along the butt region of the vortex chamber: (1) vortex systems "whiskers"; (2) area of flow toward the butt; (3) peripheral region of the flow toward the active region; (4) quasi-solid flow. $Re_c \approx 8 \cdot 10^5$, $L/d_0 = 4.4$ [87,562].

vortices have a spiral shape. When these vortices propagate from the nozzle to the end, the velocity and their energy decrease. The least energy-bearing areas of the flow are two areas. One of them has the form of a meandering annular tube (2), the flow in which is directed to the end.

The second region (3) is directed from the butt toward the active region. Such a movement is caused by the relative movement of the fluid during its flow around the whiskers moving at high speed. The wake of "whiskers" in visualization suggests that "whiskers" are a large pair of vortices rotating toward the wall. Such a complex structure of motion also causes the relative movement of fluid (3) from the butt.

4.6.2 Flow in the area of the nozzle toward the butt and toward the active area

Fig. 4.21 shows a photograph of a vortex camera with a long butt, obtained by a video camera while visualizing the powder flow. In the photo the butt is located on the right and the active zone on the left. Above the nozzle is visible cylinder (2), from which the powder flows into the vortex chamber (1). The direct arrows show the direction of the incoming and outgoing air. When watching a movie, it is seen how, when viewed from the front of the nozzle, stable large vortex structures (3) ("whiskers") are formed, diverging to the side of the nozzle along a helix. The direction of rotation of the "whiskers" is shown by curvilinear arrows. The gray color on these arrows corresponds to the movement of air in the rear part of the

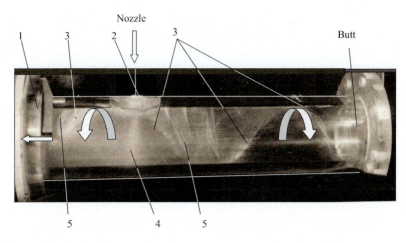

Figure 4.21 Visualization of the vortex structures in the vortex chamber using powder: (1) the hull of the vortex chamber; (2) cylinder for feeding powder; (3) large divergent vortex formations ("whiskers"); (4) small vortex structures; (5) vortex structures absorbed by the "mustache." $L/d_0 = 4.4$, Re $\approx 7.8 \times 10^5$, $\alpha = 0$ degree, $\gamma = 67$ degrees [128].

inner surface of the cylinder, and the white color of the arrow corresponds to the direction of flow along the inner surface of the cylinder in the front part. Under the nozzle, a zone is formed in the form of a curved cone, filled with small vortex structures (4) moving along the circumference of the inner surface of the cylinder. "Whiskers" (3) constantly absorb other smaller vortex structures (5) in relation to them, which flow downwards along the periphery of the inner wall of the cylinder of the vortex chamber. With the help of a video camera, the process of formation of vortex structures that are absorbed by the "whiskers" is clearly visible.

Large vortex structures ("whiskers") were observed, moving along a helical line along the inner surface of the cylinder of the vortex chamber, both toward the butt and into the active zone (on the left in the Fig. 4.21). The speed of the "whiskers" in their trajectory is 2–4 m/s, but when approaching the butt the speed decreases. The angle of the curved cone is approximately 55 degrees. The curved cone under the nozzle is uneven. Depending on the length of the butt zone, the angle of the cone region is different. So, with a short butt, the angle of taper toward the end is about 30 degrees, and toward the active zone—about 25 degrees. With a long butt, on the contrary, in the direction of the end, the angle of the cone zone is approximately 25 degrees, and in the direction of the active zone it increases and is approximately 45 degrees. Smaller vortex structures (5), "feeding" "whiskers" also move along a helix in the same directions, but the pitch of their helix is significantly smaller. Therefore, when rotating along the generator of the cylinder, these vortex structures intersect with "whiskers" that absorb them. With a long butt, the absorption ends to the end, and with a short butt, such absorption occurs quickly and the enlarged vortex structures—"whiskers"—approach the end. Vortex structures (5), as will be clear from further visualization photos, these are quasi-Taylor vortex systems.

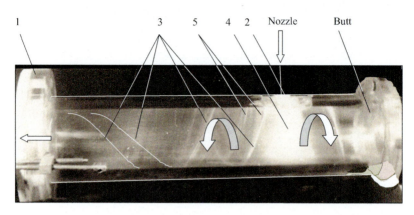

Figure 4.22 Visualization of the vortex structures in the vortex chamber with a short butt-end: (1) the hull of the vortex chamber; (2) a cylinder for feeding powder; (3) large divergent vortex formations ("whiskers"); (4) small vortex structures; (5) quasi-Taylor vortex structures absorbed by the "whiskers" [128].

Fig. 4.22 shows a photograph of flow visualization in a vortex chamber with a short butt. Comparison with the visualization of the flow around a vortex chamber with a long butt (Fig. 4.21) showed a difference in the flow pattern in the area of the butt. In addition, the short butt allows detecting flow features in the active area. Behind the nozzle in the direction of the active region, 2−3 "whiskers" are formed, which absorb quasi-Taylor vortices at a very short distance from the nozzle. The pitch of the helix of the "whiskers" moving into the active region is 2−3 times less than the pitch of the helix of the "whiskers" moving toward the butt. Further, in the direction of the active region at a distance of 2−3 internal diameter of the cylinder of the vortex chamber, the system of "whiskers" merges into one "whiskers." In this case, the pitch of the helix of such a "whiskers" becomes approximately equal (somewhat larger) to the "whiskers" pitch moving toward the butt. It can be assumed that the formation of 2−3 "whiskers" near the nozzle in the direction of the active area is associated with the peculiarity of fluid motion in the quasi-solid core.

Fig. 4.23 shows the development of the indicated vortex structures using the soot visualization method [95]. The right side of Fig. 4.23 shows cross sections of the vortex chamber along the axis of symmetry of the nozzle. In Fig. 4.23A, a vortex chamber (1) is photographed when the butt-end is on the left. The air through the incoming nozzle (2), as well as the near-wall jet, is directed forward at an angle $\gamma = 67$ degrees (angle $\theta = 90$ degrees) in accordance with the cross-section along the axis of symmetry of the nozzle located to the right. Due to the fact that the flow enters the vortex chamber at an angle $\gamma = 67$ degrees, a separation zone (5) is formed behind the nozzle on the inner surface of the vortex chamber. Large vortex structures (3) ("whiskers") are fixed. After attachment of the separation region (5) under the nozzle on the inner surface of the vortex chamber cylinder ($\theta = 90$ degrees), small vortex structures (4), resembling Görtler-Ludwig vortices, develop, which develop along the surface of the cylinder and diverge when moving away

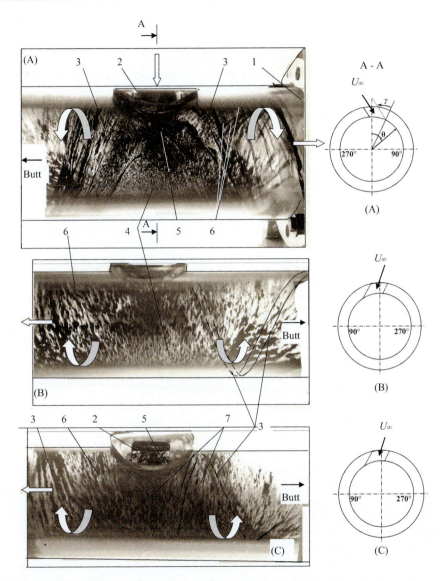

Figure 4.23 Visualization of the vortex structures in the vortex chamber using the soot method: (1) vortex chamber; (2) nozzle; (3) large divergent vortex formations ("whiskers"); (4) small vortex structures; (5) stagnant vortex area; (6) vortex structures absorbed by the "whiskers"; (7) boundaries of the primary curved wall jet. On the right is a section of the vortex chamber along the nozzle axis. (A) left butt-end; (B), (C) right butt [128].

from the nozzle (Fig. 4.23B). At the same time, they become larger and form quasi-Taylor vortex structures 6 absorbed by the "whiskers." The kinematic characteristics of these vortices are studied in detail in Refs. [101,102,126]. To the right of the nozzle in Fig. 4.23B, the direction of the front of the "whiskers" (3) is shown by dashed lines. Like the "whiskers" (3), there are recorded vortex structures (6) moving along a helical line, although they have different helical lines. They, as seen in the previous photos, are absorbed by the "mustache." These vortex structures begin to form in the lower part of the photograph in the area of the curved cone. It can be seen how the vortices 4 are enlarged—black stripes appear with a large pitch. These enlarged structures are clearly visible at the ends of the photo to the right and left (Fig. 4.23B). Since the cylinder of the vortex chamber is made of transparent organic glass, in these parts of the photograph there are fixed vortex structures on the opposite side of the inner surface of the cylinder—at $\theta > 270$ degrees. It is also seen that the dark stripes along the helix pass to the front part of the cylinder in the form of vortices (6). Right and left with respect to this near-wall jet area, a significant expansion of the jet and the formation of large sizes of vortex structures (6) were recorded. In the photo (Fig. 4.23C), the lines (7) conventionally indicate the boundaries of the "primary near-wall curvilinear jet." In the nozzle (2), vortices (5) are visible in the separation region on the wall of the cylinder behind the nozzle. When visualizing with the help of powder, light stripes are fixed on the photos. In the investigation of Görtler's stability [126], pairs of vortices were recorded. Fig. 3.43 (Section 3.7) shows the scheme of the velocity field during the formation of longitudinal vortex systems (Beni-Lin vortices) on a flat plate [612], and in Fig. 4.3 (Section 4.1) the diagram and structure of vortex systems when flowing around curvilinear plates. In both cases, pairs of vortices with mutually opposite rotation along a helix are formed. Between one pair of vortices, the velocity of rotation of the vortices is directed away from the wall. The area between these vortices is called "peak." Between the vortices, the liquid is carried to the outer region of the BL.

The adjacent pair, including one of the considered longitudinal vortices, has a rotation directed to the wall. The area between these vortices is called the "hollow." Between the vortices flow is directed to a streamlined wall. In the above photographs, bright stripes are visible—areas of the "hollows" between the rotating adjacent eddy, in which the flow is inhibited. Therefore, the powder in these places sticks to the wall. The films show that the movement of vortices is dynamic—light stripes is constantly changing shape and position. To improve the visibility of light stripes, the background is black during filming. When visualizing with kerosene, photographs show dark stripes—particles of soot, also located in the stagnant regions between the vortices. In this method, the background for photographing is made white.

4.6.3 Flow in the peripheral area of the vortex chamber

Fig. 4.24 shows a photograph of the visualization of the soot method of vortex structures on the inner surface of the vortex chamber. The flow in the incoming jet of the nozzle (Fig. 4.24A) is directed toward the observer: the cylinder of the vortex chamber is located in the same way as in Fig. 4.23C, but the butt of the cylinder is on the left, as in Fig. 4.23A wherein the nozzle axis is rotated by angle $\theta = 90$ degrees relative to the vertical axis of symmetry of the cylinder. Behind the nozzle, two tornado-like vortices are visible in relation to the generator of the cylinder. The axes of these vortices are displaced with respect to the generator of the cylinder.

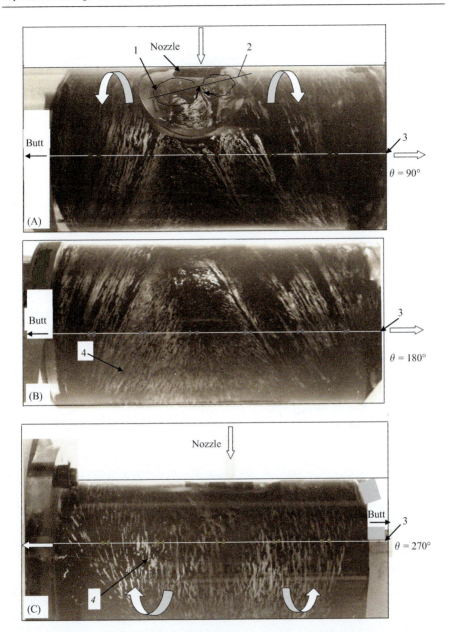

Figure 4.24 Visualization of vortex structures in the peripheral region of the vortex chamber: (1) tornado-like vortices; (2) axis of mutual arrangement of vortices; (3) longitudinal axis of symmetry of the cylinder; (4) conical vortex of the Görtler-Ludwig type [128].

The position of these vortices is nonstationary relative to each other. In Fig. 4.24B, the current is recorded when the axis of the vortex chamber is located at $\theta = 180$ degrees. One can see a cone of eddies of the Görtler-Ludwig type. In Fig. 4.24C

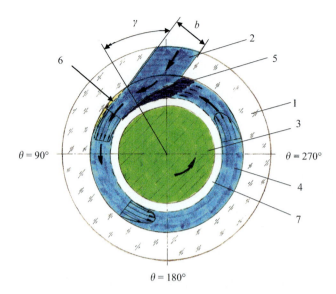

Figure 4.25 The flow pattern inside the vortex chamber: (1) the cylinder of the vortex chamber; (2) nozzle; (3) quasi-solid longitudinal axial flow; (4) "primary" near-wall jet; (5) "secondary" near-wall jet; (6) separation area zone of tornado-like vortices; (7) areas of flow to the butt [128].

($\theta = 270$ degrees), it can be seen that as these vortices move along the inner surface of the cylinder, the dimensions of these vortices become significantly larger: quasi-Taylor vortex systems were formed.

The flow structure inside the vortex chamber in the cross-section of the nozzle is shown in Fig. 4.25. The middle part of the flow in the vortex chamber is a quasi-cylindrical quasi-solid-state rotation of flow 3 from the butt-end to the active part of the vortex chamber (toward the observer). The diameter of such a flow is about half the internal diameter of the cylinder. A near-wall jet (4) is formed in the nozzle, the cross-section of which immediately after the nozzle cut decreases and presses against the inner surface of the cylinder under the action of two factors: centrifugal force and the "secondary" near-wall jet 5. In accordance with the laws of continuity and conservation of mass, the "primary" near-wall jet 4 significantly reduces the thickness and simultaneously increases the width so that its cross-sectional area is constant with regard to friction losses. The thickness of the "primary" wall jet is approximately half the width b of the nozzle. Having made one revolution, the "primary" near-wall jet 4 flows onto the stream leaving from nozzle. At the same time, it presses the flow coming out of the nozzle to the inner wall of the cylinder, and flows around this width in width, forming a "secondary" parietal jet (5). The area of the parietal jet that flows around the exit of the nozzle is conventionally called the "secondary" parietal jet (5).

Fig. 4.26 shows the scheme of the formation of vortex structures in the peripheral area of the vortex chamber. The internal diameter of the cylinder (1) of the vortex chamber was 102 mm. The cylinder wall thickness in the model experiment was

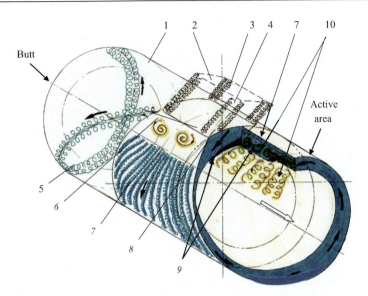

Figure 4.26 The scheme of formation of the vortex structures in the peripheral region of the vortex chamber: (1) cylinder of the vortex chamber; (2) nozzle; (3), (4) pairs of vortices in the corners of the nozzle; (5) large vortex systems "whiskers"; (6) tornado-like vortices in the tear-off zone behind the nozzle; (7) system of vortex of the Görtler-Ludwig type; (8) "primary" near-wall jet; (9) "secondary" near-wall jet; (10) Taylor-type vortex system [128].

20 mm. The nozzle (2) was made in the form of a rectangle with a peppered section of 41×25 mm^2, the corners were rounded. Rotary nozzles were made with different angles γ. As can be seen from Figure 10.19, depending on the angle γ, the length of the walls of the nozzle varies. For example, for $\gamma = 67$ degrees, the length of the wall on the right is approximately 21 mm, and on the left—36 mm. Despite the rounding, in the corners of the nozzle, there are prerequisites for the formation of pairs of vortices in the corners of the nozzle. In this case, the velocity along the walls of the nozzle increases substantially from approximately zero on the outer surface of the cylinder to a value exceeding the velocity in the active zone of the vortex chamber. For example, at a speed in the active zone of the order of 10 m/s in the lower section of the nozzle, the speed is approximately 30 m/s.

Fig. 4.26 shows pairs of vortices (3), (4) in the angular regions of the nozzle. In accordance with this picture of unsteady flow in a nozzle, vortex pairs will occur not at the beginning of the nozzle, but approximately in the middle of the thickness of the nozzle. The diameter of the vortices will vary significantly due to the acceleration of flow along the walls of the nozzle. According to Fig. 4.26, the dimensions of a pair of vortices (4) formed in the angular regions of the nozzle on the side of the nozzle adjacent to the inner surface of the cylinder will differ significantly from the dimensions of the vortices (3) that will flow out of the nozzle and develop in the inner region of the "primary" wall jets. The first of these vortex pairs (4) will be the source of vortex (5) generation ("whiskers"). In the separation area behind the nozzle, tornado-like vortices 6 are formed on the inner surface of cylinder (1). In the near-wall area of the "primary" near-wall jet, systems of Görtler-Ludwig type (7) are recorded, which increase as the wall jet moves along the cylinder surface.

Measurements showed that between the currents of the near-wall jet (4) and the quasi-solid core (3) (Fig. 4.25) in the vortex chamber area between the nozzle and the butt there is an area in which the flow in the inner near-wall area of the cylinder spreads and flows from the nozzle axis to the butt-face and in the opposite direction— to the active area. In this case, the direction of motion in this region (7) is opposite to the flow of quasi-solid-state flow (3) in the direction from the nozzle to the butt-face and coincides with the direction flow (3) in the active area. Such a movement is due to the fact that with a strong rotation of the near-wall jet in the area of the nozzle a vacuum is formed. The region of the quasi-solid flow is directed from the butt-end to the nozzle and further to the active area. In this connection, in flow (3) other conditions arise immediately after the plane of the cross-section passing through the vertical axis of symmetry of the nozzle (Fig. 4.25). In this place, the conditions of interaction with current (7) change abruptly: the direction of the flow changes from the opposite in the direction toward the butt-end to unidirectional in the direction toward the active region. Behind the axis of symmetry of the nozzle, the quasi-solid state current 3 begins to merge with flow (7) and expand in diameter. As can be seen from Fig. 4.20, the quasi-solid flow (4) has, in addition to the translational and rotational motion, and its cross-section varies in diameter due to the movement of "whiskers." Therefore, the quasi-solid flow after passing through the axis of symmetry of the nozzle has a pulsating character. This leads to the formation of a periodic system of vortices "whiskers" in the direction of the active region. As these oscillations of the diameter of the quasi-solid-state flow decay, a vortex "whiskers" structure of the same type as in the butt area is formed. Depending on the length of the active area, the vortex structure "whiskers" gradually dissipates and merges with the solid-state flow, which has a weakly screw motion. In accordance with the continuity condition, a stream is sent to the butt region to compensate for the motion of the quasi-solid part (see also Fig. 4.20). The field of flow to the butt consists of an energy-intensive high-speed flow in the area of vortices of the "whiskers" type and a substantially less high-speed flow (7) (Fig. 4.25) between the "whiskers" and the quasisolid core. A thinner layer of flow (7) is also fixed in the area of the nozzle—between the curved near-wall jet and the quasi-solid core. In this part of the stream (7) relative to the axis of symmetry of the nozzle very slowly moves to the butt-end, and the second half of the stream (7) moves toward the active zone. Therefore the cone of the near-wall jet in the area of the nozzle arises. According to Fig. 4.24, thus, there is a strong rotation of the near-wall jet layer, which from the outside moves along the cylinder, and from the inner side the cylindrical near-wall jet rotates along the practically stationary layer of liquid (7). When conducting numerous experiments of the Taylor's current, several typical options are known:

- The outer cylinder rotates, while the inner cylinder remains stationary.
- The inner cylinder rotates and the outer cylinder remains stationary.
- The outer cylinder rotates, and the inner cylinder rotates in one direction or another with a much lower speed.
- During the experiments, the inner cylinder is driven in a periodic motion relative to the longitudinal axis of the cylinder.

Thus, when analyzing the detected types of flow in a vortex chamber, there are all the necessary prerequisites for the existence of a Taylor's type flow in the nozzle area. Fig. 4.26 shows a pair of Taylor's vortices (10) arising in the "primary" (8) and the "secondary" (9) near-wall jets. Since the thickness of the "secondary" near-wall jet (9) is greater than the thickness of the "primary" near-wall jet, the dimensions of the Taylor vortices also differ in the region of these jets. Taylor' vortices absorb less energy-consuming systems of Görtler-Ludwig type (7) vortices. And already after one revolution of the near-wall jet, the visualization fixed traces in the region of the hollows from the Taylor vortex pairs. Video films clearly show how these Taylor vortex systems are absorbed by the latter when the vortices of the "mustache" type are reached, the energy of which is substantially greater, and the dimensions are comparable. Thus, the interaction of vortices occurs in accordance with the laws of interaction of vortex disturbances [112,113,247].

4.6.4 Flow near the butt

Extremely important are the regularities of the flow immediately near the butt, where the flow of vortices "whiskers" is transformed into a quasi-solid flow moving toward the active region. Fig. 4.27 shows the results of visualization of the flow near the butt, obtained using powder. A transparent plate (2) made of Plexiglas, adjoins to the cylinder of the vortex chamber (1) from the butt. This made it possible to observe the nature of the flow near the butt, on the inner surface of the cylinder, and at the butt. The vortex systems moving to the butt along the inner surface of the cylinder, when in contact with the butt, make an incomplete rotation in the angular region of the interface between the cylinder and the butt. Then, this angular flow is converted into a conical spiral flow (4), which moves away from the butt. As the distance from the butt increases, the diameter of the surface of the cone-shaped flow decreases, and the flow velocity along the spiral cone-shaped surface increases significantly. At a distance equal to approximately the radius of the

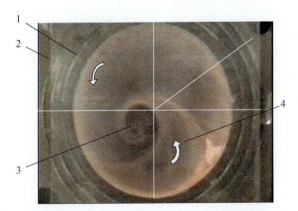

Figure 4.27 Visualization of the flow structure in the area of the butt of the vortex chamber: (1) cylinder of the vortex chamber; (2) butt transparent plate; (3) quasi-solid flow; (4) conical spiral flow near the butt [128].

cylinder, the rotational speed increases so that the powder diffuses. Therefore, in Fig. 4.27, near the center of the butt, a dark circular spot is seen, which represents the beginning of the development of quasi-solid-state flow (3) located in the region of the longitudinal axis of the cylinder. The curvilinear arrows in Fig. 4.27 show the direction of rotation of the corresponding vortex structures at the butt, and the bright lines show the axis of symmetry of the cylinder.

On the right in Fig. 40.27, there is a stagnant area in which powder accumulates. In accordance with the characteristics of the flow of a pair of vortices, we can assume that the white traces of powder reflect the flow of a pair of vortices rotating toward each other. In Fig. 4.27 two pairs of spiral vortices are fixed. There are also fuzzy traces of smaller pairs of vortices. These vortices are traces of "whiskers" rotating along the butt. Two pairs of spiral vortices are connected at a certain distance from the butt into one pair of vortices, which sharply reduces the radius of the spiral and forms a quasi-solid motion. The period of changing the picture of visualization at the butt is 0.16–0.24 s. The speed of rotation along the end is the same as in the vortex pair "whiskers," measured near the butt. The diameter of the spiral tornado-like vortex decreases up to four times in comparison with the size of the vortex at the butt and is: $d_v/d_0 = 0.22 - 0.24$.

Fig. 4.28 shows the results of visualization of the flow near the butt, obtained by the carbon black method. A transparent plate made of organic glass adjoined the cylinder of the vortex chamber from the butt. This made it possible to observe the

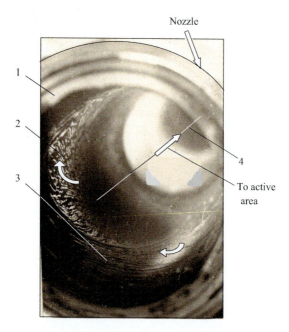

Figure 4.28 Visualization of the flow structure near the butt of the vortex chamber: (1) cylinder of the vortex chamber; (2) vortex structures on the inner surface of the cylinder near the butt; (3) vortex systems at the butt of the vortex chamber; (4) longitudinal axis of symmetry of the cylinder [128].

nature of the flow both near the butt, and at the butt. This figure shows a picture of the visualization obtained when installing a short butt on the vortex chamber. In this case, besides the "whiskers," systems of quasi-Taylor vortices formed in the "primary" near-wall jets also approach the butt. The traces of these vortices (2) are approaching to the butt along the helix, flow to the butt and form a system of vortices (3), moving along the end of the helix. Picture of these vortices (3) at the end almost coincide with the flow visualization pattern on a rotating disk [406].

Fig. 4.29 shows the results of visualization of the flow at the butt, obtained by the carbon black method. As in Fig. 4.27, the nozzle is located on the left, so that the movement of the vortices on the butt-face is directed counterclockwise. A theoretically similar flow pattern was solved in Ref. [177] near a flat disk. An exact solution of the Navier-Stokes equation is obtained, when at a large distance from the wall a liquid rotates with a constant angular velocity ω around an axis perpendicular to the disk plane. For fluid particles at a great distance from the wall, the centrifugal force and the radial pressure gradient are mutually balanced. For fluid particles near the wall, the circumferential speed due to braking is reduced, so here the centrifugal force is significantly reduced. In this case, the inward radial pressure gradient remains the same as at a large distance from the wall. As a result, an inward radial flow arises near the wall, which in turn causes, as a result of the continuity condition, an upward flow in the axial direction.

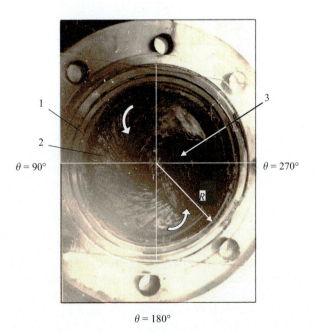

Figure 4.29 Visualization of the flow structure at the butt of the vortex chamber: (1) cylinder of the vortex chamber; (2) vortex systems at the butt of the vortex chamber; (3) zone of flow separation [128].

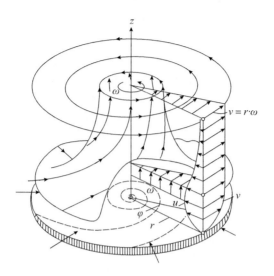

Figure 4.30 Rotational motion of fluid over a fixed base [177].

Fig. 4.30 shows the flow scheme adopted in the calculation [177]. The quantities u, v, w are the components of the velocity in the radial, circumferential and axial directions, respectively. Near the disk, the peripheral speed due to friction is inhibited. As a result, a secondary flow arises radially inside the disk. Considering the above pictures of visualization of the flow in the vortex chamber, it becomes obvious that the flow patterns in the vortex chamber and in the theoretical U.T. Bödewadt problem differ significantly. In the vortex chamber in the central axial part there is a vacuum, due to which a quasi-solid current is formed. The flow is unwound mainly by the "whiskers" with vortex systems, and their screw movement leads to the meandering of all the enclosed cylindrical flow structures inside the vortex chamber. Analysis of the video films on which features of the vortex structures were recorded showed that at the butt-end of the center of the flow separation and the formation of a tornado of such a flow is about half the radius, located approximately in the middle of the third quadrant—between $\theta = 180$ degrees and $\theta = 270$ degrees. The area of separation is nonstationary and, in addition, depends on the geometrical parameters of the nozzle and the angles of its installation.

With a long butt portion of the vortex chamber (Fig. 4.21), when approaching the butt, all quasi-Taylor's vortex systems are absorbed by vortex systems "whiskers." To the butt of the stream is transferred mainly by the "whiskers." In this case (Fig. 4.20, arrows (3)) on the inner surface of the cylinder there are areas of low velocity flow directed toward the active region (from the butt). At the same time, on the cylinder surface, visualization allowed us to fix in the BL of these areas a system of longitudinal vortices, similar to the *VII* stage of the transition BL on a flat plate (Section 3, Fig. 3.33). Disturbances similar to Λ-shaped vortices at stage *IV* of the transition BL of a flat plate were also recorded (Section 3, Fig. 3.34, item (2)). Fig. 4.31 shows photographs of such vortex structures. The shape of these vortex structures depends on the local Reynolds number, which depends, in particular,

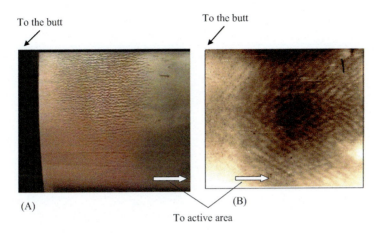

Figure 4.31 Longitudinal (A) and Λ-shaped (B) vortex structures on the inner surface of the cylinder near the butt of the vortex chamber [128].

on the velocity in the nozzle and the length of the butt portion. The flow in the region of the specified longitudinal vortex structures is directed from the butt toward the active area (shown by the arrow) in accordance with Fig. 4.20.

4.6.5 Model of the vortex structures of the vortex chamber

V.V. Babenko developed and manufactured a three-dimensional model for the development of vortex structures of a vortex chamber (Fig. 4.32) based on the above results of V.N. Turick research. The main types of vortex structures in the internal region of a vortex chamber are shown in Fig. 4.20. Four types of vortex structures are considered: (1) "whiskers" vortex systems; (2) area of flow toward the butt-end; (3) peripheral region of the flow toward the active area; (4) quasi-solid flow. The quasi-Taylor vortices (2) indicated in Fig. 10.32 is formed in the "secondary" parietal jet. Vortices (2) interact with vortices (1), arising in the angular places of the nozzle. Downstream along the inner surface of the cylinder, a local separation of the "primary" near-wall jet occurs in the region of the nozzle-cylinder interface. In the region of attachment of the separation zone to the surface of the cylinder as a result of this interaction of the vortices (1) and (2), large vortex pairs (4) "whiskers" appear. These vortex pairs (4) move along the inner surface of the cylinder along a helix to the butt-end and into the active area. As you move, the "whiskers" vortex pairs absorb quasi-Taylor's vortices, whose energy and pitch of the helix are substantially less than those of the "whiskers." The model (Fig. 4.32) shows a white line showing the longitudinal axis of symmetry of the vortex chamber, arrow (3) shows the quasi-solid flow along the longitudinal axis of symmetry of the cylinder from the butt to the active zone, arrow (5) shows a photo of the internal surface of the cylinder of the vortex chamber when the vortex structures are visualized by the soot method.

Fig. 4.33 shows a model of the shape of a quasi-solid axial flow 1 from the butt to the active area. Large arrows show the rotation of flow 1 along a helix from left to right. Points (2) show the location of the vortex pair "whiskers" in the near-wall region of flow

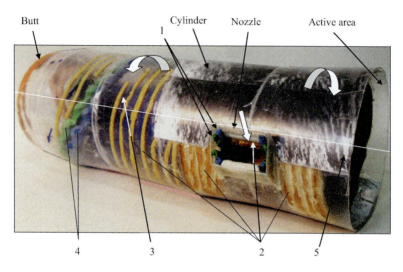

Figure 4.32 Model of the vortex structures of the vortex chamber with the butt-end part (A) and without it (B): (1) vortices in the corners of the nozzle; (2) quasi-Taylor vortices; (3) quasisolid flow along the longitudinal axis of the cylinder; (4) vortex pairs "whiskers" [128].

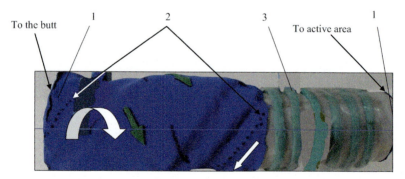

Figure 4.33 The shape of a quasi-solid cylinder located along the longitudinal axis of the vortex chamber: (1) quasi-solid flow along the longitudinal axis of the cylinder; (2) the location of the vortex pair "whiskers"; (3) quasi-Taylor's vortices [128].

around the inner surface of the cylinder. The light arrow shows the direction of movement of the "whiskers" from right to left. According to the measurement results shown in Fig. 4.20, there are two quasi-cylindrical flows between the quasi-solid flow and the "whiskers": (2) movement from nozzle to butt due to vacuum in the area of the nozzle and movement from the butt to the active area. Flow (3), shown in Fig. 4.20, is caused by the fact that the direction of flow in the quasi-solid-state region and in the "whiskers" is opposite. Since the shape of these vortex structures is helical, their mutual movement is similar to the movement of gears in the corresponding pump. This movement causes the direction of flow near the "whiskers" in the opposite direction: from the butt to the active area.

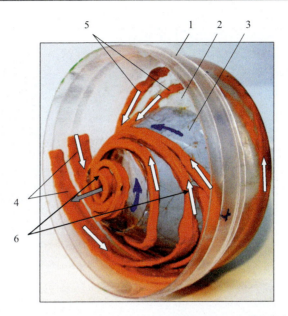

Figure 4.34 Model of the vortex structures at the butt-end when viewed from inside: (1) the cylinder of the vortex chamber; (2) butt; (3) cone of fluid motion over a fixed base [177]; (4) vortex pairs "whiskers"; (5) vortex systems rotating on the butt-end; (6) vortex systems twisting in a quasi-solid flow [128].

Fig. 4.34 shows the model of the flow in the area of the butt-end. In accordance with Fig. 4.32, two types of flow were photographed in the end area: the vortex system "whiskers" (4) and the quasi-cylindrical swirling flow (3). At the same time, the speed, and, consequently, the amount of movement in the vortices (4), is three times more than (3). In the area of the butt-end, both types of flow are connected and rotate at the butt-end as shown in Figs. 4.27–4.29. Due to the rotation near the butt-end in the area of the longitudinal axis of the cylinder of the vortex chamber, a vacuum occurs, as in the vortex systems of a tornado. In Figure 10.30 shows the results of the calculation of U.T. Bödewadt [177] for a simpler case. Fig. 4.34 shows a model of vortex structures based on video films. It is seen (shown by light arrows) that the vortex systems (5) rotating at the butt (2) of the vortex chamber (1), which are a continuation of the vortex pairs "whiskers" (4), are drawn under the action of centrifugal forces into the vortex systems (6), which move along the funnel-shaped surface (3) toward the active area (direct dark arrow) and converted into a rotating quasi-solid flow, located along the longitudinal axis of symmetry of the vortex chamber.

Fig. 4.35 shows the model of interaction of vortex systems in the region of the butt of the vortex chamber. Light arrows indicate the direction of flow in vortex systems. The vortex pairs "whiskers" (3) follow a helix to the butt (1), on which they rotate in the form of butt vortices. Vortex systems (6) are accelerated and shifted to the central part of cylinder (2), twist into a tornado-like vortex and

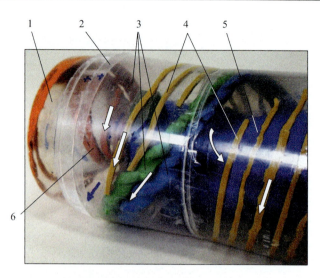

Figure 4.35 Interaction of vortex structures in the butt-end area: (1) butt; (2) vortex chamber cylinder; (3) vortex pair "whiskers"; (4) quasi-Taylor's vortices; (5) quasi-solid axial flow; (6) vortex systems twisting into a tornado-like vortex [128].

screwed along the surface of the funnel into a quasi-solid flow (5) at some distance from the butt approximately equal to the radius of the butt. As mentioned earlier, the longitudinal axis of the funnel of the tornado-like vortex 5 is offset relative to the longitudinal axis of the cylinder and is located about half the butt-end radius. This distance depends on the location of the nozzle. Trajectory velocity when screwed into a tornado, it was about 0.5 m/s. Since the radius of a tornado decreases with distance from the butt, the angular velocity increases. Then the diameter of the funnel increases, and the flow is transformed into a quasi-solid state, the angular velocity of which decreases as it moves to the nozzle.

4.7 Görtler's vortices on a concave surface behind the vortex chamber nozzle

Experimental studies of the parameters of the Görtler's vortices arising behind the vortex chamber nozzle were carried out on an aerodynamic bench (Section 4.6, Fig. 4.19) [126]. A coordinate device, in which wire thermo-anemometric sensors were fixed through appropriate holders and attachment points, was installed on an independent frame near the end of the vortex chamber. The frame was not connected to the vortex chamber. The coordinate device is equipped with a micrometer device that provides linear movement of sensors in three mutually perpendicular directions. Sensor installation was monitored using laser beams, which were directed through a lens system to produce two laser light blades. By lighting the sensor's sensitive wire at the intersection of the laser knives, the location of the hot-wire anemometer was recorded and its coordinates were calculated. One laser beam was directed along the side surface of the vortex chamber. This made it possible to obtain the coordinates of the sensor with an error not exceeding 10^{-6} m. Measurements were carried out using instrumentation (voltmeters, oscilloscopes, frequency meters, amplifiers), Pitot, Pitot-Prandtl pneumometric tubes, Disa Electronic thermal anemometric instruments, and spectral correlation analysis using a Bruel and Kjaer multichannel tape recorder. Spectral correlation analysis performed using a Bruel and Kjaer multichannel tape recorder.

The tangential angle γ of the inlet nozzle (Section 4.6, Fig. 4.19B) in the experiments varied from 32 to 90 degrees. The range of Reynolds numbers calculated by the mean air velocity in the nozzle and its equivalent diameter d_n was $Re_n = d_n U_\infty/\nu = (5-8) \times 10^4$. The distance between the nozzle and the end butt-face varied within 1.1−4.4 of the diameter of the vortex chamber. Before conducting instrumental measurements, experiments were performed on visualization of the flow structure in the vortex chamber. Visualization pictures of Görtler's vortices are given in (Sections 4.2, 4.3, 4.5, and 4.6). The error of geometric and averaged kinematic characteristics of the vortex motion and CVS that are formed in the near-wall region of the vortex chamber did not exceed 10% with a confidence of 0.95, which corresponded to a dispersion of 2σ for instrumental investigations, and 15% for visual observations.

After receiving the visual pictures, characteristic sites for instrumental measurements were identified. The measurements were performed under a nozzle (Section 4.6, Fig. 4.24) in the vicinity of $-0.15 < z/d < 0.15$ for 70 degrees $< \theta < 160$ degrees and $-0.2 < z/d < 0.2$ for 200 degrees $< \theta < 320$ degrees. In these parietal regions, there are necessary prerequisites for the existence of Görtler's vortices. In the case when the tangential angle $\gamma < 50$ degrees, the BL separates from the entrance edge of the rectangular nozzle, therefore Görtler's vortices are not formed here. Fig. 4.24 (Section 4.6) shows photographs for relations in a vortex chamber $L/d = 4.4$, $U_\infty = 37.6$ m/s, $\gamma = 67$ degrees, and $\theta = 90$ degrees. A fine mesh of washed light strips was recorded. Only in the plane of the axis of the nozzle are the strips directed parallel to axis OX. Further, they diverge, creating on the inner surface of the vortex chamber a nonuniform trapezoid with curvilinear lateral sides (Section 4.26, Fig. 4.24).

The instability of the Görtler's vortices is determined by an imbalance between the centrifugal force and the radial pressure gradient in the BL over a concave streamlined surface (Section 4.1, Fig. 4.3). The presence of the Görtler's vortices makes this flow three-dimensional. They raise the sedentary fluid from the streamlined surface, and high-speed fluid is directed to the wall, thus creating areas of flushing and ejection, according to the location of the Görtler's pairs. The result of this redistribution of fluid mass is the creation of mushroom-like structures with a strong distortion of the velocity profiles in the normal and transverse directions (Section 4.2, Fig. 4.5). When crossing the flow field, where the influence of the Görtler's vortices manifests itself, at a fixed distance from the streamlined surface, the velocity profiles acquire a periodic character (Fig. 4.36). The velocity maxima correspond to the "hollow" regions that occur between pairs of Görtler's vortices, where a high-speed fluid is directed to the streamlined surface (Section 4.1, Fig. 4.3).

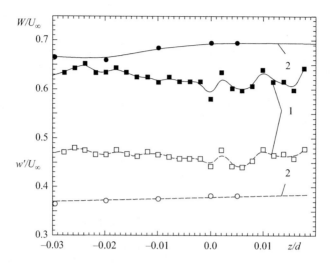

Figure 4.36 Profiles of averaged W/U_∞ (●, ■) and pulsation w'/U_∞ (○, □) transverse velocity at the entrance to the vortex chamber at $\theta = 90$ degrees for different distances from the streamlined surface: (1) $y/d = 0.02$, (2) $y/d = 0.04$ [109,126].

The minima in the velocity profiles characterize the zones of the ejection lift of a low-mobility fluid ("peaks") from the near-wall region of the BL to its outer boundary. The distances between the corresponding extremes make it possible to measure the wavelength of the Görtler's vortices, which in our experiments is $\lambda_z = (0.8-0.9)$ mm. This wavelength correlates well with the data obtained from visual experiments (Section 4.6, Fig. 4.24A), namely $\lambda_z = (0.9-1.2)$ mm for angles of 70 degrees $< \theta <$ 110 degrees. With increasing distance from the surface of the vortex chamber (Fig. 4.36), the influence of Görtler's vortices on the streamlined flow was not recorded (uniform distribution of velocity profiles). Similar features of the velocity distribution over concave curved surfaces were observed in Refs. [169,243,449,613].

Using measured average velocity profiles relative to the normal direction of the streamlined surface, as well as calculations for the Blasius BL, we obtained the BL thickness δ, extrusion thickness δ^* and impulse loss thickness δ^{**} for flow regions where Görtler's vortex wavelengths are measured. So, for $U_\infty = 37.6$ m/s, $R = 51 \times 10^{-3}$ m and 70 degrees $< \theta <$ 110 degrees, was received 0.88 mm $< \delta <$ 1.2 mm; 0.28 mm $< \delta^* <$ 0.41 mm; and 0.11 mm $< \delta^{**} <$ 0.16 mm, and for 250 degrees $< \theta <$ 290 degrees, 1.4 mm $< \delta <$ 1.8 mm; 0.47 mm $< \delta^* <$ 0.61 mm; and 0.18 mm $< \delta^{**} <$ 0.23 mm. These parameters of the BL over the streamlined curvilinear surface of the vortex chamber correspond to the Görtler's numbers $11.9 < G_o < 22.1$ for 70 degrees $< \theta <$ 110 degrees and $26.0 < G_o < 37.2$ for 250 degrees $< \theta <$ 290 degrees.

In Fig. 4.37, neutral curves and maximum gain curves are constructed based on the results obtained from studies of the formation of Görtler's vortex systems on the surface of the vortex chamber. This graph shows the calculated and experimental results of the parameters of the longitudinal vortex systems arising from the flow around a solid curvilinear surface. In Fig. 4.37, neutral curves (1) were obtained in Ref. [239] for different

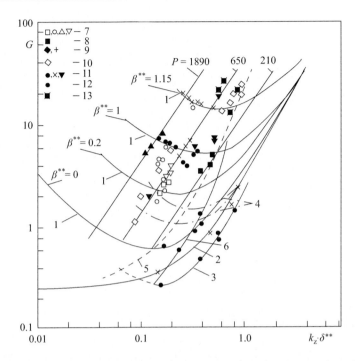

Figure 4.37 Görtler's stability diagram on concave and flat rigid surfaces and maximum amplification curves: (1) [239], (2) [243], (3) [626], (4) [515], (7) [463], (8) [466], (9) [168], (10) [433], (11) [34], (12) [613], and (13) are the results of performed studies [126]; maximum amplification curves: (5) [239], (6) [627].

normalized spatial gain factors of instability, curve (2) [243], curve (3) [626], and curves (4) [515]. The curves of maximum amplification of instability (5) and (6) are presented in Refs. [239,627]. Experimental results, denoted by dots (7), are given in [547], (8) [466], (9) [168], (10) [433], (11) [34], (12) [613], and (13) the results of performed investigations [126]. The neutral curves shown in Fig. 4.37, show that while the Blasius BL has a large instability for all disturbances with wave numbers $k_z\delta^{**} = 2\pi\delta^{**}/\lambda_z < 1.5$, the near-wall jet has a greater instability for disturbances with wave numbers $k_z\delta^{**} > 1.5$. This is due to the fact that the vortices, which correspond to the wave numbers $k_z\delta^{**} > 1.5$, are very small and concentrated near the wall. Therefore, these vortices are stable only on a small part of the flow, since they are located near the wall and are located deeper in the BL, where the instability mechanism has a greater influence. It has been established [243,247,627] that the motion of a fluid with a monotonic velocity distribution becomes unstable before the motion of a fluid with a nonmonotonic velocity profile.

Experimental instability curves for Görtler's vortices are also presented in Fig. 4.38; the technique for their preparation was proposed in Ref. [36]. Fig. 4.38 shows calculated from experimental data the values of Görtler's numbers for the concave surface of the input part of the vortex chamber, which are located in the region of instability. If we take into account that λ_z is the wavelength of a vortex pair, then the lateral scale of Görtler's

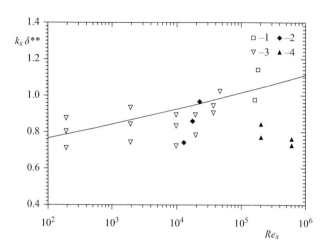

Figure 4.38 The effect of the Reynolds number on the normalized wave number of Görtler's vortices [126].

vortices will be less than the thickness of the BL. Since the given calculated results are in the zone of instability, the Görtler's vortices must develop, which leads to an increase in their scale. Therefore, when flowing near the nozzle of the vortex chamber, the scales of the Görtler's vortices increase both in the longitudinal direction along the axis OX, and in the transverse direction along the axis OZ (Section 4.6, Figs. 4.24 and 4.37). The cross-section of the Görtler's vortices obtained for such flow conditions is characterized by the ratio of their wavelength to [242,613]. The cross-section of the Görtler's vortices obtained for such flow conditions is characterized by the ratio of their wavelength to the thickness of the boundary layer and correlates with the results for rigid concave surfaces [242,613].

Fig. 4.38 shows the experimental results of [449] (1), [240] (2), [466] (3) and (4) obtained in these investigations. According to Refs. [206,243,542], the growth rate of the Görtler's vortices on a concave curved surface is very small, while the thickness of the BL grows much faster: for example, for the Blasyca flow, it is proportional to the longitudinal coordinate in step 0.5. Therefore, for the classical development of the Görtler's vortices, the value of $k_z\delta^{**}$ increases with an increase in the Reynolds number. Using the least squares method, an approximation dependence was obtained between the wave number of a pair of Görtler's vortices and the Reynolds number, namely, $k_z\delta^{**}=0.64\mathrm{Re}_x^{0.04}$, which corresponds to the data shown in Fig. 4.38. In the present experiments, the normalized wave number, on the contrary, decreases with increasing Reynolds number. This is due to the fact that in a vortex chamber with a butt-end there are oppositely directed flows that stretch the inlet stream under the nozzle and Görtler's vortices, and accordingly cause an increase in their transverse wavelength. At the same time, the growth rate of the transverse wavelength of a pair of Görtler's vortices significantly exceeds the rate of growth of the BL thickness, which differs from the classical formation of Görtler's vortices when the flow around a concave surface in the longitudinal direction.

Experimental investigations of the characteristics of the boundary layer on the analogs of the skin covers of hydrobionts

5.1 Structures of elastic surfaces

In chapters 7−10 (Part II, Experimental Investigations Flow around Bodies), verification of the discovered features of the swimming of aquatic organisms, discussed in chapters 1−6 (Part I, Experimental Hydrodynamics of Fast-Floating Aquatic Animals), was carried out with the help of technical experiments without taking into account the influence of specific features of the interaction of all aquatic organisms on the economical expenditure of energy. In the future, some specific features of the structure of organisms of hydrobionts will be simulated, in particular, in this chapter; the influence of the elastic-damping properties of the skin on the characteristics of the boundary layer will be simulated.

Models of elastic surfaces are considered in section 6.3 (Part I, Experimental Hydrodynamics of Fast-Floating Aquatic Animals), as well as in Refs. [60,139,378]. The equipment for measuring the mechanical characteristics of elastic surfaces was used the same as that presented in section 6.4 (Part I, Experimental Hydrodynamics of Fast-Floating Aquatic Animals). The standard equipment given in Refs. [60,366] was also used. The main characteristics of flexible coatings and similarity criteria are given in section 1.4 (Part I, Experimental Hydrodynamics of Fast-Floating Aquatic Animals), as well as in Refs. [16,60,139,378].

For the first time, the drag reduction of an axisymmetric model with an elastic coating was experimentally investigated in Refs. [382,383]. Elastic coatings were designed and manufactured by analogy with the outer layers of the skin of dolphins, when flown around which a significant drag reduction was obtained (57%). In accordance with the level of development of hydromechanics at the time, M. Kramer explained the effect obtained by stabilizing Tollmien−Schlichting waves, and called his approach a method of stabilizing the laminar boundary layer distributed damping. All variants of elastic coatings were made in the form of composite materials, and the most successful was a composite with longitudinal internal ribs. Subsequently, most of the papers took into account ideas [166] about the occurrence of three types of waves on the surface of elastic plates under the influence of the incident flow.

Initially, investigations of hydrodynamic stability on elastic surfaces were performed. The most interesting and important results were obtained in [282,325,471].

The investigations of hydrodynamic stability on elastic coatings were carried out by Carpenter and his students [111,196,197,424], and in Refs. [410,411]. The experiments were carried out on single-layer coatings made of a layer of polyurethane foam or porous rubber, outside of which was located moiré film. Gad el Hak M. [249–252,509] also carried out research in this direction. In Refs. [249–252,509], experimental investigations were carried out in the flow around a gel-like material, which has no physical and practical interest.

Investigations of the turbulent boundary layer in the flow around elastic surfaces were also performed in Refs. [11,60,139,171–176,235,237,238,253,275,277,302, 321,322,3327,360,361,378,382,404,405,416,417,441,535,586,587,589–593]. In Refs. [410,411], an elastic plate in the form of polyurethane foam was investigated, and in Ref. [237] experiments were performed using a thermal anemometer on an elastic plate made of a layer of polyurethane foam, on the outer surface of which there is a thin layer of silicone rubber coated with moiré film. The most advanced theory of the turbulent boundary layer in the flow around compliant surfaces was developed at the Institute of Hydromechanics of the National Academy of Sciences of Ukraine (Ukraine, Kiev) G.A. Voropaev [247,589,593]. A rather complete list of studies performed in this direction is given in Refs. [60,139,197,235,247,378].

In Refs. [404,405], experimental investigations of the structure of a turbulent boundary layer in a flow past a passive pliable plate using holography and smoke imaging in a wind tunnel are performed. The scales of λ^+ transversal vortex structures and other kinematic characteristics of the turbulent boundary layer are given. Similar results were published much earlier in Refs. [60,322]. In addition to studying the kinematic characteristics of the boundary layer in the flow around elastic coatings, it is also of interest to investigate the influence of elastic surfaces on the field of pressure pulsations [171,302,303,334].

Based on the aforementioned brief review, it becomes obvious that, along with simple designs of elastic coatings, it is necessary to develop active coatings of a more complex structure.

During evolution, the structure of hydrobiont skin covers became optimal for each range of Reynolds numbers calculated for characteristic swimming speeds. Based on the results of hydrobionics investigations (Part I, Experimental Hydrodynamics of Fast-Floating Aquatic Animals, chapters 2, 4–6), initial data were obtained for determining the structural and mechanical characteristics of elastic materials when conducting model experiments. In addition, the hydrobionic approach made it possible to understand the structure of the skin of hydrobionts, which was developed as a result of evolution when interacting with the environment.

The theoretical analysis of the flow on elastic surfaces gave some recommendations for choosing the type of material [166]. The stability of the laminar boundary layer in the flow above the elastic surface is achieved if the surface rigidity is high. Moreover, the correspondence condition is satisfied for the propagation velocity of the surface wave on the elastic surface c_o and the velocity of the instability wave from the point of view of the Tollmien–Schlichting stability, corresponding to the second branch of the neutral curve. If the rigidity of the elastic surface is small, the velocity c_o must be large enough to prevent the appearance of class B waves

[166], and the disturbance decrement is small to prevent the development of Reynolds stress due to class A waves, according to Benjamin's terminology. However, it is almost impossible to find an elastic material that complies with such theoretical recommendations. The choice of the most dangerous wave of Tollmien−Schlichting is also in doubt. The results of an experimental investigation of hydrodynamic stability on elastic plates are presented in Refs. [17,19−23,26,378]. The main recommendations are that the phase velocity of the surface wave arising in an elastic coating under the action of the pulsation velocities of the boundary layer should approximately correspond to the phase velocity of the main energy-carrying disturbances in the boundary layer with an appropriate flow regime. The energy of pressure pulsations in the boundary layer must be sufficient for oscillations in an elastic material to appear. If the pulsating energy of the boundary layer is insufficient, the elastic surface behaves like a rigid surface. If this energy is large, depending on the mechanical characteristics of the elastic material, large amplitude oscillations will appear on its surface. The amplitude of oscillations in an elastic material should be small and not exceed the size of the maximum allowable roughness.

There is considerable experience in the use of elastic materials to absorb vibrations and noise. The physical character of the interaction of disturbances and deformations caused by them in an elastic material is different compared with the flow in the boundary layer of elastomers. However, the results of experiments in which elastomers were used to absorb vibrations, provide some recommendations on the choice of the design of elastic materials: elastomers must have a multilayer structure and longitudinal ordering.

To carry out experimental investigations of hydrodynamic stability when flowing around elastic surfaces, V.V. Babenko developed a design of elastic plates with allowance for small and large flow velocities necessary for investigations laminar, transitional and turbulent boundary layers. Fig. 5.1 shows the scheme of the structure of two types of elastic plates for conducting investigation at low flow velocities. The length of these plates was somewhat less than the length of the working section (3 m) of the hydrodynamic stand, in order to facilitate their

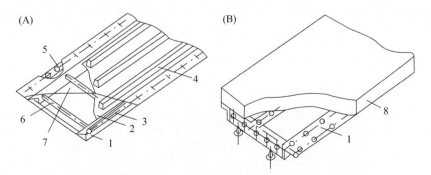

Figure 5.1 Design of the membrane (A) and viscoelastic (B) surfaces: (1) frame; (2) duralumin plate; (3), (5) pressure plate, (4) polyurethane foam strips, (6) frame rods, (7) membrane, and (8) foamed polyurethane sheet [378].

installation in the working part. The width of the plates is 1—2 mm less than the width of the working part, thereby creating a gap at the edges of the plates in order to remove the corner boundary layer along the length of the working part. The plate (Fig. 5.1A) is made in the form of a frame consisting of two longitudinal corners connected by rods installed in a checkerboard pattern so that they can be equipped with a duralumin plate. This made it possible to provide variable distances from the plate to the elastic membrane. A membrane is stretched on the frame, which is fixed on four sides in clamping plates with clamping screws. On one side, the membrane is rigidly fixed on the channel to the frame, and on the other three sides, the membrane is fixed to the frame with these clamping screws so that the tension of the membrane can be changed. At different distances between the duralumin plate and the membrane provides a different thickness of water or air located under the membrane.

This design allows you to explore the so-called simple and complex membrane surfaces. In the second case, three strips made of polyurethane foam are installed between the membrane and the plate. The strips prevent membrane vibrations (strips width 12 mm and height 11 mm). The design allows you to change the size of these strips, their material, and quantity. An elastic plate made of polyurethane foam was attached to the frame (Fig. 5.1B), covered with a thin elastic film on the outside. On the elastic plate on top of it was mounted a frame made of three duralumin corners (Fig. 5.2).

Both designs of elastic plates have adjustment screws that allow the plates to be installed horizontally in the working section and flush with the surface of the lower walls of the confuser and diffuser in the region of their connection with the plate located in the working section. Such constructions make it possible to create a uniform horizontal elastic surface in the working section without folds and without tension, in order to vary the thickness and material of the elastic plates in a wide range during the tests. Theoretical models of elastic plates are given in Refs. [60,139].

One of the important mechanical characteristics is the tension of the surface layer of an elastic material or membrane. Theoretically, you can determine the tension with the help of a stresses sensor installed on the surface of an elastic material. However, the layer of glue, which is attached to the strain gage on the elastic membrane, causes significant errors in the measurement results. Babenko developed a device (Fig. 5.3) [24], which allows determining the dependence of tension on elasticity, measured by the developed instrument for measuring elasticity (Part I,

Figure 5.2 Photograph of a membrane (1) and visco-elastic surfaces: without a border frame (2) and with a frame (3).

Figure 5.3 Photograph of the device for determining the dependence of elasticity on the magnitude of the membrane tension.

Experimental Hydrodynamics of Fast-Floating Aquatic Animals, section 6.4, figs. 13–15). The device is made in the form of a rigid frame with a length of 1 m and a width equal to the width of the elastic coating. On the one hand, the membrane is firmly fixed with a plate fixed to the frame. On the other three sides, the membrane was fixed in clamping plates of the same design as in the first type membrane coating (Fig. 5.1A). Various weights were attached to these clamping plates, evenly spaced along the length, acting per unit length of the membrane. A device for measuring elasticity was used to measure the elasticity of plates in various places along the length of the plate, and the tension per unit length of the membrane was calculated. The load values varied, and a calibration curve for tension and load was obtained. Similarly, on this device, you can determine similar dependencies for monolithic and laminated composite elastic materials in the case of, when their thickness is small.

Experimental investigations of the boundary layer characteristics on these structures of elastic coatings allowed us to obtain new results, both on rigid and elastic plates. Visualization and measurement of quantitative characteristics showed significant differences in the structure of the flow near the elastic boundary compared with a rigid plate.

Further studies were needed when changing the flow rate and viscosity in a wide range of parameters. To solve this problem, a complex of experimental installations and new designs of elastic plates was developed [60,139]. The structures of elastic surfaces shown in Figs. 5.1 and 5.2 are simple and have no practical significance. But these constructions are convenient for comparison with theoretical models of elastic materials. Unfortunately, such designs have a narrow range of frequency characteristics (relaxation times) and have no prospects, since with increasing speed the range of energy-carrying frequencies in the spectrum of disturbing flow movement significantly expands, and the external load leads to a significant deformation of the elastic material.

The principles of developing new designs of elastic plates were based on the following main considerations. Elastic surfaces should be monolithic and composite. It is known that each type of monolithic elastic material has a spectrum of natural frequencies, in the range of which the elastomer absorbs the energy of external oscillations most intensively. Therefore composite materials expand the spectrum of intensely absorbed oscillations. The design of elastic coatings should be based on the physics of coherent vortex structures in the near-wall regions of the boundary layer. The nature and characteristics of the development of coherent vortex structures in the transition boundary layer (Chapter 3, Experimental Investigations of the Characteristics of the Boundary layer on a Smooth Rigid Plate) are investigated. Currently, sufficient material has been accumulated to understand the development of coherent vortex structures also in the turbulent boundary layer [60,139]. To determine the influence of elastic surfaces on the structure of a turbulent flow, it is necessary to design appropriate elastic surfaces. Therefore it is necessary to have monolithic elastic plates with different mechanical characteristics and various designs of composite elastic plates.

Based on the results of studies of the interaction of disturbances in the boundary layer [53,139,247] and the results of the bionic approach (Part I, Experimental Hydrodynamics of Fast-Floating Aquatic Animals), it became apparent that the presence of a certain structure of the elastic surface intensely and purposefully affects the structure of the turbulent flow and other flow regimes streams. In this regard, by analogy with the structure of the skin of fast-swimming hydrobionts, in designing elastic materials, certain orderliness was provided in the longitudinal and transverse directions (Part I, Experimental Hydrodynamics of Fast-Floating Aquatic Animals, section 4.3). The ideas about the waveguide character of the boundary layer and the mechanism of the complex waveguide (Section 3.8) were the basis for the development of structures for elastic coatings. By analogy with the structure of the skin of dolphins and swordfish, in the design of elastic coatings, active control systems were applied by changing the temperature and pressure inside the elastomers. The skin of hydrobionts is equipped with control sensors that allow you to have an automatic control system for the mechanical and geometric characteristics of the skin. One of the basic principles underlying the design of elastic coatings is that such elastic coatings significantly change the flow structure. When designing extended coatings, it is necessary to take into account the distribution of various environmental energy loads along the length of the coating (Part I, Experimental Hydrodynamics of Fast-Floating Aquatic Animals, section 2.3, fig. 2.4). It is also necessary to take into account that during real body movement, external perturbations and perturbations acting on the body, acting from below—from the body on which the elastic coating is fixed, will affect the elastic coating. In this regard, it is necessary to appropriately change the design of the elastic coating and its characteristics along the body. It is necessary to take into account that the characteristics of the main flow are also important when flowing around an elastic coating. This means that the flow acting on the subsequent sections of the coating has a structure that was modified during the flow past the previous sections. Therefore the developed elastic coatings cannot be designed in accordance with the schemes and methods of flow when flowing around a rigid surface.

When considering such a complex multipurpose task, it is necessary to conduct experimental investigations of the same elastic plates in various experimental installations and in a wide range of flow velocities. The design of the elastic coating should effectively attenuate disturbances acting from the outside of the flow and from the side of the rigid body. The elastic coating must have a system for regulating its mechanical and geometrical characteristics and allow the installation of control sensors inside the coating. V.V. Babenko has developed various design options for such a coating. It is also necessary to develop in the future not only adjustable elastic coatings, but also, by analogy with hydrobionts, active elastic coatings with an automatic control system for their dynamic characteristics.

V.I. Korobov and V.V. Babenko [139,356] jointly developed a standard panel design for testing elastic coatings on various experimental installations. Fig. 5.4 shows a photograph of the underside of this panel. On this panel, on the front side, rigid and various types of elastic plates with different structure and thickness were installed. Hard or elastic plates (3) are glued to duralumin plates, which are screwed to stiffeners (2) with screws (4), located at the intersection of stiffeners. On the front side of the panel, the test plate is additionally pressed against the panel (1) by frame (5) using a system of pairs of adjustment screws consisting of mounting screws (6) and clamping screws (7). These screws make it possible to achieve the same plane of the test plate and the outer plane of the frame so that the gap between them is practically absent.

Fig. 5.5 shows the basic frame of the resistance balance designed by V.I. Korobov consisting of two levers connected by a jumper. Attached to this frame are four screws with coated covers. The panel is installed flush with the upper bottom of the working area with backlash. For this purpose, at the end of the working section of the hydrodynamic installation (from 2.0 to 2.5 m), a section was cut out for installing the measuring panel—insert (Section 3.7, Fig. 3.41).

Fig. 5.6 shows the mounting scheme of the elastic surface (5) to the panel (4) and the installation of the panel in the working section of the hydrodynamic stand [139,356]. The frame of the panel (20) with four screws (19) is attached to two brackets of the strain gage (3), which are fixed by means of spring plates (11) tangential flow stress and allow you to determine the friction on the plate using

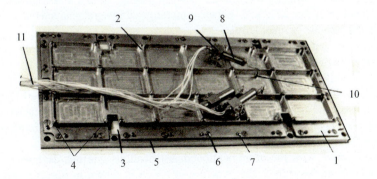

Figure 5.4 Photograph of the panel structure for fixing the elastic surface: (1) panel, (2) stiffening ribs, (3) test plate, (4) screws, (5) fringing, (6) fixing screws, (7) clamping screws, (8) sensor, (9) sensor corps, (10) moving axis of the sensor, and (11) wires.

Figure 5.5 Photograph of the strain gage suspension in the working section of the hydrodynamic stand: (1) suspension brackets of the main frame of the strain gage, (2) design of the lower part of the working section, (3) surface of the second bottom made of organic glass, and (4) the working section of the hydrodynamic stand.

mechanotron or sensitive strain gage (6). Thus on the elastic plate, the kinematic characteristics were simultaneously measured using thermo-anemometer and integral characteristics using a strain gage device. On the same elastic plates, M.V. Kanarsky, together with V.V. Babenko, carried out experimental investigations of the characteristics of a turbulent boundary layer on the same strain gage device in a wind tunnel, and V.I. Korobov, together with V.V. Babenko investigated integral characteristics in high-speed hydrodynamic channel [139].

Integral measurements on a strain gage device in various experimental installations were obtained using strain gauges glued on four elastic plates (11) of a strain gage. The width of the gap A the butt-end surface of the window and the front edge of the insert was 0.05–0.08 mm, the width of the gap C—0.1–0.15 mm. In the region of the strain gage panel, the kinematic and integral characteristics of the boundary layer over various elastic plates were investigated simultaneously for different flow regimes.

In connection with the aforementioned justifications, various designs of elastic plates have been developed, the structure of which is shown in Fig. 5.7. All elastic

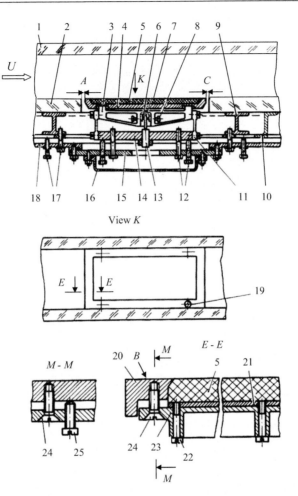

Figure 5.6 Scheme of placement of the strain gage on the hydrodynamic stand: (1) working area cover, (2) lower part (second bottom) of the working section, (3) two strain gage brackets (see (1) in Figs. 5.4 and 5.5) panel for mounting rigid and elastic surfaces, (5) elastic plate, (6) strain gage, (7) movable pin of mehanotron, (8) adjusting screws, (9) I-shaped profile of the tensometric section of the second bottom, (10) cross-section of the longitudinal beam of the second bottom, (11) four elastic plates of the strain gage, (12) four pairs of adjustment screws, (13) mechatronic corps, (14) the main plate of the strain gage, (15) the hatch cover of the strain gage insert, (16) the cap to ensure hermetic, (17) stop-spacer adjustment screws of the second bottom, (18) the first bottom of the working section, (19) screws for mounting the brackets (3) to the panel (4), (20) the pressure frame of the elastic surface (see 5 in Figs. 5.4), (21) is the substrate for gluing the elastic surface, (22) are screws for fastening the substrate to the panel, 23 is the panel, (24) is the clamping thrust screws, and (25) is the spacer screws of the adjustment pairs [139,356].

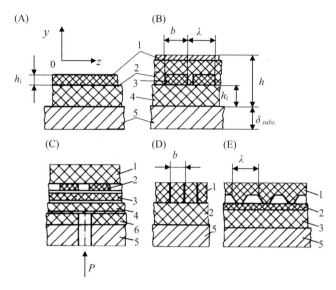

Figure 5.7 Schemes of composite elastic plates cross-sections: (1)–(4) and (6) elastomer layers; (5) metallic plate.

plates were glued to a thin duralumin plate (5). For some plates, layer (1) was made of a thin rubber film (Fig. 5.7A, B), layers or longitudinal strips (3) were made of electrically insulated conductive graphite fabric. In the plates in Fig. 5.7C, layer (2) was made from longitudinal elastic bands, and in variant (D), the external layer (1) was made of thin longitudinal glued together elastic material bands.

In plates of option (E), the outer layer (1) is made in the form of "inverted riblets" glued to layer (2). In this case, the layer of inverted riblets was installed in some coatings along the stream, and in others, across the stream. At low flow velocities, one or two outer layers interacted with the flow. At high velocities, the underlying layers also interacted with the flow. In addition, the lower layers played an important role as absorbers to reduce vibrations affecting the slab from below and caused by the vibration of the unit or base in the laboratory. The middle layers also absorbed various external disturbances of great intensity.

In the investigation of hydrodynamic stability in accordance with Figs. 5.1 and 5.2, various versions of elastic coatings were made, the geometrical parameters and mechanical characteristics of which are given in Tables 5.1–5.3. The data in the "measurement location" columns was defined as follows. As a result of experimental investigations of hydrodynamic stability on various types of flexible coatings, neutral curves were constructed. For each type of coating along these curves, the Reynolds number of loss stability was determined (interpolation ally for some cases). Then, Reynolds numbers and boundary layer thicknesses were calculated. The coordinates x_1 were determined from the results of calculations graphically by the boundary layer thickness from the

Table 5.1 The measurement location and dimensionless parameters of the elastic plates.

No	$x_1 \cdot 10^2$ m,	$U_\infty \cdot 10^2$ m/s,	$\delta \cdot 10^2$ m,	Measurement location Re_{δ}^*	Re_δ	$\alpha\delta^*$	$\alpha\delta$	$Ca \cdot 10^3$	K_E	$K_T \cdot 10^{-5}$
1	7.8	11.0	0.44	160	480	0.48	1.44	0.314	145	7.16
	14.8	11.0	0.61	220	660	0.42	1.26	0.314	145	7.16
	2.0	9.1	0.266	80	240	0.4	1.2	0.216	225	8.57
2	48.0	11.0	0.99	360	1080	0.4	1.2	0.389	117	3.26
3	52.0	11.7	1.06	410	1230	0.45	1.35	0.595	73	1.69
	61.0	10.4	1.19	410	1230	0.35	1.05	0.47	104	1.69
	31.5	7.8	1.05	270	810	0.59	1.77	0.265	245	2.23
4	67.6	13.0	1.19	515	1545	0.42	1.26	0.625	62	1.9
	166.0	13.0	1.87	800	2400	0.41	1.23	0.625	62	1.9
5	66.0	13.6	1.15	515	1545	0.42	1.26	0.89	48	1.1
	124.0	13.9	1.55	710	2130	0.39	1.17	0.93	39	1.07
6	45.6	13.0	0.98	420	1260	0.4	1.2	0.338	1.15	
									1.22	
	45.6	13.0	0.98	420	1260	0.4	1.2	0.483	0.8	1.22
7	116.0	13.4	1.54	680	2040	0.45	1.35	0.36	1.14	1.18
	116.0	13.4	1.54	680	2040	0.45	1.35	0.5	0.82	1.18
8	43.6	13.6	0.935	420	1260	0.45	1.35	0.272	1.36	1.61
	45.6	13.0	0.98	420	1260	0.45	1.35	0.445	0.46	1.67
9	101.0	12.0	1.52	600	1800	0.5	1.5	2.32	0.036	0
	168.0	14.9	1.75	860	2580	0.58	1.74	3.6	0.018	0
10	46.0	11.7	1.04	400	1200	0.5	1.5	0.195	0.44	0
11	52.0	11.7	1.06	410	1230	0.45	1.35	1.37	0.063	0
	55.0	9.8	1.24	400	1200	0.45	1.35	9.6	0.106	0

Table 5.2 Mechanical characteristics of elastic plates.

No	Measurement series	Elasticity modulus $E_{Sh.c.units}$	Elasticity modulus $E \cdot 10^{-4}$ N/m²	Tension T N/m	Dencity ρ kg/m³	Tickness $t' \cdot 10^{-4}$ N m	Oscillation mass M, kg/m
1	B1–B5	9.5	1.92	79	950	1	0.095
	B6–B10	9.5	1.92	79	950	1	0.095
	B11–B15	9.5	1.92	79	950	1	0.095
2	B16–B20	8.0	1.55	36	950	1	0.095
3	B21–B25	5	1.15	17.5	950	1	0.095
	B26–B34	5	1.45	17.5	950	1	0.095
	B27–B35	5	1.15	17.5	950	1	0.095
4	B36–B42, Measurement location	7.0	1.35	25	950	1	0.095
5	Increased stability BT43–BT44, Measurement location	7.0	1.35	25	950	1	0.095
		4.0	1.04	15	950	1	0.095
6	Increased stability P1–P4 along the lanes foam rubber	4.0	1.04	15	950	1	0.095
		12.5	2.5	–	950	100	0.095
7	over the membrane P5–P8 along the lanes foam rubber	8.0	1.75	16	950	1	0.095
		12.5	2.5	–	950	100	0.095
8	over the membrane P9–P12 along the lanes foam rubber	8.0	1.8	16	950	1	0.095
		15	3.4	–	950	100	0.095
	P13–P16 over the membrane	10	1.9	22	950	1	0.095
9	P17–P	2	0.31	0	63	500	0.0063
	P21–P24	2	0.31	0	63	500	0.0063
10	P25–P28	13.5	3.5	0	950	501	0.095
11	P29–P33	3	0.5	0	60	310	0.006
	P34–P36	3	0.5	0	60	310	0.006

Table 5.3 Dimensionless mechanical and kinematic parameters of elastic plates.

No	Oscillating mass parameter K_M	Density coefficient $m \cdot 10^2$	Limit frequency parameter K_ω	Dimensionless frequency $\omega_0 \cdot 10^{-3}$	Damping parameter K_η	Damping factor d	Viscosity (damping) $\eta \cdot 10^{-3}$, N·s/m³	Wave velocity in the membrane C_m, m/s	Dimensionless velocity in the membrane $C_0 = C_m/U_\infty$	Dimensionless group speed $C_0 \cdot 10^{-3}$
1	10.4	2.16	3.74	1.8	0.95	38	4.17	28.8	26.2	1.28
	10.4	1.56	3.74	2.47	0.995	39	4.28	28.8	26.2	1.97
	8.57	3.58	5.46	1.31	0.885	43	3.91	28.8	316	1.13
2	10.4	0.96	3.36	3.63	1.0	35	3.85	19.5	177	3.02
3	11.0	0.896	2.58	3.18	1.0	28.3	3.31	13.6	124	2.34
	9.8	0.8	3.26	4.0	1.0	31.9	3.32	13.6	131	3.81
	7.36	0.905	5.57	4.52	0.99	41.2	3.22	13.6	174	2.58
4	12.2	0.8	2.25	3.44	1.0	27.4	3.56	16.2	125	2.72
	12.2	0.508	2.25	5.4	1.0	27.4	3.56	16.2	125	4.39
5	12.8	0.826	1.94	3.0	1.0	24.8	3.37	12.6	92.5	2.38
	13.1	0.613	1.73	3.68	1.01	22.7	3.15	12.6	90.6	3.12
6	12.2	0.972	0.306	0.386	0.96	3.76	0.48	13.0	100	0.336
	12.2	0.972	0.258	0.325	0.934	3.13	0.407	13.0	100	0.288
7	12.6	0.617	0.3	0.72	0.836	3.73	0.5	13.0	97	0.542
	12.6	0.617	0.255	0.614	0.833	3.22	0.43	13.0	97	0.464
8	12.8	1.02	0.326	0.412	0.94	4.18	0.57	15.2	112	0.324
	12.2	0.972	0.194	0.244	0.814	2.38	0.31	15.2	117	0.223
9	0.75	0.625	0.219	0.394	1.0	0.163	0.0195	0	0	0.262
	0.93	0.543	0.143	0.488	0.755	0.133	0.0197	0	0	0.281
10	11.0	0.914	0.2	0.24	1.0	2.2	0.258	0	0	0.16
11	0.7	0.896	0.3	0.369	1.65	0.21	0.0246	0	0	0.165
	0.58	0.766	0.447	0.535	2.0	0.26	0.0254	0	0	0.199

photographs of the velocity profiles. This takes into account the accumulation of the thickness of the boundary layer on the walls of the confuser of the hydrodynamic stand. Wave numbers α were measured directly during the experiments. A values α were larger than expected. This is explained by the fact that the maximum value of α is taken, which corresponds to the Reynolds number of loss stability, and that when flowing around flexible surfaces, the value of α is greater than that of flowing around hard surfaces. The remaining values in the table were determined in accordance with the measurement methods described in [378].

The content of measurements series of experiments in accordance with their sequential numbering in the table is as follows.

1. B1–B5: investigation of hydrodynamic stability on a simple membrane coating with a thickness h of the water layer under the membrane equal to 7 cm. B6–B10: the same measurements, but with $h = 1$ cm. B11–B15: the same measurements, as in the previous series, but with a reduced velocity of the main flow.
2. B16–B20: investigation of hydrodynamic stability on a simple membrane coating, when there is a layer of air 2 cm thick under the membrane, and a layer of water equal to 5 cm was under the layer of air.
3. B21–B25: investigation is the same as in the series B1–B5, but the tension of the membranes is reduced, B26–B34: investigation are the same as in the previous series, but with $h = 1$ cm; B27–B35: repetition of previous studies with reduced velocity.
4. B36–B42: investigations are the same as in the series B26–B34, but with an increase of the tension in the membrane.
5. BT43–BT44: repetition of the previous series of experiments, but water with elevated temperature is poured under the membrane.
6. P1–P4: investigations of hydrodynamic stability on a complex membrane coating, when polyurethane foam strips were shifted together and located along the longitudinal axis of symmetry of the coating.
7. P5–P8: repetition of the previous series of measurements, but the strips of polyurethane foam were moved apart.
8. P9–P12: the same investigations as in the previous series, but with increasing tension membranes. P13–P16: repetition of the previous series of experiments, but a disturbing motion of a nonsinusoidal character.
9. P17–P20: investigations of hydrodynamic stability on a viscoelastic coating made of polyurethane foam sheet with a nonsinusoidal disturbing movement. P21–P24: repetition of the previous series of measurements in a sinusoidal disturbing motion.
10. P25–P28: the same investigations as in the previous series, but polyurethane foam is covered plastic film without tension.
11. P29–P33: repetition of a series of measurements P21–P24 with a flow around a sheet of polyurethane foam having a smaller thickness and greater hardness, P34–P36: the same investigations as in the previous series, but with a decrease in the velocity of the main flow.

To carry out experimental investigations on the inset in the working area (Section 3.7, Fig. 3.41), V.V. Babenko designed and manufactured various versions of passive and adjustable elastic plates, the geometrical parameters of which, in accordance with Fig. 5.7, are given in Tables 5.4 and 5.5, in which geometric parameters of elastic plates,

Table 5.4 Geometric parameters of elastic plates of group I.

Plate number	Dimensionless thickness of elastic plates ($10^2\, h/l$)	Quantity of layers	Layer number	Dimensionless thickness of i-layer ($10^3\, h_i/l$)	Material of i-layer	Plate design features, notes
1, 1a	2	2	1	0.12	RF	Solid sheet of rubber film
			2	19.9	FP-1	Sheet of polyurethane foam in Fig. 5.7A
2, 2a	2	2	1	0.12	RF	SAA
			2	19.9	FP-2	SAA, Rigidity above
3, 3a	1.8	4	1	0.12	RF	SAA
			2	6	FP-2	SAA
			3	2	CCF	LS with width $b = 2h$ and pitch $\lambda = 2.5h$; is pasted with RF
			4	12	FP-2	CC according to Fig. 5.7B
4, 4a	1.8	4	1	0.12	RF	CC
			2	12	FP-2	SAA
			3	2	CCF	LS $b = 3.7\, h$; $\lambda = 4.5\, h$; is pasted w/RF
			4	6	FP-2	CC, SAA as plate 3, 3a
5, 5a	1.2	3	2	6	FP-2	CC pasted w/RF
			3	2	CCF	CC according to Fig. 5.7B
			4	4	FP-2	CC
6, 6a	1.6	3	2	7	FE-1	LS $b = 2\, h$; $\lambda = 2.5\, h$; is pasted w/ RF
			3	2	CCF	CC according to Fig. 5.7B
			4	7	FE-1	

(*Continued*)

Table 5.4 (Continued)

Plate number	Dimensionless thickness of elastic plates ($10^2\,h/l$)	Quantity of layers	Layer number	Dimensionless thickness of i-layer ($10^3\,h_i/l$)	Material of i-layer	Plate design features, notes
7,7a	1.6	3	2	7	FL-1	CC
			3	2	CCF	CC pasted w/RF
			4	7	FL-1	CC according to Fig. 5.7B
8, 8a	1.8	4	1	6	F-2	CC
			2	2	FP-2	$b = 1.2h; \lambda = 2h$
			4	8	CCF	CC pasted w/RF
			6	8	FL-1	CC according to Fig. 5.7B
9,9a	2.6	5	1	10	FE-2	CC
			2	4	FP-2	$b = 1.2h; \lambda = 2h$
			3	2	FP-2	CC
			4	2	CCF	CC; layer 3, 4 are pasted w/RF
10, 10a	10, 10a	1.2	5	8	FL-1	CC according to Fig. 5.7B
			2	1	8	FE-1 For plate 10 $b = 0.85\,h$; for 10a: CC according to Fig. 5.7B
11, 11a	1.2	3	2	4	FE-1 CC	
			1	1	Rough film CC, roughness in one direction	
			2	5	FP-2	CC
			4	6	FP-2	CC according to Fig. 5.7B

respectively, of group I (research in a water flow) and group II (research in a wind tunnel). Part of the plates of the second group was also tested in the water flow.

In Tables 5.4 and 5.5, the notation is taken:

- RF: rubber film.
- FP: foam polyurethane.
- CCF: synthetic fabric (carbonaceous conductive).
- FE: foam elast.
- FL: foam latex.
- PU: polyurethane.
- CC: continuous coating.
- SAA: same as above.

In the column "Design features" Table 5.4 for elastic plates of group I below are the following clarifications:

- 1, 1a – layer 1 – rubber film with a thickness (on all plates) – 0.006 mm (Fig. 5.7A), layer 2 – a sheet of old foam polyurethane with a thickness of 10 mm.
- 2, 2a – layer 1 – rubber film (Fig. 5.7A), layer 2 – sheet of new FP with a thickness of 10 mm.
- 3, 3a – layer 1 – rubber film (Fig. 5.7B), layer 2 – sheet of new FP with a thickness of 3 mm and layer 4–6 mm, layer 3 – strips of conductive fabric with width $b = 20$ mm, and with a step $\lambda = 40$ mm, connected at the ends by transverse strips of the same material.
- 4, 4a – layer 1 – rubber film (Fig. 5.7B), layer 2 – sheet of new FP – 6 mm and layer 4–3 mm, layer 3 – strips of conductive fabric with a width of $b = 30$ mm, and with a step $\lambda = 50$ mm, connected at the ends by transverse strips of the same material.
- 5, 5a – layer 1 is missing (Fig. 5.7), layer 2 – sheet of new FP with thickness 3 mm, layer 4–2 mm, layer 3 – a continuous sheet of conductive fabric with a thickness of 1 mm.
- 6, 6a – layer 1 is missing (Fig. 5.7B), layer 2 and layer 4 – new material – FE blue 3 mm thick, 3 – the same as that of 3, 3a.
- 7, 7a – layer 1 is absent (Fig. 5.7B), layer 2 and layer 4 – a sheet of white foam latex 4 mm thickness, layer 3 – a continuous sheet of conductive fabric 1 mm thick.
- 8, 8a – layer 1 – a sheet of yellow foam latex 4 mm (Fig. 5.7C), layer 2 – strips of new foam polyurethane with a thickness of 0.8 mm, $b = 10$ mm, $\lambda = 15$ mm, layer 3 – continuous sheet of conductive fabric with a thickness of 1 mm, layer 4 is missing, layer 6 – sheet white foam latex 4 mm thick.
- 9, 9a – layer 1 – sheet of gray foam latex 5 mm thickness (Fig. 5.7C), layer 2 – strips of new foam polyurethane with a thickness of 0.8 mm, $b = 10$ mm, $\lambda = 15$ mm, layer 3 – continuous sheet of new foam polyurethane with a thickness of 1 mm, layer 4 – conductive fabric with a thickness of 1 mm, layer 6 – a sheet of white foam latex thickness of 4 mm.
- 10, 10a – layer 1 – glued longitudinal strips of blue foam elast thickness of 4 mm (Fig. 5.7D) layer 2 – a continuous sheet of new foam polyurethane thickness of 2 mm, for 10 – in $b = 10$ mm, for 10a – $b = 5$ mm.
- 11, 11a – layer 1 and layer 2 – a continuous sheet of new foam polyurethane with a thickness of 3 mm (Fig. 5.7D), a layer of shark skin is pasted on top layer 1.

In the column "Design features" Table 5.5 for elastic plates of group II the following clarifications are given:

- 1, 1a – layer 1 – a sheet of blue foam elast with a thickness of 10 mm (1) and 6 mm (11a) (Fig. 5.7A).

Table 5.5 Geometric parameters of elastic plates of group II.

Plate number	Dimensionless thickness of elastic plates ($10^2\,h/l$)	Quantity of layers	Layer number	Dimensionless thickness of i-layer layer ($10^3\,h_f/l$)	Material of i-layer	Plate design features, notes
1	2	1	2	20	FE-3	CC
1a	1.2	1	2	12	FE-3	CC according to Fig. 5.7A
2, 2a				Same as in Table 5.4		
3	2.2	3	1	8	PU-1	CC, bugles and hollows are oriented transverse to flow, $\lambda = 1.8 \cdot 10^{-3}$ m, according to Fig. 7Bb. Layer 2 and 3 the same as for plate 3 and 3a in Table 5.4
			2	2	CCF	
			3	3	12	FP-2
5, 5a				Same as in Table 5.4		
6, 6a	Same as in Table 5.4	The outer surface follows to the form of the strips of CCF				
7, 7a				Same as in Table 5.4		
8a	Same as 8; 8a in Table 5.4, but in place of two upper layers the PU-2 layer is glued on the CCF, dimensions of the layer are the same as by PU-1 of plate 3, bugles/hollows are transverse					
9	Same as 9 in Table 5.4, but in place of three upper layers the CC of FE-3 is glued, $h_f/l_E = 8 \cdot 10^{-3}$					
9a	Same as 9a in Table 5.4					
10, 10a				Same as in Table 5.4		
11	1.6	3	1	8	PU-1	CC, bugles-hollows are oriented along the flow, $\lambda = 1.7 \cdot 10^{-3}$ m, $b = \lambda = 0.21\,h$
			2	2	CCF	CC tested w/RF
			3	6	FL-1	CC according to Fig. 5.7D
11a	Same as plate 11, but the third layer — CC of FPPU-3 bulges/hollows are oriented along the flow					

- 3 — layer 1 — is absent (Fig. 5.7B), instead of layer 2 there is a sheet of polyurethane with an outer smooth surface, and the bottom surface of layer 2 has a corrugation, like "riblets" (layer 2 — inverted riblets), layer thickness 2—4 mm, the height of the riblets is 0.6 mm, the pitch of the riblets is 5 pieces per 9 mm ($\lambda = 1.8$ mm), layer 2 is set so that riblets are located across the stream.
- 8, 8a — instead of the top layer (yellow latex), soft polyurethane is glued — the same as for plate 3.
- 9 — the same as in Table 5.4; 9a — the same as for plate 7 in Table. 5.4, but the top layer is made of blue latex, and the bottom layer is made of white latex.
- 11 — (Table 5.4) layer 1 is made of solid polyurethane 4 mm thick with inverted riblets (riblets height -0.4 mm, riblets pitch $-\lambda = 2$ mm), layer 1 was set so that riblets were located along the stream; 11a is the same as the plate 11, but the top layer is made of medium-hard polyurethane.

All plates, with the exception of 3 and 8a in Table 5.5, have inner bands oriented along the stream. Each plate was made in two copies to conduct of an investigation of the integral characteristics in the high-speed hydrodynamic channel, indicated by the corresponding number and number with indices "a." The first group of elastic plates was investigated on a hydrodynamic stand of small turbulence [378] and in various installations in a water flow. Later on, these plates were modified (Table 5.5), which was investigated mainly in the air flow. Some plates of this group were also investigated in the water flow.

Figs 5.8–5.10 are photographs of some elastic plates made using special tools. Plate 1 in Fig. 5.8 is made according to Fig. 5.7C from latex—a material with an average density, a small number of closed pores and a smooth outer surface. A layer of white FE is glued onto the duralumin plate 5 (Fig. 5.7C), on which a thin rubber film with a slight preliminary stretch is glued on top. On top of this film is glued a continuous layer of CCF, on which a thin rubber film is also glued on top

Figure 5.8 Photograph of various types of elastic plates, the regulation of the mechanical characteristics of which is carried out with the help of: (1) pressure and heating of longitudinal conducting strips; (2), (3) heating of longitudinal conducting bands; (4) the outer layer is made of longitudinal strips of FE; (3) FE-3 elastic plate installed on the panel with edging.

Figure 5.9 Photograph of various types of elastic plates: (1) monolithic cylinder with longitudinal conductive wires, (2) surface with longitudinal slits for injection of polymer solutions into the boundary layer, (3) nonadjustable plate "10" according to Figs. 5.7D, (4) plate with conductive wires located on its surface, and (5) with narrow conductive ribbons on its surface.

Figure 5.10 Photograph of the inner surface of inverted riblets in accordance with the outer layer of Fig. 5.7E, on the left, the cross-section of the plate.

to seal the layer of CCF. Then pasted the longitudinal strips of the new foam polyurethane, on which a continuous layer of yellow PE is glued on top. The end parts of the CCF layer were clamped in thin brass transverse grids, to which conductive wires were soldered in the middle. Fig. 5.8 on plate 1 shows the outer continuous layer of submarines and dark stripes of a layer of electrically conductive CCF with a transverse bus (dark color). Above you can see the electrically conductive wire and the elastic tube, with the help of which in the cavity between the longitudinal strips 2 (Fig. 5.7) liquid is pumped at constant or variable pressure. The numbers 8 and 8a denote this plate in Table 5.4.

Plate 2, shown in Fig. 5.8, is made according to the scheme in Fig. 5.7B from foam polyurethane and is indicated in Table 5.4 with numbers 3 and 4 (in Fig. 5.8, the outer layer is removed and the CCF is visible in dark color through a rubber film. Plate 5.8 shows plate 3 in the form in which it is installed in experimental installations. The elastic plate 3 corresponds to the number 6 and 6a, and the plate 4 to the number 10 and 10a in Table 5.4.

Detailed experimental investigations of the static and dynamic mechanical characteristics of elastic material samples from which composite elastic plates are made, as well as the mechanical properties of the plates, the geometric parameters of which are given in Tables 5.1–5.5, were carried out in [60,139]. Tables 5.6 and 5.7 show some of the results obtained for elastic plates, the design of which is given in Tables 5.4 and 5.5.

Foam elast consists of the following components: polyvinyl chloride, plasticizer, butadiene—nitrile rubber, porofor, calcium stearate, sulfur, diphenylguanidine, polyethylene, carbon black.

Polyurethane consists of the following components: polyester, diisocyanate, a crosslinking agent, a plasticizer.

Table 5.6 Mechanical characteristics of elastic plates, presented in Table 5.4.

Plate number	Density ρ, kg/m^3	Quantity pore/cm	Static elasticity modulus 10^{-4} E_{st}, MPa	Dynamic elasticity modulus 10^{-4} E_{dyn}, MP	Anisotropy coefficient $k_1 = E_{dyn} / E_{st}$
1	40	14–16	0.47–0.8	–	–
1a	40	14–16	0.43–0.9	–	–
2	42	20–25	0.3–1.4	1	1
2a	42	20–25	0.3–1.16	1.23	1
3	42	20–25	0.17–0.2	–	–
3a	42	20–25	0.37–1.1	–	–
4	42	20–25	0.65–1.0	–	–
4a wet	42	20–25	0.5–1.0	–	–
5	42	20–25	0.3–0.5	0.37	1
5a	42	20–25	0.4–0.64	0.7	1
6	107	30–40	0.57–1.6	0.536	9
6a	107	30–40	0.55–0.87	0.384	7
7	200	40–60	0.26–0.4	–	–
7a	200	40–60	0.3–0.35	3.8	12
8	300	30–40	0.3–4.6	–	–
8a	300	30–40	0.4–0.8	–	–
9	173	45–55	0.28–0.33	13.2	27
9a	173	45–55	0.4–5.8	13.2	23–33
10	410	50–65	0.98–0.4	11.2–29	7.2–11
10a	410	50–65	0.98–0.4	11–33	7.2–11
11	42	40	0.27–0.89	–	–
11a	42	40	0.4–0.86	–	–

Table 5.7 Mechanical parameters of elastic plates, presented in Table 5.5.

Plate number	Density ρ, kg/m^3	Quantity pore/cm^2	Frequency f, Hz	Phase velocity longitudinal	C, m/s transverse
Standard Plate	1190	–	639	1230	1230
5	42	20–25	78	29	29
5a	42	20–25	99	43–48	40–47
2	42	20–25	81	47	53
2a	42	20–25	83	–	48
1	107	32	214	64.3	58.6
1a	115	36	246	66.6	48.6
6	107	30–40	260	70.7	67
6a	107	30–40	247	60.4	61
9	107	32	137–153	51.7	50.5
9a	173	45–55	75	87.4	40
7a	200	40–60	74	43.6	–
8	300	30–40	65	123	–
10	410	50–55	200	52.7	84.1
10a	410	50–55	231	52.7	90
11	1050	5–10	185	61	66
11a	1045	4–9	239	49.7	68.2
3	1050	4–7	169	40	23
8a	930	4–7	233	63.2	52.7

Plate number	Static modulus of elasticity 10^{-4} E_{st}, MPa	Dynamic modulus of elasticity		Anisotropy factor	
		10^{-4} $E_{long.}$, MPa	10^{-4} $E_{trans.}$, MPa	$k_1 = E_{long.}/E_{st}$	$k_2 = E_{long.}/E_{trans}$
standard plate	1.77	1.77	1.77	1	1
5	0.35	0.353	0.353	1	1
5a	0.53	0.7–1.0	0.7–0.9	1.3–1.9	1
2	0.3–1.4	0.9	1.15	1	0.8
2a	0.3–1.2	1.23	0.98	1	1.25
1	0.49	4.3	3.6	8.85	1.2
1a	0.49	5	2.67	10	1.87
6	0.59	0.536	0.48	0.9	1.1
6a	0.59	3.76	3.9	7	0.97
9	0.49	2.8	2.7	5.7	1.04
9a	0.39–5.7	13	2.7	2.3	4.8
7a	0.35	3.82	–	11	–
8	4.5	44	–	9.8	–
10	0.35	0.98–3.9	28.4	2.8–11	0.14

(*Continued*)

Table 5.7 (Continued)

Plate number	Static modulus of elasticity $10^{-4} E_{st}$, MPa	Dynamic modulus of elasticity		Anisotropy factor	
		10^{-4} $E_{long.}$, MPa	10^{-4} $E_{trans.}$, MPa	$k_1 = E_{long.}/E_{st}$	$k_2 = E_{long.}/E_{trans}$
10a	0.35	0.98–3.9	32.4	2.8–11	0.13
11	8.8	39	45.6	4.3	0.85
11a	5.9	25.3	47.6	4.3	0.53
3	3.7	15.7	5.5	4.1	2.9
8a	7.85	36.5	25.3	4.65	1.45

5.2 Experimental investigation of coherent vortex structures in the transition boundary layer of elastic plates

Regardless of the type of disturbing motion, with a favorable selection of mechanical properties of the damping surfaces, one of the following effects was observed during the experiments:

- Increased Reynolds number of loss of stability.
- The transition zone and the Reynolds transition number increased.
- The frequency ranges of unstable oscillations of the laminar boundary layer decreased.
- The phase velocity of the disturbing motion decreased.
- The wave number increased and the growth factors decreased.
- The coordinate of the critical layer decreased; the kinetic energy of the disturbing motion decreased [378].
- Elastic plates contribute to the formation of longitudinal vortices in the boundary layer.
- Flow structure and disturbance behavior in the boundary layer change.
- The characteristics of the boundary layer, considered as a waveguide, can be controlled by changing the structure and properties of the elastic surface.

Several of these effects were observed at the same time, as they appear interrelated.

The elastic surface, made in the form of longitudinal strips having different mechanical characteristics in the transverse direction, contributes to the formation of longitudinal vortex structures in the boundary layer.

Investigations have shown that the similarity criteria used (Part I, Experimental Hydrodynamics of Fast-Floating Aquatic Animals, section 1.4, table 1.5.) correctly simulate the features of the flow over the elastic surfaces under consideration. For more complex structures of elastic plates, it is necessary to develop more complex equations of motion of such plates and, accordingly, to clarify the similarity criteria.

Considering the aforementioned, the integral result of the flow interaction with the damping surfaces can be represented as the dependence of the plate resistance coefficient on the Reynolds number (Fig. 5.11). Solid lines (1)–(3) indicate curves for a laminar, transitional and turbulent flow around a rigid flat plate. Points A, B, and C correspond to the characteristic Reynolds numbers in the flow past a rigid plate. Dashed lines show patterns resistance to damping surfaces. Reynolds number loss of stability remains almost constant during flow hard plate. Experimental investigations of the hydrodynamic stability of the flow around various types of elastic plates showed that $Re_{st.l.}$ this varies widely depending on the mechanical characteristics of the material of elastic plates and their design.

Investigations on simple membrane surfaces have shown that, with nonoptimal mechanical properties of surfaces, the hydrodynamic stability of the flow along them deteriorates. If you interpolate the results obtained in experiments BI–B5, then the Reynolds number of stability losses will be $Re_{st.l.} = 1.35 \cdot 10^4$. Point loss of stability in Fig. 5.11, under these conditions, it is designated by the letter D. With respect to point A, when flowing around a rigid surface, the number $Re_{st.l.}$ has decreased four times. Suppose that the deterioration of hydrodynamic stability leads to a decrease in the transition zone and Re_{cr1} approaches $Re_{st.l.}$. Indeed, oscillations

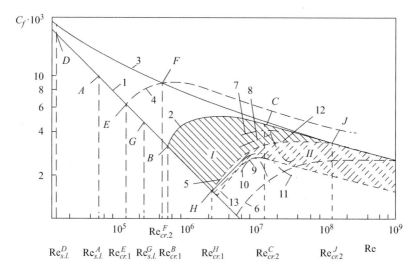

Figure 5.11 The dependence of the resistance coefficient of a longitudinally streamlined smooth plate on the Reynolds number: (1), (2), and (3) is the same as in Fig. 3.6; (4) the law of resistance of a flexible surface destabilizing the boundary layer; (5) the law of resistance of a flexible surface, which tightens the transition of the laminar boundary layer into a turbulent one; (6) the optimal law of resistance of the flexible surface when the transition is tightened [Kramer] tests: (7) is a rigid standard, (8) is a damping surface at $E' = 1.63 \cdot 10^8$ N/m^3, $d = 57\%$ and $\eta' = 10$–400 cSt, (9) $E' = 4.34 \cdot 10^8$ N/m^3, $d = 44\%$, $\eta' = 1200$ cSt; (10) $E' = 2.17 \cdot 10^8$ N/m^3, $d = 47\%$, $\eta' = 300$ cSt; (11), (12) laws of resistance when testing a pop-up model; (13) the law of resistance of the flexible surface, reducing turbulent friction.

occur in the damping surface destabilizing the boundary layer, and the oscillation amplitude of the damping surface may exceed even the allowable roughness height.

Thus the effect on the flow of a destabilizing surface may be similar to that of a rough or wavy rigid surface. We assume that the stages of the transition I–V (Section 3.6, Fig. 3.33) during a flow around the destabilizing surface will decrease and the point B corresponding to $\text{Re}_{\text{crl}}^{B}$ the rigid surface will move upstream to point E by an amount proportional to the ratio $\text{Re}_{\text{s.l.}}^{A}/\text{Re}_{\text{s.l.}}^{D}$. Then the critical Reynolds number for a flow around a destabilizing flexible surface will be $\text{Re}_{\text{crl}}^{E} = 1.25 \cdot 10^5$. With strongly destabilizing properties of a flexible surface, the resistance curve from point D to point E will shift above the solid line due to the fact that the pressure and velocity pulsations in the laminar boundary layer will be larger than when the rigid surface is flowed. The flow around such a surface from point E to point F corresponds to stage VI of the occurrence of turbulence (Section 3.6, Fig. 3.33), and its resistance [curve (4) Fig. 5.11] due to the aforementioned reasons will increase compared to the resistance of a rigid smooth plate: curves (2), (3). Depending on the degree of destabilization of the boundary layer by the flexible surface, the character of curve (4) may differ from that indicated in Fig. 5.11.

In a investigation on a damping surface with the best set of mechanical parameters for stabilizing the boundary layer, the Reynolds number of stability losses was $\text{Re}^* = 850$ and $\text{Re}_{st.l} = 2.4 \cdot 10^5$. The point corresponding to this Reynolds number is indicated in Fig. 5.11 the letter G. Due to the above effects, obtained in the investigation on flexible surfaces, we can assume that the point of stability losses will not only move downstream, but the entire transition region will increase. Suppose that when such a damping surface is flowed in comparison with a rigid surface, the stages I–V increase their length in proportion to the ratio $\frac{\text{Re}_{\text{s.l.}}^{G}}{\text{Re}_{\text{s.l.}}^{A}} = \frac{2.4 \cdot 10^5}{5.4 \cdot 10^4} = 4.45$. Then $\text{Re}_{\text{cr.1}}^{H}$ correspondingly it will increase by flowing around this flexible surface and will be $2.22 \cdot 10^6$.

We denote in Fig. 5.11 beginning of stage VI of the beginning of turbulence at the letter I flow around this damping surface, the end of this stage is letter J, and the resistance curve at this stage—numeral 5. The gain in reducing the resistance when such damping surfaces flow around compared to a hard surface can be estimated using area I, shaded in Fig. 5.11. The method of drag reducing by applying flexible surfaces is based on tightening the transition of a laminar boundary layer into a turbulent one and is called a method of distributed damping.

According to the results of the research, the coefficient of increasing the number of $\text{Re}_{\text{cr.1}}$, or the coefficient of laminarization, was 4.45. In Refs. [282,325] it is calculated that the use of flexible surfaces can increase this coefficient to 10. The resistance curve of the damping plate built in Fig. 5.11 with this coefficient is indicated by the number (6). It is clear from the calculations that this curve characterizes the maximum laminarization effect of the boundary layer due to the use of flexible surfaces. Curves (7)–(10) represent the results of the Kramer test [383,387,389], in which the resistance of the body of rotation of large elongation with various damping surfaces fixed on the cylindrical part was studied. The body of rotation was towed by a boat—a catamaran, between the hulls of which this model was installed. The best results were obtained when testing the surface indicated by curve (10). The friction resistance decreased by

59% compared with a rigid standard. This means that 83% of the length of the model was flown around the quasi-laminar flow.

The results of testing a floating rigid model of the body of rotation of the "Dolphin" model, which has a laminarized shape with contours along the profile of the NACA-66 series, are represented by curves (11), (12) [488]. This form allowed for a long distance, the body to maintain a laminar flow in the boundary layer. Curve (12) describes the test model at different speeds. It can be seen that this form of the body allows, at different speeds, to have a different length of the laminar boundary layer. Curve (11) is constructed from the results of model testing, taking into account that 60% of its surface is flowed around the laminar boundary layer at any speed.

Comparing the curves (7)–(10), (11), and (12), we can conclude that the damping surfaces reduce the hydrodynamic resistance not only by tightening the transition, but also by laminarizing the boundary layer. Curve (13) denotes the case if along curves (10) and (11) we draw an envelope and extrapolate it in both directions. This curve characterizes region II of hydrodynamic drag reduction and includes region I. The method of distributed damping virtually eliminates the mixed flow around the body if it is completely covered by a damping surface with properly selected mechanical characteristics. In this case, a significant transition to the initial part of the body occurs, and the flow structure at stages I–VI differs from the similar structure along a rigid surface. For values of x, corresponding to turbulent flow around a rigid surface, a boundary layer will be formed along the damping surface, its structure differs from the turbulent BL. Damping surfaces have optimal stabilizing properties in a limited range of flow rates. Therefore if the mechanical properties of the damping surface are constant, then as the numbers Re increase, the values of the body resistance coefficient will start to increase and can not only reach, but also exceed the values of curve (3). In order for the right branch of curve (13) to be parallel to curve (3), it is necessary with increasing Re adjust the mechanical properties of the damping surface so that its stabilizing properties are optimal all the time.

It can also be argued that the effectiveness of damping surfaces is determined not only by the correct selection of their mechanical properties, but also by the choice of an optimal design that takes into account the physical features of the flow near the damping surfaces. For this purpose, as well as in order to identify similarities and differences in the flow, at certain stages of the transition in the flow around rigid and damping surfaces, qualitative studies of the characteristics of the boundary layer in flow around various types of elastic surfaces were performed.

The development of the amplitude of the normal velocity v' along the working section of the hydrodynamic stand is presented in Fig. 5.12. Like experiments on a rigid plate (Section 3.3, Fig. 3.14), the visualization of a disturbing motion over the thickness of the boundary layer was carried out using three tellurium jets. The mechanical characteristics and the corresponding dimensionless coefficients of the elastic plate are presented in Section 5.1 in Tables 5.1–5.3 [378].

The disturbing motion was visualized in a stream over a simple membrane surface—experiments of the B1–B15 series (Section 5.1). Character of the disturbing movement in Fig. 5.12A, B was different due to the change in the velocity of the main flow. On these membrane surfaces, the zone of loss of stability was at

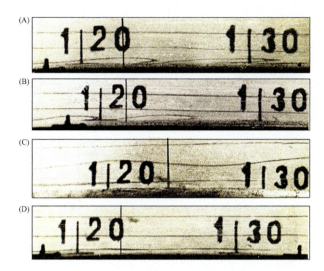

Figure 5.12 Photographs of velocity amplitudes v' when flowing around a membrane surface: oscillation frequencies near the second branch of the neutral curve: (A), (B), (D) under the membrane—water, (C) air; (A) experiments of the B3 series, $n = 1.1$ Hz, $U_\infty = 11$ cm/s, $E = 1.92 \cdot 10^4$ N/m^2, $T = 79.0$ N/m; (B) B8, 1.1 Hz, 12.4 cm/s, $1.92 \cdot 10^4$ N/m^2, 79.0 N/m; (C) B18, 0.5 Hz, 11.7 cm/s, $1.55 \cdot 10^4$ N/m^2, 36 N/m; (D) B38, 0.4 Hz, 13 cm/s, $1.35 \cdot 10^4$ N/m^2, 25 N/m.

a given x coordinate. At this point in x, all typical characteristics of the stability loss point were recorded [378]. However, compared with a rigid plate (Section 3.3, Fig. 3.14), the waveform and the size of its amplitude differed significantly: the waveform immediately became nonlinear, reached a maximum, but without folding the wave and the amplitude of oscillations significantly (3–4 times) decreased. In addition, disturbances in the flow over the membrane surfaces were observed only near the plate. It can be assumed that only the coordinate of the critical layer has changed. The only exception was the membrane surface with air located under the membrane (Fig. 5.12C). Although the tension of the film was reduced, the air under the film stretched the membrane. As a result, the disturbing motion was repelled from such a surface, and was recorded near the outer boundary of the boundary layer. When changing the tension of the membrane, the amplitude of the disturbing movement significantly decreased (Fig. 5.12D). When flow around such a surface, generated by the vibrator, almost damped at any x coordinate.

Fig. 5.13 shows graphic copies of photographs of three tellurium jets, characterizing the evolution of a disturbing motion along the boundary layer thickness as it flows around a rigid and simple membrane plate. Fig. 5.13A, B shows the development of a disturbing movement in the area of stability loss for a rigid plate: the oscillation amplitude of the oscillating plate for a rigid surface was 0.32 mm, the oscillation frequency $n = 0.74$ Hz, and for a membrane surface, 0.24 mm, $n = 0.98$ Hz. Disturbing motion in the flow past a rigid plate extends over the entire thickness of the boundary layer and is characterized by all the signs characteristic of the point of stability loss [378]. In the

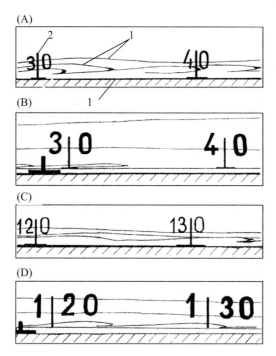

Figure 5.13 Development of disturbing movement over the thickness of the boundary layer in the flow around a rigid (A), (C) and simple membrane surface (B), (D): (1) tellurium jets, (2) distance markers along the working section, cm, (3) streamlined surface.

flow above the membrane plate, despite its shape of a neutral curve [378], which is significantly different from that of a rigid plate, the disturbing motion is concentrated near the streamlined surface. The shape, amplitude and type of the wave are different from those of a rigid plate and show that when the membrane plate is flowed oscillations become nonlinear. The disturbing movement around the damping surface is concentrated near the surface. The shape and type of the wave differ from similar waves when flowing around a hard surface. The tendency to twist the crest of the wave when flowing around the membrane surface has increased significantly. The amplitude of oscillations above the flexible surface was smaller when the characteristics of the disturbing motion corresponded to frequencies close to the first neutral frequency for the flow above a rigid surface.

Comparison of the disturbing motion in the flow around both surfaces in the area $x = 120$ cm showed that the wavelengths are longer on a rigid surface (Fig. 5.13C), the waveform is tilted, the wave crest is twisting characteristic of a rigid surface, and the disturbing motion is fixed to the entire thickness of the boundary layer. In the flow above the membrane surface (Fig. 5.13D), the wavelength λ increased compared to $x = 30$ cm, but decreased as compared to λ on a hard surface at $x = 120$ cm. The waveform also varied, but remained the same as at $x = 30$ cm. The disturbing motion was observed approximately in the same part of the boundary layer as in the flow along a rigid surface (Fig. 5.13C, D). The oscillation frequency of the wave was 0.68 Hz on a rigid and 1 Hz on a membrane surface. The disturbing motion in the boundary layer of a simple membrane surface becomes similar in shape to a disturbing motion when a rigid plate flows past a positive pressure gradient (Section 3.3, Figs. 3.15 and 3.17).

In the study of hydrodynamic stability on a solid surface, it was found that there is certain distance between the appearance of visible vibrations and the "folding" of a wave for each vibration frequency. This distance was conventionally called the stabilization zone. First, the stabilizing zone is reduced, and then increases with increasing frequency of oscillation. The stabilization zone is minimal near the second neutral oscillation and shifts against the flow toward the vibrator strip as the frequency increases. The principles of "folding" waves in a stream over membrane surfaces were essentially the same. The stabilization band characterizes the speed of growth of the disturbing movement. Thus the stabilizing strip on almost all of the tested damping coatings was smaller than when flowed around a hard surface. For example, the stabilization band was 20 cm for a rigid surface with $x = 70$ cm and an oscillation frequency of 0.74 Hz (the frequency is close to the second neutral oscillation)—the oscillations occurred at $x_{beginning} = 60$ cm and decayed at $x_{end} = 80$ cm, respectively. With a simple membrane surface flowed around the same x (B2 experiments), the stabilization band was 18 cm ($x_{beginning} = 67$ cm and $x_{end} = 85$ cm, respectively) with the oscillation frequency 0.76 Hz; 10 cm ($x_{beginning} = 70$ cm and $x_{end} = 80$ cm) at 0.96 Hz; 6 cm ($x_{beginning} = 72$ cm and $x_{end} = 78$ cm) at 1.09 Hz, close to the frequency of the second neutral oscillation. With a subsequent increase in the oscillation frequency, the stabilization band increased. On the stabilization band—see Section 3.8.

Fig. 5.14 shows photographs of the visualization of a disturbing motion in the flow past a simple membrane surface when hot water was poured under the membrane ($t = 32$ degree) and then the working section was filled with cold water

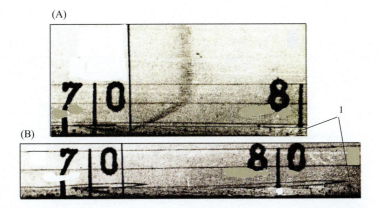

Figure 5.14 Photographs of the profiles of average velocity U (A) and amplitudes of velocity v' (B) in the flow around a simple membrane surface with heated water under the membrane: (A) BT44 experiment, $n = 0$ Hz, $U_\infty = 13.9$ cm/s, $E = 1.04 \cdot 10^4$ N/m^2, $T = 15.0$ N/m; (B) BT44 experiment, $n = 0.56$ z, $U_\infty = 13.9$ cm/s, $E = 1.04 \cdot 10^4$ N/m^2, $T = 15.0$ N/m; n is the second neutral oscillation; (1) the surface of the membrane coating.

($t = 7.3$ degree). The temperature of the water under the membrane and above it gradually leveled off. Due to the heating of the membrane, its mechanical characteristics have changed. As a result, the disturbing oscillations stabilized and concentrated in a very narrow region near the membrane surface. The waveform indicated nonlinear oscillations. Moreover, the stabilization band was not detected. After the formation of the sawmill wave, disturbances with this form of the wave moved downstream, without being deformed, as in a hardened form. The average velocity profile is typical for a boundary layer on a heated plate.

When swimming dolphins, there is a small temperature difference between the skin surface and water. The skin has specific structures that significantly reduce thermal radiation and temperature differences on the skin surfaces, and the mechanical characteristics of the skin are regulated in accordance with the coherent vortex structures of the boundary layer (Part I, Experimental Hydrodynamics of Fast-Floating Aquatic Animals, chapter 5). The character of the disturbing movement changed when the complex membrane surfaces flow around, while the characteristics of the disturbing movement when flowing around a rigid and simple membrane surface seem to be combined.

The design of membrane and viscoelastic surfaces is given in Section 5.1. A graphic copy of photographs of disturbing motion with a frequency of $n = 0.8$ Hz and a flow rate of 13.6 cm/s (experiments P11 in Tables 5.1–5.3) is shown in Fig. 5.15A, and a scheme of the behavior of the disturbing motion over these complex surfaces is shown in Fig. 5.15B. Three features of the behavior of the disturbing motion were found. The stabilization band was the same as for the flow around a hard surface. With increasing frequency of oscillation, the stabilization band, approaching to the vibrator, narrowed and then expanded. At the same time, unlike a rigid surface, a double folding of the wave was observed. Part of the crest of the wave "evolved," and the upper part, without collapsing, continued to move over it. Three features of the behavior of the disturbing motion were found. The stabilization band was the same as for the flow around a hard surface. With increasing frequency of oscillation, the stabilization band, approaching the vibrator, narrowed and then expanded. At the same time, unlike a rigid surface, a double folding of the wave was observed. Part of the wave crest "evolved," and the upper part, without

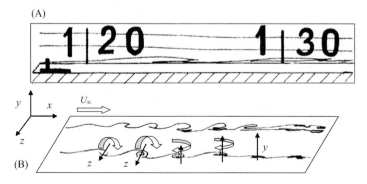

Figure 5.15 Graphic copies of a photograph of a disturbing motion (A) and a diagram of its development (B) when a complex membrane surface flows around.

collapsing, continued to move over it. The upper part of the wave crest was then also "evolved," and two stabilized ridges moved parallel to each other downstream.

The second feature was that with a further increase in the oscillation frequency, the disturbing motion did not immediately stabilize, but turned inwards so that if the axis around which the crest twisted was previously parallel to the streamlined surface of the stream, then downstream the axis became perpendicular to the surface. At such frequencies, the crest twisted so quickly, and the wavelength decreased so much that small vortices initially formed, parallel to the surface, which stabilized after turning their axis of rotation through 90 degree. In addition, as the disturbing movement went downstream, it appeared that the lower layer of fluid under the oscillating layer moved against the flow, it was so strongly inhibited, and above this layer the waves moved without increasing and without slowing down and stabilized significantly lower flow compared to a hard surface. Detailed results of investigations of the development of a disturbing motion in the flow around various types of elastic surfaces are presented in [378]. Membrane surfaces are not practical. But for such structures it is convenient to write the equation of motion and experimentally verify the value of certain mechanical characteristics. During the experiments, the thickness of the liquid under the membrane, its density, the tension of the membrane and the elastic characteristics of the membrane surfaces, the influence of nonharmonic perturbations and the temperature of the liquid under the membrane varied.

Of greatest practical interest are viscoelastic surfaces made of foam polyurethane (experiments P17−P36 in Tables 5.1−5.3). The experiments were carried out with a film placed on a sheet of polyurethane foam, without a film, as well as in the case of a dry or water-saturated layer of polyurethane foam with an outer film. Fig. 5.16

Figure 5.16 Photographs of a disturbing motion during the flow around viscoelastic surfaces: (A) experiment P23, $n = 0.83$ Hz, $U_\infty = 14.9$ cm/s, $E = 0.31 \cdot 10^4$ N/m^2, $T = 0$, thickness $t = 5$ cm; (B) P19, $n = 0.67$ Hz, $U_\infty = 12.4$ cm/s, $E = 0.31 \cdot 10^4$ N/m^2, $T = 0$, $t = 5$ cm (nonsinusoidal disturbances); (C) P27, $n = 0.83$ Hz, $U_\infty = 11.7$ cm/s, $E = 3.5 \cdot 10^4$ N/m^2, $T = 0$, $t = 5$ cm (with a membrane); (D) P35, $n = 0.83$ Hz, $U_\infty = 10.0$ cm/s, $E = 0.5 \cdot 10^4$ N/m^2, $T = 0$, $t = 3.1$ cm. The second neutral oscillation is: (A) $n = 0.83$ Hz, (B) 0.67 Hz, (C) − 0.67 Hz, (D) 0.46 Hz.

shows photographs of the disturbing motion with different variants of viscoelastic coatings. The character of the disturbing motion in the boundary layer of a viscoelastic surface also differed from the character of the flow around a rigid surface.

Fig. 5.17 shows graphic copies of a photograph of the behavior of a disturbing motion in a stream over a thick plate of polyurethane foam. Everywhere, the oscillation frequencies corresponded to the frequency of the second neutral oscillation. The oscillation amplitude is greatly reduced compared to the amplitude on a solid surface, the waveform is similar to that when a simple membrane surface flows around, but the bottom layer adjacent to this surface is strongly inhibited, as in the case of a complex membrane surface flowing (Fig. 5.15). The disturbing motion propagates in the immediate vicinity of the damping surface. Fig. 5.17A demonstrates the development of a disturbing movement in place the location of the point of stability loss over the surface of the foam polyurethane. The behavior of a nonsinusoidal disturbing movement in a stream above a sheet of foam polyurethane

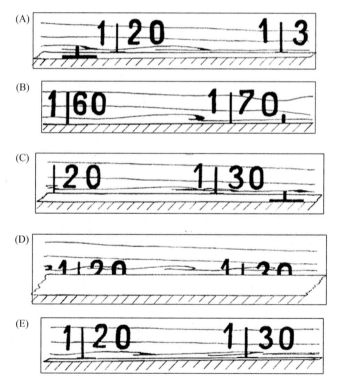

Figure 5.17 Graphic copies of a photograph of a disturbing motion in the flow around viscoelastic surfaces (FP): (A) experience P23, $n = 0.83$ Hz, $U_\infty = 14.9$ cm/s, $E = 0.31 \cdot 10^4$ N/m^2, $T = 0$, thickness $t = 5$ cm; (B) P24, 0.77 Hz, 14.9 cm/s, $0.31 \cdot 10^4$ N/m^2, 5 cm; (C) П19, 0.67 Hz, 12.4 cm/s, $0.31 \cdot 10^4$ N/m^2, 5 cm (nonsinusoidal disturbances); (D) P27, 1.4 Hz, 11.7 cm/s, $3.5 \cdot 10^4$ N/m^2, 5 cm (with a membrane); (E) P35, 1.0 Hz, 10 cm/s, $0.5 \cdot 10^4$ N/m^2, 3.1 cm.

(Figs. 5.16B and 5.17C) is the same as when a solid surface is flowed around it (Figs. 3.11 and 3.30), except that the wave crest is folded almost immediately vibrator without developing oscillations. Character folding waves is the same as on a rigid surface, and is similar to double folding waves when flowed around a complex damping surface.

The difference from double folding is that almost immediately after the vibrator, the initial oscillation in the xoy (in accordance with List of symbols) plane very quickly collapses with almost a turn into the xoz plane. And further in the xoy (in accordance with List of symbols) plane, all oscillations cease, and for some time until the oscillation in the xoz plane is fully stabilized, the oscillations continue. The behavior of the tellurium jet after double folding is easily seen when $x = 130$ cm, and the stratification jet is in the xoz plane (Fig. 5.17C).

The amplitude of oscillation of the vibrator tape with sinusoidal disturbances on viscoelastic surfaces was 0.24 mm, and with nonsinusoidal oscillations 0.3−0.4 mm. In the first series of measurements at $x = 30$ cm, the oscillations damped immediately behind the vibrator (experiment P17). In the second series, at $x = 80$ cm, the amplitude was increased to 0.7 mm (experiments P18). However, in this case too, the oscillations were sufficiently damped by the elastic coating. It was found that the fluid near the surface was severely braked at frequencies close to the frequency of the second neutral oscillation, and as it were, higher along y, it slipped over the bottom layer of liquid, so that it seemed that the boundary layer of liquid below moved against the flow. A similar pattern was observed when flowing around complex membrane surfaces. Subsequently, the oscillation amplitude of the vibrator tape was reduced to 0.4−0.5 mm, disturbing movement was concentrated directly near the surface.

The structure of the disturbing movement when the polyurethane foam surface flowed around the outside is coated with a film is shown in Fig. 5.17D. Although the character of the disturbing movement remained similar to the character of this movement in experiments on a surface not covered by the film, the frequency and amplitude of oscillations increased. The same was observed in experiments on a thin sheet of foam polyurethane, which has worse stabilizing properties than a thick sheet of foam polyurethane (Fig. 5.17E).

As in the case of a rigid surface flow around (Section 3.3, Figs. 3.8 and 3.16), measurements were made in III region of the natural transition with a flow around a simple membrane surface (Fig. 5.18). The speed of the main flow was 9 cm/s, the coordinates of the tellurium wire were $x = 96$ cm, $y = 3$ mm. It turned out that the deformation of the $U(z)$ profile in the flow above the membrane surface was about the same as when the rigid plate was flowed around, the tellurium wire $x = 180$ cm (Fig. 3.8). In both cases, tellurium wires were located from the surface in the region of the critical layer. From here, as well as from the shape of the velocity profile $U(z)$, it can be concluded that membrane oscillations occur in the flow above the surface, which lead to acceleration of the nonlinear deformation of the plane wave compared to a rigid plate. Vibrations are caused by vibrations of the vibrator tape. At the same time, compared with the flow around a rigid plate with a positive pressure gradient

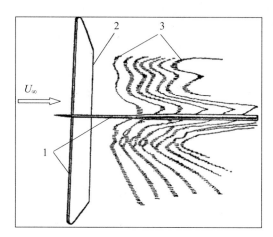

Figure 5.18 Graphic copies of the photograph of the velocity profiles in the flow around a simple membrane surface: (1) holder of tellurium wire, (2) tellurium wire, and (3) velocity profiles $U(z)$.

(Fig. 3.16), the deformation of the $U(z)$ profile in the flow above the membrane surface is much less. Görtler vortex wavelength significantly decreased in the flow above the elastic surface.

The features of the natural transition in the flow around the monolithic viscoelastic plate FE-3 were also investigated (Section 5.1, Fig. 5.8). Fig. 19 shows photographs of the velocity field $U(z)$ during a natural transition on an elastic plate made of FE-3 foam elast (mechanical characteristics are given in Tables 5.5 and 5.7).

On a rigid plate (Section 3.3, Fig. 3.9), the velocity profiles $U(z)$ are measured at different distances from the streamlined surface at $U_\infty = 6.7 \cdot 10^{-2}$ m/s and $x_{t.w.} = 1.1$ m—at the first stage of the transition. Already at the very beginning of the working section, the velocity profile $U(z)$ was formed practically over the entire thickness of the boundary layer, indicating the formation of large-scale longitudinal vortex structures.

Experimental studies of the velocity field in the flow past an elastic plate FE-3 were performed with $U_\infty = 5 \cdot 10^{-2}$ m/s and $x_{t.w.} = 2.2$ m—on the inset located at the end of the working section (Section 3.7, Fig. 3.41)—at the sixth stage of transition when a rigid plate flows around when complex vortex structures form in the boundary layer (Section 3.6, Fig. 3.33). Despite this, the flexible FE-3 plate effectively stabilized the disturbances of the boundary layer: the velocity profile $U(z)$ was plane-parallel across the entire thickness of the boundary layer (Fig. 5.19). The maximum bending of the tellurium cloud, recorded at $y = (0.01-0.05)$ δ, turned out to be insignificant compared to the flow past a rigid plate, for which the maximum velocity profile bend was at $y = 0.4$ δ. Speeds increase were less than when flowed around a rigid plate. When photographing three jets along the entire FE-3 plate, an exceptionally stable flow of them was recorded at a speed of 0.1 m/s and in the

Experimental investigations of the characteristics of the boundary layer on the analogs of the skin 271

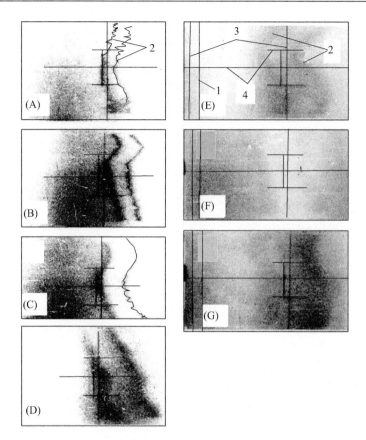

Figure 5.19 Photographs of the velocity profiles $U(z)$ at the natural transition of the boundary layer of an elastic plate FE-3 at $x_{t.w.} = 2.2$ m, $U_\infty = 5$ cm/s: (A) $y_{T.w.} = 2 \cdot 10^{-3}$ m; (B) $4 \cdot 10^{-3}$ m, (C) $6 \cdot 10^{-3}$ m; (D) $8 \cdot 10^{-3}$ m; (E) $1 \cdot 10^{-2}$ m, (F) $1.2 \cdot 10^{-2}$ m; (G) $1.6 \cdot 10^{-2}$ m [43].

case where the generators of longitudinal vortices GV3were installed (Section 4.1, Fig. 4.2), were fixed on a rigid plate with the same x nonlinear transition stages.

From this we can conclude that elastic plates with a large damping coefficient prevent rapid changes in the kinematic characteristics of the flow. This effect is similar to that found with a distributed impact on the boundary layer of a rigid ribbed surface (riblets). Thus reducing the gain of the waveguide due to the properties of its lower boundary, reduce velocity of propagation of disturbances in the waveguide (Section 3.8).

Thus taking into account the shape of tellurium jets, which characterize the development of disturbances in the flow of various elastic plates, we can conclude the following.

1. On all the plates, the disturbing motion immediately became nonlinear: the wave assumed a pointed spindle-shaped, but not sinusoidal form.

2. The disturbing motion, in contrast to the rigid plate, propagates in a very narrow conical layer with a small angle of inclination near the surface. In this case, the disturbances were concentrated along the longitudinal axis of this liquid layer, which is much closer to the wall than the critical layer in the flow past a rigid plate (standard), and below this layer the disturbances began to stabilize very quickly. The phase velocity in comparison with the standard also decreases: $c_r/u <1$.
3. If the mechanical parameters of the elastic plates differed by one or two orders of magnitude from the optimal ones, then the Reynolds number $Re^*_{st.l.}$ increased in comparison with the standard. In this case, the disturbing motion developed very slowly at the nonlinear stages of the transition. If, moreover, the parameters differed by 4 orders of magnitude, then $Re^*_{st.l.}$ became either smaller or comparable to the standard, and the disturbing motion developed very quickly to the VII stage of transition (fast alternation of the transition stages), after which the disturbances stabilized.

Experimental investigations of the flow around elastic plates were also performed by Carpenter [196–198,211].

5.3 Neutral curves of linear stability of a laminar boundary layer of elastic plates

Nonlinear Tollmien–Schlichting waves in a flow around an elastic surface were investigated in Ref. [573]. Experimental investigations of hydrodynamic stability during flow past various types of simple membrane surfaces at different velocities of the main flow have been performed [378]. At the same time, various parameters were varied: the thickness of the water layer under the membrane changed, the properties of the liquid under the membrane changed (there was air under the membrane), the membrane tension changed, the effect of combined properties on hydrodynamic stability was determined.

Below are the results of the investigation of hydrodynamic stability in the flow around various types of complex membrane surfaces [378]. Investigations on simple membrane surfaces have shown that when flown around them, the hydrodynamic stability substantially and approximately equally depends on many factors; it is very difficult to single out the advantage of any of them. Complex membrane surfaces are obtained with the complication of simple membrane surfaces (Section 5.1). The same membrane stretched on the same frame, but relied not on a layer of water or air, but on three longitudinal strips made of foam polyurethane.

In Ref. [388], the effect of elastic-damping surfaces on the reduction of friction resistance was investigated depending on the location of the strips or diaphragms on which the membrane rests with respect to the direction of the flow. The author of this work found the optimal mutual arrangement of the strips and indicated that the strips should be approximately at an angle of 15 degree to the direction of the main flow. Therefore it became necessary to check whether the mutual arrangement of the bands on which the membrane is supported affects the mechanical properties of the entire plate and whether the mechanical properties of the plate affect the hydrodynamic stability. The measurements were carried out in two cases: when the three strips supporting the membrane are shifted together in the region of the longitudinal axis of symmetry of the surface and when the strips are moved apart. In the latter case, one strip was located along the longitudinal axis of symmetry, and the other two—parallel to it at a distance of 7 cm from the axis in both directions.

Fig. 5.20A show the neutral curves plotted in the coordinates of the dimensionless frequency. The solid line indicates the results of measurements by the authors when a hard surface is flowed around it, the dashed and dotted lines indicate the first and second branches of a neutral curve when a complex membrane surface is flown around when the strips are shifted together, the dashed line is when the strips are apart. The line with two points indicates the flow around the same surface, but with increased membrane tension, and the line with crosses indicates the flow around with nonsinusoidal disturbances.

In accordance with the numbers of experiments, the characteristics of complex membrane surfaces are given in Section 5.1, Tables 5.1–5.3. The mechanical properties of the plates in both cases of the location of the strips do not differ, except for the damping properties. This apparently explains the fact that the frequency ranges of unstable oscillations (shaded under different slopes) differ from each other, and the maximum frequencies of the neutral oscillations of these areas differ little and remain smaller than in the case of a flow around a rigid wall. For various arrangements of the supporting strips of the membrane surface, velocity amplitudes v' were measured as a function of the frequency of oscillation. The amplitudes of the transverse velocities of the disturbing motion in the case when the strips are apart are smaller than when they are shifted together. Moreover, in the initial part of the plate with the striking apart strips, the amplitudes of the transverse velocities of the disturbing motion are so small that they can be neglected. In this case, the region of frequencies of unstable oscillations, limited by dashed line (3), would be closed in the region of Reynolds numbers of the order of 700. This means that the Reynolds number of instability would increase 1.7 times. Thus a change in the damping properties of the surface and the location of the supporting longitudinal strips leads to an increase in the stabilizing properties of the plate.

Fig. 5.20B, C presents the results of measuring the hydrodynamic stability in the coordinates of the dimensionless wave number and velocity of propagation of the disturbing motion. In these Figs., the dashed lines indicate the first and second branches of the neutral curves when a complex membrane surface flows around, when the longitudinal strips are shifted together, and the line with two points, when the strips are apart. Comparing both cases of the flow around a membrane surface, one can draw conclusions: the region of wave numbers of unstable oscillations is much smaller in the second case, especially if one considers that the amplitudes of oscillations in the initial part of the plate are very small and can be neglected. However, there is a region of Reynolds numbers, where in the first case the maximum values of wave numbers are greater. The growth coefficient c_i is proportional to the square of the

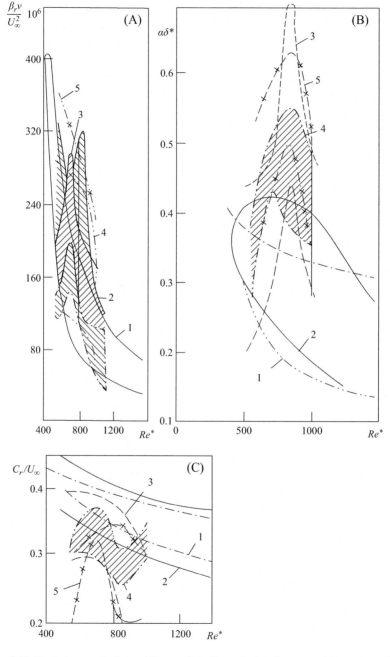

Figure 5.20 Neutral curves in the coordinates of the dimensionless frequency (A), wave number (B) and phase velocity (C) when a complex membrane surface flows around: (A): (1) measurements on hard surfaces, (2) measurements on membrane surfaces (experiments PI–P4); (3) experiments P5–P8, (4) experiments P9–P12, (5) experiments P13–P16; (B), (C): (1) measurements on rigid surfaces of Shubauer and Scramsted, (2) measurements of the authors, (3) measurements on damping surfaces (experiments PI–P4), (4) experiments P5–P8, (5) experiments P9–PI0.

wavelength of the disturbing motion λ. Therefore in those cases when the wave numbers $\alpha = 2\pi/\lambda$ are large, the growth in the disturbances occurs more slowly. When a complex membrane surface flow around with propped strips apart, the region of wave numbers of unstable oscillations (shaded in Fig. 5.20) is smaller than in the case of a flow around a rigid surface and is located in the region of large values of wave numbers. From the latter it follows that the magnitude of the growth of the disturbing motion in the flow around this type of membrane surfaces less. The same conclusions can be drawn from the analysis of the data on the propagation velocity of the disturbing motion in the flow around complex membrane surfaces. Also noteworthy are the appearance of neutral curves and the shape of the region of instability when flowing around membrane surfaces. This specificity compared with the case of a hard surface wrapping is due to the fact that at constant velocity the mechanical properties of the flexible surface are optimal for a certain coordinate value along the body.

The effect of membrane tension on the hydrodynamic stability of the flow past a plate with extended strips was also investigated. With an increase in the tension of the membrane in comparison with the previous measurements, the parameters characterizing the mechanical properties of the surface changed (experiments P9–P12 in Section 5.1, Tables 5.1–5.3).

The magnitude of the plate elasticity was measured in the area between the supporting strips and above them. In this case, the elastic parameter of the plate with the strips moved apart at low tension ($T_F = 16$ N/m, $k_T = 1.18 \cdot 10^5$) was $k_E = 1.05$, and at increased tension ($T_F = 22.0$ N/m, $k_T = 1.67 \cdot 10^5$)—$k_E = 1.56$, the modulus of elasticity also increased. Thus with an increase in the membrane tension, the elasticity of the plate and the parameter of the limiting frequency increased. From Fig. 5.20 it is clear that in this case the region of unstable oscillations increases and there is a tendency to decrease in the value of the Reynolds number of instability. The amplitude of the transverse velocities of the disturbing motion increases with increasing membrane tension.

On the other hand, the maximum frequencies of unstable oscillations generally decrease. The maximum values of the wave numbers increase, and the phase velocities decrease, which indicates an increase in hydrodynamic stability as compared with the cases of a flow around a rigid wall and simple membrane surfaces.

According to the results of measurements on a complex membrane surface (experiments P5–P8), a limit neutral curve was constructed, which limited the frequency range significantly exceeding the range of unstable extreme oscillation frequencies when flowing around simple membrane surfaces. Thus the maximum frequency value, limited by this curve, was $\frac{\beta_r \nu}{U_\infty^2} 10^6 = 1600$.

The analysis of the conducted research led to the following conclusions.

1. The stabilizing properties of complex membrane surfaces do not change so drastically depending on the mechanical properties, as in measurements performed on simple membrane surfaces.

2. The maximum frequencies of unstable oscillations when measured on complex membrane surfaces are smaller than in the case of a flow around a rigid surface.
3. In the region of small Reynolds numbers for a rigid wall (in the region of Reynolds number loss of stability) when flowing around complex membrane surfaces, the regions of maximum values of frequencies and phase velocities of the disturbing motion decrease, which is especially important, since this leads to an increase in the stability and length of the transition region boundary layer.
4. In the investigations of complex membrane surfaces, even in the case of an increase in the instability region, there are signs of a delay in the transition (an increase in the wave numbers and a decrease in the phase velocities of the disturbing motion).
5. A complex membrane surface stabilizes the laminar boundary layer better than a simple membrane surface.

In the investigations of viscoelastic surfaces, it was envisaged to assess the effect on hydrodynamic stability of the following parameters [29]: (1) mechanical properties of surfaces; (2) the thickness of the damping surface; (3) surface roughness. For the accepted measurement conditions (small values of velocities and Reynolds numbers), the mechanical properties of membrane surfaces were not optimal for stabilizing the boundary layer.

This, as well as design flaws, explains the results obtained when testing simple and complex membrane surfaces. The design and mechanical properties of surfaces made of polyurethane foam, had to improve their stabilizing properties.

The tests were carried out on a 5 cm thick polyurethane foam sheet. In order for polyurethane foam roughness not to affect the measurement results (since polyurethane foam was first tested without an outer film), a 25 cm long Plexiglas plate was installed behind it, and a flexible surface was mounted behind it. This made it possible to increase the thickness of the boundary layer to the beginning of the surface, as a result of which the roughness of polyurethane foam became permissible. Figs. 5.21 and 5.22 show the results of the study of hydrodynamic stability during the flow past a thick sheet of polyurethane foam. In Fig. 5.21, a dash-dotted line indicates a neutral curve when a polyurethane foam sheet is wrapped around it without a film, a line with crosses—in the case of wrapping around the same surface, but covered with a plastic film without tension; the dashed line is wrapping a sheet of polyurethane foam without a film during nonsinusoidal vibrations. Measurements of mechanical properties showed that the values of most of the parameters decreased significantly, although they did not reach the optimal values.

During the flow past a thick sheet of polyurethane foam without a film, the frequency range of unstable oscillations and their maximum values, as well as the range of unstable vibrations, limited by the limit neutral curve, decreased. The critical Reynolds number of instability increased in comparison with the case of a flow around a rigid wall 2.1 times.

If, at a rigid wall flow past, a point of stability loss was observed at the beginning of the working part, then when flow past a thick sheet of polyurethane foam, oscillations in the entire frequency range up to the middle of the working part were not observed, since they were completely damped near the vibrator strip. In the middle of the working part of a hydrodynamic stand, when testing on a rigid

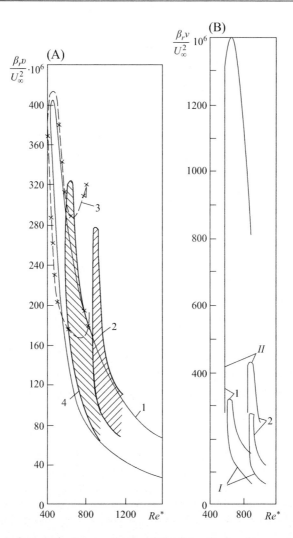

Figure 5.21 Neutral curves in the coordinates of the dimensionless frequency are ordinary (A) and limiting (B) when flowing around a viscoelastic surface. (A): (1) measurements on a hard surface, (2) measurements on a viscoelastic surface: experiments P21–P24, (3) P25–P26, (4) P17–P20; (B): ordinary (*I*) and limit (*II*) neutral curves: (1) P17–P20, (2) P21–P24.

surface, unstable frequencies are significantly lower than those recorded during tests with a polyurethane foam surface. The disturbing movement in this place has acquired all the characteristic signs corresponding to the point of loss of stability on a rigid surface (Chapter 3, Experimental Investigations of the Characteristics of the Boundary layer on a Smooth Rigid Plate, Section 3.4). Thus

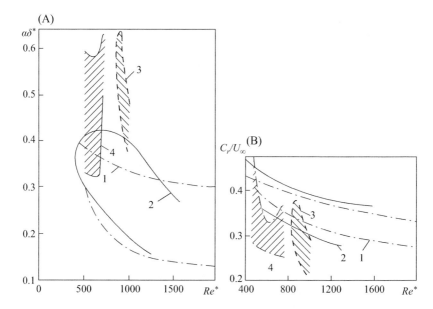

Figure 5.22 Neutral curves in the coordinates of the dimensionless wave number (A) and phase velocity (B) for a flow past a viscoelastic surface: measurements on a rigid surface: (1) Shubauer and Scrammsted [523], (2) Babenko [378]; measurements on the viscoelastic surface: (3) experiments P21−P24, and (4) experiments P25−P28.

for the first time, the phenomenon of a transfer of a loss of stability point was visually observed.

For the control test of the influence of the mechanical characteristics of the plate on the stabilizing properties, the polyurethane foam sheet was covered with a film used in testing membrane surfaces. The film was smoothed to avoid folds, but there was no tension in it. From Fig. 5.21, it can be seen that, in this case, the region of instability increased and assumed a specific shape. Hydrodynamic stability on such a plate has become worse than on a rigid wall, since the mechanical parameters have deteriorated sharply. The pressure and velocity pulsations are small. If the soft polyurethane foam effectively damped the oscillations of the boundary layer, then the film sharply increased the elasticity of the surface, and the positive effect of the interaction of the surface with the boundary layer was not fixed.

From Fig. 5.22A it follows that, although the values of $\alpha\delta^*$ are approximately the same in both cases, the areas of unstable oscillations differ sharply (in Fig. 5.22 these areas are shaded). The range of $\alpha\delta^*$ values when testing a polyurethane foam surface without a film is preferable. The speed of propagation of disturbances in the flow over polyurethane foam is significantly less (Fig. 5.22B), as well as the amplitudes of the velocity v', than in the flow over a hard surface. A number of researchers believe that the thickness of the flexible surface is important for stabilizing the boundary layer [325,382,383,387,389,471,546]. If we take into account previous studies and the role of the "substrate" in the technique of measuring polymers and rubbers, then the thickness of the flexible surface really matters in this sense. Thickness affects on rigidity, and, therefore on mechanical properties of a surface. In the following series of experiments, hydrodynamic stability was investigated on a flexible surface made of a thinner sheet of polyurethane foam (3 cm thick). This surface was harder than the previous one, and the values of its parameters were large.

As can be seen from Fig. 5.23, when flowing around a thinner sheet of polyurethane foam [curve (3)], which also has less favorable mechanical properties for stabilizing of the boundary layer, the region of unstable oscillations increased, and the Reynolds number of stability loss decreased in comparison with tests on a thicker sheet polyurethane foam [curve (2)]. As with tests on simple and complex membrane surfaces, the investigations were repeated at a lower value of velocity of the main flow [curve (4)]. In accordance with the similarity criteria, a decrease of the velocity has led to a deterioration of mechanical parameters. As was to be expected, hydrodynamic stability under these conditions deteriorated, but the amplitudes of the oscillations of the velocity v' were, nevertheless, rather small. The difference in the results of testing surfaces made of thick and thin

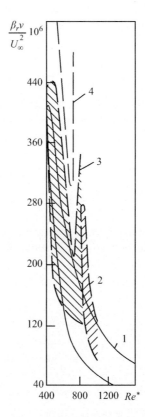

Figure 5.23 Neutral curves in the flow past viscoelastic surfaces: (1) measurements on a hard surface; measurements on viscoelastic surfaces: (2) experiments P21–P24; (3) experiments P29–P33, and (4) experiments P34–P36.

polyurethane foam is explained both by the difference in their mechanical properties and the effect of their thickness on the hydrodynamic stability. Consequently, the influence of elastic surfaces on hydrodynamic stability is also determined by their mechanical properties [20,22].

The results of experimental studies of hydrodynamic stability during nonsinusoidal perturbing motion and with heating of water under a membrane of simple membrane surfaces are also given in [378] (experiments BT43–BT44). The results of investigations performed on flexible surfaces indicate that, with the proper selection of mechanical properties and design of damping surfaces, it is possible to effectively influence the hydrodynamic stability boundary layer.

In Fig. 5.24, the results of measurements of the values of growth are graphically depicted in the form of dependences of the rate of increase on the wave number (see also Section 3.4, Fig. 3.24).

The shift of the curves characterizing the disturbing motion in the flow around the damping surfaces to the region of large values of $\alpha\delta^*$ compared to a rigid surface indicates a decrease in the value of λ, which was photographed during the experiments. Since the coefficients of increase c_i the velocities of the disturbing motion is proportional to the value of λ^2, the rate of growth decreases, and the hydrodynamic stability increases.

Not always a decrease in the value of λ is a sufficient condition for increasing stability. A decrease in λ at the same velocity amplitude v' means an increase in the oscillation wave steepness, and this can lead to the appearance of prerequisites for the premature occurrence of further transition stages in the boundary layer, or to the destruction of the wave crest and the appearance of secondary high-frequency oscillations. Therefore with an increase in $\alpha\delta^*$, a simultaneous decrease in the amplitude of the velocity v' is necessary, which was observed when tested on damping surfaces that have the best stabilizing properties. A decreases in the value of λ and, consequently, of the value of c_i indicates, moreover, the possibility of increasing the length of the transition zone of the laminar boundary layer to the turbulent one [22,309]. Using the value of c_i, it is quite simple to determine the Reynolds number of the transition with a small degree of turbulence of the main flow in the case of a flow around a rigid surface [378]. Applying the described method and knowing the characteristic values of the

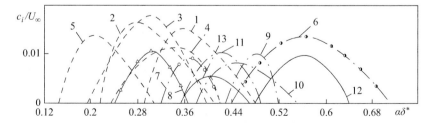

Figure 5.24 Growth rates in the flow around a rigid surface (dashed lines) with Re* equal to 676 (1), 693 (2), 786 (3), 865 (4), 1160 (5) and damping surfaces (solid lines) with Re*, equal to 710 (6), experiments B13, 690 (7), experiments B37, 484 (8), experiments B20, 700 (9), 828 (10), 1000 (11), 900 (12), and 1000 (13).

values of c_i for any damping surface, one can determine the corresponding Reynolds number. A decrease in the value of λ and, therefore the value of c_i indicates, moreover, the possibility of increasing the length of the transition zone of the laminar boundary layer to the turbulent one [22,309].

The analysis of the given data allows drawing the following conclusions. When flowing around all effective flexible surfaces, the rate of growth of the disturbing motion is less than when flowing around a rigid surface. For a flexible surface with better stabilizing properties, the value of c_i/U_∞ has decreased twice as compared to a rigid surface. When flowing around effective flexible surfaces, not only the Reynolds number of stability loss increases, but also the length of the transition region, as the oscillations increase more slowly. Unstable oscillations in the flow around flexible surfaces are shifted to the region of higher Reynolds numbers. The value of c_i depends on λ^2. Therefore if, when flowing around flexible surfaces, it is found that the value of $\alpha\delta^*$ increases, then this means that the magnitudes of growth increase accordingly.

5.4 Distribution of disturbing movements across the thickness of the laminar boundary layer of elastic plates

During the flow around various types of elastic surfaces, experimental investigations of the patterns of the distribution of characteristics and energy of the disturbing motion along the thickness of the boundary layer were carried out [29]. With the optimal location of the vibrator tape and its optimum amplitude, the amplitude distribution of the transverse pulsation velocity v' of the disturbing motion along the thickness of the boundary layer is investigated. Fig. 5.25A shows the results of investigations of hydrodynamic stability in the flow around a simple membrane surface, under which there was air, with $Re^* = 770$, $x_l = 168$ cm and $U_\infty = 11.7$ cm/s (experiments BI8, $E = 1.55 \cdot 10^4$ N/m², $T = 36$ N/m). Fig. 5.25B shows similar results for the flow around a simple membrane surface under which there was water, with $Re^* = 690$, $x_l = 122$ cm and $U_\infty = 13$ cm/s (experiments B38, $E = 1.35 \cdot 10^4$ N/m², $T = 25$ N/m). In Fig. 5.25C presents the results of measurements under the same conditions as in Fig. 5.25B, but the mechanical properties of the surface were different: $Re^* = 710$, $x_l = 163$ cm, $U_\infty = 10.4$ cm/s (experiments BI3, $E = 1.92 \cdot 10^4$ N/m², $T = 79$ N/m). In Fig. 5.25D presents the results of studies of the flow around a complex membrane surface at $Re^* = 700$, $x_l = 123$ cm, $U_\infty = 13.4$ cm/s (curve 1, experiments P6, $E = 2.5 \cdot 10^4$ N/m², $T = 16$ N/m), $Re^* = 826$, $x_l = 175$ cm (curve 2, experiments P7, $E = 1.8 \cdot 10^4$ N/m², $T = 16$ N/m), $Re^* = 1000$, $x_l = 225$ cm (curve 3, experiments P8, $E = 1.8 \cdot 10^4$ N/m², $T = 16$ N/m) and polyurethane foam plates at $U_\infty = 14.9$ cm/s, $Re^* = 900$, $x_l = 183$ cm (curve 4, experiments P23, $E = 0.31 \cdot 10^4$ N/m², $T = 0$). In Fig. 5.25, the vertical dashed line characterizes the thickness of the boundary layer $y = \delta^*$, and the x_l coordinate is determined taking into account the area of the confuser on which the boundary layer was formed.

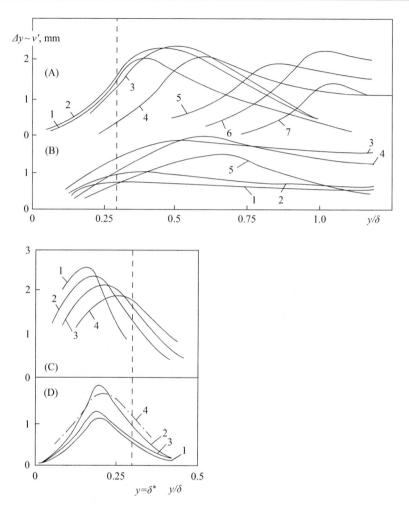

Figure 5.25 The distribution of the amplitudes of the velocity v' over the thickness of the boundary layer in the flow around a simple (A), (B), (C), a complex membrane and viscoelastic (D) surfaces at different values of the oscillation frequency. (A): (1) $(\beta_r \cdot \nu/U_\infty^2) \cdot 10^6 = 175$, (2) 259, (3) 332, (4) 387, (5) 526, (6) 680, (7) 920; (B): (1) 134, (2) 170, (3) 188, (4) 227, (5) 262; (C): (1) 475, (2) 680, (3)—875, (4) 1170.

Compared to the flow around a rigid plate, a number of features were discovered. One of them is that with an increase in the frequency of oscillations, the disturbing motion moves, as it were, to the outer boundary of the boundary layer, and at the same time the amplitudes of oscillations increase. Such a pattern was observed with poorly selected mechanical properties of simple membrane surfaces (Fig. 5.25A). On other membrane surfaces (Fig. 5.25B–D), the same picture was observed, which made it possible to conclude that upward oscillation movement with increasing oscillation frequency with increasing viscosity of the fluid under the membrane occurs more slowly.

In investigations of the patterns of distribution of disturbing motion along the thickness of the boundary layer holder with several tellurium wires allowed simultaneously to release tellurium jets at different distances from the bottom.

This made it possible to trace how oscillations develop at different coordinates $\bar{y}$ (Fig. 5.26). Solid lines show the flow around a simple membrane surface under which water was located (experiments B13), with $Re^* = 710$, $U_\infty = 10.4$ cm/s and $x_l = 163$ cm, and dash-dotted lines represent the case of a flow around a membrane surface under which there was air (experiments BI8), with $Re^* = 770$, $U_\infty = 11.7$ cm/s and $x_l = 168$ cm.

As can be seen from the Figure, for each coordinate in the boundary layer of the liquid when flowing around the membrane surface there is its second neutral oscillation. Moreover, in the direction to the outer boundary of the boundary layer, the frequency of the second neutral oscillation increases in the membrane, "resting" on water, more sharply than in the membrane, "resting" on the air. It should be noted that when a rigid surface was flowed around the entire thickness of the boundary layer, for each x there was only one value of the first and second neutral oscillations.

The data presented in Fig. 5.26 also confirm the results shown in Fig. 5.25A. Thus in Fig. 5.26 it is shown by solid lines that with an increase in the oscillation frequency to $n = 0.85$ Hz in a layer $\bar{y} = y/\delta = 0.16$ thick, the amplitude of the oscillations begins to increase. With further increase in the frequency, the oscillation amplitude in this layer decreases, but starting from $n = 1.5$ Hz, the oscillation amplitude in the layer $\bar{y} = 0.376$ begins to increase, and for $n \approx 2.25$ Hz, the oscillation amplitude in this layer becomes the same as in the layer $\bar{y} = 0.164$. With a further increase in the oscillation frequency, their amplitude in the layer $\bar{y} = 0.64$ continues to decrease and then becomes hardly noticeable, and in the layer $\bar{y} = 0.376$ increases to $n = 2.7$ Hz, after which the oscillation amplitude also begins to decrease in this layer.

During the flow around complex membrane surfaces, oscillations were observed only in the area not exceeding $\bar{y} = (0.3 - 0.4)$, and when flowing around the plate made of polyurethane foam, near its surface, in a layer of liquid not exceeding $\bar{y} = 0.2 - 0.3$ (Fig. 5.25D). The deterioration of the stabilizing properties of the

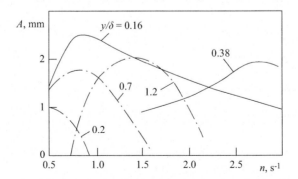

Figure 5.26 Dependences of the amplitude of velocity v' on the frequency of oscillation and the location of the place of fixation of oscillations along the thickness of the boundary layer.

damping surfaces leads to an increase in the relative thickness of the boundary layer in which a disturbing movement was observed, regardless of the device of the damping plate. The design affects only the range of variation of the magnitude $\bar{y}$ at which oscillations were observed.

The same conclusions can be made on the basis of the patterns of the distribution of the values of $\alpha\delta^*$ and c_r/U_∞ across the thickness of the boundary layer (Fig. 5.27). The flow over simple membrane surfaces is characterized by continuous (experiments B37), dash-dotted (experiments B13) and dashed (experiments B18) curves. Arabic numerals in Fig. 11.27 indicate oscillation frequencies in accordance with the numbers of experiments (Table 5.8).

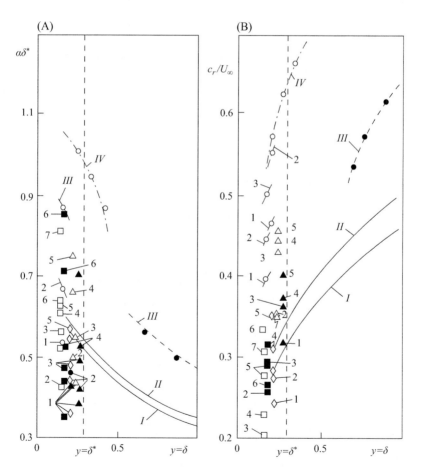

Figure 5.27 Distribution of the wave number (A) and phase velocity (B) over the thickness of the boundary layer when flowing around elastic surfaces: simple membrane—○—experiments B13, ●—experiments B18; complex membrane △—experiments P6, ◇—experiments P7, ▲—experiments P8; polyurethane—□—experiments P23—■—experiments P24; *I*, *II*—experiments B37, *III*—experiments B18, *IV*—experiments B13.

Table 5.8 Dimensionless oscillation frequencies along the thickness of the boundary layer elastic surfaces.

Experience number	Figures in Fig 5.27	Oscillation frequency $(\beta_r \cdot \nu/U_\infty^2) \cdot 10^6$	Experience number	Figures in Fig 5.27	Oscillation frequency $(\beta_r \cdot \nu/U_\infty^2) \cdot 10^6$
B37	1	27	P8	1	124
	2	262	2	150	
B13	1	300	3	168	
	2	420	4	199	
	3	615	P23	5	220
	4	876	6	286	
P18	1	80	2	117	
	1	259	3	127	
	2	332	4	158	
	3	387			
	5	196			
P6	1	220	6	238	
	2	254	7	286	
	3	340			
	4	442	P24	1	68
	5	492	2	114	
	3	140			
P7	1	110	4	168	
	2	152	5	204	
	3	168	6	227	
	4	210			
	5	248			

From Fig. 5.27 it can be seen that when the fluid flows over complex membrane and polyurethane foam surfaces, the values of $\alpha\delta^*$ and c_r/U_∞, in contrast to similar values, when a rigid plate is flown around (Section 3.5, Figs. 3.27 and 3.28) are practically independent of the thickness of the boundary layer. When flowing over simple membrane surfaces, when their mechanical properties were unfavorable for stabilization of the boundary layer, the values of $\alpha\delta^*$ and c_r/U_∞ were distributed over the thickness of the boundary layer approximately like the same as in the case of flowing a rigid plate. The difference was only in the nature of the laws and the absolute values of $\alpha\delta^*$ and c_r/U_∞.

During investigations on a hard surface, it was found that in each place along the working part, oscillations were observed in a strictly defined frequency range with the vibrator oscillation amplitude selected (Section 3.4, Fig. 3.22). These frequencies are plotted on the graphs depending on the wave number and phase velocity. From the limit values of these points, a curve is drawn (in Fig. 5.28, a dash-dotted curve) that bounds the area outside of which oscillations of any frequency in the boundary layer were not observed, and is called the limit neutral curve. Normal and limit neutral curves for damping surfaces in these coordinates were not constructed. The dashed lines in Fig. 5.28 show the dependences of $\alpha\delta^*$ and c_r/U_∞ on the oscillation frequency at different flow velocities of a rigid plate. Measured at the flow around elastic surfaces, the values of $\alpha\delta^*$ and c_r/U_∞ are plotted in Fig. 5.28 as points (Fig. 5.27).

A comparison of the data obtained for rigid and damping surfaces shows that in both cases between the wavelength and phase velocity, on the one hand, and the oscillation frequency of the disturbing motion, on the other hand, ambiguous patterns are observed, depending on the velocity of the main flow. In the case of a flow around damping surfaces, the value of $\alpha\delta^*$ generally increases, and c_r/U_∞ decreases as compared with the case of a flow around a rigid surface, that is, the oscillation spectrum in the boundary layer changes.

An increase in $\alpha\delta^*$ means a decrease in λ compared to a rigid plate. It is photographed in experiments. Since the coefficients of increase in the disturbing motion c_i are proportional to the value λ^2, the growth rate decreases, and the hydrodynamic stability of the flow increases. As it turned out, a decrease in λ is not a sufficient condition for increasing stability. It is significant that a decrease in λ at the same value of the velocity v' leads to an increase in the slope of the oscillation wave. This may lead to prerequisites for the premature development of further stages of the transition region of the boundary layer or to the destruction of the wave crest and the appearance of secondary oscillations of high-frequency character. Therefore simultaneously with increasing $\alpha\delta^*$, a decrease in the velocity amplitude v' is required. This ratio was observed on the best tested damping surfaces. A decrease in the value of λ and a decrease in the value of c_i additionally mean the possibility of increasing the length of the transition zone from the laminar boundary layer to the turbulent one. The same follows from Fig. 5.28, from which it can be seen that even with unstable oscillations, the values of c_r decrease.

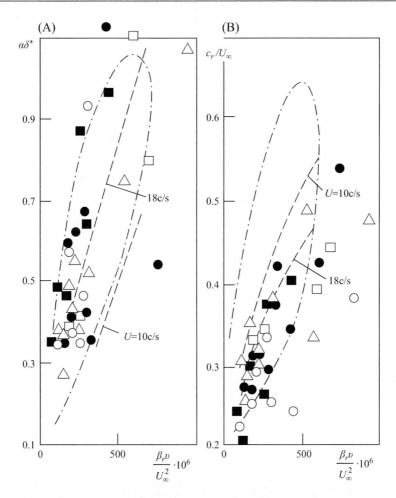

Figure 5.28 The dependence of the wave number (A) and phase velocity (B) on the frequency of disturbing oscillations in the flow around elastic surfaces. The designations are the same as in Fig. 5.27.

Direct measurements of the values of c_i were also carried out, and the dependences of the values of c_i on the wave number were constructed in the same way as in [247,378]. The corresponding graphs are given in Refs. [17,22,23,26,29,378]. The same physical features were revealed during the flow around a rigid plate. The difference was that c_i values were lower when all effective damping surfaces were flowed around. For the best tested damping surface, the c_i values are reduced by half. Using the value of c_i, the Reynolds number of the transition at low free-flow turbulence can be easily and reliably determined for the case of a flow around a rigid surface [378]. Using this technique and knowing the characteristic values of c_i for any damping surface, one can determine the corresponding Reynolds number of the transition. According to measurements on elastic plates, it was found that in the process of complex interactions the ratios of the characteristic thicknesses of the boundary layer change.

As in the case of flowing around a rigid surface, the kinetic energy of the disturbing motion was determined on the basis of the velocity distribution u' and v' over the thickness of the boundary layer (Fig. 5.29). If we compare the regularities of the velocity distribution u' over the thickness of the boundary layer for rigid [378] and damping surfaces (Fig. 5.29A), it is found that in the second case the maximum positive value of u' is smaller, the maximum negative value is larger and is somewhat closer to streamlined surface. In addition, the velocity u' in the region of negative values in the second case decays faster and closer to the outer boundary of the boundary layer.

Comparing the patterns of velocity distribution v' across the thickness of the boundary layer of rigid [378] and damping surfaces (Fig. 5.29B) shows that when flowing around damping surfaces, depending on their mechanical properties and construction, the velocity v' may be either greater or less than the corresponding values for flowed around a hard surface. In addition, when flow around damping surfaces, the maximum of v' is shifted along the thickness of the boundary layer, and its rate of decrease also changes.

It was not possible to perform a large number of simultaneous measurements of the velocities u' and v' on the damping surfaces in the same place. Therefore the

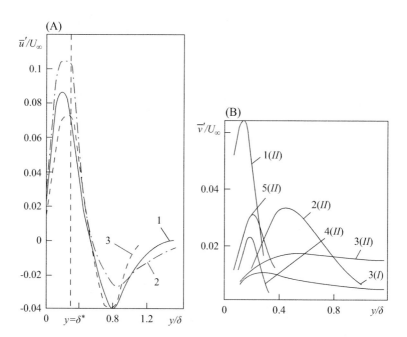

Figure 5.29 The distribution of the amplitudes of the velocities u' (A) and v' (B) on the thickness of the boundary layer when flowing around elastic surfaces: (A) (1)—Re* = 765, $(\beta_r \cdot \nu/U_\infty^2) \cdot 10^6 = 310$ (experiments B8), (2)—Re* = 795, $(\beta_r \cdot \nu/U_\infty^2) \cdot 10^6 = 230$ (experiments B3), (3)—Re* = 765, $(\beta_r \cdot \nu/U_\infty^2) \cdot 10^6 = 276$ (experiments of the P33); (B) (1)—Re* = 710, $(\beta_r \cdot \nu/U_\infty^2) \cdot 10^6 = 500$ (experiments B13), (2)—Re* = 770, $(\beta_r \cdot \nu/U_\infty^2) \cdot 10^6 = 246$ (experiments B18), (3)—Re* = 690, $(\beta_r \cdot \nu/U^2_\infty) \cdot 10^6 = 113$ (*I*) and 188 (*II*) (experiments B37), (4)—Re* = 700, $(\beta_r \cdot \nu/U_\infty^2) \cdot 10^6 = 294$ (experiments P6), (5)—Re* = 900, $(\beta_r \cdot \nu/U_\infty^2) \cdot 10^6 = 278$ (experiments P23); *I* and *II*—the first and second neutral oscillations.

constructed pattern of the distribution of the kinetic energy of a disturbing motion over the thickness of the boundary layer when flowing around damping surfaces is rather qualitative. This pattern is graphically presented in Fig. 5.30. Line (1) indicates the flow around a simple membrane surface, line (2) shows a membrane under which air was located, line (3) indicates flow past a sheet of polyurethane foam.

For a rigid surface, the maximum value of the kinetic energy was $2.4 \cdot 10^{-2}$, and its maximum was in the vicinity of $\bar{y} = 0.2$. When flowing around damping surfaces, the maximum value of the kinetic energy was within $(0.6-1.1) \cdot 10^{-2}$, and the coordinate of the maximum—at $(0.2 - 0.3)\bar{y}$. The largest values of these values corresponded to simple membrane surfaces, and the smallest—to a polyurethane foam plate.

In [519], the energy exchange between the main and disturbing motions was analyzed. It is argued that the energy of the disturbing motion changes due to the transfer of kinetic energy from the main motion to the disturbing one, as well as due to a change in pressure and dissipation. When a rigid surface is flowed around a wall, dissipation near the wall is maximum, while the critical layer ($y = \delta^*$) does not matter much for dissipation. The transition of energy from the main motion to the disturbing one, on the contrary, is maximum in the region of the critical layer, where the velocity components of the disturbing motion are maximum.

Thus when a rigid surface is flowed around, the energy of the disturbing motion strictly depends on the value of y/δ.

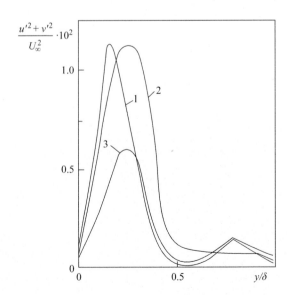

Figure 5.30 The distribution of the average kinetic energy of the disturbing motion along the thickness of the boundary layer when flowing around elastic surfaces for the second neutral oscillations: (1) Re* = 710, $(\beta_r \cdot \nu/U_\infty^2) \cdot 10^6 = 500$ (experiments BI3); (2) Re* = 770, $(\beta_r \cdot \nu/U_\infty^2) \cdot 10^6 = 246$ (experiments B18); (3) Re* = 900, $(\beta_r \cdot \nu/U_\infty^2) \cdot 10^6 = 278$ (experiments P23).

When flowing around damping surfaces, the mechanical properties and the design of the plate, which form its stabilizing properties, are decisive. As can be seen from the above facts, the maximum values of the pulsating velocities of the disturbing motion are determined by the indicated parameters and for different damping surfaces vary in thickness of the boundary layer.

With the proper selection of the mechanical properties of the damping surfaces, the liquid layer in which the kinetic energy is exchanged between the main and disturbing motions approaches the damping surface in which the energy of the disturbing motion dissipates. The energy balance is concentrated in almost one zone immediately near the surface of the damping plate.

Based on the above experimental results, we can assume that when the damping surfaces flow around, the ordinate of the critical layer changes. If we assume that this ordinate approximately coincides with the value of δ^*, as when a hard surface is flowed around it, the following conclusion can be made: when the stabilizing properties of the damping surfaces improve, the value of δ^* decreases, and when the stabilizing properties deteriorate, it increases.

It can also be assumed that when the damping surfaces flow around due to a change in the boundary conditions, the ratio between the characteristic values of the boundary layer changes.

To test this assumption, we measured the velocity profile, which revealed some difference between the Blasius profile and the resulting profiles. However, the accuracy of the measurements made does not allow us to consider this difference as a reliable confirmation of the hypothesis put forward.

As on a rigid plate, the wavelength λ_x of plane oscillations versus x_1 (Fig. 5.31) was measured at small ε in the flow around all types of simple and

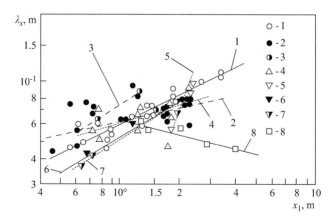

Figure 5.31 Change of the plane wavelength λ_x and longitudinal vortices λ_z along the plates: (1) rigid horizontal plate, (2) simple membrane surface, (3) the same as (2) with heating, (4) complex membrane surface, (5) polyurethane foam-1 (FP -1), open, (6) the same as (5), with the outer film, (7) the same as (6) with different mechanical properties of the polyurethane foam-2 (FP -2), (8) the regularity of λ_z in the flow past a rigid plate.

complex membrane, as well as viscoelastic plates. On simple and complex membrane plates [curves (2), (4)] patterns λ_x (x_1) differ from the standard (Section 3.6, Fig. 3.32). The large scatter of the points of curve (2) for small values of x_1 is due to membrane oscillations and the peculiarities of the development of a disturbing motion along the boundary layer thickness. The regularities λ_x (x_1) on viscoelastic plates made of polyurethane foam-1 and polyurethane foam-2 and covered outside with a membrane [curves (6), (7)] differ in their character. The heating membrane [curve (3)] has a dependency that is equidistant with it, and there is a completely different relationship for the FP-1 plate without film [curve (5)]. Thus each type of elastic plate is characterized by its own regularity in the evolution of a disturbing motion along x_1. In other words, it is shown that, due to a change in conditions at the lower boundary of the waveguide (boundary layer), the characteristics of the oscillations in the waveguide are different.

The conclusions from Fig. 5.31 become more obvious if the results are presented in a dimensionless form: $\overline{\lambda}_x = \lambda_x/\delta^*$.

The curves in Fig. 5.32 are plotted for second neutral oscillations. Although all the curves in Fig. 5.32A are smooth, it is clear that for each elastic plate there is its own type of pattern. The same curves as in Fig. 5.32A, presented depending on the Reynolds number Re* (Fig. 5.32B), have a certain minimum. Similar dependencies, but with a different characterization of dimensionlessness [54], do not possess such information ability. The values of the minima are noticeable, especially in Fig. 5.32C those constructed in normal coordinates. It is characteristic that on a rigid plate the extremum occurs at $1.35 \cdot 10^5$; for a simple membrane plate, it corresponds to a smaller Reynolds number (Re = 10^5) in the case of adverse mechanical characteristics and otherwise the same Reynolds number for a simple membrane plate or a larger Reynolds number (Re $\approx 2.35 \cdot 10^5$) for a complex membrane plate.

The patterns of development of values $\overline{\lambda}_x$ on the left and right of the extremum are different (Fig. 5.32). Since λ_x monotonously grows along x_1 (Fig. 5.31), this character of the curves in Fig. 5.32 is explained only by the fact that the growth of δ^* changes in the process of evolution of the boundary layer. In addition, this, in turn, is associated with the character of the flow at various stages of the transition. For example, the Reynolds number Re $\approx 1.35 \cdot 10^5$ for a rigid plate is in good agreement with the Reynolds number, at which the development pattern of λ_z changes (Fig. 5.31). Thus based on the shape of the curves in Fig. 5.32C, one can estimate the areas of development of the nonlinear stages of the transition.

In Section 3.6, the dependences λ_x and λ_z are presented as functions of the Reynolds number in the flow past a rigid plate. Empirical regularities $\overline{\lambda}_x$ (Re$_x$) in accordance with Fig. 5.32C have the following form [60]:

• for a rigid plate [curve (1)]

$$\overline{\lambda}_x = 7.19 \cdot 10^5 \text{Re}_x^{-1} + 3.56 \cdot 10^{-5} \text{Re}_x + 1.7 \tag{5.1}$$

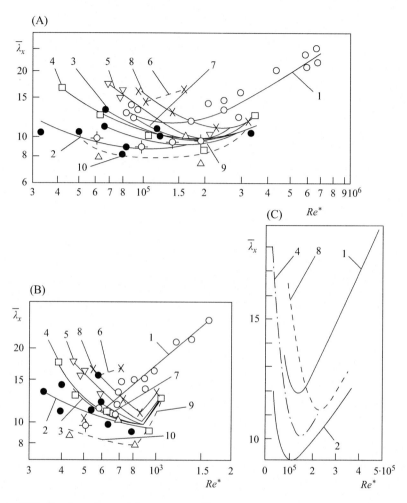

Figure 5.32 The dependence of the wavelength $\overline{\lambda}_x$ on the Reynolds number calculated using the length (A), (C) and extrusion thickness (B): (1) is a rigid plate. Simple membrane surface: (2) experiments B1–B4, (3) B8–B10, (4) B11–B14, (5) B20–B24, (6) BT43 (heated). Complex membrane surface: (7) P9–P12, (8) P5–P8. Viscoelastic surface (9) P23–P24, (10) P25–P35.

- for a simple membrane plate [curve (2)]

$$\overline{\lambda}_x = 2.28 \cdot 10^5 \mathrm{Re}_x^{-1} + 1.76 \cdot 10^{-5} \mathrm{Re}_x + 4.96 \tag{5.2}$$

curve (4)

$$\overline{\lambda}_x = 6.3 \cdot 10^5 \mathrm{Re}_x^{-1} + 2.2 \cdot 10^{-5} \mathrm{Re}_x + 2.5 \tag{5.3}$$

• for a complex membrane plate [curve (8)]

$$\overline{\lambda}_x = 17.17 \cdot 10^5 \text{Re}_x^{-1} + 3.13 \cdot 10^{-5} \text{Re}_x - 3.15 \tag{5.4}$$

When a rigid plate is flown around, the point of intersection of curves (1) and (8) (Fig. 5.31) corresponds to the second stage of the transition [60,247], when the evolution of the disturbing motion allows us to identify the deformation of the velocity front in the xz plane. The point of intersection of curves (2), (4–8) (Fig. 5.32C) shifts downstream at a distance two times larger than for a rigid plate (stabilization of the development of the I transition stage). From this point on, in the boundary layer, the development of three-dimensional deformations will begin, which will also develop slowly. Therefore the curve for an elastic plate, similar to curve (8), will begin its own beginning from this point of intersection and will be located parallel to the x-axis (Fig. 5.32A).

In Section 3.3 it is shown that the character of the development of a disturbing motion when a rigid plate is flown around is conservative with respect to any external conditions. Investigations carried out on elastic plates suggest that the character of the development of a disturbing motion essentially depends on many factors, which are determined by the properties of the bottom wall of the waveguide (boundary layer). All of them can be combined into two types: the first type consists in a substantial increase in the length along x (along Re) of individual stages of the transition [60,378]. For the second type, on the contrary, a very rapid alternation of the initial stages of transition one after another and the rapid formation of a stage of a developed system of longitudinal vortices, which is strongly extended along x, is characteristic.

When flown around a plate made of FP-1 without a film (Fig. 5.33), the fluctuations u' are smaller than those of the same plate covered with film on top. On plate 10a (Section 5.1, Table 5.4), the difference between curves (5) and (6) indicates the presence in the boundary layer of longitudinal vortex structures of a smaller scale, and the transition from a laminar flow to a turbulent one occurs more smoothly. The presence of two maxima, as well as the dependence of the curves on z for plates 11 and 11a (Section 5.1, Table 5.4), is a sign of a no stationary system of longitudinal vortices. As the elasticity of plate 11a decreases as compared with plate 11, the vortex systems become smaller in z (Fig. 5.33B). This flow pattern explains the shape of such curves.

The regularities of the longitudinal pulsation velocity shown in Fig. 5.33 are important to compare not only with similar dependencies when flow around a rigid plate (Section 3.7, Fig. 3.43), but also with measurements of the pulsating velocity of the floating dolphins boundary layer (Part I, Experimental Hydrodynamics of Fast-Floating Aquatic Animals, Section 6.1, Figs. 6.2 and 6.5). In accordance with the measurements shown in Fig. 6.2, the dependence of the longitudinal pulsation velocity on the velocity or the Reynolds number in the boundary layer of dolphins has a completely different character than in the flow past a rigid plate or a rigid dolphin model (Fig. 6.5). In addition, the dependence of the floating dolphin depends on the acceleration or deceleration modes. In the boundary layer of a rigid plate there is a maximum, indicating a jump in energy and the beginning of a turbulent flow regime. In the boundary layer of a floating dolphin, a similar dependence is

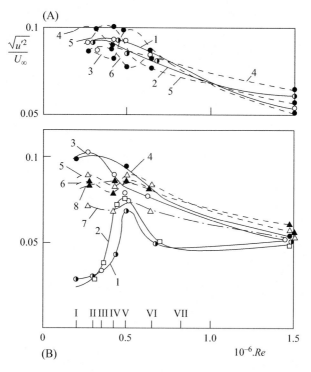

Figure 5.33 The change in the longitudinal pulsating speed at different stages of the transition when flow around a hard and elastic plates. (A): (1) rigid plate, (2) rigid plate with vibrator, (3) plate FP-1, (4) FP-1 with film, (5) plate 10a, between embedded stiffeners, (6) plate 10a, above embedded stiffeners, (7) the same as (5), but with a vibrator, (8) the same as (6), but with a vibrator; (B): (1) plate 11, $z = 0$; (2) 11, $z = 0.01$ m; (3) 11a, $z = 0$; (4) 11 a, $z = 0.01$ m; (5) 11a, $z = 0.02$ m; (6) 11a, $z = 0.04$ m.

stretched and has a smooth character, which indicates other regularities of the parameters of the boundary layer and its smooth transition. When flowing around a dolphin, there is no turbulent boundary layer, which is characteristic of a rigid surface for a boundary layer.

All the above data allow us to conclude that with the first type of transition in the boundary layer of elastic surfaces, the model of development of disturbing motion in the boundary layer over elastic plates looks different in comparison with the boundary layer of a rigid plate (Section 3.6, Fig. 3.33). In the first stage of the transition, under the influence of wall oscillations, the shape of the plane Tollmien–Schlichting wave is accelerated and greatly stretched (Section 5.2, Fig. 5.12). This stage is greatly extended by x. The second, third and fifth stages of transition are absent. A nonlinear plane wave is very quickly transformed into a stage *IV*, which, without interrupting the heads of the pin-shaped vortices, transforms into a *VI* stage of the transition, which is significantly extended in x, as in the *I* stage of the transition. The transition to turbulence takes place evolutionarily without the separation of peripheral vortices on longitudinal vortices (Section 3.6, Fig. 3.33).

The second type of transition differs from the first, since on a rigid wall, as ε increases, there is also a very rapid alternation in time of the first five stages of the transition. Then, without interrupting the heads of the hairpin-shaped vortices, there comes the VI stage of the transition, which, as in the first type of transition, is much extended in x. However, during its evolution, due to surface oscillations, the system of longitudinal vortices becomes nonstationary (Section 3.6, Fig. 3.33). The transversal form of the vortices is constantly deformed: the enlarged cross-sectional shape of an enlarged scale is distorted so that a system of longitudinal multilayer (layered) small-scale eddies arises, which, in turn, merge with each other during evolution and again form large-scale eddies with irregular cross-sectional shape. The evolution of these systems is accompanied by the appearance of peripheral vortices, the destruction of which leads to turbulence. Thus due to the oscillations of the elastic plates and their surface waves, the disturbing motion has specific features.

The above data confirm the conclusions that an adequate choice of mechanical properties of damping surfaces allows obtaining the effects listed in Section 5.2.

5.5 Laser Doppler velocity measurements of the structure of the boundary layer of elastic plates in the process of laminar-turbulent transition

Experimental investigations of the field of longitudinal velocities in the flow around various elastic plates were carried out by V.P. Ivanov using laser Doppler velocity measurement (LDVM) equipment, developed jointly with V.A. Blokhin (Section 3.7), in a hydrodynamic stand of small turbulence [378] in the region of a strain gage insert located in the end of the working area (Section 3.7, Fig. 3.41). The insert at the end of the work area was attached to a strain gage suspension (Fig. 5.6). In order not to introduce errors into the measurement results, a wedge was installed in the rear gap C between the insert and the fixed part of the bottom of the work area, which pressed the insert tightly to the bottom in the area of the front gap A and made the insert stationary.

The measurements were carried out, as in the case of a flow around a rigid plate, in the ninth section ($x = 2.43$ m) in the inset (Section 5.1, Figs. 5.4 and 5.6) with different U_∞. The analysis of the data obtained was carried out in accordance with the classification of the transition stages described in Section 3.6 (Fig. 3.33). The profiles of averaged and pulsating longitudinal velocities in the flow past a rigid plate are given in Section 3.7 (Fig. 3.44). With the help of LDVM, at large ε, longitudinal averaged and pulsating velocities were measured with the flow over plates no. 5 and no. 2 made of polyurethane foam FP-2 (Section 5.1, Fig. 5.7). Plate no. 5 on top was open, and no. 2 was on top plastered with a thin rubber film (Section 5.1, Table 5.4).

The investigation of hydrodynamic stability on the plate no. 5 was not carried out, and the investigation on the plate no. 2 showed that the stability decreased compared to the standard [20,48]. The data presented in Fig. 5.34 are consistent with these findings: the maxima of the pulsation profiles increased compared with

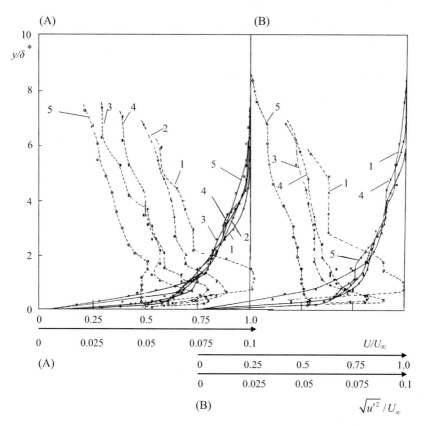

Figure 5.34 Profiles of averaged and pulsating longitudinal velocities when flowing around elastic plates no. 5 (A) and no. 2 (B). (A): (1) $U_\infty = 0.11$ m/s, (2) 0.18 m/s, (3) 0.21 m/s, (4) 0.26 m/s, (5) 0.6 m/s; (B): (1) $U_\infty = 0.1$ m/s, (3) 0.21 m/s, (4) 0.26 m/s, (5) 0.6 m/s.

the rigid plate (Section 3.7, Fig. 3.42B), especially in the initial stages of the transition. Comparison of the shape of the averaged profiles of the standard and the plate number 5 showed that at stage *II* of the transition (Section 3.7, Fig. 3.42A, curve (2) and Fig. 5.34A, curve (1) on plate no. 5 appeared in the profile an inflection characteristic of the third stage of transition at the standard. The velocity profile at the fourth stage of the transition [Fig. 5.34A, curve (2)] has an inflection located substantially lower in δ than on the standard [Fig. 3.42A, curve (4)]. At stage *V* of the transition [Fig. 5.34A, curve (3)], the profile deformation is the same as on the standard [Fig. 4.42A, curve (5)], but δ is again closer to the streamlined surface. At stage *VI* of the transition [Fig. 3.42A, curve (4); Fig. 5.42A, curve (6)], the velocity profile is smoothed on the standard, and on plate no. 5 has several bends. Finally; at the *VIII* stage of transition, on plate no. 5 [Fig. 5.34A, curve (5)], a bend in the profile in the region of $y/\delta^* = 1$.

In general, the profiles on plate no. 5 are equidistant profiles on the standard at the corresponding stages of the transition, that is, the roughness of plate no. 5 did not affect the measurement results. The character of the averaged profiles indicates the acceleration of the alternation of transition stages in the flow around the plate no. 5, with the exception of *VIII* stage of transition on which a turbulent boundary layer has not yet formed. This is also evidenced by the shape and amplitude values of the pulsation profiles of the longitudinal velocity (more informative than the profiles of the averaged velocities). Already at the second stage of transition on the plate no. 5 a maximum appears, which is characteristic of the later vortex stages of the transition. Then, in the subsequent stages of the transition, the maxima decrease and approach the surface. In accordance with the averaged profile at the *VIII* stage of transition on plate no. 5, the shape of the pulsation profile differs from the reference one; it remains a two-hump characteristic of the previous stages of the transition. Pulsation profiles on plate no. 5 have characteristic minima, indicating the organization in the boundary layer of a system of longitudinal vortices at all stages of the transition. In Section 5.2 it is shown that with a natural transition on the membrane plate, the formation of ordered longitudinal vortices spontaneously accelerates. This fact was also obtained during the flow around plate no. 5. Along with the acceleration of the formation of vortex systems, a delay in the transition of the boundary layer was found.

Fig. 5.34B shows the results of measurements on plate no. 2, the analysis of which confirms the same conclusions. The difference from the plate no. 5 is that, judging by the pulsation profiles, nonlinear stages are formed even faster. Beginning with stage *V* (earlier than on plate no. 5), a nonstationary system of longitudinal vortices was formed in the boundary layer, which persists even at stage *VIII* of the transition. Because of this, the maxima of the pulsation profiles are located even closer to the streamlined surface. At the *VIII* stage, the maxima of the pulsation profiles are smoothed (relaminarization). Thus when various types of elastic plates flow around in the initial stages of a transition, the *phenomenon* of spontaneous organization of a system of stable longitudinal vortices is discovered, leading to a significant delay in the transition and accompanied by a change in the flow structure in the boundary layer.

The tested elastic plates had a high absorption capacity. This led to the fact that a change in the character of the flow in the boundary layer was observed only with the introduction of concentrated three-dimensional disturbances with the help of VG3 longitudinal vortex generators (Section 4.1, Fig. 4.2). The height of the VG3 was $7 \cdot 10^{-3}$ m and its length was $18 \cdot 10^{-3}$ m. At the same time, even immediately behind the vortex generators, the bends of the $U(z)$ profile were insignificant compared to a rigid plate. With distance from VG3 along x, the deformation of the profile decreased even more, and the shape of the profile $U(z)$ had the form, as in the natural transition (Section 3.7, Fig. 3.44). The increase or decrease in λ_z of input three-dimensional perturbations is almost not changed the stable nature of the flow in the boundary layer.

Distributed formation of longitudinal vortex perturbations with step $\lambda_z = 0.012$ m was carried out on plate no. 1 (Table 5.9) and (Section 5.1, Fig. 5.7, Table 5.5) in the same way as on a rigid standard—thin wires of $1.2 \cdot 10^{-4}$ m in diameter were glued on the plate surface in the longitudinal direction. In this case, the vortex formation will be stimulated due to the different thickness of the boundary layer in the z direction [397,548]. The measurements showed that the velocity field before the elastic insert ($Re = 1.1 \cdot 10^5$) is the same as in the natural transition on a rigid plate (Section 3.7, Fig. 3.44). As expected, less intense distributed disturbances did not (compared to the

Table 5.9 The design of elastic plates.

No plates	Thickness elastic plates $10^2 h_j/l_e$	Number layers	No layer	Thickness ith layer $10^3 h/l_e$	Material ith layer	Constructive features and notes
1	2.	1	1	20	FE-3	Solid sheet
10A	1.2	2	1	4	FE-1	Solid sheet according to Fig. 5.7D
			2	8	PU-1	Solid sheet, ledges-hollows downstream $\lambda = 1.7 \cdot 10^{-3}$ m, $b = \lambda = 0.21 h$
11	1.6	3	1	2	CCF	Solid sheet, plastered with a thin rubber film
			2	6	FL-1	Solid sheet (Fig. 5.7E)
			3	8	PU-1	Solid sheet, ledges-hollows downstream $\lambda = 1.7 \cdot 10^{-3}$ m, $b = \lambda = 0.21 h$
11A	1.6	3	1	2	CCF	Solid sheet, plastered with a thin rubber film
			2	6	FP-2	Solid sheet, ledges- hollows downstream

effect of concentrated generator of VG3 vortices) any significant changes in the stabilization of the boundary layer: at $y = 0.01$ m and $(y/\delta = 0.5)$, the greatest deformation of the $U(z)$ profile front, as in the natural transition on an elastic plate. Some difference consisted only in the sharp-pointed form of the front of the profile and a decrease in the size of the deformation to $\lambda_z = 0.024$ m. Thus the intensity of the introduced concentrated or distributed three-dimensional disturbances is insufficient to determine the patterns of development of the nonlinear transition stages due to the damping properties of elastic plates. Therefore as on a rigid plate with a small ε, techniques were used to increase the intensity of the disturbing motion. One of them is to heat the stiffeners (wires) with an electric current. Due to the heat-insulating properties of elastomers, the heat flow was directed to the boundary layer, increasing convective currents. As shown in [60,139], the absorption capacity of plate no. 1, on which wires were attached, decreased when heated. All this was to increase the intensity of the input disturbances. However, with low wire heating (power up to 8 W), a slight deformation of the $U(z)$ profile with $\lambda_z = 0.036$ m was observed. An increase in current power to 12 W was accompanied by a deformation of the profile with $\lambda_z = 0.012$ m only near the wall to $y = 0.003$ m. The velocity profile $U(z)$ leveled off with the wall. With an increase in current power above 15 W, irregular longitudinal bundles formed in the boundary layer immediately behind the tellurium wire.

When conducting experimental studies on elastic plates, the structure of fast-swimming hydrobionts was simulated, in which longitudinal structures were found in the outer covers that interact with the vortex structures of the boundary layer (Part I, Experimental Hydrodynamics of Fast-Floating Aquatic Animals, Sections 4.3, 4.5–4.7).

Another method of introducing three-dimensional disturbances into the boundary layer was in the manufacture of elastic plates, the outer layers of which had a corresponding ordering [25,27,32,37]: with increasing U_∞, the increased dynamic loads caused a reaction in elastomers, as a result of which longitudinal disturbances "generated" into the bottom layer, as well as using heated wire. The measurements were carried out with the help of the LDVM. Fig. 5.35 shows velocity profiles for flow past a plate 10A (Table 5.9). Its outer layer was made of longitudinal strips of FE-1 foam material, glued together so that vertical adhesive membranes (more rigid than FE-1 material) were located along the flow with a step of $\lambda_z = 0.005$ m (Section 5.1, Fig. 5.7). Comparison of the obtained results is conveniently carried out with a hard (Section 3.7, Fig. 3.44), and with elastic standards (Fig. 5.34).

The main results were obtained:

1. All averaged profiles, including those in the turbulent flow regime [curves (5)], have inflection points, which indicate, as in elastic standards, on the presence in the boundary layer of longitudinal nonstationary vortex structures with a size of y on the order of $2\delta^*$ (on a hard standard, the size of y on the order of $4\delta^*$).
2. Character of the pulsation profiles, as on elastic standards, indicates the accelerated formation of longitudinal vortex systems compared with a rigid standard. If on a rigid plate the maximum pulsations u' are fixed at the V stage of the transition, then on the elastic one—at II stage of the transition. The shape of the pulsation velocity profiles is similar to those with the receptivity of three-dimensional disturbance [247], only the maxima in Fig. 5.35 are located closer to the wall. The shape and maximum values of the pulsation

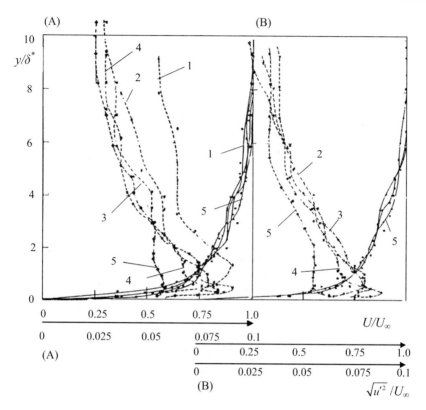

Figure 5.35 Longitudinal averaged (solid curves) and pulsation (dashed) velocities at flow around plate 10A at various stages of transition when measured between glue membranes (A) and above them (B): (1) $U_\infty = 0.11$ m/s, (2) 0.18 m/s, (3) 0.21 m/s, (4) 0.27 m/s, and (5) 0.6 m/s.

profiles vary little depending on U_∞, which indicates the rapid formation in the boundary layer of an elastic plate of longitudinal vortex structures and their preservation up to the turbulent flow regime compared to a rigid standard.

The difference in the velocity profiles when measured over glue seams and between them occurred only at $U_\infty > 0.18$ m/s, when the increased dynamic loads of the boundary layer began to cause a corresponding reaction of the streamlined surface. This allows us to conclude that the surface of the plate 10A interacts with three-dimensional disturbances of the boundary layer. However, the λ_z plate 10A is small, and this interaction manifests itself in the same way as on elastic standards.

Difficulties in studying the patterns of development of the nonlinear stages of transition to elastic plates are caused not only by the stabilizing properties of elastomers, but also by the structural and kinematic-dynamic features of the interaction [60,139]. In accordance with the structural feature of the interaction, it is necessary to determine what size of disturbances introduced into the boundary layer "from below" will interact with disturbances corresponding in size during the natural transition on monolithic elastomers. In accordance with the kinematic-dynamic interaction features, it is necessary to determine at what mechanical characteristics and at what stages of the transition the dynamic loads of the flow will cause the greatest response in the elastomer. The decrease in the maximum u' of pulsations near the wall at $U_\infty = 0.6$ m/s is due to the better dynamic interaction of the elastomer with the boundary layer, as well as the fact that the size of the longitudinal vortex structures in the viscous sublayer is smaller, therefore the interaction of the natural and inserted longitudinal vortex structures occurs more intensively in accordance with the specified interaction features.

Calculation of the parameters of the similarity of the plate 10A at low speeds indicates its high rigidity. Therefore in order to identify general patterns at the nonlinear stages of transition, experiments were continued on more compliant elastomers—plates 11 and 11A (Table 5.9). (Their Section 5.1). The pitch ordering of their outer layer was even smaller: $\lambda_z = 0.0017$ m.

Fig. 5.36 shows the profiles of averaged and pulsating longitudinal velocities on plate 11 (Table 5.9). The pitch λ_z was small; therefore measurements at $z = 0$ and $z = 0.01$ m were carried out on the same design features—over the ledge in the outer layer of elastomer. Thus the differences in the shapes of the profiles are not due to the geometric structure of the composite, but to the peculiarity of the flow in the boundary layer. At stage II of the transition [curve (1)], the shape of the averaged profile at $z = 0$ and $z = 0.01$ m has similar excesses, as well as the elastic standard (Fig. 5.34) at the same speed and the hard standard (Section 3.7, Fig. 3.44) on V transition stage. At $U_\infty = 0.18$ m/s, the excesses at both z are the largest and are located higher than those of rigid and elastic standards: the vortex system becomes more complex, a nonstationary system of vortices begins to form. The velocity profiles at $z = 0$ and $z = 0.01$ m are very different.

At $U_\infty = 0.21$ m/s, the kinks in the profiles decrease and fall below to the wall, as with a rigid standard after the V stage of the transition. With increasing U_∞ to 0.26 m/s, the size of the kinks in the profile decreases, but the nonstationary vortex system remains. At the maximum speed $U_\infty = 0.6$ m/s, the kinks remain and indicate that the size of the vortices has become smaller. A similar transformation of the shape of the averaged profile was also recorded on the plate 10A.

The general pattern of changing the shape of the pulsation profiles at different stages of the transition is the same as on plate 10A (Fig. 5.35). In both cases, the shape of the profiles is similar to the profiles when the boundary layer of a rigid plate interacts with plane perturbations at the 5th transition stage [60]. All of this, as well as the difference in the shape of the pulsation profiles at $z = 0$ and $z = 0.01$ m, indicates that in the boundary layer of plate 11, like in plate 10 A, the formation of a specific complex nonstationary vortex structure with pulsation velocities characteristic of it u' near the wall. An increase in compliance and a decrease in λ_z size on plate 11 compared to plate no 10A resulted in a decrease in the ripple value u' at the VIII stage of the transition [curves (5) of Figs. 5.35 and 5.36].

The elasticity of plate 11A was less than that of plate 11 (Table 5.10). Considering that measurements at $z = 0$, 0.01, 0.02, and 0.03 m were carried out on the same geometric irregularities, the results given in Fig. 5.37 allow us to determine the size of the vortex system by z, as well as the role of the mechanical characteristics of the elastomers in the formation of this system. The shape of the averaged and pulsation velocities at different stages of transition at plates 11 and 11A is similar.

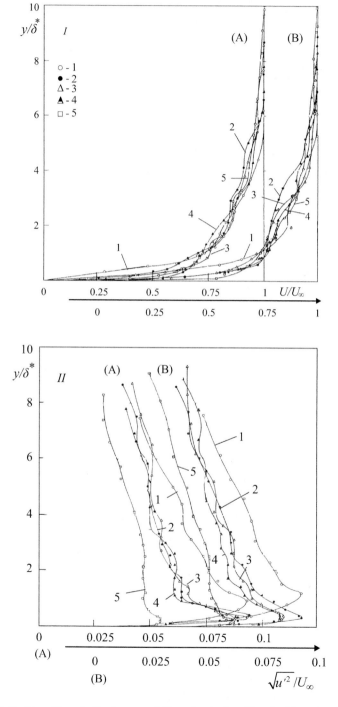

Figure 5.36 Profiles of longitudinal averaged (*I*) and pulsation (*II*) velocities when flow past plate 11 when measured along the longitudinal axis of the working section, at $z = 0$ (A) and at $z = 0.01$ m (B): (1) $U_\infty = 0.11$ m/s, (2) 0.18 m/s, (3) 0.21 m/s, (4) 0.26 m/s, and (5) 0.6 m/s.

Table 5.10 Mechanical characteristics of elastic plates according to Table 5.9.

No plates	ρ, kg/m^3	Quantity pores	f, Hz	C, m/s Longitudinal	C, m/s Transverse	Static modulus of elasticity $10^{-4} \cdot E_{st}$ (MPa)	Dynamic modulus of elasticity $E_{long.}$, MPa	Dynamic modulus of elasticity $E_{tr.}$, MPa	Anisotropy coefficient $k_1 = E_{long.}/E_{st}$	Anisotropy coefficient $k_2 = E_{long.}/E_{tr.}$
1	107	32	214	64.3	58.6	0.49	4.3	3.6	8.85	1.2
10A	410	50–55	231	52.7	90	0.98–3.9	11	32.4	2.8–11	0.3
11	1050	5–10	185	61	66	8.8	39	46.5	4.3	0.85
11A	1045	4–9	239	49.7	68.2	5.9	25.3	47.6	4.3	0.53

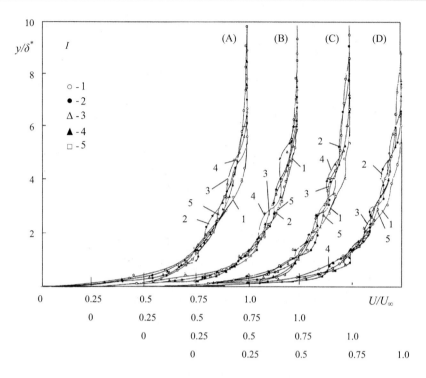

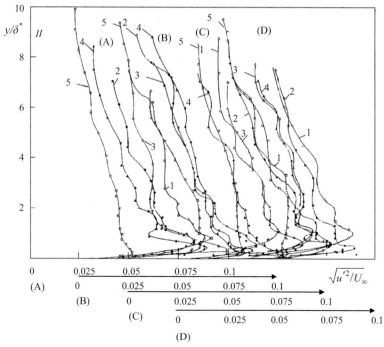

(*Continued*)

However, due to a change in the mechanical characteristics of plate 11A, the $U(y)$ profiles at stage II of the transition [curves (1)] are practically without excesses.

Based on this and on the basis of the type of pulsation profiles, it can be concluded that the plate 11A has a slower formation of the longitudinal vortex system than the plate 11. In addition, in the boundary layer of plate 11A, according to Fig. 5.37, it changes from $\lambda_z = 0.04$ m (stage II) to 0.02 m at the last stages of the transition. All other features of the flow are the same as in plate 11.

Since the kinematic-dynamic interaction of the boundary layer and elastomers manifested more intense in the turbulent flow regime ($U_\infty = 0.6$ m/s), detailed measurements were made at this speed along the strain gage in the working part of the hydrodynamic test bench (Section 3.7, Fig. 3.41) on the plates no. 5, 10A, 11, and 11A (Figs. 5.38 and 5.39). The flow along the plate no. 5 indicates a different degree of fullness averaged profile. The "double hump" of a pulsation profile usually corresponds to nonlinear transition stages.

At the same time, along plate 10A, a turbulent averaged velocity profile is formed with an inflection characteristic of longitudinal vortex systems. The shape and size of the pulsation profile corresponds to the preturbulent transition stages on a rigid standard. The same can be said about plates 11 and 11A [compare with Figs. 3.42 and 3.44 (Section 3.7) and Fig. 5.34]. The pulsation profiles of the plate 11A are smaller than those of the plate 11, when measured by z, a decrease in the profile maximum was recorded.

Thus measurements on elastic plates with an ordered structure showed that, despite their nonoptimal mechanical parameters, specific longitudinal vortex systems are formed in the boundary layer.

Their formation is accelerated in comparison with a rigid plate, and the shape and size of the pulsation profiles testifies to a transition tightening: even at $U_\infty = 0.6$ m/s. On elastic plates, the averaged and pulsating profiles are the same as those of a rigid standard in nonlinear transition stages. It is characteristic that the increased values of the pulsation velocity u' at various stages of the transition during the flow around elastic plates correspond exactly to those at stage $V-VI$ of the transition on a rigid standard, when longitudinal vortex systems are formed in the boundary layer.

The fact that the shape of the pulsation profiles when flowing around elastic plates is the same as when the investigation of receptivity to flat disturbances, as well as the type of averaged and pulsation profiles—all this indicates that the elastic surface, perceiving the ripple of the pulsations of boundary layer, oscillates with very small amplitude. In other words, a consequence of the kinematic-dynamic interaction of the boundary layer and the elastomer is the appearance of surface waves on it and the corresponding change in the kinematic characteristics of the flow. (In [60,139], the results of measurements of oscillations of the surface of

Figure 5.37 Profiles of longitudinal averaged (I) and pulsation (II) velocities with flow around a plate 11A when measured along the axis of the working section, with $z = 0$ (A), with $z = 0.01$ m (B), with $z = 0.02$ m (C), with $z = 0.03$ m (D): (1) $U_\infty = 0.13$ m/s (A), (B) and 0.11 m/s (C), (D); (2) 0.17 m/s (A), (B) and 0.18 m/s (C), (D); (3) 0.21 m/s; (4) 0.26 m/s; (5) 0.6 m/s.

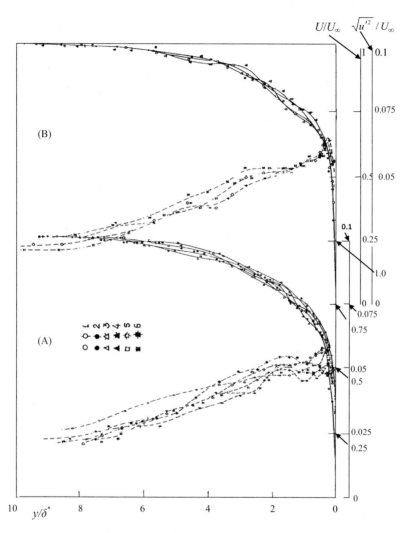

Figure 5.38 The profiles of averaged (solid) and pulsation (stroke) longitudinal velocities on the plates 5 (A) and 10 A (B) at $U_\infty = 0.6$ m/s and at $z = 0$: (1) section № I, (2) II, (3) III, (4) IV, (5) V; with $z = 0.045$: (6) section IV (section numbers see Section 3.7, Fig. 3.41).

elastomers are given). The nonstationarity of the longitudinal vortex system and the corresponding irregular and irregular shape of the cross-section of the vortices are explained by the oscillation of the surface of the elastic plates.

Hence the following signs of complex interactions. If the flow conditions and the properties of the elastomers are such that the similarity parameters are optimal, the elastic plates interact with the flow in the absorption mode, and then all changes in the boundary layer will be fixed near the surface. If the similarity parameters are

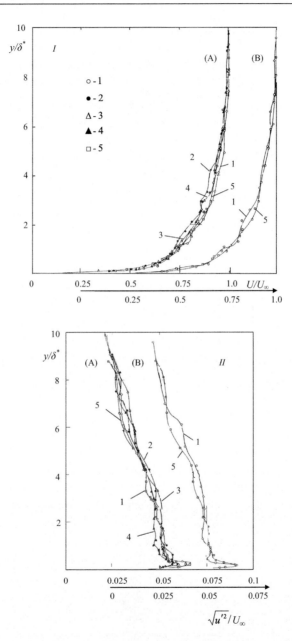

Figure 5.39 Profiles of the longitudinal averaged (*I*) and pulsation (*II*) velocities with flow around the plate 11 at $U_\infty = 0.6$ m/s. (A): $z = 0$, (1) section I, (2) section II, (3) section III, (4) section IV, (5) section V. (B): section V, (1) $z = 0.02$ m, (5) $z = 0.04$ m (for section numbers, see Section 3.7, Fig. 3.41).

far from optimal, the elastic plates operate in a resonant mode, and due to the increased amplitude of their surface waves, changes in the kinematic characteristics will be recorded across the entire thickness of the boundary layer [378].

It was advisable to find out whether the ordinate of the fluid layer, in which the disturbances (the critical layer) propagates, does not coincide with the value δ^* on elastic inserts (as is the case on a rigid plate). To do this, in the IV section at different speeds U_∞ (at different stages of the transition), using LDVM, we measured the profiles of the longitudinal averaged velocity on a rigid (Section 3.7, Fig. 3.44) and elastic plates (Figs. 5.34–5.40) and δ and δ^* are calculated from them.

The measurement results are summarized in Tables 5.11 and 5.12. On a rigid plate δ^*/δ for the laminar and turbulent boundary layers are 0.33 and 0.125,

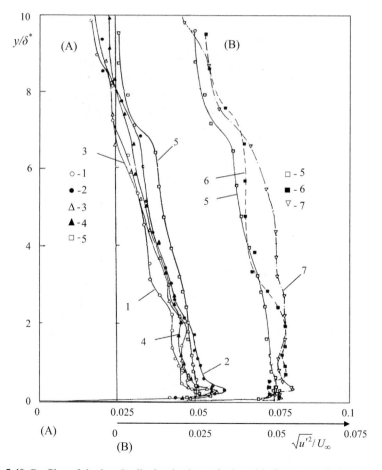

Figure 5.40 Profiles of the longitudinal pulsating velocity with flow around plate 11A at $U_\infty = 0.6$ m/s (A): $z = 0$, (1) section I, (2) section II, (3) section III, (4) section IV, (5) section V. (B): section V, (5) $z = 0$, (6) $z = 0.02$ m, (7) $z = 0.04$ m (the section numbers see Section 3.7, Fig. 3.41).

Experimental investigations of the characteristics of the boundary layer on the analogs of the skin

Table 5.11 The characteristic thickness of the boundary layer in the IV section of the insert.

U_∞, m/s	0.11	0.18	0.21	0.26	0.6	Experiments conditions
δ	2.68	2.91	3.8	4.51	6.5	
δ	14	16	24	32	48	Plate 5
δ^*/δ	0.19	0.18	0.16	0.14	0.135	$z = 0$
δ^*	2.8	–	4.49	4.1	5.25	
δ	16	–	32	28	44	Plate 2
δ^*/δ	0.18	–	0.14	0.15	0.12	$z = 0$
δ^*	3.05	4.07	3.4	($U_\infty = 0.27$)	3.41	Plate 10A, 3.87 measurements between glue membranes, $z = 0$
δ 20	32	32	32	36		
δ^*/δ	0.15	0.13	0.1	0.12	0.09	
δ^*	3.08	3.96	3.43	3.4	3.68	Measurements over
δ	20	32	36	20	40	glue membranes,
δ^*/δ	0.15	0.12	0.1	0.17	0.09	$z = 0.005$ m
δ^*	4.38	4.19	–	3.75	4.6/4.9	$z = 0$,
δ	24	32	–	32	44/14 vibrator	
δ^*/δ	0.18	0.13	–	0.12	0.1/0.11	$h_v = 3-4$ mm
y_v/δ^*	0.8	0.84	–	0.94	0.76/0.59	$h_v = 1.5$ mm
δ^*	4.51	3.74	–	4.14	3.88	
δ	24	20	–	28	32	$z = 0.005$ m
δ^*/δ	0.19	0.19	–	0.15	0.12	Vibrator
y_v/δ^*	0.58	0.7	–	0.63	0.68	
δ^*	3.22	4.83	3.7	4.61	4.85	
δ	20	32	3	32	48	Plate 11
δ^*/δ	0.16	0.15	0.1	0.14	0.1	$z = 0$
δ^*	2.62	4.66	3.87	($U_\infty = 0.27$)	3.97	3.59
δ	20	3	32	32	44	$z = 0.001$
δ^*/δ	0.13	0.15	0.12	0.11	0.09	
δ^*	4.2	4.29	–	($U_\infty = 0.27$)	3.8/4.1	4.89
δ	28	32	–	36	32/40	$z = 0$,
δ^*/δ	0.15	0.13	–	0.14	0.12/0.1	vibrator
δ^*	3.62	4.26	0.2	0.26	4.83	
δ	16	28	24	32	48	plate 11A
δ^*/δ	0.23	0.15	0.19	0.13	0.1	$z = 0$
δ^*	2.72	4.49	4.65	6.01	5.16	
δ	20	28	36	40	44	plate 11A
δ^*/δ	0.14	0.16	0.13	0.15	0.12	$z = 0.01$ m
δ^*	3.01	3.1	3.99	6.19	4.7	
δ	16	24	24	40	40	plate 11A
δ^*/δ	0.19	0.13	0.17	0.15	0.12	$z = 0.02$ m

(*Continued*)

Table 5.11 (Continued)

U_∞, m/s	0.11	0.18	0.21	0.26	0.6	Experiments conditions
δ^*	2.66	3.75	5.72	6.15	5.03	
δ	16	24	40	40	44	Plate 11A
δ^*/δ	0.17	0.16	0.14	0.15	0.11	$z = 0.03$ m
U_∞	0.11	0.16	–	0.27	–	
δ^*	4.34	3.88	–	5.16	4.92	
δ	24	28	–	32	44	$z = 0$
δ^*/δ	0.18	0.14	–	0.16	0.11	Vibrator
y_v/δ^*	0.81	0.9	–	0.68	0.71	
δ^*	–	4.14	–	4.82	–	
δ	–	28	–	32	–	$z = 0$
δ^*/δ	–	0.16	–	0.15	–	Vibrator
y_v/δ^*	–	0.8	–	0.73	–	
δ^*	–	5.71	–	5.53	–	
δ	–	28	–	36	–	$z = 0.02$ m
δ^*/δ	–	0.2	–	0.15	–	Vibrator
y_v/δ^*	–	0.61	–	0.63	–	

Table 5.12 The characteristic thickness of the boundary layer in different sections of the insert at $U_\infty = 0.6$ m/s.

Section number	I	II	III	IV	V	Plate
δ^*	4.47	4.7	5.43	4.83	5.08	
δ	44	48	48	48	48	Plate 11A
δ^*/δ	0.1	0.1	0.11	0.1	0.11	$z = 0$
δ^*	4.87	4.81	4.81	4.85	5.45	
δ	36	48	40	48	44	Plate 11
δ^*/δ	0.14	0.1	0.12	0.1	0.12	$z = 0$
δ^*	–	3.4	3.66	3.41	4.18	Plate 10A
δ	–	48	32	36	44	Measurement between
δ^*/δ	–	0.07	0.11	0.09	0.1	Glue membranes
δ^*	5.4	5.3	5.6	6.5	6.05	
δ	44	44	48	48	48	$z = 0$
δ^*/δ	0.12	0.12	0.12	0.14	0.13	

respectively. On all elastic plates in the transitional boundary layer, the δ^*/δ values are significantly less than on the hard, and with increasing speed, the δ^*/δ values decrease to $U_\infty = 0.26$ m/s, at which this value sometimes increases. On a rigid plate, a jump of this magnitude was earlier at the V stage of transition after the destruction of the heads of the hairpin-shaped vortices at $U_\infty = 0.21$ m/s. At $U_\infty = 0.11$ m/s on elastic plates $\delta^*/\delta = 0.15-0.19$, and at $U_\infty = 0.6$ m/s, 0.1–0.12 (for a rough plate, 0.13).

It turned out that during complex interactions of elastic plates with disturbances of the boundary layer, the ratios of the characteristic thicknesses of the boundary layer changed.

5.6 Experimental construction of the Görtler neutral curve in the flow around elastic curvilinear plates

The investigation of complex interactions of natural vortex structures arising in the boundary layer of curvilinear elastic plates and flowing around a cylindrical surface with longitudinal vortex systems introduced into the boundary layer is carried out in accordance with the methodology used to investigation curvilinear rigid plates (Section 4.5). In accordance with this method, the visualization of vortex structures in the boundary layer of curvilinear elastic plates was performed using the tellurium method. Then, with the help of LDVM equipment, velocity profiles were measured in various longitudinal and transversal directions of curvilinear surfaces. On streamlined elastic surfaces, generators of longitudinal vortices were installed and the above investigations were repeated [622,625–628,629].

The investigations were carried out on the same curvilinear plates (Section 4.1, Fig. 4.1), on which monolithic elastic plates of the type FE-3 were glued (Section 5.1, Table 5.4 and 5.5) [67]. The thickness of the elastic plates was 0.003 and 0.01 m, the mechanical characteristics were: instantaneous modulus of elasticity $E = 5 \cdot 10^5$ H/m^2, loss tangent tg$\varphi = 0.62$ and density 120 kg/m^3. The design of rigid curvilinear plates with glued elastic plates made it possible to set the plates flush with the confuser and the diffuser of the hydrodynamic stand using adjustment screws. Below are the results when using a rigid curvilinear plate with a radius of curvature R = 1 m. The intensity of the introduced longitudinal vortices increased due to the use of VG4 vortex generators having a height 2 times the height of VG3 (Section 4.1, Fig. 4.2).

The gluing of the elastic plate FE-3 with a thickness of 0.003 m in areas of 0.5 m $< x <$ 0.65 m and 1 m $< x <$ 1.15 m was performed not over the entire surface, but with longitudinal adhesive strips with a pitch $\lambda_z \approx 0.042$ m (when flowing around a flat plate the longitudinal structures were characterized by $\lambda_z \approx 0.045$ m (Section 3.3). This led to the rise of unglued sections of the bands by an amount of $y \approx 7 \cdot 10^{-4}$ m, to an increase in their compliance and absorption capacity due to the appearance of an air gap. Fig. 5.41 presents photographs of the visualized flow field in these areas without vortex generators (A), (B) and in their presence (C), (D). The greatest effect was obtained for the diffuser section (A), in which a separation with a random distribution of the perturbation field was observed on a rigid surface under such flow conditions. In this case, the accelerated flow over convex stripes leads to a regular z-shaped tellurium cloud, and due to the increased damping properties, not the destruction of the tellurium cloud, but the formation of longitudinal bundles occurs here. The same can be seen in the confused

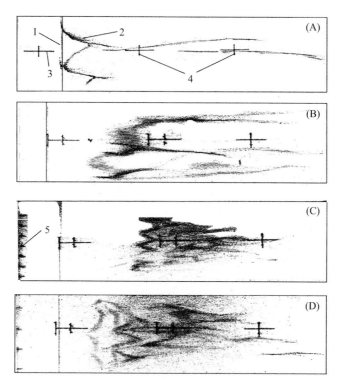

Figure 5.41 The effect of a change in z of the properties of the streamlined elastic plate FE-3 on the form $U(z)$ at $U_\infty = 0.05$ m/s and $y = 0.002$ m in diffuser: (A) $x = 0.53$ m) and confused: (B) $x = 1.1$ m plots; (C) $x = 1.1$ m, generators of eddies VG1, $\lambda_z = 0.01$ m; (D) $x = 0.05$ m, GV1, $\lambda_z = 0.02$ m. (1) tellurium wire, (2) tellurium "clouds," (3) longitudinal marks, (4) transverse marks at the bottom of the working section (distance between longitudinal marks—0.1 m and transverse marks—0.5 m), and (5) vortex generators.

section (B). Vortex generators create a more powerful disturbance field, which suppresses natural disturbances (C), (D), but exists at a much smaller distance downstream than on a rigid surface, regardless of the wavelength.

Fig. 5.42 shows the visualization of the tellurium method the vortex structure in the boundary layer when flowing around a concave curvilinear surface on which a uniform elastic material FE-3 is glued 0.01 m thick is shown. Unsteady bending of the tellurian line (A) is observed already to the concave section. In contrast to the case considered in Fig. 5.41, in the diffuser section, even with a decrease in U_∞, tellurium cloud is blurred randomly right behind the tellurium wire (B); in the confused section, the flow somewhat stabilizes, and the erosion of tellurium occurs in the form of longitudinal bundles (C).

The longitudinal vortex systems behind the vortex generators stabilize the flow in the diffuser section (D), (E), but not to the same extent as in Fig. 5.41A. At the exit from the confused section (F), (G), the forced structure is maintained at

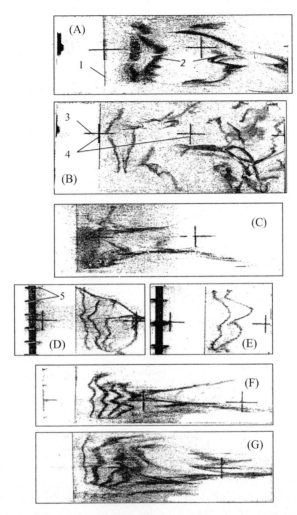

Figure 5.42 Photographs of the visualization of the $U(z)$ profiles in the flow around a uniform elastic curved plate FE-3 with $R = 1$ m at $U_\infty = 0.033$ m/s in the case of their natural occurrence at $x = 0.4$ m (A), 0.6 m (B) and 1.1 m (C), as well as under the action of the vortex generators VG1: (D), (E) $x = 0.53$ m, $\underline{x} = 0.05$ m, $y_{tw} = 0.02$ m, $\lambda_z = 0.012$ m (D) and 0.024 m (E); (F), (G) $x = 1.15$ m, $y_{tw} = 0.002$ m; (F): $\lambda_z = 0.016$ m, $\underline{x} = 0.035$ m; g: $\lambda_z = 0.02$ m, $\underline{x} = 0.05$ m. (1) tellurium wire, (2) tellurium "clouds," (3) longitudinal marks, (4) transverse marks on the bottom of the working section (distance between longitudinal marks—0.1 m and transverse marks—0.5 m), (5) vortex generators.

a greater distance than in the diffuser section, but at a smaller distance than under similar conditions on a hard surface.

With the help of the LDVM, the velocity profiles $U(y)$ were measured during the flow around elastic curved plates. Differences from a rigid curvilinear plate are more significant in the confused section: if, when a rigid curvilinear plate flows around, the $U(y)$ profile is strongly filled near the wall and has a bend point, then the elastic plates completely eliminate these features. The change in the thickness of the elastic layer from 0.003 to 0.01 m had practically no effect on the shape of the profiles. From this it follows that at the indicated value of E, the loading effect of the flow is insufficient for the effective action of the thickness of the elastic plate. A slight increase in velocity in the area following the concave area, from 0.106 to 0.12 m/s, resulted in differences between the corresponding $U(y)$ profiles when flowing around elastic plates of different thickness. The greatest difference in them is observed at $y/\delta_1 = 1.4$.

At $x = 1.35$ m, the velocity profiles $U(y)$ were also measured for flowing around these elastic curvilinear plates, when generators of VG1 longitudinal vortices with $\lambda_z = 0.02$ m were installed on streamlined surfaces. Measurements were performed at $U_\infty = 0.112$, 0.136, and 0.213 m/s at $z = 0$, $z = \lambda_z/2$, where $\lambda_z = 0.012$ m. When using a thin layer of elastomer, sharp bends were found in the velocity profiles for both values of z; for a thicker material, the effect is smoothed, the deviation from the reference profile (hard streamlined surface) caused by the generators vortices, decreased. Consequently, the disturbances introduced are intense enough to make the elastic layer work throughout its thickness. An increase in U_∞ leads to a deepening of the inflection of the profile and the classical difference between the profiles measured above the "peaks" and above the "hollows." Decreasing λ_z makes the profile above the "hollows" somewhat less filling, that is, under such flow conditions, the preferred scale is $\lambda_z = 0.02$ m.

Fig. 5.43 shows graphic copies of photographs of the flow field visualized in the xz plane along the curvilinear elastic surface of FE-3. The flow field in the region of the tellurium wire 1 was recorded at a distance from the streamlined surface, approximately equal to the displacement thickness δ_1 ($y_{T.w} = 0.006$ m). Tellurium clouds were emitted automatically when electrical pulses were applied to the wire after 0.5 s (Sections 3.1 and 3.2). When propagating downstream, the initially linear cloud deformed in accordance with the velocity field in the boundary layer, that is, it acquired the form of a $U(z)$ profile. Marks (3), (4), applied to the surface in the area of the longitudinal axis of the working section, made it possible to calculate the propagation velocity and the magnitude of the grows in disturbances along the x and z coordinates.

The various forms of tellurium clouds emitted at successive moments of time characterized the nonstationarity of the flow field above a concave elastic surface. Compared to a rigid standard (Section 4.3, Fig. 4.9), natural disturbances in the flow around an elastic curvilinear surface were characterized by a large-scale and lower growth rates downstream.

The installation of vortex generators (5) (Fig. 5.43C–H) significantly modified the flow pattern. Comparison with the similar Fig. 4.16 (Section 4.5) shows that, first, the natural disturbances on the elastic surface are distinguished by a larger structure and lower growth rates downstream while maintaining the unstable form $U(z)$. Introduction of too small perturbations on a rigid surface leads to the formation of perturbations with twice the value of λ_z (Section 4.5, Fig. 4.16), on an

Experimental investigations of the characteristics of the boundary layer on the analogs of the skin 315

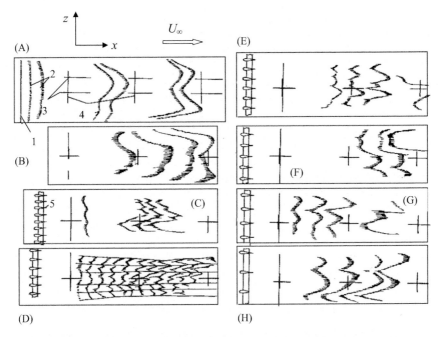

Figure 5.43 The reaction of the boundary layer of the elastic surface of FE-3 to disturbances of different scale: $R = 4$ m, $U_\infty = 0.033$ m/s; $x = 2.2$ m; $x_- = 0.05$ m; $y_{t.w} = 0.006$ m; without making disturbances on rigid (A) and elastic (B) surfaces; (C)–(H) when introducing disturbances with $\lambda_z = 0.01$ m (C); 0.012 m (D); 0.014 m (E); 0.016 m (F); 0.02 m (G); 0.028 m (H): (1) tellurium wire; (2) tellurium clouds; (3) longitudinal and (4) transverse marks on the bottom of the work area; (5) vortex generators; the distance between the transverse marks—0.5 m and the longitudinal marks—0.1 m [627].

elastic plate—to the imposition of a forced small-scale waviness on a large-scale natural structure (Fig. 5.43C). In contrast to a concave rigid surface, such an effect here manifested itself downstream from the vortex generators. A superposition of perturbations and natural perturbations introduced by generators of VG3 disturbances was observed when a flat rigid plate was flown around at $Re = 5 \cdot 10^4$, when disturbances of lesser intensity were introduced.

This shows that the growth velocity of natural disturbances is greater than that introduced with a small value of λ_z: modulation of the natural wave was observed before the formation of "longitudinal bundles" with forced λ_z began. The same is manifested when λ_z is increased to 0.02 m (Fig. 5.43C, D, E, F). Consequently, the organization of a given flow structure in the boundary layer of curved monolithic elastic surfaces, on the one hand, requires the introduction of sufficiently intense perturbations and, on the other hand, such a structure is formed at a greater distance downstream from the impact point. The boundary layer on elastic surfaces is a more inertial system than on rigid ones. The obtained data are in good agreement with the findings of [48], according to which, during a natural transition on elastic horizontal uniform plates, the formation of Benny-Lin vortices and their slow development downstream occurred more intensely than the rigid standard, and the interaction with the imposed field of three-dimensional disturbances depended from the ratio of the scale of natural and introduced disturbances and contributed to the delay of the transition.

The smoothing, damping character of the influence of elastic surfaces on the development of processes in the boundary layer makes it difficult to find clear boundaries of the areas of growth and attenuation of three-dimensional disturbances. In this case, the principle of selectivity of the boundary layer reaction with respect to the scale λ_z of the introduced disturbances does not manifest itself so clearly and makes it impossible to uniquely identify neutral disturbances, as was the case for a rigid surface (Section 4.5). From data analysis (Fig. 5.43D–G), according to the criteria adopted for a rigid curvilinear plate, it follows that such disturbances can equally be structures with $\lambda_z = 0.012 - 0.02$ m. Calculation of parameters for constructing the stability diagram gives the following values: $G = 2.1$; $\alpha_z \delta_2 = 0.5 - 0.84$. Enhanced insertion disturbances with $\lambda_z = 0.028$ m (Fig. 5.43H) are characterized by the parameters: $G = 2.1$ and $\alpha_z \delta_2 = 0.36$, and natural perturbations with $\lambda_z = 0.05$ m—by the values $G = 2.1$ and $\alpha_z \delta_2 = 0.2$. Note that natural disturbances are best stabilized superimposed with $\lambda_z = 0.012$ m (Fig. 5.43D).

A series of investigations on the same elastic surface in a cylindrical channel with a height of 0.07 m when installing VG4 vortex generators (Section 4.1) made it possible to determine several more points characterizing the formed systems of longitudinal vortices on the Görtler stability diagram. The results were used not only of visualization of the $U(z)$ profiles, but also of measurements of the $U(y)$ profiles with a laser anemometer. Only those profiles $U(y)$ that had a characteristic form for the developed system of longitudinal vortices were analyzed [622]. The parameters G and $\alpha_z \delta_2$ obtained for such conditions, as well as the last pair of these parameters for a concave wall, can be attributed to the region of maximum gain in the Görtler stability diagram. Table 5.13 shows the kinematic characteristics of the boundary layer in the cylindrical section in the presence of such disturbances. The resulting pairs of parameters are plotted as points on the grid of the Görtler stability diagram (Fig. 5.44). Despite the fact that a clear boundary between damped and increasing disturbances (neutral curve) is not defined, it can be argued that the instability region here covers a smaller range of disturbances than in the case of a rigid surface.

Calculations showed that the first and second critical Görtler numbers increased. The first critical Görtler number G_o is the minimum value of G at which the neutral development of longitudinal vortices is fixed: it characterizes the beginning of the formation of ordered longitudinal vortex systems in the boundary layer and is determined by the minimum extreme point of the Görtler neutral curve. The results of the investigation of the linear stability of the boundary layer (Section 3.4, Section 5.3) indicate an increase in G_o for elastic plates compared to hard plate, for which $G_o = 0.3 - 0.5$ (Section 4.5). According to Fig. 5.44 G_o for elastic plates increased 2–3 times.

Table 5.13 Parameters of a disturbed boundary layer in a cylindrical channel.

x, m	U_∞, m/s	$\delta \cdot 10^3$, m	$\delta_1 \cdot 10^3$, m	$\lambda_z \cdot 10^2$, m	$x_- \cdot 10^2$, m	$\langle\delta\rangle \cdot 10^3$, m	$\langle\delta_1\rangle \cdot 10^3$, m	$\langle\delta_2\rangle \cdot 10^3$, m	G	$\alpha_z \delta_2$
0.85	0.107	15	1.9	2	5	20	3	1.67	7.2	0.52
0.9	0.2	13	1.67	2	5	19	2.45	1.94	18.8	0.61
0.85	0.106	15	1.9	2	10	15	2.9	1.71	7.5	0.52
0.85	0.105	15	1.9	2.8	5	20	2.2	1.5	6.1	0.34
0.9	0.055	20	4.43	2.4	5	20	5.1	2.76	80	0.72

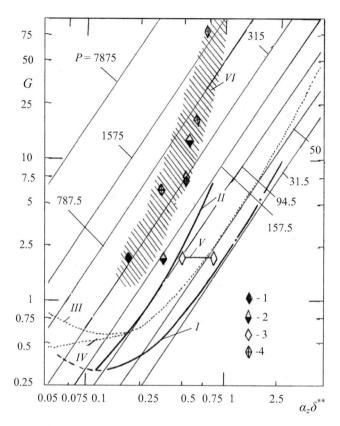

Figure 5.44 The stability of three-dimensional disturbances on the concave elastic surface of FE-3: I, II experimental neutral curve and maximum strengthening curve for a rigid surface (Section 4.5, Fig. 4.18); III, IV calculated neutral curves [270,534]; V, VI areas of neutral and maximally growing disturbances for an elastic surface; characteristics for a concave elastic surface: (1) natural disturbances; (2) increasing forced disturbances; (3) forced disturbances close to neutral; (4) increasing forced disturbances in the cylindrical channel [627].

Another important result is an increase in the second critical Görtler number G^*, which characterizes the destruction of ordered vortex systems and the transition to turbulent flow. Depending on the flow conditions for a rigid surface, $G^* = 2-8$ and $1-1.5$ [239,241,243,609] were obtained, respectively, and oscillation of longitudinal vortices along z (meandering) was fixed in [609] with $G^* = 6-7$.

In these measurements with artificial generation of vortex disturbances above a curvilinear elastic plate $G^* = 18.8$ values were obtained when a cover was installed over this plate (Section 4.1, Fig. 4.1) and thus a cylindrical channel was formed. The largest VG4 vortex generators were installed, $G^* = 80$ was obtained. For a rigid concave wall, the value $G^* = 6.3$ was fixed (Section 4.5, Fig. 4.18, Table 4.1), that is, 3–12 times less than for an elastic surface. If we also take into account the lower growth velocities of three-dimensional perturbations on the elastic surface, then it is obvious that the nonlinear transition region increases with this.

Thus the results obtained, as in the case of a flow around a flat plate (Sections 5.2−5.5), showed that the curvilinear elastic surface effectively stabilizes three-dimensional disturbances, and the length of stable longitudinal vortex systems significantly increases compared to a rigid plate. It should be noted that the value of U_∞ when conducting experiments on curved elastic surfaces was small, and the elastic modulus of the material was large and did not correspond to the optimal similarity parameters (Part I, Experimental Hydrodynamics of Fast-Floating Aquatic Animals, Sections 1.4, 6.3, 6.5−6.8). When performing optimal similarity parameters, the efficiency of correctly selected elastic materials for curved surfaces substantially optimizes the values of G_o and G^*, so that the difference from similar parameters for a rigid curved surface will be even greater.

In [50, 625] on a rigid plate, the control of longitudinal vortex systems was carried out using distributed generators of longitudinal vortices in the form of thin longitudinal wires glued to a rigid surface. These wires, to increase efficiency, were heated: these experiments show that either uniform elastic surfaces or ordered in the transversal direction (Section 5.1) can be successfully used for these purposes, since it was discovered that the properties of such plates to organize and stabilize the system of longitudinal vortices [60,139,247,378].

Accuracy and reproducibility of results in various conditions, as well as the choice of the type of streamlined surface that is most suitable for solving the problem, is achieved while observing the principles of modeling (Section 1.4). To this end, dimensionless complexes characterizing the properties of the materials used and their behavior under different conditions were compiled [60,139,338].

Inertial properties were determined by the coefficient of oscillating mass:

$$M = \rho l/\rho_o \delta_1, M^* = \rho l/\rho_o \overline{\delta}_1, \tag{5.5}$$

where δ_1 and $\overline{\delta}_1$ are the thickness of the displacement of the boundary layer during its natural development and with the introduction of controlled disturbances.

The elastic and damping properties were characterized by the Cauchy Ca and D damping parameters:

$$Ca = \rho_o U_\infty^2/2E, D = k\rho_o U_\infty^2/2\sqrt{(\overline{p'})^2}, \tag{5.6}$$

Here k is proportional to the energy absorbed by the layer of elastic material ($k = 1 - \exp[-\pi \, \text{tg}\varphi]$). The parameter D can be written as $D = \kappa/\alpha \, \tau_w$. According to [393]

$$\left(\overline{p^2}\right)^{1/2} = \alpha \tau_w. \text{ Обычно } 2 \leq \left(\overline{p^2}\right)^{1/2}/\tau_w \leq 4$$

For the material FE3 used in these experiments, the values of the similarity criteria had the following values:

$M = 0.2$; $M^* = 0.16$; $Ca = 1.25 \cdot 10^{-4}$, $D = 98.8$.

For other materials tested, these values varied within:

$M = 0.09-0.89$; $M^* = 0.16$-0.54; $C = (0.4-2.4)\,10^{-4}$; $D = 72.9-98.8$.

In this case, one should also take into account the geometric similarity of the transverse regularity λ_z of the induced system of longitudinal vortices in the case of an elastic surface with an outer layer ordered in the transversal direction—the step of this ordering. To create and maintain longitudinal vortex structures that promote the best mixing of the medium near the streamlined surface, the λ_z pitch should be selected based on recalculation of disturbance parameters that fall into the region of neutral stability (Fig. 5.44). In this case, the diffuse character of this region (in contrast to the localized curve for a rigid surface) allows us to have a certain range of acceptable values of λ_z.

5.7 Determination of the mechanical characteristics of elastomers by the method of wave and pulse propagation

To determine the dynamic modulus of elasticity by the method of propagation of waves and pulses in materials, Kanarsky developed the appropriate apparatus [51,60,139,323]. Elastic oscillations generated by the falling metal balls on the surface of the elastomer allow us to determine the internal frequency of oscillations, the parameters of the surface waves, the damping factor, the dynamic modulus of elasticity, the gain, etc. At the same time, the corresponding bounces of the elastomer can be found and elastic characteristics depending on the height of the falling balls [499,531,630]. The scheme of the device is presented in Fig. 5.45 [51,60,323,366].

Steel balls were dropped on the elastic material under investigation using the device given in Part I, Experimental Hydrodynamics of Fast-Floating Aquatic Animals, Section 6.4, Fig. 6.18. The balls were placed at the top of the tube, which was placed above the test plate. The ball (13) was held in the tube using an electromagnetic coil (12) located at the top of the tube. Pressing the start button of the electronic trigger (14) turns on the tape recorder (16), the oscilloscope (17) and after a while the electromagnetic starter (19), which opens the button (11), opens the network of the electromagnetic holder (12) connected to the power supply (10), the steel ball (13) is released, which falls on the test surface, causing elastic oscillations on the surface and inside the test material. These oscillations are recorded using piezoelectric sensors (2)–(9), (18), the signals of which are amplified at amplifier (15) and recorded on tape of tape recorder (16) at the maximum speed of drawing the tape. Block (16) is a parallel synchronous oscilloscope H-041 and magnetograph H-036. At the same time signals from the sensors are fed to the oscilloscope (17) to monitor the conduct of the experiment. The signal from the trigger unit allowed us

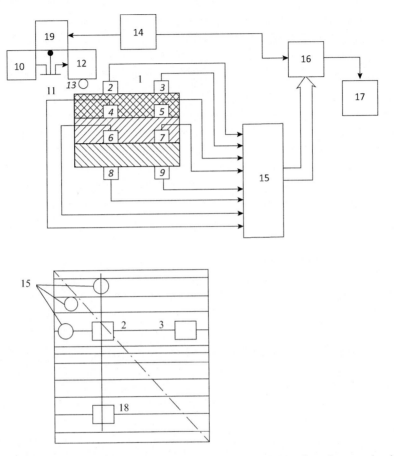

Figure 5.45 Apparatus for determining the dynamic characteristics of an elastomer by the method of wave and pulse propagation: (1) test material, (2)–(9), (18) piezoelectric sensors, (10) power supply, (11) button, (12) electromagnetic holder, (13) steel ball, (14) electronic trigger unit, (15) amplifiers, (16) tape recorder, (17) oscilloscope, and (19) electromagnetic starter [51,60,323,366].

to enable scanning of the S1-18 oscilloscope, a contour oscilloscope and a magnetograph. Kanarsky developed and manufactured the electronic part of the equipment [51,60,323,366].

To control the scanning of the oscilloscope C1-18, an electronic circuit has been developed, which uses a multivibrator as the basis. The delay for the period of the falling ball was smoothly adjusted and adjusted so as to start scanning at the moment of contact of the ball with the surface under study. Signals on the oscilloscope screen were photographed and then analyzed.

The signals reproduced from the magnetic tape were processed using an electronic time measurement unit and converted to input into a computer, or the processing was carried out using photographs obtained from the oscilloscope screen.

By damping the amplitude of oscillations, it is possible to determine the damping and nonlinear properties of the material. The dynamic modulus of elasticity and the anisotropy coefficient of the material were determined by the phase difference between different signals, the natural frequency of the material in different layers. To investigation the anisotropic properties of materials and parameters of surface waves, piezoelectric sensors were fixed outside and inside the composite at the boundaries of the layers in three perpendicular directions to each other and at certain distances from each other. The place of the ball's fall was chosen either along the diagonal of the angle, on the sides of which the sensors are located, or alternately in the longitudinal and transverse directions. Such an arrangement of piezoelectric sensors made it possible to expand the number of measurement parameters and control the parameters during testing and using elastomers. It is important to develop methods for investigation the properties of the composite and its layers, as well as to increase the efficiency of elastic materials, as well as friction drag reduction.

The method for determining the mechanical characteristics of elastic materials was developed by M.V. Kanarsky together with V.V. Babenko [60,139]. The technique consists in generating oscillations inside the material, whose parameters are measured in three mutually perpendicular directions. Thus it is possible to determine the characteristics of different layers, which determine the characteristics of the entire sample and the anisotropy in different directions.

It is known that in multilayer composite materials oscillations are concentrated near the border of adjacent layers (in waveguides) and on the surface of the composite in the form of surface Rayleigh waves and in the form of Stoneley and Lava waves between the layers (Section 3.8). The elastic dynamic characteristics of the layers of the composite depend on the parameters of the oscillations, primarily on its phase and amplitude. Therefore sensors located in this way allow us to determine and constantly monitor the dynamic characteristics of the composite and its response to external disturbances. Sensors allow you to constantly monitor the character of external oscillations and the laws of their dispersion within the material, including in waveguides. It also allows you to constantly receive information about the complex mechanical characteristics of the composite, depending on temperature and other physical parameters of external conditions, as well as changes in material properties during long-term operation. In addition, if the composite has a system for controlling mechanical characteristics, the mechanical properties of the composite can be adjusted under various conditions.

The scheme for performing experimental investigations was simplified compared with Fig. 5.45 and is shown in Fig. 5.46. In this case, the sensors were placed only on the outer surface of the plates. Plates (3) were investigated with a surface size of 530×250 mm and a thickness of 6–12 mm (Section 5.1; Figs. 5.4, 5.7 and 5.8; Table 5.4). On top of the plate were installed piezoelectric pressure pulsation sensors (1), (2). The distance between the sensors was set to 200 and 100 mm. To determine the anisotropy coefficient, the sensors were placed on the plate alternately in the longitudinal and transverse directions with respect to the plate. To excite waves on the surface, a steel ball (5) was used, which was held by the

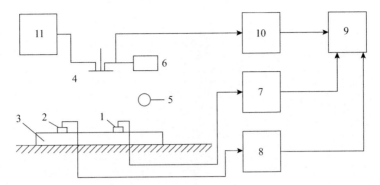

Figure 5.46 The scheme of the experimental device: (1), (2) sensors, (3) plate, (4) button, (5) ball, (6) electromagnetic coil, (7), (8) preamplifiers, (9) oscilloscope, (10) block delays, and (11) block power supply [323].

magnetic field of the electromagnetic coil (6) at a certain height H. The balls were used in various sizes. Experiments have shown the independence of the speed of wave propagation from the mass of the ball. The coil was connected to the power supply unit (11) via button (4). When the button was pressed, the current was interrupted and the ball fell to the surface the test material, initiating an elastic wave in the material. The peculiarity of the method was that the glass tube inside which the ball fell on the surface of the elastic plate did not touch the plate so as not to deform the surface wave shape. Elastic waves that occur on the surface of the plates when the balls fall, propagate at a speed that depends on the dynamic elastic modulus of the material. The excited waves were perceived first by the sensor (1) and after a while by the sensor (2). The signals from the sensors through the preamplifiers (7), (8) were fed to a dual-beam oscilloscope.

To start the oscilloscope sweep, a signal from the button was applied to the delay circuit (10), which was necessary to start the sweep after the ball touched the plate. Smooth adjustment of the start-up delay allowed to record on the oscilloscope screen the beginning of the propagation of elastic waves in the material. The waves that appeared in the material were recorded on an oscilloscope screen and recorded with a camera, and then the results were processed. Knowing the oscilloscope scanning speed and measuring the distance between the signals recorded on the photographic film obtained from two sensors, the time of the passage of a wave excited in the material from one sensor to another was determined.

Fig. 5.47 shows a photograph of an elastic wave excited in a standard made of acrylic plastic [Fig. 5.48, curves (1), (2)]. These measurements were reference when analyzing the propagation of oscillations on elastic plates. After the ball fell and the oscillation propagated up to the first sensor, the largest oscillation amplitude A on the studied surfaces depended on the frequency of its own fluctuations in the material or composite.

Thus in Fig. 5.47, the first front of the surface waves and the oscillations following it are recorded. It is seen that the character of the oscillation is different from

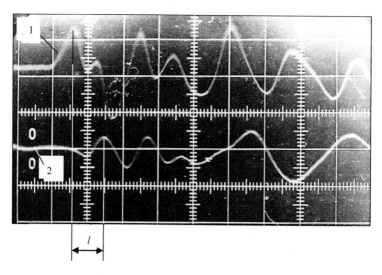

Figure 5.47 Photograph from the oscilloscope screen of oscillations on a solid surface (reference): (1) data of the first sensor, (2) data of the second sensor, (*l*) distance between the wave fronts recorded by the first and second sensors.

the harmonic. This is due to the presence of not only the spectrum of disturbances, but also the presence of scratches and irregularities on the surface of the standard, causing distortion of the waveform. The amplitude and character of the oscillations recorded by both sensors determine the attenuation value. Oscillations detected by the first sensor weakly fade out, and those recorded by the second sensor are similar in shape to the first with slightly smaller amplitude.

From the known distance between the sensors and the measured time of passage of the wave between them, the propagation velocity of the elastic wave in the material was determined.

The velocity of propagation of an elastic wave is related to the dynamic modulus of elasticity E by the dependence

$$E = \rho c^2 \tag{5.7}$$

where ρ is the density of the material; c is the propagation velocity of the elastic wave in the material. The dynamic modulus of elasticity was calculated by the formula (5.7) (Table 5.14), the numbers of the investigated elastic plates are given in Table 5.5. The geometrical parameters of the studied plates are given in Tables 5.4 and 5.5, and the mechanical characteristics of elastic plates are given in Tables 5.6 and 5.7, and also in Ref. [60,139]. Table 5.14 shows the results of measured mechanical characteristics of elastic plates given in [323].

Analyzing the obtained results, it should be noted that the materials used to investigation the effect of damping on the turbulent boundary layer are characterized by a wide range of values of the dynamic modulus of elasticity.

Experimental investigations of the characteristics of the boundary layer on the analogs of the skin 325

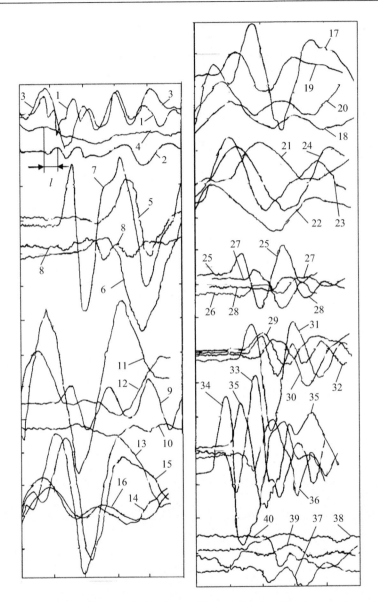

Figure 5.48 Surface oscillations of elastomers. The notation for the curves is shown in Table 5.15.

The plates 10, 10a under study were made of longitudinal glued strips, and plates 3, 8a, 11, 11a had a ribbed surface on the side of gluing them to the base, and besides, in plates 11, 11a, ribbing was longitudinal (along the stream), and in plates 3, 8a—transverse.

Table 5.14 Dynamic modulus of elasticity, calculated from Table 5.7.

Plate number	ρ, kg/m^3	c, m/s longitudinal	c, m/s transverse	E, N/m^2 longitudinal	E, N/m^2 transverse	k_2	Number of pores on 10^{-2} m
2	42	62	62	$1.61 \cdot 10^5$	$1.61 \cdot 10^5$	1	26
2a	42	66	66	$1.83 \cdot 10^5$	$1.83 \cdot 10^5$	1	26
5	42	67	67	$1.88 \cdot 10^5$	$1.88 \cdot 10^5$	1	32
1	107	74	74	$5.86 \cdot 10^5$	$5.86 \cdot 10^5$	1	36
1a	115	85	85	$8.31 \cdot 10^5$	$8.31 \cdot 10^5$	1	31
6a	107	146	146	$2.28 \cdot 10^6$	$2.28 \cdot 10^6$	1	31
6	107	160	160	$2.74 \cdot 10^6$	$2.74 \cdot 10^6$	1	3
10	410	266	200	$2.9 \cdot 10^7$	$1.64 \cdot 10^7$	1.77	34
10a	410	200	160	$1.64 \cdot 10^7$	$1.05 \cdot 10^7$	1.56	34
11a	1045	194	162	$3.93 \cdot 10^7$	$2.74 \cdot 10^7$	1.43	6
8a	930	183	200	$3.11 \cdot 10^7$	$3.72 \cdot 10^7$	0.84	8
11	1050	190	173	$3.79 \cdot 10^7$	$3.14 \cdot 10^7$	1.21	6
3	1050	152	177	$2.43 \cdot 10^7$	$3.29 \cdot 10^7$	0.74	6

Research results confirm that along the direction of the strips or edges, the speed of propagation of waves and, accordingly, the dynamic modulus of elasticity is higher than with the propagation of waves across the strips and edges. Obviously, with a decrease in the number of pores per centimeter and a decrease in the pore sizes of the dynamic modulus of elasticity should increase, which was confirmed experimentally.

Knowing the scanning speed of the electron beam of the oscilloscope and the distance L between two sensors mounted on the surface, determines the velocity of propagation of the elastic wave. Fig. 11.48 shows graphic copies of photographs of the shape of surface waves on the surfaces under study. In Table 5.15, balls with a diameter d were positioned from the surface of the plates under study at a distance H. In the experiments, balls were also dropped onto a plastic rod supported on the plate under study plate with longitudinal arrangement vibration sensors (curves 33–40).

According to formula (5.7), the dynamic modulus of elasticity of a material can also be calculated by the formula:

$$E = \rho(L^2 v^2 / l^2) \tag{5.8}$$

Table 5.15 Conditions for performing experiments (to Fig. 5.48).

Curve number	Plate number	$H \cdot 10$, m	$d \cdot 10^3$, m	Sensors arrangement
1, 2	Standard	1.3	2	Longitudinal
3, 4	5	1.3	10	Longitudinal
5, 6	5	1.45	10	Longitudinal
7, 8	2	1.3	7	Longitudinal
9, 10	2	1.45	10	Transversal
11, 12	1	1.3	10	Longitudinal
13, 14	1	1.3	8	Longitudinal
15, 16	1	1.3	7	Longitudinal
17, 18	1	1.3	6	Longitudinal
19, 20	1	1.3	7	Longitudinal
21, 22	1	1.45	10	Transversal
23, 24	6a	1.3	7	Longitudinal
25, 26	10	1.3	7	Longitudinal
27, 28	10	1.45	7	Longitudinal
29, 30	10a	1.45	10	Transversal
31, 32	11a	1.3	10	Longitudinal
33, 34	11a	1.3	10	Longitudinal with a rod
35, 36	11a	1.3	10	Longitudinal with a rod
37, 38	11a	1.3	8	Same, $h = 70$ mm, water
39, 40	11	1.3	10	Same, $h = 70$ mm, oil

The anisotropy coefficient of the properties of materials K_1 can be determined by the formula:

$$\kappa_1 = E_{dynam.}/E_{static.} \tag{5.9}$$

The anisotropy coefficient K_2 is calculated according to the longitudinal and transverse flow directions:

$$k_2 = E_{long.}/E_{transv.} \tag{5.10}$$

As can be seen from the oscillograms of oscillations, after the third period on the rigid plate, the amplitude distortion begins, caused by nonlinear effects. A completely different form and character of the development of surface waves on elastic plates. Although the oscillations are distorted, their development in time is much better characterized by an exponential law.

To generate surface waves on a standard, it was enough to use a ball with a diameter of $d = 2 \cdot 10^{-3}$ m, and record on an oscilloscope with a gain of $K_{ampflic.} = 10$.

Since elastomers absorb disturbances, these parameters had to be increased. The gain is graded in decibels, so a change in $K_{ampfl.}$ from 10 to 20 corresponds to a signal gain of 3.16 times, and a change from 10 to 30 means a gain of 10 times. To determine the influence of the disturbance energy on the amplitude of the oscillations, experiments were conducted with different balls, the parameters of which are given in Table. 5.16.

The elastic plates were divided into two groups. In the first group there were plates with an outer layer made of a material with a large number of pores (flexible), and in the second group there were plates with a small number of pores (elastic).

Fig. 5.48 (curves 3–6) shows the propagation of oscillations on the surface of the plate 5 (Fig. 5.7, Tables 5.4 and 5.5). The outer layer of the plate is an open sheet of FPU-2, which intensively dampens oscillations. The amplitude of oscillations A is about 15 times less than on the reference plate. The second sensor recorded only very small oscillations that could be detected only with an increase in the disturbance energy (mgH) [curves (5), (6)]. A plate glued with a rubber film showed an increase in elastic parameters. Thus the amplitude A recorded by both sensors on plate 2 increased in compared with the amplitude of plate 5 by 1.2 times [compare curves (5), (6) with (7), (8)].

Table 5.16 Range of parameters of balls.

Parameter			Range of balls			
$10^3 \cdot d$ (m)	2.0	5.3	6.0	7.0	7.9	9.9
$10^3 \rho$ (kg/m^3)	7.8	10.1	7.6	7.8	7.7	7.77
$10^3 m$ (kg)	0.33	0.6	0.88	1.4	2.0	3.96

On a plate 2 made of a monolithic homogeneous material glued on top with a thin rubber film, the fluctuation parameters were measured in the longitudinal and transverse directions. The amplitude A in the transverse direction is 4 times less than in the longitudinal direction. This means that the longitudinal energy of the directed disturbance propagates more intensively. Therefore the design of the plate allows you to orient the disturbance in a certain direction.

More rigid plates (1, 1a, and 9, Table 5.5) are made of the same material FE-3, but of different thicknesses. Oscillations of the surface of the plate 1, which has the greatest thickness, depend on the intensity of the load [curves (11)−(18) Fig. 5.48]. The shape of the oscillations remains the same for all masses of the ball. With a decrease in the disturbances energy, the amplitude of fluctuations A decreases. In all cases, due to the gain coefficient $K_{ampi.}$, the value A of the plate 1 is 2.6 times larger than that of the plate 2. On the plate 1, the development of oscillations in the longitudinal and transverse directions was also checked [curves (19)−(22)]. Considering that the values of d and H in the second case are larger, the anisotropy of properties is characteristic of plate 1, although the plate was also monolithic. In the longitudinal direction, the value of A is 2−3 times larger than in the transverse direction. In the latter case, the oscillations are fixed in antiphase, which is due to a more significant attenuation of the wave at a distance L.

Fig. 5.49 shows a photograph of the development of oscillations on the surface of the elastic plate 1 in the longitudinal direction. The elastic layer of the plate 1a is thinner than that of the plate 1. Therefore the plate 1a has a higher stiffness and the oscillation amplitude is greater. The outer layer of the plate 9 has the same material as the plates 1 and 1a, but its thickness is less. The stiffness of the layer is the largest, but the amplitude A is the lowest since the plate 9 is composite (plates 1 and 1a are single-layer plates). Anisotropy of properties is absent.

The elasticity of the plates 6, 6a, 10, and 10a is even higher. Curves (23), (24) in Fig. 5.48 represent oscillations on the surface of plate 6a. The value A of plate 6a

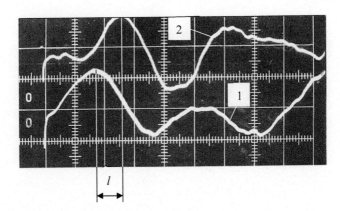

Figure 5.49 Photograph from the screen of an oscilloscope on an elastic plate 1: (1) data of the first sensor, (2) data of the second sensor, l, distance between the wave fronts recorded by the first and second sensors [curves (19), (20) Fig. 5.48].

is 1.7 times lower than on plate 1 because of its composite structure. The anisotropy properties of the plates 6 and 6a are provided by the structure of their inner layers. The anisotropy of the plates 10 and 10a is ensured by the structure of their outer layers made of strips glued to each other from the same material from which the plates 6 and 6a are made.

The strips were rotated 90 degree on plates 10 and 10a (the properties of porous materials depend on the direction of foaming). With the propagation of oscillations in the transverse direction, their amplitude decreases, and the phase shift increases up to 180 degree. Plate 10 has the most pronounced anisotropy of its properties [curves (25)–(28)]. The amplitude of the oscillations on the surface of the plate 10 in the longitudinal direction is 2 times less than on the plate 1, and in the transverse direction, the amplitude of the oscillations is 2.7 times less than in the longitudinal direction. The pitch of the longitudinal strips in the outer layer of the plate 10a is 2 times smaller than on the plate 10. The value A of the plate 10a is slightly smaller in the longitudinal direction and 1.5 times smaller in the transverse direction than on the plate 10.

Fig. 5.50 shows a photograph from the screen of the oscilloscope of the development of oscillations on the surface of the elastic plate 10 in the transverse direction.

Plates 7, 7a, and 8, the outer layer of which is made of FE material, are characterized by rapid damping of oscillations. Of greatest practical interest is the group of plates made of low porous material (polyurethane). These are plates 11, 11a, 3, and 8a (Tables 5.4 and 5.5). Plates 11 and 11a have a longitudinally ordered inner surface, and plates 3 and 8a have a transverse ordered inner surface. Fig. 5.48 shows the surface oscillations of the plate 11a.

Among the compliant plates of the first group, the maximum amplitude was for plate 1. On plate 11a, the oscillation amplitude decreased by 1.6 times compared to plate 1. The first group of compliant plates includes plates made or having PU foam material in the structure. Other plates conditionally belong to the second group of elastic plates.

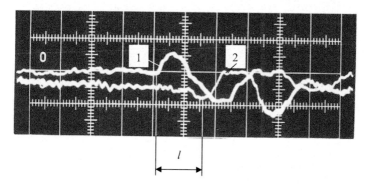

Figure 5.50 Photograph from the screen of the oscilloscope on the elastic plate 10: (1) data of the first sensor, (2) data of the second sensor, l, distance between the wave fronts recorded by the first and second sensors [curves (27), (28) Fig. 5.48].

Experiments were also carried out when the ball did not fall on the elastic plate, but on a special rod mounted vertically on elastic plates with rubber braces. The rod is made light—from a thin tube with tips made of organic glass. Despite this, the oscillations amplitude of plate 11a in this experiment increased by 1.25 times. When the plate 11a was heated [curves (35), (36)], the value of A decreased 1.3 times [curves (33), (34), and (35, 36)].

Fig. 5.51 shows a picture of the development of oscillations on the surface of the elastic plate 11a in the longitudinal direction.

To compare the obtained data with the results of testing elastic plates in a hydrodynamic channel, the plates were mounted in a cuvette filled with water or another viscous liquid. The thickness of the liquid layer above the plate was 7 cm. The oscillations characteristics changed, and the value of A decreased by 8.5 times. The oscillations parameters of the plate 11 were the same as on the plate 11a. A decrease in fluctuations in this case indicates the damping properties of a viscous fluid (oil) located above the elastic plate (Fig. 5.52). When photographing oscillations on the surface of elastic plates from the screen of an oscilloscope, due to the limited area of the monitor screen, it is impossible to track the dynamics of the development of oscillations in time. This is conveniently performed using a loop oscilloscope (Fig. 5.53).

The oscillogram was used to determine the frequency of own oscillations of the elastic surface and the logarithmic decrement of attenuation. Free oscillations of an elastic surface can be expressed by the equation:

$$y = Ae^{at}\sin\left(\beta t + \varphi_o\right) \tag{5.11}$$

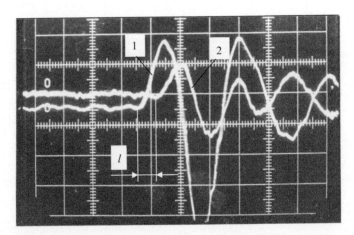

Figure 5.51 The screen shot of the oscilloscope on the elastic plate 11a: (1) data of the first sensor, (2) data of the second sensor, l, distance between the wave fronts recorded by the first and second sensors [curves (31), (32), Fig. 5.48].

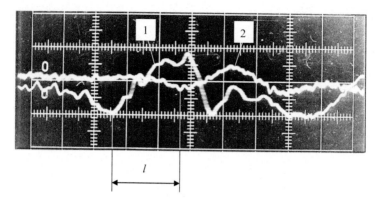

Figure 5.52 A screen photograph of an oscilloscope on an elastic plate 11: (1) data of the first sensor, (2) data of the second sensor, l, distance between the wave fronts recorded by the first and second sensors [curves (39), (40) Fig. 5.48].

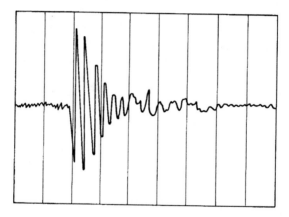

Figure 5.53 Oscillogram of surface oscillations obtained using a loop oscilloscope. Vertical lines are time stamps.

As follows from the character of the oscillograms, the oscillations of the elastic plates are nonlinear and damped. Such oscillations lead to the appearance of an added mass of liquid, which begins to change with a phase delay in comparison with the oscillations of elastic plates. This also leads to a decrease in the amplitude of oscillations of the elastic plates.

Table 5.17 shows the mechanical parameters of the elastic plates, calculated in accordance with the measurements according to the above measurement method [60, 139].

It is interesting to compare the above mechanical parameters of elastic plates with similar data obtained by measuring living dolphins on the surface of the body. So, for example, in Part I, Experimental Hydrodynamics of Fast-Floating Aquatic Animals, Section 6.5, Fig. 5.6 shows the distribution of the skin elastic modulus over the body of a dolphin dolphin. The measurements were performed by Babenko with the help of a compact elasticity meter developed by him (Part I, Experimental

Table 5.17 Mechanical characteristics of elastic plates.

Plate no.	Density ρ, kg/m^3	Amount pore/cm	Frequency f, Hz	Phase longitudinal	Velocity C, m/s transversal
Standard plate	1190	–	639	1230	1230
5	42	20–25	78	29	29
5a	42	20–25	99	43–48	40–47
2	42	20–25	81	47	53
2a	42	20–25	83	–	48
1	107	32	214	64.3	58.6
1a	115	36	246	66.6	48.6
6	107	30–40	260	70.7	67
6a	107	30–40	247	60.4	61
9	107	32	137–153	51.7	50.5
9a	173	45–55	75	87.4	40
7a	200	40–60	74	43.6	–
8	300	30–40	65	123	–
10	410	50–55	200	52.7	84.1
10a	410	50–55	231	52.7	90
11	1050	5–10	185	61	66
11a	1045	4–9	239	49.7	68.2
3	1050	4–7	169	40	23
8a	930	4–7	233	63.2	52.7

Plate number	Static modulus of elasticity 10^{-4} E_{st}, MPa	Dynamic modulus of elasticity		Anisotropy coefficient	
		10^{-4} $E_{long.}$, MPa	10^{-4} $E_{trans.}$, MP	$k_1 = E_{long\ dyn}/E_{st}$	$k_2 = E_{long.}/E_{trans.}$
Standard plate	1.77	1.77	1.77	1	1
5	0.35	0.353	0.353	1	1
5a	0.53	0.7–1.0	0.7–0.9	1.3–1.9	1
2	0.3–1.4	0.9	1.15	1	0.8
2a	0.3–1.2	1.23	0.98	1	1.25
1	0.49	4.3	3.6	8.85	1.2
1a	0.49	5	2.67	10	1.87
6	0.59	0.536	0.48	0.9	1.1
6a	0.59	3.76	3.9	7	0.97
9	0.49	2.8	2.7	5.7	1.04
9a	0.39–5.7	13	2.7	2.3–3.3	4.8
7a	0.35	3.82	–	11	–
8	4.5	44	–	9.8	–

(*Continued*)

Table 5.17 (Continued)

Plate number	Static modulus of elasticity 10^{-4} E_{st}, MPa	Dynamic modulus of elasticity		Anisotropy coefficient	
		10^{-4} $E_{long.}$, MPa	10^{-4} $E_{trans.}$, MP	$k_1 = E_{long\ dyn}/E_{st}$	$k_2 = E_{long.}/E_{trans.}$
10	0.35	0.98–3.9	28.4	2.8–11	0.14
10a	0.35	0.98–3.9	32.4	2.8–11	0.13
11	8.8	39	45.6	4.3	0.85
11a	5.9	25.3	47.6	4.3	0.53
3	3.7	15.7	5.5	4.1	2.9
8a	7.85	36.5	25.3	4.65	1.45

Hydrodynamics of Fast-Floating Aquatic Animals, section 6.4, fig. 6.13) on the surface of the body of various species of dolphins. As shown in Part I, Experimental Hydrodynamics of Fast-Floating Aquatic Animals, sections 6.5 and 6.6, the elasticity of the skin along the body varies depending on the type of dolphin and their swimming speed. In section 6.5, the elastic modulus $E \cdot 10^{-4}$, N/m² for dolphin dolphins varied between 8.38–10.1, for bottlenose dolphins—7.93–9.62, and for porpoise—4.55–7.06. According to the data in Table 5.16, the static modulus of elasticity of the plates 11, 11a, 3, and 8a was closest to the measurements of the static modulus of elasticity on dolphin skin.

Part I, Experimental Hydrodynamics of Fast-Floating Aquatic Animals, Section 6.9 in Fig. 6.35 shows the distribution of the phase velocity of forced oscillations on the surface of the skin of a bottlenose dolphin (see Part I, Experimental Hydrodynamics of Fast-Floating Aquatic Animals, S.M. Kidun). The phase velocity c of the plates 11, 11a, 3, and 8a is an order of magnitude smaller than on the surface of the dolphin skin. Fig. 6.36 (Part I, Experimental Hydrodynamics of Fast-Floating Aquatic Animals) shows photographs of oscillograms of fluctuations in the surface of a dolphin's skin during slow swimming and at rest. The amplitude of the oscillations on the surface of the dolphin's skin is constant, while the forced oscillations of the elastic plates are damped (Fig. 5.53).

5.8 The investigation of oscillations on the surface of elastic plates during flow around a water stream

In Section 5.7, the results of experimental investigations of the development of oscillations on the surface of elastic plates are presented. In this case, the elastic plates were in the air, and oscillations in the plates occurred when steel balls of various weights fell on their surface. Below are the results of an experimental investigation on the same elastic plates installed in a hydrodynamic stand of small turbulence [378] in the region of the tens metric insert located at the end of the working section (Section 3.7, Fig. 3.41). The aim of the investigation was to study the pulsation characteristics on the surface of elastic plates caused by pulsations of the boundary layer. It is known that the amplitude of oscillations of the elastic surface is less than 20 μm [8, 217,235,323,603,630]. This creates certain difficulties in measuring fluctuations in the surface of an elastic plate. For measurements with high accuracy, the optical method was chosen, the main advantage of which is non-contact determination of the amplitude of surface motion [217,229,230,630].

The measurement method is as follows: a spot of light was projected onto the measured surface from the light source through focusing lenses. Light reflected from the surface was perceived by a photo detector. Depending on the distance, from the light source and receiver to the surface, the spot is defocused and a different amount of light energy enters the receiver.

A device that implements this method was developed by the M.V. Kanarsky [323] (Fig. 5.54).

The light from the DC power supply (1) and the light source (2) through the light guide (3) and the focusing device (4) falls on the test surface (14) mounted on the bottom of the working section of the hydrodynamic stand. The reflected light enters the receiving part (7) and through the optical fiber (8) enters the photo electronic multiplier (9). The signal in the form of voltage is supplied through the amplifier (10) to a DC voltmeter (11), a RMS meter (12), and a tape recorder (13). A photoelectric multiplier was used in this measuring device PMT-2, a DC voltmeter and RMS values of DISA (55D30 and 55D35), as well as an MP-1 tape recorder.

The focusing device (4) and the receiving part (7) are made in the form of a tube with a diameter of 10 mm. Lenses are mounted in the focusing and receiving tubes for

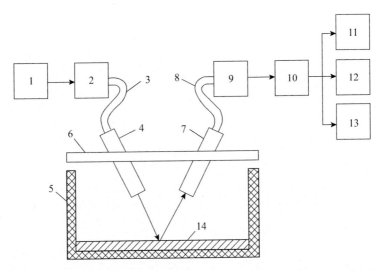

Figure 5.54 Scheme of a device for measuring vibrations on the surface of an elastic plate [323].

focusing the light beam on the test surface and the light reflected from it on the end surface of the light guide. To increase the accuracy of measurements, a thin mirror elastic foil with a diameter of 1 mm was glued on the test surface. The focusing device and the receiving part were fixed on one holder (6), which was connected to the coordinate device necessary for calibration of the measuring device. The holder (6) was not installed on the hydrodynamic stand and had an independent mount. This avoided the influence of vibrations of the design of the hydrodynamic stand on the readings of the measuring equipment. The coordinate device was connected to a Measure, which measured the displacement of the holder together with the focusing and receiving lenses mounted on the holder, with an accuracy of 1 μm. The calibration function of the output voltage relative to the distance from the surface was linear.

At the same time, pressure pulsations on elastic plates were investigated using pressure sensors. The signal processing technique is given in Ref. [324,536]. In elastic plates, vertical holes with a diameter of ∼1 mm were made. Pressure pulsation sensors were fixed from below to elastic plates on the body of the strain gage insert instead of mechanotronic devices (Section 5.1, Fig. 5.4). The measurements were carried out on the same plates as in the air flow (Section 5.1, Table 5.4).

The insert at the end of the working section was mounted on a strain gage suspension (Section 5.1, Fig. 5.6). The strain gage suspension of the insert allows it to move in the longitudinal direction. In order not to introduce errors in the measurement results due to possible fluctuations of the insert in the longitudinal direction, a wedge was installed in the back gap B between the insert and the fixed part of the bottom of the working section, which pushed the insert tightly to the bottom in the region of the front gap A and made the insert stationary. Investigations of fluctuations on the surface of elastic plates caused by the flow of water were carried out on two types of plates: compliant 2a and elastic 11 (Section 5.1, Table 5.5). The measurements were carried out on a hydrodynamic stand at a flow velocity of 0.6 m/s in section III (Section 3.7, Fig. 3.41). The design of the hydrodynamic stand is given in Section 3.7.

To distinguish the level of interference, the output signal from the displacement meter was also recorded during the flow around the standart rigid plate and at zero flow velocity, including with an idle pump. Each plate was measured three times. Fluctuations in the surface displacement in the form of stress were recorded on an MP-1 tape recorder. The measurement results of the RMS values of surface oscillations are presented in Table 5.18.

Table 5.18 Amplitudes of oscillations of the elastic plate surface.

Plate number	$\sqrt{a'^2}$, μm Measurement number			$\left\langle \sqrt{a'^2} \right\rangle$, μm
	1	2	3	
2a	12	17	15	14.7
11	18	21	25	21.3

The oscillation amplitude of plate 2a (Section 5.1, Table 5.5, FPU with a film) turned out to be less than that of plate 11 (PU). This can be explained by the resonance properties of the plate 11, which has higher values of the own frequency of oscillations and the elastic modulus (Section 5.7, Tables 5.14 and 5.16).

In theoretical investigations of the interaction between flow and elastic plates, a complex stiffness coefficient was calculated that relates the intensity of pressure fluctuations to the amplitude of surface vibrations [235,559]:

$$Z = \frac{\sqrt{\overline{p'^2}}}{\sqrt{\overline{a'^2}}} \tag{5.12}$$

where

$$Z = S - iarc - m\alpha^2 c^2 + T\alpha^2 \tag{5.13}$$

Based on (5.12), we can estimate the magnitude of fluctuations of surface displacements [323]. To do this, you can apply the formula [513,536,521,605,606]:

$$\frac{2\sqrt{\overline{p'^2}}}{\rho U_\infty^2} = d_\tau \lambda \tag{5.14}$$

where d_τ is the proportionality coefficient (for a smooth surface, $d_\tau = 2.1$, and λ is determined by the formula:

$$\lambda = \frac{2\tau_0/\rho}{U_\infty^2} \frac{2\overline{u'v'}}{U_\infty^2} \tag{5.15}$$

In accordance with the graphs of the distribution of the shear stress over the thickness of the boundary layer during flow around the standard and elastic plates given in Refs. [60,139], we take, respectively, for a rigid plate $\lambda = 0.004$ and for an elastic plate $\lambda = 0.0032$. Then, for $U_\infty = 0.6$ m/s, the value calculated by formula (11.14) is 1.5 N/m² for a rigid plate and 1.2 N/m² for an elastic plate.

The approximate values of the mean square amplitudes of the surface displacement calculated by (5.12) for plates 2a and 11 based on their dynamic elastic moduli were 8.1 and 0.5 μm, respectively.

A good agreement was obtained between calculation and experiment for plate 2a. For plate 11, the calculation differs significantly from experiment.

In [166] another expression for compliance is presented:

$$Z^{-1} = \left[\sigma\omega^2 - T\alpha^2 - S\alpha^2 - \rho_w g - \sigma_s(\rho_w, \alpha, c, H, \nu)\right]^{-1} \tag{5.16}$$

A detailed analysis of the boundary conditions was performed in Refs. [33,60,589]. Obviously, the pressure fluctuations of the boundary layer and the frequency properties of the streamlined surface influence the amplitude of the oscillations of the surface of the elastic plate. The receptibility of the boundary layer is also of great importance, which consists, in particular, in the fact that under the influence of rather weak pressure fluctuations in the flow, a resonant increase in the amplitude of the surface displacement or, on the contrary, its substantial decrease can be observed.

The assumption of the resonant character of the interaction of surface fluctuations and oscillations of the boundary layer is confirmed by the values of f and f^+ (Section 5.7, Table 5.16, [60,139]) and the investigation of spectral characteristics of these oscillations (Fig. 5.55). Fig. 5.55A shows the oscillations spectra of the surface of a rigid plate made of plexiglass (standard). When the pump is running and the valve is closed [$U = 0$, curve (2)], an increase in relative amplitudes is observed

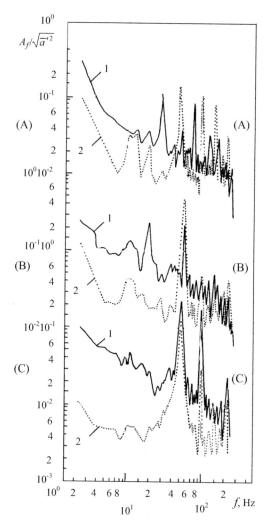

Figure 5.55 Oscillation spectra of the hard surface (A) and elastic plates 11 (B) and 2a (C); (1) $U_\infty = 0.6$ m/s, (2) $U_\infty = 0$ (the pump is running, the valve is closed) [60,139].

at frequencies of 12 and 20 Hz, which can be caused by oscillations of the entire strain gage suspension.

The increase in the relative amplitudes of oscillations at a frequency of 12 Hz is noticeable on elastic plates, but it is not so large—damping properties plates, especially compliant plate 2a (Fig. 5.55C).

When the pump is running and the valve is closed, an increase in the relative amplitudes of curves (2) at a frequency of 50 Hz is noticeable, caused by network interference. There is also a slight increase in amplitude at the second harmonic of the network pick-up—100 Hz. When the pump is running and open valve curves (2) characterize the vibration of the plates. When a plate flows around a stream under the influence of pressure pulsations, oscillations begin to interact, caused by vibration of the base of the elastic plates and pressure pulsations of the boundary layer. In this case, the spectra of surface oscillations are reconstructed on the plates [curves (1)]. On a hard plate noted above although vibration and network-induced bursts of pulsations are preserved, their contribution to the energy of the displacement spectrum decreases. Due to the superposition of the indicated oscillations, bursts at 30 and 75 Hz appear on curve 1. On the whole, the energy of oscillations of the plate surface caused by the flow increases compared with the case of no flow around [curve (2)].

The elastic plate 11 in terms of its mechanical characteristics remains, as in the case of air flow around it, rigid at a speed of $U_\infty = 0.6$ m/s. Therefore the general character of curve (1) relative to curve (2) remains the same as that of the standard. The difference is that due to the resonant increase in the amplitude of the oscillations (Table 5.17), the energy of surface oscillations during flow around the plate 11 increases more intensively than in the standard over a wide range of oscillations frequencies. Thus the pressure fluctuations of the turbulent boundary layer [curve (1)] significantly increase the amplitude of the surface oscillations.

According to the similarity criteria, plate 2a is more effective at low flow velocities; therefore its vibration-absorbing properties are better [curve (2), Fig. 5.55C]. Since the compliance of plate 2a and its oscillation amplitude is higher than that of a rigid plate, the differences between curves (1) and (2) are also greater than that of a standard. Due to the high vibration-absorbing properties, all peaks on curve (1) are smoothed out.

A comparison of the data is shown in Fig. 5.56. It is seen that the energy of surface oscillations is greatest at plate 11. Its oscillation spectrum [curve (2)] has a classical shape typical of pulsations longitudinal velocity in the boundary layer [curve (5)]. At low frequencies, the oscillation spectrum of plate 2a [curve (3)] is smaller than that of a rigid standard [curve (1)]. In the other part of the frequency range, these curves almost coincide. The vibration-absorbing properties of the plate 2a in the absence of flow [curve (4)] remain constant in almost the entire frequency range. Probably, this is precisely why the minimum friction coefficients were found when a stream 2a flows around a plate in a hydrodynamic stand with a low degree of turbulence in the main stream.

Thus taking into account the kinematic characteristics of the boundary layer above the elastic plates and the data given in Figs. 5.55 and 5.56, we can assume that the process of interaction of oscillation damping surfaces with the flow is

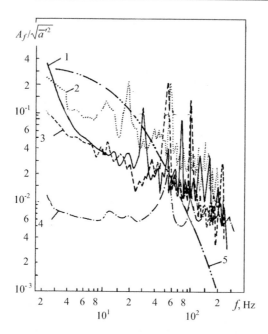

Figure 5.56 Comparison of the oscillation spectra of the hard (1) surface and elastic plates 11 (2) and 2a (3) at $U_\infty = 0.6$ m/s, (4) plate 2a at $U_\infty = 0$ (the pump is running, the valve is closed), (5) spectrum of longitudinal fluctuations in the speed of plate 11 [60,139].

complex character. It can be assumed that the main mechanism of interaction of the boundary layer with compliant plates is the absorption of pulsating energy [355–360,586,587,589,590]. When flowing around elastic plates, the main mechanism is the effects of preexplosive modulation [192–194].

From a comparison, curves (1), (2) in Fig. 5.55 show that the pressure fluctuations of the surface are significantly affected by the pressure pulsations of the turbulent boundary layer, which correlate well with the longitudinal pulsation velocities [227,228,605] and therefore with pulsations near-wall shear [228,229]. Therefore it is interesting to compare the spectra of surface fluctuations with the spectra of pressure pulsations (Fig. 5.57). In Ref. [208], a technique was proposed for measuring the pressure fluctuations of a rigid plate according to the data of motion pulsations. Pressure pulsation sensors in these experiments were less sensitive to vibrations. Therefore bursts at vibration frequencies of 12 and 20 Hz are negligible, and the main bursts on a hard standard are caused by interference from the power line. Comparing series "A" of Figs. 5.55 and 5.57, we can conclude that pressure pulsations correlate well with movement pulsations. Curve (3) in Fig. 5.57 (the pump does not work) is metrological. It is located above all other curves, because in the absence of disturbances the denominator of the ratio $p(f)/\sqrt{\overline{p'^2}}$ decreases. When flowing around elastic plates, these values correlate weaker [compare curves (1), (4) of series B]. Instead of two vibration bursts at frequencies of 12 and 20 Hz, when measuring pressure pulsations sensors, elastic bursts were recorded at frequencies of 15–17 Hz.

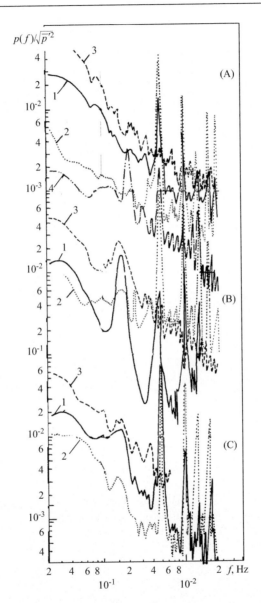

Figure 5.57 Spectra of pressure pulsations on rigid (A) and elastic plates 11 (B) and 2a (C): (1) $U_\infty = 0.6$ m/s; (2), (3) $U_\infty = 0$; (2) the pump is working, (3) the pump is not working, (4) the spectrum of fluctuations of the surface of the plate 11 (dimension see Fig. 5.56).

Comparison of the spectra of pressure pulsations on rigid and elastic plates is shown in Fig. 5.58. If in Fig. 5.56 the energy of the spectrum of pulsations on the surface of plate 11 is greatest, then in Fig. 5.58 the energy of pressure pulsations on this plate is the smallest. This contradiction is explained by the fact that compliant plate 2a only absorbs

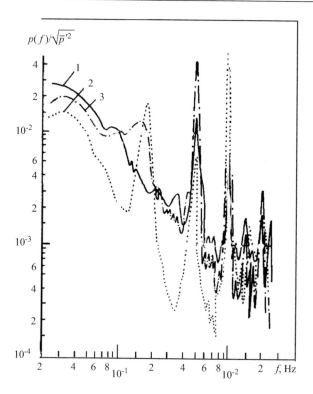

Figure 5.58 Comparison of the spectra of pressure pulsations on rigid (1) and elastic plates 11 (2) and 2a (3) at $U_\infty = 0.6$ m/s.

transverse pulsations of speed and pressure. Its amplitude of surface fluctuations is smaller than the plate 11, and the vibration-absorbing characteristics are higher.

Pressure pulsations reflect the nature of the structure of the boundary layer, which changes in the flow on elastic surfaces. The spectra of pressure pulsations better characterize the change in the kinematic characteristics of the boundary layer on elastic surfaces. It follows that there were significant changes in the structure of the boundary layer during the flow around the elastic plate 11.

The plate 11 generates three-dimensional distributions in the form of longitudinal vortex systems into the boundary layer from below. It was found that this leads to an increase in the fluctuations of the surface of the plate 11 in comparison with the plate 2a. In addition, this leads to an increase in the transverse components of the oscillation velocity, a decrease in the anisotropy of turbulence, and, consequently, to a decrease in pressure pulsations. This is in good agreement with the measurements of the pulsation velocities of the boundary layer and the kinetic energy profiles [60,139] and corresponds to the formulas [605,606]:

$$\sqrt{\overline{p'^2}}/q = 0.0035 \qquad (5.17)$$

and [393]:

$$\sqrt{\overline{p'^2}} = \alpha \tau_w \qquad (5.18)$$

$$2 \leq \sqrt{\overline{p'^2}}/\tau \leq 4 \qquad (5.19)$$

The best normalization of the spectra of wall pressure fluctuations is given in Ref. [228]. The decrease in pressure fluctuations during flow around elastic plates at low frequencies (Fig. 5.58) is consistent with similar measurements in elastic pipes [217].

In Refs. [47,60,139], the results of experimental studies of the characteristics of boundary layers at various stages of the transition, including a turbulent boundary layer during flow around various types of simple and multilayer elastic surfaces, are presented. It was shown that the longitudinal vortex structures in the viscous sub layer of the turbulent boundary layer during flow around elastic surfaces significantly increased compared with similar parameters when flowing around a rigid surface.

In Refs. [404,405], the structure of a near-surface flow with zero pressure gradient of a turbulent boundary layer of a flat plate with a single-layer viscoelastic compliant surface was visualized using the hydrogen bubble method. Compliant materials were made by mixing silicone elastomer with silicone oil. Stream visualization experiments fixed low-speed parietal longitudinal strips with an increased transversal interval between the strips and extended spatial coherence compared with similar results obtained on a solid surface. At low Reynolds numbers, an intermittent relaminarization phenomenon was observed for the concrete surface under study. Apparently, the observed changes in the structure of the near-wall flow above the compliant surface are due to the stable interaction between the surface and the turbulent flow. To better understand the physics of the interaction between a turbulent boundary layer and a passive flexible surface, optical holographic interferometry and LDVM were used to obtain the main parameters of the turbulent boundary layers and associated with the flow displacements flexible surfaces.

Fig. 5.59 shows a set of all data available from the values of the average dimensionless gap between the near-wall strips, using the hydrogen bubble method for different Re values. The existing observed values of λ^+ are in good agreement with the data given in [345]. An average value of $\lambda^+ = 95$ was found for $y^+ = 2-6$. Fig. 5.59 also clearly shows that low-speed strips are essentially invariant with Re_θ.

Fig. 5.60 shows the change in the transverse interval between the near-wall strips with increasing distance above the surface. The results show that λ^+ increases with increasing y^+ (for $y^+ > 6$). At $6 < y^+ < 30$, the strip begins to merge, which leads to an increase in the distance between the strips.

Apparently, the merger events are the result of small stretched and raised vortex loops generated in the near-wall region. These vortex loops interact strongly with each other when they move away from the wall. This fusion and interaction appears at $10 < y^+ < 30$, that is, in the buffer zone region, which determines the main production of turbulence in a flat turbulent boundary layer.

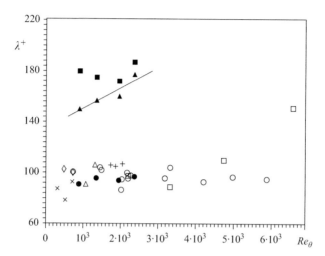

Figure 5.59 Dependences of the average dimensionless transverse interval between the near-wall rows of strips with the Reynolds number: △ Kline et al., [345], $y^+ = 4$; □ Gupta et al. [280], $y^+ = 2 \div 7$; + Lee et al., [405]; × Thompson (1976); ◇ Nakagawa & Nezu (1981), $y^+ = 4$; ○ Smith & Metzler (1983), $y^+ = 5$. Presented measurements: ● hard surface, $y^+ = 2-6$; ▲ flexible surface, $y^+ = 2-7$ (based on u^* obtained from measurements of average speed on a hard surface); ■ flexible surface (based on u^* obtained from measurements of average speed on a flexible surface).

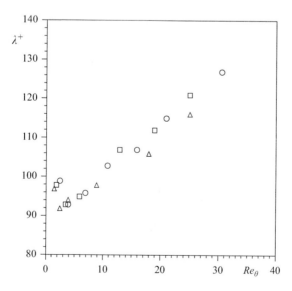

Figure 5.60 The average dimensionless wall distance between the strips as a function of the distance from the streamlined surface: ○ $Re_\theta = 2347$; □ $Re_\theta = 1952$; △ $Re_\theta = 1348$.

5.9 Physical substantiation of the interaction mechanism of the flow with an elastic surface

In most papers interaction of flow and an elastic surface are considered as static or quasi-static process. The field of external loadings is represented in the form of bending mode of pressure. With this simplified approach, many features of interaction are not included in consideration.

To develop and describe a complete rheological model of a material, which is exposed to a flow, it is necessary to consider real structure of an elastic surface and distribution of outer loading.

The basis for the development of schemes for elastic surfaces laid the foundations of the structural and kinematic-dynamic interaction of the boundary layer with the surface [35,53,56,60,67,80,139,247], similarity criteria (Part I, Experimental Hydrodynamics of Fast-Floating Aquatic Animals, Section 1.4, Table 1.5, [60,139]) and patterns of interaction of disturbances in the boundary layer (the problem of receptivity) [53,139,247].

The structural principle is to satisfy the correspondence between the structures of the disturbing motion of the boundary layer (or its characteristic regions) and the elastic plate. This means that if in the critical layer of the boundary layer at the initial stages of the transition there is an intense plane wave, then it is desirable that the outer layer of the elastomer has such a structure that, under the action of the pressure field of the boundary layer, will generate into the boundary layer a similar two-dimensional disturbing motion interacting with natural flat wave of the boundary layer.

The kinematic-dynamic principle is that the mechanical properties of all layers or the surface layer of a composite elastic plate should ensure kinematic and dynamic compliance with the characteristics of the boundary layer. In this case, the energy of a two-dimensional wave in the critical layer of the boundary layer at the initial stages of the transition must have energy high enough to cause deformation of the outer layer of the elastomer. The frequency of the driving force of the boundary layer must correspond to the natural frequency of this layer (or several layers) of the elastomer. In this case, the external load leads to the generation of vibrations in the elastomer. These fluctuations, with a certain degree of compliance with the driving force phase, will dampen the external load. In the case of the waveguide structure of an elastomer, under the action of an external load, surface waves will arise in the elastomer and on the wall-liquid interface, to maintain which the energy of the two-dimensional wave of the boundary layer will be selected and the energy of the boundary layer will be damped.

The principle of agreement is that the first two principles should be followed along the length of the surface in each place. In other words, when the structure and properties of the disturbing motion in the boundary layer change, the structure and properties of the elastomer (or its outer layer) must change.

With a large length of the streamlined plate and a small ε along the plate in the boundary layer, all stages of the transition are formed. These stages have a certain length and structure of the flow (Section 3.6, Fig. 3.33). Then at the zero stage of transition up to the area $Re_{l.st}$ (Section 3.4, Fig. 3.20) all oscillations must attenuate, so there is no need to use elastic coatings. But the experiments showed oscillation before $Re_{l.st}$, so the monolith elastic plate has to be placed. After this region, the surface has to be divided into sections in accordance with the stages of transition. At stage I, the elastic surface structure must be conformed to two-dimensional sinus waves. Because there is a spectrum of flat fluctuations in a boundary layer, ordering in an elastomer should be determined from the condition of conformity to second neutral fluctuation. After the location area $Re_{l.st}$ the elastic surface should be divided into sections in accordance with the transition stages. At stage I (Section 3.6, Fig. 3.33), the elastic surface must correspond to two-dimensional sinusoidal waves of the boundary layer.

The elastic surface section in the region of forming longitudinal vortices in a boundary layer must satisfy the structure, dimensions and directions of these vortices. At the turbulent boundary layer stage, the elastic surface structure must satisfy the viscous sublayer flow structure (Section 3.6, Fig. 3.40).

When designing an elastic coating, it is necessary to take into account that the kinematic-dynamic characteristics of a boundary layer at different stages of its formation vary significantly. For example, at the initial stages of transition of at least five initial stages, where a two-dimensional wave exists, a quasi-static approach can be applied and a simplified model of elastomer can be used. At the following stages of the transition and in the case of a turbulent boundary layer, the interaction between the flow and the elastic surface occurs in a wide frequency range under dynamic loading. In this case, the surface model must be statistical, for example, taking into account equations of the type (3.11)–(3.13) or equations given in Refs. [60,139,586]. At the same time, it is necessary to take into account such processes as "bursts" from a viscous sublayer, "falling" on a streamlined surface from the outer region of the boundary layer of "lumps" of a liquid, etc. (Section 3.6, Fig. 3.40). In addition, it is important to take into account the results obtained in Refs. [53,139,247] of an experimental investigation of the process of interaction in the boundary layer of various disturbances introduced with natural disturbances of the boundary layer (the problem of receptivity).

The foregoing approach to selecting elastic coatings in real conditions can be explained on an example of the flow over a wing (Fig. 5.61). The dashed lines indicate the boundaries of the sections of the unregulated uniform elastic surface mounted on the wing profile; points show the modified distributions of loading parameters and new borders of sections after selection of optimum characteristics elastomers for them. The Roman numerals designate numbers corresponding sections. Borders of sections are determined from following requirements:

- Within each section, gradients of parameters of the loading should be constant or small.
- In area jointing of sections, parameters should not contain singularities.

The same is valid for bodies of revolution.

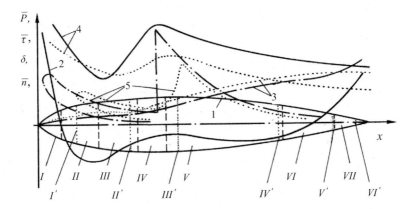

Figure 5.61 Standard distributions of loadings along a wing: (1) outer contours of a wing, (2) distribution of pressure $\bar{p}$ on x axis, (3) boundary layer thickness δ, (4) shear stress $\bar{\tau}$, (5) frequency ranges of laminar and turbulent fluctuations $\bar{n}$ in the boundary layer.

Presented in Fig. 5.61 patterns show that the elastic surface along the wing chord is subjected to different static and dynamic load actions. With constant velocity flow over a wing in section I, the pressure gradient $\bar{p}$ is maximal, that is, pressure sharply decreases, but, remaining positive, it compresses the elastomer. Gradient of shear stress $\bar{\tau}$ is as maximal, which means quick changing of shear load in elastomer. In the same section intensive high-frequency velocity fluctuations take place. Because of the maximum gradient of δ and of its minimal value, the heat transfer gradient is maximal. Thus in section I the elastic surface must be rather stiff to be stable to the action of high loads and their gradients. With this, the high-frequency velocity and pressure fluctuations are being dumped. The positive effect can be amplified by dividing section I into subsections.

In section II the aforementioned loads and their gradients are altered not so intensively and do not reverse their sign; the region of unstable oscillation frequencies and the frequencies themselves decrease in this section. Because of this, the mechanical properties of the elastomer must differ from that in the section I. For example, the compliance of the surface must be increased in comparison with that in section I. The $\bar{p}$ load, in contrast with that in section I, acts in separating. At the end of section II, the gradients of magnitudes $\bar{p}$ and $\bar{\tau}$ become positive, and the sign of the gradient changes, that is, the load on the separation decreases.

In section III the gradient $\bar{p}$ positive and gradient $\bar{\tau}$ is changed sign that means that loading on separation decreases. The range and frequencies of velocity fluctuations in the boundary layer tend to a minimum, as well.

In section IV the $\bar{p}$ gradient practically becomes equal to zero, that is, the elastomer is subjected to shear loads due to $\bar{\tau}$. With this, the stress $\bar{\tau}$ becomes maximal and its gradient reverses the sign. Besides that, due to transition of laminar boundary layer into turbulent boundary layer the region, the range of load $\bar{n}$ sharply widens and the law of growth of δ and heat transfer coefficient modify.

In section V all loads vary weakly and they can be assumed constant. In sections VI and VII, the load $\bar{p}$ increases, that is, the separating load increases. Besides these loads, we have to add the vibration loads, caused by nonsteady motion and by vibration of the base and the loads, caused by modification of the environment properties and flow regime.

At the first stage of the selection of the mechanical characteristics and design of the elastic coating, these considerations are guided by the considered sections along the body length, and the length of the sections is determined on the basis of the load parameters for a rigid body. Based on the results of the experimental studies given in Sections 5.2–5.8, it becomes apparent that some of the loading parameters change at flow around the body with an elastic coating. Conventionally, the modification of load parameters in Fig. 5.61 is shown in dots. With this change in load parameters will depend on the effectiveness of the selected elastic coating. In accordance with the new patterns of distribution of load parameters along the body with an elastic coating, the length of the sections will increase, as shown in Fig. 5.61 by dots and Roman numerals with strokes. This method of selection of elastic coatings depends on the specific geometric parameters of the body and the speed of its movement.

In Part I (Experimental Hydrodynamics of Fast-Floating Aquatic Animals), experimental investigations have shown that high-speed hydrobionts automatically regulate the characteristics of their body and skin. Various active methods for controlling the characteristics of the boundary layer and reducing resistance are considered. Therefore the third stage in the design of elastic coatings is the simulation of active and passive boundary layer control methods based on hydrobiological investigations. Some results of the simulation of phenomena detected in high-speed hydrobionts are considered in Sections 5.2–5.8. Based on the simulation results, it is possible to design effective elastic coatings to drag reduction at various speeds.

According to ideas of work [165], the complex interactions of disturbances in a boundary layer can be presented as a multivariate system with arbitrary input processes and one output process (Fig. 5.62). All kinds of outer and inner disturbances with input processes X_i, describable by frequency spectrum H_I, are summed up in an elastic surface with outer disturbance N. As a result, at exit the outer frequency response of elastomer is obtained.

The input process x_1 is a frequency load of the entire elastic surface as a single plate. The load x_1 is caused by nonstationary flow in real hydroaeromechanic problems. Fluctuation of an elastomer occurs under influence of the load x_1 regularly distributed and acting from above.

The input process X_2 is a frequency of load of the entire elastic surface as a single plate due to vibrations and fluctuations of the basis—that surface, to which the elastic coating is attached. Loading X_2 acts on the elastomer from below and can be either uniform along the plate or with various laws of distribution along the plate.

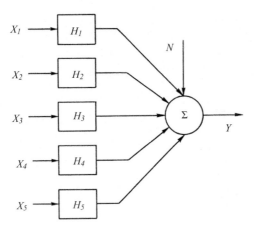

Figure 5.62 Multivariate system with arbitrary input processes and one output process [83].

The input process X_3 is the frequency of quasistatic loads, caused by distribution $\bar{p}$ and $\bar{\tau}$ along the body, under which influence normal and shear loads appear in each section, in addition to vibrating loads X_2. The loads X_3 have a certain distribution law along the plate and act from above.

The input process X_4 is local dynamic frequency loads due to fluctuations of pressure and velocities in the boundary layer. The pressure fluctuations from above determine the sign-alternating normal stresses and the velocity fluctuations induce shear stresses.

The input process X_5 is a frequency of thermal loading acting from above and from below. The properties of elastomers are substantially dependent on the temperature. Therefore energy dissipation of fluctuations and vibrations, depending on the thermodynamic parameters of complex composite plates, result in a change of temperature of composites. The heat flux direction is downward from the elastomer to the base and upward to the boundary layer. This heat flux is not uniform in different directions because the elastomer consists of sections and because of different intensity of dissipative energy in the transition stages, and so on. In addition, the elastomer is influenced by thermodynamic fields of the boundary layer and of the base. Thus there are differences in the thermodynamic regime of the flow over different parts; hence, there are additional energies in each section, for example, changed elastomer rigidity and viscosity, and so its frequency characteristics change.

The disturbance N is caused by variable external conditions and flow conditions. If values X_i are expressed in terms of their components, the scheme in Fig. 5.62 would be more complicated. The multivariate system, such as presented in Fig. 5.62, is solved by means of Fourier transformation [165]. In reality, the interaction process is very complex. Therefore in the presented scheme the elastic surface acts as an integrator of different disturbance spectrum. Because of summation of all disturbances getting into an elastic surface, one-frequency process of oscillation of an elastic surface occurs at exit (see Section 1.12). This process interacts with input processes, that is, a feedback takes place. For example, a feedback modifies the parameters of base vibrations, the loads $\bar{p}$, $\bar{\tau}$, and $\bar{n}$.

Simplified models of real complex interactions are solved by means of calculation of the conditional spectral density and a spectral matrix [165]. Thus an elastic surface at complex interactions is always subjected to complex load condition. Simultaneously there are various kinds of loadings—compressing, stretching, and shear, unidirectional, sign-variable, static, and dynamic loads.

It is practically impossible to construct the model including all these loadings. Therefore it is obviously not possible to develop mathematical model for the description of a real picture of interaction of a flow with an elastic surface at all kinds of loadings. However, different simplified approaches are possible. In particular, we can use a superposition method. At any interaction, an elastic surface must be hydraulically smooth, monolith, composite, with appropriate inner structure, and must be controllable to adjust it for optimal interaction.

The hydrobionics approach is promising for understanding the processes of interaction of elastic coatings with a flow, and for designing various characteristics and structures of drag reducing elastic materials.

5.10 Boundary layer when flow around a controlled elastic plate

It is known that the mechanical properties of elastomers are temperature dependent. Therefore in Refs. [16,17,32,37,51,59,60,139,378] it was proposed to change the temperature of the elastomer to control the coherent vortex structures in the boundary layer of the elastic surface.

If we consider the interaction of elastomers with the flow from the standpoint of receptivity by the boundary layer of various disturbances (Section 3.9) or from the standpoint of a complex waveguide (Section 3.8), then the regulation of stiffness or viscosity of the elastomers through heating or cooling allows changing the interaction process of disturbances in the boundary layer. It is known that hydrobionts use this method reflexively—the principle of thermoregulation (Part I, Experimental Hydrodynamics of Fast-Floating Aquatic Animals, Sections 1.2, 5.4, 5.5).

Depending on the design of the composite elastic plate (Section 5.1), by heating it can change its stiffness and damping properties. In this case, it is possible to change the receptivity to external disturbances depending on their type and waveguide properties of the material. To check these positions, we used the thermoanemometric apparatus to measure the kinematic and dynamic characteristics of the boundary layer over elastic plates of various designs in water and air flow when the elastomers are heated.

Experiments in water were performed on a hydrodynamic stand with regulated turbulence (Section 3.1) [17,59,60,139,378]. Elastic plate 6 in accordance with the Tables 5.4 and 5.5 (Section 5.1) was placed in the inset (Fig. 5.6, Section 5.1) at the end of the work area (Section 5.1, Fig. 5; Section 3.7, Figs. 3.42 and 3.43). Composite plate 6 (Fig. 5.7B) consisted of two layers of foam elast FE-1 (thickness of each layer was $4 \cdot 10^{-3}$ m), one of which was glued to an aluminum plate with a thickness of $2 \cdot 10^{-3}$ m. Outside of the bottom glued layer FE was glued with a thin rubber film on which the longitudinal strips of conductive synthetic fabric CCF were laid. The end parts of these strips CCF were clamped in transverse narrow strips made of thin brass meshes. Conductive wires were soldered to these brass strips. A thin rubber film was glued to the strips of CCF, and the outer FE-1 layer was glued over, the outer surface of which was hydraulically smooth. When electric current was passed through the CCF, the elastic plate was heated. The layers of FE-1 have good thermal insulation properties; therefore the plate was kept warm. The mechanical properties of the elastic plate changed upon heating. Since the CCF layer is made of longitudinal strips, the places of the outer layer of PE-1, located above the CCF layers, were heated more than the neighboring places located between the ST strips. Therefore in the transverse direction in the outer layer of the elastic plate, the mechanical characteristics varied unevenly. This accelerated the formation of longitudinal vortex structures above the elastic surface.

To control the temperature during the experiments, two holders with MT-54 microthermopairs fixed on them were installed in the region of the trailing edge of the elastic plate (temperature accuracy was 0.01 degree). One sensor measured the temperature of the elastic plate, and the second was set along the longitudinal axis of the flow and monitored the temperature of the water during the experiments. The voltage was 19.2 V and the current was 5.9 A. The water temperature did not change, and the temperature of the outer surface of the elastomer increased by 0.1 degree. Therefore the change in the kinematic characteristics of the boundary layer occurred only due to a change in the mechanical properties of the elastic plate.

The measurements were carried out along the channel axis in section IIIa (Section 3.7, Fig. 3.42) at a distance of 0.485 m from the front edge of the insert or 2.485 m from the beginning of the working section at the main flow velocity of 0.06 m/s. The transverse correlation coefficients of the longitudinal pulsation velocity were measured using two thermoanemometer sensors. One sensor measured the field of longitudinal velocity and spectral characteristics. The transverse correlation coefficients and spectral characteristics are given in Refs. [60,139].

Fig. 5.63 shows typical profiles of the longitudinal averaged velocity in semi-dimensional coordinates. The main change occurred in the near-wall region and at the boundary between the buffer region and the core of the turbulent boundary layer. The velocity profile in the near-wall region when heated the elastic plate becomes less filled, which indicates a decrease in friction. During measurements, it is necessary to take into account the change in the thermal conductivity

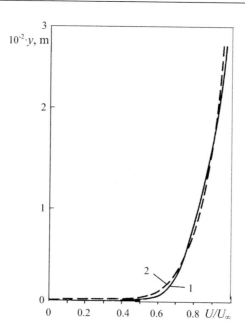

Figure 5.63 Profile of the longitudinal averaged velocity when flow around the elastic plate 6 without heating (1) and with heating (2) [59,60,139].

of the elastic wall and the heat loss of the thermoanemometer sensor near the surface.

Fig. 5.64 shows the profiles of the averaged velocity measured at the flow around the elastic plate 6 (Section 5.1, Tables 5.4 and 5.5), with different heating parameters. Curve (1) shows the velocity profile in the viscous sublayer $u^+ = y^+$, curve (2) shows the law $u^+ = 2.5\ln y + 5.5$, which is characteristic of the turbulent core. One of the laws of flow in the boundary layer when applying aqueous solutions of the polymer in the buffer layer (3) and the turbulent core (4) can be written as:

$$u^+ = 11.7\ln y^+ - 17$$
$$u^+ = 2.5\ln y^+ + 5.5 + \Delta B \tag{5.20}$$

The results of measurements on a rigid plate [curve (5), $u_\tau = 3 \cdot 10^{-2}$ m/s] are in good agreement with the known data and lie slightly below curve (2), as is usually the case when flow around rough plates. The obtained law [curve (5)] can be explained by an increase in the level of turbulence at the end of the working section at a speed of 0.6 m/s. Curve (5) is a reference when analyzing the results of the flow around an elastic plate.

The hydrodynamic stand is a closed hydrodynamic tunnel in which the heat exchanger in the return channel did not work during the experiments, as a result of

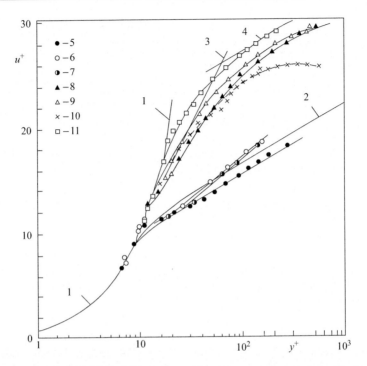

Figure 5.64 The influence of the duration of heating of the elastic plate 6 on the change of its stabilizing properties: the velocity profile on a rigid plate [200] in the viscous sublayer (1) and in the core (2) of the turbulent boundary layer; the law of flow in the buffer layer (3) and in the core of the turbulent boundary layer (4) in the flow of the polymer solution. The rigid plate (5) and elastic plates (6)–(11) without heating (6), (8), (10) and with heating (7), (9), (11) [59,60,139].

which the temperature of the water gradually increased by 1.1 degree during one series of experiments during the experiments. At the beginning of the experiment on an elastic plate [curve (6)], the water temperature was 20.4 degree, and at the end 21.5 degree. In experiments with heating an elastic plate [curve (7)], the temperature corresponded to 21.5 and 22.5 degree. With this increase in temperature, the mechanical characteristics of the elastic plate changed, which did not affect the characteristics of the boundary layer.

The difference in the velocity profile from the reference one is insignificant, but the slope of the curve has changed (the value of the Karman constant $\varkappa$ has changed), which indicates a thickening of the viscous sublayer. For curve (6), $u_\tau = 2.7 \cdot 10^{-2}$ m/s, and for curve (7), $u_\tau = 2.65 \cdot 10^{-2}$ m/s. To assess the effect of the sensor holder configuration on the measurement results, subsequent measurements [curves (8), (9)] were performed with a complex holder. The initial temperature was 23.7 degree, and the final temperature was 24.7 degree. For curve (8), $u_\tau = 1.9 \cdot 10^{-2}$ m/s, and for curve (9), $u_\tau = 1.8 \cdot 10^{-2}$ m/s. This series of measurements was carried out immediately after the previous series. During this time, the elastic plate is well heated; its compliance has increased, so the result is comparable to that obtained by injection of polymer solutions into the boundary layer.

To test the results of the first series, the experiments were repeated, but the initial the temperature was lower and was 18 degree, and at the end of the experiments it was 19.3 degree for the plate without heating and 20.7 degree for heating experiments. These experiments gave the most significant results [curves (10), (11)] in comparison with the rigid plate and the heated plate. The shape of the curves indicates a thickening of the viscous sublayer. For curve (10), $u_\tau = 2.08 \cdot 10^{-2}$ m/s, and for curve (11), $u_\tau = 1.8 \cdot 10^{-2}$ m/s.

Despite the identical experimental conditions, significant differences are found with repeated testing. This means that in the initial experiment the elastic plate was well heated, therefore in the region of the heating strips, the properties of the elastic material changed significantly. As a result, the properties of the material in the transverse direction changed with a step equal to the distance between the heating strips. Thus in the boundary layer, stationary systems of longitudinal vortices with λ_z, corresponding to λ_z of the natural boundary layer at the measurement site, were induced. According to the principles of receptivity [53,247], this led to the effective interaction of natural longitudinal vortex systems and longitudinal vortices introduced into the boundary layer by means of a heated elastic surface. When heated, the efficiency of the elastic plate increased in all cases.

The data obtained correlate with measurements of the pulsation velocities (Fig. 5.65). The conditions of the experiments are the same as in series 1−3 in Fig. 5.64. As shown in Fig. 5.65, when the elastic plate was flow around the pulsation velocity in the 1st series only slightly changed when the plate was heated, and in the 2nd and 3nd series the pulsation velocity decreased. In all cases, the maximum of the pulsating velocity when flowing around an elastic plate is located closer to the wall and its magnitude is smaller compared with the maximum on a rigid plate. These data are similar to those in the laminar boundary layer (Sections 5.2, 5.4, and 5.5) [378].

These series of experiments were repeated when conducting experiments in a wind tunnel. The results of experiments in the air flow in accordance with Table 5.5 (Section 5.1) with elastic plates 11a (PU) and 9 (FE) are given in [60,139]. Measurements in air were carried out in the same section of the insert IIIa, as in water. At the same time, in the wind tunnel this section was located at a distance of 1.4 m from the beginning of the working section with a working section length of 3 m. The measurements were carried out at the main flow velocity of 9−18 m/s. The profiles of the longitudinal averaged velocities are given in [60,139]. The thickness of the viscous sublayer of the corresponding plate 9 in section IIIa increased slightly, compared with a rigid plate. In this case, the additive constant ΔB when heated increased. Heating compliance plate 9 led to an increase in its elasticity. In the air flow, the elastic properties of the

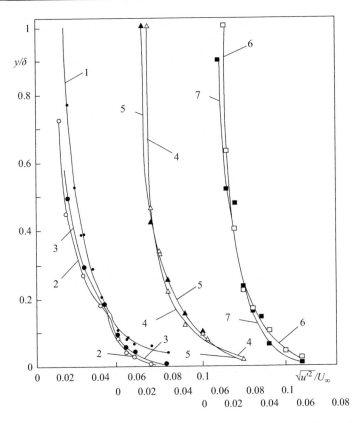

Figure 5.65 The profiles of the longitudinal pulsation velocity when the rigid plate (1) and the elastic plate flow around without heating (2), (4), (6) and with heating (3), (5), (7) [59,60,139].

plate without heating were not optimal, so the heating could not greatly affect the results of the experiments.

The heating of the elastic plate 11a increased its compliance, therefore the efficiency of this plate when heated increased. The thickness of the viscous sublayer, as well as on plate 9, did not change, but the value of $\delta_{lam.}$ was greater than on a rigid plate. The ΔB value of the plate 11a was maximum when heated. According to [48,60,139], when flow around the plates 9 and 11a, the most pronounced S-shaped velocity profiles in the near-wall region (as well as the flow velocity in the boundary layer) were recorded. It follows that, although the efficiency of plate 11a is better than that of plate 9, but the effect on heating turned out to be maximum compared with the standard in both cases. The pulsation velocity profiles of plate 9 with heating were the same as on a rigid plate [48,60,139]. The maximum of the pulsation velocity on the plate 11a increased and was located higher in y than on a rigid plate. Analyzing the flow in the near-wall layer [48,60,139], it can be seen that in both cases (on both plates) the longitudinal pulsation velocities increased throughout the entire layer. S-shaped profiles on both plates increased significantly. This indicates a decrease in friction and shear pulsations near the wall.

Fig. 5.66 presents the profiles of the pulsation velocity when the elastic plates are heated as they flow around air and water flows. The elastic plate 11a and the compliance plate 9 were examined in the air flow. In both cases, the distribution of the pulsation velocity across the boundary layer changed: the maximum of the pulsation velocities decreased and was located closer to the streamlined surface, which agrees well with the visualization data in the laminar boundary layer [139,378]. In contrast, the pulsation speeds in the buffer layer and in the core are larger than on a rigid plate; the lower the flow rate, the greater the difference. This can be explained by the oscillation of the elastic surface. When water flows around elastic plates, the attenuation of surface oscillations leads to a decrease in the difference between the rigid and elastic plates (Fig. 5.65).

When the elastic plate 11a is heated (Fig. 5.66B), the level of the longitudinal pulsation velocity throughout the entire boundary layer thickness decreased both in comparison with the absence of heating and in comparison with the flow around a rigid plate. The maximum of pulsation velocities increased in the near-wall region and its vertical coordinate decreased.

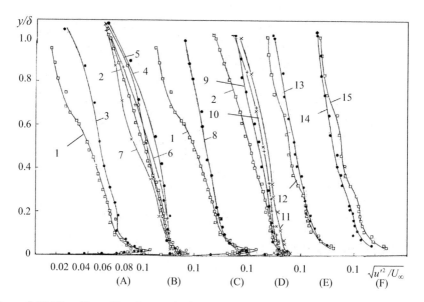

Figure 5.66 The effect of heating an elastic plate on the distribution of the pulsation velocity over the boundary layer thickness in air (A), (B), (C), (D), and water (E) and (F) flows: (1), (2) rigid plate, III section, $U_\infty = 10$ and 17 m/s. Plate 11a, section III, $U_\infty = 9.9$ m/s (3), 17.43 m/s (4), 18.23 m/s (5); section IIIa: without heating, 19.38 m/s (6), with heating, 19.45 m/s (7). Plate 9, section III, $U_\infty = 9.765$ m/s (8), 15.92 m/s (9); section IIIa: without heating, 16.26 m/s (10), with heating, 16.14 m/s (11). Plate 6 (water flow), section IIIa, $U_\infty = 0.6$ m/s, without heating (12), (15) and heated for 15 min (13) and 30 min (14) [59].

The heating of the plate 9 slightly affected the distribution of the pulsation velocity over the thickness of the boundary layer. A similar pattern was observed when the water flow over the plate 6 with a brief heating. An increase in the duration of heating increases the efficiency of the elastic plate, while the pulsating longitudinal velocity decreases throughout the entire thickness of the boundary layer.

In Refs. [60,139,324], the spectral densities of the longitudinal pulsating velocity are measured, measured with air flow around heated elastic plates 11a and 9 in section IIIa.

Heating the elastic plate 11a leads to a better interaction with the flow than in the absence of heating. As a result, the energy of high-frequency pulsations increased, and the energy of low-frequency energy-carrying pulsations decreased. In this case, a maximum appeared at small wave numbers, as well as dips in the spectral curve, which indicates the formation of longitudinal vortex systems in the near-wall region of the boundary layer. The same picture was observed when testing the same plate in a water stream. The change in π-parameters (Part I, Experimental Hydrodynamics of Fast-Floating Aquatic Animals, section 1.4) and the ratio between the densities of the liquid and the plate improved their interaction and allowed to determine the characteristic features of the formation of longitudinal vortices [60,139,247]. The plate 9 became more elastic when heated, therefore the interaction of the plate with the boundary layer in the air flow deteriorated. As a result, the spectral properties during heating remained almost the same as those without heating, although the shape of the spectral curves changed similarly to the shape of the spectral curves in the flow past the plate 11a.

These data indicate the possibility of controlling the mechanical and stabilizing properties of elastic plates by heating. Results depend on many factors and require careful analysis.

As was shown, it is possible to control the characteristics of the boundary layer when flowing around elastic surfaces by generating longitudinal vortex disturbances in the boundary layer using special generators of longitudinal vortices or a special design of elastic plates. Another method of controlling the boundary layer is to change the mechanical characteristics of elastic plates in transverse directions, as well as by controlling the temperature of an elastic material. The design of elastic plates allows you to control the pressure and density of liquids that fill the cavity of the coating inside the plate (Section 5.1, Figs. 5.7 and 5.8) [27,32,378]. By changing the pressure inside the coating, you can control the rigidity of the coating, to withstand environmental pressure. The density of the liquid filling the coating affects the characteristics of the distribution of pulsations in the coating caused by the pulsating field of the boundary layer (complex waveguide).

The combination of all these methods allows you to effectively control the flow characteristics when flow around an elastic surface. Another effective method is the creation of pulsations in an elastic coating, for example, using pulsating pressure inside cavities in composite coatings. All this allows you to control the complex waveguide, representing the boundary layer, and an elastic plate with a given structure.

Experimental investigations of friction drag

6.1 Control methods for coherent vortex structures of the boundary layer

Control of the regularities of the formation of coherent vortex structures (CVS) in the boundary layer and the determination of the mutual influence of the CVS in the near-wall and outer regions of the boundary layer are actual tasks. The CVS of the transition boundary layer are shown in Section 3.6 in Fig. 3.33, and the turbulent boundary layer in Fig. 3.40. In real conditions of flow, various disturbances will act on the boundary layer with its external and internal boundaries. These disturbances are called degrading factors in comparison with idealized conditions of flow around a flat plate by an unperturbed flow. In 1980, a hypothesis was put forward [35] that in the viscous sublayer of the turbulent boundary layer, the development of coherent structures occurs approximately the same as in the transition boundary layer, but under the influence of degrading factors. Since measurements in a viscous sublayer are very difficult, taking into account this hypothesis, it is sufficient to study the development of coherent structures in the transition boundary layer when exposed to degrading factors, for example, in the form of increased turbulence of the main flow and vibration of the streamlined surface. Indeed, reinforced vortex disturbances contained in the outer region and in the core of the turbulent boundary layer act on the viscous sublayer from the outside. To model this picture, it is necessary to create increased turbulence of the main flow in the transition boundary layer or to disturb the outer boundary of the boundary layer with flat or three-dimensional finite disturbances. On the other hand, constant emissions of a inhibited fluid from a viscous sublayer, as it were, create additional disturbances from the lower boundary of the viscous sublayer. This can be modeled in a transition boundary layer by creating vibrations of a streamlined boundary. This fundamental idea was explored in Chapter 5, Experimental investigations of the characteristics of the boundary layer on the analogs of the skin covers of hydrobionts. In addition, three regions of CVS are formed in the turbulent boundary layer through its thickness. These are large vortices in the outer region, ordered vortex systems in the buffer region, and systems of vortices in a viscous sublayer.

In addition to the idea of the structural character of the development of disturbances at various stages of the transition, the development of control methods for CVS is also based on the idea of influencing the frequency characteristics of ordered vortex structures. A particle of fluid has a dualism of properties when moving: a particle moves along a certain trajectory and has a spectrum of oscillations during movement, among which one can distinguish certain energy-carrying frequencies for each stage of the transition or for each type of CVS.

In accordance with these two approaches, methods were developed to influence CVS. The main factors (the third approach or the principle of interaction) is that the impact energy must be of the same order as the energy and size of the disturbing motion at the appropriate stage of the transition.

Table 6.1 shows the developed methods for control the CVS, consisting of seven groups: mechanical (I), dynamic (II), kinematic (III), electrical (IV), acoustic (V), combined (VI), and active methods with a feedback system (VII) [57,64,66,71,74,91,100,103,104,132,139].

Consider mechanical methods. The shape of the streamlined surface can vary either stationary or depending on the flow velocity. Stationary plates are installed, for example, on the surface of the profile. Oscillating, for example, in the transversal direction, the plates should contribute to the organization or damping of the corresponding CVS of the boundary layer. Part of the surface may fluctuate, for example, the nose.

The transverse ledges or grooves are intended for the organization of transverse vortices with an axis of rotation directed in the transverse direction.

Longitudinal ledges or grooves are designed to organize longitudinal vortices. Various riblets have been investigated most extensively for this purpose.

Longitudinal stationary vortex generators differ from riblets in discreteness of longitudinal size and variety of forms. Longitudinal oscillating vortex generators are distinguished by the possibility of dynamic effects on the boundary layer.

Ordered roughness is designed for the same purposes: the organization of longitudinal or transverse vortices.

Table 6.1 Control methods of CVS boundary layer.

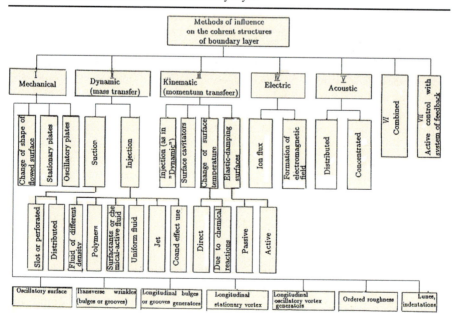

Holes: this is a new little-studied method of influence on the vortex structure of the boundary layer. Depending on the task, the wells can be installed with a longitudinal or transverse sequence.

Changing the shape of a streamlined surface can be either cyclic or stationary.

Stationary plates can be installed in the boundary layer so as to break up large-scale vortices. An example of such plates can be Large-Eddy-Break-up device (LEBU), that is, rings set equidistantly to the streamlined cylindrical surface.

Oscillating plates are, for example, oscillating LEBUs.

Dynamic methods are based on the principle of mass transfer and are distinguished from the known ones in that they are applied to a part of the washed surface. These methods include slotted, perforated, or distributed suction of the boundary layer.

Injection or blowing into the boundary layer is also carried out in discrete areas. Injection can be carried out by blowing into the boundary layer of a heterogeneous or homogeneous fluid compared to the incoming flow, aqueous solutions of polymers, surface-active substances, and also a chemically active fluid. The injection method can be different, in particular, jet or using the Coanda effect.

Kinematic methods are based on the principle of transfer of momentum. These include the injection, as in the previous case, the installation of discrete surface cavitators, and control of the temperature of the washed surface by heating or due to chemical reactions with the main flow liquid.

A separate class is a single-layer or multilayered elastic-damping surfaces, isotropic or anisotropic, passive or active.

Electric methods include a method of establishing an ion flow in a boundary layer, as well as the formation of an electromagnetic field around a streamlined body in the case of a weakly conducting main flow liquid.

Acoustic methods can be distributed or concentrated. These methods are based on the principle of organizing an additional pulsation specially directed pressure field. An example of artificial laminarization of the boundary layer is sound irradiation [326].

The most promising are combined methods in which a combination of the various methods considered are used simultaneously, and also (or) in combination with active control methods.

Some of the methods considered were experimentally investigated in previous sections.

In the future, the proposed mechanisms for influencing CVS will be investigated.

6.2 Review of experimental investigations of integral characteristics of elastic plates

The first report on the theory of fluid flow in elastic pipes was made at a meeting of the physicotechnical section of the Society of Natural Scientists at Kazan University Russia on May 14, 1883 by Professor I. Gromeko [275].

The results of investigations of complex interactions of the skin of hydrobionts with the boundary layer, performed in part I, have shown that this interaction depends on many parameters of the external environment, the way of swimming, on the mechanical characteristics of their skin and the interaction of body systems during movement.

To date, a large number of theoretical and experimental investigations of the boundary layer have accumulated during the flow around various elastic plates without taking into account the features of aquatic organisms swimming noted in part I. Examples of some such studies are given in Refs. [60,139,247,378]. Below will be considered the results of experimental investigations in the flow around various types of elastic coatings.

Most investigations have been performed on membrane surfaces, which are described by the simple Voigt-Kelvin model [60,139,247,378,499,531] and are easily amenable to theoretical analysis. Other studies have investigated smooth monolithic plates, for which theoretical models can also be constructed. And only two studies carried out on composite elastic plates.

It should be noted that theoretical solutions of the problems of complex interactions of composite covers—analogues of the skin of hydrobionts with a real flow—are still not known.

In experimental investigations conflicting data were obtained due to three factors: incorrect measurement techniques, no experiment planning, and the dependence of the efficiency of elastic plates on many factors.

Based on observations of dolphin swimming, M.O. Kramer for the first time experimentally investigated the interaction of a flow with an elastic surface [382–390]. During the rotation of two coaxial cylinders, one of which was plastered with an elastic surface, he obtained an increase in revolutions of the pasted-up cylinder by 20% compared with a rigid standard. Determining the coefficient of friction by this method is very problematic. However, these experiments served as the basis for the continuation of experimental investigations by both Kramer MO himself and other researchers.

At the lowest numbers of Re, the investigation were performed in Ref. [327]. An open rectangular working section with dimensions of $0.6 \times 0.14 \times 0.07$ m was installed in the open-type hydrodynamic tunnel. Above the working section, a notch 0.406×0.086 m was cut, in which the insert 0.394×0.074 m was mounted as a sealed box fixed along its ends along the longitudinal axis on flat springs with strain gauges. The depth of the box is 0.042 m. The box was filled with either air or olive oil and outside it was covered with a film of natural rubber 0.15 or 0.3 mm thick, or neoprene rubber 2.5 mm thick. The tension varied and the mechanical characteristics of the elastic plate were determined. The flow velocity was measured up to 20 cm/s. In the range of Reynolds numbers $Re = 10^3 - 10^5$, the resistance coefficients of an elastic plate were measured depending on its mechanical properties. Some of the results obtained in Ref. [327] are given in Ref. [60]. The character of the curves due to the large gaps between the insert and the working area (6 mm). The scatter of points on the standard is less than on elastic plates. The magnitude of the friction was so small that the design of the strain gauges did not allow for high accuracy measurements. The conditions of flow around the elastic plate were unfavorable due to these gaps. At its leading edge, a pressure drop was formed, which formed forces commensurate with the amount of friction on the plate. The pressure distribution along the plate was measured, which led to the formation of a standing wave on the plate. All this led not only to difficulties in determining the actual efficiency of elastic plates, but also to an increase in measurement errors. So, when testing the best plate (in the article: curve (2) triangles) and worst (in the article: curve (3) squares) it is very difficult to average the measurement points, and the error has the same order as the measurements of the useful signal. Although this work revealed the dependence of the efficiency of an elastic plate on its mechanical characteristics, it was not possible to obtain unambiguous dependencies. To obtain a reliable result, it is necessary to change the design of the elastic plate, methods of fixing it on the insert, reduce the gaps between the insert and the working section, increase the range of Re numbers.

The same experiments in the water flow were performed in Ref. [546] (Fig. 6.1). For towing tests in the towing tank with dimensions of $80 \times 8 \times 3$ m were made 2 plates. The maximum towing speed was 3 m/s, and the Reynolds number was $6 \cdot 10^5 - 1 \cdot 10^7$. At the same time, the influence of the elastic surface on the transition of the laminar boundary layer to the turbulent one (the first measurement series) and to the turbulent flow (the second measurement series) was investigated. In the first series of measurements, the plate was towed in a vertical position and was only partially immersed in water, and in the second series, the plate was completely immersed and towed in a horizontal position. The design of the plates was different in both cases. In the first series, sheets of porous rubber 5, 6, or 10 mm thick were glued onto a dural plate with a thickness of 6 mm on both sides.

Edges of the same thickness as the rubber layer were fixed on the ends on a solid plate base. The width of the edging was 50 mm, with the front edging ogival form. A 70 mm wide edging was attached to the plate from above, and there was no edging on the bottom. Then the whole plate was covered with a 0.2 mm vinyl film. The plate dimensions are 1×2 m. The reference rigid plate consisted only of solid aluminum sheet 6 mm thick.

The second plate was different in shape from the first and was made of acrylic sheet 3 mm thick. On this sheet, at a distance of 50 mm from its front edge, sheets of porous rubber 5 or 10 mm thick were glued on both sides. The front end of the rubber was cut on the body to a length of 30 mm from its beginning. The whole plate was also covered with the same film. Rubber was soaked with air or water.

Fig. 6.1 shows some results of the second series of measurements. Almost all the results were negative, only in the first series of experiments the elastic surface 6 mm thick slightly reduced the resistance. In spite of the fact that, in comparison with the above works, the experiments correctly investigated an elastic plate, the following methodical errors were made. Both plates were structurally different from each other, with the front edge of the second plate creating large pressure gradients. It was not taken into account, and the pressure distribution along the plate was not measured. It is obvious that the pressure distribution along the plate led to large pressure gradients in the region of the leading edge, which led to tear-off loads in

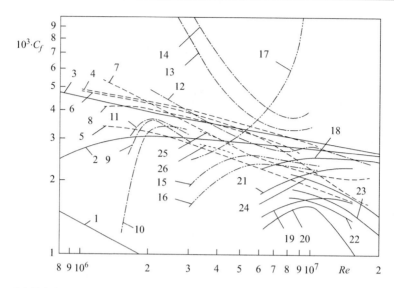

Figure 6.1 Friction drag coefficients on rigid and elastic plates: (1), (2), (3) friction resistance at in laminar, transitional, and turbulent flows in a boundary layer over a rigid plate [520,521]; measurements [172]: (4) hard standard, (5) polyurethane foam (FP) with a porosity of 16 pores/cm, soaked with water and covered with polyvinyl chloride film (PVC) 0.064 mm thick with longitudinal tension $T_x = 35$ N/m, (6) the same, as (5), but $T_x = 23$ N/m, (7) FP 10 pores/cm impregnated with water, $T_x = 31$ N/m, (8) the same as (7), but the FP is dry, everywhere $T_z = 54$ N/m; measurements [253]: (9) standard, (10) FP with a thickness of 6.4 mm, covered with a film "melinex" with a thickness of 0.127 mm, (11) the same as (9), but covered with a film "melinex"; measurements [546]: (12) standard, (13) porous rubber 5 mm thick, impregnated with water and covered with PVC film 0.2 mm thick, (14) the same as (13), but the thickness of porous rubber is 10 mm; measurements [382]: (15) standard, (16) elastic surface with elasticity $E = 10.3$ N/cm^2 and a ball impact absorption coefficient $\eta = 0.7$, (17) the same as (16), but $E = 10.3$ N/cm^2, $\eta = 0.88$; [387]: (18) standard, (19) – (21) columnar coating: (19) $E = 1100$ N/cm^2, (20) $E = 550$ N/cm^2, (21) $E = 410$ N/cm^2; [389,390] optimal coatings: (22) columnar, (23) ribbed, inside of which is a fluid with a viscosity of 7500 cCt; (24) the same as (23), but the viscosity is 10,000 cCt; measurements [416,417]:(25) standard, (26) membrane surface.

the front part of the elastic plates. This led to deformation front parts of elastic plates and, as a result, to negative results. The front edge of the elastic plates must be specially secured. In addition, elastic plates in shape and thickness differed from their rigid standards. Measuring systems of resistance force based on dynamometry gave errors commensurate with the resistance value of the plate. The mechanical characteristics of the elastic plates were not optimal for the selected range of towing speeds. Towing studies of elastic surfaces were performed most correctly in Ref. [387], and elastic surfaces that were optimal for towing speeds were tested. Due to the wrong measurement technique, most researchers were unable to repeat Kramer's experiments. Unlike other authors, Kramer studied the structure of the outer layers of the skin of dolphins [389] and patented the design of their coatings, which, in his opinion, modeled the structure of the outer layer of dolphins' skin [384–386,388,390]. The investigations of elastic coatings Kramer carried out on an axisymmetric cylinder of large elongation, towed under the bottom of a double-hulled speedboat. The model was located under the boat in the area between the keels in a kind of niche, stabilizing the flow.

In the first series of experiments (Fig. 6.1, curves (16) and (17)), soft materials were investigated.

It was recorded as a decrease and an increase in friction due to the low strength and elastomer detachment. In the second series of experiments, complex composite column coatings were investigated with variations in the mechanical characteristics of the composites. Very encouraging results were obtained—friction resistance decreased to 57%. However, the strength of these coatings at high speeds was unsatisfactory, as a result of which a third group of ribbed coatings was developed. The effectiveness of these coatings has decreased, but remained fairly high. The laws obtained in the experiments of Kramer are of the same type and characterize, as it were, the inhibition of the transition of the boundary layer. The shape of the curves is similar to series (9)–(11), curve (3). Curve (20), in which C_f after the maximum decreases significantly, is doubtful. To date, the results of experiments Kramer remain one of the best. For a long time no one could get similar results. There is still no theory and physical explanation of the results. Kramer first tested composite elastomers and removed most of the shortcomings of the above techniques. However, there are still doubts about the effectiveness obtained. These doubts are due to the fact that the tests were conducted not in the laboratory, but in full-scale. The visualization of the flow conditions of the model located between the two hulls of the boat and the flow structure formed during the flow around the hulls at different speeds is not performed. The vibration characteristics of the model were not measured at different speeds.

In a review [304], data from studies performed in Ref. [470] are presented. The measurements were carried out on a pop-up model with a length of 0.915 m, nose section which was plastered over with a surface that mimics the coating of Kramer. Received increase in resistance in comparison with the standard. Similar results were recorded in the investigation in a hydrodynamic tunnel [510]. It remains unknown what were the conditions and methods of measurement.

Numerous investigations were carried out in the United States by the Blik group [172–176,421]. Both a significant (up to 60%) [421] and a smaller decrease in friction resistance (up to 37%) [535] on the same plates were obtained. However, the above-mentioned shortcomings of the majority of authors cast doubts on the authenticity of these numbers since the materials of elastomers (FP), their structure and mechanical properties do not allow obtaining a significant reduction of friction. The results obtained by the authors can be obtained only in a narrow range of speeds or on composite plates.

These deficiencies were partially eliminated when tested in a wind tunnel [172]. The gaps between the insert and the cut in the working section (1.6 mm) were reduced and the range of Re numbers at which measurements were made was increased. However, the design of the elastic surface and the method of its fastening on the insert did not change and, moreover, the fastening of the insert on the strainer deteriorated, and the gaps remained sufficiently large. Since the entire insert was fixed on one rod—strainer, despite the available compensating device for setting the insert surface at a zero angle of attack, the measurement was influenced by the moment characteristics of the insert. The plate design remained very sensitive to the pressure distribution along the insert. The working section of the wind tunnel had dimensions of 1.2 × 0.5 × 0.35 m, and the inserts—0.66 × 0.2 m, which was a box, 0.13 m deep, filled with various damping fluids. A polyurethane foam plate could also be placed in the box, and a polyvinyl film 0.064mm thick stretched over the box, the transverse tension of which was constant ($T_z = 54$ N/m), and the longitudinal tension T_x varied. Fig. 6.1 shows the resistance values of the tested plates obtained in Ref. [172]. The best results allowed reducing the resistance by 25%–33% compared with a rigid standard. However, due to the fact that the measurements were carried out in the air, despite the increase in the Re numbers, the resistance force was small, therefore the measurement errors were of the same order as the measured quantity, and the results obtained are rather qualitative character.

The measurement procedures described in Refs. [173,327] do not allow checking whether the stabilizing properties of elastic plates depend on their mechanical characteristics, as shown, for example, in Refs. [282,325]. In addition, when flowing around membrane surfaces, depending on the velocity, a vacuum occurs over the elastic surface, which deforms and tears the membrane from the frame. To eliminate this drawback, you must either balance the pressure inside the elastic surface with the pressure of the flow, or experience a monolithic surface firmly connected to the insert.

In Ref. [253], an attempt was made to eliminate the noted deficiencies. Two plates and an aerodynamic profile were made, on which elastic surfaces were glued. Investigations were conducted in a wind tunnel. Fig. 6.1 shows some results of measuring the resistance of one of the plates 1.22 × 0.84 m in size. The plate was made of 2 aluminum sheets 6.4 mm thick each. The nasal edge of the plate was made in the form of an ellipsoid with semiaxes 50.8 × 6.4 mm, and the rear edge was blunt. A turbulizer with a diameter of 0.356 mm was installed in the nose of the plate. In one sheet of aluminum, a 1 × 0.3 m cut-out was made, in which a sheet of polyurethane foam was placed. The appearance of the curves shows that the diameter of the turbulizer is insufficient for the formation of a turbulent boundary layer when flowing around the plate. A thin film of "melinex" practically did not change the characteristics of the boundary layer. A sheet of polyurethane foam placed in the cut-out of the plate and covered with the same film made it possible to reduce the resistance of the plate by about 20%. The area of the elastic surface was only 1/3 of one side of the plate. Despite the achieved positive result, the method and place of fixing the elastic surface on the plate and the method of measuring the resistance value of the elastic insert remained unsatisfactory. In Refs. [416,417], the experiments of Blick were thoroughly checked: the reduction of friction did not exceed 10%.

Further investigations in this area were carried out in the United States by a group of researchers led by Denis Bushnel [192–194,235,237,238,603]. A detailed review of the investigations performed by this group on various types of elastic plates was performed in Refs. [192,238,603]. The most significant results are:

- A hypothesis was put forward on the role of elastic plates in the pre-explosion modulation of burstings from a viscous sublayer [192];
- Some data on the regulation of elastic plates by heating [235];
- The effect is obtained in the form of $C_{fel.}/C_{fhard} = 0.4$ [235];
- Recorded maximum reduction of friction in the case when the natural frequency of oscillation of the membrane coating is approximately half the frequency $f_p = U_\infty/2\pi\delta$, where f_p is the characteristic frequency of the turbulent boundary layer, corresponding to the maximum power in ripple spectrum of the boundary layer.

In Ref. [441], detailed investigations have been carried out on the reduction of resistance in the flow around elastic plates. In reviews [193,194,439] it is argued that elastic surfaces are promising for reducing friction. Their main drawback is noted—the mood for a narrow range of speed.

In Refs. [60,139], based on the analysis of our own work and the investigations of other authors, the features of the two mechanisms of the influence of elastic plates on friction resistance are shown: one of them causes a transition to be delayed, and the second adds a decrease in turbulent friction to the transition. Character of curves (9)−(1) is explained by the action of the first mechanism, and the Kramer curves (15)−(24) are explained by the action of the second mechanism. Character of some of the other curves in Fig. 6.1 is also explained by the influence of elastic plates on the turbulent boundary layer (for this, a turbulizer was installed on the models).

In addition to the known methods of controlling the boundary layer, for example, using suction or injection, recently, new methods for controlling the boundary layer have been widely developed: injection of polymer solutions or fine air bubbles into the boundary layer, the use of elastic-damping coatings and others. In addition to the above-mentioned shortcomings of the considered studies on the reduction of friction resistance by elastic coatings, we note the general shortcomings of all investigations, including modern ones, in the field of methods for controlling the characteristics of the boundary layer, in particular, to reduce friction.

In order to study theoretically or experimentally any of the methods for controlling the characteristics of the boundary layer, it is necessary to take into account the results of investigations of CVS found in the boundary layer of flat or curvilinear rigid plates. Known results in this direction are systematized and supplemented by the results of their own experimental research: Chapter 3, Experimental investigations of the characteristics of the boundary layer on a smooth rigid plate, and Chapter 4, Experimental investigations of the characteristics of the boundary layer on a smooth rigid curved plate [60,139,378]. Developed models for the development of CVS in a laminar, transitional (Section 3.6, Fig. 3.33) and turbulent (Section 3.6, Fig. 3.40) boundary layers.

For successful application of elastic coatings, it is necessary to experimentally investigate the characteristics of the laminar, transitional, and turbulent boundary layers in the flow around various types of elastic flat and curved plates. Such investigations were carried out in Chapter 5, Experimental investigations of the characteristics of the boundary layer on the analogs of the skin covers of hydrobionts, and in Refs. [60,139,378]. When developing a methodology for studying the flow of elastic plates, it is necessary to take into account the results obtained for the development of CVS in the boundary layer of elastic plates.

In Refs. [53,56,591], Section 3.9, and others, the problem of receptivity by a boundary layer of various disturbances is investigated. In particular, the interaction of the introduced disturbances in the boundary layer with the forming natural CVS of the boundary layer is investigated. These interaction mechanisms also need to be considered when planning experimental investigations of the flow around elastic plates.

When conducting investigations on the control of the CVS of the boundary layer and drag reducing, it is necessary to take into account the results of investigations of fast-swimming hydrobionts: Part I (Experimental Hydrodynamics of Fast-Floating Aquatic Animals) and [15,28,42,43,45,55,62,63,65,71,72,76,77,98,111, 115,118,121,123,139,579]. Analysis of the investigations cited in Part I showed that the structure of the skin of high-speed hydrobionts and the mechanical characteristics of their skin depends on the speed of their swimming and change along the body. All hydrobiont systems are interconnected and interact in the process of movement in order to achieve optimal swimming speeds and minimize energy losses [98].

To obtain reliable experimental data on friction resistance, it is necessary for each velocity range to develop a methodology for conducting research and conduct investigations in various experimental installations designed for a specific range of flow around of the model. It should also be borne in mind that the experimental investigations performed in Chapter 5, Experimental investigations of the characteristics of the boundary layer on the analogs of the skin covers of hydrobionts, as well as in Refs. [60,139], showed that the characteristics of the boundary layer in the flow around of elastic coatings differ significantly from similar results in the flow around hard surfaces. Part I shows that, in hydrobionts, the boundary layer when flown around their bodies is significantly different from the case of the flow around rigid bodies.

Theoretical investigations of the turbulent layer on an elastic surface were performed by G.A. Voropaev [586,587,589−593] and etc.

6.3 Experimental complex for investigations of friction drag

At present, various theoretical methods for calculating the turbulent boundary layer are being intensively developed. One of the results of these calculations is the determination of turbulent friction. In Refs. [60,139] a list of various semiempirical methods for calculating the turbulent friction drag is given. Despite the variety of theoretical and experimental methods for determining friction resistance on smooth flat plates, under new boundary conditions, direct measurement of friction resistance on strain-gauge dynamometers is still reliable.

The development of numerical methods and computer technologies contributes to the solution of many complex problems. However, in solving new nonlinear and nonstationary problems, the physical experiment remains as relevant and irreplaceable when the physical concepts of the processes occurring are unknown. Therefore, an experimental analog of computational methods for solving new problems was implemented. Such experimental complex is presented in Fig. 6.2. It consists of the setup *I* (the hydrodynamic stand of small turbulence) for physical

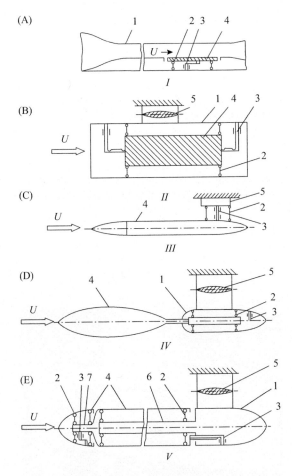

Figure 6.2 Experimental complex for the investigation of the boundary layer: *I V* types of strain gauges. Placement of strain-gauge inserts in the hydrodynamic stand (A) and wing (B): (1) hydrodynamic stand and wing, (2) elastic suspensions, (3) strain gauge, (4) test surfaces, (5) knife-pylon; (C)–(E) schemles of tensor dynamometers of longitudinal streamlined cylinders: (1) cowl, (2)–(5) the same as in (A) and (B); (6), (7) rods [60].

research of the boundary layer at low Reynolds numbers, and of installations $II-V$ (strain dynamometers) for towing tests in a wide range of Reynolds numbers.

The hydrodynamic stand of small turbulence is given in Section 3.1. Installation II is designed as a thin axisymmetric wing and allows investigate in the towing tank on the same inset (Section 5.1, Fig. 5.4) the same plates (Section 5.1, Figs. 5.7–5.10), which were examined at installation I (Section 5.1, Figs. 5.5 and 5.6). In towing tests, it is impossible to design a drag balance, which would be equally sensitive at various ranges of towing speeds. Therefore, it is necessary either to have several drag balances, or to take into account that one drag balance will produce increased errors at other ranges of towing speeds. The possibility to test the same surfaces in installations I and II (Fig. 6.3) allows increasing accuracy of measurements in a wide range of Reynolds numbers.

Installations III, IV allow to investigate the same surfaces, as in installations I, II, in 2D approximation. On these installations, it is also possible to investigate various spatial problems of a boundary layer. On installation III (Fig. 6.4), a tensometer is placed in a knife-pylon. On installations IV, V, a tensometer is placed in the tail cowls. It reduces the measurements error caused by the moments from asymmetrical loadings. The installation V allows separate measuring the loads on the nose, middle, and tail parts. Tests on installations II-V can be made in both direct and reversed motion.

V.V. Babenko developed the idea of creating an experimental complex with unified units and principles of operation, features of developed strain-gauge diaphragms in models of a body of revolution and separation of resistance measurements into separate parts along the body of rotation on a strain-gauge V. He designed and participated in the manufacture of a hydrodynamic stand and model III (Fig. 6.2A and C). V.I. Korobov, under the scientific guidance of V.V. Babenko, upgraded the hydrodynamic stand, and also developed the design of strain gauges and models II, IV, V (Fig. 6.2B–E).

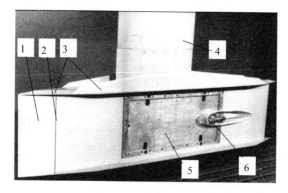

Figure 6.3 Photograph of the vertical axisymmetric wing for towing plates (strain-gauge dynamometer II): (1) wing, (2) wire turbulence promoter, (3) end vertical plates, (4) knife-pylon, (5) test surface, (6) sensor for measuring elasticity and vibrations (Part I, Section 6.4, Figs. 6.15–6.17) [357].

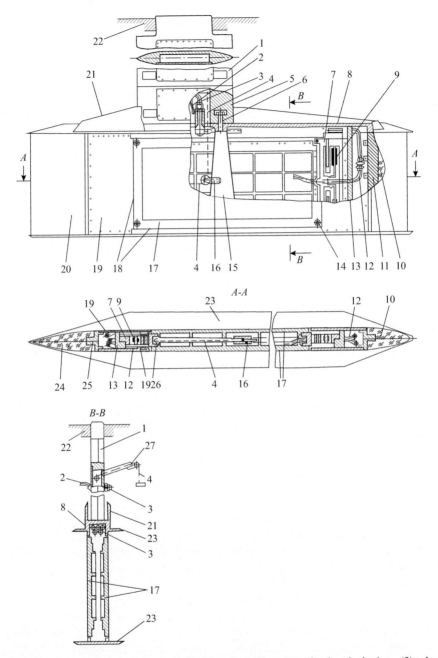

Figure 6.4 Wing design—strain gauge *II*: (1) main knife-pylon, (2) electrical wires, (3) tubes for injections of liquids, (4) cable of the calibration device of the strain gauge, (5) knife-pylon mounted on the wing, (6) pin for mounting the calibration device, (7) strain gauges, (8) top power profiled profile, (9) resistive-strain sensor, (10) rear plexiglass cowl, (11) rear leg of the power frame, (12) pipelines, (13) power profile for fixing the strain gauges, (14) frame fastening of the panel, (15) frame calibration device, (16) roller, (17) panels for examining plates, (18) gaps between the panel and the body of the strain gauge *II*, (19) front removable plate, (20) front cowl, (21) vertical washers, (22) frame of the towing trolley of high-speed towing tank, (23) horizontal power washer, (24) rear cowl, (25) front power strut, (26) hole for the calibration cable, (27) tube calibration device [357].

During the development of the experimental complex, the unification of the greatest number of nodes was provided, which made it possible to reduce the magnitude of systematic errors, to facilitate and unify the tests, and also to increase the interchangeability of the nodes and the possibility of their application for various experiments.

Experience shows that it is impossible to create a universal installation, in which many boundary layer problems could be investigated in a wide range of Reynolds numbers. The developed experimental complex made it possible to plan the investigations of a wide class of physical experiments, both of the boundary layer, and other problems of hydromechanics. The experimental complex in combination with the devices presented in this chapter (Part I) and in Chapters 3–5 (Part II), is a peculiar system of machines, devices, and mechanisms, which made it possible to develop a new, more economical technology for conducting experimental investigations and to obtain qualitatively new results.

The device of the hydrodynamic stand of small turbulence is given in Section 3.1. The data obtained in the experiments on the hydrodynamic stand formed the basis for the development of the methodology for the subsequent investigations presented below. At the end of the working section there is a strain-gauge insert containing a unified bracket attached in the middle to the strain gage, which, in turn, is attached to the first bottom of the working section. The removable panel is screwed to the four legs of the bracket. The measurement in the channel was carried out using a strain-gauge dynamometer I. To reduce the influence of moments on the results of friction measurement, the strain-gauge attachment structure was modified: the bracket was attached to the plate using four elastic plates acting as kinematic elements of the strain gauge at the ends, and the plate was attached to the strain-gauge beam. The same strain gauge I was used in measurements in a wind tunnel [60,139].

To extend the range of Reynolds numbers during measurements, a strain gauge II was designed in the form of an axisymmetric wing 1 (Fig. 6.3). Due to the small wingspan, it was equipped with end horizontal and vertical washers (3). On the back of the surface (5) on the wing (1), a spring sensor and displacement (6) are installed in the fairing. A wire turbulence promoter (2) is installed on the forward fairing of the wing.

The power frame of a vertical axisymmetric wing (Fig. 6.4) consists of two powerful horizontal washers (23) profiled with stiffness edges and force racks (11), (13), (26). Horizontal force washers (23) perform the function of preventing fluid from flowing from the ends of the wing of a small extension. Vertical washers (21) perform the hydrodynamic function of stabilizing the influence of large vortices formed in the corners between the pylon-knife and the upper washer (23). In addition, the washers (21) provide additional rigidity of the washers (23). The racks (11), (25) are attached to their outer edges and horizontal power washers (23) forebody (20) and rear (10) interchangeable cowls. Symmetric strain-gauge beams (7) are attached to the power profiles (13). Uniform removable panels (17) are attached to strain-gauge beams on two sides for examining the plates, which were also mounted on the strain gauge I. Outside of the strain gauge (7) are flush with the panels (17) are covered with removable plates 19. The design of the wing has systems for injection or suction of liquid or bubble solutions. The design of the wing is made in this way (3), which allows you to enter a liquid or microbubble solutions through the front gap (18) between the insert and the wing and suck the liquid through the rear gap of the insert. There is a gap along the perimeter of the panel (17) between its front surfaces and the corresponding details of the wing surface. The front clearance is $(1.0-1.2) \cdot 10^{-4}$ m, the rear clearance is $(1.2-1.8) \cdot 10^{-4}$ m, side gaps $(0.8-1.6) \cdot 10^{-4}$ m. A calibration device (4), (6), (15), and (16) is built in the wing, allowing calibration of strain gauges (9) on the model at any time without placing the wing in the dry dock of the high-speed hydrodynamic channel.

To the upper power washer (23) is mounted a knife-pylon 5, connected with a unified knife-pylon (1), mounted on the frame of the towing cart of the high-speed towing tank (22) or in the cavitation pipe. In pylon 1 there are all highways: electrical wires (2), tubes for supplying liquids (3) and a cable for calibration of (4) strain gauges (7). Knives-pylons (1), (5) are equipped with removable plate guards for easy access to the specified highways.

The strain gauge *III* consists of a power knife with front and rear cowls (Fig. 6.2). In the front cowls there are pipelines, and in the rear electrical wires of the strainer and the calibration device tube. In these investigations, a longitudinally streamlined cylinder (4) was investigated on a strain gauge, which, through an intermediate short knife-pylon, was attached to a strainer installed in the main pylon-knife. The cylindrical part of model consists of several cylinders, which are screwed to each other. It allows changing the model length over a wide range. The calibration device has the same unified principle of action: it allows making calibration during tests without disassembling the setup.

The strain gauge *IV* (Fig. 6.2) consists of a power body—tail cowl (1) (diameter 0.1 m), which fastened to the unified knife-pylon (5). In contrast to strain-gauge dynamometers *I—III*, elastic suspensions and strain gauges were made of separate blocks. The elastic suspension have a special design and are made in the form of annular membranes (2), which are installed at a sufficient distance from each other. They are fixed hollow rod (pipe), having a seating surface. On this rod are installed various test surfaces (4), having various shapes and surfaces. In the end of the case in a sealed chamber is a strain gauge (3). This layout is convenient in operation and allows you to improve the measurement accuracy.

The strain gauge *IV* is fastened to the power section of the towing trolley of the towing tank or in the cavitation tunnel by means a unified pylon-knife consisting of a short (5) and a long (6) parts (Fig. 6.5). The short part (5) includes a dynamometer with a strain gauge, and the long part after connecting with the short part is attached to the towing trolley (8) of the high-speed towing tank (9). Both parts of the knife are connected by locks closed by hatches (2). They allow you to remove the power body—tail cowl (4) with a strain gauge located in it. The tail fairing (4) is rigidly connected to the short part of the knife-pylon (5). It allows removing the strain dynamometer (4) and short part of the knife (5) for adjustment and alignment, not disrupting the alignment of installation of the strain dynamometer in high-speed

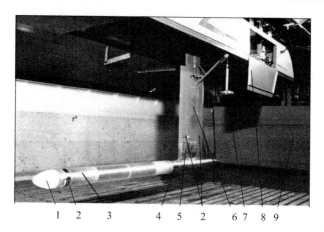

Figure 6.5 Photograph of a model of a body rotating with a diameter of 100 mm (strain gauge *IV*): (1) forward cowl, (2) cylindrical mounting hatches, (3) cylindrical part of the model, (4) power case–tail cowl, (5) short part of knife-pylon, (6) long part of the unified knife-pylon, (7) calibration system, (8) towing carriage, (9) towing tank dock, (10) hatches [357].

towing tank. In the pylon-knife there are cavities in which there are communications for connecting strain gauges, tubes for injecting models of various liquids into the boundary layer and components of the calibration system (7), which has the same unified components and operating principles shown in Fig. 6.4.

In the power case (4), membrane ring suspensions with an axial cylinder fixed on them are located. At the same time, the cylindrical part of the model, consisting of three cylinders (3) and the front cowl (1) of various shapes, is mounted on the seats on the cylinder. Cylindrical dashboards (2) cover the junction of the forward cowl with cylinder. Depending on length of the rod, it is possible to mount certain number of cylinders (3). In this case the model length changes. The length of the strain-gauge model *III* is changed in the same way.

The most complex and perfect is strain dynamometer *V* (Fig. 6.6). It consists of the same unified units and details, the strain dynamometer *IV* is assembled of. The difference is that diameter of the strain dynamometer *V* is increased by 1.75 times. It has allowed in the case (1) (Fig. 6.2) to fix rigidly the system of hollow rods (6), (7). The external pipe (4) leans on those rods through the system of membranous diaphragms (2). On this pipe, the investigated surface and nasal cowl are installed. The developed design of the strain dynamometer *V* has allowed having three independent surfaces on the model length, in central cylindrical parts, nasal and tail parts. Each of these surfaces has an independent strain dynamometer. It allows investigating hydrodynamic characteristics of the model simultaneously on the specified parts and on the entire model.

The investigation of tensor dynamometers *II*, *IV*, *V* was conducted in a high-speed towing tank designed and developed by N.V. Shaibo at the Institute of Hydromechanics of the National Academy of Sciences of Ukraine, Kiyv (Fig. 6.7). In the high-speed towing tank [161] there are two towing trolley:

Experimental investigations of friction drag

Figure 6.6 Photograph of a model of a rotating body with a diameter of 175 mm (strain-gauge dynamometer V): (1) body of the front strain gauge, (2) cylindrical shield of the mounting flap, (3) cylindrical part of the model, (4) power body of the strain gauge of the central part of the model, (5) back part of the model (back strain gauge), (6) the long part of the unified knife-pylon, (7) calibration system, (8) towing carriage, (9) pipelines for supplying the injected fluid, (10) blocking for increasing the stability of the model [357].

Figure 6.7 Photograph of high-speed towing tank of the Institute of Hydromechanics of the National Academy of Sciences of Ukraine [161].

- Low-speed towing trolley driven by an endless cable-cable has a towing speed of up to 6 m/s.
- A high-speed towing trolley driven by a linear electric motor has a towing speed of up to 25 m/s and an acceleration of up to 7.0 m/s.

The dimensions of the towing tank are $140.0 \times 4.0 \times 1.8$ m.

Experimental investigations on strain gauges *II*, *IV*, and *V* on a high-speed towing trolley were performed by V.I. Korobov. Investigations on the strain-gauge dynamometer *I*, *III*, and *IV* and the processing of the obtained data were carried out (with the installation of the short cylindrical part) V.V. Babenko together with V.I. Korobov.

6.4 Experimental investigations of friction drag on elastic plates

The design of the strain-gauge insert (Section 5.1, Fig. 5.6) and the results of an experimental investigation in a hydrodynamic stand of small turbulence of one type of elastic plate (Section 3.1, Figs. 3.1 and 3.2), located on the inset, are given in [357].

Fig. 6.8 shows the results of an experimental investigation of the resistance of an insert on which an elastic plate 4a is installed, whose structure is shown in Fig. 5.7(B) (Section 5.1), and its geometrical parameters are given in Table 5.4

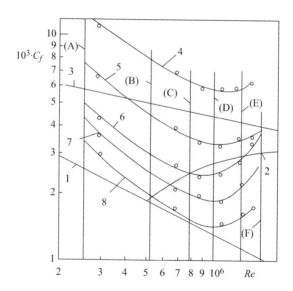

Figure 6.8 The dependence of the friction drag coefficient on the number Re on the insert of the working section of the hydrodynamic stand of small turbulence: (1)–(3) are the calculated dependences for the laminar, transition, and turbulent boundary layers, respectively [292,568]; Experimental data: (4) rigid insert (standard) with a turbulizer and a 1.5 mm front slit, (5) a standard without a turbulizer and with a 1.5 mm front slit, (6) the same as (5), but a 0.5 mm front slit, (7) polyurethane foam with an outer film (plate 4a according to Table 5.4, Section 5.1), (8) is the same as (7), but when the plate is heated (Fig. 5.7, Section 5.1), and (A)–(E) are indicated in the text.

(Section 5.1). In accordance with Fig. 5.7B, layer (1) of plate 4a is a continuous thin sheet of rubber film with a thickness of 0.06 mm, layer (2) is a sheet of new polyurethane foam 6 mm thick and layer (4) is a sheet of new polyurethane foam 3 mm thick, layer (3) is conductive synthetic fabric (CCF) with a thickness of 0.4 mm in the form of longitudinal strips with a width of = 30 mm, arranged with a step $\lambda = 50$ mm. Layer (3) on all sides plastered with rubber film for sealing. A general view of this plate is shown in Fig. 5.8, position (2) (Section 5.1), in which this plate has no outer layer FP. The area of the elastic layer is 72% of the area of the insert with the outer metal edging (Fig. 5.8, position (3) Section 5.1).

The purpose of these investigations is to check the basic concepts of the receptivity problem [53,247] and to determine the effect of gaps around the insert on the results of investigations of elastic plates: as shown in Section 6.2, the gaps can significantly affect the C_f value.

The method of carrying out these measurements was as follows. Before the experiments and after they were carried out, the strain gauge was calibrated when the pump was not running. Then the pump was turned on and the speed flow in the area of the insert was determined. After reaching a constant speed, measurements of the friction resistance of the insert were carried out. Then, using a valve, the speed flow increased and, at a constant value of the speed, sequential measurements of the friction resistance of the insert were carried out.

In Fig. 6.8, the first measurements of the friction resistance were carried out at $U_\infty = 0.1$ m/s direct (A) (Re $= 2.59 \cdot 10^5$), second measurements $U_\infty = 0.2$ m/s direct (B) (Re $= 5.17 \cdot 10^5$), third measurements $U_\infty = 0.3$ m/s straight line (C) (Re $= 7.75 \cdot 10^5$), fourth measurements $U_\infty = 0.4$ m/s straight line (D) (Re $= 1.033 \cdot 10^6$), fifth measurements $U_\infty = 0.5$ m/s straight line (E) (Re $= 1.293 \cdot 10^6$), sixth measurements $U_\infty = 0.6$ m/s straight line (F) (Re $= 1.55 \cdot 10^6$). The last point on curve (4) (between lines (E) and (F)) corresponds to Re $= 1.33 \cdot 10^6$. The numbers Re are not calculated from the geometrical data of the length of the working section (Section 3.7, Fig. 3.41), but taking into account the virtual beginning of the boundary layer at each flow rate. The virtual coordinates origin was determined experimentally by visualizing the velocity profile of the boundary layer using a tellurium method. The velocity profile was used to determine the thickness of the boundary layer, from which the virtual plate length and the Reynolds number were calculated. The measurements showed that the virtual beginning of the length of the working section is located on the surface of the confuser (5) of the hydrodynamic stand, according to Fig. 3.1 (Section 3.1).

Curve (4) in Fig. 6.8 was obtained when a wire turbulator inserted in the nose of the insert, located along the front edge of the insert and with a front slot gap of 1.5 mm. The obtained regularity of curve (4) characterizes the friction resistance of the insert when it is flown around by a turbulent boundary layer. The resulting pattern of curve (4) indicates that the turbulator with the specified the size of the front gap of the gap between the insert and the bottom surface of the working section creates an additional vortex flow around the insert surface so that they obtained C_f (Re) pattern is significantly higher than the curve (3).

To determine the effect of the front slot gap on the results of friction resistance measurements, measurements were repeated in the absence of a turbulizer curve (5) of Fig. 6.8. The shape of the (5) C_f (Re) curve is slightly different from curve (4). This gap size plays the role of a turbulizer. Consequently, with such a gap, tear-off structures are ejected from the slot into the boundary layer of the insert, which form on the front side of the notch in the bottom for insertion. Curve (5) is located in the region of curve (3) corresponding to the friction resistance at a turbulent boundary layer.

Curve (6) in Fig. 6.8 was obtained with a reduced gap (0.5 mm) of the front insert gap three times. Its pattern is determined as follows. On the abscissa in Fig. 6.8 Roman numerals mark the Reynolds numbers at which the transition stages of the boundary layer are formed. In Fig. 3.33 (Section 3.6) the photograph shows a layout of CVS that form during the transition of a laminar boundary layer to a turbulent flow around a rigid smooth plate, and Fig. 3.40 shows the development scheme of the CVS at the corresponding transition stages in the boundary layer of a rigid smooth plate, including the turbulent boundary layer. Thus, in accordance with the layout of the CVS transition (Fig. 3.33) in Fig. 6.8 according to curve (6), corresponding to a rigid standard without a turbulator and with an insert front slot equal to 0.5 mm at a speed of 0.1 m/s, the first point corresponds to the CVS mode of forming and growth the amplitude of Tollmien–Schlichting waves (Fig. 3.33A and B). As the speed increases to $U_\infty = 0.2$ m/s (line (B) in Fig. 6.8), the *III* and *IV* stages of the transition form in boundary layer, in which the Tollmien-Schlichting waves deform and form "hairpin-shaped" (Λ-shaped) vortices (Fig. 3.33B and C), in which the friction resistance decreases. With an increase in the flow around of the insert to $U_\infty = 0.3$ m/s (line (C) Fig. 6.8), the *V* stage of the transition formed in the insert boundary layer, where separation occurs Λ-shaped vortices (Fig. 3.33D). After the line (C) in Fig. 6.8 the *VI* stage of transition was formed (Fig. 3.33D), where the heads of Λ-shaped vortices separation. The head sections of these vortices are beginning take shape. The longitudinal side parts of the Λ-shaped vortices are connected. This leads to the formation of a longitudinal vortex system with mutually opposite rotation of neighboring vortices. This process ends at the *VII* stage of transition (straight line (D) in Fig. 6.8). At the same time, the systems of longitudinal vortices in the boundary layer (Fig. 3.33E) were fully formed; the friction coefficient in Fig. 6.8 is minimal and corresponds to pattern (2). With a further increase in the flow velocity, the longitudinal vortices in the boundary layer begin to meander—their trajectories move from a straight line of movement to zigzag in the XOZ plane (Fig. 6.8, line (E)), which leads to an increase in friction. Line (F) in Fig. 6.8 corresponds to the formation of vertical vortices in the boundary layer due to the meandering of longitudinal vortices (Fig. 3.33E). With a further increase in the velocity of the main flow in the boundary layer, a turbulent boundary layer is formed in the insertion region (Fig. 3.40), and curve (6) coincides with curve (3).

To determine the kinematic structure of the boundary layer with the regularity of curve (6), were measured the profiles of the longitudinal velocities using the LDVM (Section 3.7, Fig. 3.44). These velocity profiles were measured in the ninth section of the insert ($x = 2.43$ m from the beginning of the working section – Fig. 3.42) with: $U_\infty = 0.09$ m/s (1), 0.12 (2), 0.15 (3), 0.18 (4), 0.21 (5), 0.27 (6), 0.35 (7), 0.4 (8) and 0.6 m/s (9). The obtained profiles at the indicated speeds U_∞ correspond to: (1), (2) *I* stage of transition according to Fig. 3.33A (Section 3.6); (3) *II, III* stages of transition (Fig. 3.33B); (4) *IV* stage of transition (Fig. 3.33C); (5), (6) *V, VI* stages of transition (Fig. 3.33F); (7), (8) *VII, VIII* stages of transition (Fig. 3.33G), curve (9) turbulent boundary layer. Fig. 3.44 shows the results of measurements using LDMV of averaged profiles of the longitudinal velocity of the boundary layer in the ninth section according to Fig. 3.42 at each stage of transition for different values of z, whose coordinates are determined to study characteristic sections of the vortex system in the boundary layer of a rigid plate during a natural transition according to Fig. 3.33 (Section 3.6). The results obtained correspond to the layout of the development of the CVS in the boundary layer (Section 3.6, Fig. 3.33).

To carry out investigations of the friction resistance on the insert with elastic surfaces placed on it, physical investigations of the boundary layer structure were previously performed. Fig. 3.31 (Section 3.6) [34] shows a diagram of the formation of three-dimensional deformation on the crest of a plane Tollmien–Schlichting wave. Confirmation of the occurrence of a vortex motion already at the initial stage of the transition is photographs of the streamlines from the side and from above (Section 3.3, Figs. 3.6, 3.9). The laws governing the development of the Tollmien – Schlichting plane wave and Benny – Lin vortices along the length of the working section of the hydrodynamic stand were examined (Section 3.6, Fig. 3.32). The obtained regularities on a logarithmic scale are presented in the form of straight lines. The small-scale wave recorded at the beginning of the working section along z has a length λ_z very close to λ_x. As it develops along the x axis, the wavelength λ_z increases, as does λ_x, but, when $Re = (1.1-1.35) \cdot 10^5$, it decreases, while λ_x continues to increase (the length $\lambda_z \approx 6$ cm when $Re \approx 10^5$ and $\lambda_z \approx 4$ cm at $Re \approx 5 \cdot 10^5$). According to Fig. 3.9 (Section 3.3) $\lambda_z \approx 6$ cm at $x \approx 1.3$ m $U_\infty = 6.7 \cdot 10^{-2}$ m/s ($x_{tel.w.} = 1.1$ m). Visualization showed that, as the boundary layer develops, an axisymmetric system of longitudinal vortices with a wavelength $\lambda_z \approx 4.8 \cdot 10^{-2}$ m in the region of the insert in the hydrodynamic stand is formed throughout its thickness.

The interaction of natural longitudinal vortex structures of the specified scale λ_z, formed along the length of the working section, with introduced disturbances from the side of the streamlined wall was investigated. For this purpose, an elastic plate *4a* was designed and manufactured by V.V. Babenko (Section 5.1, Fig. 5.7B, Table 5.4), layer (3) of which has a pitch $\lambda_z = 5 \cdot 10^{-2}$ m in the longitudinal direction. Longitudinal layers CCF in layer (3) of an elastic plate *4a* were plastered with a thin rubber film, so when the glue dried out, these layers of CCF had a stiffness substantially greater than the outer layers (1) and (2), which damped the disturbances of the boundary layer and interacted with the boundary layer in accordance with the provisions given in Section 3.8. At the same time, due to the stiffness of the CCF of layer (3), longitudinal vortex disturbances with a wavelength $\lambda_z = 5 \cdot 10^{-2}$ m were introduced into the boundary layer from the elastic insert in contrast to the local longitudinal vortex generators given in Section 4.1 in Fig. 4.2. Fig. 4.16 (Section 4.5) shows photographs of the formation of longitudinal vortex systems using vortex generators shown in Fig. 4.2.

One of the positions of the receptivity problem [53,247] is that the interaction of natural disturbances in the boundary layer with disturbances introduced into the boundary layer will be if their geometrical parameters are identical (in the present experiments this is the wavelength λ_z).

Curve (7) of Fig. 6.8 is obtained by flow around the elastic plate 4a located on the inset. The flow around the elastic plate made it possible to reduce the friction resistance. Although the main material of the elastic plate was made of polyurethane foam, however, the outer rubber film and the inner layers CCF pasted over with this film made the plate 4a rigid for the indicated flow around. Therefore, a decrease in resistance was obtained due to the formation with the help of plate 4a in the boundary layer of a system of longitudinal vortex disturbances interacting with natural similar structures that form in the boundary layer (Section 3.6, Fig. 3.33).

Curve (8) of Fig. 6.8 was obtained by flow around an elastic plate 4a on the inset in the case when current was passed through a conductive, electrically insulated synthetic fabric (CCT). The longitudinal strips of the CCF were heated and the surrounding areas of the elastic surface were also heated. The results of experimental investigations of the mechanical characteristics of Section 5.1, Table 5.4 show the structure of the elastic plates, and Table 5.6 shows their mechanical characteristics. Based on the data in these tables, it can be seen that changes in the mechanical characteristics of the PU foam layer and the layers of CCF strips pasted over with a rubber layer when heated will be significantly different. Therefore, the pliability of the plate 4a increased as a whole when the CCF layer was heated. In addition, the longitudinal layers of the CCF will heat up more intensively than the entire plate 4a, which will lead to a significant difference in the mechanical characteristics of the plate 4a in the transversal direction. Heating of the layers of CCF leads to more intensive generation of longitudinal vortex structures into the boundary layer by plate 4a. As a result, the friction resistance of the plate 4a decreased (curve (8)) compared with the absence of its heating (curve (7)).

The magnitude of the friction drag reduction in accordance with the test measurements given in Fig. 6.8 depends on a number of factors:

- It is necessary to follow the correct methodology for performing experimental investigations, in particular, it is necessary to take into account the Reynolds numbers for which certain vortex structures exist in the boundary layer.
- When conducting experiments, the clearances around the measured areas should be as minimal as possible.
- Design and execution of elastic surfaces should be carried out in accordance with the basic similarity criteria (Part I, Section 4.1, Table 1.5) and the provisions of the receptivity problem [53,247], without which uniform elastic plates will always be effective only in a narrow range of Reynolds numbers.
- It is also necessary to take into account the provisions given in Section 5.9.

At low values of Re, elastic plates were tested using a strain gauge I in the range of Reynolds numbers $2.5 \cdot 10^5 - 1.55 \cdot 10^6$ (Fig. 6.8). To extend the range of Re numbers, the same and other elastic plates were investigated on strain gauge II (Figs. 6.2 and 6.3) in the range of Reynolds numbers $1.8 \cdot 10^6 - 2.0 \cdot 10^7$. Fig. 6.9 shows the results of experimental investigations of the effect elastic plates on the

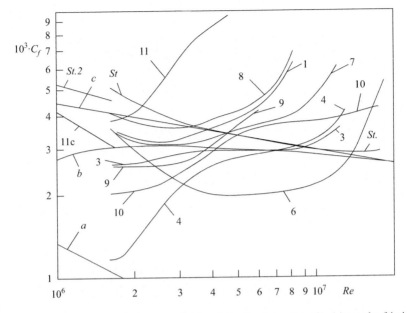

Figure 6.9 Friction drag coefficients on rigid and elastic plates: (*a*), (*b*), (*c*) are the friction resistance of a rigid plate under laminar, transitional, and turbulent boundary layers [520,521]. Measurements in the water flow: St. rigid plate (standart), the numbers of the curves correspond to the numbers of the plates given in Section 5.1, Table 5.4. Measurements in a wind tunnel: St.2 rigid plate, (11c) plate 11 (Section 5.1, Table 5.5).

friction resistance when placing the plates on the strain gauge *II*. The design of elastic plates is given in Section 5.1 in Figure 5.7, and the features of their structure are in Tables 5.4 and 5.5, respectively, as well as in the text of Section 5.1.

Fig. 6.10 shows the structure and the main characteristics of the plates indicated in Figs. 6.9 and 5.7:

- Plate 1: a continuous sheet of polyurethane foam with a thickness of 10 mm, glued to a rigid plate and pasted on the outside with a thin rubber film.
- Plate 3: composite, in which the outer layer is a rubber film glued with glue to a layer of new (more rigid) 3 mm thick polyurethane foam, under which are located longitudinal strips of synthetic conductive fabric (CCF) 20 mm wide with 40 mm pitch, glued to a base made of the same CCF with a thickness of 6 mm and glued to an aluminum plate.
- Plate 4: composite—the same sandwich as plate 3, only the thickness of layers made of polyurethane foam is vertically changed to the opposite position, the outer film is glued with leukonat (plates 3 and 4 are shown in Fig. 5.8 with position (2)).
- Plate 6: composite—layer 1 is absent (Fig. 5.7B), layer 2 and layer 4 are new blue foam-elast (FE) material 4 mm thick, layer 3 is the same as for plates 3, 3*a* (plate 6 is shown in Fig. 5.8 position (3)).

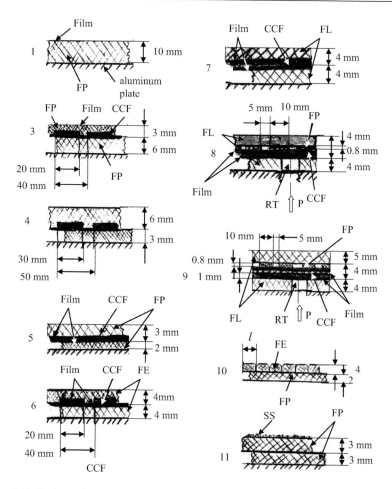

Figure 6.10 Design schemes elastic plates.

- Plate 7: composite—layer (1) is absent (Fig. 5.7B), layer (2) and layer (4) are a sheet of white foam-latex (FL) 4 mm thick, layer (3) is a continuous sheet of synthetic conductive fabric (CCF) 1 mm thick, glued with leuconat.
- Plate 8: composite—layer (1)—a sheet of yellow latex foam (FL) 4 mm thick (Fig. 5.7(C)), layer (2)—strips of new polyurethane foam with a thickness of 0.8 mm $b = 10$ mm, $\lambda = 15$ mm, layer (3)—a solid sheet of CCF with a thickness of 1 mm, layer (4) is absent, layer (6) is a sheet of white latex foam 4 mm thick, inside which is fixed a rubber tube (RT), (plate 8 is shown in Fig. 5.8 position (1)).
- Plate 9: composite—layer (1) a sheet of gray latex foam with a thickness of 5 mm (Fig. 5.7 C), layer (2) strips of new FP with a thickness of 0.8 mm $b = 10$ mm, $\lambda = 15$ mm, layer (3) a continuous sheet of new FP with a thickness of 1 mm, layer (4) conductive fabric (CCF) with a thickness of 1 mm, layer (6) a sheet of white latex foam 4 mm thick, inside which is fixed a RT.
- Plate 10: composite layer (1) glued longitudinal strips of blue FE $l = 4-5$ mm wide (Fig. 5.7D) for plate 10a and $l = 10$ mm for plate 10, layer (2) a continuous sheet of new

FP with a thickness of 2 mm [plate 10 is shown in Fig. 5.8 by the position (4), and Fig. 5.9 by the position (3)].
- Plate 11: composite layer (1) and layer (2) a continuous sheet of new FP with a thickness of 3 mm (Fig. 5.7D, the outer layer of shark skin (SS) is pasted on top of layer 1).

The elastic surface can affect various stages of the transition, including the turbulent boundary layer [60,139,378]. All elastic plates, depending on their structure, increase the length of the transition region and, consequently, the lower critical Reynolds number increases (Fig. 6.8). At the stage of development of longitudinal vortices in the boundary layer, the transition to turbulence (from the lower Reynolds number to the upper one) occurs faster than in accordance with the hypothesis [35]. This is due to the fact that the structural principle of interaction has not been fulfilled, with the exception of plate 10 (Fig. 6.8). The effect on the turbulent boundary layer with large numbers of Re was not observed, with the exception of (6), which satisfied the structural and dynamic principles of the interaction of elastic coatings with the flow, resulting in a friction coefficient C_f in the range Re = $1.7 \cdot 10^6 - 1.2 \cdot 10^7$ was significantly less than the standard (curve St.).

All plates showed a significant increase in C_f, starting with a certain Re value for each plate. Fig. 5.59, Section 5.9 shows standard load distributions along a symmetrical rigid wing profile. Figs. 6.2–6.4 show the wing design—strain gauge *II*, designed for testing elastic plates with a vertical wing in a high-speed hydrodynamic towing tank at towing speeds significantly exceeding the flow rate of the same elastic plates with small turbulence (Fig. 6.8). When designing the strain gauge *II*, the load distribution shown in Fig. 5.59 (Section 5.9) was taken into account. Therefore, the standard insert was located on the wing in the middle of the wing in such a way that the vertical axis of symmetry of the insert was shifted to the stern with respect to the vertical axis of symmetry of the wing. In this case, the insert was located, as far as possible, in the region of the zero gradient of the pressure distribution along the x axis (Fig. 5.59). However, the elastic plate was located in the region of a constant negative pressure distribution along the insert. Therefore, when towing on the surface of the elastic plates, a separation load acted depending on the towing speed of the strain gauge *II*. This leads to deformation of the elastic plate, and in some cases to the detachment of the front edge of the elastic plate from the rigid plate, to which all composite elastic plates were glued. Therefore, almost all tested elastic plates with an increase in towing rate, the drag coefficient increased.

In addition, when analyzing the results obtained, other types of loads should be taken into account, as shown in Fig. 5.61 (Section 5.9). So the distribution of tangential stress on the wall correlates with the location of the spectra of velocity pulsations at various stages of transition along elastic plates. In accordance with the distribution of tangential stress along the wing (Section 5.9, Fig. 5.61 curve (4)), the second maximum value $\bar{\tau}$ can be located near the front edge of the elastic plate located on the wing insert. As a result, shear stress will act along the elastic plate, trying to "pull" the elastic plate along the stream, and the maximum $\bar{\tau}$ in the region of the leading edge will contribute, like the pressure distribution, to tear the elastic coating from the rigid base. Therefore, as shown by the results of experiments in Fig. 6.9, when increasing the speed, special attention should be paid to the method of fastening the elastic plate in the region of the leading edge. In addition to the distribution of tangential stress along the wing, ripple of tangential stress acts on the surface of the elastic plate, which can be determined according to the formula [393]: $(\overline{p^2})^{1/2} = \alpha \tau_w$.

Another factor affecting the efficiency of the interaction of elastic plates with the boundary layer at different flow around is the thickness of the boundary layer δ and the frequency ranges of laminar and turbulent pulsations $\overline{n}$ in the boundary layer.

Analysis of the skin structure of dolphins (Part I, Sections 4.1, 4.3, and 5.2) showed that there are special devices in the skin structure of dolphins that prevent the skin from tearing or deforming during fast swimming. In addition, surface periodic oscillations appeared in the skin for ejection of inhibited fluid from the near-wall region of the boundary layer (Part I, Sections 6.1 and 6.9). As an explanation of the selective character of the influence of the elastic surface on the turbulent boundary layer, the hypothesis of frequency interaction is proposed. It is known that any elastomer absorbs mechanical vibrations. The degree of such absorption depends on the frequency of applied loads (Sections 5.7—5.9). Comparing the frequency of the spectral function of the energy of turbulent pulsations and the spectral function of the energy dissipation coefficient in the elastomer [60,139]; you can determine the effectiveness of the elastomer.

The essence of the proposed approach from integral positions and taking into account the dynamic viscoelastic properties of elastic surfaces is shown in Fig. 6.11 [60,139,356,361]. The energy spectra (I–III) of the pulsation load, perceived by the elastic surface at different Re numbers, are scaled to the same level. The change of the spectral function of pulsations in the flow on elastic plates compared with the spectra on a rigid wall confirms the existence of a mechanism for selecting and redistributing the turbulent energy of the flow by a viscoelastic wall.

Case I corresponds to the measurement of spectral characteristics at low flow speeds. In these experiments, the upper limit of the energy part of the spectrum does not exceed 100 Hz and cuts off the ascending branch of the absorption

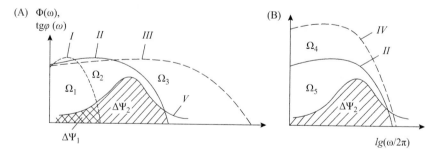

Figure 6.11 Frequency dependences of the energy absorption coefficient $tg\varphi(\omega)$ for elastic plates. The spectra $\Psi(\omega)$ in the boundary layer of the plate under different flow conditions: I–III (A), and at different levels of perceived energy for one flow regime IV, V (b); I–III flow regime with $Re_I < Re_{II} < Re_{III}$; $\Omega_1, \Omega_2, \Omega_3, \Omega_4$ are spectral density domains; $\Delta\Psi$ energy dissipated in an elastic plate; $\Delta\Psi_1, \Delta\Psi_2$ are the regions of the pulsating energy of the boundary layer absorbed by the plate in various flow regimes [60,139,356,361].

coefficient (curve (2), Fig. 6.11). Therefore, in the experiments on towing the plates, it was stated that the elastic wall reduced the energy of high-frequency fluctuations de greases more strongly than the energy of low-frequency fluctuations.

With widening of the frequency range of fluctuation loads (with increase in flow velocity), when the value of $tg\varphi(\omega)$ lies within the range of energy-carrying frequencies, that is, $\omega/2\pi|_{tg\varphi=\max} \ll U/\delta_T$, the distribution of the spectral function of fluctuations near a pliable wall should change significantly. This is confirmed by dips in the pulsation spectra of the longitudinal velocity near the viscoelastic boundary [60,139]. In this frequency range, the extraction of fluctuation energy from the flow is strongest. It corresponds to the case *III* in Fig. 6.11.

We can assume that the value of integral effect ξ is in proportion to the part of fluctuation energy dissipated in the elastic wall. Integrating the spectrum of longitudinal fluctuations in the turbulent boundary layer with respect to all frequency region ($0 < \omega/2\pi|_{tg\varphi(\omega)=\max} < U/\delta_T$), It can be concluded that the maximum of the hydrodynamic effect will be in this flow mode, when the relative area $\Delta\Psi_i$ "cut" by the curve $tg\varphi(\omega)$ for a particular elastomer from the area Ω_i of the pulse load spectrum on the rigid wall $\Phi(\omega)$ will also be maximum, i.e.

$$\xi = \xi_{\max}, \text{ when } d\left(\frac{\Delta\psi(\omega)}{\Omega(\omega) - \Delta\psi(\omega)}\right)\Big/d\omega = 0 \tag{6.1}$$

On the basis of (6.1), we can see from Fig. 6.10 that maximal effect should be in the case *II*: $\xi_I < \xi_{II}$, $\xi_{III} < \xi_{II}$ Hence $\xi_{II} = \xi_{\max}$.

The idea of energy exchange between the turbulent boundary layer and the viscoelastic boundary as a necessary condition provides for the deformation capacity of the pliable plates—the elastic surface makes forced oscillations in accordance with the superimposed spectrum of pressure pulsations. The resulting waviness of the elastic surface can be considered as some equivalent roughness.

As the flow speeds increases, the pressure pulsations on the wall increase and, as a result, the amplitude of oscillation of the elastic boundary increases. If the oscillation amplitude exceeds the level of the equivalent wavy roughness in the boundary layer of an elastic plate, the balance of turbulent energy in the opposite direction changes. In this case, the generation of the pulsating energy due to the dynamic surface roughness beginning to manifest itself due to the oscillation of the elastic layer will exceed the damping energy of the oscillations by the elastic surface. Considering the element of the compliant wall under the loading $p'(Re)$, we have

$$\Delta h = \frac{\sqrt{\langle p'^2 \rangle}}{c} \tag{6.2}$$

where $c = E'/h$. Using the known condition [254,467,468,521] for the allowable roughness height for hydraulically smooth surface

$$Re_{k(accept.)} = 100 \tag{6.3}$$

we have $k_{(accept.)} = \Delta h_{(accept.)} = \Delta y_{(accept.)} = 100\nu/U$

Then the minimal allowable rigidity of elastic plate to hold the hydraulically smooth (static) surface can be calculated as follows:

$$C_{min} = 0.005 \rho U^3 \nu^{-1} \mathrm{Re}^{-0.3}. \tag{6.4}$$

The average value of pressure pulsations over the entire energy-carrying frequency band was determined from the expression [228,489]: $\sqrt{\langle p'^2 \rangle} = 0.5 \rho U^2 \mathrm{Re}^{-0.3}$

Knowing the rigidity of an elastic plate C_{min} determined from the formula (6.4) and elastomer thickness, it is possible to calculate the theoretical module of elasticity E'_{min}, at which the plate under investigation remains hydraulically smooth. The ratios E'_{min}/E'_{meas} were determined for all tested plates when their efficiency dropped to zero. The values of E'_{meas} were taken from Tables Section 5.7. For all elastic plates, with an error of order 3%, we obtained:

$$E'_{min}/E'_{measured} = 1.5 \tag{6.5}$$

The obtained result can be used as a limit criterion for rigidity when selecting the elastic walls intended to reduce turbulent friction.

Optimum mechanical characteristics of an elastic plate under certain conditions of application can appear insufficiently stiff to obstruct the occurrence of a wavy roughness on its surface. According to the hydrobionics approach, the design of such a plate should provide an arrangement, for example, of longitudinal rigid internal walls interfering shear deformations of the elastic plate surface.

Consider the effectiveness of elastic plates according to the degree of increase in their effectiveness. Plate 11 (Fig. 6.10) is made of a single layer of old polyurethane foam (more flexible than the new polyurethane foam), outside of which the outer layer of SS is glued so that the inclination of the scales is located along the stream. According to the structure of the skin of sharks (Part I, Section 4.5), their skin is functionally similar to the skin of dolphins, in particular, the outer layer is smooth, the scales are recessed into the mucous layer and form, with their scallops, longitudinal ordered structures stabilizing the CVS of the boundary layer. In the experiments, a dried outer layer of SS was used, in which the scallops formed a rigid longitudinal structure, mistakenly modeled in various scientific investigations by the so-called riblets. Experiments have shown (Fig. 6.9) that plate 11 has the lowest efficiency compared to the standard (St.) in a very narrow range of Reynolds numbers—$\mathrm{Re} = 1.6 \cdot 10^6 - 2.1 \cdot 10^6$, when the effect was provided by the influence of riblets on reducing C_f. As the towing speed increased, the boundary layer became thinner, the scallops protruded from the boundary layer, forming a roughness, which led to an increase in C_f compared to the standard, and with an increase in velocity the negative effect increased.

In plate 8 (Fig. 6.10), the efficiency (Fig. 6.9) compared with St. is fixed in a wider range of Reynolds numbers as compared with plate 11—$\mathrm{Re} = 1.6 \cdot 10^6 - 3.2 \cdot 10^6$. The design of this plate has a theoretically correct structure, since the second from above layer has longitudinal strips, the width and pitch of which turned out to be suboptimal with a given range of towing speeds. In addition, the outer layer (FL) had elasticity, which did not allow the formation of longitudinal vortex structures in the boundary layer under the influence of the second layer from the top, and the outer layer in the middle had an adhesive joint due to the small size of the elastic plates. This joint increased the roughness of the plate 8 as the towing speed increased.

In plate 1, the magnitude and range of efficiency (Fig. 6.9) increased compared with St. (Re = $1.6 \cdot 10^6 - 3.9 \cdot 10^6$). The plate consisted of a layer of old, more compliance polyurethane foam, outside of which a thin rubber film was glued. At the same time, the elasticity of plate 1 increased and its compliance with the load parameters decreased. With an increase in towing speeds under the action of load parameters, the surface of plate 1 was deformed, which led to an increase in resistance compared to the standard.

At plate 7, the efficiency value was almost the same as that of plate 1, while the efficiency range slightly increased (Re = $1.6 \cdot 10^6 - 4.7 \cdot 10^6$). The design of plate 7 was ineffective, because on the outside, there was a hard layer for a given speed range, a FL, under which there was a continuous layer of CCF glued over with a thin rubber film.

The plate 9 on the outside had the same FL layer as the plate 7, but the CCF layer below was made of longitudinal strips. Therefore, the efficiency range of plate 9 remains the same as that of plate 7, the efficiency value has doubled.

In plate 10, the efficiency value has increased even more—about two times as compared with plate 9 with the same range of efficiency. As the towing speed increased, only this plate had no surface deformation, like other elastic plates. This is explained by the design of the plate 10 and the material from which the outer layer (FE) is made—the outer surface of this material was hydraulically smooth, and the longitudinal layers formed longitudinal vortex structures in the boundary layer.

In plate 3, the efficiency value decreased—approximately two times as compared with plate 10, but at the same time the efficiency range increased (Re = $1.6 \cdot 10^6 - 9 \cdot 10^6$). This is explained by the design and material of the sandwich plate 3: the outer layer is malleable (new foam), which "skips" the effect on the boundary layer below the longitudinal plates arranged, made of CCF.

The greatest decrease in C_f was obtained when towing tests in the speed channel of plate 4 in the range of Reynolds numbers Re = $1.6 \cdot 10^6 - 3 \cdot 10^6$ (Fig. 6.9 curve (4), the standard – St. curve). In the range Re = $3 \cdot 10^6 - 9 \cdot 10^7$, the efficiency of plate 4 is inferior only to plate 6. The design of plate 4 is the same as that of plate 3; the difference lies in the different thickness of the outer and underlying layers made of polyurethane foam. The large thickness of the outer layer of the plate 4 in combination with the longitudinal rows of the CCF strips made it possible to more effectively influence the CVS of the boundary layer in the specified range of Re numbers.

In experimental investigations of the same plate in a hydrodynamic stand of small turbulence in the range of Reynolds numbers Re = $2.6 \cdot 10^5 - 1.6 \cdot 10^6$ (Fig. 6.7, curve (7)), the decrease in C_f was not so significant compared with the standard (Fig. 6.7, curve (6)). Taking into account all the factors enumerated for the interaction of elastic plates with flow parameters, plate 4 turned out to be quite effective in the specified range of Re numbers in accordance with its design. Control tests of the same plate in the air flow did not reveal a decrease in C_f due to the unfavorable mechanical properties of plate 4 when flowing around the air flow (see the list of π — parameters in Section 1.4, Part I).

In plate 6, the range of efficiency in comparison with St. is the largest (Re = $1.6 \cdot 10^6 - 1.5 \cdot 10^7$). In addition, in a wide range of Re = $3 \cdot 10^6 - 1.5 \cdot 10^7$, the efficiency value of plate 6 is the highest compared to all tested elastic plates. This is due to the fact that the material of the elastic plate (FE) had a smooth outer surface, and the underlying layer consisted of longitudinal strips CCF. In the manufacture of plate 6, the outer surface of the sandwich looked like longitudinal flat bumps and hollows by analogy with riblets. In this range of vertical wing towing speeds, plate 6 best met all of the above conditions for the interaction of elastic plates with flow characteristics.

The dependences obtained indicate the selective character of the influence of the elastic surface on the boundary layer. Within towing speeds, the curves have minima for different values of Re numbers. The results obtained differ significantly from those obtained by other researchers, where the optimum has the opposite sign (Section 6.1).

Most curves indicate that the design of elastic plates and the material of the composite structure of the plates affect the transition boundary layer. The physical mechanism of influence is described in detail in [60,139,378]. However, when analyzing towing tests, it is necessary to pay attention, first of all, to the correct fastening of elastic plates on metal substrates. At low towing speeds, elastomer composites materials work more efficiently and interact better with the boundary layer. As the towing speed and load parameters increase, elastomers stretch in bulk deformation, their parameters change and interaction with the boundary layer ceases (plates 6 and 10 are an exception). With further increase in the speed of towing under the action of load parameters, elastomer deformation occurs, which leads to an increase in friction resistance. Sometimes this leads to the separation of the elastomer from its substrate in the region of the leading edge. Thus, the resistance of the plate increases significantly.

Experiments have shown that glued outer rubber film reduces the roughness of elastic materials made of polyurethane foam. At the same time, this film does not allow the underlying layers of elastomers to interact with the boundary layer due to its rigidity on the glue base. Therefore, the most effective material was FE.

Unfortunately, towing tests were not carried out by heating a layer made of CCF, as was done in investigations in a hydrodynamic stand of small turbulence (Fig. 6.8). Heating the CCF layer allowed us to "tune-up" the mechanical characteristics of the elastic plate 4 to the optimum range and significantly increase its efficiency.

In [60,139], experimental investigations of the kinematic characteristics of the turbulent boundary layer with flow around the same elastic plates in a wind tunnel are performed. The flow velocities varied around two speeds—10 m/s and 17 m/s. At the same time, the frictional resistances of the elastic plates were fixed. Fig. 6.8 shows the $C_f(Re)$ patterns for plate 11. Table 6.2 shows the measured drag coefficients for elastic plates. During 1995—1997 Institute of Hydromechanics of the National Academy of

Table 6.2 Friction drag coefficients of the plates in airflow.

Plate number	10^{-6} Re	$10^3 C_f$	$f_1 = U_\infty/\delta$, s^{-1}	$f_2 = u/2\pi$, s^{-1}
Rigid plate (Standart)	0.99	5.2	404.1	85.4
	1.6	4.6	723.1	114.8
5	1.0	5.2	297.6	67.02
	1.66	5.08	499.4	96.8
2a	1.02	5.0	258.2	69.0
	1.6	4.42	429.9	104.4
9	0.95	4.0	410.8	70.4
	1.57	4.1	739.4	109.2
3	0.95	4.58	336.4	69.0
	1.5	3.8	544.9	113.9
10a	0.85	4.95	262.9	69.7
	1.5	4.76	594.2	112.0
8a	0.95	4.6	339.2	64.9
	1.49	4.2	541.1	112.7
11a	1.03	4.9	406.8	72.3
	1.52	4.51	703.9	120.5
11	0.95	4.68	341.3	67.6
	1.57	4.5	580.5	114.0

Sciences of Ukraine performed works under a contract with Cortana Corporation (United States). Based on the results of experimental investigations given in Fig. 6.8, as well as the results given in [60,139] (the monograph [60] was translated into English at Cortana Corporation), we concluded that the most effective for reducing friction resistance in the water flow were plates made of polyurethane (Section 11.1, Fig. 5.10), the bottom surface of which is made in the form of inverted riblets. The mechanical characteristics of such materials are given in Table 5.7.

To validate the results we obtained, G.A. Voropaev, an employee of The Institute of Hydromechanics, confidentially made plates of PU, which were transferred to The Pennsylvania State University, Applied Research Laboratory. Experimental investigations of these plates in the cavitation tube were performed. Unfortunately, the results were not published, but positive results were obtained in a certain range of speeds.

6.5 Drag of longitudinally streamlined cylinders

It is known that the integral characteristics of the boundary layer in the flow around plates and longitudinally streamlined cylinders practically do not differ if measurements are made on cylinders of large elongation. In order to eliminate the noted deficiencies identified during the testing of elastic plates in Section 6.4, experimental investigations were carried out on longitudinally streamlined cylinders—on strain gauges *IV* and *V* (Section 6.3, Figs. 6.2, 6.5, and 6.6). In the factory, metal cylindrical inserts were covered with elastomers with the help of specially designed devices. To obtain reliable results V.I. Korobov developed designs of strain-gauge devices for measuring friction resistance on longitudinally streamlined cylinders [359].

The Fig. 6.12 shows the structures of strain gauges and the character of their loading and deformation (the author of the structures is V.I. Korobov). With the

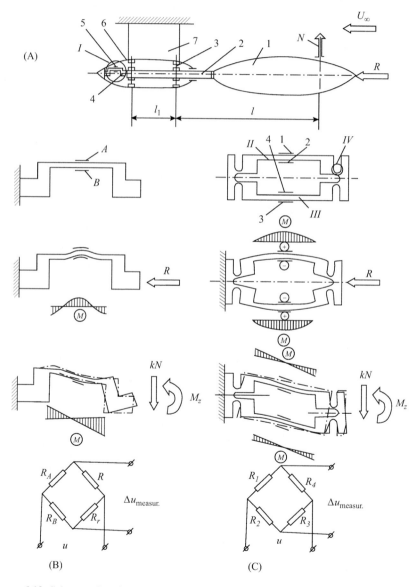

Figure 6.12 Scheme of strain-gauge device. Designations are given in the text.

chosen scheme of strain gauges, strain gauge (5) is not located in pylon (7), as in strain gauge *III*, but in the tail section (6) of the model of axisymmetric cylinder (1). In particular, model (1) ends with cylinder (2), on which two elastic round membranes (3) are fixed. Tension gauge (4) is fixed at the end cylinder (2) and at the end part of the fairing (5). Water resistance R causes longitudinal movement of cylinder (2) and longitudinal loading of tension beam (5), on the lateral sides of which are pasted tenzo-resistors A and B (in Fig. 6.12B) and tenzo-resistors (1)−(4) (in Fig. 6.12C). When compressing the tension beam deformed their thin sides. Below in Fig. 6.12 shows the corresponding load diagrams. In the process of towing a model in a high-speed towing tank, the lateral force N can also act, arising either when the model is not sufficiently accurate in the longitudinal direction or when the round membranes *3* are not sufficiently rigid. Below in Fig. 6.12 the deformation schemes of strain gauges caused by the moment M_z from the lateral force N are shown.

The choice of the design of internal hydrodynamic weights (6) is determined by the measured components, their ratio and limiting values, as well as the geometric dimensions of the bodies being tested. To eliminate the influence of supporting devices on the characteristics of the boundary layer, the test body (1) is usually placed on console (2) upstream from pylon (7) (Fig. 6.12A). In this case, to increase the accuracy of the experiment, it is important to exclude the mutual influence of the measured force components. It can be solved by proper selection of the elastic geometry (kinematic (3) and measuring (4)) of the strain-gauge elements, as well as through the implementation of the principle of electromechanical compensation (placement and activation of sensors (5)) based on the analytical properties of the measuring bridges.

The main aspects of the methodology for solving this problem are shown in Fig. 6.12 and Fig. 6.13 (position (8)). Thus, the influence of the transverse component N on the measurement accuracy of the longitudinal component R is substantially less, and the sensitivity of measurement R is higher in the case shown in Fig. 6.12C than in the case shown in Fig. 6.12B.

In addition to the known sources of errors caused by the method (instrument) of measurements, it is necessary to identify systematic errors and introduce appropriate corrections, which is necessary to compare experiments performed on different installations. In all cases, test measurements of the characteristics of the boundary layer are the data of experiments on the plate. However, a number of investigations on flat plates are difficult for technological reasons, whereas the use of cylindrical surfaces for this purpose greatly simplifies the design of experimental devices and the carrying out of experiments.

Investigations, as in the strain gauge box II (Sections 6.3 and 6.4), were carried out in a high-speed towing tank (Section 6.3, Fig. 6.7). Since the water in the high-speed towing tank had a limited depth, preliminary investigations were carried out to determine the effect on the hydrodynamic characteristics of the geometric parameters models: elongation, cylinder diameter and shape of the nasal tip. The investigations results and features of the measurement technique on models are given in

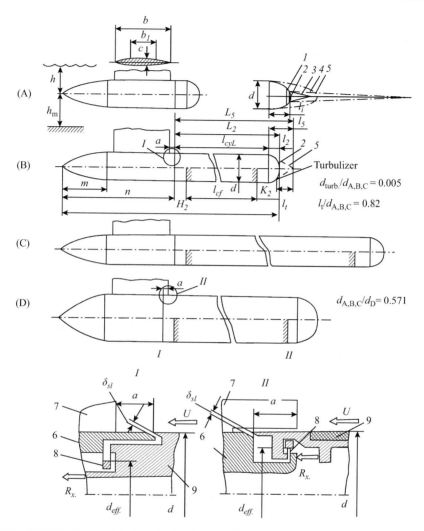

Figure 6.13 The layout of the longitudinally streamlined cylinders with respect to hydrodynamic weights (notation given in the text) [60].

[358]. These investigations allowed us to obtain the dependences of C_F (Re) when testing on strain gauges *IV*, *V* with rigid surfaces.

The geometrical parameters of the investigated cylinders and their arrangement with respect to hydrodynamic weights are maintained in accordance with the recommendations [269,532]. The cylinder parameters are shown in Table 6.3 and in Fig. 6.13. The letters denote a series of tests (cylinder lengthening), the numbers 1–5—the number (shape) of the nasal tip. The distances from the longitudinal axis of the cylinder to the water surface (h-depth) and the bottom of the pool (h_m) when towing it are given in Table 6.4. When conducting experimental investigations with

Table 6.3 Geometric parameters of models according to Fig. 6.13.

(A)

Test series and model designation	l_{cyl}/d	l_{cyl}/L_i	L_2/d	l_2/d	L_5/d	l_5/d	l_{ef}/d	k_2/d	H_2/d
$A_{1,2,3,4}$	0.84	0.535	1.57	0.73	-	-	-	-	9.09
A_5	0.84	0.328	-	-	-	1.72	-	-	-
$B_{1,2,3,4}$	6.45	0.898	7.18	0.73	-	-	5.07	1.39	14.7
B_5	6.45	0.789	-	-	8.17	1.72	5.07	2.38	-
$C_{1,2,3,4}$	11.25	0.838	12.0	0.73	-	-	9.87	1.39	19.5
C_5	11.25	0.865	-	-	13.0	1.72	9.87	2.38	-
D	5.65	0.916	6.07	0.514	-	-	4.92	0.754	10.1

(B)

Test series and model	n/d	m/d	a/d	a/c	b/d	b/c	b_1/c	c/d
A, B, C	7.52	2.82	0.40	1.11	3.80	10.56	4.4	0.36
D	4.0	1.83	0.23	1.11	2.17	10.56	4.45	0.206

(C)

I	1	2	3	4	5
l_i/d	0.55	0.73	1.30	4.50	1.72

This table shows the relation of the length of the probationer part of the model of the axisymmetric cylinder (it is shown in Fig. 6.13 by the vertical shaded lines) to its diameter.

Table 6.4 The relative embedding models during towing in a high-speed towing tank.

Series of and model designation	h/d	h_m/d	h/L_2	h/L_5	h_m/L_2	h_m/L_5	$10^2 S_{ef}/S$
$A_{1,2,3,4}$	4.5	4.7	2.87	-	3.0	-	12.02
A_5	4.5	4.7	-	1.76	-	1.84	8.4
$B_{1,2,3,4}$	4.5	4.7	0.03	-	0.655	-	2.56
B_5	4.5	4.7	-	0.55	-	0.575	2.35
$C_{1,2,3,4}$	4.5	4.7	0.375	-	0.39	-	1.53
C_5	4.5	4.7	-	0.346	-	0.36	1.45
D	2.44	2.9	0.4	0.48	-	-	2.73

model D, the value of h_m was ~0.465 m, and the diameter of the model was 0.175 m, the diameter of the other models was 0.1 m. Since the experiments were carried out for a long time, the depth of the models was varied. The maximum water depth in the high-speed towing tank was 1.8 m.

Fig. 6.13 below shows the design solutions for the formation of a slit δ_{sl} between the model 6 hull, fixed on the knife—pylon (7), and the movable part of the longitudinal cylinders, on the outer surface of which elastic cylinders (9) were fastened. The outer slot of the joint of the model hull parts in the region of δ_{sl} is inclined to reduce the effect of separation of the flow from the outer surface of the cylinder and the resulting suction force.

When conducting experiments, it is necessary to find out the influence of a number of factors when evaluating, for example, a component of viscosity resistance (friction resistance) according to the results of measurements on longitudinally streamlined cylinders. These factors are: the layout of the cylinders with respect to the body of the scales and supporting devices, their elongation and the shape of the front tip, the role of the kinematic junction gap.

In Fig. 6.13, the geometric parameters of the models investigated are shown in dimensionless coordinates. For a better understanding of the measurement results, Fig. 6.14 shows the scheme of the investigated models with specific dimensions. Series 1–5 characterizes the shape of the nasal parts of cylindrical models. Models have some uniform parts.

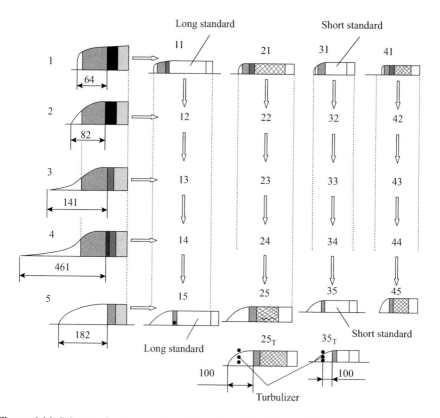

Figure 6.14 Scheme of variants of investigated models of longitudinal axisymmetric cylinders. Designations are given in the text.

So on the left in Fig. 6.14 in the 1–5 series, the first part (the light contour) denotes parts that are screwed onto the same standardized nasal part of the model, consisting of an elliptical truncated cylinder (dark contour). The first continuous cylindrical section (black contour) and the second cylindrical section (gray contour) adjoin the truncated cylinder, which consists of two removable half-rings, which are necessary for mounting the entire axisymmetric cylindrical model. Series 1–5 is clearly visible on the photo of the model – Fig. 6.5 (Section 6.3), where it is indicated by the positions (1), (2).

Series 11–15 characterizes the shape of long cylindrical models with various shapes of nasal parts and a hard surface. Series 21–25 is the same as the series 11–15, but the cylindrical part of the models has an elastic surface. Series 31–35 is the same as the series 11–15, but the cylindrical part of the models is short. Series 4–45 is the same as the series 31–35, but the cylindrical part of the models has an elastic surface. The 25_T and 35_T series are the same as the 25 and 35 series, but boundary layer turbulators are installed on the nose parts of the models.

Fig. 6.15 shows the geometric parameters of the models of longitudinally streamlined cylinders with a diameter of 175 mm (A) and the tail section of a model with a diameter of 100 mm (B). Fig. 6.15A shows a principled scheme of the model, the kinematic scheme of which is given in Fig. 6.2 (tensiometer V) and a photograph of the model is given in Fig. 6.6 (Section 6.3), which shows that the tail cylindrical part of model (4) is rigidly attached to the knife-pylon (6) with the help of powerful bolts. In Fig. 6.6, a pair of hatches for bolt installation is visible on the knife.

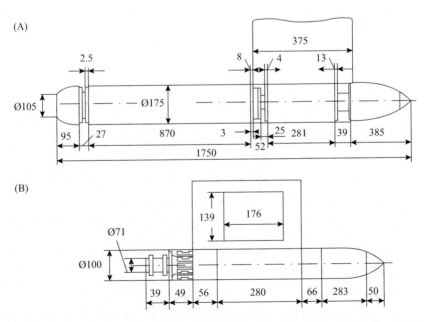

Figure 6.15 Model parameters of a longitudinally streamlined cylinder with a diameter of 175 mm (A) and the tail of a model with a diameter of 100 mm (B).

Sideways in the middle of this cylindrical part there is a cylindrical shield (2) performing assembly works.

In the central part of the power corps of the strain gauge (4), a pipe is rigidly fixed along the longitudinal axis of symmetry (Section 6.3, Fig. 6.2). On this pipe, the outer cylinder and the head part of the model are mounted on circular membranes. Fig. 6.15A shows three places along the longitudinal axis with dimensions of 27 mm, 52 mm and 39 mm for installation of two cylindrical half rings. These half rings close the places for installation work when preparing the model for testing. The dimensions shown in Fig. 6.15 are necessary for calculating the drag coefficient when carrying out towing experiments, in addition, using these dimensions, according to Tables 6.3 and 6.4, can to obtain model sizes and calculate model towing speeds.

shows the dimensions of the tail section of the model, the kinematic diagram of which is given in Fig. 6.2, Section 6.3 (strain gauge *IV*) and the photograph of the model is given in Fig. 6.5 (Section 6.3). In this model, round membranes are fixed in the tail section of the model, on which a short pylon-knife with a mounting flap, 139 × 176 mm in size, is rigidly fixed. This short knife of the model is attached to a standard towing knife. In front of the tail section of the model (Fig. 6.15B) there is a section of 49 mm for fastening the outer cylinder and a section of 39 mm for installing additional cylinders—in Fig. 6.5 (Section 6.3) there are four identical longitudinal sections. The front parts of various configurations (Fig. 6.14, series 1—5) are screwed onto longitudinal cylinders, as shown, for example, in Fig. 6.15 and the front shape is set with a size of 95 mm.

In Ref. [521], a formula is given for calculating the drag coefficient c_f of a longitudinally streamlined body:

$$c_f = \frac{2W}{\frac{1}{2}\rho U_\infty^2 S}, \tag{6.6}$$

where W is the resistance of the model, ρ is the density, and S is the wetted surface.

Fig. 6.16 shows examples of calculating the geometric parameters of the models needed to calculate the resistance coefficients. The value of W was determined when conducting towing tests. The following notation is used: $S_{II} = S_j + S_0$, where S_j is the area of the model's spin-on tip, $_0$ is the curvilinear standard section of the model's nose; S_{h-r} surface area of the cylindrical half-ring of the mounting plate of the model; S_{ad} the surface area of the cylindrical plot (adapter) for testing series 1—5; $S_{l.s.}$ the area of the cylindrical section of the long standard (hard surface); $S_{t.s.}$ the area of the tail section of the model; $S_{s.s.}$ the area of the cylindrical section of the short standard; $S_{r.s.}$ is the area of the ring section of the cylindrical surface; $S_{e.s.l.m.}$ the area of the elastic surface of the cylindrical section of the long model; $S_{e.s.s.m.}$ the area of the elastic surface of the cylindrical section of the short model; $S_{h.f.}$ the square of the heel face of the nasal surface of the model with the tip removed.

Below are given the results of the calculation of the corresponding surfaces:test series 1—5 (Fig. 6.14) $S_0 = 0.01826$ m^2; $S_1 = 0.0202$ m^2 with $l_1 = 64$ mm;

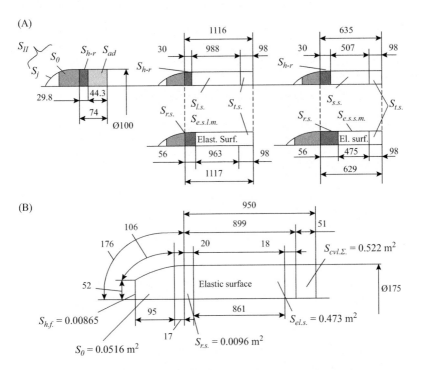

Figure 6.16 The scheme for calculating the areas of the washed surface of models towed in a high-speed towing tank. Explanation of symbols given in the text.

$S_2 = 0.0243$ m² with $l_2 = 82$ mm; $S_3 = 0.0252$ m² with $l_3 = 141$ mm; $S_4 = 0.0332$ m² with $l_4 = 461$ mm;
$S_5 = 0.0442$ m² with $l_5 = 182.5$ mm;
$S_{h-r} = 0.0094$ m²; $S_{r.s} = 0.0176$ m²; $S_{ad} = 0.0139$ m²; $S_{t.s.} = 0.0308$ m²; $S_{l.s.} = 0.31$ m²; $S_{s.s.} = 0.159$ m²; $S_{e.s.l.m.} = 0.306$ m²; $S_{e.s.s.m.} = 0.151$ m².

Relationships for models with a diameter of 100 mm (Figs. 6.14–6.16A).

$$S_1 = S_I + S_{h-r} + S_{ad} = S_I + S_\alpha = 0.0436 \text{ m}^2 \quad l_1 = 138 \text{ mm};$$

$$S_2 = S_{II} + S_\alpha = 0.0477 \text{ m}^2 \quad l_2 = 156 \text{ mm};$$

$$S_3 = S_{III} + S_\alpha = 0.0486 \text{ m}^2 \quad l_3 = 215 \text{ mm};$$

$$S_4 = S_{IV} + S_\alpha = 0.0565 \text{ m}^2 \quad l_4 = 535 \text{ mm};$$

$$S_5 = S_V + S_\alpha = 0.0675 \text{ m}^2 \quad l_5 = 2565 \text{ mm};$$

$$S_{11} = S_I + S_{h-r} + S_{l.s.} + S_{t,s.} = S_I + S_\beta = 0.371 \text{ m}^2 \quad l_{11} = 1180 \text{ mm};$$

$S_{12} = S_{II} + S_\beta = 0.375$ m^2 $l_{12} = 1198$ mm;

$S_{13} = S_{III} + S_\beta = 0.376$ m^2 $l_{13} = 1257$ mm;

$S_{14} = S_{IV} + S_\beta = 0.384$ m^2 $l_{14} = 1577$ mm;

$S_{15} = S_V + S_\beta = 0.395$ m^2 $l_{15} = 1299$ mm;

$S_{21} = S_I + S_{r.s.} + S_{e.s.l.m.} + S_{t.s.} = S_I + S_\gamma = 0.374$ m^2 $l_{21} = 1181$ mm;

$S_{22} = S_{II} + S_\gamma = 0.378$ m^2 $l_{22} = 1199$ mm;

$S_{23} = S_{III} + S_\gamma = 0.379$ m^2 $l_{23} = 1258$ mm;

$S_{24} = S_{IV} + S_\gamma = 0.387$ m^2 $l_{24} = 1578$ mm;

$S_{25} = S_V + S_\gamma = 0.398$ m^2 $l_{25} = 1300$ mm;

$S_{31} = S_I + S_\eta = 0.22$ m^2 $l_{31} = 699$ mm;

$S_{32} = S_{II} + S_\eta = 0.224$ m^2 $l_{32} = 717$ mm;

$S_{33} = S_{III} + S_\eta = 0.225$ m^2 $l_{33} = 776$ mm;

$S_{34} = S_{IV} + S_\eta = 0.233$ м^2 $l_{34} = 1096$ mm;

$S_{35} = S_V + S_\eta = 0.244$ m^2 $l_{35} = 818$ mm;

$S_{41} = S_I + S_\zeta = 0.22$ m^2 $l_{41} = 693$ mm;

$S_{42} = S_{II} + S_\zeta = 0.223$ m^2 $l_{42} = 711$ mm;

$S_{43} = S_{III} + S_\zeta = 0.224$ м^2 $l_{43} = 77$ mm;

$S_{44} = S_{IV} + S_\zeta = 0.232$ m^2 $l_{44} = 1090$ mm;

$S_{45} = S_V + S_\zeta = 0.243$ m^2 $l_{45} = 812$ mm;

Relationships for models with a diameter of 175 mm (Fig. 6.16B):
$S_{noses} = S_{h.f.} + S_0 = S_{r.s.} = 0.0697$ m^2, $S_{cylind.\Sigma} = 0.522$ m^2, $S = 0.592$ m^2, $\rho = 101.8$ kg$\cdot$s^2/m^4;

$S_{\text{nose s.}} \cdot \rho/2 = 3.087 \text{ kg} \cdot \text{s}^2/\text{m}^2$; $S_{\text{cylind.}\Sigma} \cdot \rho/2 = 26.57 \text{ kg} \cdot \text{s}^2/\text{m}^2$; $S \cdot \rho/2 = 29.66 \text{ kg} \cdot \text{s}^2/\text{m}^2$;

$$\overline{S}_{e.s.} = \frac{S_{\text{onlye.s.}}}{S} = 0.812, \quad \overline{S}_{\text{cyl.e.s.}} = \frac{S_{\text{onlye.s.}}}{S_{\text{cylind.}}} = 0.906.$$

The water temperature during measurements was $t^o = 14.5-19$ degree, and the value of kinematic viscosity was $\nu = 1.1463 \cdot 10^{-6}$ m²/s ($t^o = 15$ degrees) and $\nu = 1.036 \cdot 10^{-6}$ m²/s ($t^o = 19$ degrees); $\nu_{\text{aver.}} = 1.088 \cdot 10^{-6}$ m²/s; $l/\nu = 1.035 \cdot 10^{-6}$ m²/s.

During the research, the elongation of the longitudinally streamlined cylinders was changed, and the variation of the longitudinal distribution of pressure was carried out within small limits through the use of interchangeable nose extremities of various shapes.

The drag of the cylinders can be written as the sum of the components:

$$C_x = C_{F\text{cyl}} + C_{xw} + C_{xb},$$

where $C_{F\text{cyl}}$, C_{xw} and C_{xb} are the viscous, wave, and bottom resistance, respectively.

1. The nominal viscosity resistance is $C_{F\text{cyl}} = C_{F0} + C_{Pv}$, where C_{F0} is the friction resistance of the equivalent plate, C_{Pv} is the viscous pressure resistance. For test series B, C and D (Table 6.3, Fig. 6.14), according to [480] with $L/d > 4$. $C_{F\text{cyl}} = (1 + k_{\text{form}}) \cdot C_{F0}$, where k_{form} is the coefficient of influence of the surface curvature of the axisymmetric cylinder, which mainly depends on the relative elongation L/d.
2. Bottom drag is due to the design feature of the hydrodynamic balance of the experimental installations EI-I (A) and EI-II (D) and depends on the value of the pressure coefficient $C_{P\text{crack}}$ in the region of the crack of the kinematic decoupling of the tensor suspension, which is determined by the distribution of pressure along the body of the entire installation, as well as influenced by the pylon:

$$C_{xb} = C_{P\text{crack}}(S_{\textit{eff}}/S) = (C_{P\text{crack.cyl.}} + C_{P\text{crack.pylon}})(S_{\textit{eff}}/S) \tag{6.7}$$

where $S_{\textit{eff}} = \pi/4 \, (d^2 - d_{\text{ef}}^2)$ is the cross-sectional area at which the pressure drop "works," $d_{\textit{eff}}$—see Fig. 6.13, S is the wetted surface of the cylinder. Fig. 6.13, section I, is shown kinematic decoupling scheme on the front support (Fig. 6.13, position (8) dual diaphragm) on the installation EI-I (A–C), section II: for the installation EU-II (test series D).
3. The characteristic drag C_{xw} of immersed cylinders depends on many parameters, the main of which are [480]: the shape of the hull (lengthening L/d), the shape of the contours, the ratio of the displacement (δ_k); Froude number Fr; the influence of the basin boundaries—the relative depth of the model (h/L, h/d), and the distance to the bottom (h_{mod}/L, h_{mod}/d).
4. It should be noted that the sterned (curvilinear) extremities of strain-gauge cylinders (Fig. 6.13) reduce the influence of the wave component, as well as the viscosity resistance of the pressure, since there is a unseparation flow around the cylinders.
5. When towing tests in the towing tank of longitudinally streamlined cylinders (B, C, D, Table 6.3 and series 11–45 Fig. 6.14), the main component is the viscosity component.

Fig. 6.17 shows the results of experimental investigations of the coefficients of drag C_x of longitudinally streamlined cylinders in accordance with Fig. 6.14.

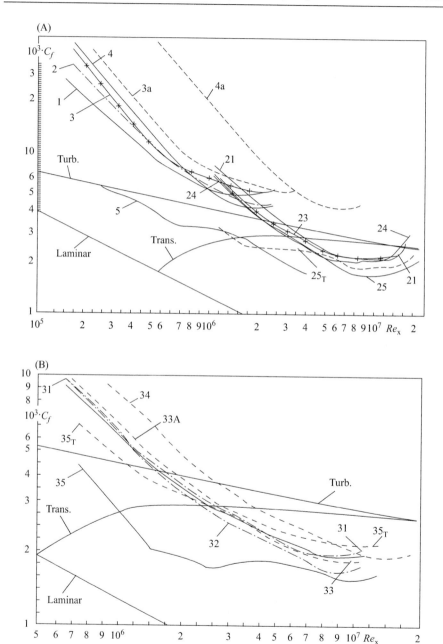

Figure 6.17 Dependencies of drag coefficients C_x of longitudinally streamlined cylinders on Reynolds number Re_x in accordance with Fig. 6.14: (A) a series of experiments 1–5 and 21–25, (B) a series of experiments 31–35, (C) a series of experiments 11–15 (Fig. 6.14).

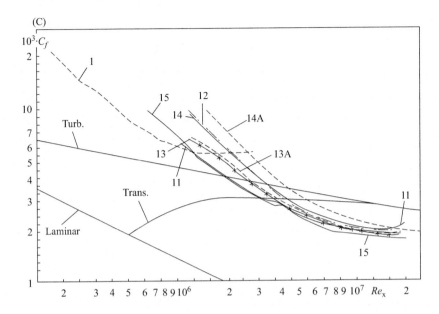

Figure 6.17 (Continued)

Reynolds numbers were calculated by the length l_i of the strain-measuring parts of the models. The curves with the designation of the series with the letter "a" were obtained by calculating the Reynolds number along the entire length of the model (Fig. 6.15A), and the curves with the letter "T" when installed in the head part of the turbulizers model according to the scheme of Fig. 6.14. For cylinders of small elongation (a series of experiments A in Table 6.4 and Fig. 6.14, series 1–5 in Fig. 6.15), the wave component is practically absent, since for all modifications 1–5 $h/d > 3$ ($h/d \approx 5$), $h/L_i > 1.1$ [480] and the number $Fr > 1.4$. Since the elongation of the cylindrical surface is small compared with the curvilinear front part, in this series of experiments, C_{P_v} is an essential additive to friction resistance, and with variations in the geometry of the nasal tip, C_{xb} has a strong influence.

Modifications of the front extremity of a semielliptic form ((2), Figs. 6.13 and 6.14) with a small ratio of semiaxes (1 to 1.46), as well as due to a thin nasal needle (3), (4), lead to a significantly smaller redistribution of pressure in the vicinity of the cowl (decrease peak discharges [374,375,632]). As for the strain gauge section, the elongation $\lambda_2 = L_2/d$ is small, it is also near the slot, compared with the use of the cowl (5) ($\delta_{k5} \ll \delta_{k2}$) with a large ratio of semiaxes ($a/b = 1/3.44$). In the latter case, the drag of the cylinder of the 5i series differs significantly from the 1_i–4_i series (Figs. 6.13 and 6.14), mainly due to C_{xb} and to a lesser extent due to the difference in the C_{P_v} coefficient due to the difference in elongation.

The model with a nasal tip in the experiments of series 1 has the smallest drag (S_1 is the smallest) in comparison with the series of experiments 2–4 (Fig. 6.17A). The drag of the model with a nasal tip in the form of a longer nasal needle (series 4) to $Re = 7 \cdot 10^5$ had the greatest drag coefficients due to the fact that it is washed area S_4 was the greatest.

Despite this, with $Re = 10^6 - 2.5 \cdot 10^6$ in series 4, the smallest C_x was obtained compared with series 1–3. The drag of the model in series 5 had significantly less drag in the whole range of Re numbers compared to series 1–4.

In Fig. 6.17A, models of rigid cylinders of small elongation (series 1–4), despite the significantly smaller surfaces being washed, had significantly larger C_x values, compared to other cylinder lengths (Fig. 6.17B and C).

Fig. 6.17B shows the dependences C_x (Re) in the investigation of short rigid cylinders of the 31–35 series (Fig. 6.13). Curves of series 31–35 are obtained in the range of numbers $Re = 7 \cdot 10^5 - 1.4 \cdot 10^7$ and, as it were, continue the regularities of series 1–5, obtained in the range $Re = 1.4 \cdot 10^5 - 2.5 \cdot 10^6$. The different forms of the nasal contours of short rigid cylinders of the 31–35 series had practically no effect on the C_x (Re) dependences, in contrast to the 1–5 series. Only the model with a long xiphoid tip increased the drag of the model in the whole range of Reynolds numbers, and the model with the short ogival tip (series 32) had less drag in the range $Re = 1.6 \cdot 10^6 - 1.4 \cdot 10^7$ compared to series 31, 33, 34. In the range $Re = 1.6 \cdot 10^6 - 1.4 \cdot 10^7$ the drag of the cylinders of the series 31–34 with various forms of the nasal contours was less than the drag of the rigid plate in the case of a turbulent boundary layer.

As in the series of measurements 1–5, the drag of a short longitudinally streamlined cylinder with a long ogival tip (series 35) turned out to be significantly less than with other forms of nose tips, in the whole range $Re = 1.75 \cdot 10^5 - 1.4 \cdot 10^7$. In this case, the drag was less than that of the plate in the transition and turbulent boundary layers. When a turbulizer was installed in the nose of this model, the resistance of the model of the 35_T series was greater, but over a long Re distance the C value was less than when the rigid plate was wrapped around a turbulent boundary layer. This can be explained by the small length of the strain-gauge insertion of short cylinders and the small size of the turbulizer.

The calculation of the C_x(Re) dependence of the model of the 33 A cylinder over the entire length of the model slightly increased the drag of the short cylinder, since the ratio of the total length of the cylinder to the length of the tensiometric part of the cylinder decreased compared to the series 1–5.

In Fig. 6.17C, the C_x(Re) dependences are located when investigated long rigid cylinders of the 11–15 series (Fig. 6.14). For comparison, in this Figure, the drag curve of a cylinder of series 1 is plotted. Curves of series 11–15 are obtained in the range of numbers $Re = 6 \cdot 10^5 - 2 \cdot 10^7$. The different shape of the nasal contours of the long rigid cylinders of the 11–15 series had practically no effect on the C_x(Re) dependences, unlike the 1–5 series. The model with a long xiphoid tip (series 14) and a short ogival tip (series 12) increased the drag of the model in the range of Reynolds numbers $Re = 10^6 - 4 \cdot 10^6$. The model with a long ogival tip (series 15), although it had less drag compared with tips of series 12, 13, was significantly less than that of series 5 and 35 cylinders. When calculating drag, taking into account the entire length of the model (series 13A and 14A), C_x increases, as in the series of experiments with long cylinders (33A series), but significantly less compared to the 3A and 4A series.

In the range of $Re = 1.8 \cdot 10^6 - 2 \cdot 10^7$ the drag of cylinders with various shapes nasal contours (series 11–15) was less than the drag of a rigid plate with a turbulent boundary layer.

Comparison of $C_x(\mathrm{Re})$ dependence was performed when conducting towing tests of short and long longitudinally streamlined cylinders. Thus, the dependences of $C_x(\mathrm{Re})$ cylinders with a dead-end (series 31 and 11) in the range $\mathrm{Re} = 6 \cdot 10^5 - 2 \cdot 10^6$ almost coincided, and as Re increases to $1.1 \cdot 10^7$ C_x, the short cylinder is smaller. For cylinders with short ogival (series 32 and 12) and short xiphoid (series 33 and 13) tips in the entire range of numbers Re, the drag of a short cylinder is less than that of a long cylinder. For cylinders with a long xiphoid tip (series 34 and 14) up to $\mathrm{Re} = 3 \cdot 10^6$, the drag of a short cylinder is slightly less than that of a long cylinder, and as Re increases, the drag of both cylinders practically coincides. And only for cylinders with long ogival tips (series 35 and 15) the drag of a short cylinder is substantially less than that of a long cylinder.

Changing the geometry of the front parts of the cylinders leads to a redistribution of pressure in the area of the curvilinear forming. On the cylindrical part, however, the difference in the distribution of overpressure decreases rapidly with increasing distance from the nose. Therefore, for cylinders of greater elongation, the difference in drag between modifications due to the component C_{xb} should be smaller, which is observed in experiments.

It is possible to estimate the C_{xb} value in these experiments using the graph of the measurement error of the static pressure of an axially symmetric air pressure receiver (APR) depending on the relative location of its receiving holes [269], which is caused by the pressure distribution over its body (distance from the nose) and the influence of the holder. So, for cases B and C (Fig. 6.13) or respectively for short (series 31–35) and long (series 11–15) cylinders ($\lambda = 8-13$) $C_{P\mathrm{slot \cdot cyl}} \approx + 0.003$; $a/d = 0.4$, $C_{P\mathrm{slot\ pylon}} \approx -0.053$. Then by the formula (6.6) $C_{P\mathrm{slot}} = -0.05$. For series B, $S_\mathrm{eff}/S = 0.024$ (Table 6.3) and $C_{xb} \approx -1.2 \cdot 10^{-3}$; for $CS_\mathrm{eff}/S = 0.015$ and $C_{xb} = -0.75 \cdot 10^{-3}$. In the case of D (Fig. 6.13), a/b is such that $C_{P\mathrm{slot\ pylon}} \approx 0$, and $C_{P\mathrm{slot}} \approx 0.3$ ($\bar{l}_{u_l} = L_2 \approx 6.1\ d$); $S_\mathrm{eff}/S = 2.725 \cdot 10^{-2}$; then $C_{x \cdot \mathrm{cyl.tang}} = C_{x \cdot \mathrm{meas}} - 0.817 \cdot 10^{-3}$. The calculated dependence of the coefficient of cylinder drag, taking into account the amendment to the bottom drag in the latter case given in [60,139].

As can be seen, with a certain fixed number Re, in the middle of the towing speed interval (where the C_{xw} component and the surface roughness do not affect), the corresponding experimental dependencies (Fig. 6.17) differ from the law of turbulent friction for the plate (Turbulent line) by the indicated values, which agrees with the evaluation of the component drag.

However, with increasing cylinder elongation in certain modes, the wave resistance affects the results obtained—with an increase of λ_2 (Section 6.3, Fig. 6.2) the relative depth decreases—h/L, h_m/L (Table 6.4). So, for a short cylinder (series 35, Fig. 6.17B)

at $\mathrm{Re} = 1.5 \cdot 10^6$ $\mathrm{Fr}_L = 0.706$;
at $\mathrm{Re} = 3.0 \cdot 10^6$ $\mathrm{Fr}_L = 1.412$;

For a long cylinder (series 15, Fig. 6.17C)

at $\mathrm{Re} = 2.4 \cdot 10^6$ $\mathrm{Fr}_L = 0.56$;
at $\mathrm{Re} = 4.8 \cdot 10^6$ $\mathrm{Fr}_L = 1.12$;

The change in the geometric parameters of the cylinders (λ, the shape of the nasal tip, δ_k) more influenced on the component C_{xb} than on the other components of their drag (C_{Pv}, C_{xw}).

It is preferable to eliminate such a source of systematic error, for example, by means of constructive detuning by sealing off the decoupling cavities with the help of a bellows. In this case, restrictions are imposed on its dimensions (according to S_{eff}/S) from the condition of ensuring the permissible measurement error. This error can be ignored in a comparative experiment.

The maximum relative standard error of C_x did not exceed 2% in each interval of towing speeds.

6.6 Friction drags of elastic longitudinally streamlined cylinders

Fig. 6.18 shows the dependences of $C_x(Re)$ in the investigation of short elastic cylinders (series 41–45 of Fig. 6.14) in the range of numbers $Re = 6 \cdot 10^5 - 1.4 \cdot 10^7$. In contrast to the series of studies given in Section 6.5 at flow around a short elastic cylinder with dead-end nasal shape (series 41) and short xiphoid shape (series 43), the $C_x(Re)$ dependences indicated that the cylinder drag was greatest in the range $Re = 6 \cdot 10^6 - 4 \cdot 10^6$. In this range of Reynolds numbers, the drag of a cylinder with a short ogival tip (series 42) had less drag compared to series 41, 43 and even less in

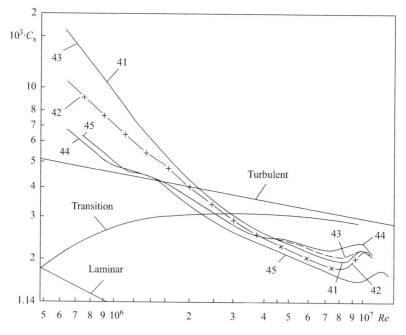

Figure 6.18 The dependences of the drag coefficients C_x on the Reynolds number Re_x of longitudinally streamlined short elastic cylinders (Fig. 6.14, Section 6.5)—a series of experiments 41–45.

series 44, 45. As in previous experiments, the cylinder with a large ogival tip had the least drag (series 45). In the range $Re = 1.6 \cdot 10^6 - 1.6 \cdot 10^7$, the drag of an elastic cylinder with various shapes of the bows of the 41–45 series was less than the drag of a rigid plate in a turbulent boundary layer.

In Section 6.5 in Fig. 6.17A, the dependences of $C_x(Re)$ are shown when examining long elastic cylinders (series 21–25 Fig. 6.14) in the range of numbers $Re = 10^6 - 2 \cdot 10^7$ and, as it were, continue the regularities of series 1–5, obtained with $Re = 1.4 \cdot 10^5 - 2.5 \cdot 10^6$. The range of Re numbers has increased in comparison with the series 41–45 due to the increase in the length of the cylinders. The difference in the shape of the nasal contours of a long model with an elastic coating had practically no effect on the dependences $C_x(Re)$ when testing long elastic cylinders (series 21–24). When flowing around a long elastic cylinder with a dead-end nasal shape (series 21), the $C_x(Re)$ dependences indicated that the drag of the cylinder was greatest in the range of $Re = 1.1 \cdot 10^6 - 6 \cdot 10^6$. In the range of $Re = 2 \cdot 10^6 - 2 \cdot 10^7$, the drag of a long elastic cylinder with various forms of the nasal contours of the 21–25 series was less than the drag of a rigid plate with a turbulent boundary layer. In series 25, as in series 45, the drag with the ogival shape of the nasal contours was minimal. When installed on a long cylinder turbulizer (curve 25_T), the elastic surface began to interact with the pulsations of the turbulent boundary layer. As a result, the drag of the elastic cylinder in the 25_T series was consistently less than a flat plate with a turbulent boundary layer and with a transition boundary layer in the whole range of Re numbers.

Comparison of the drag of short elastic cylinders with short rigid cylinders (standards) showed:

- With a blunt nasal shape (series 41 and 31) in the range $Re = 6 \cdot 10^5 - 3.3 \cdot 10^6$, the drag of the elastic cylinder was greater than that of the standard, and with a further increase in towing speed, the drag of the elastic cylinder became *less* than the drag of the standard.
- With a nasal form in the form of a *long xiphoid needle* (series 44 and 34) in the range $Re = 6 \cdot 10^5 - 1.6 \cdot 10^6$ the drag of the elastic cylinder was *less* than that of the standard, then the drag was the same, and in the range $Re = 4 \cdot 10^6 - 1.1 \cdot 10^7$ the elastic cylinder resistance has become greater than that of the standard.
- In series 42, 43, and 45, the drag of the elastic cylinder was greater than that of the standard in the entire range of Re numbers.

Comparison of the drag of long elastic cylinders with long rigid cylinders (standards) showed that drag only an elastic cylinder with a long xiphoid tip (series 24) was *less* than drag the standard in the entire range of Re numbers.

V.V. Babenko developed the design of elastic cylinders, as well as elastic plates (Section 5.1), and the technology of their manufacture. Partially elastic cylinders were made by hand on a special polishing table, while elastic plates of a given thickness were made using special devises. An elastic substrate was glued to metal cylinders of a given diameter in the same way as in the manufacture of plates (Section 5.1, Table 5.15). On top of the substrate, a layer of CCF was glued onto which a layer of FE or FL was glued. In this way, two short and one long cylinders were made with the parameters: a metal cylinder diameter of 88 mm, a substrate layer −2 mm, a CCF layer—0.5 mm, an outer FE layer—4 mm. Thus, the elastic layer thickness δ was

6 mm, the length of short cylinders—516 mm and the long cylinder—996 mm. In addition, one long cylinder was made with the following parameters: the diameter of the metal cylinder is 75 mm, the thickness of the elastic sandwich is 12.5 mm: the substrate is 6 mm, CCF is 0.5 mm, the outer layer of FE is 6 mm, the length of the cylinder is 996 mm. Two short cylinders were also manufactured: one with a metal cylinder diameter of 80 mm, an elastic sandwich thickness of 10 mm: a substrate of 6 mm, a CCF of 0.5 mm, an outer layer of 4 mm, a cylinder length of 288 mm, and a second short cylinder with parameters: the diameter of the metal cylinder is 72 mm, the thickness of the elastic sandwich is 14 mm: the substrate is 8 mm, CCF is 0.5 mm, the outer layer is 4 mm, the length of the cylinder is 276 mm.

The main disadvantages of this manufacturing technology are the presence of a step in the front end of the splicing of an elastic cylinder with the cylinder head and the presence of a longitudinal glue seam of an elastic sandwich. Especially unfavorable was the strength of the adhesive joint in the front end—at high speeds, the negative pressure that occurs when the cylinder flows around caused an increase in the thickness at the end of the elastic coating and even flaking of the coating. To eliminate this drawback, a special form of the head of the model and methods to increase the strength of fixing the elastic cover on a metal cylinder were designed. In addition, were made in the factory, cast elastic cylinders made of polyurethane, which were fixed on a metal cylinder at a high temperature for the manufacture of PU composition. The elastic cylinders were mounted on strain gauges (Section 6.5, Fig. 6.15). The following uniform elastic cylinders with an outer layer of PU were manufactured: long cylinders 6 mm and 14.5 mm thick with elasticity: 13–15 Shore units and thickness of 12 mm with elasticity − 35–40 Shore units, as well as short cylinders with a thickness of 6 mm and 20 mm with elasticity—20 Shore units. In Table 6.5 shows the corresponding parameters of the models. Elastic coatings are made of PU, with the exception of "Manual assembly"—FE material. Designations

Table 6.5 Geometric parameters of elastic cylinders.

(A) Long cylinders						
№	l, mm	L, mm	d, mm	S, m^2	δ, mm	E_w, arbit.un.
1	958	1294	101	0.401	6	15
2	958	1294	101	0.401	15	15
3	961	1297	101	0.402	12	40
4	950	12864	101	0.398	6	Manual assembly
				$S_{st.} = 0.393$		
(B) Short cylinders						
№	l, mm	L, mm	d, mm	S, m^2	δ, mm	E_w, arbit.un.
1	477	813	99	0.238	20	15
2	481	817	101	0.246	6	Manual assembly
3	470	806	99	0.238	6	40
				$S_{st.} = 0.2435$		

in the Table 6.5: l—length of the elastic coating, L—length of the entire model, d—diameter of the model, S—surface area of the entire model, δ is the thickness of the elastic coating, E_{sh} is the elasticity measured by Shore elastomer in arbitrary units. All elastic coatings are made of polyurethane, except for the line "Manual assembly," where the elastic coating is made by hand from FE.

6.6.1 Long cylinders

6.6.1.1 Short cylinders

Table 6.6 shows the parameters of samples of elastic materials from which elastic cylinders are made (Table 6.5). Table 5.7 (Section 5.1) shows the results of measuring the elasticity of elastic plates, measured using standard Shore hardness testers (Fig. 6.19) in arbitrary units and using the developed elasticometer (Part I, Section 6.4, Fig. 6.13). As an example, a plate made of polyurethane foam, on the surface of which a polyvinyl film was located, had elasticity, measured by a Shore hardness tester, 13.5 arbitrary units, and simultaneously measured by an elastic meter—0.05 MPa. From Table 6.6 it can be seen that the elasticity of samples of elastic coatings made of PU was approximately 15 times greater than the elasticity

Table 6.6 Parameters of samples of elastic materials.

Species of elastomer	Density (ρ, kg/m³)	Elastic modulus (E, MPa)
ПЭ-3	107	0.49
ПЭ-1	129	0.56
ПУ-3А	1050	8.8
ПУ-3Б	1050	9.0
ПУ-3В	1050	9.5

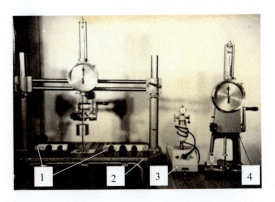

Figure 6.19 Standard instruments for measuring the static characteristics of elastomers: (1) samples of elastomers to investigation their mechanical characteristics; devices: (2) VN-5704, (3) VN-5705, (4) VN-5404.

of samples made of FE. The elasticity of the elastic cylinder in Table 6.5 it is given in arbitrary Shor units, since the elasticity of elastic materials was determined in the factory conditions. Given the above ratios, we can express the elasticity of elastic cylinders in MPa. The elastic cylinders made by hand were mounted on a strain gauge IV (Section 6.3, Figs. 6.2 and 6.5; Section 6.5, Fig. 6.13 option B).

Elastic cylindrical surfaces made of monolithic PU, FL materials and various FE variants were investigated (Section 5.1, Tables 5.4 and 5.5). Investigations were conducted in a high-speed towing tank (Section 6.3, Fig. 6.7). The calibration measurements of strain gauges carried out before and after each series were obtained as linear dependencies.

Fig. 6.20 shows the dependence of the drag force of the cylindrical insert model on the towing speed (2−18.5 m/s). Curve (4) (dash-dotted) defines measurements on a rigid standard. The square points (3) located in the region of the standard corresponded to an elastic cylinder made of FE. Curve (1) and round dots characterize an elastic plate

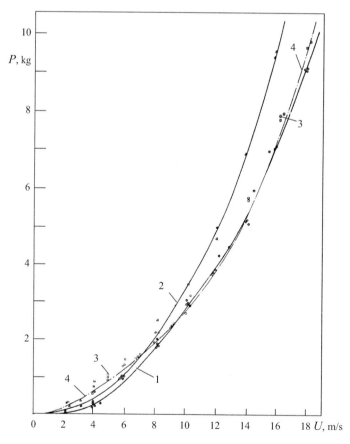

Figure 6.20 Dependence of the drag force of the short model on the towing speed: (1) elastic cylinder made of PU $\delta = 20$ mm, (2) elastic cylinder made of PU $\delta = 6$ mm, (3) squares characterizing an elastic cylinder made of FE, (4) rigid standard.

made of PU with a thickness of 20 mm. Curve (2) and triangles correspond to a cylinder made of the same material with a thickness of 6 mm. The effectiveness of the elastic surface (curve (1)) was significantly higher than the standard to a towing speed of 9.5 m/s, and then the surface drag was the same as that of the standard.

A thinner PU surface (curve (2)) up to a towing speed of 7 m/s was more effective than a standard, but less than curve (1) and, subsequently, its drag was significantly greater than the standard. The elastic cylinder made of FE had greater drag in the entire range of towing speeds.

The results obtained in a dimensionless form are presented in Fig. 6.21. Curve (4) corresponds to the drag of a reference cylinder made of plexiglass (dash-dotted curve in Fig. 6.20). Curve (5) corresponds to curve (1) in Fig. 6.20, and curve (6) to curve (2) in Fig. 6.20. It turned out that the same material, depending on the thickness, has maximum efficiency with different Reynolds numbers.

With a greater thickness of the elastomer, its stiffness becomes less and, therefore, it must be more effective with lower Reynolds numbers. It was expected that a thinner coating would be more effective with large Reynolds numbers. However, although the minimum of curve (6) (Fig. 6.21) was shifted to a larger Reynolds number, the coating was not effective because of the poor attachment of the elastomer to the metal pipe.

Composite elastic coatings were installed on a long model of strain gauge IV (Section 6.3, Figs. 6.2 and 6.5; Section 6.5, Fig. 6.13 option B). Fig. 6.22 shows a photograph of long elastic cylinders for towing tests on strain gauge IV. Cylinders

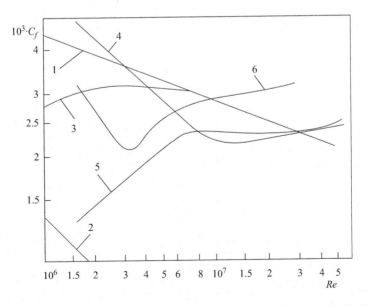

Figure 6.21 The dependence of the drag coefficient on the Reynolds number in the flow around elastic cylinders: (1)–(3) resistance of a smooth rigid plate with laminar, transitional, and turbulent boundary layers; (4) rigid standard cylinder, (5) elastic cylinder made of PU $\delta = 20$ mm, (6) elastic cylinder made of PU $\delta = 6$ mm.

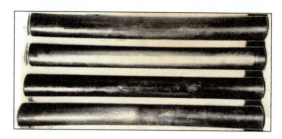

Figure 6.22 Photograph of long elastic cylinders designed for towing tests on strain gauge IV (Section 6.3, Fig. 6.2).

are located in a box on special lodges. The photograph shows a slight optical distortion of the cylinder shape around the edges. The color of the coating varies depending on the material and its buildings. A sample of one coverage is shown in Fig. 5.10 (Section 5.1).

Like the short cylinders, the diameter of the model with an elastic coating was the same (0.1 m), but the length of the cylindrical part was almost twice as large.

Fig. 6.23A presents the results of repeated series of tests of the drag of elastic surfaces in dimensional form. Curve (1) (dash-dotted line) corresponds to a rigid cylinder (standard), curve (2) corresponds to a coating made of monolithic elastic material PE 6 mm thick, and curve (3) that is a coating like PU 12 mm thick with elasticity 40 Shore units (solid material). The elastic cylinder made of FE (curve 2) did not differ from the standard to $U = 13.3$ m/s, and with an increase in speed, the drag of the cylinder was significantly higher than the standard due to the low strength and resistance to shear in FE. The elastic cylinder, made of PU, had a drag less than that of the standard, in the entire range of U, and in the range of $U = 1-12$ m/s was significantly more efficient than the standard.

The Fig. 6.23B shows the results of testing the drag of elastic cylinders made of PU in dimensional form. Elastic cylinders had the same elasticity 15 Shore units (soft material), but were of different thickness: 15 mm (large thickness) curves (2) and (3), and 6 mm (small thickness) curve (4). During testing, it was found that the front edge of the elastic cylinder was poorly fixed curve (2). After fixing the front edge of this elastic cylinder, the measurements performed are indicated by curve (3). With poor fixation of the elastomer near the front edge, the efficiency of the elastic cylinder (curve (2)) was in the range of towing speeds $U = 5-10.8$ m/s, and with increasing towing speed, the drag of this cylinder was significantly larger than the standart (curve (1)). After fixing the elastomer in the area of the leading edge, the velocity range with positive efficiency increased and amounted to $U = 5-13.7$ m/s (curve (3)). The cylinder with a soft PU and a thin layer (curve (4)) was less effective than the same, but thicker PU, with $U = 2-9$ m/s, and with increasing U the drag of this elastic cylinder was greater than the standard.

Fig. 6.24 shows the results in dimensionless form. Curve (4) corresponds to the drag of the long reference cylinder, and (5) the short reference. Some of the patterns shown in Fig. 6.24, correspond to Table 6.4 curves (10), (11), and curve (12)

Experimental investigations of friction drag

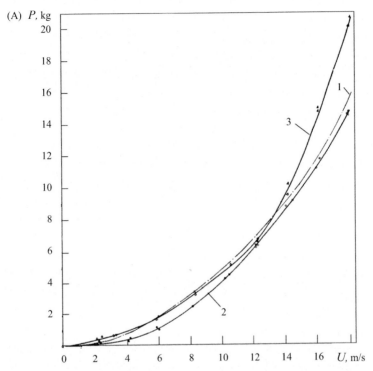

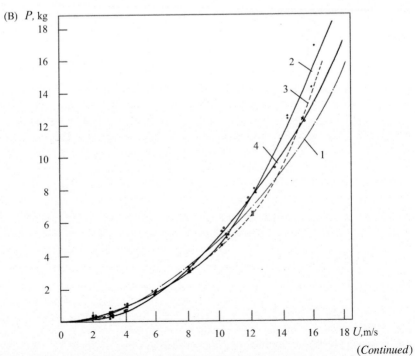

(*Continued*)

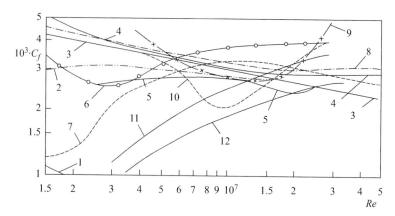

Figure 6.24 Dependence of the drag coefficient on the Reyolds number: (1)–(3) the same as in Figs. 6.20, (4) a long rigid cylinder (standard), (5) a short rigid cylinder (standard); short cylinders: (6) PU, $\delta = 6$ mm, $E = 20$ Shore units, (7) PU, $\delta = 20$ mm, $E = 20$ Shore units, (8) FL (foam latex); long cylinders: (9) FL, $\delta = 6$ mm, (10) PU, $\delta = 14.5$ mm, $E = 15$ Shore units, (11) PU, $\delta = 6$ mm, $E = 15$ Shore units, (12) PU, $\delta = 12$ mm, $E = 40$ Shore units.

corresponds to curve (3) in Fig. 6.23A. Curves (8), (9) correspond to cylinders made of FL. The drag of the short cylinder (curve (8)) with respect to the short standard (curve (5)) was higher for all numbers Re. The coating made of elastic material FL on a long cylinder (curve (9)) had a glued seam located across the flow in the middle of the cylinder, since the sheets FL were short.

Despite this, when Re $= 5.5 \cdot 10^6 - 2 \cdot 10^7$, the drag of a long cylinder with FL in relation to a long standard (curve (4)) in this range of numbers Re was less than a standard. This is primarily due to the pressure distribution along the cylinders at various towing speeds and the values of other loading parameters.

The remaining cylinders were made of the most perspective material (PU) and were effective for different ranges of Reynolds numbers. The short elastic cylinder (curve (6)) with Re $= 1.5 \cdot 10^6 - 5.5 \cdot 10^6$ was less effective; with an increase in the speed of towing, its drag was higher than that of the standard. The maximum efficiency was at Re $= 2.8 \cdot 10^6$. Approximately in the same range of numbers Re had a significantly greater efficiency as well as short elastic cylinder with other parameters δ and E (curve (7)). In this case, the maximum efficiency was at Re $= 1.5 \cdot 10^6 - 2.1 \cdot 10^6$. The second feature of this cylinder was that at Re $= 8 \cdot 10^6 - 3 \cdot 10^7$ its drag slightly exceeded the drag of the standard, and at $8 \cdot 10^6 - 3 \cdot 10^7$ the drag decreased again in relation to the standard. The elastic long

Figure 6.23 Dependence of the drag force of a long model on the towing speed on a strain gauge IV (Section 6.3, Fig. 6.2). (A): (1) standard, (2) elastic cylinder FE $\delta = 6$ mm, (3) elastic cylinder PU $\delta = 12$ mm, $E = 40$ Shore units; (B): (1) standard, (2) PU $\delta = 15$ mm, $E = 15$ Shore units (bad sealing of the leading edge), (3) the same as (2) (good sealing of the leading edge), (4) PU $\delta = 6$ mm, $E = 15$ Shore units.

cylinder (curve (10)) was effective in approximately the same pattern as the cylinder, characterized by curve (6), but in a different range of towing speeds—with $Re = 3.2 \cdot 10^6 - 1.7 \cdot 10^7$. Elastic cylinders (curves (11), (12)) had efficiency similar to curve (7), but for large numbers Re: for curve (11) in the range $Re = 3 \cdot 10^6 - 1.5 \cdot 10^7$, for curve (12) in the range $Re = 3.4 \cdot 10^6 - 1.5 \cdot 10^7$. But curve (11) at $Re = 1.5 \cdot 10^7$ drag became higher than that of the standard, and curve (12) was effective for all numbers Re. The patterns of curves (7), (11) and (12) were similar and, as it were, equidistantly shifted toward large numbers Re. In addition, these patterns were similar to curve (2) of the transition boundary layer when a rigid plate flowed around. This indicated that, in the boundary layer of elastic cylinders, the structure of characteristic CVSs changed, which led to a delay in the transition of the boundary layer (Chapter 5: Experimental investigations of the characteristics of the boundary layer on the analogs of the skin covers of hydrobionts).

The effect of elongation of an elastic cylinder on its efficiency has been investigated [358,360]. Although the cylinders were not long, the thickness of the boundary layer changed along the length of the cylinder. The region of the energy frequencies of the spectrum of turbulent pressure fluctuations should also vary according to Section 6.4. Therefore, if the mechanical characteristics of the elastomer are constant in length, only in a certain place in x will the greatest coincidence of the range of frequencies of fluctuations of the elastomer with maximum values of $tg\varphi$ with the same range of energy-carrying frequencies of the turbulent boundary layer (dynamic disturbance interaction principle). Therefore, two cylinders of different lengths, but with the same elastic coatings, will have different efficiencies. The overall efficiency of a long cylinder should be higher than that of a short one, since an elastic material effectively interacts with the boundary layer over a larger relative length of the elastic cylinder.

All elastic cylindrical inserts (PU) were manufactured in the industry by a specially developed method and contained certain chemical components. The thickness of the PU, the length of the cylinder and the composition of the PU material varied. It is known that PU has an unlimited set of structures, depending on its components and recipes. We used such composition of components, which provided the necessary durability of fastening on the metal substrate, and such mechanical characteristics, which allowed fixing the greatest effect under the given test conditions. Particular attention was paid to the profiling of the metal substrate for fixing PU on it, especially in places where PU was fixed in the area of transverse edges. These were necessary to prevent the coating from separating from the substrate when towing cylinders at a higher speed. The surface of the metal cylinders had a longitudinal ribbed regular structure or transverse cutting. In the manufacture of elastic cylinders, the inner surface of the PU had either the shape of riblets (such elastic surfaces were called "inverted riblets") or the shape of transverse ledges- grooves. The inner surface of the PU is shown in Fig. 5.10 (Section 5.1). Such an internal surface was chosen on the basis of the provisions of the problem of interaction of CVS in the boundary layer [53,247].

Fig. 6.25 shows a photograph of the strain gauge V (Section 6.3, Figs. 6.2 and 6.6; Section 6.5 Fig. 6.13) with an elastic cylindrical insert. In all

Figure 6.25 Photograph of the strain gauge V with an elastic cylindrical insert, the surface of which is made of PU.

subsequent tests, the turbulizer was installed in the bow parts of the models. This allowed us to obtain a turbulent boundary layer on models in the entire range of towing speeds. The measurements were carried out using a strain gauge V to assess the influence of the mechanical characteristics of the elastomer on its frequency properties.

The results are presented in Fig. 6.26. The characteristics of the elastomers are listed in Table 6.7. As can be seen from Table 6.5, the hardness of the elastomers increases from PU A to PU B, as the thickness decreases. The results show that in series A and B (Table 6.7), the efficiency of elastomers is practically independent of their thickness. The difference in their mechanical characteristics also did not affect the efficiency. The difference is found on a harder elastomer in series B, in which the efficiency depends on the hardness of the elastomer or the frequency range of the loss tangent $tg\varphi$.

All tested elastomers showed a decrease in drag over a wide range of Reynolds numbers. The decrease in drag did not exceed 35%. The positive effect (Fig. 6.26), obtained by testing the elastic coatings on the strain gauge B (Section 6.5, Fig. 6.13), and indicates the correct choice of the mechanical characteristics of the elastomer in the manufacture of elastic cylinders. In addition, the result was due to the high quality of the outer surface of the elastic cylinders and the correct design of the metal substrate. It provided strong adhesion of the elastomer to the metal substrate. Therefore, higher loads with an increased towing speed could not deform or separate the elastomer from the substrate. It is very important to ensure a strong fastening of the elastomer in the area of connection with the transverse ends of the metal body of the model. The only exception was cylinder (9) (PU-3B), the elastic coating of which was separated from the substrate at high-speed towing.

The integral efficiency coefficients are shown in Fig. 6.26B:

$$\xi(x) = (C_{Frigid} - C_{Felastic})C_{Frigid}^{-1}, \tag{6.8}$$

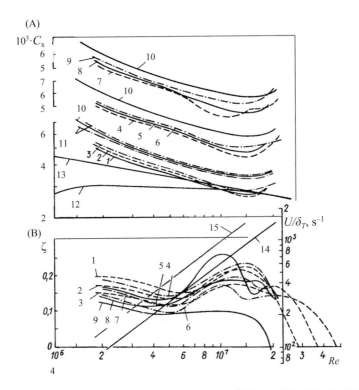

Figure 6.26 The coefficient of friction drag (A) and the efficiency (B) of elastic (1)–(9) and rigid (10) cylinders depending on the Reynolds number. The designation of the curves is given in Table 6.7. Curves (11) denote the root-mean-square error in the determination of the friction coefficient, (12), (13) is the friction resistance of a rigid plate at the transition and turbulent boundary layer [292, 566 and 568], (14), (15) U/δ_T (Re) dependence in the middle and at the end of elastic insert [358,360].

where $C_f = 2\tau_w/\rho U_\infty^2$ —the dependence of the range of energy-carrying pulsations of the turbulent boundary layer, $U/\delta_{\text{turbulent}}$ on the Reynolds number. Indexes F_{rigid} and F_{elastic} designate corresponding tests for rigid and elastic cylinders. Thus, the thickness of the boundary layer at $x = l_i + 0.5 l_{\text{elastic}}$ was calculated by the formula:

$$\delta_T(x) = 0.37 x\, \text{Re}_x^{-2}, \tag{6.9}$$

The condition of maximum efficiency (6.1) (Section 6.4) can be simplified. With sufficient accuracy for engineering estimates, we can assume that the maximum of friction reduction of the given elastic insert will occur in this flow mode, when the upper limit of the range of energy-carrying frequencies becomes approximately equal to (or a little greater) to the frequency corresponding with the a maximum of the elastomer absorption coefficient:

$$U/\delta_T \approx \omega/2\pi \big| tg\varphi(\omega) = \max \tag{6.10}$$

Table 6.7 The thickness of the elastic coating cylinder.

Number of the elastic cylinder	Elastomer	h/L
1	PU—3A	0.01062
2	PU—3A	0.00531
3	PU—3A	0.00265
4	PU—3B	0.01062
5	PU—3B	0.00531
6	PU—3B	0.00265
7	PU—3C	0.01062
8	PU—3C	0.00531
9	PU—3C	0.00265

The recorded maxima of the effect are correlated with the dynamic properties of the material and the frequency of the fluctuation load.

The revealed effect can be demonstrated on the elastic cylinder (4) (Table 6.7), made of PU-3B polyurethane. As the velocity increases up to $Re = 5 \cdot 10^6$ and the range of energy-carrying fluctuation loads expands to 400 Hz at $x_{elastic} = l_{05}$ (curve (15) in Fig. 6.26), Young modulus $E'(\omega)$ and, consequently, the wall rigidity of the wall increases. Consequently, the relative level of energy susceptibility decreases. Since the loss coefficient of the PU-3B material remains constant as ω increases in a given frequency range, the relative part of the turbulent energy dissipated into the elastic wall decreases. Therefore, the integral effect ξ (Re) decreases as $Re \to 5 \cdot 10^6$.

At frequencies that corresponded to the flow regime at $5 \cdot 10^6 < Re < 1.4 \cdot 10^7$, the growth rate of $E'(\omega)$ remained the same (or fairly reduced), and the loss factor in the material $tg\varphi(\omega)$ increased dramatically. At the same time, the spectral density of the frequencies corresponding to the maximum absorption coefficient increased. Thus, the hydrodynamic effect should increase when integrated over the entire frequency range. This is observed in the tests. Thus, the maximum effect corresponds in frequency to the maximum of loss coefficient. On the one hand (Fig. 6.26, curves (4), (15)) ξ (Re) = max at $Re_L \approx 1.4 \cdot 10^7$ and $[U/\delta_T(Re)]_{x=l(0.5)} = 1400$ Hz; on the other hand, ξ (Re) grows and reaches a maximum of about 800 Hz (see Fig. 2.55, curve (5) in [139]). As can be seen from the graph $tg\varphi(\omega)$ (Fig. 6.11, Section 6.4), the range of energy-carrying part of the spectrum of fluctuating loads in this case overlaps the bell-shaped frequency dependence of the loss coefficient of the PU-3B elastomer, similar to case *II* shown in Fig. 6.11A (Section 6.4). At greater frequencies, the function $E'(\omega)$ reaches a constant value, and $tg(\omega)$ decreases. With the corresponding flow regime ($Re > 1.4 \cdot 10^7$), ξ (Re) also begins to decrease. This corresponds to the diagram in Fig. 6.11A (Section 6.4) case *III*.

From the experimental results it can be concluded that the integral effect of elastic inserts made of the same material is proportional to the geometrical parameters characterizing their rigidity and the level of turbulent energy absorbed by the material. For example, $h_4 > h_5 > h_6$ for PU-3B on the cylinders $4 \div 6$ (Table 6.7), consequently $(\xi_{max})_4 > (\xi_{max})_5 > (\xi_{max})_6$ (Fig. 6.26). A similar correspondence is observed for the PU-3A and PU-3C modifications.

In Section 6.5, examples are given of calculating the geometric parameters of the models necessary to obtain dependencies in a dimensionless form. In particular, the obtained ratio for the strain gauges V:

$$\overline{S}_{cyl.e.s.} = \frac{S_{onlye.s.}}{S_{cylind.}} = 0.906$$

This ratio is obtained taking into account the fact that the area of the washed surface of the elastic insert is less than the area of the standard rigid insert. Therefore, the measured efficiency of elastic cylinders will be greater given this ratio.

The tensometer V (Section 6.3, Figs. 6.2 and 6.6) constructively has the ability to measure independently the drag of the head part and the cylindrical part, as well as the drag of the conscientiously head part and the cylindrical part. Fig. 6.27 shows the results of measurement on the strain gauge V of the drag of the head part (curve (1)) and together the head and cylindrical part (curve (2)) depending on the speed of towing the model. It can be seen that the drag of the head part is substantially greater than the drag of the head and cylindrical parts of the model. It can be seen that the drag of the head part is substantially greater than the drag of the head and cylindrical parts of the model. When measuring the joint head and cylindrical parts

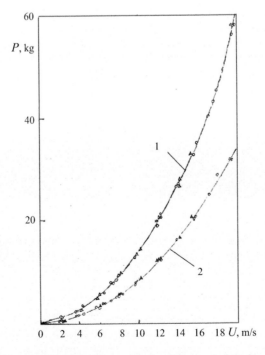

Figure 6.27 Dependence of the drag force on the towing speed of the head part (1) and together the head part and the cylindrical section (2) of the longitudinally streamlined cylinder of the strain gauge V.

there is no the crack between the head and the entire model. In addition, all drag components considered in Section 6.5 are reduced. The similar character of $C_x(\text{Re})$ models depending on the elongation is considered in dimensionless form in Fig. 6.17 (Section 6.5). Similar results of experimental investigations of longitudinally streamlined cylinders A, B, C are given (Section 6.5, Fig. 6.13).

Based on the results of an experimental study of the physical mechanisms of the formation of a CVS in the boundary layer (Section 3.6, Figs. 3.33 and 3.40), V.V. Babenko developed and manufactured a perfect metal base for fixing elastic materials on it, including PU (Fig. 6.28). The results of experimental investigations of the interaction of the CVS during the flow around elastic surfaces (Chapter 5: Experimental investigations of the characteristics of the boundary layer on the analogs of the skin covers of hydrobionts) showed that the CVS of the boundary layer changes as compared to the flow around a rigid surface when flowing around homogeneous elastic surfaces. The flat disturbances of the boundary layer in the flow around an elastic surface are quickly transformed into longitudinal CVS, which become more stable compared to the flow around a rigid surface. The same picture is observed in the turbulent boundary layer.

In connection with the change in the structure of the CVS and the order of their alternation during the flow around elastic surfaces, a change in the surface structure of the metal cylinder was designed to fix an elastic material on it. In Fig. 6.28B, the area of the metal surface of the model in the nose part (2) has a longitudinal structure a small step for firm fixing of the elastic surface and counteracting the separation loads that occur during the transition from the hard surface of the tip (1) to the pliable elastic surface. Section (3) has a transverse structure of fine step to increase the strength of fixing the elastic surface and to interact with flat disturbances of the boundary layer. Experiments have shown

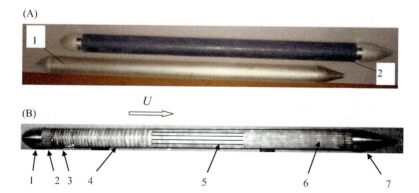

Figure 6.28 Photographs of the design of a rigid longitudinally streamlined cylinder (1) and a cylinder with an elastic coating of the central part (2) with a diameter of 5 cm. (A), as well as a rigid profiled base of the cylinder (B) for fixing the elastic surface: (1) nasal part of model, (2) longitudinal ledge-deepening, (3) transverse ledge-deepening of small step, (4) transverse ledge-deepening of large step, (5) longitudinal ledge-deepening of large step, (6) longitudinal ledge-deepening of small step, (7) tail part of the model.

that the T−S wavelength in the boundary layer of an elastic surface increases in comparison with the flow around a rigid surface. Therefore, section (4) has a larger and increasing step compared to section (3). Such a structure of the elastic surface, which will be fixed on the metal base of the model, will lead to the rapid formation of longitudinal CVS in the boundary layer of the elastic surface. Therefore, in section (5), longitudinal ledge-deepening of large step in the transversal direction, which allow an elastic surface in this place to stabilize longitudinal CVSs in the transition boundary layer. During the formation of a turbulent boundary layer, section (6) has made longitudinal ledge-deepening small steps in the transversal direction, which allow the elastic surface in this place to stabilize the longitudinal GVS in the viscous sublayer of the turbulent boundary layer. Fig. 6.27A shows a photograph of a rigid cylinder (1) and the same diameter of a cylinder with an elastic surface (2), which was applied to a rigid base by R. Bannash. The drag of both cylinders was investigated by R. Bannash in the central hydrodynamic laboratory of Berlin (Germany).

The technology shown in Fig. 6.28 allows the manufacture of optimal elastic PU coatings for a given range of speeds.

The results of experimental investigations of the drag of cylinders, the elastic coating of which was applied to the smooth metal surface of the cylinders, showed that for each set of parameters of elastic surfaces there is a narrow range of Reynolds numbers in which the efficiency of elastomers is maximum. It follows that the elastic surface must be manufactured with varying characteristics along the length of the cylinders (sectioning the elastic surface of the cylinders). I.V. Shcherbina developed equipment and technology for manufacturing continuous elastic cylinders from FE material with given cylinder diameters and distribution of the mechanical characteristics of the material along the length.

From the analysis of numerous experimental studies of the drag of elastic cylinders (here are only some of the results) it is possible to draw the following conclusions:

1. The results obtained on strain gauge IV, variant (B) and (C) (short and long cylinders), in general, showed more stable results compared to investigations on the same elastic materials when towing them on strain gauge II. The effectiveness of the materials investigated was better when tested on a strain gauge IV.
2. The best results are obtained on a homogeneous material, which is more drag to separation loads at high towing speeds.
3. Control of the mechanical properties of materials by heating a conductive fabric inside the coating during towage did not give a positive result.
4. This indicates that the smoothness of the coating and its durability were more important, as the boundary layer became thinner as the towing speed increased, and the separation loads increased significantly. Therefore, a small degree of roughness becomes greater than the permissible roughness. With an increase in towing speed, the coating at the end of the cylinder was separated, which significantly increased the drag.
5. Tests on short and long cylindrical strain gauges did not allow reaching a greater Reynolds number and determine the effectiveness of the coating with a fully developed turbulent boundary layer on elastic surfaces.

6. On all tested variants of elastic coatings, the shape of the $C_x(Re)$ dependences was predominantly equidistant to the regularities of the transition boundary layer in the flow past a rigid plate. Thus, the corresponding critical Reynolds numbers were significantly shifted by large Reynolds numbers. This means that even imperfect coating manufacturing techniques and insufficiently effective coating design significantly stabilized the boundary layer and drag reduction.

6.7 Influence of polymer additives on the friction drag of elastic cylinders

Chapters 1−6 (Part I: Experimental Hydrodynamics of Fast-Floating Aquatic Animals) showed that all fast-floating hydrobionts have a complex adaptations for fast and energy-efficient swimming in the aquatic environment, while hydrobionts have combined methods for drag reduction worked through the evolutionary process. In this chapter and Chapter 5, Experimental investigations of the characteristics of the boundary layer on the analogs of the skin covers of hydrobionts (Part II: Experimental Investigations Flow around Bodies), combined methods of drag reduction are investigated: the influence of the xiphoid nasal form in combination with an elastic cover and the formation of longitudinal or transverse disturbances in the boundary layer using predetermined structures of elastic coatings.

In 1976−1977 V.V. Babenko and V.I. Korobov carried out experimental investigations of the injection of water solutions of polymers on the model of a longitudinal streamlined rigid cylinder developed by V.V. Babenko with interchangeable nose tips, which allow injection of polymer solutions through one and two nasal slits. In this case, the tips were made in an ogival form and in the form of a short and long xiphoid tip [139,378].

Numerous investigations of the effect of polymer water solutions on the drag reduction of plates and longitudinally streamlined cylinders have been carried out in various countries. The most detailed review of the techniques for preparing polymer solutions and their injection methods, as well as the results obtained, are given in the reports by K.J. Moore (Cortana Corporation, US) [452,453] and in US patent [450,451], and also the results of the injection investigations are given polymer solutions through three slits, according to the patent [83,89,90,92,93,105−107]. The configuration of a pre-gap chamber for polymer injection using the Coanda effect was proposed and experimentally investigated by V.N. Vovk and A.I. Tsyganyuk [557,594,596].

A large contribution to the investigation of various methods of drag reduction was made at the University of Pennsylvania (The Pennsylvania State University, Applied research Laboratory) [190,244−246]. Experimental investigations of a turbulent boundary layer modified by the injection of polymer solutions, which reduce friction resistance, through a slit into the near-wall region of the flow are performed in Ref. [245]. The velocity profiles were measured with a two-component laser velocity meter at the main flow rate of 4.5 m/s with the injection of the polymer solution, the injection of water and without injection. The concentration of the injected polymer solutions was 500 and 1025 wppm. The results of the distribution of the concentration profiles of polymer solutions across the thickness of the boundary layer were obtained using laser fluorescent technology. The investigations were performed on a flat plate placed in the working part of the hydrodynamic tube. Friction resistance was also measured on the strain gauge part of the plate during injection of polymer solutions. At a distance of 353 mm from the nozzle of the plate there was a slit for the injection of polymer solutions. The measurements were carried out at points located at a distance of 50.8 mm, 129 mm, 232 mm, and 384 mm from the injection slot. Integral friction resistance is made in the area located at a distance of 19 mm to 337 mm from the slit. The reduction of resistance was investigated at various values of the solution concentration and polymer consumption, as well as at different speeds of the main flow (4.5, 8.14, 13.7, and 18.3 m/s). The resulting effect of drag reduction was 10%-55%.

Profiles of longitudinal and transverse pulsation and averaged velocities, as well as Reynolds stresses and correlation coefficients, were measured depending on the same parameters. The data obtained in [245,246] are qualitatively similar to the previously obtained results of systematic investigations of the kinematic characteristics of the turbulent boundary layer in the flow around various elastic plates [60,78,139,321,322,324].

In Ref. [190], experimental investigations of the spectral characteristics of the pulsations of the wall pressure and the integral characteristics of the friction resistance during the flow of pure water and modified with the addition of polymer solutions around the body were performed. The experiments were performed on a cylindrical longitudinal streamlined body of rotation fixed in a cavitation tube with a diameter of 0.3048 m. The body of rotation had a diameter of 89 mm and a length of 632 mm. The nasal part of the model had a dead-end. At a distance of 46 mm from the beginning, a wire turbulator was installed on the model. The central part of the cylinder with a length of 237 mm was mounted on annular diaphragms attached to the inner tube, which was rigidly fixed to the end part of the model. The front slit of the strain gauge central part was located at a distance of 196 mm from the beginning of the model. Air bubbles or polymer solutions were fed through a porous surface fixed in front of the front slit of the strain gauge. The front slit had a width of 0.127 mm, and the rear slit—0.254 mm. The pressure pulsations were measured by piezoelectric pressure sensors with a diameter of 0.254 cm installed in the center of the balance axis and mounted flush with the model surface. In about a few minutes, polymer solutions with concentrations of 1, 5, 10, and 20 weight parts per million (wppm) were injected at a speed of 5 m/s. The obtained regularities of the spectral characteristics of pressure pulsations are qualitatively similar to the spectral characteristics of the longitudinal pulsation velocity when flowing around various elastic plates, measured with a thermal anemometer in a wind tunnel [60,139,324].

The above-mentioned similarities of the obtained experimental results were taken into account in the investigation of the flow around elastic surfaces and when a rigid surface of water solutions of polymers was injected into the boundary layer. Experimental investigations of the combined method of drag reduction were performed during the injection of polymer solutions into the boundary layer of elastic surfaces [27,30,60,84,85,91,100,133–135,139].

In [378], it was hypothesized that the boundary layer, as a body with some elasticity, has a range of resonant frequencies bounded by a neutral curve. In accordance with this hypothesis, the energy frequency range characterizes the resonant frequency region of the turbulent boundary layer. The dissipative properties of the fluid affect the region depending on the flow velocity and flow regime in the boundary layer. Therefore, the polymer solution should change the region of natural oscillations of the boundary layer and the condition of resonant interaction with the elastic surface. In Section 3.8, the boundary layer is represented as a nonuniform asymmetrical wave-guide.

Measurements were made of the friction resistance on a rigid (standard) and elastic cylinder surface during the injection of polymer solutions into the boundary layer of the model through one nasal slit. The investigation was carried out subject to the relevant similarity criteria (geometric, Reynolds number, etc.) and under the assumption that the radius of the transverse curvature of the cylinder surface r was greater than the thickness of the boundary layer (δ) [521]: $\delta/r < 1$.

Hydrodynamic friction measurements of longitudinally streamlined rigid and elastic cylinders with the injection of polymer solutions were carried out using a strain gauge V (Section 6.3, Figs. 6.2 and 6.6), on model D (Section 6.5, Figs. 6.13 and 6.15A, Section 6.6, Fig. 6.25), with towing in a high-speed hydrodynamic towing tank (Section 6.3, Fig. 6.7) in the speed range $U_\infty = 2.0$–22.0 m/s. Outer diameter of the cylinder $d = 0.175$ m, elongation $\lambda_{cyl.} = L_{cyl.}/d = 5.65$, and the elongation of the cylinder with a nose cowl $\lambda = 6.07$. The scheme of the experimental setup is shown in Fig. 6.29. Using pylon (8) model G with a modernized nose section hung from a towing cart. A cylinder with an investigation solid (13) or elastic (5) surfaces was fixed on a strain-gauge suspension, the kinematic elements of which are elastic

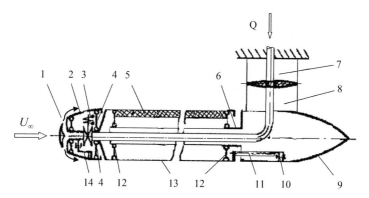

Figure 6.29 Scheme of the experimental setup: (1) annular crack, (2) nose cowl, (3) solenoid valve, (4), (12) elastic annular membranes, (5) elastic cylinder, (6) pipe, (7) tube for feeding polymer solution, (8) knife-pylon, (9) model body, (10) strain-gauge sensor, (11) rod, (13) hard cylinder, (14) strain gauge of the nose section of the model [363].

annular membranes (12) mounted on pipe (6), which is part of body model (9). The kinematic decoupling between cylinders (5) or (13) and the body (9) is made at the expense of a gap of $8.0 \cdot 10^{-5}$ m. The force of hydrodynamic friction resistance acting on the surface of the test cylinder (5) or (13) with a nose cowl (2) is transmitted through the rod (11) to a strain-gauge force sensor (10).

The injection of water polymer solutions into the boundary layer was carried out tangentially to the streamlined surface through the nose annular crack (1) with a size of $3.0 \cdot 10^{-4}$ m. The edge of the slit (1) is located from the front critical point of the nose fairing along its generatrix at a distance of $0.19d$. The angle between the direction of injection and the axis of the cylinder was 20 degrees. Due to the Coanda effect, the jet adhered to the surface of the nose cowl. The flow speed of the polymer was varied by varying the pressure in the line (7). Controlling the release of water or water solutions of the polymer was carried out using an electromagnetic shut-off valve (3). The technique and technique of the experiment are described in [363,364].

The results of experimental investigations on a rigid standard in the form of a metal cylinder with a polished outer surface are shown in Fig. 6.30. The parameter $\delta/r = 0.165$. In Fig. 6.30, the boundaries of the errors in measuring the friction coefficient $\sigma_{\Delta R\text{-}1}$ and $\sigma_{\Delta R\text{-}2}$, respectively, are plotted in two ranges of towing speeds ($10^6 < \text{Re}_1 < 1.2 \cdot 10^7$; $1.2 \cdot 10^7 < \text{Re}_2 < 2 \cdot 10^7$). The measurement in two speed ranges was made with the aim of improving the accuracy of measurement of errors and is due to the different gain of the measuring recording equipment.

When the Reynolds numbers are less than $5 \cdot 10^6$, there is a slight bend in the experimental curve due to an increase in wave resistance due to the influence of the free surface of the basin as it approaches the critical value of the Froude number

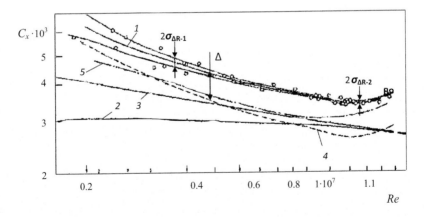

Figure 6.30 The drag coefficient of a streamlined cylinder with a solid surface (1) at various Reynolds numbers [363]; $\sigma_{\Delta R}$ is the error in measuring the resistance coefficient; (4) taking into account the amendment Δ on the bottom resistance; (5) the same as (1), but with the injection of the polymer solution; the law of change in the coefficient of consumption of the polymer by the numbers Re corresponds to the curve B in Fig. 6.31; (2), (3) is the law of resistance of a rigid plate under transition and turbulent boundary layers [521].

($Fr \approx 0.5$ at the beginning of the towing speed range; model depth $h_m/d = 2.4$, where h_m—the distance from the axis of the model to the surface of the water).

In the range of Reynolds numbers $5 \cdot 10^6 - 1.5 \cdot 10^7$, the experimental dependence of the friction coefficient of a standard rigid cylinder, taking into account the calculated correction Δ for the value of its bottom resistance in the zone of kinematic decoupling of the strain-gauge suspension (curve (4)), coincides quite well with the corresponding dependence for a smooth rigid plate. When the Reynolds numbers are greater than $1.5 \cdot 10^7$, the resistance coefficient of the reference rigid surface begins to increase compared with the known value for a smooth plate in a turbulent flow (curve (3)), which is due to the appearance of roughness with increasing flow velocity. Given that these experiments are based on a comparative experiment in testing geometrically identical rigid and elastic surfaces, introducing corrections for these factors, as well as taking into account the systematic error Δ, is not necessary and it suffices to compare the hydrodynamic effect ξ in each individual case (or test series).

The elastic cylinder was made of polyurethane PU-3B with a thickness of $3 \cdot 10^{-3}$ m ($h/L = 0.00265$ in accordance with Fig. 6.13, Section 6.5 and Table 6.7, Section 6.6) and density $\rho = 1250$ kg/m³. The outer diameter of an elastic cylinder is equal to the diameter of a rigid standard—$d = 0.175$ m. The results of measurements of the dynamic viscoelastic characteristics of an elastomer in the range $0 < \omega < 320$ s^{-1} are given in [60,139,378]. The static and dynamic modules of elasticity of an elastomer are $1.6 \cdot 10^3$ and $5 \cdot 10^3$ kPa, respectively. The coefficient of mechanical losses at frequencies up to 100 s^{-1} was about 0.53, and at frequencies up to 300 s^{-1} increased to a value of 0.7.

An water solution of polyethylene oxide (POE-WSR-301) with a molecular weight $M_\omega = 4 \cdot 10^6$ and a weight concentration of 10^{-3} (1000 wppm) was used as high-molecular additives.

Two series of experiments A and B were performed on rigid and elastic surfaces (Fig. 6.31). They differed in the amount of polymer solution injected into the boundary layer. The value of the volumetric flow coefficient C_q dependent on the number Re is presented in Fig. 6.31.

Volume flow coefficients were determined by the formula $C_q = Q/U_\infty S$, where $Q = V/t$ is the volume flow rate of the polymer solution; U_∞—towing speed; S and L are, respectively, the area of the wetted surface and the length of the cylinder, V is the volume of the injected fluid, and t is time of injection; $Re_L = U_\infty L/\nu$, where ν is the kinematic viscosity coefficient.

In a comparative experiment with the injection of polymer solutions, it is very important to maintain a constant pressure in the polymer injection system. However, during the experiments it was not possible of regulating the pressure with sufficient accuracy. When the volumetric flow rates of the fluid obtained in the experiments were determined depending on the Reynolds number, a certain violation of these data was obtained. Curves A and B average these flow rates simultaneously for rigid and elastic cylinders. If you set the average curves separately for each type of cylinder, the spread of points will be small. In this case, two curves A and B could be obtained. The average value of all points led to an increase in the

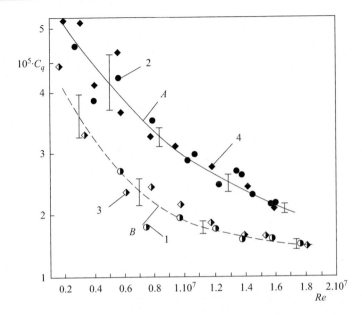

Figure 6.31 The coefficient of volumetric flow rate of the polymer solution, depending on the Reynolds number: A, B—the flow rate of the polymer solution; (1), (2) rigid and (3), (4) elastic cylinders [363].

spread, which is shown in Fig. 6.31 vertical solid lines. It was found that the greater the towing speed, the smaller the spread of points. Therefore, direct comparisons of the resistance curves of different cylinders should be with a certain measure of stock, since at each point of the resistance curves the costs vary in size depending on the towing speed.

Despite the stated disadvantage, it is possible to analyze the resistance curves shown in Fig. 6.32. You can determine the flow rate of the polymer solution for each Reynolds number in Fig. 6.31. When flowing around an elastic cylinder without injection of a polymer solution (curve (6), Fig. 6.32), a decrease in resistance with respect to a rigid standard was obtained. The magnitude of the resistance decrease is less than in accordance with Fig. 6.26 than in accordance with Fig. 6.26, due to the fact that in this series of experiments the front annular slit additionally turbulizes the boundary layer, up to the separation of the boundary layer. With the injection of polymer solutions onto the surface of a rigid cylinder (4), (5), the reduction of drag was almost the same as on an elastic cylinder without injection of polymers (6). Change expense of the polymer solutions when towing a rigid cylinder did not have a strong effect on the drag reduction (compare curves (4) and (5)). Injection of the polymer solution on the elastic cylinder (7), (8) significantly increases the effect. With a larger value of polymer expense (8), the effect increased compared with a rigid cylinder (4) in a wide range of Reynolds numbers.

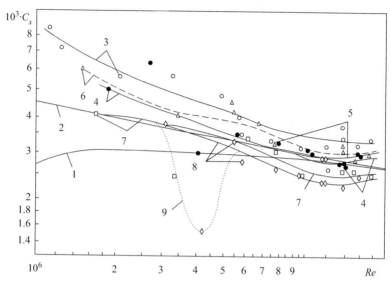

Figure 6.32 Combined method of drag reduction. Coefficient of drag: (1), (2) flat rigid plate at the transition and turbulent boundary layer; (3) rigid cylinder (standard); (4), (5) rigid cylinder at injection of polymer solutions with flow rates A and B (Fig. 6.31); (6) elastic cylinder without injection of polymer solutions; (7), (8) elastic cylinder with injection of polymer solutions with rates A and B. (9) abnormal drag reduction. Numbers and corresponding points denote corresponding curves [363].

During the experiments, an abnormal drag reduction (9) was found during the experiments on elastic cylinders (7), (8). Unfortunately, these experiments were not repeated. A similar case was recorded during experiments on elastic plates in a wind tunnel [60,139]. A similar pattern was also observed during the water flow around elastic plates (for example, Section 6.4, Fig. 6.9, curve (4)). During the experiments in the wind tunnel was found abnormally low friction resistance. At a certain flow velocity, the elastic plate began to make longitudinal oscillations with a high frequency. The plate was fixed on four flat springs, on which strain gauges were glued. Moreover, the width of the flat springs was located in the transverse direction, which allowed fixing the friction resistance. Apparently, the natural frequency of the springs coincided with the frequencies of the longitudinal pulsating velocity, which led to resonant longitudinal oscillations of some elastic plates, with the flow around which the longitudinal pulsations of the boundary layer were amplified.

On the basis of experimental studies of live dolphins, a hypothesis was put forward (part I), according to which dolphins automatically regulate the tension of the skin muscle, which contributes to the resonance interaction with the flow. This leads to the interaction of the pulsations of the boundary layer and the self-controlled frequency of oscillation of the dolphin skin. As a result, the skin vibrates with a resonant frequency with little or no energy. Fluctuation of the skin occurs at a frequency that controls the frequency of bursting from a viscous sublayer. This is also the reason for the abnormal low resistance that was found in our model experiments.

The values of $\xi(Re)$ in the flow around rigid and elastic cylinders in a water flow during injection of low concentration polymer solutions are shown in Fig. 6.33 (curves (1)–(5)), where $\xi(Re) = (C_{xrig.} - C_{xi})/C_{xrig.}$ is the coefficient of relative change of friction depending on Reynolds numbers; $C_{xrig.}$- coefficient of friction of a hard surface; C_{xi}- coefficient of friction during injection of polymer solutions. Curves (6), (7) in Fig. 6.33 show the relative mean-square measurement error $\sigma_{\Delta C_x}/C_x$. Curves (1), (3) correspond to the coefficient of flow rate $C_q(Re)$ according to the law "B" (Fig. 6.31), and curves (2), (4) correspond to the law "A." The designation of points in Fig. 6.33 is the same as in Fig. 6.31.

As can be seen from the results of the experiment, the injection of a small amount ($C_q \approx (2-5) \cdot 10^{-5}$) of low concentration polymer solution reduces the friction drag. The greater the amount of injected polymer, the more drags reduction. This is clearly seen when $Re > 1 \cdot 10^7$, where the coefficients in "A" and "B" are clearly different:

$$C_{qA} > C_{qB}\xi_2 > \xi_1 \tag{6.11}$$

The indexes at ξ denote the curve number in Fig. 62.33. On the elastic surface, the effect of polymer solutions on friction drag becomes even clearer:

$$\xi_4 > \xi_3 \text{ and } \xi_4 > \xi_2 \tag{6.12}$$

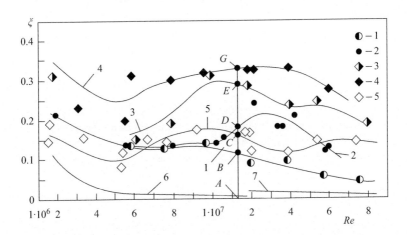

Figure 6.33 Coefficient of drag reduction ξ (Re) depending on the Reynolds number: at injected polymers of polymer solutions in the boundary layer of rigid (1), (2) and elastic (3), (4) cylinder surfaces; (5) without injection of polymer solutions ($C_q = 0$) on an elastic surface. For curves (1), (3), the law of supplying PEOs corresponds to the flow rate $C_q(Re)$ according to pattern B, for curves (2), (4)—according to the law A (Fig. 6.30); (6), (7) relative mean-square measurement error in the experiments of the drag coefficient for Reynolds numbers $\sigma_{\Delta C_x}/C_x = f(Re)$ [363].

Fixed that the effects of the elastic surface and the polymer are summed. In Fig. 6.33, the vertical line shows the letters that determine the magnitude of the efficiency for a given Reynolds number. From Fig. 6.33, we obtain $AC + AB \approx AE$ and $AC + AD \approx AG$.

The complex effect on the boundary layer of the elastic surface and polymer solutions is complex character. Conducting special studies or identifying flow mechanisms is necessary for various combined methods of reducing resistance. With various combined methods of drag reduction, it is necessary to perform special investigations to determine the flow mechanisms in the boundary layer.

6.8 Engineering method for the selection of elastic plates to drag reduction

As shown in Section 5.9, a complex of loads from above (Fig. 5.61) and from below acts on elastic surfaces under real flow conditions. Such a complex effect should be analyzed using the spectral characteristics at the output of the multiparameter system (Fig. 5.62).

The interaction of an elastic surface with a flow causing such a complex of loading can be determined using numerical calculations. However, such calculations can be performed only under corresponding boundary conditions, which are ambiguous because the action of loads on elastomers occurs in an interconnected manner and is determined by the conditions of body motion with an elastic surface. Let us consider some particular cases of the effect of loads on elastic surfaces.

One of the important loads is the spectral characteristics of velocity pulsations and pressure pulsations of the boundary layer. Based on the experimental data obtained in the flow around various types of elastic plates, it is known [60,139] that the characteristics of the boundary layer in the flow around elastic plates differ from the spectral characteristics of the boundary layer in the flow around a rigid standard at all frequencies of the boundary layer.

Let as consider the spectral characteristics of the longitudinal pulsation velocity in the boundary layer of rigid and elastic plates. Laminar flow regime exists when flowing around any surface. You can calculate the frequency of unstable oscillations by the formulas:

$$x = 0.388 \frac{\nu \mathrm{Re}_*^2}{U_\infty}, \tag{6.13}$$

$$n = 0.159 \frac{\beta_r \nu}{U_\infty^2} \frac{U^2}{\nu}. \tag{6.14}$$

The results of the investigations performed and their analysis are presented in [378]:

1. The region of unstable oscillations depends on the value of U_∞.
2. Elastic plates significantly reduce this area.
3. At $U_\infty > 1$ m/s, there is a narrow area on the plate along x, over which a wide range of unstable oscillations is formed in the boundary layer.

Although the laminar section has a small extent along x, a wide spectrum of unstable fluctuations arises in this area. In the transitional boundary layer, the spectrum of these oscillations gradually expands and becomes filled in the turbulent boundary layer. The elastic surface with corresponding choice of its design and mechanical characteristics significantly reduces the spectrum of unstable fluctuations at all stages of the transition, including the turbulent boundary layer.

Fig. 6.34 shows the specific power of the pressure pulsation spectra calculated by the formula [391]:

$$S_p(\omega) = \frac{4b^2\gamma^2\tau^2\alpha U_c}{\pi(\alpha^2 U_c^2 + \omega^2)}, \qquad (6.15)$$

where $b = 0.3$; $U_c = 0.8 \cdot U_\infty$; $\gamma = 4.34$; $\alpha = 2/\delta$; δ is the boundary layer thickness.

The spectra were calculated over the plate as it flows around it at velocity of 5 m/s at different x coordinates [28,465]. It can be seen that the power-density spectra of energy-carrying frequencies are located in the low-frequency range. With increasing coordinate x, the range of energy-carrying frequencies decreases and greater part of the energy of pressure pulsations of the boundary layer is concentrated in a narrow range of low frequencies.

Fig. 6.35 presents the areas of unstable pulsations of the laminar boundary layer in the flow around rigid and elastic plates (Sections 3.4 and 5.3) [378]. In the investigation of hydrodynamic stability, the maximum amplitudes of the pulsations are analyzed. The limit neutral curves denote the range (spectrum) of frequencies that are detected using tellurium-method. In fact, the spectrum of pulsations in the laminar boundary layer in the field of the neutral curve will be even larger. In Fig. 6.35,

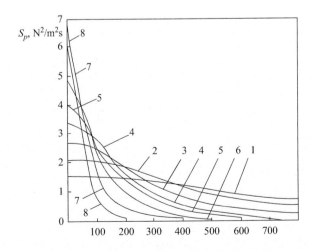

Figure 6.34 Power-density spectra of energy-carrying pressure pulsations when at a flows around of a rigid plate with a velocity of 5 m/s depending on various values x: (1) 0.05 m; (2) 0.1 m; (3) 0.2 m; (4) 0.4 m; (5) 0.6 m; (6) 0.8 m; (7) 1.4 m; (8) 2.0 m [465].

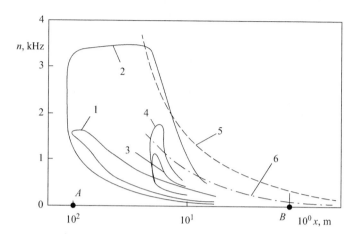

Figure 6.35 Neutral (1), (3) and limiting neutral (2), (4) curves of hydrodynamic stability flat rigid (1), (2) and elastic (3), (4) plates and power-density spectrum of pressure pulsations of a turbulent boundary layer of a rigid plate at velocity of 5 m/s. Power-density spectrum: (5) 90% of the full range, (6) 75% of the full range. A—point of loss of stability, B—point of conditional beginning of the turbulent boundary layer.

dashed and dash-dotted curves limit the spectra obtained according to Fig. 6.34. Curve (5) represents 90% of the spectrum of the full range of pressure pulsations and curve (6) – 75% of the full range. Qualitatively, this corresponds to the limit (curves (3), (4)) and normal neutral curves (curves (1), (2)). It can be seen that the law of dependence of the range of the fluctuations spectrum in laminar and turbulent boundary layers in the flow on a rigid plate with reard to x has a similar character.

Maximum ranges of fluctuation frequencies are in the area of stability loss and the beginning of development of the turbulent boundary layer. Between these characteristic values of x and further along the plate, the range of fluctuation frequencies is essentially narrowed.

In a wind tunnel with low turbulence intensity, V.A. Tetyanko [550] has performed experimental research on spectral characteristics on a rigid plate in laminar, transitional, and turbulent boundary layers. Part of his data is put on plots of spectra at receptivity of various disturbances by a boundary layer [247]. According to the spectral curves he obtained, the spectrum of low-frequency fluctuations is narrower in a laminar boundary layer, and is not full in the high-frequency area relative that in turbulent boundary layer. In the process of passage through various stages of development to transitional boundary layer (with growth x), the high-frequency part of the spectrum was and more and more filled. In comparison with the spectrum of turbulent boundary layer, in low-frequency area the spectrum essentially differs from the turbulent spectrum, there is obviously expressed symmetric maximum in the spectral curves being expressed, in comparison with the turbulent spectrum, the spectral curve of which has no maxima and tends to 1 at reduction of fluctuations frequency.

The spectral curves obtained in Ref. [550] are well agreed with the data in Figs. 6.34 and 6.35. According to the model of development of disturbances in transitional boundary layer (Section 3.6 Figs. 3.33 and 3.40), in the process of deformation of disturbances in a boundary layer from linear to three-dimensional form, the high-frequency part of spectrum is more and more filled. The results of an experimental research of spectral curves at receptivity of the boundary layer on a rigid plate to three-dimensional disturbances presented in Ref. [247] have shown the formation of characteristic maxima of spectral curves at low frequencies and valleys in the high-frequency part of the spectrum. A similar picture is obtained in [550]. In Refs. [60,139,247], the spectral dependences of the longitudinal pulsation velocity in the flow of elastic plates around the water flow and in the receptivity of the boundary layer of elastic plates of three-dimensional perturbations are given. At low frequencies, a statistically significant picture of the formation of maxima was found. Such data indicate the formation of a narrower spectrum in the flow of elastic plates, and the maxima of the spectra at low frequencies characterize the presence of longitudinal vortex structures in the boundary layer.

These data and results in Fig. 6.35 are the basis for the following hypothesis: it is possible to influence the characteristics of a boundary layer provided that it is possible to reduce essentially areas of instability and the initial section of development of the turbulent boundary layer. Such an assumption is confirmed by the results of physical experiments partially presented in the previous chapters.

Based on this hypothesis, it is possible to assume that dolphin and other high-speed hydrobionts produced mechanisms for the realization of the revealed laws. Thus, it is possibly that dolphins are capable of reducing the range of unstable fluctuations and quickly generate quasi-turbulent boundary layer with a narrow spectrum of fluctuations by means of the specific skin structure. Regulating the skin's muscles tension, dolphins may adjust fluctuations of the skin to achieve resonance with the narrow spectrum of fluctuations of the boundary layer (Part I, Sections 5.2 and 6.9). Along with the formation of pre-burst modulation mechanism mentioned before, this allows dolphins to regulate the resonant work of the skin coverings as a complex wave-guide. Fluctuations directed by the skin wave-guide are dissipated or thrown into the flow from the end part of the waveguide (the tail part of the body) due to the work of the fin mover without the wave reflection on the end face of the wave guide.

The same mechanism of the organization of quasi-turbulent boundary layer with narrow range of low-frequency energy-carrying fluctuations is utilized by high-speed fishes. Dolphins use this mechanism due to drastic reduction of the spectrum of neutral frequencies of flat fluctuations and fast passage of other stages of transition with the purpose of formation of quasi-turbulent boundary layer. The swordfish use such mechanism due to xiphoid tip, which allows the generating of the necessary quasi-turbulent boundary layer along its body. Swordfish skin has the same functions, as high-speed dolphins, including a complex wave-guide. These topics are considered in the following chapter in more detail.

The first problem in the engineering method of selection of elastic plates is the choice of materials, which decrease the amplitude and spectrum of disturbance oscillations as much as possible. Thus, a basis of the first task is definition of necessary mechanical characteristics of the materials, of which a simple or composite elastic plate with optimal stabilizing properties should be made, simulating the structure and functioning of the skin of fast-floating hydrobionts in the best possible way.

The second problem is the development of methods for damping unstable oscillations of the laminar boundary layer in the widest range of frequencies of unstable oscillations—in the instability region ($Re_{loss\ stability}$, Sections 3.9 and 5.2).

Third task is development of a corresponding design of an elastic plate. To achive this task, it is necessary to consider the boundary layer on an elastic plate and a design of an elastic plate from positions of a complex wave-guide (Section 3.8).

As the surface of elastic plates interacts with the field of fluctuation velocities indirectly and it directly interacts with pressure fluctuations and near-wall shear of the boundary layer, we shall further consider the spectrum of turbulent pressure fluctuations. Besides, it is justified by that fact that velocity fluctuations correlate with pressure fluctuations:

$$\sqrt{\overline{p'^2}} = 0.5 \rho U_\infty^2 \, Re^{-0.3} \tag{6.16}$$

Calculations of pressure fluctuation spectrum can be carried out according to [29,36]. Data in this paper are well agreed with the experimental data in [277,605,606].

In the range of x from the expected beginning of transitional boundary layer (in the area of the beginning of curves (5), (6), Fig. 6.35) calculations of the range of pressure fluctuation frequencies must be done according to curve (5), in view of 90% of a full range of pulsations of pressure. From $x = 0.8$ m, after point B it is possible to consider the range of frequencies of pressure fluctuations on curve (6).

Therefore, according to the principles presented in Part I and to the above-presented calculation results, the simplest method of selection of elastomers is the following. By means of equipment and methods described in [60,139], the frequencies of elastomers are determined: the frequency of natural oscillations and frequency range, in which the elastomer strongly absorbs mechanical oscillations.

The pressure fluctuation spectrum and the location of regions with the widest range of frequencies of the disturbance motion are then determined. According to this calculations and results of the investigation done in previous chapters, the elastomer composite with the natural frequency and $tg\varphi$ range is selected which best correspond to the calculated pressure fluctuation spectra. This composite is mounted in the place of the body surface with the most corresponded frequencies. Taking into account the absorption of pressure fluctuation by the first composite, the correction is made and then the second composite (next along the body length) is placed with other properties (frequency characteristics), corresponding frequencies of fluctuation loading in the following region. Since the elastic composites are being selected for certain free-stream velocity range, the mechanical properties of an elastomer (and, therefore, frequency characteristics) must be controlled in accordance with the requirement of the greatest coincidence of the natural frequencies of the composite and the ranges of the characteristic frequencies of the boundary layer at the x location of the composite.

As well as a correspondence in frequency characteristics, it is necessary to reach the energy correspondence. For this purpose, the root-mean-square values of pressure fluctuations are determined using measurements of Reynolds stresses on elastic plates [60,139] and the formula (6.16). Using the value of dissipative vortex scale near the wall [184,299,300,435,436,521,533,549,597] and its distance from the wall [184,300] ($y/\delta = 0.15-0.25$), it is possible to calculate the pressure fluctuation energy on the rigid and elastic plates. Comparing the energy that can be absorbed by the elastic composite with the energy of turbulent pressure fluctuations, composites are selected that satisfy maximum correspondence of the above-mentioned energies, that is, maximum response of the elastic surface to disturbances generated by the flow are implemented. Only in case of conformity of the frequency range and energy of fluctuations of disturbances movement of the boundary layer and natural elastomer fluctuations will there be the best absorption and redistribution of energy of external disturbances.

Other π-parameters (Part I, Section 1.4) should be taken into account to provide optimal work of the composite elastic coating.

The engineering method for the selection of the elastic plates must be verified by the experiments below (Sections 6.4 and 6.6). The selected in such way elastic plates must be checked by measuring the friction drag. It is necessary to provide a mechanism for control the mechanical parameters of the composite elastic plates for different velocities of flow.

It has been shown above how to define and optimize mechanical characteristics of the elastomers in view of obligatory sectioning along the body length. Manufacturing a material with the specified change of mechanical characteristics along the length, width, and through the thickness of a coating would be optimum. We shall now consider the principles of designing elastic coatings, for which it is necessary to consider the following research findings:It is necessary to approach an optimum design of high-speed hydrobiont skin coverings and their specific functioning:

- It is necessary to approach the optimum design of high-speed hydrobiont skin coverings and their specific functioning.
- It is necessary to know principles of a complex wave-guide.
- It is necessary to apply results of complex interactions of various disturbances in a boundary layer.
- It is necessary to apply main principles of functioning of hydrobionts.

Let us consider these features consistently.

1. The structure of dolphin's skin coverings (Part I, Section 4.3) is such that there is an ordered system of dermal papillae in the outer layer. When swimming velocity increase, the blood system of these papillae increases blood pressure in these capillaries which leads to a change in the shape of the papillae. At the increase blood filling, according to increasing energy of pressure fluctuations of the boundary layer, the fluctuating mass

of skin coverings in the outer layer changes. Thus, the parity of energy of shaking force and the caused oscillatory energy of unit fluctuating mass is kept optimum. The temperature of the skin outer layer simultaneously increases and its mechanical characteristics change.

2. The structure of a dolphin's skin outer layers represents the classical sample of a waveguide. When speed of motion is increased, according to principles of a complex waveguide (Part I, Section 3.8) changes of the structure and blood filling of the outer layer of dolphin's skin automatically change characteristics of the skin covering as a wave-guide. Indeed, at increased speed of motion the phase velocity of fluctuations in the skin caused by the fluctuations of pressure and the boundary layer shear increases. Hence, it is necessary to change the impedance of such a wave-guide.

3. Experimental investigations of interaction of various disturbances in a boundary layer [247] have shown that there are certain laws of interaction of various disturbances. In particular, introducing flat disturbances into a boundary layer in the area of three-dimensional disturbances in the boundary layer of the plate, there is very intensive interaction of these disturbances leading to accelerated development of stable three-dimensional disturbances. Furthermore, these conclusions are obtained for various types of interaction of disturbances generated in a boundary layer by means of elastic plates. The results confirm that transversal microfolds on the surface of dolphin skin cause quasi-planar fluctuations in a boundary layer with a certain frequency, which interact with 3D disturbances generated from below in the boundary layer due to the longitudinal ordered structure of the outer layers of the skin. Interaction of these two kinds of disturbances leads to the formation of steady three-dimensional disturbances in the boundary layer, which develop in a complex wave-guide: "boundary layer—skin coverings."

4. Periodically, at certain velocity of motion, the skin muscles contract that causes such tension in the outer skin covering that intensive fluctuations of the boundary layer cause self-oscillation of the outer layer of the skin (Part I, Section 6.9) and essential reduction of friction drag (Part II, Section 6.7, Fig. 6.32).

Realization of the considered features of functioning of high-speed hydrobionts is implemented in the form of the developed designs of elastic coverings (Part II, Section 5.1). In Part I, Section 4.3, Chapter 6, the structure and mechanical characteristics of dolphin's skin coverings is described. In Section 6.9, Part I results of research of parameters of fluctuation of dolphins skin coverings at various modes of swimming are represented. Based on dimensionless π—parameters (Section 1.4, Part I), one can be define initial data for designing elastic coatings for technical objects. However, each concrete technical device with an external flow or internal flow demands the account of specific features of the flow structure.

In addition to the above features of the interaction of coatings with the flow, the technological features of fixing coatings on moving bodies and the interaction of coatings not only with disturbances of the boundary layer, but also with other loading parameters are also important (Part II, Section 5.9, Fig. 5.61). Experimentally investigated a large number of variants of monolithic and combined, passive and controlled elastic coatings. The most promising materials are PU and FE elastomers, the recipes of which are given in Section 5.1. I.V. Shcherbina (Kiyv, Ukraine) developed the FE and a device for its manufacture in the form of a continuous elastic pipe of a given diameter. In this case, the necessary distribution of mechanical characteristics and the structure of the elastomer can be specified along the length of the pipe.

However, each specific technical device, depending on the external or internal flow, requires an individual approach to the development of the formulation and design of an elastic coating.

Combined methods of drag reduction

7.1 Combined methods of drag reduction

In Ref. [560], problems of controlling turbulence and relamin/arization were considered. The use of combined methods of drag reduction is promising (Section 6.1, Table 6.1). It is assumed that each method of influencing on the coherent vortex structure (CVS) of the boundary layer affects the specific features of the physical process of the development of disturbing motion. In other words, each method mainly affects certain characteristics of the development of the disturbance. The simultaneous use of several methods of influencing on the CVS contributes to the use of the advantages of each individual method, and in the aggregate, the effectiveness will increase. If both or several methods affect the same characteristics of the disturbing motion, then the efficiency increases due to the increased effect on a certain characteristic of the disturbing motion. For the optimal application of the combined method, it is necessary to determine what characteristics of the development of the disturbing motion are affected by each individual method of controlling the CVS. In hydrobionts, only combined methods of drag reduction are used.

The following combined drag reduction methods were investigated:

1. The combination of the shape of the nasal contours of a hard surface, in particular, various xiphoid tips and injection into the boundary layer of polymer solutions. In this case, longitudinal vortex disturbances were formed in the boundary layer.
2. Injection into the boundary layer of a rigid model of polymer solutions from two slits with the simultaneous formation of two longitudinal vortex systems.
3. Injection into the boundary layer through three slits of various solutions in combination with three systems of longitudinal vortex structures.
4. Elastic surfaces in combination with the formation of longitudinal vortex structures and two-dimensional disturbances in the boundary layer [624]. In this case, the generation of disturbances when placed on an elastic surface and inside its rigid structures for the formation of vortices is investigated. In addition, disturbances were generated using a special design of the elastic surface, as well as during heating and pressure control of individual elements of the elastic surface. The joint effect on the resistance of the elastic surface of its curvature is studied.
5. The same, but with the formation of a static electric field around the model.
6. The combination of elastic surfaces, injection into the boundary layer of polymer solutions, the formation of longitudinal vortex systems in the boundary layer and the use of nasal contours in the form of xiphoid tips.
7. Various methods of injection of gas bubbles into the boundary layer using thin slots, conductive wires, strips, or diaphragms placed along or across the flow.

Some options for combined methods of drag reduction in accordance with the list are discussed later.

7.1.1 Flow around elastic surfaces

On the basis of the hydrobionic approach (Part I), various versions of elastic surfaces were designed and manufactured (Section 5.1). Along with homogeneous monolithic surfaces, multilayer anisotropic elastic surfaces were investigated to simulate the skin structure of high-speed hydrobionts. In addition, various types of vortex generators and conductive heated elements were placed inside and outside the elastic plates. Inside some plates were placed longitudinal and transverse cavities with a liquid, the pressure of which was regulated statically or dynamically. Such elastic surfaces made it possible to combine two or more CVS control methods. All this made it possible to regulate the characteristics of the boundary layer by regulating the mechanical characteristics of elastic surfaces and the formation of various types of CVS in specified combinations. The placement of conductive elements on elastic plates made it possible to generate longitudinal layers of microbubbles (MB) in the boundary layer. The characteristics of the boundary layer of such surfaces were investigated in various hydrodynamic installations and in a wind tunnel [60,139,624]. Almost all investigated elastomers reduced hydrodynamic resistance to 35% in a wide range of Reynolds numbers.

7.1.2 Injection of polymer solutions when flowing around elastic surfaces generating longitudinal vortex structures

The outer layer of the elastic plate is made of permeable elastic material. The internal cavity of the coating is filled with a polymer solution with a concentration of 2000 ppm. In accordance with the pressure distribution along the streamlined body, a small part of the polymer solution will flow to the outer surface of the elastomer. The design of the composite elastic coating contributes to the formation of longitudinal CVS in the boundary layer. This design simulates the skin of a swordfish. A similar combined method can be implemented by injection into the boundary layer of the elastic surface of polymer solutions through the nasal slit. Experimental investigations have increased the efficiency of this method to 40%–60%.

7.1.3 The influence of the shape of the nasal contours on the efficiency of injection of polymer solutions into the boundary layer

Experimental investigations of the flow around a model with various types of tips without injection showed that the model with a ogival tip has the least resistance. In the model with xiphoid tips, during injection of polymer solutions, the resistance decreased more than with the ogival tip. The effectiveness of the results obtained is greatest for a model with a long xiphoid tip. The xiphoid tips are effective over a wider range of

Reynolds numbers. At certain values of the concentration and specific consumption of polymers, the formation of nasal thrust was recorded in a model with xiphoid tips.

The influence of various forms of the shape of the nasal contours was also investigated on longitudinally streamlined cylinders with large sizes. The length of the cylinders varied over a wide range, and the cylinder diameter was 0.1 and 0.175 m.

7.1.4 The effect of a static electric field on the characteristics of the boundary layer

A cylindrical model of organic glass was made, which had interchangeable tail and nose fairings. These fairings were made of organic glass and brass. Insulated wires were connected to the fairings made of brass, which were connected to a direct current source. As a result, a static electric field was formed around the model, whose characteristics under certain conditions influenced the efficiency of injection of polymer solutions.

It is known that water molecules can make up various liquid crystalline structures, the so-called clusters, which differ significantly from each other. So, for example, clusters of sea water differ from clusters of fresh water. Therefore the effectiveness of polymer solutions in seawater decreases. It is also known that the most favorable water clusters for the human body are obtained by defrosting water. A photograph of such a cluster is given in Ref. [139]. It can be assumed that such clusters will be most effective when polymers dissolve in water. Effective polymer solutions can be obtained by using thawed water or by creating an electromagnetic field whose lines of force are directed along the streamlined body.

7.1.5 Injection of polymer solutions from two slots

The design of the model had a bow in which one or two transverse slots were formed. In this case, a continuous veil of a parietal water jet was injected from the first slit. A near-wall jet of polymer solutions was generated from the second slit in the form of a system of longitudinal vortices. It was possible to generate a system of longitudinal vortices during the injection of solutions and from the first gap. The design of the bow made it possible to prestretch the polymer molecules, which increased the efficiency of polymer solutions. The formation of one or two systems of longitudinal vortices in injected near-wall jets increased the stability of near-wall jets and reduced the diffusion of polymers along the thickness of the boundary layer.

7.1.6 Injection of polymer solutions from one nasal slit when flowing around a cylinder with an elastic surface

The results of the performed experiments are given in Section 6.7. The models had a rigid surface and various types of elastic surfaces. The length of the first model varied and for both models the shape of the bow. It was also possible to form various sizes of longitudinal CVS in injected near-wall jets with polymer solutions. The main conclusion is that combined methods can increase the effectiveness of drag reduction. The results obtained depend on many parameters, for example, those listed above in this method of drag reduction. Therefore the effect of each parameter on the decrease in resistance must be investigated separately and independently. And then examine the effect of each additional parameter in sequence.

7.1.7 Various methods of injection into the boundary layer of microbubbles of gas

In Ref. [75], the results of measuring the resistance of a model with a short xiphoid tip are presented. In a highly aerated flow, the resistance of this model is significantly reduced. The effect of reducing drag is maintained for almost all Reynolds numbers.

7.1.8 Injection of polymer solutions through three slits

The possibility of reducing friction resistance in the presence of a viscosity gradient across the boundary layer is theoretically investigated. This approach is implemented using the model given in Ref. [139].

In a discussion with K.J. Moore (Cortana Corporation, United States) of scientific problems and, in particular, the role of cracks in sharks, the idea arose to develop multilayer injection systems in the boundary layer of various liquids. In Refs. [83,89,450,451], justification is given and devices for injecting various solutions into the boundary layer, including microbubbles, through three slots or through an arbitrary number of slots are given.

An important task for increasing the efficiency of such methods as polymer solutions and MB is to reduce the diffusion of these solutions across the boundary layer. The idea of a three-layer or multilayer injection is not only to provide a viscosity gradient across the boundary layer, but also to reduce the diffusion of these solutions. In point 4 above, one of the methods for realizing this problem is given when the diffusion of polymer solutions is limited by an electrostatic field. In point 5, this problem is solved by injection through the first slit of water and through the second slit of polymer solutions or through the first slit of high concentration polymer solutions, and through the second slit of low-concentration solutions. Providing the appropriate size of the longitudinal vortices leads to a significant decrease in the diffusion of the solution injected from the second slot.

In US patents, a similar idea is realized when water is injected through the first slit, a polymer solution through the second, and an aqueous solution of MB through the third. In this case, the diffusion of MB is prevented by the layer of polymer solutions located above, the diffusion of which is prevented by the located layer of injected water. This combined method reduced the resistance by 60% [453].

A promising is the combined method of drag reduction, which differs from the previous one in that the streamlined body surface, is covered with an optimal elastic material, and the nose of the body is made in the form of a xiphoid tip having appropriate dimensions for a selected range of motion speeds.

7.2 Experimental investigations of bodies with xiphoid tips

The interaction of high-speed hydrobionts with the flow is of interest from the point of view of: (1) determining the laws of physical interaction with the environment; (2) the search for mechanisms for the economical use of energy; and (3) to develop new technologies. Concerning focuses on studying features body systems and their interaction with the flow during the movement of aquatic animals, aimed at reducing energy consumption. Since the movement is realized in the aquatic environment, when these systems are taken into account, the influence of the biosphere energy on the organism is taken into account. Showing methods drag reduction to body movement.

The structural features of the outer covers of hydrobionts were developed under the influence of the hydrodynamic and physical fields of the environment, in particular, under the influence of the structure of the boundary layer characteristic of each speed range with which the corresponding groups of hydrobionts swim. At the same time, it is also interconnected during evolution, specific structures of the outer covers developed, hydrobionts change the structure of the boundary layer and form such a boundary layer in order to reduce hydrodynamic resistance and thereby reduce the effect of the environment on the skin, in particular on the skin receptor apparatus. Taking into account that the biosphere of rapidly swimming hydrobionts is the same, the mechanisms of the laws of interaction with the flow should also be identical in terms of characteristic dimensionless motion parameters (Reynolds, Strouhal number, etc.). Adaptations in the structure of the body and skin of hydrobionts may be different.

Representatives of the fourth high-speed group of fish, the sailing and xiphoid fish, swim at the highest speeds in water [381,483−485], which consist of four groups depending on the range of characteristic swimming speeds. Their speed reaches 35 m/s, and the Reynolds number is 1×10^8, so this group of fish is of particular interest. For the first time, the hydrodynamic functions of their gill apparatus were noticed in Refs. [476,477]. It has been suggested that the specific structure of the gill mucus-forming apparatus improves flow around the body [204,367−372,404]. Studies of the body shape of swordfish were performed [9,204,374,375,378,381,407,483−485,502,632]. The features of the shape of the caudal fin [371,484] and the features of the thermal regulation of the body [372] are considered. Some hydrodynamic features of the body shape of cartilaginous fish are systematized in Refs. [381,483−485]. The shape and features of the location of the scales on the surface of the body of a young swordfish are described [477].

A detailed investigation of the skin in various parts of the body made it possible to describe a number of new designs in the skin of swordfish [367−372,476]. However, these features were considered without regard to the basic principles of hydrobionics [62,65,72,77,85,115]. This did not allow to sufficiently generalize their distinguishing features and hydrodynamic functions. Table 7.1 is a list of works in which the morphological, hydrobionic, and hydrodynamic features of swordfish are investigated.

Table 7.1 Types of swordfish research.

Author	Name of journal	Year	Area of investigation
Morphology			
V.V. Ovchinnikov	Biophysics, 11, No. 1	1966	Body structure
A.P. Koval	Bionika, 11	1977	Skin structure
A.P. Koval	Bionika, 12	1978	The structure of the skin and gills
A.P. Koval	Bionika, 14	1980	Skin structure
A.P. Koval	Bionika, 16	1982	Skin structure
A.P. Koval, S.V. Butuzov	Bionika, 24	1990	Tail fin structure
A.P. Koval, A.A. Koshov-sky	Bionika, 25	1992	Heat exchanger structure
Hydrobionics			
S.V. Pershin, Chernyshov, O.B., Kozlov L.F.	Bionika, 10	1976	Hydrobionic skin parameters
Protasov V.R., Staroselskaya A.G.	Hydrodynamic Features of Fishes	1978	Theoretical drawing of the body structure
Pershin S.V.	Bionika, 12	1978	Hydrobionic body parameters
Babenko V.V., Yurchenko N.F.	Biophysics, 25, №2	1980	Hydrobionic skin parameters
Babenko V.V., Yurchenko N.F.	Hydrodynamic Questions in Bionics,	1983	Hydrobionic skin parameters
Babenko V.V., Koval A.P.	Bionika, 23	1989	Hydrobionic skin parameters
Babenko V.V.	Bionika, 25	1992	Hydrobionic skin parameters and tail fin
Problem of drag reduction			
Kozlov L.F., Leonenko I.V.	Reports AN Ukraine	1971	Sphere flow around (experiment)
Kozlov L.F., Leonenko I.V.	Bionika, 7	1973	The flow around the swordfish model (theory, experiment)
Zolotov S.S., Khodorkovsky Yu.S.	Bionika, 7	1973	The flow around the swordfish model (theory, experiment)
Aleev Yu.G., Leonenko I.V.	Bionika, 8	1974	The flow around the swordfish model (experiment)
Amfilokhiev V.B., Zolotov S.S., Ivlev Yu.P.	Bionika, 14	1980	The flow around the swordfish model (theory)

(*Continued*)

Table 7.1 (Continued)

Author	Name of journal	Year	Area of investigation
Babenko V.V., Koval A.P.	Bionika, 16	1982	Flow around the model with injection (experiment)
Voropaev G.A., Kozlov L.F., Leonenko I.V.	Bionika, 16	1982	Flow around the model (theory)
Babenko V.V.	Bionika, 17	1983	Flow around the model with injection (experiment)
Bandyopadhyay, P.R.	AIAA Journal	1989	Drag reduction of the body with a nasal protrusion (theory)
Korobov V.I.	Bionika, 26	1993	Flow around a model with an elastic surface and with polymer injection (experiment)

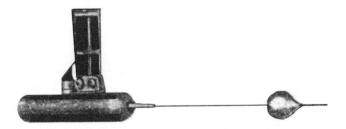

Figure 7.1 Photograph of a strain gage dynamometer with a sphere.

The length of the swordfish (*Xiphias gladius*) at the age of 5 to 6 years is 1.4–1.7 m, and the length of the xiphoid tip is 35%–45% of the body length. The speed of swordfish in extreme situations reaches 130 km/h. In connection with high swimming speeds, an assumption arose about the hydrodynamic role of the xiphoid tip. The first experimental investigations of the flow around bodies with xiphoid tips were performed by Leonenko et al. [5,374,375,378].

Test experiments are best carried out on canonical a body, which was chosen as a sphere, the resistance of which is the standard in many investigations. Fig. 7.1 shows a photograph of a strain gage dynamometer with an installed sphere with a xiphoid tip.

The sphere has a diameter of 70 mm and is made of plastic. The surface area of the standard sphere was 154 cm^2, and with a tip—197 cm^2. A wire ring turbulator was fixed at the equator of a sphere. The xiphoid tip had a length of 140 mm and a width of 30 mm. The base thickness was 7 mm. The tip was made of aluminum and screwed onto the sphere with by means of stud-bold.

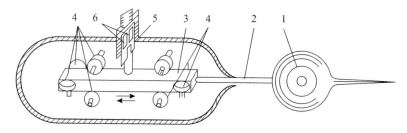

Figure 7.2 Kinematic scheme of a strain gage dynamometer: (1) sphere, (2) rod holder, (3) parallelepiped, (4) bearings, (5) measuring spring, and (6) strain gages.

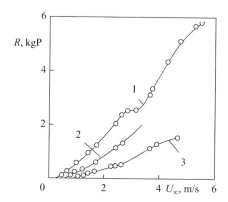

Figure 7.3 Dependence of the hydrodynamic drag of the sphere on the flow velocity: (1) reference sphere, (2) sphere with a turbulent ring, and (3) sphere with a xiphoid tip [374].

The strain gage dynamometer scheme is shown in Fig. 7.2. The experiments were carried out in a hydrodynamic towing tank. Range of working speeds—2–12 m/s. Towing the model was carried out when the ball was deepened 0.5 m from the free surface. The interval between runs was 20 min, the water temperature in the pool was 15°C. In tests, the drag of the sphere was measured for such cases:

- Standard sphere without turbulizer.
- Sphere with the turbulizing ring.
- Sphere with the xiphoid tip.

The measurement results are shown in Figs. 7.3 and 7.4. In Fig. 7.4, when calculating the Reynolds number for curve (3), the diameter of the ball was taken as a characteristic scale and for curve (4) the largest size equal to the sum of the diameter of the sphere and the length of the xiphoid tip. The xiphoid tip on the sphere like a turbulent ring significantly (2–2.5 times) reduces the hydrodynamic resistance. In both cases, the effect is explained by the displacement of the separation point of the boundary layer to the stern of the sphere. For well streamlined bodies (bodies of revolution with high elongation), the main component of the hydrodynamic resistance when moving under water is the friction resistance (80%–90%). The results obtained suggest that the

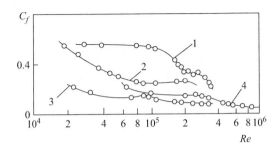

Figure 7.4 Dependence of the sphere drag coefficient on the Reynolds number: (1) standard sphere, (2) sphere with a turbulent ring, (3), and (4) sphere with a xiphoid tip [374].

Figure 7.5 Installation for towing a model with a xiphoid tip [375].

xiphoid tip will also affect the resistance of the model of the body of revolution. Experimental investigations were performed on a model, a photograph of which is shown in Fig. 7.5. The model of the body of rotation was fixed on the same tensometric dynamometer and towed in the same towing tank.

The model of the rotation body made of aluminum had the following parameters: length without tip—72.8 cm, length with tip—109.5 cm, diameter—10 cm, wetted surface without tip—2019 cm^2 and with tip—2129 cm^2, diameter of the wire turbulator—0.2 cm, the length of the turbulator is 15.9 cm, the distance of the turbulator from the front cut of the model is 2.5 cm. The model was made in the form of a cylinder with bow and stern ogival forms. The length of the cylindrical part was 27.3 cm. The nasal extremity of the model with the tip removed was in the form of a semiellipsoid of revolution with a half-axis ratio of 1–4. The aft end is made in accordance with the coordinates given in Refs. [375,378].

Before towing tests, experimental investigations of the pressure distribution along the model were carried out. For this purpose, drainage holes were made in the lateral surface of the model, connected by tubes with a pressure gage. On the model without a tip, 16 drainage holes functioned, and with a tip, 20.

The experiments were carried out in an open-type hydrodynamic pipe, the working section of which was made in the form of a channel with dimensions of

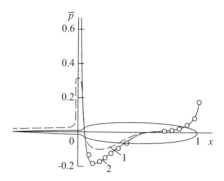

Figure 7.6 The influence of the xiphoid tip on the pressure distribution along rotation body: (1) without the tip and (2) with the tip [375].

$0.5 \times 0.6 \times 1.5$ m at a speed of 2.5 m/s. The unevenness of the velocity field in the working part was not more than 2%, the degree of turbulence of the flow was not more than 3%. The pressure measurement results are presented in a standard way in the form of a dimensionless pressure coefficient:

$$\bar{p} = 2(p_i - p_{st})/\rho U^2 \tag{7.1}$$

where p_i is the pressure at the measured point, p_{st} is the static pressure, and U is the flow velocity. Based on the results of the tests, a pressure distribution is constructed along the model without and with a tip (Fig. 7.6). The above results showed that for these contours of the body of revolution, the pressure distribution changes insignificantly with a tip.

However, for the first time, it was possible to detect a significant difference in the pressure distribution along the model with xiphoid tip. The pressure distribution on the model with the tip became more uniform, the pressure maximum at the beginning of the model decreased twice and the pressure gradients along the model decreased. This is important for living organisms.

The results of towing tests are shown in Fig. 7.7. The model was towed in a speed range of 4.6–10.4 m/s. The xiphoid tip was the same in shape as the swordfish. Detailed data on the results of towing tests are given in the tables in Refs. [375,378]. Curve (1) indicates the test results of the model without a tip, but with an installed turbulator in the form of a wire ring. Tests without turbulator were not carried out. Since the tip surface was smooth and did not have a turbulator installed on it, a natural transition in the boundary layer occurred during the flow around the model with the tip, depending on the towing speed.

For comparison with the towing results of the model with the xiphoid tip [curve (3)] with the towing data of the model without the tip [curve (1)], the turbulator's own resistance, which was determined by the formula [378], was subtracted from the obtained model resistance data:

$$R_{turb.} = 0.62\rho(U^2/2) \cdot d \cdot l \tag{7.2}$$

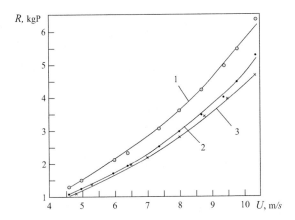

Figure 7.7 The dependence of drag of the body of rotation on the towing speed: (1) without a tip with a turbulent ring, (2) without a tip but with own resistance of the turbulator subtraction, and (3) with the tip [378].

Figure 7.8 Photograph of various forms of models and their bow parts [5].

where ρ is the density of water at temperature during the tests, d and l are the diameter and length of the wire turbulator, respectively. A comparison of curves (2) and (3) shows that the installation of the xiphoid tip led to a drag reduction on the model by about 7%–8%, despite the increase in the wetted surface of the model with the tip by about 5.5%. Moreover, the efficiency increases with large Reynolds numbers, when the boundary layer on the model with a tip is turbulized.

I.V. Leonenko conducted tests with models having different shapes of the bow, and also performed an investigation of the pressure distribution on the model, which, unlike the previous one (Fig. 7.5) was made in accordance with the shape of the body of a swordfish [5]. The prototype for constructing the theoretical drawing was a swordfish 1.5 m long. Fig. 7.8 shows photographs with different shapes of the bow of the model, and Fig. 7.9 shows a theoretical drawing of the shape of a swordfish with the location of the drainage holes.

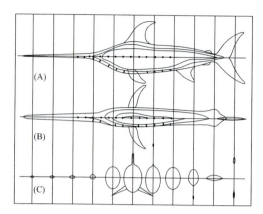

Figure 7.9 Theoretical drawing of swordfish and the location of drainage holes on the model: (A) side view, (B) top view, and (C) cross-sections [5].

Figure 7.10 Swordfish model installed in a water tunnel [5].

In accordance with this theoretical drawing, a model of swordfish was made (Fig. 7.10), the length of which with an attached tip was 1.05 m. The highest hull height (without dorsal fin) was 0.15 m, and the thickness was 0.1 m. The maximum cross-sectional area of the model was 0.0118 m^2, and the wetted surface was 0.314 m^2.

Drainage holes were made with a diameter of 2.5 mm and were located along the side (10 holes) and the lower generatrix of the model (12 holes). Drainage holes by means of rubber tubes, as usual, were attached to the battery of piezometric tubes.

The pressure head was measured using a pitot tube. Zero calibration was performed by immersing the model and the pitot tube in a suspended bath with water. The water level was at the level of the free surface in a moving stream. The pressure distribution along this model was measured in the same water tunnel at a flow

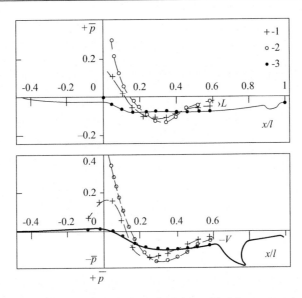

Figure 7.11 Pressure distribution along the swordfish model: L is measurement along the lateral meridional section, V is measurement along the lower vertical section, (1) with the xiphoid tip, (2) without the tip, and (3) location of drainage holes [5].

velocity of 2.5 m/s. Speed head measured using a pitot tube. Zero calibration was performed by immersing the model and the pitot tube in a suspended bath with water. The water level was at the level of the free surface in a moving stream. The pressure distribution along this model was measured in the same water tunnel at a flow velocity of 2.5 m/s.

Fig. 7.11 shows the results of measuring the pressure distribution along this model. Compared with previous measurements on a cylindrical model (Fig. 7.6), the influence of the xiphoid tip on the pressure distribution along the body is much stronger. The same conclusions obtained on this model: the xiphoid tip leads to a significant decrease in the pressure peak in the nose of the model, a smoother distribution of pressure along the body, and to a decrease in pressure gradients. The peak pressure moves to the nose of the tip. Its absolute value is less—the same as the area of the wetted surface on the tip. The pressure along the tip model is less. All this leads, according to the authors of the experiments, to earlier turbulization of the boundary layer and to a decrease in the total value of the friction drag. This is confirmed by the corresponding measurements shown in Fig. 7.7.

This measurement cycle was carried out starting in 1969. At that time, the principles of hydrobionics were not yet discovered, so the mechanism and purpose of the xiphoid tip were not understood. The authors conclude that the effects he discovered do not explain the high swimming speeds of swordfish. Unfortunately, the experimental technique was not properly developed, since the functions of the tip were not known. In particular, as a result of the investigations, the authors claim that the xiphoid tip accelerates the turbulence of the boundary layer. It was not taken into account that the upper surface of the tip of the swordfish is substantially rough and clearly has the function of a turbulizer. When conducting experiments on a sphere, the tip length was twice as large as the diameter of the sphere, although the authors note that swordfish have a tip length of 35%–45%. Therefore it would be advisable to use either a rough tip on the sphere or put a turbulent ring on it. Moreover, the authors rightly indicate the value of the Reynolds numbers, corresponding to the speeds of movement of the swordfish. With such Reynolds numbers, the boundary layer is always turbulent. The same disadvantages relate to experiments on models of a body of revolution. In the experiments, the tips were not rough and turbulizers on the tips were not used. Other conditions of hydrodynamic modeling were not fulfilled. In particular, everyone knows that a swordfish swims at high speeds with a constantly open mouth, so that most of its body is surrounded by a parietal stream flowing from the gill slit. What is the role of the xiphoid tip under such conditions of movement, remains unexplored. Despite these shortcomings, I.V. Leonenko was the first to obtain new results on the features of the flow around bodies with xiphoid tips.

In 1969–70, theoretical and experimental investigations of the flow around a body of revolution with a nasal needle were performed at the Department of Hydromechanics of the Leningrad Ship-Building Institute [9,632]. Experimental investigations were performed in a wind tunnel using a model of a body of revolution with a length of 1.8 m with a tip and 1.26 m without a tip, a maximum diameter of 0.304 m, and a tip length of 0.54 m, which amounted to 33% of the body length with a tip.

The results of measurements made using a hot-wire anemometer are shown in Figs. 7.12 and 7.13. Measurements showed that the dimensionless length of the laminar section on the model with a needle is less than in the absence of a needle in the entire range of Reynolds numbers investigated. The distribution of momentum thickness on both models at a flow velocity of 20 m/s for the model with a needle $Re = 2.4 \times 10^6$ and without a needle $Re = 1.7 \times 10^6$ showed that the momentum thickness on the model with a tip is greater, starting from $\bar{x} = 2$.

Viscous drag of the models was determined by the method of impulses from measurements of the velocity head in the flow and the aerodynamic wake behind

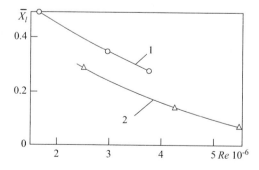

Figure 7.12 Dependence of the length of the laminar section of the boundary layer on the Reynolds number: (1) without a nasal needle and (2) with a nasal needle [9,632].

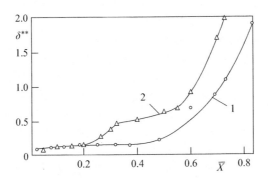

Figure 7.13 Change in momentum thickness along the length of the body: (1) without nasal needles ($Re = 1.7 \cdot 10^6$) and (2) with a nasal needle ($Re = 2.4 \cdot 10^6$) [632].

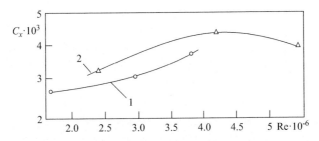

Figure 7.14 The dependence of the viscosity coefficient of the model on the Reynolds number: (1) without a nasal needle and (2) with a nasal needle [632].

the body. The measurement results are shown in Fig. 7.14 for the entire range of measurements at $Re = (1.7-5.4) \times 10^6$.

The experimental points for both models were obtained at the same free stream speeds, but in Fig. 7.14 the points obtained are spaced due to different Reynolds numbers calculated along the length of the model. In the velocity range under study, the viscosity coefficients of the model with the needle turned out to be larger than those of the model without the needle. This is due to the fact that in this range of Reynolds numbers, the viscosity drag is significantly affected by the length of the laminar sections of the boundary layer (Fig. 7.12). From Fig. 7.14 it follows that at $Re > 5 \times 10^6$, there is a tendency for C_x to decrease for the model with a needle, while for a model without a needle, C_x increases and will be larger than for a model with a needle.

7.3 Theoretical investigations of bodies with xiphoid tips

Originally in theoretical calculations, it was assumed that the shape of the body does not essentially influence the distribution of shear stress along the body, and a mixed scheme of flow in a boundary layer was adopted. It was considered that the development of a boundary layer occurs naturally from the laminar flow regime to

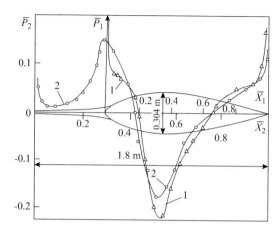

Figure 7.15 Distribution of the pressure coefficient along the length of the body: (1) without the xiphoid tip, $L_1 = 1.26$ m, $x_1 = x/L_1$ and (2) with the xiphoid tip, $L_2 = 1.8$ m, $x_2 = x/L_2$ [632].

the turbulent one. It was also assumed that the displacement of the maximum values of the tangential stress due to the presence of the xiphoid tip in the area of the tip, whose area is small, will lead to a reduction of total drag of the body. In accordance with these assumptions, the pressure distribution on the model of a body of revolution with and without a tip (Fig. 7.15) was calculated in [632] under the assumption of potential flow. Shown are also the shape of the outline of the swordfish and xiphoid tip, on which experimental investigations of the boundary layer parameters were performed (Section 7.2). The pressure distribution obtained theoretically is in qualitative agreement with similar results, according to experimental data [5] (Section 7.2, Fig. 7.11).

For the rotation body shown in Fig. 7.15, a simplified calculation of the friction resistance was proposed in [632] under the assumption that the friction resistance of the model will be similar to the resistance of an equivalent flat plate whose width varies in length according to the law $r(x)$ for the corresponding model. The length of the laminar sections was taken according to Fig. 7.12. According to the results of such a simplified calculation, it was found that up to the number $Re = 3.2 \times 10^6$, a model without a tip has a lower resistance coefficient, and for large Reynolds numbers, the resistance coefficient of a model with a tip becomes smaller.

In Ref. [375], the resistance of a body with a tip was calculated according to the same initial formula by which the calculation was carried out in Ref. [632], and the calculation scheme was selected taking into account the laminar section of the boundary layer. Since the tip of the swordfish is known to be rough, and its speed is large, it was assumed in the calculations [375] that the distribution law of tangential resistance along the body corresponds to the turbulent flow around the body. The integral of the resistance of the body is calculated for the turbulent distribution of the tangential resistance along the body. Theoretical dependences of the effectiveness of drag reducing in the presence of a tip are obtained, depending on the dimensionless length and dimensionless surface area of the tip. For all the

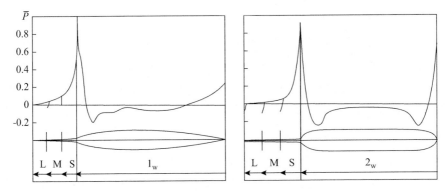

Figure 7.16 Pressure distribution along bodies [9].

parameters considered in the calculations, the tip reduces the resistance in the turbulent flow regime in the boundary layer compared to the absence of the tip. These data are consistent with Leonenko's experiments cited in (Section 7.2, Fig. 7.7). However, as in Leonenko's experiments, the proposed calculation scheme does not take into account the peculiarities of swimming swordfish.

In the calculations of [9], the assumption was made as in [632]: when flowing around the body, it has a laminar section, a transition point, and a turbulent section without separation. For the calculation, two forms of the body of revolution, shown in Fig. 7.16: 1_w: airship-like type without a tip, 2_w: with a long cylindrical insert and ellipsoidal ends without a tip. Three tip sizes were calculated for each shape of the model: S: short, M: medium and L: long. All the parameters necessary for the calculation in dimensionless form are given in [9]. The calculation was carried out on a computer using the Van Drist method, which is convenient for calculating elongated bodies and has no restrictions in the region of the junction of the tip and the main body associated with surface bends. The results of calculations of the pressure distribution on the bodies of revolution, unfortunately, do not coincide with both the simplified calculation shown in Fig. 7.16 and the experimental data shown in Figs. 7.6 and 7.11 (Section 7.2).

Fig. 7.17 shows the calculations of the drag coefficients of the models taking into account changes in the Reynolds numbers and the value When calculating the drag coefficient of the bodies of revolution shown in Fig. 7.16, theoretical dependences of the drag coefficient and the relative change in the friction resistance are obtained. of the wetted surface in the presence of a tip.

Curves without a tip are located between the laminar and turbulent laws of friction for a flat smooth plate and at $Re = 5 \times 10^6$ they are located along the curve for the transition boundary layer, and at $Re = 7 \times 10^7$ they are located almost in the region of turbulent friction for a smooth flat plate.

These reference curves indicate that a mistake was made in the calculation results, since it is known from experimental results that the friction curves for a body of revolution are always higher than the regularities for a flat plate. In addition, for the plate, the transition begins at $Re = 5 \times 10^5$, while, according to the

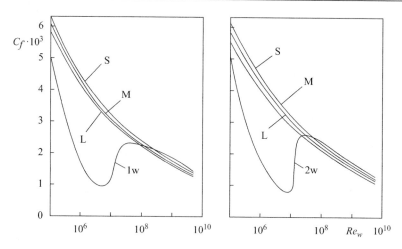

Figure 7.17 Dependence of the coefficient of resistance of models on the Reynolds number [9].

calculation for the reference body of revolution, the transition begins an order of magnitude later in the Reynolds numbers. The authors do not explain why due to the flow around the reference body of revolution this transition is delayed. The curves for the body of revolution with tips are located in the region of the pattern of turbulent friction for a flat plate, which also indicates an error in the calculations. Considering that the calculations had a comparative character, interesting results are obtained indicating that for the assumptions made, the drag of a body of revolution without tips to $Re = 7 \times 10^7$ is smaller than that of body with a tip. With a further increase in Reynolds numbers, the tips lead to a drag reduction, compared with a standard body of revolution. The effect increases with increasing tip length.

In Ref. [588], the potential flow around axisymmetric bodies was calculated in the presence of inflection points on the contour line. When developing a program of numerical calculations, much attention was paid to the features of conjugation of the xiphoid tip with the body of revolution, which has a spherical shape or the shape of ellipsoids of revolution. In addition, the program made it possible to calculate the corresponding characteristics for other forms of the body of revolution.

The rotation ellipsoid had an axis ratio of 1:4 with a tapered tip. The cone with a solution of 5 degrees has a spherical rounding with a radius of 1/10 midsection of the ellipse. In the calculations, the following parameters were changed: the length of the tip was 0.5; 0.25; 0.1, fillet radius 1; 0.5; 0.3 and a nose radius of 0.025 times the length of the ellipsoid of revolution, and the flow velocity is 10, 20, and 30 m/s. For the sphere, the following parameters were changed: tip length 1; 0.5; 0.3 and mating radius 1; 0.7; 0.35 sphere radius. At the indicated parameter values, the distributions of velocities and pressures along the contours of the bodies were calculated. In Fig. 7.18 shows the result of calculating the pressure distribution when flowing around an ellipsoid of revolution. The abscissa shows the dimensionless coordinate S along the contour of the body of revolution.

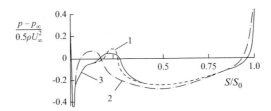

Figure 7.18 The distribution of excess pressure along the ellipsoid of revolution with a tip [588].

Curve (1) corresponds to a long tip with a radius of conjugation equal to 0.5 of the longitudinal axis of the ellipse, curve (2) with a half-length of the tip, and curve (3) with the same tip as for curve (1), but with a pairing radius equal to the longitudinal length of the ellipsoid. The most favorable is the pressure distribution corresponding to curve (3), when both the length of the tip and the pairing radius of the tip with the body of revolution are maximum from the selected calculation parameters. The maximum length of the tip in the calculations approximately corresponds to the relative size of the swordfish tip. At the same time, the calculations confirmed the main positive results of the tip according to the experimental data shown in Fig. 7.11 (Section 7.2): the pressure peak decreases in the area of the where the tip and the ellipsoid joint, the pressure gradients along the length significantly decrease, and the efficiency of the long tip is higher than that of the short tip.

Calculations made it possible to determine that for small elongation bodies (sphere), the use of a tip reduces resistance by 8%–10%, for an ellipsoid of medium elongation by 6%–8% and for bodies of high elongation by 4%–5%. These results are consistent with the basic experimental data of I.V. Leonenko (Section 5.2) and make it possible to determine that the body of rotation of small elongation has the greatest efficiency from the use of the tip. According to calculations, the short tip turned out to be more effective for medium and large elongation bodies. It is also determined that a larger radius of pairing of the tip with the body is preferable. Thus in the joint calculation of the potential flow and the turbulent boundary layer (TBL), it is possible to choose the optimal tip lengths and the radius of pairing for continuous flow with the lowest tangential stresses throughout the body.

For many years, the problem of flow on curved surfaces has been investigated. In relation to this problem of the influence of the tip on the resistance of the body, the role of curvature in the interface between the body and the tip from the position of potential flow was studied in Ref. [588]. However, it is known that longitudinal vortex systems such as Görtler vortices are always formed on a curved concave surface. Until now, the role of these vortex systems has not been investigated. Bandyopadhyay [151] also investigated the effect of curvature in the interface between the body and the tip on the pressure distribution along the body. His work consists of two parts. In the first part, he performed a detailed review of the known theoretical results, which he compared with the corresponding experiments around a flat plate having sections with a convex or concave surface. Initially, various options for the location of a convex surface area behind, in front of and in the middle of a flat surface were considered. At the same time, the length of the convexity site in dimensional form was also analyzed. The analysis showed that in all cases the convex surface leads to a decrease in the drag coefficient, which increases slightly after the convex section. Work is also systematized in which similar combinations of the location of a concave portion behind a flat surface or in some other combinations of various options for surface areas are investigated. In this case, in the region of concave surface sections, the friction coefficient increased. These conclusions are consistent with the known features of the flow around the confuser and diffuser and correspond to the Bernoulli equation. Of interest are cases where the length of the site of convexity and concavity was small. In these cases, the patterns of change in the coefficient of friction along the length were similar: in both cases, in the region of irregularity, the coefficient of friction decreased and then increased. Behind the convex section, this increase was small, and at the end of the concave section—significant. But in the future, the friction coefficients changed slightly. In general, based on this analysis, we can conclude that the convexity favorably reduces the coefficient of friction, and the concavity leads to an increase in the coefficient of friction. It is noted in the work that in concave sections the development of vortex systems of the Taylor–Görtler type is quite possible.

Based on a detailed analysis of the features of the flow around convex and concave sections on a flat surface, we consider the problem of flow around a cylindrical body of revolution with nasal cylindrical and cone-shaped tips. Using a Fortran numerical program for the inviscid subsonic flow around axisymmetric bodies, Bandyopadhyay [151] performed the calculations for these two types of tips. Fig. 7.19 shows the calculation results for a cylindrical tip, and Fig. 7.20 for a conical tip. Bandyopadhyay [151] analyzes the results: Fig. 7.19 shows the application of the D.M. Bushnell concept [194] to a cylindrical body. The following characteristic features of the nasal part of the body should be noted: (1) the purpose of the

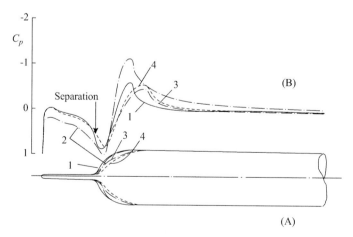

Figure 7.19 Numerical calculation of the surface pressure distribution for a cylinder with single-stage geometry of the cylindrical nose. The calculated separation point in all four geometries is indicated at $M = 0.5$ [151].

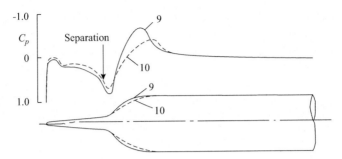

Figure 7.20 Numerical calculation of the surface pressure distribution for a cylinder with single-stage nose geometry with gradually increasing lengths. The calculated separation location in both geometries is indicated at $M = 0.5$ [151].

small diameter of the front of the body is to increase the TBL, which can "absorb" viscous resistance with convex curvature of the body.

The dimensions of the front body, its diameter and length should be small in order to maintain low resistance; (2) the inevitable concave region between the front body and the convex region should be short for the following reasons: (a) to avoid the growth of Taylor–Görtler-lice vortices, (b) to reduce integrated turbulence amplification, and (c) to utilize the asymmetric effect of a convex − concave surface curvature; (3) the surface length of the convex region in terms of δ should be as large as possible, and δ/R on the convex surface should be >0.05; analysis of available convex curvature data shows that an asymptotic state is reached for $\delta/R>0.05$, and relaxation is slower with increase in $\Delta s/\delta$ [151]; (4) according to the concept, the levels of near-wall shear stress will now be lower than the equilibrium levels, and a decrease can be expected viscous resistance due to relaxation of flow over the main cylinder.

The above is a single-stage nasal body. Fig. 7.19 also shows the distribution of surface pressure. All calculations of the nasal bodies were performed at $M_{ref} = 0.5$. In this case, the reference speed, compressibility effects are present, but the shock waves do not appear at the nose-cylinder junction. The corresponding Reynolds number based on the diameter of the main cylinder is $U \cdot d/\nu = 5 \times 10^6$. The boundary layer calculations were performed using these inviscid pressure distributions. There are two potentially troublesome areas of the pressure gradient: one is the strong negative pressure gradient near the fore body/concave junction; the other is the strong acceleration followed by a negative region of the pressure gradient at the transition to the master cylinder. Calculations showed that in the worst case, the first leads to separation, and later a shock wave and/or separation forms. The separation location (zero wall shears) is indicated in Figs. 7.19 and 7.20. A parametric study was then conducted to control the pressure gradient in these two areas.

In Ref. [151], tangent surfaces intersect at the joints of curvature. The dimensions of the nose—body are dimensionless using the main diameter of the cylinder d. The parameters studied for a body with nasal contours, shown in Fig. 7.19, represent the diameter and length of the body, the ratio of the concave to convex radii of curvature, and the location of the concave/convex coincidence point (determined by the turning angle φ), which determines the lengths of the curved areas. Fig. 7.19 shows the changes in the pressure distribution due to changes in the parameters of the curvature, keeping the fore body unchanged. However, none of these geometries can prevent separation near a plane/concave transition.

Changes in the length of the forebody did not prevent separation. To alleviate adverse pressure gradients, the forebody was slightly widened (5 degrees). Nasal body pressure distributions with this additional parameter are shown in Fig. 7.20. Although the taper of the nasal body reduced the negative pressure gradient, it was still not sufficient to prevent separation. There are several ways to reduce the probability of separation of this boundary layer. One of them is the so-called reverse design approach of Zedan and Dalton. According to this approach, it is desirable that the necessary pressure distribution that is probably prevent separation, should first be chosen from the experiment. In Figs. 7.19 and 7.20, such a distribution can be obtained by smoothing the dip in the neighborhood of separation. Then the body geometry, which could create such a distribution, could be calculated back. Boundary layer calculations can then be performed to verify that separation has indeed been prevented. However, this approach has not been made, suggesting that it is likely to redefine the geometry substantially in such a way that separation is prevented. Geometric functions, such as δ/R and $\Delta s/\delta_i$ of convex curvature and other concepts of reducing viscous resistance, which are the main focus of this work, are likely to be investigated in the future.

The systematic calculations of the influence of the curvature radius of the joining surface on the pressure distribution along the body with a tip, shown in Figs. 7.19 and 7.20 are in good agreement both with similar theoretical calculations shown in Figs. 7.15 and 7.16, and with the experimental data shown in Figs. 7.6 and 7.11 (Section 7.2). In Ref. [151], calculations were performed for bodies moving in air. But since his calculations were performed for small Mach numbers, the results shown in Fig. 7.20, curve (10), which are closest to the results of experiments on a model of a swordfish body, whose nasal tip has a large radius of conjugation with its body and shape, are especially important similar to that accepted in the calculations.

In Mach research, the features of the shape of the nasal parts of various bodies on their hydrodynamic characteristics are investigated. For example, in Ref. [210] the pressure distribution on the elliptic nose of the profile and cylinders was studied. Theoretical investigations of the influence of xiphoid tips located on bodies of revolution on the characteristics of elongated bodies have shown that tips change some characteristics of bodies depending on a number of parameters. The relative length of the tip and the radius of joining of the tip with the body turned out to be significant. It was revealed that in the presence of a tip, the pressure distribution along the body changes significantly. Tips lead at certain Reynolds numbers to a decrease in resistance. The circle of problems that have not yet been adequately studied is determined.

The experimental and theoretical investigations of the characteristics of elongated bodies of revolution have shown that there is no purposeful and interconnected study of this problem. Technological approaches did not solve the task. At the same time, the hydrobionic approach made it possible to determine the range of tasks necessary to investigation the problem of the flow around elongated bodies with tips [123]. Note, for example, that the body shape of swordfish and other xyphoid fish determined the value of the conjugation radii generated as a result of evolution. This allows us to determine the values of these radii for theoretical and experimental studies on models. The analysis of the considered results showed that it is necessary to carry out experimental and theoretical studies based on the hydrobionic approach. It is known that in all high-speed fish, due to the gill apparatus behind the gill slit, there is fluid injection into the boundary layer (parts I Sections 4.5 and 4.6).

7.4 Brief review of the problem of injection of polymer solutions into the boundary layer

The effect of high molecular weight polymers on drag reduction has been extensively studied since the discovery of this phenomenon by B.A. Thoms [555] in 1948. Due to the ability of polymers solutions to reduce turbulent shear stress, the dependence of the spectral pressure amplitude on the wall of the boundary layer on shear stress was investigated. The polymer solutions can suppress noise and vibration, characteristic of an unsteady boundary layer. One of the first to investigation the drag coefficient of a sphere at high Reynolds numbers in a polymer solution flow [604].

A method of control of TBL using high molecular weight polymer additives of low concentration, which give the liquid non-Newtonian properties, is an effective way to influence the hydrodynamic processes in the boundary layer. Such additives, in particular, reduce the level of turbulence and increase the thickness of the laminar sublayer, which leads to a significant decrease in the friction resistance of the streamlined surface. The effect of drag reduction is manifested when polymer additives are present in the buffer zone between the viscous sublayer and the turbulent core.

Exists several methods of injection of polymer solutions in TBL. One of them is the use of soluble coatings applied to the streamlined surface. Gradually dissolving, they report "non-Newtonian" properties to the flow in the boundary layer. In Ref. [336], a 0.85 m long Plexiglas plate was investigated, which was coated with 1.5% polyoxyethylene (WSR-301) and dried. Then the plate was towed at a Reynolds number $Re = 2 \times 10^6$. Initially, a decrease in friction of 10% was recorded. After 25 s, the resistance increased to the value that was before coating the plate with a solution. The result shows the effect of polymers in a thin boundary layer near the plate.

In Ref. [273], a cylindrical model with stabilizers in the aft part 3.97 m long and 0.46 m in diameter was investigated. The maximum Reynolds number in the experiment was $Re = 1.2 \times 10^6$. The nasal part of the model had two forms: well streamlined and with a nasal cross-section. The nasal tip with blunt contours provides a turbulent flow along the body. In a special way, the prepared chemical substance in the required amount was applied to the nose fairing. The required amount of substance was determined during experiments, which showed that the coating thickness should be about 3.2 mm. Drag reduction of the model by approximately 18% is noted up to $Re = 1.2 \times 10^6$ in fresh water. When conducting experiments in sea water, the total drag reduction in model was 16%. It is indicated that the calculated ratio of the friction resistance to the total resistance is 59%. This low ratio is due to the high value of the shape resistance and the influence of inductive resistance. Full drag reduction by 18% in fresh water corresponds to a decrease in the friction resistance of the shell by approximately 30%, and a decrease in full drag by 16% during experiments in sea water corresponds to a decrease in friction resistance of the hull by 27%. These results are based on the assumption that such coatings only reduce frictional resistance.

A porous surface (distributed supply of polymer solutions [6]) is also used for injection of polymer solutions into TBLs. The results of testing a porous surface with a uniform polymer supply over the entire surface of both sides of a plate 1.8×0.36 m in size are given in Ref. [337]. In experiments, a solution of polyoxyethylene (POE) of concentrations 3×10^4 and a flow rate of 0.0057 and 0.025 cm^3/(s·cm^2) was injected through a porous plate.

The most common way of injected polymer additives into the boundary layer is to inject a polymer solution through a slot into the near-wall region of TBL [274,320,338]. In Ref. [423], a drag reduction of a flat plate 0.457 m long, which was towed in a channel, was investigated. The polymer was injected through slots on each side of the plate near its front edge. To determine the amount of friction reduction on the plate, a wake jet was investigated. The maximum drag reduction to 50% was obtained by slow injection of a POE solution with a concentration of 5×10^{-5} ppm. High rates of injection of the solution onto the plate and high concentration significantly reduced the effect. The decrease in the friction resistance of a 2.5-meter-long plate with the injection of the POE solution from the slot near the leading edge was measured in Ref. [498]. The decrease in the coefficient of friction was 50% ($Re = 2 \times 10^6$) at a concentration of 2×10^{-4} ppm.

In Ref. [218], a decrease in the friction resistance of a cylindrical body with a diameter of 0.127 m and a length of 1.27 m when a WSR-301 polyoxyethylene solution with a concentration of 1×10^{-3} ppm was injected from the hole near the nose when towing ($Re = 2.9 \times 10^6$). A 23% decrease resistance of a model was noted. It is noted that at these speeds it is necessary to turbulence the boundary layer and ensure turbulent flow around in the absence of polymer injection.

In Ref. [223], the resistance of a sphere during flow around a solution of polymers was investigated.

In Ref. [544], data were obtained on the resistance of a plate measuring 1.945×1.04 m at its movement in a circular towing tank. The tested plate on each side had 50 slots for injection the polymer solution. The plate speed varied to 4 m/s. When pure water or a polymer solution was injected through all slots, with an increase in flow rate, the plate resistance decreased. The weight concentration of the polymer solution was $C < 2 \times 10^{-5}$ ppm. The maximum value of friction reduction ($\approx 50\%$) was achieved when the solution was injected with a concentration of 2×10^{-5} ppm. When a solution of a higher concentration was injected, the effect of reducing friction was observed only at low discharges. The results of experiments with injection of a polymer solution through only nine slots, selected in such a way as to ensure an almost uniform concentration of the solution in the wall region of the boundary layer, are also presented. It is shown that with a nonuniform injection of the solution along the plate length, the magnitude of the reduction in friction can be reduced by 20% compared to uniform injection (with equal weight flow discharges of the polymer substance).

In the case when the polymer solution is injected through the slot into the TBL of the body streamlined by the Newtonian fluid, due to turbulent diffusion, an alternating concentration profile $C(x, y)$ is formed downstream of the slot. Moreover, the character of diffusion, on the one hand, depends on the ability of the flow to turbulent mixing, and on the other hand, directly affects this ability. The mutual conditionality of these phenomena makes the diffusion of active impurities in turbulent flows more difficult.

The qualitative side of the process of impurity diffusion in the TBL is characterized by the presence of four regions downstream behind the slot [498]. In the region closest to the slot, diffusion propagation of the impurity to the upper boundary of the viscous sublayer occurs. In the second intermediate zone, the diffusion layer is immersed in the buffer region of the boundary layer. The layer thickness in this region is determined by the dependence $\delta_d = 0.076 \cdot x^{0.8}$, where x is the distance from the slot downstream. The third zone is transitional, where the diffusion layer gradually fills almost the entire TBL. Here

$$\delta_d/\delta = 0.64 \tag{7.3}$$

In the fourth zone, the diffusion layer merges with the dynamic boundary layer, after which diffusion into the main flow begins. The concentration of polymer solutions in the final region is usually small, and their effect on the flow gradually decreases with increasing x coordinate. Due to significant concentrations of polymer additives in the near-wall region of the TBL, the main effects on the turbulence parameters occur.

In Ref. [581], the diffusion processes of a solution of Polyethylene oxide (PEO) or POLYOX WSR-301 in the intermediate region of the boundary layer of a flat plate were studied. The experiments were carried out in a TBL on the wall of a working section 1 m long of a hydrodynamic tunnel with a cross section of 0.15×0.075 m. The velocity at the outer boundary of the boundary layer varied from 2 to 12 m/s. The Reynolds number over the displacement thickness Re^* varied in the range $Re^* = 6 \times 10^3 - 6 - 10^4$. The authors introduced the longitudinal diffusion scale L using the relation

$$C_W(L) = C_0 e^{-1} \tag{7.4}$$

where C_0 is the initial polymer concentration upon exiting from slot, C_W is the polymer concentration on the wall after exiting the slot. The law of concentration drop on the wall at $0 < x/L < L$ is close to exponential:

$$\frac{C_W}{C_0} = \exp\left(-\frac{\alpha x}{L} - \beta\right) \tag{7.5}$$

moreover, for the WSR-301 solution, $\alpha = 0.7$, $\beta = 0.3$, and x is the distance from the slot downstream. The scale of L depends on the flow rate of the solution q and the concentration of the solution C_0, and depends on their product, that is, on the resulting amount of the polymer substance. This dependence is such that for small $q \cdot C_0$ the scale of L grows approximately proportionally to $q \cdot C_0$, and for large $q \cdot C_0$ it is almost independent of this quantity. The dimensionless shape of the diffusion scale has the form L/h (h is the characteristic particle size of the molecules). Another form of representation may be $L\mu D/kT$ that follows from the relation $D = kT/6\pi\mu h$, connecting the molecular diffusion coefficient D with the size h of diffusing particles, viscosity μ and temperature T of the medium (K is the Boltzmann constant). At present, the values of D and h for polymer particles are not known with sufficient accuracy, therefore the scale L in the formula (7.15) is given in dimensional form:

$$\frac{L\mu D}{KT} = \varphi\left(\frac{qC_0}{\mu}\right) \tag{7.6}$$

where qC_0/μ is the dimensionless argument of the longitudinal scale L.

A feature of the results obtained is their independence from the velocity U_0 flow around the plate and, therefore from the intensity of turbulent diffusion in the outer parts of the TBL.

It was found that the concentration distribution in the boundary layer (with the exception of the small near-wall region) is determined by the value of the parameter $\frac{qC_0}{\rho U_0 x}$, where ρ is the density of the medium. This parameter is the ratio of two parameters:

$$\frac{qC_0}{\rho U_0 x} = \frac{qC_0/\mu}{U_0 x/\nu}. \tag{7.7}$$

Near the wall, only the dimensionless flow rate $q \cdot C_0/\mu$ is decisive, and the Reynolds number plays a role in the outer part of the boundary layer.

A formula is obtained for describing the concentrations across the boundary layer in the intermediate region [580]:

$$\frac{C(x,y)}{C_W} = 0,1\left(\frac{y}{\Omega}\right)^{-4/3} \tag{7.8}$$

where Ω is a certain conditional thickness of the diffusion layer.

The investigation [580] was continued in Ref. [581]. The experiments were carried out in the boundary layer on the flat wall of the working section of the hydrodynamic tunnel with a cross section of 0.36×0.15 m² and a length of 5 m. The velocity at the outer boundary of the boundary layer varied from 2.5 to 10 m/s, with a change in the Reynolds numbers in the displacement thickness Re_{δ^*} from 4×10^3 up to 1.1×10^5. The experiments used aqueous solutions of polyethylene oxide WSR-301 and P-31 and a copolymer (based on acrylamide P-20–77). The dimensions of the exit slot were 0.4×150 and 0.8×130 mm². The slope of the slit and the streamlined surface ensured that the solution was supplied at an angle $\varphi = 20$ degrees (in experiments [580] this angle was equal to 7 degrees). As in [580], the longitudinal diffusion scale is introduced—the distance from the slit at which the concentration of diffusing impurities on the wall decreases e times. A dimensionless quantity $k = \frac{L\rho V_0}{qC_0}$ is also introduced (ρ is the density of the solvent), which serves as a measure of the polymer's ability to weaken the processes of turbulent exchange. It also introduces $k_0 = k/k_{stand}$, where $k = k_{stand} = 6.25 \times 10^6$ for WSR-301. The value of k_0 is determined either by the formula

$$k_0 = \frac{\rho}{C_0} 10^{\frac{S_T - 2,49}{0,35}} \qquad (7.9)$$

or according to the formula

$$k_0 = \frac{x\rho U_0}{qC_0} 10^{\frac{S_T - 3.08}{0.35}} \qquad (7.10)$$

where x is the distance from the slit downstream. It is indicated that in the formula describing the concentration distribution in the transition zone $\beta = 1 - \alpha$, the exponent α depends on the angle φ of the slope of the slit to the streamlined surface and varies from $\alpha = 1.8$ for $\varphi = 20$ degrees to $\alpha = 0.7$ for $\varphi \leq 7$ degrees. It is concluded that at $\varphi \leq 7$ degrees the dependence of the diffusion scale L on the flow velocity disappears, while at $\varphi = 20$ degrees this dependence is clearly observed. A universal dependence is obtained that describes the wall concentration for large x/L values, taking into account the difference in diffusion properties of different polymers:

$$\frac{k_0 C_W}{\rho} = 4,16 \cdot 10^{-2} \frac{x}{L}. \qquad (7.11)$$

It is concluded that in the intermediate region of polymer diffusion in the TBL, their concentration decreases along the streamlined surface according to the exponential law. Formula (7.10) takes into account the influence on the diffusion of the initial concentration of C_0 and (through the scale L) the influence of the specific flow rate of the polymer solution q, its efficiency k_0 and the flow velocity U_0. In the final region of diffusion of polymers in the boundary layer, their concentration decreases along the wall according to a power law, which does not depend on the initial concentration of C_0, but takes into account the total polymer consumption $q \cdot C_0$, its efficiency, and flow velocity U_0 through L and k_0.

In Refs. [614,615], the diffusion of the WSR-301 polymer solution was investigated by injecting it through a slot located at an angle of 7 degrees to the surface of the plate, which is one of the walls of a rectangular channel 1.12 m long and with a section of 0.38 × 01 m. The channel velocity was constant—2.62 m/s. The width of the slit was 0.56 mm. To calculate the local resistance on the plate in a pure liquid, the formula was used:

$$C_f = 0.059 \cdot Re_x^{-1/5} \qquad (7.12)$$

where $Re_{x'} = \frac{U_0 x'}{\nu}$. Here U_0 is the velocity in the channel, x' is the distance from the transition point of the laminar boundary layer to the turbulent one. To determine the thickness of the boundary layer δ, the thickness of the viscous sublayer δ_s and the thickness of the buffer zone δ_b the formulas apply:

$$\delta = \frac{V_*}{x'_*} = 0.377 Re_{x'}^{-1/5}; \quad \delta_s \frac{V_*}{\nu} = 11.6; \quad \delta_b \frac{V_*}{\nu} = 30.$$

$$V_* = \sqrt{\tau_w/\rho}, \quad \tau_w = \rho \cdot \nu \left(\frac{\partial U}{\partial y}\right)_w, \quad C_f = \frac{2\tau_w}{\rho U_\infty^2}.$$

In Ref. [614], it is assumed that the diffusion boundary layer is divided into four regions along the length: in the first, the diffusion layer is located in the laminar sublayer. The thickness of the diffusion layer is the distance normal to the wall at which the mass concentration decreases by half its initial value on the wall; the second intermediate region is characterized by the fact that the diffusion layer is immersed in the boundary layer, thicker than the laminar sublayer. The thickness of the diffusion layer is $\delta_d = 0.076 \cdot x^{0.8}$, where x is the distance from the beginning of the diffusion boundary layer; in the third transition region, the surrounding flow begins to influence the growth of the diffusion layer; fourth final area—the diffusion of the mass of the polymer solution expands beyond the boundary layer. The thickness of the diffusion layer continues remain a constant part of the thickness of the boundary layer:

$$\delta_d/\delta = 0.64. \tag{7.13}$$

The parietal concentration of C_W in the border zone is determined by the formula:

$$C_W = \frac{26{,}2 G_x^{-0{,}9}}{U_0} \tag{7.14}$$

and in the final area:

$$C_W = \frac{(G_x/0.55)}{U_0 \delta} = 1.81 G_T \tag{7.15}$$

where G_x is the flow rate of the polymer solution injected through the slit. The formula for the distribution of polymer concentration across the boundary layer in the final region is given:

$$\frac{C(x,y)}{C_W} = \frac{1}{\exp(0.695(y/\delta)^{2.15})}. \tag{7.16}$$

In Ref. [498], as in Ref. [581], four regions of the diffusion boundary layer are considered, which are determined by the same scheme. The diffusion of polymer additives in the initial 1st diffusion region is considered. For this, a diffusion scale L is introduced, which shows the distance for which the concentration maximum decreases e times from its initial value. Using this scale, parietal concentration can be expressed as follows:

$$C_W(x) = C_0 \exp(-x/L_0), \tag{7.17}$$

where

$$L_0 \approx 0.1 \cdot u_c h^2 / D \tag{7.18}$$

or

$$L_0/h \approx 0.1 (u_c h/v)(v/D) \tag{7.19}$$

Here x is the distance from the beginning of the slit, u_c is the injection velocity, C_0 is the initial concentration of the solution, h is the thickness of the laminar sublayer, D is the diffusion coefficient for the viscous sublayer, when $h \cdot V_*/v = 5$ and $u_c/V_* = 5$,

$$L_0/h \approx 2.5 \cdot v/D. \tag{7.20}$$

The scale L_0 in the region of the viscous sublayer, where $y^+ = y \cdot V_*/v < 2.5$, is determined by the value of the coefficient D, which can be calculated by the formula:

$$D/v = 0.00032 (y \cdot V_*/v)^4 \tag{7.21}$$

In this case, v/D can reach about 10^2. This suggests that diffusion from the viscous sublayer is not laminar (molecular), but many times more intense, although the velocity profile value hardly differs from the laminar velocity distribution:

$$U/V_* = y \cdot V_*/v. \tag{7.22}$$

Therefore further, at high polymer flow rates, L_0 can be of the order of only a few centimeters.

In the same work, the diffusion process in the near-wall region in the intermediate zone is considered. To do this, a statistical approach is used, with the help of which the model of the diffusion of an ensemble of certain particles dumped into the logarithmic part of the boundary layer is studied, that is, the flow of a polymer solution through a slit is simulated:

$$C_W(x) = \frac{Q}{1.45 \bar{z} U(\bar{z})} \tag{7.23}$$

where $\bar{z}$ is the average value of particles in a given cross-section behind the place of injection of the polymer solution. Value $\bar{z} = 0.76 l_{lg}$, and $l_{lg} = 0.076 x^{0.8}$ in the case when the diffusion layer is immersed in a logarithmic sublayer, that is, $y/\delta < 0.15$.

In Ref. [430], the pattern of distribution of polymer additives along the wall was investigated. The process of diffusion of the polymer WSR-301 is investigated on a flat plate 600 mm long, placed in a channel with a water velocity of 10.65 m/s. The polymer solution was injected into the boundary layer using a special slit device providing a zero angle between the direction of the jet and the plane in which the plate surface lies, which ensured minimal flow disturbances in the region of the boundary layer. Samples of the solution were taken from the side of the wall. The dimensionless thickness of the selected layer was

$$y^+ = (2q_s/v)^{1/2} = 9.27 \tag{7.24}$$

where q_s is the flow rate of the selected solution.

The dimensionless distance is introduced to describe the near-wall concentration $(u_c/U_\infty)^{-1.5} x/S$, where u_c is the injection velocity, U_∞ is the incident flow velocity, x is the distance from the slit, and S is the width of the slit.

The concentration on the wall was calculated by the formula:

$$C_W(x) = 10.79 (u_c/U_\infty)^{1.74} (3/x)^{1.16} C_o^{2.16}. \tag{7.25}$$

The length of the first (initial) zone is determined by the formula:

$$\ell_1 = (2/3)\tau_W/(\mu D_m)\delta_D^3 \tag{7.26}$$

where δ_D is the thickness of the diffusion sublayer, D_m is the molecular diffusion coefficient; or when using the dimensionless thickness of the diffusion sublayer:

$$\ell = (2/3)(v/D_m)(\delta_D^+)^3 \cdot v/V_* \tag{7.27}$$

The relationship of the diffusion layer with a viscous sublayer in the first region is determined by the formula:

$$\delta_D^+ = (D_m/v)^{1/4} (\delta_S^+)^{3/4}, \tag{7.28}$$

where δ_S^+ is the dimensionless thickness of the viscous sublayer. In this case

$$\ell = (2/3)(v/D_m)^{1/4} (\delta_S^+)^{9/4} v/V_*. \tag{7.29}$$

The dimensionless thicknesses of the viscous sublayer and the diffusion sublayer are respectively determined by the formulas:

$$\left. \begin{array}{l} \delta^+ = \delta V_*/v \\ \delta_s^+ = \delta_s V_*/v \end{array} \right\}. \tag{7.30}$$

In Ref. [435], equations of nonlocal matter transport in a turbulent diffusion layer are presented.

In Refs. [576,577], the relationship between the longitudinal distribution of polymers in the near-wall region and injection parameters was studied. The injection parameters include: the initial concentration of the polymer C_0, the relative injection rate $\lambda = u_i/U_\infty$, the incident flow velocity U_∞, and the slit width s. In the experiment, a solution of polyethylene oxide was used. The free-stream velocity was 5.8, 6.6, 7.9, and 10.5 m/s, the slit width was 0.3 and 0.15 mm, $\lambda = 0.4$; 0.5; 0.8; 1.0; the value of the initial concentration is 1×10^{-4}; 2×10^{-4}; 5×10^{-4}; 6×10^{-4}. As a result of the measurements and analysis of the data obtained, the dependence of the concentration of polymer solutions in the near-wall region of the boundary layer on the values of U_0, C_0, $x = x/s$ (x is the distance from the gap downstream) and Reynolds numbers within $1.2 \times 10^7 - 2.1 \times 10^7$ as:

$$\overline{C_W} = C_0 \exp\{-2,3[(M+N)Re_\alpha + \varphi(\lambda)]\overline{x}^{0,75}\} \tag{7.31}$$

here $M = 8.7^{-13}$; $N = -2.3 \cdot 10^{-13}$; $\varphi(\lambda) = 4.9 \cdot 10^7$.

In Refs. [398,399,615], empirical formulas are given for calculating the concentration in various sections of the boundary layer containing, as a parameter, the injection intensity and hydrodynamic characteristics. The experiments were carried out in a hydrodynamic tunnel. Free free-stream velocity $U_{max} = 6$ m/s. For the experiments, we used a polymer (anionic acrylic) with a molecular weight of 10^7-10^8, which has a high resistance to destruction. Injection of the polymer solution passed through a slot in the plate. Downstream, according to the authors, there are zones of polymer diffusion. In the first parietal region, the concentration gradient is very large; therefore it is impossible to accurately determine the value of the time-averaged near-wall concentration $\overline{C}_W$. In the fourth parietal region, the distribution is described with good accuracy concentration according to the formula:

$$\frac{\overline{C}}{C_W} = \exp\left[-0,693\left(\frac{y}{\eta}\right)^\alpha\right] \tag{7.32}$$

where y is the distance normal to the surface, η is the distance at which $\overline{C}/C_W = 0.5$. The constant α depends on the determination of the location (x coordinate) behind the slit downstream and the ratio of the velocities u_{max}/u_s, the concentration of the injected solution C, and the width of the slit η. Velocity u_i is the injection velocity

$$\alpha = f\left(\frac{u_{max}}{u_s}; \frac{x}{h}; C_W\right).$$

The value α takes the value 1.7 for the intermediate region and 2.15 for the final region. Near-wall concentration is determined by the following empirical equation:

$$\overline{C}_W = 0.81 \cdot 10^8 \frac{u_s h}{u_{max} x} C_0 \tag{7.33}$$

where

$$\frac{u_s h}{u_{max}} \overline{C}_W \leq 3.5 \cdot 10^{-8}.$$

It is assumed that by increasing the amount of polymer solution injected into the boundary layer, the initial zone will be expanded in comparison with the intermediate zones. This is important, as the main part the solution will be flooded in a viscous sublayer, where turbulence and molecular diffusion will be low.

The effect of polymer additives of variable concentration on the velocity distribution in the boundary layer was studied in a number of works. So in Refs. [498,614] a relation is given for determining the velocity in the inner part of the boundary layer under the influence of polymer additives. In Refs. [577,578], the effect of polymer additives of variable concentration on the flow characteristics in a TBL was studied.

In Refs. [311,438], the resistance of a flat plate was calculated upon injection of a polymer solution into a boundary layer. In Ref. [311], a simplified idea is used of the possibility of replacing the actual distribution of the concentration of polymer additives along the normal to the plate surface with an effective uniform distribution. The concentration distribution on the wall is given by the formula:

$$C_W(x)\delta(x)U_o = (\alpha_1 q)/(\rho_p q) \tag{7.34}$$

where x is the distance from the beginning of the plate, q, ρ_p is the specific (per unit width of the plate) weight flow rate and density of the polymer solution, α_1 is a dimensionless parameter that characterizes the heterogeneity of the concentration distribution and the longitudinal component of the average velocity in the cross section of the boundary layer. From the experiments it follows that α_1 changes slightly: $\alpha_1 = 2 \pm 0.2$. In the calculations, the value of α_1 was taken equal to 2. It was assumed that polymer additives did not change the magnitude of the velocity defect in the outer region of the boundary layer on the plate. To solve the problem, an integral method is used in which the form of the transverse velocity profile is assumed to be known:

$$u^+ = 1/x \cdot ln(\eta) + B + W(\eta/\eta_\delta) \cdot \Pi/x \tag{7.35}$$

$$u^+ = u/V_*; \quad \eta = V_* y/v; \quad \eta_\delta = V_* \delta/v$$

where $x = 0.41$, B is the additive constant, Π is the Coles parameter ($\Pi = 0.55$), W is the "wake function":

$$W(\eta/\eta_\delta) = 1 - cos(\pi\eta/\eta_\delta). \tag{7.36}$$

Formula (7.36) is unsuitable in the immediate vicinity of the wall. The effect of polymer solutions is taken into account by changing the value of B. In the case of a pure liquid flow, $B = 4.9$.

To determine the friction, the integral relation is used:

$$d(\delta_2^* u^+)/d\eta = 1 \qquad (7.37)$$

$\delta_2 = \delta_2 \, V_*/\nu$; $\eta = U_o x/\nu$; $u^+ = u^+(\eta_\delta) = U_o/V_*$, which is numerically solved together with Eq. (7.36) and the relation for u^+ for $y = \delta$ ($\eta = \eta_\delta$):

$$u^+ = x^{-1}\ln(\eta_\delta) + B + 2\prod x^{-1} \qquad (7.38)$$

under the corresponding initial boundary conditions. As a result of the solution, u_1^+, η_δ, and C_W are determined and the dependence of the local friction coefficient C_f on the dimensionless longitudinal coordinate r is found:

$$C_f = 2\left[u^+(r)\right]^{-2}. \qquad (7.39)$$

Coefficient full friction resistance is determined by the formula:

$$C_f(R) = \frac{2\eta_\delta V^*}{2}\left(\frac{G_1}{u^+} - \frac{G_2}{V^{*2}}\right) \qquad (7.40)$$

where $Re = U_o L/\nu$; $G_1 = 1 + \Pi/x$; $G_2 = (2 + \Pi/\pi + 3\Pi^2/2)/x^2$.

It follows from the calculations that for a fixed flow rate of the polymer solution injected into the boundary layer at the nose edge of the plate, there exists a finite range of Reynolds numbers Re, in which there is a decrease in turbulent friction. The width of this range is greater, the greater the specific consumption of the polymer.

The appearance of the effect of drag reducing of the plate at injecting a polymer solution of a variable concentration is due, according to the authors of [311,438], to the same reasons as in a homogeneous solution. With increasing Reynolds number, the effect of influence increases, and then gradually decreases. The decrease in the effect, and then its complete disappearance, is due to the fact that with an increase in the Reynolds number, the velocity of flow around the plate U_o also increases, while the thickness of the boundary layer decreases. However, a decrease in δ occurs much more slowly. In this regard, the product $U_o \delta$ (y_p, y_o) increases, and the concentration of the polymer solution in any section of the boundary layer decreases. A decrease in concentration in turn leads to a decrease in the effect of polymer molecules on turbulent friction.

In Refs. [7,149,154,248,256,306,308,340,435,469,493,533,538,543,578,582–584,594] various theoretical methods for calculating the boundary layer upon injection of polymer solutions into the boundary layer are developed. To date, a large number of theoretical models have been developed with the help of which various sides of the process of flowing aqueous solutions of polymers are calculated. In Ref. [469], a theoretical model of the flow of polymer solutions in the boundary layer was developed taking into account longitudinal vortex coherent structures. In Ref. [493], the physical mechanism of the development of coils into long chains of polymer molecules was investigated. It is known that the effectiveness of the effects of polymer solutions on reducing the pulsation velocities of the boundary layer and reducing friction depends on the degree of "unfolding" of the macromolecules.

It is known that polymer solutions that reduce flow resistance reduce pressure fluctuations on the wall. In Ref. [190], pressure fluctuations on the wall and integral friction were measured for a TBL, which was modified by the addition of a polymer solution that reduces friction resistance. The measurements were carried out on an axisymmetric model equipped with an insulated cylindrical section with a balance of resistance and placed in a working section of a cavitation tunnel with a diameter of 0.3048 m installed in ARL Penn. State. The data were obtained at a main flow velocity of 10.7 m/s with pure water and with the addition of polymer to water at concentrations of 1, 5, 10, and 20 weight parts per million (ppm). The dimension lessness of the frequency spectra of pressure fluctuations on the wall with traditional external, internal, and mixed flow variables did not allow us to properly evaluate the obtained data. It was found that the mean square pressure fluctuations on the wall linearly change with shear stress. The addition of polymer had little effect on the characteristic timeline of the flow.

In Refs. [214,244], the pressure fluctuations on the wall in a TBL were investigated, which was modified by the addition of a polymer solution that reduces the friction resistance compared to pure water flow. Friction resistance measurements were carried out in the same place under appropriate test conditions. Unmeasuring the frequency spectra of pressure fluctuations on the wall with traditional external and internal flow variables could not adequately analyze the data obtained. It was found that the mean square pressure fluctuations on the wall linearly change with shear stresses. The addition of a polymer solution had little effect on the characteristic timeline of the flow. These properties were used to develop a spectrum of dimensionless pressure fluctuations on the wall.

Over the past decades, pressure fluctuations on the wall caused by a TBL have been investigated. The results obtained were of great practical importance, since pressure fluctuations are a source of noise and vibration in engineering applications, from sonar systems to underwater vehicles.

In Refs. [306,398], the effect of polymer solutions on the mechanism of turbulent friction and drag reduction was experimentally investigated.

In Ref. [256], a drag reduction in a large-diameter pipe was studied. In Ref. [538], an experimental investigation of the effect of injection of a polymer solution on an extended flat plate were carried out in a closed-type hydrodynamic tunnel with a working section 3.6 m long and a cross-section 0.36 m wide and 0.13 m high at an average flow velocity of 0.5 m/s. A plate was installed at the top of the working section. A turbulator in the form of a wire with a diameter of 0.6 mm was placed on the plate in the bow over the entire width. This led to a complete TBL. The injection of polymer solutions with a concentration of 1000 ppm into the boundary layer was carried out through a transverse slit located at a distance of 147 mm from the beginning of the plate. Behind the slit, investigations were performed at distances from the slit $x = 0.254, 343, 483, 597, 737, 1168, 1651,$ and 2108 mm. Using a CCD camera and a pulsating YAG laser, PIV measurements were performed in these sections, which made it possible to construct the longitudinal velocity profiles in universal coordinates, as well as the Reynolds stress in each indicated section. All this made it possible to determine the dynamics of the effect of injection of polymer solutions on the characteristics of the boundary layer with distance from the location of the slit.

Similar investigations were performed in the cavitation tunnel ARL Renn. State [486]. The investigations were performed on a plate with a length of 50.8 cm and a width of 25.4 cm, in which a window with dimensions of 31.75 × 15.24 cm was placed. Three mini-balances with measuring points located at a distance of 107, 221, and 335 mm from the slit for injection of the polymer located at the beginning of the window. Solutions of WSR301 polymers with a concentration of 100, 200, and 500 ppm, as well as N60K at 100, 200, 500 and 1000 ppm were injected into the boundary layer. The drag reduction coefficients were measured on three balances at various polymer concentrations and flow rates of 6, 12, and 18 m/s. Depending on the rates of polymer solutions, the maximum decrease in resistance was 60%. A field of pulsation velocity vectors is constructed over the thickness of the boundary layer as a function of x.

Most researchers studied the technical side of the problem. At the same time, it is known that many aquatic animals have a biopolymer cover on the surface of the skin. This cover has a different structure and different location along the body depending on the swimming speed and lifestyle of aquatic animals (part I Chapter 4, Experimental investigations of the characteristics boundary layer on a smooth rigid curved plate, and Chapter 5, Experimental investigations of the characteristics of the boundary layer on the analogs of the skin covers of hydrobionts). In Refs. [257,305,373,583], various versions of biopolymers and some results of a drag reduction when they are injected into a stream are considered. The effect of fish mucus on a drag reduction of models and fish was also investigated [374].

Despite the huge amount of research in this area, to date, there are still many outstanding questions. Obviously, during the diffusion of polymer molecules in the boundary layer, it remains unclear how to reduce this diffusion and thereby increase the extent along the body of the effective effect of polymer solutions on friction. It is clear that for maximum efficiency polymer molecules should concentrate on the outer boundary of the viscous sublayer. However, there is no clear idea of how to retain polymer molecules in this region of the boundary layer for longer. Investigations were carried out with a slotted and distributed injection of aqueous polymer solutions along the body, as well as with a constant and variable concentration of polymer solutions. However, the questions of the conditions of injection of polymer solutions through the slit, the effects of the slit on the development of disturbances in the boundary layer, the dependence of injection along the body on the shape of the body, and other factors remain unclear. The main question remains unclear: what is the mechanism of action of polymer molecules on the characteristics of the boundary layer and on the physical mechanism of friction reduction. There is a hypothesis that polymer macromolecules deployed under the action of shear stress (Reynolds stresses) propagate in the boundary layer in the form of extended filaments. These filaments are united in associates in such a way that an elastic grid forms, as it were, in the boundary layer, damping vortex disturbances in the boundary layer. At the same time, it remains unclear how quickly a ball can be turned into threads, how to make associates stable formations. The question of the development of CVS in the TBL of aqueous polymer solutions remains completely unexplored. In studying this problem, the basic laws of interaction in the boundary layer of various CVS were not taken into account.

7.5 Experimental equipment and investigation methodology for the combined method of drag reduction

The search for new ways to save energy and methods to drag reduction remains an urgent task of scientific investigations. In Refs. [28,33,43,378], a new method was proposed for drag reduction by controlling CVS using the combined method. The developed methods for control the CVS [57,64,66,71,74,91,104–107,114,119,120,132–135,139] do not affect the entire flow, but only it's ordered vortex structures (CVS). A combined method of drag reduction was found in all high-speed hydrobionts (Part I) [15,43,111,121,123,378]. In particular, for the first time, a combined method was proposed simultaneously acting on the CVS of the boundary layer using elastic coatings and injection of polymer solutions into the boundary layer [25,27,32,59]. Other combined drag reduction methods have been investigated [30,43,55,61,118,363–365,378].

The method of controlling a, TBL using high molecular weight polymer solutions of low concentration, which give liquids non-Newtonian properties, is an effective way to influence hydrodynamic processes in the boundary layer. In particular, such additives reduce the level of turbulence and increase the thickness of the laminar sublayer, which leads to a significant decrease in friction resistance. The effect of drag reduction is also manifested in the buffer zone between the viscous sublayer and the turbulent core.

There are several methods for injecting a polymer solution into a TBL. One way is to use soluble coatings on streamlined objects. Porous surfaces are also used to inject polymer solutions in a TBL. The results of testing a porous surface with uniform injection of polymer solutions on the surfaces of both sides of the plate are given in Ref. [337]. The device of the porous surface is given in Ref. [27].

The most common method for injecting polymer additives into the boundary layer is to inject of the polymer solution through slits. There are a large number of works devoted to the problem of injection polymer solutions through the slit. A number of investigations in this direction have been published in conference proceedings [452,486]. It was also proposed to carry out the injection of various liquids through several gaps [83,84,89,90,92,93,100,103–107,139]. The combined method of drag reduction was also investigated in Ref. [216]. The combined method of drag reduction using the xiphoid tip consists of the simultaneous use of several methods of drag reduction, in particular:

- Using polymer solutions.
- Using the xiphoid tip.
- By means of formation in the boundary layer of the CVS in the form of a system of longitudinal vortices.
- With adjustable elastic coatings.

The following are the results of an experimental investigation of the combined method, which includes the first three methods. The parameters of the first two methods vary, and the parameters of the third method in this work are fixed—the size of the system of longitudinal vortices in the transverse direction is identical in all experiments.

7.5.1 Experimental equipment and measurement techniques

To study the effect of injection of polymer solutions on the characteristics of the boundary layer, V.V. Babenko designed a model of a body of revolution and devices for conducting experiments. In Fig. 7.21 shows a scheme of a model made of organic glass.

The model consists of a tail part, which is made of a cone (1) and a body (2), to which a fairing (3), having the shape of a symmetrical wing profile, is attached. Using two screws (4), the casing (2) is attached to the strain gage (5) mounted inside the pylon (6). The casing (2) is attached with a threaded connection to the cylindrical part of model (7), consisting of two connected sections. You can install one or both sections. When installing two sections, the length of the model with an ogive tip (OT) is 0.415 m, diameter—0.04 m, elongation 10.4. Accordingly, the elongation of models with short sword-shaped tip (SXT) and long sword-shaped tip (LXT) was 11.1 and 11.9. In the front, on the cylindrical section, the nose of model (8) is screwed on, in the front of which one axial and eight inclined peripheral holes are drilled. Conical (9) and nose (10) fairings are mounted on the cylindrical part in front, so that one or two coaxial slots of adjustable width can be fixed. The parameters of the front slit: width—0.8 mm, diameter—29 mm. The nose fairing (10) was made of various shapes and, like the tail fairing, made of plexiglass or metal (brass). Knife (6) has three tubes with a diameter of 10 mm, which are tightly

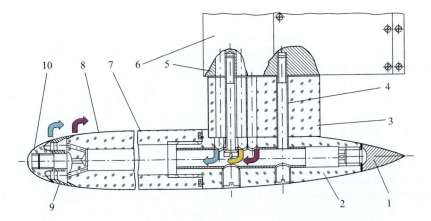

Figure 7.21 Scheme of the model for injection of polymer solutions into the boundary layer: (1) tail fairing, (2) model body, (3) fairing, (4) screw, (5) strain gage, (6) pylon-knife, (7) cylindrical part of the model, (8) nose part of the model, (9) conical fairing, and (10) nose fairing [378].

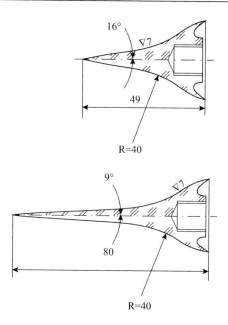

Figure 7.22 Scheme of the nasal xiphoid fairings of two types—short and long.

connected to the holes inside the model. The device of the model provides a uniform fluid or with a different density in the model.

Fig. 7.21 shows the injection scheme of liquids of different densities into the model. The device allows you to mix these fluids inside the model before they enter the slots or provides an independent flow of different liquids into each slot.

In Fig. 7.21, an ogive nose cone is installed on the model. Fig. 7.22 shows the scheme of two types of nasal xiphoid-shaped fairings—short and long. The xipoid-shaped tips are made of plexiglass or brass with a front taper of 9 and 16 degrees. In addition, the nose part of the tips was smoothly matched with the stern, the contours of which on all the tips were the same. All turbulators were square cross-section with a side of 1.5 mm and mated with the inner surface of the nose surface of the model. Turbulizers were installed in all series of experiments. A turbulator with an outer diameter of 20 mm at a distance of 6 mm from the beginning of the model was installed on the ogival tip (OT). Correspondingly, on the short SXT, the outer diameter of the turbulator was 11 mm and was installed at a distance of 25 mm from the beginning of the tip, and on the LXT, respectively, 7.5 and 31 mm. When analyzing the results obtained, it must be taken into account that the intrinsic resistances of the turbulators were not determined and were not subtracted from the total resistance of the model.

The designs of all nasal tips and the front cylindrical part make it possible to inject liquid into the boundary layer in which a system of longitudinal vortices was generated. The length of the ogive and two sword-shaped (xiphoid) tips was 0.02, 0.05, and 0.08 m, and the wetted surface area of the models was 465, 468, and 471 cm^2, respectively. The range of Reynolds numbers was $Re = 3 \times 10^5 - 1.5 \times 10^6$.

Combined methods of drag reduction 475

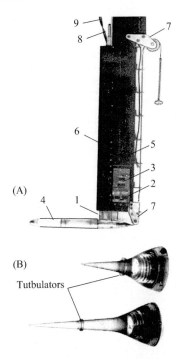

Figure 7.23 Photograph of the model of the body of revolution (A) and the nasal xiphoid tips of the model (B). In (A): (1) fairing; (2) strain gage; (3) resistor element (displacement sensors); (4) model; (5) knife—pylon; (6) front fairing of the knife-pylon; (7) calibration device; (8) a tube for supplying a polymer solution, and (9) electric wires.

Fig. 7.23 shows photographs of the model and xiphoid tips in the working position. Model (4) using a cowl (1) is attached to a strain gage (2), on which the deformations sensor is installed strain gage) (3). The area of the washed cowl surface (1) was 65 cm^2. The strain gage is fixed in the knife-pylon (5) with the front (6) and the tail fairings (the tail fairing is missing in the photo). Calibration equipment (7) is located in the tail of the knife-pylon, which allows calibration before and after the experiment without removing the model. It consists of a removable lower roller, a fixed upper roller, a cable mounted on the rear of the model and passing through these rollers, a loading platform is attached to the cable at the top.

In the nose fairing (6) installed tubes (8) for injection of fluid through the nose model slits and electrical wires connecting the strain gage with amplifier deformation. In addition, it is possible to place DC electric wires (9) inside the model, attached to the nose and tail fairings, when fairings made of brass are installed on the model. The design of the knife (5) made it possible to measure the resistance of the model (4) with the fairing (1), the washed area of which was much smaller than the washed area of the model. Knowing the drag coefficient of the profile that fairing (1) had at different Reynolds numbers, one can determine the resistance value of fairing (1) and subtract it from the total resistance of the model with fairing.

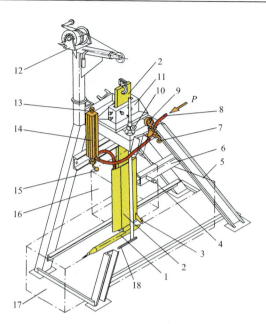

Figure 7.24 Experimental installation for investigation the influence of polymer solutions on the model resistance: (1) model, (2) pylon-knife, (3) calibration device, (4) limit stops for the stability of the pylon-knife, (5) frame, (6) channel for ensuring the strength of the device, (7) an electric crane for supplying air under pressure, (8) compressed air line, (9) manometer, (10) tester for moving a pitot tube, (11) knife fixing unit, (12) crane-beam with a winch, (13) plug, (14) capacity, (15) electric crane polymer solution in the model, (16) shaft for lips models in the working section, (17) the working section of the hydrodynamic pipe, and (18) Pitot tube.

Experimental investigations were carried out in a closed-type hydrodynamic tunnel [504]. The length of the working area is 1.8 m and its cross-section is 0.4 × 0.4 m. The side walls of the working area are made of silicate glass. The maximum flow velocity is 3 m/s. To eliminate the influence of vibration of the body of the hydrodynamic tunnel on the measurement results, the model was mounted on a specially made frame designed by V.V. Babenko (Fig. 7.24). The frame is made of powerful channels. The model was placed in the working section of the hydrodynamic pipe and fixed using a crane-beam developed by V.V. Babenko and used in experiments on the hydrodynamic stand of small turbulence (Section 3.1). To eliminate the transmission of vibrations from the engine of the hydrodynamic tunnel, the frame base was mounted on rubber sheets, and the frame was additionally attached by two channels to the laboratory wall.

Model (1) is attached to a knife-pylon (2), which is suspended from a crane-beam (12) and then the model with a knife-pylon is lowered into the shaft 16 and then into the working section (17). The upper edge of the knife-pylon (2) is fixed in the block (11). The model is set in a horizontal position along the longitudinal axis of symmetry of the working area. The knife-pylon in the middle part of the knife is fixed by restrictive stops (4) installed in the beam (6) of the frame (5). The stops (4) are necessary to eliminate the vibration of the model caused by the flow.

The experimental technique was as follows. At various flow speeds, the model resistance value was determined using a strain gage. The container (8), the pressure in the system was controlled using a manometer (9). Then, the valve (15) opened, (14) was filled with water and the plug (13) closed. The valve (7) opened, supplying pressure through the pipe supplying pressure to the container (14), at the same time, the electric stopwatch was turned on. After exiting the fixed volume of liquid from the tank, the valve (15) was closed. The liquid discharge time was determined using an electric stopwatch. The resistance of the model at various flow speeds and various water flow velocity's was determined by a strain gage.

Measurement of the model drag was similarly measured by injection through the nasal slots of other types of liquid (polymer solutions of various concentrations) and by installing two xiphoid tips. Before each experiment with polymer solutions, the loading container (14) was washed with water. After each series of experiments, water was discharged in a closed hydrodynamic tunnel and pure water was filled. At the beginning and at the end of each series of measurements, the strain gage was calibrated to check the quality of the strain gage and the absence of the effect of polymer solutions on the characteristics of the main flow. The system of fluid injection into the boundary layer was developed by V.I. Korobov, and experimental investigations were carried out jointly by V.V. Babenko and V.I. Korobov

In addition, the result of the injection of liquid into still water was recorded. Based on the calibration lines according to the readings of the strain gage, the model resistance force was determined, and the dependences of this force on the towing speed were built. Drag force R of the model is determined by the well-known formula (Section 6.5, formula 6.6):

$$R = C_f \cdot qS \qquad (7.41)$$

where C_f is the model drag coefficient, $q = 0.5 \, \rho U^2$ is the hydrodynamic pressure, S is the washed surface area of the model.

Subsequently, the results obtained were constructed in the form of dependences on Re. Fig. 7.23B shows the specific form of turbulators. In Ref. [363] and Section 6.5, a methodology for taking into account various drag components is described in detail. As comparative experiments, these additional drag components were not subtracted from the total drag in order to obtain only the drag values on the cylindrical part of the model. The size of the turbulator can be estimated using the method described in Ref. [378]. The size of the fairing intrinsic drag is 14% of the total drag. If we subtract these values from the obtained curves during the injection of polymer solutions, then their efficiency will increase by 14%. When direct current was applied to the nose (10) and end (1) fairings of the model (Fig. 7.21), the effect of the electrostatic field on the stretching of polymer macromolecules and their efficiency was determined, as well as the influence of MB formed at a certain voltage and current on the drag reduction.

7.5.2 Drag of the model without injection of polymer solutions

Fig. 7.25 shows the results of experiments performed on a model with different types of tips without and when water is injected through slits into the boundary layer. Curves (1)–(3) denote the standard dependences of the drag of a longitudinal streamlined flat plate on the number Re for laminar, transitional, and TBL.

Curve (4) denotes the corresponding dependence of the drag of the model with a OT without a turbulizer and with closed nasal slots, curve (5) is the same as (4),

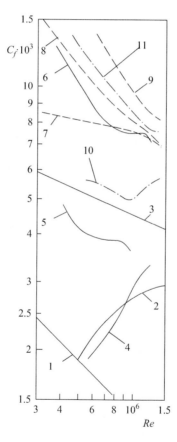

Figure 7.25 Dependence of the resistance coefficient on the Reynolds number of a model with a different shape of the bow: (1)–(3) standard dependencies of the resistance of a longitudinal streamlined flat plate with laminar, transitional and turbulent boundary layers. Resistance of the model with ogival tip: (4) without turbulizer and with closed nasal slots; (5) with a turbulizer and closed nasal slots; (6) with a turbulizer and open nasal slots. Resistance of the model with short sword-shaped tip (dashed lines): (7) without a turbulizer and with open nasal slots when flowing around a highly *aerated* stream; (8) without turbulizer with open slots; (9) with a turbulizer and open slots. Resistance of the model with long sword-shaped tip (dashed-dotted lines); (10) without turbulizer with open slots; (11) with a turbulizer and open slots [378].

but with a turbulizer, and curve (6) is with a turbulizer and open nasal slots. The dashed lines show the data for the model with KMN: (7) without a turbulizer and with open slots in a highly aerated stream, (8) the same, but in a deaerated stream, (9) the same as (8), but with a turbulator. The dashed-dotted lines show the results of test measurements of the model with SXT: (10) without a turbulator with open slots, (11) the same as (10), but with a turbulator.

A model with OT in the further test will be called the "standard" model. Its drag without the turbulizer and with closed slots indicates that the shape and roughness of its surface are such that its drag corresponds to the transition boundary layer on a plate at corresponding numbers Re. If you subtract the drag of the fairing on which the model is mounted, the drag of the standard model will be less than shown by curve (4). Curve (5) indicates that the turbulator on the standard model is not effective enough. However, the combined turbulizer with open nasal slots allows for a turbulent flow in the boundary layer of the model (curve (6), standard).

It is known that the xiphoid tips are effective in the turbulent flow regime; therefore test studies have not been conducted without a turbulizer and with closed slots. The drag of the models with xiphoid tips with a turbulator on the standard model was not effective enough. However, the combination of a turbulizer with open nasal slots allows for a more turbulent flow in the boundary layer than in the standard. Thus the growth rate of drag slows as the length of the xiphoid tip increases. Based on these results, it can be argued that the xiphoid tips in these test conditions can reduce or increase drag compared to the standard.

The growth rate of drag decreased with increasing length of the xiphoid tip. There may be optimal xip tip sizes for drag reduction. With a significant size of the tip, its wetted surface increases, this neutralizes the positive effect. The shape of the cross-section of the tip, as well as the complex effect of drag reduction, possibly plays a certain role. During the experiments, the device of the turbulator on the xiphoid tips was chosen so that when determining the *Re* numbers, the effective lengths of the model with different tips did not differ much. The effective length of the model was determined from the beginning of the turbulator. Given the roughness of the tip of the swordfish, in the future it would be advisable to check the effectiveness of the results obtained by placing the turbulator at the beginning of the xiphoid tip.

Curves (6) (OT), (9) (SXT), and (11) (LXT) denote the results of model experiments with different tips, but under the same experimental conditions. A model with an OT had less drag in the range of experiments Reynolds numbers, and a model with a long xiphoid tip had less drag than with a short tip.

Comparing curves (7) and (8), it is seen that the tip efficiency increases significantly in *highly aerated* flows. A comparison of curves 6 and (7) shows that the model with a less efficient short tip in the aerated stream has less drag than the standard model in a weakly aerated stream.

To determine the scale effect, a series of similar investigations was performed on larger models—Section 6.5. Three types of models with a diameter of 100 mm were investigated. The first view is the head of the model: OT 82 mm long, SXT 141 mm long and LXT 461 mm long. The second type is a short cylindrical part

635 mm long, which is connected to the indicated nose parts of the model, the total length, respectively, was 717, 776, and 1096 mm. The third type is a long cylindrical part 1116 mm long, which is connected to the indicated nose parts of the model, the total length, respectively, was 1198, 1257, and 1577 mm. The elongations of each species, respectively, amounted to: 0.82, 1.41, 4.61; 7.2, 7.8, 11; 12, 12.6, 15.8. In the model investigated above (Fig. 7.21), the elongations are 10.4, 11.1, and 11.9, respectively, which is closer to the third type of experiments given in Section 6.5. In Section 6.5, five series of experiments were investigated (Fig. 6.14), of which, for comparison, Fig. 6.17A shows the results of measurements of the first type (the head part of the model), from which it is clear that the model with the bow OT has the least drag (the wetted surface is the least). The drag of the model with LXT $Re = 7 \cdot 10^5$ had the highest drag coefficients due to the fact that its washed area was the largest. Despite this, at $Re = 10^6 – 2.5 \cdot 10^6$, the lowest C_x in the series with LXT were obtained in comparison with the series 1–3.

Fig. 6.17C shows the $C_x(Re)$ dependences in the investigation of long rigid cylinders of the 11–15 series (Fig. 6.14). The different shape of the nose contours of long rigid cylinders of the 11–15 series practically did not affect the $C_x(Re)$ dependences, in contrast to the 1–5 series. The model with a long xiphoid tip (series 14) and a short OT (series 12) increased the drag of the model in the range of Reynolds numbers $Re = 10^6 – 4 \cdot 10^6$.

Thus experiments on models with large sizes, although allowed us to obtain new physical results, but basically confirmed that the xiphoid tips do not reduce the drag of the model compared to the nose in the form of OT in the absence of fluid injection through the nasal slots.

7.6 Drag of the model with the ogival tip and at an injection of polymer solutions into the boundary layer

Figs. 7.21–7.24 (Section 7.5) show the design of the model for investigation the injection of fluid solutions through the nasal slots for various nasal parts of the model. Fig. 7.26 shows the hydrodynamic drag of a model with an OT type at various flow rates and concentrations of fluids introduced through the nasal slots. Only one front slit was used in the experiments, since the second slit was located close to the first and, when only one type of liquid was injected, the efficiency of injection through the second slit was negligible.

The dashed lines indicate the results of the first series of measurements on the standard model. Curve (4) denotes the dependence of drag on the Reynolds number of the model when a turbulator was installed in the nose part and the slots for liquid injection were open. No fluid was injected through the slits. Curve (5) denotes the law of drag of this model when water was injected into the stream through an open nasal slit. The average water flow rate through the slit was $Q_{av} = 57.6 \text{ cm}^3/\text{s}$. To

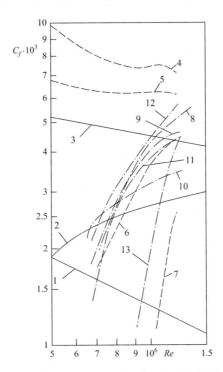

Figure 7.26 Effect of injection of polymer solutions on the drag of a model with ogival tip: (1)–(3) is the same as in Fig. 7.25; (4) with a turbulator, open slots and without fluid injection; water injection at an average flow rate: (5) Q_{av} = 57.6 cm³/s, (8) 71 cm³/s; polymer solution injection at an average flow rate: (6) 0.1%, 84 cm³/s, (7) 0.1%, 100 cm³/s, (9) 0.05%, 54 cm³/s, (10) 0.1%, 59 cm³/s, (11) 0.1 %, 49 cm³/s, (12) 0.15%, 54 cm³/s, and (13) 0.1%, 80 cm³/s [139].

identify the dependence of C_f on u_{sl}/U, the values of the C_μ coefficient of momentum were determined:

$$C_\mu = 2Qu_{sl.}/\rho U^2 S = 2C_q \rho_c u_{sl.} \rho U \tag{7.42}$$

where ρ and ρ_c are the density of the injected fluid and flow, respectively. Given that in this experiment, $\rho \approx \rho_c$:

$$C_\mu = 2C_q u_{sl.}/U \tag{7.43}$$

The average water flow rates through the slots in each series of measurements are presented in Table 7.2.

The pressure in the solution injection system was not controlled to ensure the injection rate in each series of experiments so as to be such that $u_{sl}/U \approx 1$. Constant pressure was applied to the injection system, and the flow rate in the hydrodynamic pipe was varied. The air was pumped into a special container, from

Table 7.2 Parameters of water injection through the slit.

N	U, m/s	u_{sl}, m/s	u_{sl}/U	Q, cm³/s	$Cq \cdot 10^3$	$C_\mu \cdot 10^3$
1	1.19	1.61	1.335	55.7	1.07	2.9
2	1.29	1.69	1.31	58.5	1.01	2.65
3	1.82	1.66	0.92	57.4	0.71	1.31
4	2.52	1.67	0.67	57.6	0.51	0.68

Table 7.3 Parameters of injection of polymer solutions.

N	U, m/s	u_{sl}, m/s	u_{sl}/U	Q, cm³/s	$Cq \cdot 10^3$	$C_\mu \cdot 10^3$
1	1.04	2.16	2.08	74.7	1.61	6.7
2	2.02	2.68	1.33	92.6	1.03	2.74
3	2.48	2.38	0.96	82.0	0.74	1.421
4	2.90	2.38	0.82	82.0	0.631	1.03

which, when its pressure reducing valve was opened, it entered the system for injection of fluid through the slot under pressure (in Fig. 7.24, Section 7.5, position (8) shows the compressed air line). Further experiments were carried out according to the method described above. The pressure in the first series of experiments was about 2 atmospheres, and with an increase in flow rate—3 atmospheres. Therefore in each series of experiments with fluid injection, the rate of fluid injection from the slit and the flow rate were approximately the same. To obtain optimal results on drag reduction, it was necessary that the ratio $u_{sl}/U \approx 1$ be satisfied. If $u_{sl}/U > 1$, then the positive effect is more it is determined not by a decrease in the friction resistance in the boundary layer, but by the magnitude of the thrust impulse that occurs during fluid injection at high flow rates through the nasal slit. Therefore in a series of experiments 3 in Table 7.2, this ratio approaches 1, and other results are not optimal for reducing friction resistance in the boundary layer. Thus the Q values correspond to the average of all series of these experiments in Table 7.2. All results obtained should be analyzed based on these reasons and the corresponding Tables. For example, the value of curve (4) in Fig. 7.26 at $Re = 0.9 \times 10^6$ is optimal, other values on the right and left on this curve are not optimal. Therefore comparing other curves, it is necessary to take into account the corresponding values given in the Tables. Comparison of curves (4) and (5) shows that drag of the model decreases at the injection of water.

Curve (6) denotes the law of drag of the model during injection through the nasal slots of an aqueous solution of polyoxyethylene (PEO) at a concentration of 1000 ppm. The average consumption of PEO solution in each series of measurements is given in Table 7.3. The total average fluid flow rate for curve 6 was $Q_{av} = 84$ cm³/s. The average values of the flow velocity in most cases of measurements were approximately the same as in previous experiments (Table 7.3).

Pressure in the injection system of POE solutions was approximately the same as in the previous experiment. However, due to drag reduction in the injection system of polymer solutions, the flow rates Q and the $u_{sl.}$ velocity injection increased.

It is known that polymer solutions change the flow rate in the injection system; it should be taken into account when calculating in specific technological devices and planning such experiments. Already from these data, it is possible to approximately evaluate the effectiveness of the use of polymer solutions for drag reduction. It can be seen that the injection rate from the slot increased by about 40%. The optimal results for drag reduction performed in this series of experiments will be at a flow velocity of 2.48 m/s (series No. 3, Table 7.3). In the previous experiment (Table 7.2), the optimal value was at a flow velocity of 1.66 m/s (series No. 3, Table 7.2). Comparing curves (5) and (6), it can be seen that with an increase in the flow rate of the polymer solution, the drag of the model compared with water injection decreases significantly.

Curve (7) denotes the law of drag under the same experimental conditions as in the case indicated by curve (6), but the flow rate of the POE solution was increased (Table 7.4). The total average flow rate of the liquid for curve (7) was $Q_{av} = 100$ cm³/s.

In fact, the flow rate increased even more and amounted to $Q_{av} = 110$ cm³/s, except for the last series of experiments in which the flow rate decreased sharply. Given this correction when comparing curves (6) and (7), it can be seen that at the same concentration of the polymer solution, but with an increase in flow rate of about 50%, the model drag decreases significantly.

To verify the reliability of the results obtained, hydrodynamic drag of the standard model was measured again. Thus flow rates and concentration of polymer solutions changed. The results of these measurements are presented in Fig. 7.26 by dashed-dotted lines. Curve (8) denotes the law of model drag for water injection through the nasal slots with the average flow rate of $Q_{av} = 71$ cm³/s, curve (9) for the injection of a polyox solution with a concentration of 500 ppm with an average flow rate of $Q_{av} = 54$ cm³/s (Table 7.4), curve (10) when injecting of a preconditioned polyox solution before the experiment with $C = 1000$ ppm and $Q_{av} = 59$ cm³/s (Table 7.5), curve (11) when introducing the same concentration of a solution of polyox prepared 7 days before the experiment with $Q_{av} = 49$ cm³/s (Table 7.6), (12) with $C = 1500$ ppm and $Q_{av} = 54$ cm³/s (Table 7.7), q with the injection of a *synthetic solution Wallpaper glue* with $C = 1000$ ppm and $Q_{av} = 80$ cm³/s. The corresponding average flow rates for each flow speed in the latter case are given in Table 7.9 (Tables 7.8 and 7.9).

Table 7.4 Parameters of injection of polymer solutions.

N	U, m/s	$u_{sl.}$, m/s	$u_{sl.}/U$	Q, cm³/s	$Cq \cdot 10^3$	$C_\mu \cdot 10^3$
1	1.0	3.16	3.16	109	2.241	14
2	2.06	3.16	1.54	109	1.193	3.7
3	2.48	3.22	1.3	111	1.0	2.6
4	3.02	2.06	0.68	71	0.53	0.72

Table 7.5 Parameters of injection of polymer solutions.

N	U, m/s	$u_{sl.}$, m/s	$u_{sl.}/U$	Q, cm³/s	$Cq \cdot 10^3$	$C_\mu \cdot 10^3$
1	1.4	1.57	1.12	54	0.86	1.94
2	2.08	1.58	0.76	53.9	0.58	0.88
3	2.8	1.57	0.65	54.1	0.5	0.66
4	3.0	1.56	0.53	54	0.4	0.425

Table 7.6 Parameters of injection of polymer solutions.

N	U, m/s	$u_{sl.}$, m/s	$u_{sl.}/U$	Q, cm³/s	$Cq \cdot 10^3$	$C_\mu \cdot 10^3$
1	0.98	1.72	1.76	59.1	1.35	4.75
2	2.2	1.71	0.78	58.9	0.6	0.93
3	2.62	1.7	0.65	58.8	0.5	0.665
4	3.0	1.65	0.55	59	0.44	0.485

Table 7.7 Parameters of injection of polymer solutions.

N	U, m/s	$u_{sl.}$, m/s	$u_{sl.}/U$	Q, cm³/s	$Cq \cdot 10^3$	$C_\mu \cdot 10^3$
1	1.0	1.412	0.71	48	1.08	1.54
2	2.24	1.43	0.64	48	0.48	0.61
3	2.68	1.42	0.53	51	0.43	0.45
4	3.04	1.43	0.47	48.8	0.36	0.34

Table 7.8 Parameters of injection of polymer solutions.

N	U, m/s	$u_{sl.}$, m/s	$u_{sl.}/U$	Q cm³/s	$Cq \cdot 10^3$	$C_\mu \cdot 10^3$
1	1.0	1.56	1.56	54.1	1.21	3.8
2	2.0	1.56	0.78	54.1	0.6	1.02
3	2.5	1.56	0.63	53.9	0.485	0.61
4	3.0	1.58	0.53	54.0	0.4	0.425

Table 7.9 Parameters of injection of synthetic glue solutions.

N	U, m/s	$u_{sl.}$, m/s	$u_{sl.}/U$	Q, cm³/s	$Cq \cdot 10^3$	$C_\mu \cdot 10^3$
1	1.08	2.14	1.98	74	1.54	6.15
2	2.05	2.44	1.19	84	0.92	2.2
3	2.48	2.52	1.02	87	0.78	1.6
4	2.98	2.2	0.74	76	0.57	0.85

The preparation of polymer solutions, as a rule, is carried out with prolonged mechanical stirring of the solution with a rubber screw. Previously, when POE powder is poured into water, a phase of its swelling occurs when water molecules are combined with POE molecules. During mixing in a liquid, a uniform distribution of POE (WOX 301 polyox) and a gradual unfolding of its molecules are obtained. The processes of obtaining expanded POE molecules before injection were considered in a number of works [452,493]. In Refs. [88,119,120,133−135] a device was developed for the preparation of polymer solutions. If the POE is well prepared for use, its effectiveness increases. If POE is not immediately applied to the body after preparation, its molecules begin to coagulate, so its effectiveness decreases. To verify this position, a series of experiments were presented, the results of which are shown by curves (10), (11). It can be seen that the injection of the polymer solution prepared before the experiments [curve (10)] showed better results compared to POE solutions, which were used a week after its preparation. The polymer solution prepared before the experiments [curve (10)] at high Reynolds numbers was effective, and at lower Reynolds numbers it was less effective compared to the long-prepared POE solution of the same concentration [curve (11)].

Curve (9) (C = 500 ppm and Q_{av} = 54 cm^3/s) illustrates the injection of the polymer solution at a low concentration and the same flow rate as with the injection of water (curve 5, Q_{av} = 57.6 cm^3/s). It can be seen that the injection of the low-concentration polymer solution is significantly more effective than the injection of water. With an increase in the concentration of the polymer solution [curves (10), (11)], its efficiency increases [compared to curve (9)]. However, an even greater increase in concentration at the same flow rate reduced the effect [curve (12)]. The increase in concentration and flow rate [curves (6), (7), and (13)] made it possible to obtain maximum efficiency in the investigation on this model.

Curves (8), (6), and (13) characterize the experiment with approximately the same flow rates of the injected liquid. When polymer solutions were injected, the drag reduction was significantly higher than when water was injected into the boundary layer. However, during water injection [curve (8)], approximately the same effects were obtained as during injection of a polymer solution with a lower flow rate [curve (12)]. In this regard, it was suggested that, due to the small size of the model and the large flow rates of injected fluids through the nasal slots, one of the known schemes for creating hydrojet thrust is implemented [532]. To verify this assumption, when conducting a series with water injection, measurements were made in the mooring mode (i.e., in calm water), liquid was injected through the nasal slots. Indeed, the design of the models made it possible to realize thrust during the injection of polymer solutions through the nasal slots. Thus with the introduction of polymers, a double mechanism was realized—the creation of thrust and friction drag reduction in the boundary layer due to the specific properties of polymers. It would be worth highlighting each of the components of the positive effect of fluid injection to reduce hydrodynamic drag. To do this, during each series of measurements on the model in the mooring mode (i.e., in calm water), liquid was injected through the nasal slots and the magnitude of the generated thrust was recorded on the oscilloscope. With these measurements taken into account, water injection through the nasal slots at a low flow rate [curve (5)] at low Re numbers does not affect the drag reduction, and when the Re numbers increase, it affects, slightly reducing the resistance to 10%. Exactly the same picture is observed with increasing water flow through the slots [curve (8)].

To determine the reliability of the investigations conducted, tests were conducted with solutions of *synthetic wallpaper glue* [curve (13)], the effectiveness of which was comparable with the effectiveness of POE! When injecting POE solutions, measurements in the mooring mode were not performed, and when injecting the synthetic wallpaper glue solution, measurements in the mooring mode showed approximately the same picture as when injecting water. It is quite simple to determine the role of polymer solutions in reducing the resistance by subtracting the corresponding data in the mooring mode for synthetic glue or water.

It is known that the scheme for implementing thrust by injecting fluid through the nasal slots is quite attractive. However, the scheme has a significant drawback, since the parietal stream of liquid in the nasal part of the body significantly increases the speed near the surface of the body, which leads to a significant increase in friction resistance. Therefore such a scheme did not find application in technology. The results showed that with this method of implementing waterjet thrust using polymer solutions, friction is significantly reduced and, therefore such a scheme can be promising in the future.

Comparing the obtained results with the data of [582], we can conclude the following. In Ref. [582], the POE solution was injected through four slits located in the first third of the body length, and this determined the effect of liquid injection on the boundary layer. Comparison with water injection through the slits was not carried out.

In addition, a number of other conclusions can be drawn from the measurement results. At the same flow rate, the best result was obtained in experiments with a polymer solution at a concentration of 1000 ppm [curve (10)], which indicates the effect of liquid injection on the boundary layer. At the same time, the resistance of the model became approximately the same as that of the model without a turbulizer and with closed slots [Fig. 7.25, curve (4)].

During the experiments, low values of the model resistance were obtained. Certain friction coefficients are significantly less than curves (1) in Fig. 7.26. It can be assumed that the shape of the curves located below curve (1) characterize the influence of the hydrojet formed by this method. However, at increased speeds (Re number), apparently, on all the curves, the contribution of the thrust value to the drag reduction decreased. To maintain the thrust level at a certain level at $Re > 1 \times 10^6$, any additional measures must be applied, for example, to increase the thrust momentum, form the next jet downstream or ballast the jet, that is, increase the concentration of polymer solutions, which is not always economically feasible.

Injection of polymer solutions and synthetic wallpaper glue gives a better result compared to water injection in general and especially at low Reynolds numbers, and the greatest decrease in model resistance is achieved by injection of polymer and synthetic glue [curves (7) and (13)] with a concentration of 1000 ppm (part per million) and large flow rate (respectively 100 and 80 cm^3/s).

At the same low flow rates [curves (9)–(12)] and different concentrations, the best result was obtained in an experiment with a polymer solution at a concentration of 1000 ppm [curve (10)]. Thus the model drag becomes approximately the same as the model without a turbulizer and with closed slots [curve (4), Fig. 7.25]. An increase in the liquid flow rate through the slots gives a greater effect than the choice of the optimal concentration of the liquid solution.

Injection of polymer solutions through nasal slots has led to new patterns for the drag when flowing around a model compared to water injection.

When flowing around a model with an ogival tip (Fig. 7.21, Section 7.5), the structure of the boundary layer was visualized by injecting water and a polymer solution from the slit (Fig. 7.27). Fig. 7.27A, B shows a photograph of the visualization of the flow around the model during injection into the boundary layer of water from the first slot (10), the second slot was closed. The cylindrical part

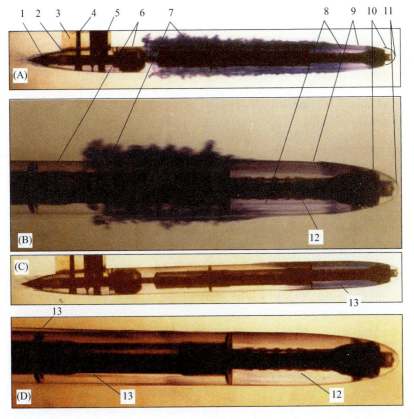

Figure 7.27 (1) cone-shaped fairing of the tail of the model, (2) body of the model, (3) fairing, (4) fixing screw, (5) tubes for supplying fluid, (6) cylindrical part of the model, (7) turbulent boundary layer, (8) longitudinal vortex, (9) the nasal part of the model, (10) the nasal front slit for fluid injection, (11) the nasal ogival fairing, (12) the supply channel, and (13) flow structure during the injection of polymer solutions.

of model (6) consisted of two cylindrical sections connected by a threaded connection and connected in the same way to the case of model (2) this can be clearly seen in Fig. 7.27D. The case of model (2) was attached through a cowl (3) to a tensometric beam located in the knife-pylon.

Tubes for supplying fluid (5) were located in the fairing (3). In Fig. 7.27A, B the structure of the boundary layer of the model is visible. When flowing around the nasal part of model (9), it is seen that the boundary layer is tinted in the form of longitudinal stripes. This shows that the design of the model made it possible to form a system of longitudinal vortices inside the slot chamber. Fig. 3.43 (Section 3.7) shows that if there exist longitudinal vortices in the boundary layer (transition stages V–VII according to Fig. 3.33, Section 3.6) that rotate in pairs, vertical layers of liquid appear between the longitudinal vortices, directed up or down. When in a cross-section a pair of longitudinal vortices rotates toward each other, moreover, the right vortex rotates clockwise and the left vortex rotates counterclockwise, then the vertical layer of liquid located between them moves upward from the streamlined wall (Peak). If, on the contrary, the right vortex rotates counterclockwise, and the left one rotates clockwise, then the vertical layer of fluid located between them moves toward the streamlined wall (Hollow). If the boundary layer in the water flow is visualized using MB of air or using tinted liquid, then in the first case ("Peak"), the dye or MB rush into the faster layers of the liquid and diffuse. Therefore such fluid layers cannot be visualized using these visualization methods. In the second case ("Hollows"), the dye or MB rush to the wall to inhibited fluid layers and accumulate near the wall between a pair of longitudinal vortices. In this case, such fluid layers in the "Hollows" are easily visualized using these visualization methods. In Fig. 7.27A, B these fluid layers are clearly visible in the "Hollows," the distance between which in the cross-section of the model determines the wavelength in the transverse direction λ_z of the longitudinal vortices (8) (Fig. 7.27). As these pairs of longitudinal vortices develop at the VII stage of the transition (Fig. 3.33, Section 3.6) due to an increase in the energy of the vortices, it leads to their zigzag development. And since the size of the vortices and their energy are large (occupy the entire region δ), the process of oscillation of the longitudinal vortices in the z direction passes with a large increase. Therefore at the end of the flow around the nose part (9) of the model (Fig. 7.27), it is seen that the longitudinal tinted strips thicken sharply at a short distance and soon a TBL (7) appears behind the nose (9) at the beginning of the cylindrical part (6) visualization of the TBL when flowing around a rigid plate is also shown Figs. 3.34B and 3.37 (Section 3.6).

When polymer solutions were injected through the anterior nasal slit located in the ogival nasal fairing (11) (Fig. 7.27), the structure of the CVS of the boundary layer at different stages of the transition changed significantly. Unfortunately, the corresponding investigations of the flow structure at various stages of the transition of the boundary layer were not performed using the tellurium method, as described in Chapters 3–5.

On the basis of the investigations performed when flowing around various types of elastic surfaces, in Section 5.4, an analysis is made of the features of the development of disturbing motion in the boundary layer of various elastic plates. The features of the characteristics of the disturbing motion during flow around elastic surfaces are listed that allow us to conclude that with the right choice of mechanical properties of damping surfaces, you can get at least one of the following effects:

- Increase in the Reynolds number of stability loss.
- Increase in the length of the transition zone and the value of the Reynolds number of the transition.
- Reduction of the frequency range of unstable oscillations in the laminar boundary layer.
- Reduction of the phase velocity of the disturbing motion.
- Increase in the wave number and a decrease coefficients gain.
- Reduction of the vertical coordinates of the critical layer.
- Reduction of the kinetic energy of the disturbing motion.

The results of hydrobionic investigations of various types of fast-floating aquatic organisms (Parts I Chapter 4, Experimental investigations of the characteristics boundary layer on a smooth rigid curved plate, and Chapter 5, Experimental investigations of the characteristics of the boundary layer on the analogs of the skin covers of hydrobionts) showed that various species of marine animals have various devices for economical energy consumption and effective drag reduction. In particular, high-speed fish types (sharks and xiphoid—xiphias) do not have methods for drag reduction, which have been studied in the art for the injection of polymer solutions. Their outer skin covers has a layer of mucus, which has similar properties and methods of reducing resistance, as in high-speed species of cetaceans. Slime-forming cells on the surface of the skin and gill slits play a role in reducing the resistance, which they tried to implement in technology by applying coatings to the body made by coating with polymer solutions of high concentration.

The results of investigation of the integral characteristics during flow around elastic surfaces (Chapter 6, Experimental investigations of friction drag) and during the injection of polymer solutions (Chapter 7, Combined methods of drag reduction) suggest that even during the injection of polymer solutions through the model slit, one or all of the features described above in the study of elastic surfaces can exist.

In Fig. 7.27C, D, when injecting polymer solutions on a model with an ogival shape of the nose, the visualization of the nose of model (9) (Fig. 7.27D) did not reveal any streaks characteristic of water injection—a continuous veil characteristic of the laminar boundary layer is visible. In Fig. 7.27D, as the polymer solution propagates along the model, the structures of disturbances wave such as a T.-Sh. wavelength with a long wavelength is also fixed on the remaining surface of the model.

The inner surface of the supply channel (12) is made in the form of a wave surface with a certain waveform [500]. This contributed to a decrease in the resistance in this channel, as well as to preliminary stretching of the polymer macromolecules, which increased the efficiency of the effect of the polymer solution on disturbances of the boundary layer.

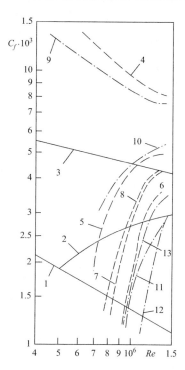

Figure 7.28 Effect of injection of polymer solutions on the drag of the model with short sword-shaped tip (dashed lines) and long sword-shaped tip (dash-dotted lines): (1)–(3) the same as in Fig. 7.26; (4), (9) with a turbulator, open slots and without liquid injection; (5), (10) during injection of water with an average flow rate at $Q_{av} = 51$ cm^3/s; injection of PEO solution and average flow rate: (6) 0.05%, 47 cm^3/s; (7) 0.1%, 34 cm^3/s; (8) 0.15%, cm^3/s; (11) 0.05%, 50 cm^3/s; (12) 0.1 %, 45 cm^3/s; (13) 0.15%, 38 cm^3/s [139].

7.7 Drag of the model with xiphoid tips and injection of polymer solutions into the boundary layer

Experimental investigations were carried out when two types of xiphoid tips were installed in the nose of the model, similar to those given in Section 7.6, Fig. 7.26. The results of these investigations are presented in Fig. 7.28 [68,139]. The dashed lines indicate the laws of the model drag with SXT and the dot-dashed lines indicate the drag with LXT. As on the standard model, mooring tests were carried out to determine the effect of fluid injection on thrust.

It was found that the injection of water through the nasal slots of the model with both xiphoid tips does not affect the reduction of hydrodynamic drag. However, the thrust in mooring mode was found to be greater than for the standard model.

Comparison of the obtained measurement results on models with xiphoid tips with similar results on the standard model allowed the following conclusions.

Despite the fact that the flow rates of injected polymer solutions on models with xiphoid tips were significantly lower than those on the standard model, their drag was less. Only when injecting a solution of *synthetic glue*, the drag of the standard is commensurate with the drag of models with xiphoid tips.

The character of the dependences of the drag model on the Reynolds number is approximately the same for all the models. The drag of the model with a long xiphoid tip was the smallest, and the best result, like that of the standard, was obtained at a polymer solution concentration of 1000 ppm (parts per million). An exception was the results when testing the model with KMN, where the best result was obtained at a polymer solution concentration of 500 ppm.

7.7.1 Friction drag for the injection of polymer solutions into the boundary layer

Given that the water injection had the same result for any shape of the tip, you can compare the results obtained by subtracting the value of the thrust arising from the injection of fluid through the nasal slit during mooring tests. Such a comparison is of a qualitative nature, since thrust was not measured during mooring tests during the injection of polymer solutions. In order to balance the absence of such measurements, the values of the drag coefficient of the model located below curve 1 in all graphs were not taken into account in the calculations. The force value determined during the injection of a solution of synthetic glue during mooring tests was subtracted from the data shown in Figs. 7.25 and 7.26. This clearly reduces the efficiency in the injection of polymer solutions on models with SXT and LXT, but allowed us to determine the main differences compared to the standard tip. Mooring tests do not allow reliable estimation of the value of the friction resistance, since the jet is injected into a stationary fluid. This is the case of underwater near-wall jets. The kinematic characteristics and structure of such a jet differ from the corresponding parameters of the wall jet injected into the boundary layer. However, the data shown in Fig. 7.28 allow a qualitative analysis of the characteristics of the problem under consideration.

Analyzing all the data of an experimental investigation of fluid injection through the nasal slots, it is obvious that after the ratio $u_{uu}/U \approx 1$ is fulfilled, in all the figures, the curves become equidistant to the standard curves (1)–(3) in the absence of fluid injection. This indicates that the influence of the thrust of the nasal jet on the friction resistance at such values of the parameter of the injected jet becomes insignificant in comparison with the influence of the wall jets. We make the assumption that for a given ratio, the difference between the drag coefficients on the control curve and during the injection of polymer solutions is constant over the entire range of Reynolds numbers. Then you can draw the resistance curves during polymer injection, at the same distance from the reference curve in the entire range of Reynolds numbers.

Resistance data are analyzed below when subtracting mooring tests.

Fig. 7.29 shows the calculated resistance data of a model with OH during injection of polymer solutions. These results are interesting for comparison with the data

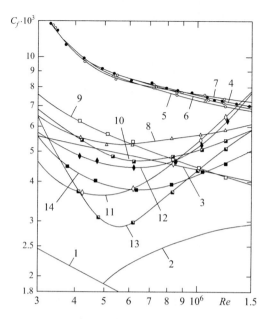

Figure 7.29 The effect of a POE solution injection on the drag of the model with ogival tip: (1)–(3) is the same as in Fig. 7.26; (4) with a turbulator, opened slots and without fluid injection (standard); water injection at an average flow rate: (5) $Q_{av} = 57.6$ cm^3/s; (6) 71 cm^3/s; 7–73 cm^3/s; injection of POE solution with a concentration and average flow rate: (8) 0.05%, 54 cm^3/s; (9) 0.1%, 59 cm^3/s; (10) 0.1%, 49 cm^3/s; (11) 0.15%, 54 cm^3/s; (12) 0.1%, 80 cm^3/s; and (13) 0.1%, 84 cm^3/s; (14) 0.1%, 100 cm^3/s.

shown in Fig. 7.26. It turned out [Fig. 7.29, curve (5)] that water injection does not affect the drag reduction at small Re and slightly decreases the drag at higher Re. The exact same conclusions were drawn with increasing water flow, as shown in Fig. 7.26, curve (8), and in Fig. 7.29, curves (6) and (7).

The data shown in Fig. 7.29 allowed us to draw a number of additional conclusions. Almost all the curves have extrema obtained at $u_{sl}/U \approx 1$. The shape of the curves to the left of the extremum is approximately the same distance from the standard curve (4). With increasing Re, the efficiency of injection of the polymer solution in the model with OT decreases. The only exception is curve (9) in Fig. 7.29 [Fig. 7.26, curve (10)], which characterizes the effect of the newly prepared polymer solution. With the growth of Re, its efficiency even increases. An increase in concentration [Fig. 7.29, curve (11) and, accordingly, Fig. 7.26, curve (12)], and in some cases, polymer consumption [Fig. 7.29, curve (13) and, accordingly, Fig. 7.26, curve (6)] led to the fact that the drag increases compared with the standard at $Re = 1.4 \times 10^6$. The effect was observed for the entire range of Re numbers upon injection of a polymer solution with a high concentration and flow rate [Fig. 7.29, curve (13) and, accordingly, Fig. 7.26, curve (7)]. In the range of numbers $Re = (3.5-9.2) \times 10^6$, the injection of a solution of *synthetic glue* had maximum efficiency and competed with the injection of a solution of a polymer in

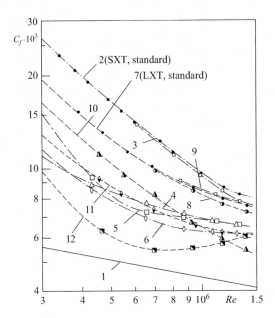

Figure 7.30 The effect of POE injection on the drag of the model with short sword-shaped tip and long sword-shaped tip: (1) is the same as (3) in Fig. 7.28; model with short sword-shaped tip: (2) with a turbulator, open slots and without injection (standard); (3) water injection at an average flow rate $Q_{av} = 51$ cm^3/s; injection of POE solution with a concentration and average flow rate: (4) 0.05%, 47 cm^3/s; (5) 0.1%, 34 cm^3/s; (6) 0.15%, 28 cm^3/s; model with long sword-shaped tip: (7) with a turbulator, open slots and without injection (standard); water injection at an average flow rate: (8) $Q_{av} = 51$ cm^3/s, (9) 70 cm^3/s; POE with concentration and average consumption: (10) 0.05%, 50 cm^3/s; (11) 0.1%, 45 cm^3/s; and (12) 0.15%, 38 cm^3/s [139].

almost the entire range of *Re* numbers [Fig. 7.29, curve (12) and, respectively, in Fig. 7.26, curve (13)].

Fig. 7.30 shows the same calculations for polymers injected on the model with SXT and LXT tips. As in the case of the model with OT, the data in Fig. 7.30 allow us to draw additional conclusions, compared to Fig. 7.28. Injection through a slot of water did not affect the model drag. Unlike OT, most curves do not have an extremum and maximum efficiency is observed in a wide range of Reynolds numbers.

Over the entire range of *Re* numbers, the considered shape of the curves is somewhat equidistant for the control curves (2) and (7). For large *Re* numbers, the efficiency of the injected polymer solutions on the model with SXT and LXT also decreases, but remains positive at any Re. Thus the effectiveness of the xiphoid tips is much higher, despite the fact that Q_{av} is significantly less than on models with OT. Therefore the efficiency doubled at small *Re* values and even more at large Re. In the case of OT, when the efficiency decreases with increasing *Re* and even in many cases disappears; the use of xiphoid tips allows maintaining high efficiency for any Re. The only exception is the model with OT, curve (9) (0.1%, 59 cm^3/s) in Fig. 7.29, when the efficiency is maintained at any number of Re. Similar results were obtained in tests with LXT, curve (10) (0.05%, 50 cm^3/s), but the concentration of the polymer solution is 2 times lower.

The efficiency of a polymer solution injection increases with the increase of xiphoid tip and depends less on Q_{av} [Fig. 7.30, curves (5) and (11)]. Tests with the model with SXT and LXT showed that injection of the polymer solution is most effective at a concentration of $C = 0.15\%$ [curves (6), (12)], but since $Re = 1.2 \times 10^6$, the maximum effect is achieved at $C = 0.05\%$ [Fig. 7.30, curve (10)]. In contrast to OT tests, the difference between the speeds of flow and injected liquid did not very much affect the efficiency of injection of the polymer solution, which was shifted to the region of higher Re values.

The experimental investigations presented here made it possible to determine the effectiveness of one type of the combined method of drag reduction, as well as the hydrodynamic value of the xiphoid tip. Results of the investigation showed that the use of the xiphoid tip, injection of polymer solutions into the boundary layer and the formation of longitudinal vortex systems in the boundary layer is an effective combined method of drag reduction. Unfortunately, in these tests it was not possible to use tips of sufficient length. But this investigation showed that the drag reduction increases with increasing tip length.

The design of the model made it possible to create the optimal longitudinal vortex size for only one flow velocity. Obviously, to increase the efficiency of the combined method of drag reduction, it is necessary to create longitudinal vortex systems similar to natural longitudinal vortices of the boundary layer for the corresponding experimental conditions. For this, it would be necessary to produce additional model details and conduct a separate research cycle. It is known that the size of longitudinal vortex systems in a TBL near a rigid flat plate depends on the flow velocity. In Refs. [60,139,378] it was experimentally determined that the size of the longitudinal vortices in the boundary layer changes during the flow around an elastic surface as compared with a flow around a rigid surface. These results are necessary to control the size of such vortex structures. During the injection of polymer solutions in the boundary layer, CVS are formed, which, as in the case of flow around elastic surfaces, differ from the CVS of the boundary layer of a rigid surface.

The maximum efficiency of this combined method was obtained when flowing around a model with LXT at $C = 0.05\%$ and $Q_{av} = 50$ cm^3/s [curve (10)]. It is also economically viable to use a solution of *cheap synthetic glue*.

In Section 4.6 (part I), structural features of the skin of the swordfish are given, and in Section 5.2 (part I), the interaction of the CVS of the boundary layer with specific structures of the skin of the swordfish is given. Swordfish developed combined control methods for the CVS of the boundary layer during fast swimming. In particular, the skin of the swordfish is elastic with external longitudinal micro folds, and these folds are covered with mucous membrane so that the skin of the swordfish is hydraulically smooth. In addition, a specific gill apparatus produces mucous cells, the mucus of which is dissolved by the oncoming flow and injected into the boundary layer of the body.

The above results showed that the method of injection of polymer solutions in the presence of the xiphoid nose of the model is effective, and with increasing length of the xiphoid tip, the efficiency increases.

In Section 6.6, another feature of the swordfish is modeled, namely a combination of an elastic surface and a xiphoid tip. Have been investigated various versions of the nose of the models, in particular, SXT and LXT, mounted on short and long cylinders, moreover, the diameter and length of the cylinders are significantly larger than in experiments with the injection of polymer solutions. In experiments with elastic cylinders, the advantage of the xiphoid tip is fixed only at large Re numbers. When flowing around a short elastic cylinder with a short xiphoid tip (Section 6.5, Fig. 6.14 series 43), the $C_x(Re)$ dependences showed (Section 6.6, Fig. 6.18 series 43) that the cylinder drag in the range of numbers $Re = 1.6 \times 10^6 - 1.6 \times 10^7$ was less than the resistance rigid plate with a TBL. Fig. 6.17A (Section 6.5, Fig. 6.14, series 21−25) shows the dependences $C_x(Re)$ when investigated long elastic cylinders in the range of numbers $Re = 10^6 - 2 \times 10^7$. Comparison of the drag of long elastic cylinders with long rigid cylinders (standards) showed that only an elastic cylinder with a long xiphoid tip (series 24) had less resistance than a standard in the entire range of Re numbers.

Unfortunately, a more complete modeling of the features flow around of swordfish was not performed, namely, the flow of an elastic cylinder model with a long xiphoid tip by injection of polymer solutions through the nasal gap was not investigated. The range of lengths of the xiphoid tips has not been investigated in order to determine the optimal length of the xiphoid tip.

In Section 6.7, the effect of polymer additives on the friction resistance of elastic cylinders is investigated. Injection of polymer solutions was performed on model V (Section 6.3, Fig 6.2E, Section 6.7, Fig 6.29) with a diameter of 175 mm and a cylindrical length of 0.635 and 1.116 m. The tips of OT, SXT, and LXT had a length of 82, 141, and 461 mm. The resistance of only the indicated parts of the models was measured. Therefore the elongations for short models were 7.2, 7.8, and 11, respectively, and for the long model, 12, 12.6, and 15.8. The elongations of the model during the injection of polymers on a rigid model were 10.4, 11.1, and 11.9, respectively.

Ring slot was located in the nose of the model when investigated the injection of polymer solutions on models with elastic surfaces located on the cylindrical part of the models, so polymer solutions first flow around the head rigid part of the model and then hit the cylindrical part with an elastic surface placed on it. It was recorded that the effects of the elastic surface and injection of polymer solutions are summed up, moreover, the drag reduction on the elastic cylinder was greater than on the rigid cylinder. Investigations on a rigid model at the injection of polymers were performed in the range of Reynolds numbers $Re = 0.5 - 1.5 \times 10^6$, and at flow around elastic cylinders during the injection of polymer solutions at $Re = 0.1 \times 10^6 - 2 \times 10^7$.

The sizes of the models in these experiments on the small model (Fig. 7.21, Section 7.5) and the large model (Sections 6.5−6.7), respectively, and the ranges of Reynolds numbers were significantly different, however, both methods of reducing the resistance with xiphoid tips and injection of polymer solutions were effective. The combined method in the form of elastic surfaces together with xiphoid tips and injection of polymer solutions significantly increased the effect of drag reduction compared to the experiments individually by each method, as well as with experiments on flowing around a hard surface during injection of polymer solutions.

Based on the results of experimental investigations shown in Figs. 7.25–7.30 (Sections 7.5–7.7) during the injection of polymer solutions for each Re number, the drag reduction efficiency, expressed in%, is determined by the formula:

$$\Delta C_f, \% = 100 \cdot (C_f - C_{f\text{inj.}})/C_f \qquad (7.44)$$

where C_f is the coefficient of drag without injection and $C_{f\ \text{inj.}}$ is with injection. Fig. 7.31 shows the results of efficiency during the injection of polymer solutions onto a model with OT, and in Fig. 7.32, with SXT and LXT [62,77].

According to Fig. 7.31, for a model with OT ΔC_f is a function of the concentration C of the injected polymer solution. The best results correspond to curve 5 at $C = 0.1\%$ and the average flow rate $Q_{av} = 59$ cm^3/s: the drag reduction is 40% in the entire range of the investigated Re numbers. The highest efficiency [curve (10)] at $C = 0.15\%$ and $Q_{av} = 54$ cm^3/s (low flow rates) was recorded at $Re = 4.5 \times 10^5$. With an increase in the Re number, the efficiency decreases to zero at $Re = 1.5 \times 10^6$.

The same results were obtained by injecting a solution of *synthetic glue* [curve (9)], but at the highest flow rate $Q_{av} = 84$ cm^3/s. At low flow rates and lower concentrations of the injected liquid [curves (4), (6)], although the efficiency decreases, it remains on the order of 20% at $Re > 1.5 \times 10^6$. With an increase in flow rates [curves (7), (8)], efficiency increases. The difference depends not only on the average flow rate, but also on the preparation time of the solution at the

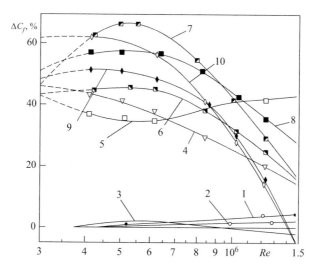

Figure 7.31 Efficiency of injection of polymer solutions of a model with ogival tip: (1)–(3) water injection at average flow rates $Q_{av} = 57.6$; 71; 73 cm^3/s; PEO injection concentration and average consumption: (4) 0.05%, 54 cm^3/s; (5) 0.1%, 59 cm^3/s; (6) 0.1%, 49 cm^3/s; (7) 0.1%, 100 cm^3/s; (8) 0.1%, 80 cm^3/s; (9) 0.1%, 84 cm^3/s; and (10) 0.15%, 54 cm^3/s [139].

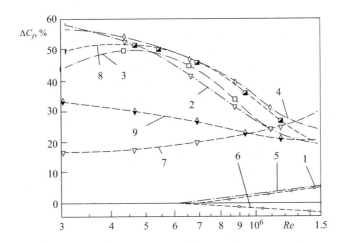

Figure 7.32 Efficiency of injection of polymer solutions of the model with short sword-shaped tip and long sword-shaped tip. Model with short sword-shaped tip: (1) water injection at an average flow rate $Q_{av} = 51$ cm^3/s; injection of PEO solutions concentration and average consumption: (2) 0.05%, 47 cm^3/s; (3) 0.1%, 34 cm^3/s; (4) 0.15%, 28 cm^3/s. Model with long sword-shaped tip: (5), (6) water injection at an average flow rate $Q_{av} = 51$ and 70 cm^3/s; injection of PEO solutions concentration and average consumption: (7) 0.05%, 50 cm^3/s; (8) 0.1%, 45 cm^3/s; and (9) 0.15%, 38 cm^3/s [139].

same concentration. The preparation of the polymer solution immediately before use is preferable in comparison with its preparation 7 days before the experiments.

In the model with SXT, the change in ΔC_f was practically independent of C (Fig. 7.32), while in the model with LXT, ΔC_f was a function of C. The decrease in C_f for all variants of nasal tips is most significant at $C = 0.1\%$, although the best results were obtained for the model with LXT starting at $Re = 1.4 \times 10^6$ at $C = 0.05\%$ [curve (7)]. Despite the smallest Q, the best results were obtained for a cylinder with SXT, and in some cases with LXT [curve (8)]. With increasing Re, ΔC_f decreases. It was impossible to conduct investigations at $Re > 1.5 \times 10^6$. This decrease was due to a change in the ratio u_{sl}/U. With a decrease in Re numbers, the ratio becomes optimal and commensurate with 1. With the growth of Re, u_{sl}/U became nonoptimal, since u_{sl} remained practically unchanged with a change in U.

Comparison of the results of the model's effectiveness with OT and with xiphoid tips allowed us to obtain the following main conclusions. The efficiency of drag reduction in the model with OT turned out to be higher to the numbers $Re = 1.4 \times 10^6$. As Re numbers increase, the efficiency of the model with OT essentially decreases, and in some cases, the drag increases. The maximum efficiency was in the range of numbers $Re = 4 \times 10^5 - 10^6$.

Curves of efficiency of the model with xiphoid tips had no expressed extremes in comparison with the model with OT. Thus high efficiency was observed in the whole range of investigated Re numbers, which remained almost constant, in contrast to the model with OT, and at $Re \geq 1.5 \times 10^6$. The advantages of the model with xiphoid tips were that, up to the numbers $Re = 10^6$, a large nasal thrust was also formed.

In Sections 7.2 and 7.3, and 7.5−7.7, the results of investigations by various authors of the hydrodynamic value of the xiphoid tip to drag reduction are given, on the basis of which can do the main conclusions:

- For a body of revolution of a cylindrical, spindle-shaped or other shape, having a xiphoid tip of various lengths, may slightly about 8%, decrease or increase drag.
- The measurement showed that in the presence of a xiphoid tip, the pressure distribution over the body changes significantly in the area of the tip location, while pressure gradients decrease.
- A xiphoid tip mounted on a poorly streamlined body (a sphere) reduces drag by 2−3 times.
- Water injection through ring slots located in the nose part of a cylindrical body of revolution with a xiphoidal tip, where the maximum pressure distribution is located, creates a slight additional thrust.
- If polymer solutions or other similar liquids are injected through this slot, depending on some parameters, a significant additional thrust arises, which remains significant in the absence of xiphoid tips, but at lower Reynolds numbers.
- The presence of xiphoid tips has a double function: with lower Re numbers, significant nasal thrust arises, and with an increase Re numbers the additional thrust disappears, but a constant drag reduction of the order of 20%−30% is maintained under the chosen experimental conditions.

The last conclusion was obtained by considering the results obtained with the injection of polymer solutions, when only the drag reduction component was considered.

7.8 Physical mechanism of the influence of xiphoid tip on drag reduction

The results of theoretical and experimental investigations so far have not led to an understanding of the physical mechanism of the impact of the xiphoid tip on drag reduction. It is known that there are various components of the resistance of a body moving in a fluid. The greatest contribution to the overall drag under certain traffic conditions is made by the shape resistance. Therefore when moving on the surface of ships, the optimal effect is obtained by the optimal shape of the body, for example, to reduce wave resistance. When bodies move under water, friction resistance is also of great importance. In any case, however, it is first necessary to determine the optimal shape of the body, which gives the greatest effect.

Based on some features of the movement of high-speed hydrobionts (Part I, Section 2.7) and the results obtained in the previous sections, we propose a hypothesis on the physical mechanism of the influence of the xiphoid tip on drag reduction.

There are two main methods of movement of hydrobionts: undulation and scombroid [331,485,488]. The first method of movement is characteristic of slowly swimming animals and is characterized by a large constant in amplitude propagating propulsive wave along the body. The second way of movement is characteristic of high-speed hydrobionts, in which a propulsive wave with small variable amplitude and a high frequency moves along the body, and the body moves only through the tail fin (part I, Chapter 2, Modeling of a waving fin mover).

However, the theory of movement of aquatic animals still does not take into account some features, in particular, the operation of the gill apparatus, the gill flaps of which are mobile and regulate the flow rate of fluid injected through a developed gill slit or system slits. In Ref. [477], a hypothesis was put forward on the analogy of the gill apparatus and flap. Indeed, when flowing around the head of fast-swimming fish with a specific body structure [477,502], a separation of the boundary layer can occur, as well as on flaps at large angles of attack [436,501]. In this section, the role of the gill apparatus in the scombroid mode of motion is experimentally investigated, and a hydrodynamic approach to the investigation of the mechanism of flow around aquatic animals from the standpoint of the laws of separated flows is proposed.

Fig. 7.33 shows photographs of visualization using the tellurium method of a flow of the cylinders located across a stream. Visualization using the tellurium method [378] made it possible to detect the difference in the flow around the cylinder not only depending on Re, but also on d. Fig. 7.33 shows the streamlines around the cylinder with $d = 6$ mm at $Re = 1 \times 10^3$ and at $d = 15$ mm at $Re = 1.8 \times 10^3$ and 2.5×10^3.

Fig. 7.34 shows graphical copies of the flow around an infinite cylinder located across the stream according to Dryden's measurements [201–203]. The Reynolds numbers in Fig. 7.34 are calculated from the cylinder diameter d. At $Re < 1$, the flow around the cylinder is continuous, and at $Re > 10$, the laminar flow separation

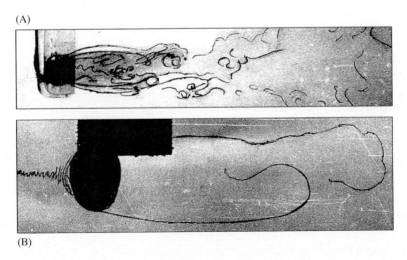

Figure 7.33 Photograph of the visualization by the tellurium method of flow around cylinders with a diameter of 6 mm (A) and 15 mm (B) located across the flow [378].

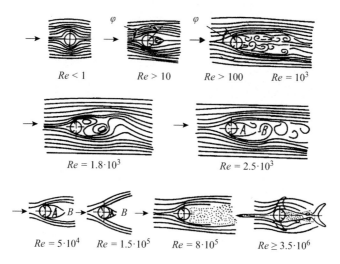

Figure 7.34 Flow around an infinite cylinder at different Reynolds numbers [45].

off with the formation of a zone of weak vorticity. The dashes inside the cylinder and the angle φ indicate the separation zones. A further increase in Re leads to the formation of two symmetric vortices, which increase in size with the growth of Re and begin to come off at $Re > 100$, began to separate, forming the Karman street. At $Re = 1.8 \cdot 10^3$ a Karman street was formed immediately behind the cylinder, and with an increase in the Reynolds number it shifts to a more remote region B, and zone A is formed near the cylinder, free of Karman vortices and filled with return flows. For smaller d, even at $Re = Re = 1 \times 10^3$, a flow pattern is obtained that similar that at $Re = 2.5 \times 10^3$. The difference was in the vortex size and their intensity. These results are consistent with measurements of the dependence of the bottom pressure coefficient on the diameter of the streamlined cylinder [512]. An increase in the Reynolds number leads to an increase in the size of zone A, which then decreases, while the size of zone B is constantly increasing. At $Re > 2 \times 10^5$, zone A practically disappears, and the Karman street is destroyed and behind the cylinder the area of turbulent flow ($Re = 8 \times 10^5$) forms.

In the range $Re = 10^3 - 10^5$, laminar flow separation occurs at $\varphi = 80-85$ degrees. The transition of the laminar flow regime to turbulent in the boundary layer occurs beyond the separation zone. However, if at $Re > 10$ the area of transition was beyond the attachment point of the separated streamlines, then with an increase in the number of Re it moves against the flow. In the range $Re = 2 \times 10^5 - 3.5 \times 10^6$, the separation of the flow remains laminar, but the transition immediately after the separation. Then, the boundary layer joins the cylinder and repeated separation of the TBL occurs at $\varphi = 110$ degrees, so on the surface of the cylinder between these angles φ a "bubble" forms in increasing the bottom pressure (decreases the negative pressure behind the cylinder) and increases the drag coefficient C_f.

At $Re > Re > 3.5 \times 10^6$, the "bubble" disappears, area B decreases significantly due to the fact that the transition in the laminar boundary layer occurs in front of the separation zone. The flow in the boundary layer, as well as the separation, becomes turbulent. At $Re = 10$ and 10^6, the separation of the boundary layer occurs at the same φ, but the resistance coefficients in both cases differ by two orders of magnitude [501] due to the difference in the bottom pressure coefficients [512]. It is known that C_f is minimal for a cylinder for $Re > 5 \times 10^5$.

The hydrodynamic characteristics of various species of aquatic animals are given in Ref. [481]. Given the principle of receptor regulation (Part I, Section 1.2 [81]), the evolution of aquatic animals can be represented from the standpoint of separated flows. The energy potential of the protozoans did not allow them to overcome the boundary of the Reynolds numbers calculated from the length of the body $Re_L = 1$, and, therefore the form of their bodies consist of either spherical or cylindrical elements. Plankton moves at $Re_L = 10^0 - 10^2$. From Fig. 7.34 and [187,512] it follows that in this case the rarefaction and degree of turbulence increase behind the cylindrical body. Increased loads in accordance with [180] led to lengthening of the body. Indeed, the results of experiments in which was vortex shedding is eliminated by placing a thin partition along the longitudinal axis of the cross section of the cylinder showed a significant decrease of bottom pressure and C_f [512].

If we calculate the Re numbers calculated by the cylinder diameter in Fig. 7.34 and the Re numbers determined by length, we get the Re numbers at which various species of aquatic animals move. Fish move at $Re_L = 10^3 - 10^7$, and individual species move at $Re_L \approx 10^8$. Comparison of the streamlines when flowing around the cylinder (Fig. 7.34) and the forms of the fish's body [502] showed that the area occupied at these Re numbers by a cylinder inhibited by a liquid in front of it and zones A, B, is similar to the form of the body of fish. For an illustration, at $Re \geq 3.5 \times 10^5$, Fig. 7.34 shows the contours of the body of a swordfish on an appropriate scale. Apparently, the nonstationary flows around and the presence of the Karman Street behind poorly streamlined bodies during evolution in order to reduce energy losses and overcome the velocity barrier led to the appearance of flexural-oscillatory motion of the body or its tail in accordance with the speeds and numbers Re, that is, led either to an undulating or to a scombroid way of movement, and the form of the body [502] became similar to the modern best technical profiles [436,501].

It is known that during cavitation in the flow around the body behind the cavitator (behind the nasal obstruction of various shapes), a separation arises within which the moving body is located. Thus the drag becomes minimal and is determined only by the drag of the cavitator. This principle of motion actually corresponds to the indicated hypothesis, when in aquatic animals during evolution to reduce drag the body was located in the separation region behind the nasal part of its body.

At $Re_L \approx 10^8$, the evolution of the body form is again determined by separated flows. With such Re numbers, xiphoid fish move: swordfish, sailfish, marlin and spearfish. Fig. 2.8 (Part I Section 2.3) shows photographs of some species of fast-swimming xiphias (xiphias) fish [139]. When analyzing these photographs, it must be taken into account that they are made on dead aquatic animals. Therefore their body form does not quite correspond to living aquatic animals. In addition, the scale of these photographs must be taken into account. The swordfish sample had a total length of 1.6 m, marlin: about 2.5 m, sailboat—about 2 m and tuna—0.85 m.

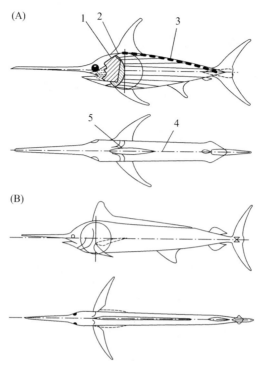

Figure 7.35 Scheme of body forms of swordfish (A) and marlin (B): (1) gill cover; (2) crypts on gill covers for mucus generation (Part I Section 4.6, Fig. 4.47); (3) ampoules in the skin (Part I Section 4.6, Fig. 4.45); (4) ampoule slits (Part I Section 4.6, Fig. 4.44); and (5) slits of gills [139].

Fig. 7.35 shows the body form of swordfish [477] and marlin [502]. Given the structural features, the body of these fish can be represented in the form of a small-sized cylinder located across the flow with rounded ends (its cross-section is indicated by a circle). As indicated above, the nasal and tail portions of the fish body occupy stagnant areas relative to this cylinder. The number Re, calculated from the diameter of such a cylinder, was 1.2×10^7 for swordfish.

It was shown in [45,378] that a xiphoid tip on a sphere acts as a turbulator: the C_f of the sphere at $Re = 10^6$ decreased by 2.5 times with the tip and reached values comparable with C_f of the aerodynamic profiles. This shows that the xiphoid rostrum in combination with a steep tip allows you to turbulize the body boundary layer and move the separation point to $\varphi = 110$ degrees (Figs. 3.34 and 35).

To prevent of separation in the confusor part of the body in fish, the specific structure and functions of the gill apparatus [368,477] and the external covers (part I Section 4.6, Fig. 4.39) [55] are used. In particular, it is possible to note the release of mucus inside the gills and immediately behind them [368], which reduces the flow resistance and accelerates its movement inside the gills. The ampoule zone [367] has the same function, preventing separation behind the dorsal fin.

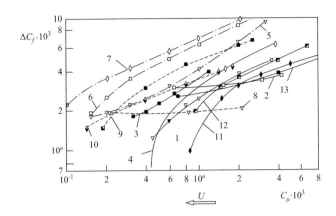

Figure 7.36 Coefficient of models drag as a function of C_μ. Efficiency of the injection of polymer solutions of model with ogival tip: (1) C = 0.05%, Q = 54 cm³/s; (2) 0.1%, 59 cm³/s; (3) 0.1%, 49 cm³/s; (4) 0.15%, 54 cm³/s; (11) 0.1%, 80 cm³/s (synthetic adhesive for wallpaper); (12) 0.1%, 84 cm³/s; (13) 0.1%,100 cm³/s. Model with short sword-shaped tip: (5) 0.05%, 47 cm³/s; (6) 0.1%, 34 cm³/s; (7) 0.15%, 28 cm³/s. Model with LX: (8) 0.05%, 50 cm³/s; (9) 0.1%, 45 cm³/s; and (10) 0.15%, 38 cm³/s.

Another hydrodynamic function of the gill apparatus is to regulate the flow injection through the gill slit depending on the curvature of the body during its oscillation, which forms a longitudinal pressure gradient that affects the stabilization of the boundary layer [621] and the body. At high speeds movement, only the tail part of the body oscillates [331,379,419] and then the jet injected from gills prevent separation from the end parts of the "cylinder" (Fig. 7.34), reduces the role of end effects when flowing around the "cylinder," and carries the function of ending washers known in aviation for the reduction of inductive resistance. Then such a "cylinder" can be regarded as infinite. These properties of the gill apparatus were tested experimentally on a longitudinal streamlined cylinder with a removable nose in the form of an ogival, long and short xiphoid tips (OT, SXT, LXT).

When blowing or suctioning the boundary layer on the wings in the calculation methods, the so-called coefficient of momentum of the blown jet C_μ is widely used [201–203,320]. This coefficient also characterizes the optimal flow rate of a jet and determines the effective speed of a jet at the exit of the slit. To identify the dependence of ΔC_f on u_{sl}/U, the obtained data were determined in the depending on the coefficient of quantity of movement (Fig. 7.36):

$$C_\mu = 2Qu_{sl.}\rho_{sl.}/\rho U^2 S_o = 2C_q\rho_{sl.}u_{sl.}/\rho U \qquad (7.45)$$

where S_o is the area of the cylinder washed by the injected jet; ρ and $\rho_{sl.}$ are the density of the main stream and the injected jet, respectively; $C_q = Q/US_o$ is flow rates coefficient. Considering that in the experiments $\rho \approx \rho_{sl.}$, expression (55) has the form:

$$C_\mu = 2C_q u_{sl.}/U.$$

It is known [436,501,557] that the injected jet is adjacent to the curved surface of the flap (Coanda effect). This leads to an increase in local velocities and introduces additional kinetic energy into the boundary layer, which neutralizes the positive pressure gradient on the flap and ensures its continuous (unseparated) flow. The wing lift coefficient increases rapidly at $C_\mu = 0.01-0.2$, and at $C_\mu = 0.5$, the effect of injected jet significantly decreases [320,501,557]. To maintain the optimum value of the C_μ, the slit is placed in front of the region of maximum vacuum on the flap to ensure sufficient mixing of the jet with the boundary layer.

A similar phenomenon was found in aquatic animals. For example, according to tests in aerohydrodynamic installations [378,632], the gill slit of a swordfish is located in front of the maximum section of the body (Fig. 7.35) in the region of maximum discharge (negative pressure) (Section 7.2, Figs. 7.9 and 7.11). Since the gill apparatus is designed to reduce C_f, and not to influence on lift force (like flaps), the range of optimal C_μ values should be defined.

In Fig. 7.36, the arrow shows the direction of increasing U values. The quantity ΔC_f is determined not in relative form, as in Figs. 7.29 and 7.30 (Section 7.7), but in absolute form:

$$\Delta C_f = C_f - C_{f \cdot inj.} \qquad (7.46)$$

Continuous lines indicate the measurement results for the model with OT, dash-dotted lines for the model with SXT and dashed lines for the LXT. It can be seen that the largest increase in C_f was obtained for lower U, for $u_{sl}/U \approx 1$ in a range $C_\mu = 1-10$. The values of ΔC_f decrease at $c_\mu = 0.1-1$, when $u_{sl}/U < 1$. For comparison with the results on aircraft flaps and to determine the optimal values of C_μ, measurements of ΔC_f should be performed at large values of U for $u_{sl}/U \approx 1$.

The results shown in Fig. 7.36 confirm the main conclusions of Figs. 7.29 and 7.30 (Section 7.7). In addition, SXT and LXT allow one to obtain large values of ΔC_f in the entire range of C_μ compared to OT. The measurements showed the role of the gill apparatus of fast-swimming aquatic animals, the advantages of SXT and LXT compared to OH.

In comparison with the other conclusions obtained by analyzing the data shown in the previous figures, the results shown in Fig. 7.36 showed:

- Xiphoid tips provide efficiency with lower values of C_μ coefficients compared to OT. Thus the values of the coefficients C_μ become closer to the known optimal values during the injection of wall jets on the flaps.
- At the same values of C_μ, the absolute efficiency of ΔC_f in comparison with to the relative efficiency (in percent) turned out to be significantly higher than that of the model with OT in the entire range of measured values of the coefficients C_μ.

Aquatic animals are known to swim at different speeds. Adaptations of the basic devices developed during evolution determine cruising swimming speeds when animals can move for a long time at such speeds. Under extreme swimming conditions, animals move for a short time at the highest possible speeds. In addition, there are slow motion modes. We analyze the developed adaptations for cruising speeds. Obviously, the shape of the fish obtained in the photographs corresponds to the proposed hypothesis: high-speed fish developed such adaptations when, during movement, their body remains in the trace field for a poorly streamlined body. Thus the body is located in the separation zone (negative pressure) due to the fact that additional thrust is realized, and the surface of the body is not flowed around by a high-speed flow, but borders on a section of this flow and a retarded track. Friction resistance during such a flow will be less than when moving in the main stream. The operation of the tail fin mover in an undisturbed flow gives the main thrust. At the same time, a receptor system should work to control the optimal speed of movement. If the speed increases, compared with the optimum, the body leaves the region of the inhibited flow and falls into the conditions of the unperturbed flow, when the resistance to motion increases significantly. Similarly, with a decrease in the speed of movement, the separation zone for a poorly streamlined body becomes isolated on the body. Thus fish, in accordance with the Karman path, increase the amplitude and decrease the oscillation frequency of the tail mover in order to get into the mode of movement with the least losses. Tuna floats at such speeds, which corresponds to our model experiment when flowing around a model with OT. They use effective nasal thrust and a drag reduction when injection solutions of biopolymers up to Reynolds numbers corresponding to our model experiment $Re = 1.5 \times 10^6$.

Xyphoid fish swim at much higher speeds, but their movement mode also complies with the flow conditions shown in Fig. 7.34. Such fish are characterized by a flat upper part of the head surface from the tip to the beginning of the vertical fin, which is a continuation of the flat tip. This is clearly seen in Fig. 2.8A (Part I Section 2.3), a top view. Therefore it is actually necessary to consider the length of the tip to the beginning of the vertical fin. In this regard, in one experiment, the ratios of the body and the xiphoid tip were not fulfilled, and all the known experimental results did not allow obtaining the maximum effect from the use of the xiphoid tip.

The body of the swordfish, when viewed from above, has a rectangular shape, which allows you to increase the moment of application of force from the muscles to the fin. The gill plates are flat and the lines of their trailing edge are straight, so the injected jet simultaneously affects the body and fins. The upper surface of the xiphoid tip is rough and turbulizes the boundary layer. Therefore in our experiments, turbulator in the form of a ring were installed on the tips. Turbulization of the boundary layer shifts the distribution area of the maximum tangential stresses along the body to the tip region. At the same time, this confirms the hypothesis put forward, since the use of a tip on a sphere reduced drag by 2.5 times. The bottom surface of the tip of the swordfish is smooth. At high speeds developed by the swordfish, the lower smooth surface of the tip can play the role of a gliding surface.

It is characteristic of these fish that at high speeds of movement their lateral fins are pressed against the body, which significantly reduces the overall resistance of the body. In the lower part of the body there are three long thin fins that seem to glide along the surface of the separation region according to the scheme shown in

Fig. 7.34. At high speeds, they are also gliding surfaces, and the body with minimal resistance moves uniformly due to the stabilization of the body using the sliding lower surface of the nasal tip and these three thin stabilizers that are located in front of the body. This also confirms the hypothesis.

When regulating the width of the gill slits simultaneously from two sides of the body, the injection speed can be changed and therefore the effectiveness of drag reduction and the magnitude of the nasal trust. By regulating the width of the slit on one side, it is possible to receive the necessary moment when maneuvering in the horizontal plane of movement. If you slightly change the angle of attack of the xiphoid tip, it is possible to receive the necessary moment for maneuvering in the vertical plane of movement. Three thin stabilizers in the lower part of the body (Fig. 2.8, part I Section 2.3) play the same role and at the same time can dampen the magnitude of the moment that appears on the tip when the angle of attack changes. When maneuvering in the horizontal plane or when moving at low speeds, the body of the fish bends either in one direction (maneuvering), or alternately in different directions. With this movement, the width of the gill slit decreases from the concave side of the body, and increases from the opposite convex side. On the concave side of the body, all the prerequisites arise for the formation of Görtler vortices in the boundary layer, which stabilize on the elastic surface of the skin and reduce the friction resistance (Part I Section 5.3, Fig. 5.16). From the opposite convex side, the liquid injected from the gill slits contributes, thanks to the Coanda effect, to continuous flow around the body. These features of the flow around the body of aquatic animals during maneuvering reduce drag and improve maneuvering characteristics. From the opposite convex side, the liquid injected from the gill slits contributes, thanks to the Coanda effect, to continuous flow around the body. These features of the flow around the body of aquatic animals during maneuvering reduce drag and improve maneuvering characteristics.

The body form of the swordfish is such that the center of gravity and the center of pressure of the body are located close to each other. All this allows you to regulating the speed of the swordfish and have the necessary devices for maneuvering during movement. At high speeds, the sailing fish (Fig. 2.8, Part I Section 2.3) reveals its vertical fin in the form of a sail, which also drag reduction. However, the role of the sail is still not well investigated. V.V. Babenko suggested using a similar form of sail for planning the body in water [494–496].

7.9 Experimental investigations of the interaction of the boundary layer with the injected semilimited jet

Despite the variety of results obtained (Section 7.4), the characteristics of the flow around a plate and curved surfaces in the presence of near-wall jets remain poorly studied. Below are the results of an investigation of the influence of a near-wall jet injected from a slot on the kinematic characteristics of the boundary layer of a longitudinally streamlined body, as well as its interaction with a flooded stream in

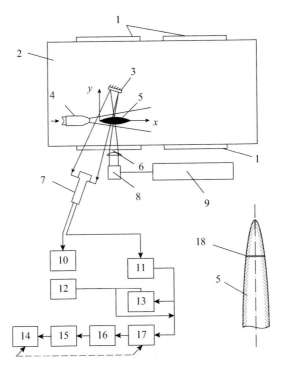

Figure 7.37 Scheme of the experimental installation for investigation of the kinematics characteristics of an injected near-wall jet: (1) glass windows; (2) water tank; (3) mirror; (4) conic caps for formation of a submerged jet; (5) ellipsoids of revolution; (6) focusing lens; (7) photograph receiver; (8) beam splitter unit; (9) laser; (10) power supply unit of the photodetector; (11) amplifier; (12) generator of calibration marks; (13) frequency spectrum analyzer; (14) self-recording potentiometer; (15) integrator; (16) quadratic voltmeter; (17) selective amplifier; and (18) slot for injection of a polymer solution [313,314].

which the body under investigation is located. Experimental investigations were performed by V.P. Ivanov [46,313,314].

In the investigation of the small model (Sections 7.6–7.8), the optimal values of C, Q, and C_μ were determined. For a better understanding of the data obtained and the mechanism of interaction of the semilimited jet with the previously formed boundary layer, we measured the profiles of the longitudinal averaged and pulsating velocities when flowing around an ellipsoid of revolution using a laser Doppler anemometer (LDA) [46,139,313,314]. A description of the experimental setup and experimental conditions are given in Refs. [313,314], and its scheme is shown in Fig. 7.37.

The research was conducted in the tank with dimensions $4.6 \times 1.5 \times 1.5$ m, in which a submerged jet were created flowing with constant speed $U_0 = 3.33$ m/s from a conic nozzle with the angle 15 degrees diameter of the outlet $d_0 = 1.24$ cm. An ellipsoid with axes $x_0 = 20$ cm and $y_0 = 2$ cm was inserted in the jet on distance of 8 cm from the nozzle outlet. The ellipsoid surface area was $S = 98$ cm^2. There

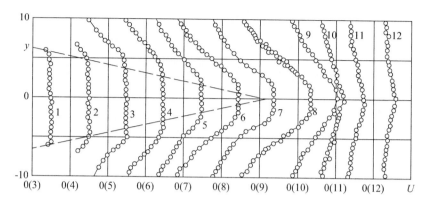

Figure 7.38 Speed profiles in a flooded jet [314].

was a 4 cm long cylindrical segment in the middle part of the. The major ellipsoid axis was aligned with the symmetry axis of the jet. The ellipsoid was fastened on the tail part to vertical fairing of an elliptic cross-section, inside of which there was a channel for feeding a liquid. A liquid was injected through a 0.08 cm wide circular slot cut through under 22.5 degrees to the major axis of the ellipsoid at distance of 2 cm from the nose parts of the ellipsoid. The slot area was 0.21 cm². A liquid was injected through the slot by means a free-flow device allowing regulating the flow rate through the slot within the range of $1-15$ cm³/s. In the experiments, we used solutions of polyox WSR - 301 with concentrations of 0.005, 0.01, and 0.015 weight percent. Polymer solutions were prepared as highly concentrated, kept for 2 days and diluted to the required concentration before the experiments. After the concentration in the main reservoir reached 0.0001%, the water in it was completely renewed.

The profiles of the velocity of the flooded jet were measured in the absence of an ellipsoid of revolution in Sections (1)–(12) corresponding to the x coordinates, spaced from the nozzle exit at distances of 2, 10, 20, 30, 40, 50, 60, 70, 100, 160, 250, and 350 mm (Fig. 7.38). The speed of the jet at the exit of the nozzle (4) (Fig. 7.37) was 2.45 m/s. The x-axis shows the origin of the velocity curves in each section, one division of the grid along the x-axis corresponds to 1 m/s. The y axis represents the distance in mm from the longitudinal axis of the jet. The dashed lines show the potential core of the jet, inside which the velocity is $U = 2.45$ m/s. The length of the initial section is 4.7 d_0, which is consistent with similar data given in [2]. The forward front of the longitudinal velocity gradually decreased in diameter with increasing x and closed at $x = 160$ mm, so that a cone was formed with a parabolic generatrix of the form of its outer surface. The diameter of the cone at the nozzle exit is equal to the internal diameter of the nozzle $d_0 = 12.4$ mm. Up to a distance $x = 30$ mm, the jet diameter decreased slightly.

The ellipsoid was mounted in a flooded jet at a distance of 80 mm from the nozzle exit. At this point, the diameter of the direct front profile of the longitudinal velocity profile of the submerged jet was 8 mm. Since the ellipsoid had a sharp nasal shape, the ellipsoid was still located in the region of the direct front of the velocity profile of the flooded jet, which enveloped the ellipsoid with a shock profile of longitudinal velocity. Thus the experimental conditions simulated a plane-parallel flow around an ellipsoid, the thickness of which slightly changed along the ellipsoid so that the thickness of the plane-parallel flow was equidistant to the variable diameter of the ellipsoid and amounted to 4 mm at the beginning of the ellipsoid. Under these experimental conditions, the flooded jet changed its shape, increasing in diameter, which equidistantly increases along the length of the diameter of the ellipsoid to its mid-section. These data must be taken into account when analyzing velocity profiles measured on the surface of an ellipsoid along x.

Analysis of velocity profiles in a dimensional form on an ellipsoid with an open slot, but without jet injection, showed that in section (1), the profile has the shape characteristic of a flooded jet flowing from the nozzle [314]. Due to the inhibitory effect of the ellipsoid on the outflow velocity of the flooded jet in the region of its axis of symmetry, the value $U/U_0 = 0.96$. In sections (2)–(7) (Fig. 7.38), the shape of the velocity profile becomes characteristic of the flow of a semilimited jet [608]. In section 3 (Fig. 7.38) (at a distance of 0.5 cm behind the slot), the area of the maximum velocity is somewhat stretched along y. In sections (8) and (9) (Fig. 7.38), the velocity profile becomes characteristic of a uniform flow around the body (as in the model).

In a dimensionless form, these results are shown in Fig. 7.39. The numbers indicate the numbers of cross-sections along the ellipsoid in which the kinematic characteristics of the boundary layer were measured. The first section was located at a distance of 4 mm against the flow, and sections (2)–(9), respectively, at distances of 10, 25, 35, 55, 75, 95, 115, 140 mm downstream from the nose part of the ellipsoid. The curves in Fig. 7.39A correspond to: c (solid curves)—without fluid injection from the slot; d (dashed curves)—with water injection at $Q = 13$ cm^3/s; e (dash-dotted curves)—with POE injection at $Q = 5$ cm^3/s and $C = 0.005\%$. In Fig. 7.39 b: g (solid curves)—POE injection at $Q = 5$ cm^3/s and $C = 0.005\%$; h (dashed curves)—POE injection at $Q = 4.5$ cm^3/s and $C = 0.015\%$; i (dash-dotted curves)—POE injection at $Q = 15$ cm^3/s and $C = 0.01\%$.

In section (1), the profile has a form for a flooded jet flowing from the nozzle. In the second section, the effect of the displacement of the flooded jet to the periphery is caused by the placement of a rotation ellipsoid in the jet, as well as the formation of a boundary layer in the presence of a negative pressure gradient in the nose of the ellipsoid. In subsequent sections, the influence of the negative pressure gradient decreases, as does the maximum velocity at the outer boundary of the boundary layer. Starting from the (7) section, a positive pressure gradient in the tail of the ellipsoid begins to influence. Water injection only changes the velocity profile in sections (3) and (4). In the same place, the effect of changes in the flow rate of water through the slot was found. In the absence of water injection, a thin laminar boundary layer is formed on the ellipsoid. The semilimited jet injected through the slot forms a displacement layer, while the thickness of the boundary layer increases. Fig. 7.39 shows that this helps to reduce the shear stress on the wall. An increase in flow rate slightly reduces the thickness of the boundary layer, but at the same time, the velocity in the boundary layer decreases and τ decreases even more on the wall.

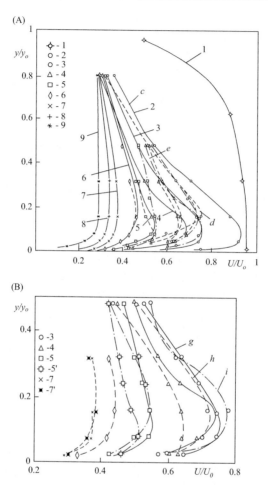

Figure 7.39 Longitudinal averaged velocity profiles measured along an ellipsoid of revolution. The numbers indicate the measurement locations at various x coordinates. (A): c (solid curves)—without fluid injection from the slit; d (dashed curves)—with water injection at $Q = 13$ cm^3/s; e (dash-dotted curves)—with POE injection at $Q = 5$ cm^3/s and $C = 0.005\%$. (B): g (solid curves)—POE injection at $Q = 5$ cm^3/s and $C = 0.005\%$; h (dashed curves) at $Q = 4.5$ cm^3/s and $C = 0.015\%$; i (dash-dotted curves) at $Q = 15$ cm^3/s and $C = 0.01\%$ [314].

A semilimited jet injected through a slit can be represented as an elastic film, since the velocity of the injected jet is greater than the local velocity of the boundary layer. Then the injected liquid will have greater elasticity than the free-stream. In this case, the effect of elastic damping of the transverse pulsating velocities of the boundary layer will be manifested until the moment when the jet velocity becomes equal to the local velocity across the thickness of the boundary layer (Fig. 7.39).

The influence of the polymer solution on the boundary layer is different from the influence of water injection. The polymer solution increases the damping of the transverse pulsation velocities and the extent of the influence along the length of the boundary layer. In this case, the thickness of the boundary layer increases in comparison with the standard and, accordingly, decreases τ. The polymer solution also influence on the flooded jet more than the injection of water. A small displacement layer is formed upon injection of the polymer solution, and the flow rate of the flooded jet increases more than upon injection of water.

Fig. 7.39B shows the results of investigation of the profiles of the longitudinal averaged velocity during injection of a polymer solution with different flow rates Q and concentrations C in sections (3)–(5) and (7). The numbers 5′ and 7′ indicate the series of experiments when the velocity profiles without injection were measured in the same sections immediately after the experiments with injection. Unlike water injection, the influence of the polymer solution on the boundary layer persists until the (7) section.

An increase in the solution concentration at the same flow rates (g, h curves) led to a decrease in τ on the wall, a slight increase in the boundary layer thickness δ and a decrease in the flow rate of the flooded jet Q. An increase in Q at the same concentration (h, i curves) increased δ and τ, therefore under these experimental conditions, it is more expedient to increase the concentration of solution polymer C.

Fig. 7.40 shows the values of the longitudinal pulsation velocity for sections whose designations correspond to the same sections as in Fig. 7.38A. In the first section of Fig. 7.40A, a characteristic distribution of the longitudinal pulsation velocity u'/U_∞ for the flooded jet is obtained, which changes due to the formation of a boundary layer on the ellipsoid. In sections 2–6, the values u'/U_∞ become the same as in Ref. [608] for a hear-wall jet. The difference is that in the boundary layer the ellipsoids u'/U_∞ increased downstream, while decreasing at the periphery of the flooded jet.

Water injection at low flow rates $Q = 4.8$ cm³/s [Fig. 7.40B, curve (1)] led to the fact that the maximum value of the pulsation velocity was observed at large y/y_o coordinates than the standard, and, starting from the outer boundary of the boundary layer (at $y/y_o \approx 0.15$) and up to $y/y_o = 0.4$ the value u'/U_∞ decreased. These differences, as in the measurement of velocity profiles, were observed in sections (3) and (4).

An increase in water flow rates to $Q = 13$ cm³/s [dashed curves (2)] led to the fact that the values u'/U_∞ decreased compared to the standard in a larger range of y/y_o—from 0.05 to 0.4. In addition, the maximum values of pulsation velocities near the wall decreased, and these effects persisted up to section (7). POE injection at low flow rates $Q = 5$ cm³/s and $C = 0.005\%$ [Fig. 7.40B, curves (2)] compared to the standard (Fig. 7.40A) significantly reduced the value u'/U_∞ in a wider range of y/y_o from 0.05 to 0.04, and compared with water near the wall and further to $y/y_o = 1$. An increase in the concentration of the polymer solution to 0.015% at the same flow rate [Fig. 7.40C, curve (2), (3)] led to a decrease u'/U_∞ not only over the entire thickness of the boundary layer, but also to preserve this effect in a flooded jet over a larger the length of the ellipsoid is up to the 8th section [Fig. 7.40C, curve (1)]. If the flow rate also increased, then immediately after the

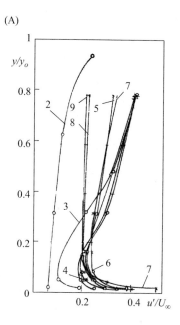
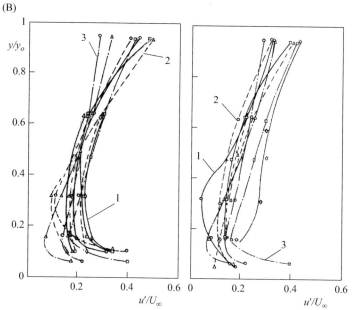

Figure 7.40 Profiles longitudinal fluctuation velocity measured along the ellipsoid of rotation: (A) without injection of liquid from the slot, (B) with injection of water, (C) with injection of polymer solutions. (A) the designations of the curves are the same as in Fig. 7.39A; (B) water injection: (1) $Q = 4.8$ cm^3/s, (2) $Q = 13$ cm^3/s, POE injection: (3) $Q = 5$ cm^3/s, $C = 0.005\%$; (C) POE injection: (1) $Q = 15$ cm^3/s, $C = 0.01\%$; (2) $Q = 4.5$ cm^3/s, $C = 0.015\%$; and (3) $Q = 5$ cm^3/s, $C = 0.005\%$ [314].

slot the value u'/U_∞ significantly decreased (at the outer boundary δ), and in other sections along the length of the ellipsoid the effect slightly decreased compared to growth C, which is consistent with the measurement of velocity profiles (squares and crosses indicate the distribution u'/U_∞ without POE solution injection).

Injection of water contributes to a decrease in the shear stresses on the wall, but since its effect on the averaged and pulsating velocity field rapidly decreases behind the slot, the integral characteristics of the boundary layer practically did not differ from the reference ones. POE solutions reduce transverse [229] and longitudinal pulsation velocities, as well as shear stresses, which leads to an increase in the length of the region of influence of POE on the ellipsoid boundary layer and, due to diffusion, affects the region located outside the external boundary δ.

Another problem was also investigated: the effect of a wall jet injected on an ellipsoid on the characteristics of an immersed jet. The results of the obtained dimensionless parameters and comparison with the data of other authors are presented in Fig. 7.41. Since the shape of the nose of the ellipsoid has a large radius of curvature, the derived laws for increasing the distance from the maximum average speeds along the ellipsoid are similar to the data of other authors for the flow on a flat plate (Fig. 7.41A). Data were also obtained for changing the maximum value of the average velocity along the ellipsoid (Fig. 7.41B). A decrease in the maximum velocity in an ellipsoid corresponds to the laws characteristic of a near-wall jet during transverse flow around a cylinder. It is likely that the energy loss of the immersed jet in our experiment increased due to the influence of the shape of the ellipsoid. In experiments by other authors for a hear-wall jet on a flat plate or cylinder, the energy loss in the jet was caused only by losses due to friction on the streamlined surface.

In experiments, the boundary layer on the ellipsoid developed naturally way. Therefore in the field of measuring the flow parameters, the transition to turbulence occurred only in the last sections along the ellipsoid in which the measurements were performed. It is known that polymer solutions are effective in the turbulent flow regime under tension of polymer macromolecules. Therefore the efficiency of polymer solutions injected on an ellipsoid was slightly different compared to water injection. However, the effect of polymer solutions in the cross-section of the submerged jet also appears in this case.

The following conclusions can be made. When flowing over an ellipsoid, the influence of polymer solutions on the kinematic characteristics of the boundary layer extended to the entire thickness. This led, in particular, to the fact that the form of the profiles of the longitudinal averaged and pulsating velocities (Figs. 7.39 and 7.40) becomes the same as during the flow around the plate. In addition, the velocity of the immersed jet, determined by the profile of the average velocity, increased during the flow around the ellipsoid because the polymer solutions injected through the slot on the ellipsoid extended their influence not only on the characteristics of the boundary layer, but also on the external flow of the flooded jet flowing from the nozzle. These results for the flow around the ellipsoid made it possible to better understand the injection mechanism of polymer solutions in the model. Obviously, the sword-shaped tips change the pressure

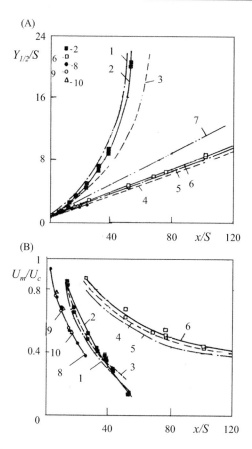

Figure 7.41 Dependence of thickness (A) and maximum speed (B) of near-wall jets on the longitudinal coordinate. (A): laws on a curved wall: (1) Fekete (see in [139]), (2) Wilson, Goldstein [608], (3) Newman [463]; laws on a flat wall: (4) Kruka, Eskinezi (see in [139]), (5) Mayer et al. (see in [139]), (6) Wilson, Goldstein [608], (7) flat free jet — Nakagushi (see in [139]); measurements on an ellipsoid: (8) without injection, (9) water injection with $Q = 13$ cm^3/s, (10) injection of a polymer solution with $Q = 15$ cm^3/s and $C = 0.01\%$. (B): laws on a curvilinear wall: (1)–(3) the same as for laws on a flat wall: (4) Wilson[607], other designations are the same as for a [139].

distribution along the model; therefore the diffusion of the polymer solution over the thickness of the boundary layer decreases. Change the flow velocity of a flooded jet flowing around an ellipsoid indicates that loss in flow decreased. If we apply these results to the flow results on the model, it becomes clearer the appearance of additional nasal traction on the models with an increased flow rates of polymer solutions, as well as an increase in the efficiency of the sword-shaped tips. On an ellipsoid, polymer solutions increase the range of influence on the flow compared to water injections. The same effects are observed when flowing around the model (Sections 7.5–7.7).

7.10 Hydrodynamic peculiarities of the skin structure and body of the swordfish

The results of a study of the structural features of the skin of swordfish showed a significant difference from the skin structure of other species of high-speed fish. An analysis of the data given in Table 7.1 (Section 7.2) allows us to consider some of the hydrodynamic features of swordfish swimming. In hydromechanics, the resistance of immersed axisymmetric cylindrical bodies consists of friction, shape and wave drag. Friction resistance can be 80% of the total resistance of the body. In accordance with the principle of interconnectedness (Part I, Section 1.2) [85,111], when analyzing any system of an organism of hydrobionts, it is necessary to take into account the influence of other systems of the organism on this system. Below are some hydrobionic features of the structure of the skin and organs of swordfish.

In Section 4.6 (part I) and Section 7.8, a diagram of the body structure and some features of the structure of the skin of swordfish are given. Chapters 1–7 present the results of numerous theoretical and experimental investigations on hydrobionic modeling of the functioning of hydrobiont organism systems. In particular, the role of the sword-shaped tip for drag reduction of an axisymmetric body has been experimentally investigated.

7.10.1 Features of fluid injection through gill slits

Experimental investigations have shown that the flow rate of fluid injection through the gill slit in the region of the trailing edge should be commensurate with the velocity of the main flow at this point [247,378]. For this, the device of the gill apparatus should be such that the flow rate during the flow inside the gills does not decrease. The gill apparatus contains cells in the form of holes that generate mucus, which reduces resistance. In addition, the structure of the gill apparatus has a structure due to which the liquid flows from the gill slit in the form of a system of longitudinal vortices. In the boundary layer, the disturbing motion has a similar structure of vortices, which interacts with the introduced vortices from the gills. This stabilizes the disturbing motion in the boundary layer and leads to a drag reduction [247].

The gill apparatus contains crypts that form mucus – Fig. 4.47A (Part I Section 4.6). This helps to equalize the values of the above velocities in the region of the trailing edge of the gill slits, reduces the pressure and shear jump, and increases the stability of the longitudinal vortices created in the gill apparatus. Features of the branchial apparatus are modeled in [83,89,92,93,105–107].

An important feature is the formation of systems in the gill apparatus longitudinal coherent vortices when flowing around three-dimensional deepening's (Part I, Section 4.6, Figs. 4.44 and 4.47A). Such a system of vortices not only stabilizes disturbances in the boundary layer and eliminates shear stresses at the edge of the gill slit, but also facilitates the preliminary preparation of mucus solutions, which consists in stretching the macromolecules in the longitudinal direction. Due to this, solutions of biopolymers become more effective. The mechanism for the formation of longitudinal vortices during the flow around the cavities will be given below.

The third feature is the problem of the interaction of disturbances [53,247]. In the conjugation region of the sword-shaped tip with the head part of the body (Section 7.8, Fig. 7.35A) there is a curved concave section on which longitudinal Görtler vortices are formed. The interaction of the Görtler vortices occurs beyond the edge of the gill cover with the longitudinal vortices generated in the gill apparatus.

7.10.2 Structure of the disturbing movement in the boundary layer of the swordfish skin

The structure of the hydrobionts skin developed under the influence of the disturbing motion of the boundary layer at characteristic swimming speeds. To determine the type and structure of the disturbing motion, we studied the laws of the formation and features of the development of disturbances in the boundary layer on a rigid plate with different velocities and flow regimes [60,139,247,378]. A model for the development of disturbing motion at different stages of its development is shown in Figs. 3.33 and 3.40 [60,139,247].

In Refs. [60,139,247], the features of the development of disturbing motion in the boundary layer on various types of elastic plates imitating the structure of dolphin skin were investigated. It was revealed that the shape and characteristics of disturbances during flow around elastic plates significantly differ from the development of disturbances in the boundary layer during flow around a rigid plate. The more fully dolphin skin is modeled, the greater the difference from disturbances when flowing around a hard surface.

Comparison of the skin of dolphins, sharks and other high-speed hydrobionts made it possible to reveal both differences and general patterns in their structure. This indicates that the structure of their skin coatings developed under the influence of the same environmental force (Part I). The hydrobionts have developed identical mechanisms of receptivity of disturbances of the boundary layer when swimming in the same medium. However, different hydrobionts have a different structure of such receptivity mechanisms. In each shark, the scales are equipped with blood vessels; the scaly cover is elastic and covered with mucus on the outside. Therefore their skin is similar to the skin of dolphins and has elastic-damping properties. With an increase in swimming speed, the structure of the outer cover changes in accordance with the force impact of the environmental. Thus high-speed fish, such as tuna, have a partially scaly cover and a partially elastic cover, the same as dolphins. Scales are located in a zone of stronger environmental influence. In addition, the skin with the highest-speed hydrobionts (swordfish and sailboats) does not have scales, but is elastic damping, like cetaceans. Transversal structures are found in the skin of all high-speed hydrobionts. On the surface of the skin of the dolphin are micro folds. On the surface of the skin of sharks and other fish, similar micro folds formed by crests of scales. Swordfish skin has transverse channels. At the same time, there are structural formations in the skin in the longitudinal direction: dermal ridges and longitudinal rows of the papillary layer of the epidermis in dolphin skin, longitudinal rows of shark crests, and longitudinal folds on the surface of swordfish.

To determine the hydrodynamic role of these microstructures, corresponding calculations were performed [58]. It turned out that during the oscillatory motion of the body during swimming, these transverse structures generate wave disturbances in the boundary layer of the same type as the Tollmien-Schlichting (T.-Sh.) waves. The calculation results for various hydrobionts and various characteristic velocities of motion were placed on the diagram of the neutral curve in the coordinates of the dimensionless fluctuation frequency of the given waves and Reynolds numbers (Part I Section 5.3, Fig. 5.13). For all considered hydrobionts, the disturbances generated by the skin are located in the stability region on the neutral curve diagram [44].

The stability calculations of longitudinal vortex disturbances arising in the boundary layer under the influence of longitudinal micro folds of hydrobionts skin are carried out. When the body oscillates, areas of convexity and concavity are formed on its surface. Calculations of the characteristics of longitudinal vortex systems thus formed in the boundary layer showed that for all hydrobionts such disturbances are located in the stability region on the Görtler neutral curve diagram (Part I, Section 5.3, Fig. 5.16).

7.10.3 Mechanisms of economical consumption of mucus of the skin

Swordfish mucus has the function of drag reduction similarly to polymer solutions, which have long been studied in technology. The difference is that the mucus formed on the surface of the skin of the swordfish is held in longitudinal folds due to the formation of longitudinal vortices on the surface of the body. It is known that in the "Hollows" between adjacent pairs of vortices, the flow is directed to the wall (Fig. 3.43, Section 3.7), as a result a stagnant longitudinal region is formed and diffusion in the boundary layer of aqueous solutions of mucus is significantly reduced. In addition, in the cross-section of the longitudinal vortices there is a large zone with velocity profiles having bends, where the friction on the wall is minimal. Thus stagnant areas are formed where the mucus is not washed away by the flow, which leads to an economical consumption of the mucous substance. Liquid particles move in longitudinal vortices along a helical line. Therefore in places where the flow goes from the surface of the body, the mucous substance is as if stretched by longitudinal vortex structures. The mucus gradually diffuses along the thickness of the boundary layer and, most importantly, remains in the near-wall region. Tests with polymer solutions have shown that polymers effectively grad reduction at low solution concentrations if they are in the laminar sublayer of a TBL.

The process of the formation of mucus and its injection to the surface of the body from the cells forming the mucus is preserved over time. With prolonged and high-speed swimming, the formation of mucus only in the surface layer of the skin is not enough to drag reduction. In this regard, there is an additional system in the skin canals with diverting canals, where the mucous substance develops and collects (part I, Section 4.6, Figs. 4.44 and 4.49, Section 5, Fig. 5.11). Calculations show that the area of the channels of the epithelium, where mucus develops, is comparable to the area of the epidermis and can produce the same amount of mucus.

The mucous substance that collects in the canal cavities enters the skin surface through the pores (part I, Section 4.6, Fig. 4.49, Section 5.2, Fig. 5.11) and is consumed more economically than the mucus located in the epidermis. This is due to the fact that the channel systems of the left and right sides of the fish body are separated from each other. When the body bends, the openings of the pores are reduced on the concave side, so the mucus is consumed on the opposite side. Thus the pore system opens on each side of the body sequentially with the body fluctuation frequency. The special arrangement of channels and pores also plays a role to minimize mucus consumption. The channels are perpendicular to the longitudinal axis of the body and have a small number of anastomoses (cross vessels) between them. This allows you to more economically redistribute the mucous environment in the channels, since the mucus goes to the surface of the skin under the influence of pressure distribution along the body with fluctuations in the movement of its tail.

The pores are distributed fairly evenly over the body (Part I, Section 4.6, Table 4.6), the only exception being on the dorsal fin, where they are located almost twice as often. An important role is played by blood vessels in the skin, which affect the thermodynamic function of the produced mucus. It is known that the degree of dilution of mucus in water depends significantly on temperature. In this regard, the location of the channels (11), (12) in the first layer of the circulatory system and the ampoule (20) between the first and second layers of the circulatory system (Part I, Section 4.6, Fig. 4.49) are very important for the production of mucus. The circulatory system in the skin of sharks (Part I Section 4.5, Fig. 4.34) has the same significance for the production of aqueous solutions of mucus as in swordfish.

7.10.4 Combined drag reduction method

One of the main results of hydrobionic research is the discovered combined methods of drag reduction [15,25,60,139,378]. The previous chapters present the results of experimental investigations of various variants of the combined method of drag reduction. Next, we will consider another version of the combined method of drag reduction, which was discovered when analyzing the structure of the skin of the swordfish (Part I, Section 4.6). However, the physical mechanism of this method has not yet been sufficiently investigated. In particular, we proposed to consider the basis of such a physical mechanism—the method of interaction of various disturbances in the boundary layer. The results of this approach are partially discussed in previous chapters. The combined method of drag reduction also consists in the fact that the circulatory system simultaneously affects the production of aqueous solutions of mucus and the optimization of the mechanical characteristics of the skin.

7.10.5 Methods of stabilization of vortex disturbances

It is known that when flowing around technical objects in angular conjugations, longitudinal large vortex or pairs of vortex arise, as, for example, shown in an experimental

Combined methods of drag reduction

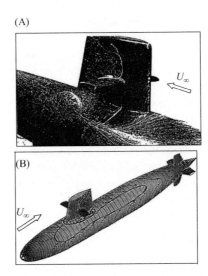

Figure 7.42 Visualization of the vortices in the angular conjugations of the bodies of revolution (A) and the scheme of the paths of the marked particles (B) (Faraokhi, see [139]).

study of the Faraokhi model—see [139] (Fig. 7.42). It is known that when flowing around the angular conjugation of two planes, pairs of large longitudinal vortices are formed. Such large vortex systems increase the drag of the body and reduce the efficiency of the propulsion system. The mucus developing in the epidermis and in the skin channels is not sufficient to stabilize these disturbances. In the skin of the swordfish on the sides and behind the vertical fin, in addition to a large number of pores, there are longitudinal folds (part I, Section 4.6, Fig. 4.43), in the channels between which slots form. Oval holes are located in these slots (part I, Section 4.6, Figs. 4.44 and 4.49), in which there are at least three ducts (vertical channels) connected directly to ampoules located in the skin under the holes (part I, Section 4.6, Fig. 4.49) Thus mucus from the ampoules flows through the ducts in the holes and accumulates in the skin slots. Therefore in places of the body where large vortices are formed when swimming at high speeds, there is a mechanism for effective influence on the stabilization of these disturbances (and the elimination of pain disturbances caused by them). With increasing swimming speed in ampoules, a greater amount of biopolymer is produced. When maneuvering, the size of large vortex systems in the area and behind the vertical fin increases. When the body bends, additional mucus is squeezed out of the ampoules. This is an example of economical, automatically controlled mucus production, when dissolved biopolymer solutions are produced in sea water.

In addition to the indicated longitudinal elliptical cavities (Part I Section 4.6, Fig. 4.44), in the skin in the region of gill slits and on the inside of the gill covers of swordfish there are skin crypt structures in the form of systems of longitudinal rows of oval cavities, located one after another (Part I Section 4.6, Fig. 4.47).

To understand the hydrobionic significance of these cavities, numerous experimental investigations have been performed on the flow of various types of cavities, both in a hydrodynamic tunnel and in the flow of air around a cavities [61,79,81,82,117, 127,129,130,131,138,141,459,460,507,508,567,568]. The methodology and results of experimental investigations of the flow around various cavities are presented in detail in these works. In addition, the patterns of flow around transverse semicylindrical cavities were investigated [110,112,124,125,564–566,569–571]. The results of experimental investigations of the flow around spherical and cylindrical cavities were applied in American patents, the list of which is given in the literature.

Investigations of the flow around a water stream were performed by V.P. Musienko and V.V. Babenko at the Institute of Physical and Technical Problems of Energy of the Lithuanian Academy of Sciences with the support of Professor A.A. Pyadishus. The measurements were carried out in a closed hydrodynamic tunnel. The working area is made of plexiglass and has dimensions: length—1.0 m, cross section—0.2 × 0.02 m. At the bottom of the working area there was a rectangular cutout in which a smooth plate was installed flush with the bottom. The smooth rectangular plate had a hole with a diameter of 0.1 m for installation of a round insert, in which there were also two holes for installing two bushings with the studied forms of cavities. The diameter of the bushings for the cavities was 0.03 m. The round insert could rotate around its vertical axis of symmetry and be mounted at different angles in the horizontal plane with respect to the incoming flow. Thus the location of the two cavities relative to the direction of the main stream was changed. This made it possible to visualize the flow around either two cavities at the same time, or their interaction during installation one after the other. To investigation the effect of changes in the geometry of the cavities a special design was used with a flexible membrane connected by a rod to the shaft of a stepper motor with an electronic control system [459]. This design allowed either to change the shape of the cavity in accordance with the specified parameters of the cavity, or to create vibrations of the flexible surface of the cavity, which allowed generating emissions from the cavity with a given frequency.

Fig. 7.43 shows photographs (top view) of the flow around the cavities (cavity 1), the diameter of the intersection of which with the flat surface of the bottom of the working area was $d = 1.4 \cdot 10^{-2}$ m, and the depth of the cavity $h = 0.3 \cdot 10^{-2}$ m [130]. The photographs were taken from a movie obtained with a camcorder. The flow velocity was $U_\infty = 5 \cdot 10^{-2}$ m/s. The dimensionless parameter of the depth of the cavity was $h/d = 0.214$, the Reynolds number $Re_d = 6.93 \cdot 10^2$, $Re_x = 4.87 \cdot 10^4$. The cavity is shallow and streamlined in a laminar flow. Visualization was carried out using colored multicolored jets (Sections 2.7 and 3.9). To obtain injet streams with a specific gravity comparable to the specific gravity of water, dyes based on pH indicators—methyl violet and methyl red—were used. The red jet (4) was fed along the axis of the working section using a tube with a diameter of 0.8×10^{-3} m, which made it possible to vary the distance of the jet from the streamlined plate within $y = (0.2-1.0) \times 10^{-2}$ m and, thus visualize the flow pattern along the entire thickness of the boundary layer.

Jets of violet color (3) were fed through three holes, 1 mm in diameter, located on a horizontal flat surface. The distance between the holes is 7.5×10^{-3} m. The height of the container with colored liquid was selected for each value of the

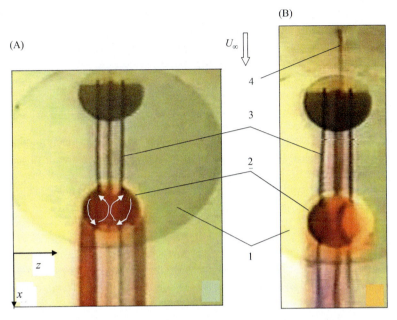

Figure 7.43 Photographs of water flow around a cavity with a diameter of $d = 1.4 \times 10^{-2}$ m, depth $h = 0.3 \times 10^{-2}$ m, $U_\infty = 5 \times 10^{-2}$ m/s (A) and $7.5 \cdot 10^{-2}$ m/s (B): (1) the bottom of the working area, (2) cavity, (3) jet of violet, and (4) jet of red [130].

average speed in the working section to ensure a uniform and continuous flow of three streams and to allow identification of the flow structure in the cavity. The average jet in the cavity region was more diffuse due to the increase in the velocity of flow around the cavity in its middle. Two side jets in the region of the leading edge of the cavity are separated: a part of the stream bends toward the center of the cavity due to a decrease in pressure along the longitudinal axis of the hole. The other part moves along the periphery of the cavity. Partially the lateral jets, having reached the trailing edge of the cavity, begin to flow in the opposite direction (upstream) toward the front edge of the cavity (shown by white arrows). This occurs, in particular, due to the flow effect in the funnel. As a result, in the lateral regions on the surface of the cavity, the jets rotate along the generatrix of the surface of the cavity against the flow. Measurement of the pressure distribution of other authors along the longitudinal axis of the cavity showed that the pressure increases asymmetrically along the longitudinal axial generatrix of the cavity and becomes greatest when the flow leaves the cavity. In accordance with this effect, the funnel-shaped flow in the cavity consists of two symmetrical funnels. Right and left jets due to increasing pressure along the axis of symmetry of the cavity are returned back against the direction of the main stream. As a result, two regions of a spindle-shaped vortex flow with a vertical axis of symmetry directed perpendicular to the surface of the cavity are formed on the sides of the cavity. Since the surface of the cavity is a part of the surface of the sphere, the axes of these vortices will not be constant in time and space. Thus in Fig. 7.39 on the sides of the cavity are visible two asymmetric elliptical regions of fluid rotation. The large axes of these ellipses are directed quasi-parallel to the direction of flow. On the sides of the cavity, liquid from the surface of the spindle-shaped vortices and from the surface of the bottom of the cavity flows from the trailing edge of the hole with two longitudinal strips in the form of vortex longitudinal vortices structures. The rotation in the longitudinal direction is due to the fact that, at the periphery of the cavity, the out flowing fluid has a speed greater than inside the longitudinal strips.

Thus due to the rotation of the spindle-shaped vortices inside the cavity, the incoming flow is pushed to the periphery of the cavity, and the stream flows out of the cavity at an angle to the diameter of the cavity contour in two strips, so that the distance between the contours of the strips is larger than the diameter of the cavity (Fig. 7.43A). With increasing velocity of the flow around the cavity (Fig. 7.43B), only the lateral regions of the spindle-shaped vortices are fixed. As a result of the increase in the velocity circulation, a liquid dye is absent in the middle of the vortices due to its leaching. As a result of this, only two lateral jets are recorded, flowing from the cavity from the region of slow rotation of the spindle-shaped vortices. The jets flowing out of the cavity from this region of vorticity become more diffuse and have the form indicative of the formation of a pair of longitudinal vortices behind the cavity. The third jet located in the middle is the jet *4* injected from the tube located above the streamlined plate.

Fig. 7.44 shows the results of visualization of the flow around a symmetric cavity [cavity (13) [138]], whose dimensions were $h = 0.5 \times 10^{-2}$ m and $d = 2.0 \times 10^{-2}$ m ($h/d = 0.25$) (experiments were carried out in December 1990). The direction of the velocity circulation inside the well is shown by small arrows. The size of this cavity and the flow rate are increased compared to the flow around the cavity shown in Fig. 7.43.

Visualization was carried out by injection of tinted jets from a sleeve located upstream. In the middle of the frames, jet (4) is visible, which is located at a

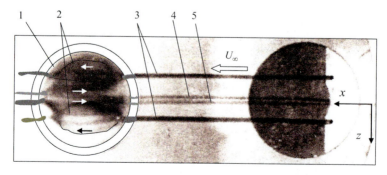

Figure 7.44 Photographs of visualization of the development of the flow pattern in a cavity at a flow velocity $U_\infty = 10.3 \times 10^{-2}$ m, $h = 0.5 \times 10^{-2}$ m and $d = 2.0 \times 10^{-2}$ m:
(1) spherical segmented cavity, (2) elongated spindle-shaped vortices, (3) side violet jets, (4) red jet located above the plate, and (5) violet jet along the longitudinal axis of the cavity.

distance of $5 \cdot 10^{-3}$ m from the bottom surface. It is seen that the trickle injected from this tube is uniform during flow along the entire frame. At a distance of $3.0 \cdot 10^{-2}$ m from the hole for the axial tube in the circular insert located upstream, three holes are made transverse to the stream. The jets injected from these three holes also move to the well without disturbance. However, the cavity affects the flow and against the flow. Therefore the photographs show that when approaching the cavity, the extreme jets bend toward each other due to the distribution of pressure inside the cavity.

Visualizing jets fall onto cavity (1) not at the edges, but at a distance of about 1/3 of the diameter of the cavity. When a jet enters in a cavity under the influence of the "funnel" effect, the flow occurs along the periphery of the cavity as shown by small arrows. When approaching the trailing edge of the cavity, the stream partially flows out of the cavity near and along the longitudinal axis of the cavity, and partially returns against the stream (shown by white small arrows). A pair of peripheral spindle-shaped vortices (2) with a vertical axis of rotation is formed in the cavity. An exchange of current takes place between the formed pair of these vortices. Between a pair of longitudinal peripheral elliptical vortices in a cavity (13) near the leading edge, a weak elliptical transverse vortex is formed with an axis of rotation located along the z axis. The distance between the vortices (2) is less than the diameter of the hole—$\lambda_z < d$. As the flow velocity increases, the velocity circulation in the longitudinal elliptical vortices increases, therefore the visualizing paint is washed out more intensively from the vortex formations. Longitudinal elliptical vortex structures do not have stagnant regions with slowly rotating paint. The surfaces of these vortex structures are flow around in thin jets of visualizing paint, and the longitudinal ellipses of the vortex structures themselves become more elongated along the flow. A weak transverse vortex between longitudinal elliptical vortices is practically washed out and is shifted in the form of a narrow region toward the outlet of the flow from the cavity (the posterior edge of the cavity). As a result, the flow flows out from cavity also in the form of a pair of longitudinal vortices. Due to their increased intensity, the paint identifying these vortices is significantly washed out. In this case, the distance between this pair of vortices decreases substantially and is approximately half the diameter of the cavity: $\lambda_z \approx 0.5\ d$. The corresponding Reynolds numbers at speeds $U_\infty = 4.4 \cdot 10^{-2}$ m/s; $7.5 \cdot 10^{-2}$ m/s and $10.3 \cdot 10^{-2}$ m/s were $Re_d = 8.7 \cdot 10^2$, $Re_x = 4.28 \cdot 10^4$; $Re_d = 1.48 \cdot 10^3$, $Re_x = 8.85 \cdot 10^4$; $Re_d = 2.04 \cdot 10^3$, $Re_x = 1.31 \cdot 10^5$.

In the considered flow regimes of the cavity, the physical picture of the flow inside the cavity is as follows. Imagine that a cavity intersects an infinite number of vertical *XOY* planes that are equidistant in the transverse direction from the edge of the cavity to the central section in the cavity. In each *XOY* plane, with the exception of sections passing through the edges of the cavity, the depth of the curve, which is the line of intersection of the *XOY* plane and the surface of the cavity, will increase to the maximum value in the center section of the cavity. If we compile the Bernoulli equation in each *XOY* plane, then it turns out that toward the center of the cavity the pressure constantly decreases and the flow velocity along the generatrix of the cavity increases. Thus a component of the fluid flow velocity appears in the transverse direction from the lateral edges of the cavity to the center. If we project each other the indicated vertical planes of XOY cross-sections, then we get a family of equidistant curve lines of intersection of the XOY plane and the surface of the cavity. Each curve is described by a part of the corresponding circle centered at the same point. Thus a family of lines with different radii of curvature is obtained so that the zero radius of curvature corresponds to the extreme section, and the largest radius of curvature corresponds to the section passing through the longitudinal axis of symmetry of the cavity. It is known that the smaller the radius of curvature, the greater the Görtler instability of the flow [609]. Based on the results of measurements of the kinematic parameters of the flow in the vicinity of the cavity using a laser anemometer, the Görtler parameters were calculated. The results showed that all parameters are in the instability region of the Görtler diagram, but the intensity of the vortices at the periphery of the hole is higher than along the axis of symmetry.

All this leads to a curvature of the stream lines toward the center of the cavity and to the appearance of longitudinal swirls along each stream line. If we imagine the flow around the cavity in the first flow regime (at low speed), then in the region of the central part of the cavity the flow will smoothly flow around the surface of the cavity. When flowing around the side surfaces of the cavity, the flow will bend along the side surfaces, which will lead to a flow type in the funnel (the pressure along the longitudinal axis of the cavity is the smallest), with the simultaneous appearance of curvilinear Görtler type vortices. As a result of this, in Figs. 7.43 and 7.44, the formation of a pair of longitudinal spindle-shaped vortex flows with a vertical axis of rotation is recorded inside the cavity.

An increase in the flow velocity in the hydrodynamic tunnel leads to the formation of the subsequent stages of the transition up to the turbulent flow regime in the boundary layer of the bottom of the working section in which the cavity is installed. With increasing velocity, it is not possible to visualize the flow into the cavity in the water stream by tinted jets. In this regard, similar investigations in the air flow were later performed. V.N. Turik and V.V. Babenko developed a small wind tunnel based on the design of the vortex chamber (Section 4.6, Fig. 4.19), in which, in particular, the flow around cavities was investigated at high flow rates and Reynolds numbers (Fig. 7.45) [95,97,101,102,108–111,563].

The working section (4) of circular cross-section was made of transparent polished organic glass. In the axial horizontal plane inside the working area, a horizontal flat plate was installed on which an insert with bushings was placed, in which cavities of various shapes were mounted. The air entered the working area (4) through the confuser (1), made V.N. Turik in accordance with the configuration profile Vitoshinsky. The confuser and damper (2), as well as the pipe behind the working section, were connected to the working section (4) using docking washers (3), which were attached to the frame (8) of the wind tunnel with the help of rigid tape metal clamps. The diffuser 6 was connected to the fan (7), which was connected to the return section located above the working section (not shown in Fig. 7.45). The return section exited from the laboratory out. A device with a micro-coordinate for mounting sensors of a hot-wire anemometer is mounted above the working section.

Figure 7.45 Small wind tunnel: (1) confuser with Vitoshinsky profile, (2) soothing section, (3) connecting unit, (4) working section, (5) microcoordinate, (6) diffuser, (7) fan, and (8) frame [563,564].

To visualize the flow structure inside and in the vicinity of the cavity, V.V. Babenko and V.A. Blokhin developed a smoke imaging device. For this, a cylindrical plug made of fluoroplastic was made, mounted flush with the plate, in which the cavity was placed. A deepening with a depth equal to the diameter of the nichrome wire installed in this groove is made along the longitudinal axis of the plug. An electrical voltage was applied to the wire. There was also a hole in the back of the wire through which motor oil was supplied to the wire surface. When the voltage was applied to the wire, the oil was heated, and a smoke stream flowed from the wire, visualizing the flow in the cavity and in its vicinity. The sizes of the hemispherical cavity were $h = 0.9 \times 10^{-2}$ m and $d = 2.0 \times 10^{-2}$ m ($h/d = 0.45$). Compared to previous experiments, when a cavity was investigated in the form of a segment of a sphere, the results of an almost hemispherical cavity, the edges of which were rounded, will be given below. The cavity and the cylindrical plug were mounted flush along the longitudinal axis of the plate with a size of 0.65×0.1 m. This plate was installed horizontally along the longitudinal axis of the working section of the small wind tunnel with an internal diameter of 0.1 m. The center of the hemispherical cavity was located at a distance of 0.5 m from the front edge of the plate. The experiments were carried out in the range of numbers $Re_x = (0.9-5.2) \times 10^5$. The air flow rate was 3 m/s, and the Reynolds numbers were $Re_d = 0.4 \times 10^4$ and $Re_x = 1 \times 10^5$, respectively. The indicated Reynolds numbers are of the same order as in the experiments with a stream of water flowing around a cavity with a velocity of $U_\infty = 10 \times 10^{-2}$ m/s, but the flow velocity was much higher.

The experimental results recorded using a video camera was decoded by the same method as in the case of a stream of water flowing around a cavity. Fig. 7.46 shows an example of a movie flow around a cavity at a flow velocity of 3 m/s. The numbers correspond in each series to four consecutive frames in the selected series of the visualization process. Each frame follows after 0.04 s. Fig. 7.46A shows a typical picture of the flow around a hemispherical cavity when photographing from

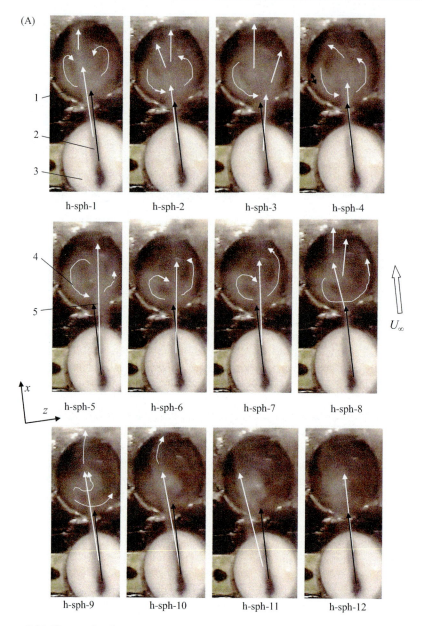

Figure 7.46 Cinema development flow patterns in a hemispherical cavity at an air flow velocity of $U_\infty = 3$ m/s, $h = 0.9 \cdot 10^{-2}$ m, $d = 2.0 \cdot 10^{-2}$ m. (A) top view, (B) side view at an angle: (1) cavity, (2) smoke stream on the surface of the plate in front of the cavity, (3) fluoroplastic plug, (4) the direction of movement of the vortex in the cavity, and (5) the direction of movement of the smoke jet.

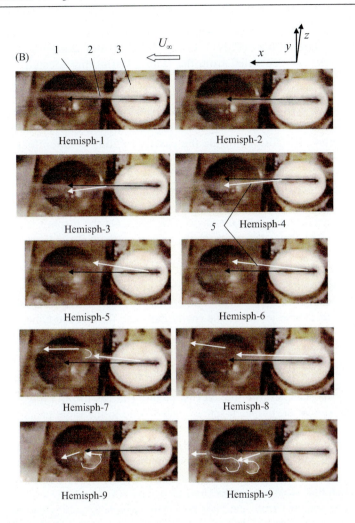

Figure 7.46 Continued

above. Light curved arrows show the direction of flow inside the cavity. Light straight arrows indicate the flow from the smoke generator on top of the cavity. The black arrows indicate the longitudinal axis of symmetry of the cavity—the direction of the *x*-axis.

From the first series of visualization of the process of ejection of swirling fluid from a cavity, two frames h-sph-1 and h-sph-2 flowing around a hemispherical well (h-sph) illustrate changes in the flow structure (light curved arrows) after the previous discharge of flow from the cavity. On the h-sph-1 frame (hemisphere), it is seen that a pair of vortices located on the sides relative to the longitudinal vertical plane XOY of the cavity rotates in the cavity. When air flows around a hemispherical cavity, this rotation is due to the effect of a funnel-shaped flow, which was recorded during the flow around various cavities of a water stream (Figs. 7.43 and 7.44). At the same time, one can see how the smoke jet generated from the fluoroplastic plug located downstream (light straight arrows), starting from the first frame, moves to the right relative to the flow direction. In the h-sph-1 frame, the stream of smoke above the cavity is still located to the left of the black arrow, but at the end of the cavity the stream of smoke bends to the right on the edge of the cavity. On the h-sph-2 frame, a vortex was formed inside the cavity, rotating counterclockwise under the influence of the beginning of the ejection of liquid from the cavity to the right.

In this case, the smoke jet moves to the right and a "fan" begins to form at the exit of the cavity—two small light arrows illustrate the solution of the "fan" in which the ejected liquid from the cavity is located. On the h-sph-3 frame, a vortex is rapidly ejected from the cavity in the region of the right half of the cavity: a wide "fan" is formed (two white arrows on the right). The ejection from the cavity draws a vortex from the cavity (on the left is a curved arrow).

A large white arrow indicates the direction of smoke movement above the cavity during the ejection. In frame h-sph-4, after the vortex is ejected from the cavity, the direction of the smoke jet above the cavity moves to the left and coincides with the x-axis. Smoke from the upper smoke jet begins to be sucked into the cavity, in which the vortex begins to rotate counterclockwise.

All smoke generated by nichrome wire is sucked into the cavity and is not observed.

A cycle was selected from the cinema picture (frames h-sph-5-h-sph-8,), similar to the cycle shown on frames h-sph-1-h-sph-4. Due to the rapid change in the frequency of ejections from the cavity, there is no mirror repetition of the entire process. The ejection from the cavity is longer—the main ejection is observed on the h-sph-7 frame, and on the h-sph-8 frame, the ejection ends. In frames h-sph-5-h-sph-8, the vortex deformation inside the cavity is clearly visible under the influence of an ejection that sucks the vortex from the cavity.

A cycle is selected from the cinema picture, which illustrates the ejection of a vortex from the hole on the left side of the cavity (frames h-sph-9-h-sph-12). One can see how the cavity is rapidly filled with smoke and how the direction of the smoke jet above the hole changes and goes to the right. On the h-sph-11 frame, an outlier from the cavity to the left of the direction of the longitudinal axis of symmetry of the cavity (black arrow) is recorded. After the vortex is ejected from the left side of the cavity, the smoke jet (straight white arrow) returns to the right and coincides with the longitudinal axis of symmetry of the cavity (straight black arrow)—frame h-sph-12. Thus the cycle of alternating ejections from the right and left sides of the cavity is unsteady.

Fig. 7.46B shows sample frames for visualizing the flow around a cavity that do not correspond to the frames shown in Fig. 7.46A. Photographs of the cavities are taken from the side and top at a certain angle. At the selected illumination, the vortex structures inside the cavity are not visible in the photographs, but the behavior of the smoke jet above the cavity is clearly visible. In the Hemisph-1 frame, a white jet of smoke moves above the cavity to the right of the direction of the longitudinal axis of symmetry of the cavity (black arrow)—in the direction of the OZ axis. The direction of flow (from right to left) corresponds to the x-axis. The y axis is vertical, and the z axis is to the right in the transverse direction from the flow direction.

On the Hemisph-2 (Hemisphere) frame, after 0.04 s, the smoke jet moving to the left coincided with the direction of the longitudinal axis of symmetry of the cavity (black arrow). After another 0.04 s in the Hemisph-3 frame, the smoke jet expands over the cavity and takes the form of a "fan" in the XOZ plane—this frame coincides with the h-sph-11 frame (Fig. 7.46A). After 0.04 s on the Hemisph-4 frame, the smoke jet continues to move in the form of a "fan" in the XOZ plane, as in the previous frame, but the direction of the "fan" shifts even more to the left of the black arrow. The frames Hemisph-3 and Hemisph-4 indicate that during the indicated time (0.08 s) a liquid is ejected from the cavity on the left side of the direction of the longitudinal axis of symmetry (with respect to the direction of the velocity of the unperturbed flow).

The next four frames (Hemisph-5—Hemisph-8) demonstrate how formed and liquid is ejected from the cavity to the right side of the direction of the undisturbed flow velocity.

Frames Hemisph-9 and Hemisph-10 illustrate the beginning of the formation of the ejection of vortices from the cavity again on the left side of the direction of velocity of the undisturbed flow.

The kinogram shown in Fig. 7.46B also illustrates the process of alternating ejection of vortices from the right and left sides of the cavity. This process of alternating ejection of vortices with a high frequency of alternation from the extreme lateral sides of the cavity leads to the formation of a longitudinal pair of vortices in the wake of the cavity in the same way as in Figs. 7.43 and 7.44. The intensity and shape of these vortices in the wake behind the cavity depends on the shape and size of the cavity, on the velocity of the main flow and on the interaction of the vortices injected by the cavity with the CVS existing in the boundary layer in front of the cavity.

Fig. 7.47 shows a visualization of the structure of the water flow in two types of cavities of various shapes: (A) spherical cavity with a diameter of $d = 1.5 \times 10^{-2}$ m, a depth of $h = 0.5 \times 10^{-2}$ m and (B) oval cavity with sharp edges, $h = 0.6 \times 10^{-2}$ m, axle lengths 3×10^{-2} m and 2.2×10^{-2} m.

Fig. 7.47A at the top shows a side view of the spherical cavity (1), the results of an experimental investigation of the flow around which are shown in Fig. 7.43 [130]. Below in Fig. 7.47A is a top view of this cavity, which is installed in the plug (5) mounted at the bottom of the working section of the hydrodynamic pipe. In the center of the bottom of this hole, the pressure will be the smallest. As can be seen from Fig. 7.43A, inside this cavity a pair of spindle-shaped vortices elongated along the stream rotates. If polymer solutions are injected into the cavity (2) located in the region of the center of the bottom of the cavity or in the region of the center of rotation of the spindle-shaped vortices (Fig. 7.47A), then they will be captured and rotated in the spindle-shaped vortices (Fig. 7.43A). The indicated location of the injection hole was chosen due to the location of the lowest pressure: due to the

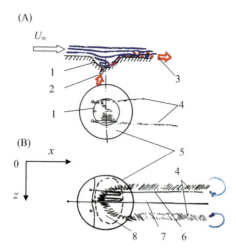

Figure 7.47 Scheme visualization of the flow structure in cavities of various shapes: (A) spherical cavity, diameter $d = 1.5 \times 10^{-2}$ m, depth $h = 0.5 \times 10^{-2}$ m; (B) oval cavity with sharp edges, $h = 0.6 \times 10^{-2}$ m, axis lengths 3×10^{-2} m and 2.2×10^{-2} m. (1) spherical cavity, (2) injection of polymer solutions or microbubbles, (3) polymer solutions or microbubbles in the boundary layer, (4) pair of longitudinal vortices, (5) plug for placement of the cavity, (6) visualization jet flowing from the plug, (7) visualization jet flowing from the tube in the boundary layer, and (8) oval cavity.

above method of vertical flat sections along the cavity and the decision of the Bernoulli equation for each section, as well as due to the circulation of fluid particles in spindle-shaped vortices.

The smallest pressure will "draw" the injected liquid from the cavity and rotate it in spindle-shaped vortices—this will stretch the macromolecules of the polymers and increase their efficiency. Behind the cavities, a pair of longitudinal vortices (4) is formed, the movement of particles in which passes along helical trajectories, which also contributes to the stretching of polymer macromolecules, a decrease in the diffusion of solutions in the boundary layer (3), and a decrease in friction resistance. The same effects will also occur when microbubbles of gas or air are injected into a cavity. Fig. 7.47B shows a scheme of the formation of vortices in oval cavity (8) [141].

The above results allow us to understand the mechanisms of interaction with the flow of vortex structures generated in the mucus-producing cavities located in the gill apparatus (part I Section 4.6, Fig. 4.47) and longitudinal oval cavities (part I Section 4.6, Fig. 4.44) located in skin folds in the region of the vertical fin of swordfish (Part I Section 4.6, Fig. 4.49).

Morphological devices for reducing two resistance components (shape and friction), as well as for reducing the vortex resistance caused by the presence of protruding parts on the body (vertical and lateral fins), were considered above. Drag of the tail part of the body of hydrobionts during active swimming is minimal, since the tail part of the body and the tail fin are the movers that not only create thrust, but also create suction force and, thus alternately from the sides reduce the resistance in the tail part of the stem.

The hydrobionic approach reveals unknown structural features skin hydrobionts and offer new ideas on the problem of drag reduction.

1. A comparative analysis showed that there are identical adaptations for drag reduction and saving energy in the structure of the skin and body of high-speed hydrobionts, as well as individual formations created in accordance with the characteristic swimming speeds.
2. New features of the structure and hydrodynamic significance of the xiphoid tip, as well as previously unknown systems for the production and transportation of mucus in the skin of swordfish, are shown. New structures have been discovered that indicate the specific role of mucus.
3. Complex of methods detected aimed at the economical use of mucus and the optimal distribution of mucus solutions over the thickness of the boundary layer.
4. Based on hydrobionic modeling, some of the detected features of the body and skin of swordfish are verified experimentally.
5. The hydrodynamic significance of micro folds on the surface of the swordfish skin and the role of skin thickness have been theoretically and experimentally studied. The area of the tail fin mover of swordfish is compared with the area of other species hydrobionts.
6. Based on a hydrodynamic analysis of some structural features of the swordfish body, a method for the interaction of various disturbances in the boundary layer and a combined method for drag reduction have been developed and experimentally investigated.
7. On the basis of the hydrobionic approach, some technical applications of the revealed structural features of the skin and the body of the swordfish have been developed (part III).

In the future, it is necessary to investigate some features revealed by the interaction of the skin of hydrobionts with the vortex structures of the boundary layer. It should be determined:

- What is the physics of the mechanism of the combined method of drag reduction.
- What is the role of vibration of the skin of hydrobionts.
- The physical basis of the boundary layer as a nonlinear waveguide.
- Interaction of disturbances in the boundary layer.
- What is the physics of the flow over various three-dimensional cavities.

7.11 Investigation of the influences of microbubbles injection into the boundary layer on the drag reduction (short review)

Table 6.1 (Section 6.1) describes the control methods of the boundary layer CVS. A large number of investigations have been performed on the effect of injection of gas MB into the boundary layer on the characteristics of the boundary layer and drag reduction. For the first time in this direction, investigations were performed, which are given in Refs. [178,180,395,448]. In Ref. [180], a list of investigations on the problems of cavitation, the transition of a boundary layer to a turbulent one, suction of the boundary layer, and the use of biopolymers and polymer solutions to drag reduction, performed at a number of institutes of the Siberian Branch of the USSR Academy of Sciences, is also given. In Refs. [178,180], the results of an investigation of bubble gas saturation of near-wall flows are presented. Extensive experimental investigations of the characteristics of a TBL with a distributed injection of gas through a permeable surface and one through narrow perforation bands and slits [178,448]. It was found that such characteristics as the intensity of pressure and friction pulsations on the wall depend on the concentration of bubbles in the near-wall zone of the boundary layer, where the velocity profile is described by a logarithmic law. A characteristic parameter in this case is the maximum value of the volume concentration of bubbles φ_{max}. The dependence on the gas injection velocity and the velocity of the unperturbed flow is investigated. Upon reaching the limiting value $\varphi_{max} \approx 0.75$, which corresponds to the dense packing of the spheres, almost complete displacement of the flow from the wall takes place.

The dependence of the coefficient of friction on the wall $\overline{c}_f(\varphi_{max})$: $\overline{c}_f = \overline{c}_0[1 - (\varphi_{max} - \varphi_o)]^2$ (Fig. 7.48) is established. It was found that there is a certain threshold concentration value $\varphi_{max} = \varphi_o \approx 0.1$, below which gas saturation does not affect the friction coefficient. When a certain value of φ_{max} is reached for a given flow velocity, instability of the bubble layer appears, which is expressed by minima in the indicated dependence. This phenomenon is accompanied by a significant increase in low-frequency disturbances. The condition for its occurrence is determined by the hydrodynamic stability criterion [395]:

$$K = \rho_2^{0.5} v_{2cr}/[g\sigma(\rho_1 - \rho_2)]^{0.25},$$

where ρ_1 and ρ_2 are the densities of the heavy and light phases, g is the acceleration of gravity, σ is the surface tension coefficient, and v_{2cr} is the critical value of the characteristic velocity of the light phase.

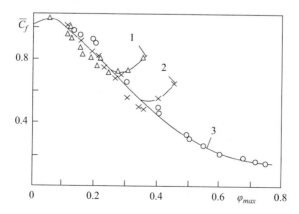

Figure 7.48 The dependence of the friction coefficient on the wall $\overline{c}_f$ on the threshold concentration φ_{max} at the flow velocity $U = 4.36$ m/s (1), 6.55 m/s (2), and 10.9 m/s (3) [180].

During the investigation, there was still no clear understanding of the development of CVS in the boundary layer. Therefore the authors associated the results with concentration bubbles and flow rate. This did not give prospects for the practical application of this method to drag reduction.

In [618], a simplified theoretical model for the flow around a plate during injection of air bubbles into a TBL was proposed. The results obtained are compared with experiment during injection of bubbles through a transverse slit with a span of 150 mm, a width of 5 mm and an angle to the flow of 20 degrees, located at a distance of 0.7 m from the beginning of the plate. Investigation was conducted in a cavitation tunnel, the working section of which was square with a side of 600 mm. The plate was located at the top. Bubbles were large; a mixture of water and bubbles was prepared in advance. Consumption ranged from 35 to 200 L/min. When injecting a bubble medium with a flow rate of 200 L/min, the drag reduction was 70% at $x = 1$ m, 60% at $x = 1.5$ m, 50% at $x = 3$ m. When injecting with a flow of 100 L/min, drag reduction was on average 20% to $x = 2$ m, and with a larger x—10%. When injected with a flow rate of 35 L/min, drag reduction was not recorded. No connection with the structure of the boundary layer was investigated.

In 1998 at the international conference "International Symposium on Seawater Drag Reduction, Newport, Rhode Island, USA" this problem was reviewed on section "Turbulent Drag Reduction. Methods: Microbubble." In Refs. [425–427], experimental investigations were performed and data were obtained on the drag reduction upon injection of MB into the boundary layer. In a cross-section in the channel in a turbulent flow, the near-wall concentration of bubbles near the wall was expressed through α_w, and the relative average value of the concentration of bubbles through α_m. This is primarily due to the fact that α_w includes most of the bubbles involved in the process of drag reduction, compatible with the general idea that the phenomena depend on the internal region of the flow. When taking into account the diameter of the bubble and the flow velocity, the near-wall concentration gives the frequency of the passage of bubbles ω_o. Accordingly, the bubbles are depicted as discrete bodies passing through a fixed point on the wall at local speed. The results show good scaling with this frequency, normalized by the internal variables ω_o^+. It has been shown that under some assumptions a similar scaling is consistent with the known results [425] obtained using a hot-wire anemometer.

Investigations in the same direction were carried out in Refs. [278,279, 289,290,292,317,353]. The experiments [278] were performed in a narrow, flat channel 10 mm thick. A pair of sintered plastic porous plates mounted flush with the upper and lower walls could create small air bubbles when dry compressed air was blown through them. The diameter of the bubbles was determined from a photograph taken through a transparent acrylic top wall. Bubble concentration profiles for shear location measurements were obtained using a probe with a tapered inlet.

The decrease in friction resistance in a TBL by introducing small gas bubbles into the flow has been the subject of many investigations [446,447,619].

A clear understanding of the physical mechanism of this method of drag reduction is necessary so that this method can be effectively applied in shipbuilding.

In Refs. [178–181] it is stated that gas-bubble saturation of near-wall fluid flows as an effective method of drag reduction has already passed the stage of laboratory tests and can now be recommended for practical use on ships.

The key issue for this method of drag reduction is the development of an optimal method for saturating the boundary layer of water with air MB. So far, three methods of gas injection into the near-wall layer of the fluid have become widespread:

- Distributed gas injection through a permeable coating.
- Slotted injection of a gas-liquid mixture.
- Gas injection through a slot under a stream of water [432].

The first impressive results in reducing friction due to saturation with gas MB of the water boundary layer were obtained using the first method. The largest number of works in this direction also involves the use of permeable surfaces. However, the possibilities of this method have not yet been fully discovered, and the application to real objects is faced with some problems.

In Ref. [181], the experiments were carried out in two series: when flowing around a flat plate and when towing an axisymmetric aluminum model. The plate had a length of 910 mm, a width of 350 mm and a thickness of 60 mm. A porous flat sheet (350 × 100 mm in size) was installed on the side of the plate at a distance of 300 mm from the leading edge. An air chamber was placed inside the plate body, coated on the outside with a porous material, which was divided into six equal parts located along the flow. Pressure-controlled air flow was injected to each of these sections through special pipelines. The permeable coating was a perforated element with microspheres, usually directed to a streamlined surface. The sheet thickness was 1.2 mm, and the slots between the stacked sheets were about 30 μm. Downstream (50 mm) behind this porous region, an impenetrable measuring plate of 55 × 80 mm was installed. Both permeable and measuring plates were mounted on spring suspensions, which made it possible to measure the integral friction forces on each plate. A flat plate was tested at a basic flow velocity of $U_\infty = 3$ m/s. The axisymmetric model had a diameter of 175 mm, its length was 1750 mm with a long cylindrical part, which was about 75% of the total length. Behind the head part (length—100 mm) were installed four removable rings made of permeable or impermeable material. The thickness of the sheets was 0.8 mm and the gap between the cells was 30 μm. The model was investigated when mounted on a special scooter while it was moving at a speed of 15 m/s.

In Ref. [618], experiments with the injection of MB were carried out in a closed hydrodynamic tunnel. A decrease in friction drag of up to 40% was obtained. The coefficient of local void was measured in two ways: one way using a suction tube in the work area, and the other way when counting the number of bubbles from photographs. The results show that the local void ratio near the wall is the dominant factor in reducing friction.

It is known that MB injected into the boundary layer on a solid surface significantly reduce the friction drag. But the energy required for injection is not nominal, and the reduction in net resistance is difficult to obtain when it is used on full-scale ships. Therefore it is necessary to reduce the amount of air and/or increase the drag reduction by studying the mechanism of reducing drag.

Experimental and numerical investigations of microbubble injection have been carried out in Japan. A group of authors studied the mechanism and scale effect during the injection of MB in order to apply the results to full-scale vessels. A small hydrodynamic tunnel was built specifically designed for microbubble research. At the end of the working section, water entered the degassing tank, in which the introduced bubbles were removed due to buoyancy, which allowed continuous testing. The test section (working section) had dimensions: 0.1 m wide, 0.015 m high and 3 m long. Bubbles are formed by injecting air through a porous metal plate with a nominal pore radius of 10 μm. The plate is located at a distance of 1.038 m from the start of the test section, where the flow is fully developed. In three neighboring places located at a distance of 0.5 m from each other, various measurements are possible. The amount of air introduced is represented by the average void coefficient $\overline{\alpha}_a$:

$$\overline{\alpha}_a \equiv \frac{Q_a}{Q_a + Q_w}$$

where Q_a is the air flow rate, Q_w is the water flow rate.

Microbubble pictures were taken using a high-definition CCD camera. A YAG laser was used as a light source. The light knife was placed at a distance from the vertical wall of 30 mm from right to left in cross-section. The upper end of each photograph corresponds to the upper wall of the cross-section of the test section. The vertical length of the photograph corresponds to 10 mm. The size of the bubbles is generally less than 1 mm in diameter, although it depends on the flow velocity. Friction resistance was measured using a floating disk with a diameter of 10 mm, located in the upper wall of the working area under various flow conditions with or without bubbles. The local void coefficient was measured in two ways.

Further investigations are needed to clarify the mechanism for reducing friction drag by MB and reduce the number of bubbles needed for the practical use of the method. Experimental dependences of the drag coefficient were obtained in three places along the working section at a flow velocity of 5, 7, and 10 m/s at $\overline{\alpha}_a$ = 026, 0.053, and 0.081. At the first $\overline{\alpha}_a$ value, at a low flow velocity, the drag reduction was 10%, at the second $\overline{\alpha}_a$ value, depending on the flow velocity, the efficiency was 10%−20%, and at the third $\overline{\alpha}_a$ value, depending on the flow velocity, the efficiency was 15%−35%. Values $\overline{\alpha}_a$ were also measured at various distances from the wall and at various bubble sizes.

The main drawback of the performed investigations was that the height of the working section was small, therefore the thickness of the boundary layer and the influence of MB on the CVS of the boundary layer were not studied. The results obtained are in no way related to the characteristics of the boundary layer.

In Ref. [491] two methods of the drag reduction (injection of a polyethylene oxide polymer and MB) were combined to study the possibility of implementing a synergistic drag reduction, that is, the drag reduction is greater than the sum of each method. The tandem injection experiment was accompanied by the measurement of integral friction using a set of balancing floating elements immediately after the injection sites. Individually, each additive showed expected levels of drag reduction. The combination of the two methods showed that the levels of drag reduction exceed the individual amount of resistance reduction to 10%. The order of injection was an important factor in achieving synergy.

A number of investigations have noted a similarity between the drag reduction of polymer and microbubble additives. It was shown that the injection of polymers or MB into the boundary layer can increase the energy of the turbulent flow and reduce friction. To drag reduction, it was demonstrated that the concentration of bubbles should be maximized between the wall and $y+ = 150$ to effectively drag reduction. ($y+ = y \cdot u^*/\nu$, where y is the normal distance from the wall, u^* is the friction velocity and ν is the kinematic viscosity).

In Ref. [599], the optimal polymer injection velocity was determined to drag reduction in the channel. The importance of maximizing the concentration of the polymer in or near the buffer layer is shown. This observed similarity in the drag reduction mechanism of each method indicates the possibility of mutual reinforcement of their respective mechanisms while injecting polymer solutions and MB in a TBL. The ability to increase the drag reduction is greater than the sum of the individual components, thereby creating a synergistic effect, implies that the bubbles can lengthen the polymer molecules and/or that the polymers increase the concentration of small bubbles near the wall. In any case, this will lead to a thickening of the buffer layer and an upward shift in the logarithmic part of the velocity profile.

The investigation presented in Ref. [431] suggests that a synergistic drag reduction occurs with the simultaneous injection of MB and polymer solutions. A. Malyuga in the late 1970s conducted experiments to reduce diffusion of MB, injecting an aerated solution of polyethylene oxide into a TBL, and measured friction below the injection point. It was concluded that there is "... mutual intensification of the two methods of drag reduction." It is assumed that this effect is mainly due to the larger concentration of small diameter bubbles that they observed when the polymer solution was aerated immediately before the injection of bubbles. An aerated polymer solution was introduced through a 1.8 mm slot with a width of 8 mm. Russian authors have suggested that a polymer solution (polyethylene oxide-PEO) reduces the surface tension of the bubbles, thereby creating smaller bubbles than expected. They measured a local drag reduction up to 80% on a floating element closest to the injection layer, with a reduction decreasing downstream. A. Malyuga noted that the levels of drag reduction achieved by aeration of the polymer solution will exceed the given resistance levels, measured only by injection of MB or only polymer solutions.

The investigation used flat test geometry with two ejectors placed in tandem to determine the possibility of a synergistic drag reduction with simultaneous but separate introduction of polymer and MB. For additives, two separate ejectors were used to eliminate the uncertain characteristics of film mixing. The parameter space for this investigation included the streamwise injection order (i.e., the polymer is higher than MB and vice versa), the volumetric flow velocity for both additives. The measured values were combined by shear stress in several places downstream, as well as other values of environmental pressure, temperature and velocity. Laser anemometry was used to confirm the parameters of the base boundary layer, in addition to integrated shear stress measurements. Salt water was used as the main liquid, since the bubbles formed in salt water are approximately an order of magnitude smaller than those obtained similarly in fresh water, and the size of the bubbles can be a factor in drag reduction of MB in combination with polymers. In a previous study of the effect of bubble size, Kuklinsky showed that salt water had a significant effect on bubble size compared to fresh water. However, the combination of the polymer with salt water did not have an additional effect on the size distribution of the bubbles.

The experiments were carried out in the NUWC closed-loop hydrodynamic research tunnel [491], in which it is possible to conduct research with both salt water and fresh water. The maximum operating velocity is 7.6 m/s (25 ft/s), which is driven by a 30 hp engine. The motor controls an axial flow pump, which has four swirl blades designed to create uniform radial velocity profiles. 0.25-inch stainless steel honeycomb with 6 inches wide is located upstream of the nozzle to even out the flow and control background turbulence. The tunnel is equipped with an automatic static pressure control system that maintains pressure in the test section within ± 1 lb/sq. inch. In a 3.05 m (10 ft) long working area with a square cross-section of 0.305 m (1 ft), the investigations were carried out at large Reynolds numbers. The test section has 16 access panels or 25.4 cm (10 inches) by 61 cm (24 inches) windows, which allows up to 65% of the test section to be optically accessible. Along the test section and on the window panels, the pressure gradient in the flow was measured.

A flat plate with an elliptical leading edge was placed in the test section. The plate has three identical inserts 0.508 m long and 0.178 m wide, which can be changed. Synergies have been found for the combined injection of polymers and MB. When MB were injected upstream and polymer solutions downstream, there were clear cases of synergetics. The reverse order did not show synergetics. It is assumed that turbulent mixing on a very small scale is locally isotropic, while mixing MB and polymer solutions is irrelevant for their effect near the wall. The absence of synergistic contractions, when the injection of the polymer is located upstream from the slit for injection of MB, implies that the internal interaction from the mixing of the two additives does not have a strong effect on the resulting TBL. In this configuration (upstream polymer, MB downstream), a polymer layer introduced into the boundary layer appears to experience greater mixing when it encounters a microbubble ejector. When MB are injected, greater mixing occurs, which reduces the overall recovery of polymer resistance, since the polymer diffuses faster from the buffer layer, where it is effective. The observed synergistic contractions obtained by injecting MB upstream of the polymer slot show that when the bubble layer passes through a limited polymer solution, large contractions are possible. Since internal interactions are not important, there should be a predominant strengthening of one of the mechanisms to drag reduction of additives. Knowing that both additives work best when they are present in the area near the wall, it was suggested that the microbubble layer interferes with polymer transfer during diffusion from the wall, thereby locally increasing the relative concentration of polymer in the buffer layer.

Testing aerated polymer injection from the same slot would strengthen this hypothesis. If this hypothesis is accurate, then significant synergistic contractions would not be observed since a mixture of aerated polymer would not effectively prevent polymer diffusion. If synergistic contractions are observed, alternative hypotheses about polymers that enhance the chemistry of surfactants with MB should be considered.

All tests were carried out in salt water. According to Kuklinsky, the size of the bubble does not change with the addition of a polymer solution. Similar experiments conducted in fresh water may indicate differences in the size of the bubbles with the addition of polymer. This can change the nature of synergistic.

In Ref. [444], it was stated that bubble formation was traditionally addressed by chemical engineers for bubbles formed in a stationary fluid. Here, the emphasis is on obtaining quantitative ratios of bubble formation in a high-speed TBL and determining whether bubble splitting is a possible mechanism for absorbing turbulence energy and, therefore reducing turbulence. This investigation is conducted to examine the mechanisms that dominate during the formation of bubbles in TBL, and to establish a relationship between Q_{pore}, d_{pore}, and U_o. This investigation offers some insight into the modes of bubble size, bubble splitting, and bubble transfer in TBL. Based on the results obtained, the concept of a possible mechanism for drag reduction of MB based on the energy splitting of bubbles is presented. The total amount of turbulent kinetic energy needed to separate the bubbles in the TBL is estimated and compared with the total energy available in the TBL. It is assumed that the splitting of the bubbles is the likely main mechanism for reducing turbulence in the TBL using MB.

It was shown in Ref. [427] that the $\Delta C_f/C_f$ ratio decreases by 50% at a distance of 25δ, where δ is the thickness of the boundary layer. This observation raises the question of why this is happening and whether it can be prevented or mitigated in practice. Intuitively, we can assume that the three main mechanisms are the potential origin of the loss of the effectiveness of drag reduction of MB: an increase in the thickness of the TBL, which reduces the local fraction of voids; diffusion of bubbles from the wall due to turbulent vortex diffusion; and fusion of bubbles and subsequent distance from the wall due to greater buoyancy. The latter mechanism is expected to be significantly reduced in seawater, especially for higher speeds. The second mechanism depends on the magnitude of the turbulent vortex diffusion $u_\tau \delta$, which also depends on δ. This fact made the discussion focus on the first mechanism: how an increase in $\delta(x)$ will lead to degeneration of the efficiency of drag reduction by MB.

The hydrodynamic forces on the wall are obtained, both tangential and normal to the wall. Fundamental unknowns are the shape of the bubble, the drag force, and the lift on the bubble in the TBL. The modes of bubble sizes as spherical caps, the modes of a single bubble, intermediate, and jet disintegration are determined. It is established that the control parameter is the ratio of the gas outlet velocity to the external flow velocity. At a very low injection velocity, the tangential force balance mode determines the size of the bubbles, that is, the balance of water resistance and surface tension. At higher injection velocities, the regime of normal equilibrium forces prevails, that is, the lifting force balances the surface tension or inertia of the water surrounding the bubble. At even higher injection velocities, jet instability determines the size of the bubble. Unknown regions remain between gas jets and bubble modes. Exposure to seawater has been identified but not quantified. It was found that the effects of bubble splitting are very significant and may well be a key source of mechanisms for drag reduction by MB. The smaller the bubble, the greater the likelihood of the appearance of bubbles near the streamlined wall. Unknown are the quantitative formulation of lift, fusion of bubbles, and splitting effects. Despite the high quality of theoretical investigations, taking into account the physical processes of the interaction of MB with the boundary layer, the authors do not mention or take into account the work of their coauthor [443], in which the model of the CVS of a TBL is presented. Without taking into account the structure of TBLs, their presentation is insufficient.

In Ref. [307], experiments were conducted to reduce friction resistance by injection of a polymer solution (polyethylene oxide) or MB. For a thin axisymmetric body, two different drag reduction mechanisms were applied when measuring the total decrease in resistance. And then the amount of reduction of the friction resistance was estimated, provided that the recovery mechanisms are effective only for the friction resistance component. As a result of tests, a decrease in the drag of polymer solutions to 23% of the total resistance was observed, which corresponds to approximately 35% of the estimated friction resistance of an axisymmetric body. This result is in good agreement with the results of a flat plate test [339]. The decrease in microbubble resistance was within 1% of its total resistance. This unexpected result was completely different than when testing a flat plate [339].

Full-scale microbubble tests have been performed [353]. In Korea, the largest shipbuilding sector in the world, such investigations that recently attracted widespread attention were also performed in a cavitation tunnel with a maximum speed of 12 m/s when flowing around an axisymmetric body. The total length of the model was 1.2 m, diameter—100 mm. The model had three parts: an ellipsoidal nose 125 mm, a cylindrical part where friction resistance was measured, 850 mm, and a tail 225 mm as a combination of five types of curves. Inside the model were pipelines for injection of MB and polymer solutions. Air was injected around the circumference to a stream of water through three checkered rows of holes with a diameter of 0.5 mm on the surface of the body. The polymer solution is introduced by a pump consisting of a piston, sliding in a cylindrical tank. An effective seal ensures that the injection velocity is fully controlled by the rotation of the threaded rod, driven by a gear motor. The injection slot has a circular shape, consists of two stainless steel rings, processed in such a way that an internal annular gap is formed when they are assembled together. The slot between the two rings is 1 mm, and the slot channel is inclined at an angle of 12.5 degrees to the flow direction, allowing the introduction of polymer solutions almost tangentially to the surface of the model. The injection devices are designed so that the injection of MB and polymer solution can be performed separately or simultaneously.

First, resistance was measured without injection, then with water injection, and then with injection of MB or polymer solutions. The measurements were carried out at a flow velocity of 7 and 10 m/s. When MB are injected, the maximum effect is 1% production at a flow rate of 3 L/h and a speed of 7 m/s. The effect of the polymer solution with a drag reduction was about 23% of the total drag reduction. When taking into account the corrections of the near-wall jet effect, measured by injecting water at the previous stage of this test, the effect was approximately a 35% decrease in the friction resistance. This result agrees well with the result of a flat plate test [338].

The effect of the injection velocity of a polymer solution of two different concentrations at two different flow velocities is shown. At a flow velocity of 7 m/s, the drag reduction increases faster, since the injection velocity increases and levels off at about 7 L/min for both PEO solutions of 200 ppm and 400 ppm. And at 10 m/s, the results are relatively moderate, and are evened out at a higher injection velocity. Limiting the injection velocity appears to result in a higher polymer concentration. This is about 3 L/min for 200 ppm and about 15 L/min for a solution of 400 ppm.

The problem of microbubble injection into the boundary layer was also considered in 2005 at Symp. on Seawater drag reduction. ISSDR 2005. Busan, Korea.

In Ref. [617], the effect of bubble size on the friction resistance in a large cavitation tunnel, the working section of which was 1.18 m long and 0.22 m width and height, was experimentally investigated. A plate measuring 812.8 mm (length) and 152.4 mm (width) made of steel was installed in the working section. A panel with a size of 127 mm (width) and 101.6 mm (length) was installed in the plane of the plate, located at a distance of 152.4 mm from the front edge of the plate, which contained a slot for injection with a width of 2 mm and a span of 76.2 mm with an exit angle of the jet of 10 degrees. At a distance of 457.2 mm from the slot, a floating element measuring 152.4×101.6 mm in size was located to measure friction resistance. The obtained test results of the gas flow for injection into tap water coincided with the results of S. Deutsch [213−216]. With increasing flow rate, the resistance decreased to 20% at a flow rate of 0.002 m^3/s. The effect of various aqueous solutions with additives to stabilize bubble size was investigated. The size of the bubbles varied between 500 and 30 μm. The authors concluded that the results were insensitive to bubble size.

In Ref. [209], the results of investigations of ship models are presented, one of which provided for injection of MB through longitudinal strips made of porous steel installed in a flat bottom. Another model of the vessel consisted of sections, at the bottom of which cavitators were installed to generate air cavities. Obtained in calm water, the drag reduction was 3%−10%.

A review of investigations on this problem is presented in Ref. [474]. Experimental investigations were also carried out using a modern two-component PIV system and the results were analyzed using an innovative technique that allowed us to trace the evolution of a TBL modified by MB. Experimental investigations were performed in an open channel 25 m long, 60 cm wide and with a maximum water depth of 1 m. The maximum velocity was 1 m/s with a water depth of 60 cm. The boundary layer developed on a transparent plate 4 m long, located below the free surface at a distance of

30 cm. Two series of experiments were performed at an undisturbed flow velocity of 0.43 m/s and 0.69 m/s. In front of the test plate, two supports were installed in the channel, between which several thin wires stretched in the transverse direction were stretched. When current was transmitted, bubbles formed on these wires, which were carried into the flow, in the same way as when the boundary layer was followed by the hydrogen bubble method. In these experiments, the bubbles floated in such a way that they reached the surface of the test plate at some distance from the leading edge. Depending on the flow velocity, a typical concentration of bubbles was 0.5%−3%. In universal coordinates, the averaged and pulsating longitudinal velocity profiles are plotted. Photographs of MB in the flow and in the boundary layer were obtained, in which the thickness of the MB was 0.1 mm, as well as the typical diameter of the bubble.

In Ref. [216], a detailed analysis of investigations by other authors in this direction was performed, and the results of an experimental investigation of friction resistance under the combined action of MB and polymer solutions are presented. The investigations were carried out at the Garfield Thomas Water Tunnel ARL Penn State, which had a cylindrical working section with a diameter of 1.22 m and a length of 4.27 m. A plate 7.14 cm thick, 1.22 m wide and 3.1 m long was installed along the longitudinal axis in the working part. Buoyancy affected the transport of bubbles from the test surface. The combination of polymer and MB is injected by assembling a double injector located approximately 0.6 m from the leading edge of the plate. The distance between the two slits was 39 mm, and the width of the slits was 59.7 cm. The separation between the gas and polymer slits is approximately 4 thicknesses δ at 10.7 m/s. Drag was measured by six floating elements located on distance 0.7; 0.87; 1.09; 1.466; 1.71; and 2.5 m from the leading edge. The drag reduction coefficients were measured during the injection of polymer solutions with a concentration of 500, 1000, and 2000 ppm at various flow rates depending on the distance from the slit. The same measurements during the injection of MB showed that behind the slit the efficiency decreases all the time. A joint injection of MB and polymer solutions revealed a synergistic effect.

Theoretical works were not reviewed, since all the physical laws of the type of flow under consideration are not yet known. Experimental investigations of the effect of injection of MB into the boundary layer were also studied in Refs. [139,207,400−402,434,440,478,479,486,539,545,551,553].

Among all known scientific investigations on methods of drag reduction, unique works are known [450−453], which present the results of full-scale investigations of the combined method of drag reduction by injection of MB and/or polymer solutions. K.J. Moore has organized directed research in the United States at Pen State University for several years.

In Ref. [452], practically important questions of the organization of injection of various liquids into the boundary layer, including MB, as well as multilayer injection of various liquids, were considered. In addition, many important investigations carried out both in the former Soviet Union and in other countries are considered.

In Ref. [453], the results of full-scale investigations of injection of solutions of MB and polymers on the underwater wing of a SEA FLYER vessel are presented.

During the summer of 2005 and the spring of 2006, four two-three-week test periods were conducted on the ONR SEA FLYER technology demonstration vessel to characterize the characteristics of an advanced polymer solution drag reduction system. The only component from which the polymer was injected was a submerged wing, which at about 27 knots of speed produces about 120 tons of dynamic lift. The Reynolds number was $1.0 \times 10^8 - 1.5 \times 10^8$. Among the results demonstrated were:

- The ability to mix polymer powder on demand and thereby avoid the need to transfer a large volume of polymer suspension.
- Polymer productivity seems to have temporal as well as spatial characteristics.
- In sea water, the decrease in viscous resistance for the treated components may exceed 60%.
- The level of costs required at sea can be reduced compared to our best laboratory experiment.

Various matrices were compiled to carry out the program of experiments. In particular, one of the matrices is intended to understand the main effect of concentration, flow rate and trapped air in the form of microbubbles (MB) individually or with a polymer (bubble polymer), as well as the interaction between these three parameters. The results provide a baseline for comparison with several ejector configurations that have also been evaluated.

After performing the planned matrices and an initial analysis of the results, specific effects were investigated. Test matrices designed to study these effects were based on the results of previous matrices. The effects of MB, acceleration, maneuvering and propeller characteristics were investigated to complete the test objectives. The use of MB and trapped air (in the form of bubbles of the order of 100 μm) was the focus of the test suite. The target was to determine the advantage of increasing the polymer additive with bubbles, as well as the benefits of ejecting the bubbles separately. MB can be injected from the first ejector and/or secondary ejector. However, MB were injected from the primary ejector in combination with the polymer.

Based on the calculations of drag and the estimated thrust subtraction fraction, the propeller designer calculated a maximum speed of 28.8 knots at 475 rpm and a load of 274 tons. During the 2004 power tests, speeds of over 29 knots were achieved. In preparation for drag reduction tests, several configuration changes were made. In addition, to protect engines, propellers were usually limited to a maximum of 470 rpm during drag reduction tests. Thus a maximum speed of about 28 knots was expected without reducing polymer resistance.

When only polymer is injected over a lifting body (wing) (LB), the contribution of friction LB in the total drag of the vessel was calculated by Navatek as approximately 24% for a nominal height of 3.2 m from the surface of the water. During the tests, they were moved away from a predetermined height of movement above water (RH), so that RH varied with the speed of the vessel. With an increase in RH, the relative contribution of friction LB increases as a percentage of the total drag. For the speed range from 24 to 30 knots, over which most of the tests were to be carried out, this value was calculated in the range from 18% to 24% and amounted to approximately 22% at 28 knots.

During previous tests in a cavitation tunnel, a more than 60% reduction in friction was demonstrated over 6.2 m downstream of a similar ejector with a flow rate of 20 Q_s and 2000 wppm polyethylene oxide-309 (PEO-309). Therefore a 50 percent reduction in friction resistance was predicted for a 10-meter LB length and the ability to accommodate about a 25% increase in flow rate (17 Q_s at 3000 wppm) was programmed to achieve this level of drag reduction. Thus at a speed of 28 knots a decrease in resistance of 11% was expected, a decrease of 9% at a speed of 24 knots and up to 12% reduction at 30 knots if this speed were achievable. It has also been recognized that a significant reduction in friction resistance can affect other aspects of viscous resistance, such as a component called profile resistance, and possibly the level of interference or transition resistance that occurs at the junction of the LB and support legs. Consequently, these forecasts can also be modest.

The polymer mixing system on demand has proven practical and efficient. Polyoxyethylene oxide (PEO) 309/310 Dow worked well in seawater. The film-forming properties of PEO were evident in the marine environment. After proper injection and when mixed with seawater, the polymer continued to have a significant effect. Apparently, the polymer film acts as a proper viscoelastic coating. In particular, it is anisotropic in the direction of flow, edge effects are eliminated, and due to its subtlety, deformations caused by turbulence remain in the viscous sublayer. These experiments show that the temporary stable of the polymer provided either a more than 50% decrease in friction, or a slightly smaller decrease to a combination of friction and other components of the viscous resistance LB. The spatial and temporal stable of the PEO film reduced the flow rate of the polymer solution significantly lower than expected. This phenomenon of synergism between a viscoelastic surface and polymer displacement is well documented in [37].

With very modest the flow rate of the polymer solution injection, a number of performance parameters can be improved. These include reducing power requirements and fuel consumption, allowing you to achieve higher speeds under more comfortable driving conditions. The impact on the performance of a particular ship will be related to the geometry of the ship, power level and ship control system. Polymer injection systems can be compact and occupy the rear on existing ships. As a result, the implementation of polymer release can also reduce power, fuel storage and control requirements without any reduction in vessel speed.

The results obtained could be improved, as it became clear after publication [84]. K.J. Moore has filed a number of patents for injection into the boundary layer of various solutions. In particular, in the full-blown research, the scheme of the pre-slit chamber [452], located in the supply channel in front of the slit chamber according to the patent [450,451], was used. This scheme was first published in Ref. [596]. In Refs. [83,89,90] other forms of pre-slit chambers are given. In Refs. [450,451,596] in the inlet channel, the pre-slit chamber is located on the left with respect to the slot. A moving flow in the pre-slit chamber spins a vortex that stretches the polymer macromolecules and deflects the flow from the curved surface so that the Coanda effect is weakened before exiting the slot. In [83,89,90], a system of pre-slit chambers is located on the convex surface in front of the slit on the right, as a result of the

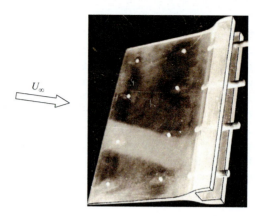

Figure 7.49 A photograph of the xiphoid nose of the hydrofoil model shown in Figs. 6.3, and 6.4 (Section 6.3).

induced rotation in these chambers enhances the Coanda effect, and the jet injected from the slit adheres more strongly to the streamlined surface. In addition, in the system of pre-slit chambers [83,89,90], CVSs are formed in the form of longitudinal vortices, which increases the efficiency of solutions flowing from the slit and, regardless of the composition of the liquid, leads to a decrease in friction resistance due to the structure of longitudinal vortices in the boundary layer.

The second factor in improving the conditions for injection of near-wall jets is the presence of xiphoid tips ([378], Section 7.7). Fig. 7.49 shows the design of the nose of the hydrofoil, which was installed during some experiments on the wing model shown in Figs. 6.2B,6.3, and 6.4 (Section 6.3). The experimental results showed a significant increase in the efficiency of polymer solutions during injection on models with xiphoid tips. This form of the nose of the wing should be installed on vessels of the type SEA FLYER, as well as other types of vessels with a dynamic principle of maintenance. To determine the effectiveness of this form of the nose of the wing (Fig. 7.49), further experimental investigations are needed.

The third factor for improving the conditions of injection of near-wall jets is to use the results of studies of the physical laws of the development of CVS in the boundary layer of rigid and elastic surfaces, as well as during the injection of polymer solutions into the boundary layer given in the previous chapters. Maltsev performed experimental studies during injection into the boundary layer of near-wall water jets and downstream MB [474]. Parietal jets contributed to the formation in the boundary layer of the CVS in the form of longitudinal vortex structures.

During the discussion with K.J. Moore, the role of the multislits structure of sharks, V.V. Babenko proposed a three-slot injection into the boundary layer, which was later issued by K.J. Moore in the form of a number of patents [83,89,90]. It was assumed that a stream of water should be injected through the first slit, a polymer solution through the second, and MB through the third slit. This order provides a significant reduction in the diffusion of liquids from the second and third slits.

In addition, it is always necessary to take into account the laws of the experimentally investigated problem of the interaction of disturbances in the boundary layer (the receptivity problem) [53,139,247].

The specific shape of the nose of the profile, resembling the shape of the nose of the wing profile shown in Fig. 7.49, implemented on the steering complex of the vessel [99]. The introduction of this form has significantly increased the maneuverability of the vessel.

7.12 Experimental investigation of the influences of bubbles injection on the drag reduction

Fig. 3.33 (Section 3.6) shows a model of CVS formed at the stages of transition of a laminar boundary layer to a turbulent one and Fig. 3.40 shows a model of the development of the CVS in the boundary layer of a rigid smooth plate at all stages of the transition, including a developed TBL. Chapters 5–7 present the results of an experimental investigation of some methods of drag reduction. In Refs. [139,247], the results of experimental investigations of the interaction in the boundary layer of various types of CVS are presented. All these results make it possible to predict the interaction of natural CVS of the boundary layer with the introduced CVS and determine the correct choice of the combined method of drag reduction for specific conditions flow around the body.

The research results presented in Section 7.11 allow us to conclude that the characteristic types of the boundary layer CVS have not yet been taken into account in the absence of any disturbances, as well as the features of the change in the CVS depending on the chosen method of drag reduction.

In Section 5.9, the physical justification of the mechanism of interaction between the flow and the elastic surface is considered, taking into account the patterns of receptivity of various disturbances by the boundary layer [53,139,247]. In developing the schemes and designs of elastic surfaces, the structural, kinematic and dynamic principles, as well as the principle of agreement the interaction of the CVS of the boundary layer with the structure and characteristics of the elastic surface are taken into account [24,35,53,69,139,247]. The dynamic and kinematic characteristics of the boundary layer cause a reaction in the elastic surface, which as a result of this action generates CVSs in the boundary layer that interact with the natural CVS of the boundary layer. The result of this interaction can be either stabilization of the development of natural CVS of the boundary layer, or their modification, or acceleration of the alternation of CVS. It will lead to, in particular, to either a significant decrease in drag or an increase in drag. The investigations of the receptivity problem made it possible to develop new control methods for natural CVS of the boundary layer (Section 6.1).

In investigated the method of drag reduction by injection of MB, it is necessary to perform a similar approach, as in the investigation of the effect of elastic coatings on the characteristics of the boundary layer. For this, it is necessary to initially take into account the development of natural CVS boundary layer in the absence of various disturbances. The results of such investigations are given in Section 3.6. Considering the models given in Section 3.6 (Figs. 3.33 and 3.40), it is further necessary to imagine what effect the introduction of MB into the boundary layer will have on these CVS. In accordance with the above principles of the receptivity problem, it is necessary to introduce MB into the boundary layer not in a continuous veil, but in a strictly ordered form, and the intensity of the introduced microbubble disturbances should correspond to the CVS intensity of the boundary layer.

Fig. 3.45 (Section 3.7) shows the dependence of the change in the maximum values of the velocity pulsations u' on the number Re. One can see an increase in the intensity of pulsations u' at the nonlinear stages of the transition and a gradual decline during the transition to a turbulent flow regime. The character of the curve is similar to that obtained in Ref. [378] when studying with a hot-wire anemometer the transition of a laminar boundary layer to a turbulent one in a wind tunnel. The data in Fig. 3.45 confirm the hypotheses [31,621] that the minimum friction coefficient corresponds to the flow patterns in the boundary layer at stages IV and VI of the transition (Section 3.6, Figs. 3.33 and 3.40; Section 5.9, Fig. 5.61). All this indicates that maintaining in the boundary layer systems of longitudinal vortex structures helps to drag reduction with minimal energy consumption.

A simplified approach to solving this problem will be given below [75,401].

Viscous drag, denoted by the coefficient of friction C_f, is a significant fraction of the drag of the vessel. It is 50%–65% of the total drag in the sea water of a surface vessel and 60%–75% of the drag for underwater vessel.

As can be seen from Section 3.11, injection air onto the surface of the hulls is one of the methods for reducing friction. Air injection can be performed in several ways:

1. Injection of bubbles obtained by electrolysis from nozzles or screens [153,440].
2. Air film from cuts in the hulls [400].
3. The formation of an air cavity [155].

The first approach is discussed below.

7.12.1 Interaction of bubbles with the boundary layer

At present, a large number of theoretical and experimental investigations of the kinematic and integral characteristics of a two-phase boundary layer have been performed, when a veil of gas bubbles is fed into the liquid. Some experiments are summarized in Table 7.10.

With the film character of the gas flow in a liquid, an assumption is made that the properties of the medium are monotonic across the boundary layer. In theoretical formulations of the problem, the assumption of uniformity of the front of the gas film in the transverse direction is used. A number of other assumptions are also used.

Another approach is also used—diffusion or bubble, in which a gas-water mixture moves near the wall. It has been shown theoretically and experimentally that

Table 7.10 Published experimental results of drag reduction during bubble injection.

Injection method	Test place	U_∞, m/s	Consumption	C'_f/C_f	Author
Electrolysis	Towing tank, body, L = 1 m	0.33–2.0	0–75 amp/s	0.7–0.95	McCormick
Bottom screen	NA	1. 4.36	1. 0%–0.3%	0.7–1.0	Barbanel
		2. 8.55	2. 0%–0.4%	0.45–1.0	
		3. 10.9	3. 0.4%–0.8%	0.2–0.4	
Top screen	0.1 m	1. 5	1. 0%–0.25%	0.75–1	Takahashi
	0.015 m	2. 7	2. 0%–0.2%	0.65–1	
	3 m lengths	3. 10	3. 0%–0.25%	0.7–1.0	
Injectors	0.6 × 0.6 m	8.0	1. 35 L/min	0.95–1	Yoshida
	Channel		2. 100 L/min	0.8–0.9	
	Slit 150 mm		3. 200 L/min	0.2–0.5	
	Width 20 degrees angle Inj.				

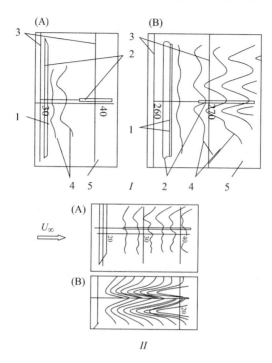

Figure 7.50 Graphic copies of photographs of tellurium lines in the *XZ* plane for $U_\infty = 10.5$ cm/s, $y/\delta \approx 0.2$ at the beginning (A) and at the end (B) of the working section for horizontal (I) and inclined (II) plates: (1) tellurium wires; (2) holder; (3) mark of the distance from the beginning of the working section (cm); (4) velocity profiles; and (5) the second bottom [378].

under certain conditions the gas saturation of the boundary layer leads to a significant decrease in the friction drag. In this case, the method of injection a gas sheet (through a porous surface or slot nozzle) and the size of the gas bubbles are essential.

Using the tellurium method was performed boundary layer investigation along the working section of the hydrodynamic stand of low turbulence at $\varepsilon \approx 0.05\%$ (Section 3.3). When photographing the $U(z)$ profiles on the side, no deformations in the vertical direction were detected—a plane jet of tellurium is visible (Fig. 3.8, Section 3.3). Fig. 7.50 shows graphical copies of photographs of $U(z)$ profiles at various x values. The relevant photographs are shown in Figs. 3.8 and 3.16 (Section 3.3). The Reynolds number of stability loss, determined experimentally, was 5.4×10^4. Photographing showed that already at $Re = 4 \times 10^4$, the $U(z)$ profile near the critical layer begins to deform (Fig. 7.50A): in the boundary layer, preconditions for the development of disturbances appeared even to the point corresponding to $Re_{los.st.}$. As the number Re increases, the distribution $U(z)$ is ordered (Fig. 6.50) and already at $Re \approx 10^5$, a stable regularity appeared with $\lambda_z = (1.5-2)\delta$ and the arrangement of "peaks" and "hollows" for certain values of z (Section 3.7, Fig. 3.43). In measurements in each series of experiments, the propagation velocity downstream of the tellurium lines is constant and depends on the location along z.

Measurements of the thickness of the boundary layer showed that in the region of the location of $Re_{los.st.}$ the $U(z)$ profile deformation was observed in a very narrow region $y/\delta = 0.2-0.3$, that is, near the critical layer, outside of which the profiles are plane-parallel. With increasing x, the region of maximum values of the nonlinear deformation of the profile front shifts upward in y, moving from $0.2(y/\delta)$ to 0.4 (y/δ), which is consistent with Klebanov's measurements [342]. At the same time, the thickness along y of this region increases.

Thus the investigations performed showed that, with a low degree of turbulence, the transition stages, although they retain the characteristic features and sequence of alternation, are very slow in their development. Nonlinear effects arise at the stage of a linear plane wave and develop in a narrow cone layer symmetrical with respect to the XZ plane of the critical layer, so that the cone expands (Section 3.6, Fig. 3.31). In this case, peaks and hollows were formed under the influence of nonlinear deformation of a plane wave caused by curvature of streamlines and a change in primary vorticity. Longitudinal vortex systems have not yet formed. In Section 3.3, Fig. 3.9 shows photographs of the velocity profile $U(z)$ at various distances of the tellurium wire from the surface of the flat bottom of the working section $y_{t.w.}$ at $U_\infty = 6.7 \times 10^{-2}$ m/s, $x_{t.w.} = 1.1$ m, and $y_{t.w.} = 3 \times 10^{-3}$, 4×10^{-3}, 5×10^{-3}, 6×10^{-3}, 7×10^{-3}, 8×10^{-3}, 9×10^{-3}, and 12×10^{-3} m. The results of visualization made it possible to develop a flow structure scheme at the initial stages of nonlinear deformation of a plane wave at large ε (Fig. 7.51).

Near the surface additional vortex pairs form, leading to a characteristic small-scale deformation of the velocity profile $U(z)$ (Section 3.3, Fig. 3.9). Taking into account the nonstationarity revealed in the experiments, we can assume that the entire system of longitudinal vortices is mobile, that is, able to change in time and space the shape, size, number of vortex pairs, their ratio and trajectory of movement. This, in turn, is reflected in the kinematic characteristics: small differences in (A) and (B) of the vortex systems (Fig. 7.47) cause a significant change in the profiles of $U(z)$.

7.12.2 The influence of the streamlined surface on the boundary layer

Currently, various mechanical devices have been developed and are used in scientific research and in technology, designed to form longitudinal and transverse vortices in the boundary layer [1,136,137,234,507]. Examples of constructive solutions can be found in scientific articles and patents. Despite their effectiveness, they have one drawback—they create additional friction drag.

One of the constructive solutions to the problem of the formation of longitudinal vortex structures in the boundary layer is a ribbed surface, the so-called "riblets." Below we consider a constructive solution that can significantly reduce the

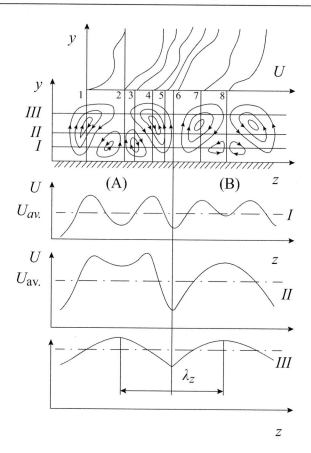

Figure 7.51 The flow structure at the initial stages of nonlinear plane wave deformation with a large degree of turbulence ε. Above: 1–8 longitudinal sections showing the velocity profile $U(y)$; (A), (B) types of longitudinal vortex structures; Bottom: I, II, III—XZ sections showing the change in velocity $U(z)$, $U_{av.}$- average speed.

additional drag and implement the recommendations of the receptivity problem. To do this, five strips of standard adhesive tape measuring $6 \cdot 10^{-5} \times 0.003 \times 0.23$ m in size were glued to the bottom of the working area along its longitudinal axis at a distance $z = 0.012$ m from each other, under which pieces of wire with a diameter of 0.12×10^{-3} m and 0.2 m long were previously laid (Fig. 7.52).

Initially, at $Re = 7 \times 10^4$, the concentrated action of such a system with a reduced strip length (0.02 m) was verified. The experimental results are shown in Fig. 7.52A: the distribution of $U(z)$ does not differ from the natural velocity profile—in the absence of a system of wires at the bottom. The velocity profiles $U(z)$ obtained by tellurium method of F.X. Wortmann with a distributed effect on the boundary layer simulating the flow around a surface with longitudinal ribs are shown in Fig. 7.52 (B–G). Ribbing of the surface began at $x = 0.9$ m from the leading edge of the bottom of the working area. It is clearly seen that the "dentation" of

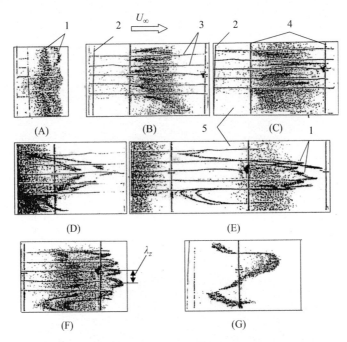

Figure 7.52 The influence of a ribbed surface regular in z on the $U(z)$ boundary layer ($\lambda_z = 0.012$ m): (A) rib height 0.002 m, $Re = 7 \times 10^4$, $y = 1.0 \times 10^{-3}$ m (rib length 0.2 m); (B) $Re = 6 \times 10^4$, $y = 1.5 \times 10^{-3}$ m (ribs length 0.2 m); (C), (D), (E) $Re = 10^5$, respectively, $y = 5 \times 10^{-4}$ m, $y = 1.5 \times 10^{-3}$ m, $y = 1.0 \times 10^{-3}$ m; (F) $Re = 1.4 \times 10^5$, $y = 1.5 \times 10^{-3}$ m; g $Re = 1.5 \times 10^5$, $y = 1.5 \times 10^{-3}$ m. (1) speed profiles; (2) tellurium wires; (3) longitudinal wires—"ribs"; (4) mark of the distance from the beginning of the working section (cm); and (5) the second bottom.

the tellurium cloud appears near the surface of the bands in the range of numbers $Re = (0.7-1.5) \times 10^5$. Moreover, the large-scale shape of the cloud remains the same both with its removal from the tellurium wire (D), (E); and when moving along the strips (F) and beyond (G)—visualization was performed 5 cm below the trailing edge of the strips. However, the rate of increase in disturbances, determined by the degree of deformation of the cloud when it moves downstream in the case of "F," is noticeably lower than in the case of "D": the overtaking part of the cloud became flatter, and the amplitude of the change in velocity along z decreased. This can be interpreted as the result of a more optimal interaction of introduced disturbances with $\lambda_z = 0.012$ m with natural disturbances with an increase in the Reynolds number. N.F. Yurchenko also investigated ribbing for other values of λ_z.

7.12.3 Test with electrolysis of bubbles

Experiments were also performed when the wires glued to the plate were not insulated and direct or pulsed current was passed through them with a voltage of 10 and

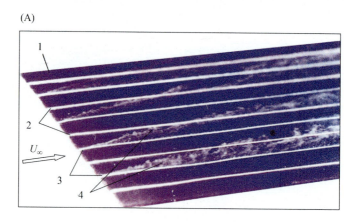

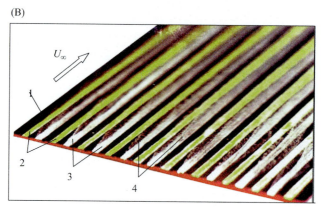

Figure 7.53 Photographs of plates for various types of combined control of the boundary layer: (A) plates with wires, (B) plates with strips; (1) elastic plates, (2) wires or metal strips, anodes, (3) the same as (2), but cathodes, and (4)—vertical layers of bubbles [84].

400 V, respectively (pulse frequency 0.1–0.5 s). The wires were glued to various plates and surfaces, including elastic ones.

Fig. 7.53 shows examples of plates on the surface of which glued wires (Fig. 7.53A) and metal strips (Fig. 7.53B). As a result of electrolysis above the wires and metal strips, vertical sheets of hydrogen bubbles were formed, moving in the longitudinal direction. Two vertical layers of bubbles were formed above the strips—along the edges of the strips.

Such longitudinal veils created intense longitudinal vortex disturbances. In addition, in the longitudinal direction along the wires, the drag of the two-phase flow sharply decreased. The longitudinal wave of the emerging MB forms a system of stabilizing barriers in the transverse direction. Large vortex structures coming from the outer boundary of the boundary layer are disintegrated in the longitudinal direction. There is a change in the kinematic balance of energy in the boundary layer. In the near-wall region, vortex structures are intensely stabilized so that stable longitudinal CVS are formed.

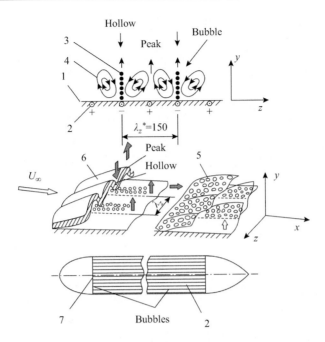

Figure 7.54 The microbubble generation scheme in the boundary layer: (1) streamlined surface, (2) electric wires, (3) vertical layers of microbubbles, (4) longitudinal vortex systems, (5) horizontal layers of microbubbles, (6) envelope surface of longitudinal vortices, and (7) transverse circular electric wire or transverse a slot for the injection of a veil of bubbles [139].

Fig. 7.54 at the top is a schema illustrating the formation of longitudinal vortex systems in the boundary layer by injection of vertical microbubble systems into the boundary layer. Below is a schema of an axisymmetric body that has a nose and tail metal surfaces, or a metal ring is mounted flush in the nose or there is a slot for the injection of MB. Injection from the bow of an axisymmetric microbubble veil stabilizes vortex systems that form at the outer boundary of the boundary layer or enter the boundary layer from its outer boundary.

Efficiency is achieved when the size of the MB does not exceed 50 μm. The layers of bubbles in Fig. 7.54 are located at a distance $\lambda_z^* = 150$, which approximately corresponds to the step of the Klein vortices [345,346]. In this case, such vertical layers either stabilize the Klein vortices existing in the TBL or form vortex systems of a given scale. Thus the size of MB no longer has such a value and may be larger. Injection of MB into the boundary layer is possible through cross-sectional slots and through longitudinal conductive elements or periodically located slits.

The distances between the wires should correspond to the λ_z of the longitudinal CVS of the boundary layer (Section 3.6, Figs. 3.33 and 3.40), and with an increase in the distance along the body or flow velocity the characteristics of the boundary layer and, accordingly, the values of λ_z of the longitudinal CVS change. As shown in Fig. 5.61 (Section 5.9), when applying the CVS control methods, the extent and characteristics of the development of the natural CVS of the boundary layer are modified. Fig. 6.28 (Section 6.6) shows that when applying any CVS control methods, it is necessary to take into account the change in the length of the structure and the nature of the CVS types. Therefore in particular, when injecting MB into the boundary layer, it is necessary to continuously or discretely change the pitch along the body length in the transverse direction of the arrangement of wires or strips in accordance with the types of CVSs shown in Figs. 3.33 and 3.40 (Section 3.6). It should be borne in mind that the use of MB will modify the shape and length of the CVS shown in Figs. 3.33 and 3.40.

It can be concluded that energy consumption is the most it is extremely important to consider that with the effective influence on the CVS of various control methods, the characteristics and structures of the CVS significantly change. This should be taken into account and significantly change the geometry of the arrangement of wires along the length of the body in the considered method in accordance with the new values of λ_z. With the help of electric strips (Fig. 7.49B), vertical veils with a significantly lower λ_z are formed in the boundary layer; therefore it is better to install the strips in the region of formation of the TBL.

It can be concluded that energy consumption is the most economical in the implementation of this method of microbubble formation. In addition, this method takes into account the features of vortex structures in the boundary layer at different flow velocities and Reynolds numbers, and does not depend on the depth of the body's motion.

To measure the effect of a two-phase flow on the characteristics of the boundary layer, experiments were also performed on a model of a body of revolution (Section 7.5, Fig. 7.21). The nose and tail parts of the model were installed made of metal (brass), which were connected to a direct current source. During the test, the voltage was 10–30 V. Under certain conditions, an annular veil of bubbles flowed from the trailing edge of the nasal fairing into the boundary layer. It was found that the drag reduction in the model depending on the current and voltage parameters reached 10%–15%. Similar experimental studies were performed by McCormick in the hydrodynamic channel (Table 7.10).

7.12.4 The combined method of forming longitudinal vortex systems in the boundary layer

In Refs. [1,136,137,234,507] the influence of devices for the concentrated formation of longitudinal vortex systems (LVS) in the boundary layer (BL) was investigated using small wings of various sizes mounted on and near a rigid surface. Chapter 4, Experimental investigations of the characteristics BL on a smooth rigid curved plate (Part I) presents the results of an investigation of the structure of the outer layers of the skin of high-speed hydrobionts, which make it possible to form the specified CVSs, including LVS, in the BL.

The distributed method of introducing LVS into BL using an elastic plate 10 was investigated (Section 5.1, Fig. 5.7D; Section 6.4, Fig. 6.10). The plate 10 is

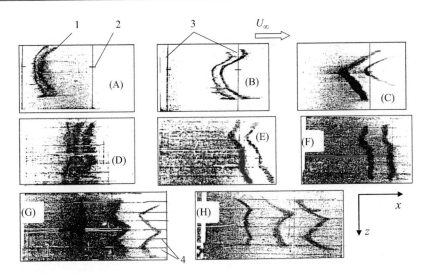

Figure 7.55 Visualization of the velocity $U(z)$ at $U_\infty = 0.05$ m/s: a rigid plate $x = 1.6$ m, $y = 0.002$ m (A), 0.004 m (B), 0.006 m (C); elastic plate 10 $x = 2.18$ m, $y = 0.002$ m (D), 0.004 m (E), 0.006 m (F), 0.008 m (G), 0.01 m (H); (1) velocity profiles $U(z)$, (2) marker of the longitudinal axis of symmetry of the working area, (3) distance markers along x (0.1 m), (4) longitudinal glue seams.

made of longitudinal elastic strips consisting of FE material (Table 5.14, Section 5.1, Figs. 5.7C—5.9). This design of elastic plates made it possible to introduce in BL longitudinal vortex structures of a given scale [69]. The introduction of LVS into BL using the plate 10 will be effective when the main points of the receptivity problem are fulfilled. It is necessary that the size and intensity of the introduced LVS be comparable with similar CVS in the natural BL.

The flow structure in the boundary layer was visualized using the Wortmann tellurium method (Fig. 7.55) in a hydrodynamic stand of small turbulence (Section 3.1, Fig. 3.1) when flowing around an elastic plate 10, the outer layer of which is glued from longitudinal strips of elastic foam material (FE) 10 mm wide.

The elastic plate was placed on the insert (Section 5.1, Figs. 5.4—5.6), which was located at the end of the working section (Section 3.7, Fig. 3.42). The velocity profiles $U(z)$ were photographed.

Tellurium clouds displaying the velocity profiles $U(z)$ were created after 0.5 s. Three tellurium wires were located at different distances from the surface. This made it possible to fix the simultaneous development of the disturbing motion in δ. Dark longitudinal stripes 4 are glue seams (longitudinal stiffeners). The first three photographs (A)—(C) visualize the flow field when flowing around the hard surface of the second bottom of the working area at $x = 1.6$ m ($Re = 1.1 \times 10^5$)—to the location of the insert with an elastic plate of 10 ($x = 2.18$ m). In the region where the insert was located, the Reynolds number was $Re = 1.4 \times 10^5$. Fig. 3.9 (Section 3.3) also shows photographs of the velocity profiles $U(z)$ when flowing around the hard bottom of the working section at $x_{t.w.} = 1.1$ m. An elastic plate with longitudinal ordering stabilized the development of disturbances (Fig. 7.55D− F) compared to flow around a rigid plate (Fig. 7.55A−C, Section 3.3, Fig. 3.9), and with increasing distance from the wall (Fig. 7.55G, H) longitudinal vortex structures with $\lambda_z \approx 38$ mm at $y = 0.008$ m were formed in the BL. When comparing with a rigid plate on which thin metal wires are glued (Fig. 7.52), it is seen that when flowing around the elastic plate 10 in the BL, small-scale dentations of the velocity profile $U(z)$ does not appear.

At a low flow velocity, the elastic plate 10 effectively stabilized the disturbing motion. To increase the intensity of the introduced LVS, longitudinal thin metal wires with a diameter of 1.2×10^{-4} m were glued on the surface of the plate 10 in the places of glued joints of the strips of the outer layer of the elastic plate (Fig. 7.56).

Current was passed through these wires, while the longitudinal regions of the elastic material, in the region of which the wires were located, were heated. Since the elastic material had good thermal insulation properties, heating wires led to a change in the mechanical characteristics of the elastomer only in the area of the wires.

On the other hand, insignificant heating led to small convective flows in the near-wall layer BL and insignificant microbubble formation due to electrolysis. In Fig. 7.56, the photographs are arranged in such a way that the photographs (A)−(I) contain photographs with the same y coordinate. The horizontal rows of photographs mean respectively: the first row without heating, the second row at a current with a voltage of 7.8 V and the third row at a voltage of 12.2 V.

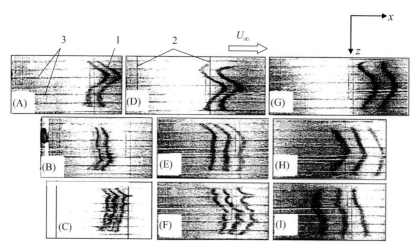

Figure 7.56 Visualization of the velocity $U(z)$ when flowing around the elastic plate 10, $U_\infty = 0.06$ m/s, $x = 2.18$ m: $y = 0.002$ m (A)−(C), 0.003 m (D)−(F), 0.005 m (G)−(I); without heating the wires (A), (F), and (G), with heating the wires at a voltage of 7.8 V (B), (E), and (H) and 12.2 V (C), (F), and (I); (1) velocity profiles $U(z)$, (2) distance markers along x (0.1 m), and (3) longitudinal wires.

Heating the wires at a voltage of 7.8 V (Fig. 7.56B, E) led to stabilization of the BL near the surface of the plate—at $y = 0.002$ and 0.003 m and, to a lesser extent, at $y = 0.005$ m (H). More intense heating at $y = 0.002$ and 0.003 m leads to the appearance of small-scale LVS near the surface (C), (F), with λ_z twice, moreover, distance between wires. With increasing distance from the wall (I) LVS was no detected: the velocity profile $U(z)$ was almost uniform.

The measurements performed, on the one hand, confirm the effectiveness of the method of interaction of disturbances in the BL (receptivity). On the other hand, it is obvious that such a method can change the waveguide characteristics of a complex waveguide (Section 3.8). In addition, these experiments are one of the variants of combined control methods for CVS (Section 6.1, Table 6.1).

7.12.5 Injection of bubbles from the upper screens

Table 7.1 summarizes some of the results of systematic testing of a plate with microbubble injection. In Ref. [153], investigations were carried out during the injection of bubbles from the lower perforated surface with round holes. The drag was measured in the hydrodynamic towing tank at towing speeds of $4 < U < 8$ m/s. The mixture of bubbles and water along the lower surface is characterized by the maximum volumetric concentration of bubbles in the BL φ_{max} measured by the laser. The total force is measured on the test plate, and the results are reduced to the coefficient of friction C'_f with injection of bubbles and C_f without injection of bubbles. The test results are characterized by the ratio C'_f/C_f on the φ_{max}. Fig. 7.57 also shows the results of measuring the resistance obtained by injecting bubbles on the upper plate in the towing channel [545].

The injection of bubbles on the upper plate leads to some differences in the drag reduction compared to the injection of bubbles on the plate located on the lower side of the working section. The difference lies in the size of the bubbles due to differences in the method of injection of bubbles, and also due to the influence of buoyancy of the bubbles. In Ref. [153] the injection of bubbles is limited by the effective contact region.

With the method of injection of bubbles on the upper wall [545], as a rule, bubbles are collected on a streamlined plate, and, as the results in Fig. 7.57, this drag reduction by 15%–20%. Fig. 7.57 also shows that this drag reduction process is more efficient for large volumes of bubble injection at higher flow rates. With these, the buoyancy of the bubbles is relatively smaller and has a smaller effect on the drag reduction process.

7.12.6 Injection of bubbles from nozzles

In Table 7.10, the upstream bubble test results are also shown. Bubbles were injected through nozzles located just below the upper wall of the working area with a cross-section of 0.6×0.6 m hydrodynamic tunnel from the University of Tokyo [618]. The diameter of the bubble from the injectors was larger than the diameter of the bubble created in the works given in Section 7.12.5. Another significant

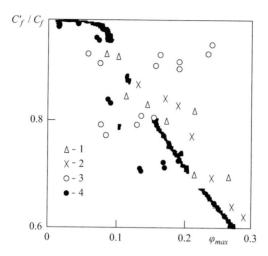

Figure 7.57 The dependence of the relative drag C_f'/C_f on the volume concentration of bubbles in the boundary layer φ_{max} at various flow rates according to [75]: (1) $U = 4.36$ m/s, (2) $U = 8.55$ m/s; and [580]: (3) $U = 5.0$ m/s, and (4) $U = 7.0$ m/s at a distance of 500 mm downstream.

difference was the orientation of the nozzles to the surface of the control plate. During these tests, the bubbles entered the stream below the plate and then collected on the upper surface as they moved downstream. The buoyancy of the bubbles helped the process through contact with the upper plate. These tests were carried out at speeds of $4 < U < 8$ m/s.

The total force was measured on the plate, and the results are presented as the ratio C_f'/C_f. The results of these tests are shown in Fig. 7.58 as the dependence of C_f'/C_f on the concentration of bubbles to water φ at different bubble rates. A comparison with the results presented in Ref. [153] indicates that the drag reduction by injection through a slit or through a porous surface is different.

The results obtained allow us to conclude:

1. The presence of bubbles in the BL has a beneficial effect on reducing energy loss due to lower surface resistance.
2. The ratio C_f'/C_f is of the order of $0.2 < C_f'/C_f < 0.85$.
3. The ratio C_f'/C_f is sensitive to the number of injected bubbles, which is expressed through φ.
4. At low velocities, the orientation of the injected bubbles can have a strong effect on the drag reduction.

The use of bubble injection has many uses for drag reduction in seawater.

7.12.7 The effect of microbubbles on the integral characteristics of axisymmetric models

In Refs. [178,440,448], the effect of gas bubbles on the integral characteristics of the BL was studied. V.V. Babenko designed a model of a body of revolution to

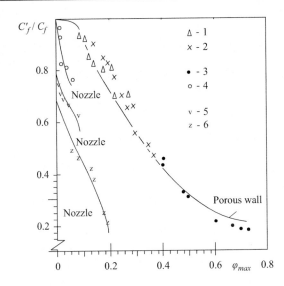

Figure 7.58 The dependence of the C'_f/C_f on the volume concentration of bubbles in the boundary layer φ_{max} at various flow rates according to [153]: (1) $U = 4.36$ m/s, (2) $U = 8.55$ m/s, (3) $U = 10.9$ m/s; and [618] at $U = 8.0$ m/s and flow rates (4) 35 L/min, (5) 100 L/min, (6) 200 L/min.

study integral characteristics for combined methods of controlling the BL (Section 6. 1, Table 6.2; Sections 7.1 and 7.5, Fig. 7.21). In Ref. [378], some obtained results of integral characteristics were presented in the study on this model. Fig. 7.25 (Section 7.5) shows the results of experiments performed on models with different types of nasal contours without and when water is injected through the nasal slots into the BL. During the experiments, it was found that the water contained MB when filling a hydrodynamic tunnel in winter. In addition, it was recorded that in a closed hydrodynamic pipe from friction on the blades of the screw there is an increase in water temperature by about 1 degrees during one hour of operation of the screw. Therefore during the measurement process, the water temperature was constantly recorded and corrections were made for the value of the kinematic viscosity coefficient when processing the results of experiments. When equalizing the temperature of cold water with the temperature of the laboratory, the water was gradually degassed. Given this, experiments were carried out, as a rule, after degassing water.

Fig. 7.25 (Section 7.5) shows the dashed lines for the model with a short xiphoid tip: curve (7) without a turbulizer and with open slots in a highly aerated stream, (8) the same, but in a deaerated stream, (9) the same as (8) but with a turbulator. During measurements in an aerated stream, measurements began at a low flow velocity and then with increasing flow velocity. During the measurements, deaeration of the flow occurred, therefore with increasing Reynolds numbers, the influence of MB on the drag reduction decreased. Probably, at a constant value of flow aeration, curve (7) should be an equidistant curve (3), while the efficiency during flow around an aerated stream will remain in the entire range of Reynolds numbers. Calculations show that, when comparing curves (7) and (8), the influence of MB dissolved in water on the drag reduction in the model at $Re = 3 \times 10^5$ is 47%, and as the flow degasses at $Re = 6 \times 10^5$, it is 20%. As shown in Section 7.5, the contribution of the profile attaching the model to the strain gage is 14% of the total measured resistance. Therefore the actual effect of aerated flow around the model was 61%. The calculation of the resistance efficiency is performed according to the formula 12.8 (Section 6.6).

Thus the results of a large number of experimental investigations carried out allow us to conclude that the injection of MB into the BL is significantly effective in drag reduction.

7.13 Modeling of disturbances development in the flow behind the ledge

In various types of shear flows, there are areas with a high concentration of shear stresses [84]. As an example of such areas, one can cite a place near the edge of the hole when a flooded stream expires, or a place near the edge of the slit through which liquid solutions are injected into the BL. The same areas are places near the edge of various ledges or deepening's on the streamlined body. Behind the trailing edge of various types of protrusions, a region of flow separation is formed, which, with an increase in speed, leads to cavitation flow around the body. A pressure jump forms on this edge, leading to the appearance of disturbances behind the edge at the boundary of the separation region and the main flow. The second region with a high concentration of shear stresses is the zone of attachment of the separation region to the wall.

The integral effects have been investigated arising in cavitation flows [199,600], for example, spectral characteristics of oscillating axisymmetric cavities, characteristics of plane flows behind a ledge on a plate [186], and on rotating bodies [291,304,601]. In Ref. [276], the influence of the thickness of the BL on the structure of the near-wall flow behind a two-dimensional protrusion was also investigated. In Ref. [186], CVCs formed behind cavitators made in the form of a disk, sphere, and hemisphere, came to life, and on a rotating rotor blade were experimentally investigated. In Ref. [186], the acoustic spectra of cavitation vortex bundles flowing down from the edge of a rotating propeller blade were measured. In Ref. [276], low-frequency quasilinear oscillations near the surface of the cavity were also measured.

In Ref. [186], the "water-air" boundary was photographed during natural cavitation and ventilated cavitation in the flow around a sphere, a truncated sphere and an paraboloid of rotation (Fig. 7.59). Visible waves were observed, similar to the T.-Sh. waves, which are deformed upon transition to nonlinear waves. At a lower flow velocity around the truncated sphere (Fig. 7.59D), T.-Sh. waves were observed at the water-air interface. Modeling of such a flow showed that T.-Sh. waves occur in

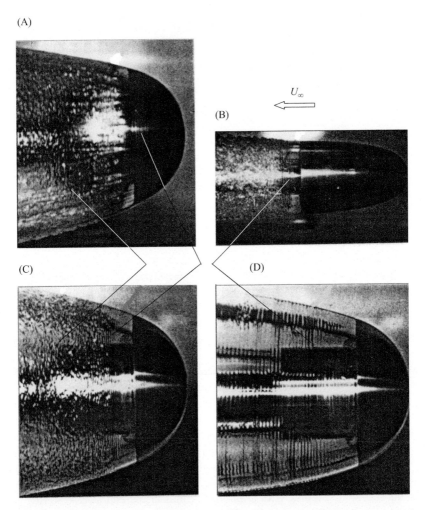

Figure 7.59 Visualization of cavitation flow around a sphere, two truncated spheres and a rotation paraboloid: (A) sphere, natural cavity $U = 35$ ft/s; (B) rotation paraboloid, ventilated cavity $U = 25$ ft/s; (C), (D) truncated spheres, ventilated cavity, $U = 20$ ft/s:
(1) Tollmien−Schlichting waves, (2) nonlinear Tollmien−Schlichting waves [186].

the zone of instability of the neutral curve (Section 3.4 of Fig. 3.20). At a higher flow velocity around the truncated spheres, the T.-Sh. waves are quickly transformed into nonlinear CVSs (Fig. 7.59B). In Fig. 7.59A when flowing around a sphere, the CVSs were photographed in the form of T.-Sh. waves, which are observed in front of the region and behind the separation region.

Fig. 7.60 shows the smoke visualization of the CVS of the BL when flowing around a sphere.

In the BL of the sphere, a T.-Sh. plane wave arises, which quickly transforms into a three-dimensional shape, and then the separation is fixed. The fixed types of

Figure 7.60 Coherent vortex structure visualization of the BL of the sphere at $Re_D \approx 10^5$.

CVSs correspond to stages I–III of the transition in the BL during flow around a flat plate (Section 3.6, Fig. 3.33). When comparing Fig. 7.60 with Fig. 7.59A, it is seen that during cavitation flow around the sphere, the classical separation is not observed behind the sphere. During cavitational flow around the sphere, a cavity of the cavity is formed, at the boundary of which in the shear layer the CVSs characteristic of the BL of a flat plate continue to develop (Section 3.6, Fig. 3.33).

In Fig. 3.34 (Section 3.6) shows smoke visualization of the development of CVS in the BL of axisymmetric bodies [188,347]. When comparing the results [188,347] with the visualization of the CVS on the surface of the cavity (Fig. 7.59), it can be seen that during the cavitation flow around the CVS on the surface of the cavity, they develop in the same way as in the case of continuous flow in the BL. The greatest similarity is seen when visualizing the surface of the cavity behind the paraboloid of revolution, which has a longer surface (Fig. 7.59B). When comparing Fig. 7.59C, D, it can be seen that at a lower flow velocity (D) the CVS s in the form of T.-Sh. waves develop longer than at an increased speed (C). In Ref. [186], photographs of the surface of the ventilated cavity behind cavitators in the form of a disk at $U = 35$ ft/s, a disk with a number of pins at the end at $U = 30$ ft/s, and a disk with a ring at the end at $U = 30$ ft/s are also presented. The pins and ring mounted on the end face of the cavitator do not significantly affect the shape and smoothness of the cavity. With all forms of cavitators, a region of large shear stress arises in the flow at the boundary of two media: water (above-outside) and vapor-gas medium (below-inside) of the cavity of the cavity.

Fig. 7.61 shows a photograph of a circular axisymmetric jet of water injected into the atmosphere [304]. According to the terminology of the author, waves of primary and secondary instability in the flow behind the nozzle are photographed. The approximate primary wave spacing is 0.046 cm. The arrows indicate areas of noticeable secondary waves. The development of disturbances at the boundary between two media is well expressed. Behind the nozzle, disturbances develop in the form of T.-Sh. waves, then they are deformed into other forms of disturbances motion, in the same way as the development of disturbances in the transition boundary layer on the plate is given (Section 3.6, Fig. 3.33). In Ref. [304], a large

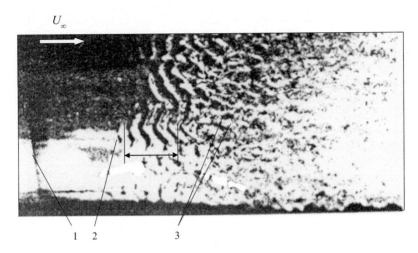

Figure 7.61 The formation of coherent vortex structures at the boundary of an immersed jet: (1) tip of the nozzle, (2) formation of the Tollmien−Schlichting wave, and (3) system of longitudinal vortices. The arrows indicate the deformation zone of the Tollmien−Schlichting wave [304].

shear stress arises at the boundary of two media: water (bottom-inside)—air (top-outside), and in investigations [186] as if in reverse motion: water (top-outside) and a vapor-gas medium (bottom-inside).

An analysis of investigations of almost all known types of movements in which various types of CVS and methods of their control were recorded was performed [114,139]. In all types of separated flows in the wake, certain characteristic types of CVS are recorded, which differ from the patterns of development of the CVS of the BL (Section 3.6, Figs. 3.33 and 3.40). This is because only at the beginning of the development of the BL in the Blasius profile (Section 3.3, Fig. 3.7) there are large shear stresses between the fluid layers along the thickness of the BL. In Section 3.6, a detailed physical justification is given for the occurrence of T.-Sh. waves both in the BL and in other types of flow (Figs. 7.59−7.61).

In Ref. [149], the results of numerical simulation for a given shear in a stream are presented. The results of the studies showed that the CVSs formed in the shear plane at the boundary of two fluid layers consist of transverse vortex structures. In the process of development, the vortex line develops along the flow and extends in the form of loops emerging from the shear plane. The loops are similar to "hairpin" vortices formed at the IV stage of the transition of the BL (Section 3.6, Fig. 3.33C), as well as at the CVS, of various shear flows shown in Figs. 7.59−7.61 and in Refs. [113,139]. Thus the CVS types considered for the transition BL (Section 3.6 of Fig. 3.33) are a reflection of the shear flow and are typical for various cases of shear flows.

Currently, a large number of investigations have been performed in various problems of cavity formation, which are of practical importance. In Ref. [186], patterns of the development of wave instability of disturbances arising during the formation of a cavity behind cavitators were investigated in a model experiment. Another area (cavity closure) also has important scientific and practical significance.

In Ref. [195], the instability of a partial cavity caused by the development of a reverse jet was investigated. A detailed visualization of the cavity behavior made it possible to identify the instability region of the return jet, which leads to the classical cavitation cloud. The surrounding regimes, in particular, the special case of thin cavities, which do not fluctuate in length but exhibit a jet of periodic behavior, are also investigated.

It is shown that two parameters are most important when analyzing the instability of the return jet: a negative pressure gradient and the thickness of the cavity compared to the thickness of the return jet.

The attached cavities are more or less nonstationary, and in some cases their length can undergo significant oscillations, and they are considered unstable. The term cavitation instability is used for situations where a stable cavity does not actually exist. Cavitation instabilities are of interest in the field of turbomachines, as they can cause abnormal dynamic behavior, as well as noise and erosion.

It is proposed to distinguish a priori two main classes of instabilities for an attached cavitation cavity: internal instabilities and system instabilities. This classification simply refers to the origin of instability. If it occurs in the cavity itself, the instability has its own character. If this is due to the interaction between the cavity and other cavities (as is the case with a turbomachine) or other components of the hydraulic system, it depends on the system.

Several researchers have described the principle of instability of the return jet. A rather accurate description of the flow at the lower end of the cavity was given in Ref. [195]. Closure of the cavity occurs where the external flow is again attached to the wall (Fig. 7.62).

The falling stream is divided into two parts flowing parallel to the wall: return, which moves against the stream and causes the cavity to separate; another stream relaxes on the wall.

However, it is clear that the return jet cannot exist permanently, otherwise the cavity will be filled with liquid. Consequently, the periods of development of the return jet, which tend to fill the cavity, are accompanied by periods of emptying,

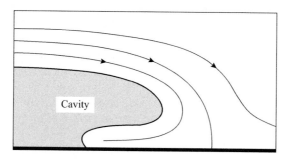

Figure 7.62 Schematic representation of relapse of the return jet in the closed area of the attached cavitation cavity [195].

and entrainment of the two-phase mixture downstream. A simple dimensional analysis leads to the conclusion that the oscillation frequency f is of the order of l/V_∞. In fact, the Strouhal number $S = f \cdot l/V_\infty$ of this instability based on the maximum cavity length l was in the range of about 0.25−0.4 in many experiments [329].

The instability of the return jet [212] is investigated during the flow around two-dimensional NACA profiles. The mechanism of cloud cavitation generation in the region of the return jet was investigated in detail in Ref. [328]. They had a simple but effective idea to set an obstacle on the profile in order to stop the return flow, and it was actually shown that this prevents the formation of cloud cavitation.

Fig. 7.63A shows the scheme of cavitation flow around a ledge in the planar approximation [186]. This scheme was used by C. Brennen in theoretical calculations of the hydrodynamic stability of plane waves recorded using high-speed filming (Fig. 7.59). The measurement results were presented in the form of the dependence of the dimensionless wavelength D/λ and dimensionless distance X_1/D on the Reynolds number $U_T D/\nu \cdot 10^{-5}$, calculated from the diameter D of the bodies investigated in the cavitation tunnel. The wavelengths on the surface of the cavity were measured during the flow around three types of sphere and OT. It was found that the wavelengths λ and X_1/D decrease with increasing Reynolds number.

The calculations of hydrodynamic stability and qualitative comparison with the results of experimental studies in accordance with Fig. 7.63A were performed in [186]. The comparison was made using the dependences of the dimensionless wave number $\alpha_r = 2\pi/\lambda$ and the dimensionless frequency $\gamma = 2\pi f \delta_2/U_\infty$ on the Reynolds number calculated from the momentum thickness δ_2. The flow behind three spheres and the ogive (a rotation paraboloid) tip was analyzed (Fig. 7.59). The frequency characteristics of the disturbing motion differed significantly from the neutral curve given in the corresponding coordinates for the Blasius profile obtained by calculation [414] (Section 3.4, Fig. 3.20). The experimental values of the dimensionless frequency during cavitation flow around three-dimensional tips were significantly larger than the neutral Lin curve. Thus in the zone of stability loss, the frequencies during cavitation flow were 2.2 times higher than during the flow around a two-dimensional plate. At $Re_{\delta 2} = 300$ and 400, the difference increased by 3 and 3.45 times. These results are due to a specific BL during cavitation flow around three-dimensional bodies. It should be noted that theoretical calculations were based on certain assumptions [186].

Fig. 7.63B shows schema of the experimental installation for modeling the flow around a ledge (Section 3.1, Figs. 3.1 and 3.2; Section 5.1, Figs. 5.1−5.3). Experimental investigations were carried out in a two-dimensional representation— during flow around a plate. Behind the ledge was a membrane surface, the construction of which consisted of a metal frame fitted with an elastic membrane (Section 5.1, Fig. 5.3). The metal transverse rods (1) were fixed in the frame, on which the metal plate (2) was located. The distance of the membrane from the bottom of the working section was 7 cm, and the height of the plate (2) was adjusted by moving the rods (1). Due to this, the air thickness under the membrane changed. In addition, in the experiments, only water was located under the film, the thickness of which was regulated by changing the location of the plate (2). The design of the

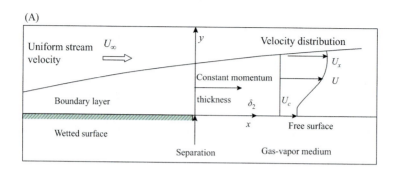

(A)

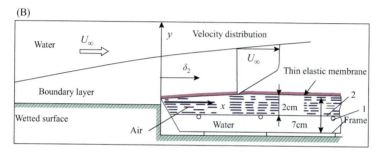

(B)

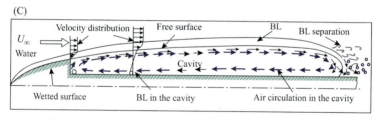

(C)

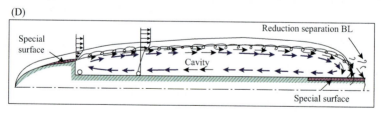

(D)

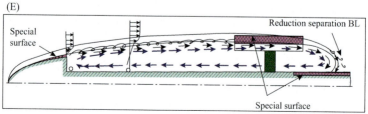

(E)

(Continued)

membrane surface also made it possible to control the tension, and, consequently, the elasticity of the elastic membrane.

A layer of a high-speed fluid boundary adjacent to a gas-vapor medium (Fig. 7.63A) can be represented as a thin elastic film, and a gas-vapor medium cavities—in the form of air under the membrane (Fig. 7.63B). By adjusting the tension of the membrane surface, it was possible to simulate the velocity of cavitation flow around the ledge and the pressure in the vapor-gas medium.

Experimental investigations of hydrodynamic stability with this representation were carried out in the hydrodynamic stand of small turbulence using the tellurium method (Section 3.4). The design of the hydrodynamic stand and the measurement procedure are given in (Sections 3.1 and 3.2). In the working section of the stand 3 m long, various types of bottom were installed (Section 5.1, Figs. 5.1 and 5.2). In the experiments, there was water under the film (series of experiments B1–15, B21–42), air (series of experiments B16–20), or heated water (series of experiments VT43.44).

In Section 5.3 and [378] the results of experimental investigations of hydrodynamic stability during flow around various types of elastic-damping surfaces at various velocities of the main flow are presented. In this case, various parameters varied: the thickness of the water layer under the membrane changed, the properties of the liquid under the membrane (there was air under the membrane), the tension of the membrane, the effect of the combined properties on the hydrodynamic stability was determined. In particular, neutral curves are presented when flowing around rigid and various types of elastic plates. Table 7.11 shows the mechanical characteristics of elastic plates in these experiments (Table 7.11).

Fig. 7.64 shows the results of experimental investigations of hydrodynamic stability during flow around membrane surfaces. It was found that the instability region increased substantially, and the number of Reynolds stability loss decreased. The instability characteristics depend on the thickness of the water layer h under the membrane. The area of instability was smaller with a smaller thickness h. At the same time, dimensionless wave numbers and phase velocities are less dependent on h, although the regularities remain the same (Fig. 7.64B, C). In this case, the maximum values of these quantities turned out to be higher than the maximum values when flowing around a rigid plate and are shifted toward large Reynolds numbers. When conducting experiments B1–B5 at approximately the same values of the Reynolds numbers as in the Brennen experiments [186], the frequencies of unstable oscillations increased in comparison with the flow around a hard surface (Section 3.4, Fig. 3.20), respectively, in 1.4; 2.1 and 1.7 times.

Figure 7.63 Schemes flow around a ledge: (A) calculation scheme of cavitation flow around a ledge [387], (B) modeling of the flow behind the ledge during experimental investigations of membrane surfaces [139,378]: (1) transverse rods, (2) metal plate, (C) scheme of formation of boundary layers on boundary two environments, (D) flow control scheme over an elastic special surface in the area of cavity closing, and (E) scheme of energy recycling using various special surfaces [139].

Table 7.11 Mechanical characteristics of elastic plates.

Numbers of experiences	$E \cdot 10^{-4}$ N/m^2	T N/m	ρ_M kg/m^2	$t_1 \cdot 10^4$ m	$h \cdot 10^2$ m	$U_\infty \cdot 10^2$ m/s
B1–B5					7	11.0
B6–B10	1.92	79.0	950	1	1	11.0
B11–B15	1.55	36.0	950	1	1	11.0
B16–B20	1.15	17.5	950	1	2 (air) + 5 (water)	10.5
B21–B25	1.35	25.0	950	1	7	10.5
B26–B34	1.04	15.0	950	1	1	10.4
B27–B35					1	7.8
B36–B39					5	11.0
B40					1	10.1
B41–B42						9.1
BT43–BT44						11.0

The ambiguity of the correspondence of the maximum values of the indicated quantities to Reynolds numbers was found. A marked increase in stability is also evidenced by a markedly increased phase velocity and graphs of the dependences of the amplitudes of the transverse velocities of the disturbing motion on the oscillation frequency [378]. Compared to the rigid standard, the maximum values of the amplitudes of the transverse component of the velocity of the disturbing motion increased, and the attenuation coefficient decreased.

In a series of experiments B16–B20, under the membrane was a layer of air 2 cm thick, and under the air was a layer of water 5 cm thick. A decrease in the tension in the membrane and the presence of air led to an increase in stability and a decrease in the range of unstable oscillations. In this series of experiments, the frequencies of unstable oscillations increased in comparison with the flow around a rigid surface (Section 3.4, Fig. 3.20), respectively in 1.2; 1.3 and 1.5 times [378].

The so-called limit neutral curves were also constructed (Section 3.4, Fig. 3.21) [378]. In the experiments, the entire range of observed oscillations identified using the tellurium method was recorded. The envelope of which made it possible to apply the limiting neutral curve. It turned out that this curve recorded a range of oscillations, which was 2–3 times higher than the range of oscillations bounded by the usual neutral curve [378]. The range of ordinary neutral oscillations when flowing around a membrane surface is 1.5 times the range of these oscillations when flowing around a rigid plate. If we interpolate the neutral curves in Fig. 7.64 for membrane surfaces on the left of the Reynolds number, the critical Reynolds number of stability loss is significantly less than for the neutral curve when flowing around a rigid plate. These data coincide with the results of measurements [186], in which the experimental points of unstable pulsations are also significantly shifted to the left of the Reynolds number of stability loss for a rigid plate.

The above results showed that the parameters of the disturbing motion significantly depend on the velocity of the main flow, the step size, the tension of the film, which simulates the speed of cavitation flow, the properties of the liquid above and below the step. As in the experiments [186], the frequency of unstable oscillations increased significantly compared with the classical data when flowing around a rigid plate. All other characteristics of the disturbing motion have also changed, which leads to the accelerated development of the CVS at the boundary of the gas-liquid shear layer. The development of the CVS behind the cavitator in the region of a high concentration of shear stresses at the cavity boundary actually occurs in the same sequence as in Fig. 3.33 (Section 3.6).

Let us consider the flow around a cavitator from the point of view of the *receptivity problem* [53,139,247], as was done in Section 7.12. The movement of the liquid medium behind the cavitator (upper arrows) causes the corresponding gas movement in the cavity (lower arrows), as shown by the arrows in Fig. 7.63C. In this case, it is possible to imagine the motion of a vapor-gas medium near the boundary with the liquid along a moving wall, which, due to friction, captures and turns a vapor-gas medium into motion. Thus at the liquid — vapor boundary, two boundary layers interact, as it were: the outer BL of liquid that appeared on the cavitator, as shown in the C. Brennen scheme [186] (Fig. 7.63A), and continues to

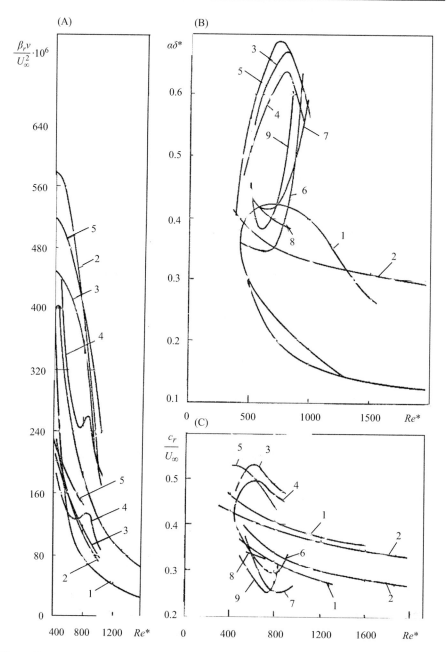

Figure 7.64 Neutral curves in coordinates of dimensionless frequency (A), wave number (B) and phase speed (C) at a longitudinal flow membranous surfaces: (A) measurements on rigid [curve (1)] and membranous surfaces at $h = 7$ cm [curve (2), series of experiences B1–B5],

(*Continued*)

move along the vapor − gas medium, and the inner BL in a gas-vapor medium caused by a running wall (liquid). Oscillations in the BL of liquid at the liquid-gas boundary that arise behind the edge of the cavitator will cause oscillations in the BL of the vapor-gas medium. These fluctuations will interact with each other. Such an interaction leads to an accelerated development of disturbances along the boundary of two media.

In the area where the cavity joins the body, the flow bifurcates: partially, the fluid joins the solid surface and continues to move along the flow, and partially goes against the flow in the form of a reverse jet (Fig. 7.62). The vapor-gas medium can partially splash out along the stream, and when the cavity is closed, a circulation movement occurs in the cavity against the stream near the streamlined boundary (Fig. 7.63C).

In Refs. [53,247], the interaction of various disturbances in the BL was experimentally investigated. It is convenient to present real disturbances with a complex form by combinations of simpler wave or vortex motions of different types, frequencies, and amplitudes. Thus the concept of receptivity includes the process of interaction in the BL of various disturbances and the result of such an interaction.

The results of an investigation of the laws of the flow of the BL at each stage of the transition under the action of a finite plane or vortex disturbance made it possible to draw two main conclusions. First, the process of interaction of disturbances in the BL can be represented as a superposition of "frozen," so-called natural structures of disturbing motion at each stage of the transition (in the absence of introduced disturbances) (Section 3.6, Fig. 3.33) with the structures of introduced disturbances. Second, the interaction of the natural structures of the disturbing motion of the BL with external disturbances has a resonant character depending on the type, energy and amplitude-frequency-wave characteristics of these structures. These conclusions apply not only to the transitional, but also to the TBL (Section 3.6, Fig. 3.40), as well as to the interactions of other types of disturbances.

The peculiarity of receptivity is that for the interaction of natural and introduced disturbances, it is necessary to create vortex structures similar to those that exist in the flow without introducing disturbances, as shown in Figs. 5.9 and 5.11 (Part I Section 5.2). Another feature of the receptivity is that it is necessary to create certain artificial vortex structures that can be easily damped and similar to structures that need to be stabilized. Based on these principles, two methods have been developed to stabilize the body in the cavity. Both of them are based on the idea of the formation of the CVS of the BL on the surface of the cavity with the initial conditions on the cavitator, as well as on the idea of gliding the body stabilizer.

The first method involves the construction of the outer surface of the cavitator so that predetermined planar or three-dimensional microvortices on its surface are

---

$h = 1$ cm [curve (3), series of experiences B6−B10], $h = 2$ cm [curve (4), series of experiences B16−B20], $h = 1$ cm [curve (5), series of experiences B11−B20]; (B), (C) measurements [378] [curve (1)] and Schubauer and Scramsted [curve (2) on a rigid plate]; measurement on membranous surfaces [378]: (3) experiences B1-B5, (4) B6−B10, (5) B16−B20, (6) B26−B34, (7) B36−B42, (8) BT43−BT44, and (9) B21−B25.

generated. These disturbances will subsequently develop in the shear layer at the liquid-gas interface. In the end part of the cavity there is a corresponding stabilizer design on the outer surface of the body, which will perceive disturbances. At the same time, the stabilizers glide along the surface of the cavity and are equipped with both known and new combined methods for drag reduction of planing surfaces. For example, it is necessary to organize some vortex structures (CVS) on the cavitator, as shown in Fig. 7.63D. Such vortex structures can be created, for example, using special elastic surfaces (Section 5.1, Figs. 5.7–5.10). The CVS generated by this surface will be transported downstream along the boundary of the two media. In the field of attachment of the cavity to the streamlined surface, it is necessary to apply the CVS control methods, which are given in Section 6.1 of Table 6.1 for effective stabilization of disturbances generated on the cavitator. For this purpose, for example, an elastic surface having a structure similar to that of a cavitator can be placed on the surface of the stabilizer, taking into account the development of disturbances when moving to the stabilizer. However, the geometric and other parameters of these special surfaces will differ, since depending on the length of the cavity, the parameters of the disturbing motion introduced on the cavitator will change as it develops along the flow.

Such methods CVS interaction in the BL of high-speed aquatic animals is considered in Part I. The indicated method of stabilization of the disturbances developing at the boundary of two environments allows one to reduce the deformation of the cavity shape, stabilizing the disturbances on the cavity surface and disturbances arising when the cavity is reattached to the body, and also prevent the formation of a reverse jet.

Two problems along with others are important at cavitation flows:

- It is necessary to stabilize the oscillation of the cavity relative to the streamlined body and, therefore to stabilize the movement of the body itself in the environment.
- At high speeds of motions, at which cavitation flows arises, greater energy is expended, which is also necessary for future to maintenance of the cavitation flow regime. In this regard, it is important to investigate the methods of utilization of expended energy.

Fig. 7.63D shows a method for stabilizing disturbances introduced on the cavitator and damping in the reattachment area, which promotes to an increase in the cavity length. In the reattachment zone, the flow bifurcates—a partially reverse movement of the liquid and a partial emission of the vapor-gas mixture downstream are observed (Fig. 7.62). With the correct design of the elastic coating in the area of the cavity closure, there is a partial utilization of the expended energy in the return flow area. The elastic coating installed in the area of the cavity closure has a different structure compared to the elastic coating placed on the cavitator, and helps to stabilize the flow in the area of the cavity closure. This allows you to reduce the loss of the vapor-gas medium from the cavity.

The application of the receptivity method allows increasing the utilization of the spent energy. The following methods are proposed for this. In the first case, the nasal cavitator has such an elastic coating design that not only allows you to generate the specified CVS, but also to twist the flow on the cavity surface in a spiral. Then the spiral-shaped

CVSs formed on the cavity surface increase the stabilization of the development of these disturbances, the shape of the cavity is stabilized, and a gyroscopic effect arises. As a result, the stability of body movement during cavitation flow increases. In this case, the elastic surface located in the back of the body below (Fig. 7.63E) has a completely different structure compared to the considered coating structure in the previous method. The new coating design stabilizes the beveled CVS s coming from the surface of the cavity, utilizes energy and increases the gyroscopic moment formed on the cavitator. The shape of the extended stabilizer can be cylindrical or in the form of discrete parts (skis) so as to glide along the inner surface of the cavity. The stabilizer can change the angle of attack and the distance from the streamlined body. The stabilizer surface can be rigidly fixed or articulated—so as to withstand the optimum angle of attack for gliding. The design of the elastic surface on the stabilizer is chosen the same as on the surface of the body. This increases the energy recovery efficiency and the magnitude of the gyroscopic moment.

It is possible to apply other designs for the formation of the gyroscopic moment (Fig. 7.63E). For example, the cavitator can be articulated mounted on the body and rotate around the longitudinal axis of the body under the action of the gyroscopic moment arising on the cavitator. A stabilizers or part of their surfaces can be made to rotate around the longitudinal axis of the body.

In Ref. [186] it was shown that in the presence of polymer solutions in the liquid on cavitators made in the form of a sphere, it increases the angle of separation of the flow during cavitation flow. At the same time, when flowing around a cavitator made in the form of a cylinder, it did not affect the flow separation angle. Polymer solutions can be introduced through slots on the surface of the cavitator or in the region of its end edge, as well as on the surface of the stabilizers. In addition, instead of or simultaneously with the polymer solution, a multilayer injection scheme for liquid solutions can also be used [83,88−90,92,93,105−107]. To apply the indicated methods of stabilization of the CVS and the body during its cavitation flow around, a new form of a gliding surface similar to fish scales proposed by V.V. Moroz ([139], Figs. 2.41 and 2.42) can be used. Recently, V.V. Moroz and V.A. Kochin have been carrying out fundamental experimental investigations of various problems of cavitation flow around bodies of various shapes near the surface of the water at different angles of attack of cavitators, speed, and the number of cavitations [351,352]. The experiments are carried out at the high-speed hydrodynamic towing tank of the Institute of Hydromechanics of the National Academy of Sciences of Ukraine, Kiev, Ukraine [161] (Section 6.3, Fig. 6.7).

Fig. 7.65 shows photographs of the shape of caverns formed at a depth of $H = 10$ cm behind a cone and a cavitator of complex shape. The main parameters used to represent the results of the experiment: α angle of attack of the cavitator, γ half the angle of the opening of the cone, d diameter of the cavity in the stream, H depth of the center of the separated flow (longitudinal axis of rod 3), U_∞ constant speed of movement and $\sigma = (P_\infty - P_c)/\frac{\rho U_\infty^2}{2}$ is the cavitation number corresponding to the depth H. Here $(P_\infty - P_c)$ is the pressure difference in the external flow and in the cavity, ρ is the density of water.

In Fig. 7.59, the "water-air" boundary is photographed during natural cavitation and ventilated cavitation [186]. In the experiments V.V. Moroz and V.A. Kochin

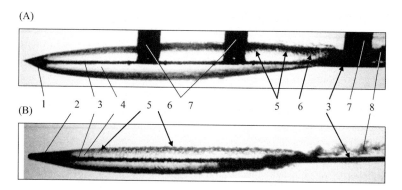

Figure 7.65 Photograph of cavitation flow around a body when moving at a depth of $H = 10$ cm and $U_\infty = 7.75$ m/s behind the cone (A) at $\gamma = 20$ degrees, diameter $d = 5$ cm, $\alpha = 5$ degrees, $\sigma = 0.062$ and behind a complex cavitator (B) with the front a part with a diameter of $d_n = 0.9$ cm, conjugated to the conical rear part with a half-angle of opening $\gamma = 10$ degrees, $d = 5$ cm, $\alpha = 2.5$ degrees, $\sigma = 0.075$: (1) cone, (2) cavitator of complex shape, (3) rod for supporting the cavitator, (4) cavity shape, (5) waves of Tollmien–Schlichting type, (6) Coherent vortex structure nonlinear in the shear flow, (7) pylon-knife to support the rod, and (8) vapor-gas mixture when the cavity collapses [352].

(Fig. 7.65) photographed the cavity forms, including behind the cavern closure region. In Fig. 7.65A, in the water-air shear layer behind the second supporting knife (7), the CVSs are clearly visible in the form of T.-Sh. waves (5), which are then transformed into nonlinear T.-Sh. waves (6), as in the experiments [186], but before closing the cavity. In Ref. [352], a photograph is presented similar to Fig. 7.65A, but at $U_\infty = 9.75$ m/s and $\sigma = 0.039$. With an increase in the speed of movement, the cavity length decreases and closes between the second and third supporting knives (7). At the same time, CVSs in the form of T.-Sh. waves are visible only behind the first supporting knife (7), which then transform very quickly in the form of nonlinear T.-Sh. waves (6), as in the experiments [186].

Fig. 7.65B shows a photograph of the cavity behind the cavitator of complex shape: the head part is made in the form of a sphere with which the conical back part is mated with the same parameters as the cavitator in the form of a cone (Fig. 7.65A). The experiments were carried out at different angles of attack of the cavitator and flow velocity. The photographs of the surfaces of cavities presented in [352] are similar to the photograph shown in Fig. 7.59B [186]. In experiments [186] on the surface of the cavity for a cavitator in the form of rotation paraboloid photographed a short part of the smooth shape of the cavity was photographed, behind which the CVSs were formed in the form of T.-Sh. waves of the order of two or three wavelengths, and then nonlinear forms of the CVS. In Fig. 7.65B, for the cavitator of complex shape at $\alpha = 2.5$ degrees, the shape of the cavity is photographed, on the surface of which the CVS develops similarly in the experiments [186]: behind the cavitator at a short distance from its trailing edge, CVSs are formed in the form of T.-Sh. waves, which at a distance, approximately equal to the length of the cavitator, begin to transform into nonlinear structures of the CVS. In

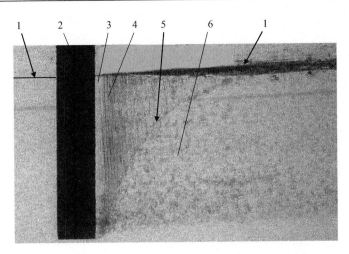

Figure 7.66 Photograph of the shape of the cavitation cavity behind the vertical strut at $Fr_H = 3$: (1) water surface, (2) pylon-knife, (3) cavern, (4) Coherent vortex structure of wave type Tollmien–Schlichting, (5) cavity boundary, and (6) vortex wake behind the cavern [352].

Ref. [352] other photographs are given, similar to those shown in Fig. 7.65B: one at $U_\infty = 7.75$ m/s, $\alpha = 0$ degrees and $\sigma = 0.074$; another at $U_\infty = 9.75$ m/s, $\alpha = 2.5$ degrees and $\sigma = 0.058$. In the first case, the CVS of the T.-Sh. wave type is practically absent—the surface of the cavity has the structure of nonlinear CVS, and in the second case, the length of the CVS of the T.-Sh. wave type is very short, as in the experiments [186].

V.V. Moroz and V.A. Kochin also performed experimental investigations of the shape of the cavity behind a vertical rack of finite length with a super-cavitating profile. The main dimensions of the rack model were (Fig. 7.66): the length of the rack model is 300 mm, the width is 50 mm, and the cross-section of the rack is made in the form of a cone with a rear end width of 5 mm.

The model of the vertical rack was mounted on the tense-dynamometer of the towing trolley and tested in the range of towing speeds of 3–9.7 m/s. The tests were carried out at immersion depths of 100, 150, and 200 mm. Fig. 7.66 shows a photograph of a cavitation cavity (3), which is formed behind a vertical rack at a towing speed of 4.25 m/s. The photograph shows that the cavitation cavity on the free surface is completely open, its shape in the vertical plane is close to triangular. In the immediate vicinity of the vertical strut, the surface of the cavitation cavity *3* is not disturbed, as in Fig. 7.59B. This area is observed in the photograph in the form of a light strip immediately behind the vertical pylon-knife (2). Then, wave disturbances arise on the surface of the cavitation cavity, which develop along the flow and increase in length and amplitude. These disturbances have the form of the type of T.–Sh. waves, as in Fig. 7.55 in experiments [186]. Behind the trailing edge of the cavitation cavity, vortex disturbances are recorded on the surface of the cavity, which carry air particles into the vortex wake behind the vertical strut.

Conclusion 8

In Part I, a hydrobionic approach am developed for solving hydromechanical problems. The structural features of the body systems and their interaction during the movement of fast-swimming hydrobionts (dolphins, sharks, swordfish, and penguins) aimed at drag reduction and reducing energy consumption have been studied and systematized. The influence of the variable shape and protruding parts of the body during movement, as well as the nonlinearity and non-stationary motion of hydrobionts on the drag to movement is investigated. The principles of the movement of hydrobionts are formulated. Using the developed special equipment, devices and techniques, experimental investigations on living dolphins were carried out. All this made it possible to detect previously unknown specific features of hydrobionts to reduce movement drag and minimize energy expenditure in the process of movement [98].

In the second real book (Part II), hydrodynamic verification of some of the discovered features of the functioning of fast-floating aquatic organisms was performed. Special devices, experimental facilities, and experimental techniques for carrying out experiments on hydrobiont analogs have been developed and manufactured, which made it possible to perform numerous physical and integral investigations, as well as develop new theoretical generalizations. The systematization of coherent vortex structures (CVS) in various problems of hydromechanics is carried out and shows the general laws of the formation of CVS. The development of the CVS in the boundary layer at various Reynolds numbers is investigated. Corresponding model of the boundary layer CVS during flow around a flat plate are constructed. Designed and systematized control methods of CVS. The interaction of various CVS in shear flows and some control methods of CVS when flowing around a plate, in particular using special designs of elastic surfaces, using various protrusions—generators of longitudinal CVS, when flowing around 21 configurations of three-dimensional cavities and their interactions, etc. were investigated. An experimental installation was made setup for studying unsteady effects when flowing around canonical bodies of various shapes, including profiles, while setting the shape of their oscillations in transversal and transverse directions, and at the same time when the rotational oscillations. The investigations were performed in the water flow using visualization and strain measurement.

A vortex chamber was designed and manufactured in collaboration with Professor V.N. Turik. The methodology for conducting experimental investigations in a vortex chamber was developed. The experiments were performed using visualization methods, a set of DISA equipment and a laser beam. Based on the results obtained, a model for the development of CVS in a vortex chamber was first constructed. Various methods for controlling the CVS in vortex chambers based on the method of interaction of various types of CVS have been developed and experimentally investigated.

New phenomena and patterns discovered in Parts I and II made it possible to develop technical solutions in the form of discoveries, patents and designs. Some of the fundamental scientific results presented in Parts I and II were verified in full-scale experiments and put into practice. Part III provides numerous examples of the implementation in various fields of technique and technology:

- Variants of underwater gliders with additive body shape, as well as elements of individual parts of gliders of nontraditional form.
- New forms of nose contours of the hull, wing profiles, and steering systems.
- Elastic coating design for the canonical form of the underwater glider.
- Gliders with unconventional propulsion and ocean energy storage methods.
- Various options for a hybrid body shape for movement under water.
- Microbubble generators for combined drag reduction methods.
- New design solutions for controlling the CVS and heat transfer.
- Designs of vortex chambers and others.

Using the conceptual approach given in Part II, the following were developed:

- Methods of controlling pressure distributions along the body.
- Adaptive wing and hull shapes.
- New principle of maneuvering.
- A new method for controlling the electromagnetic field of a body of revolution using a damping surface.
- A new method for regulating the thermal conductivity of the apparatus body.
- The method of electrification of polymer molecules and their retention by electromagnetic fields.
- The method of organizing the vortex flow during polymer injection and preliminary stretching of macromolecules.
- The method of action of polymer solutions on large vortex systems at the junction angles of two bodies.
- Active, multilayer damping coatings.
- Methods of impact on flow separation.
- A new method for stabilizing the body during cavitation flow (Part II Chapter 13.13, 6.71 and others.

Here are some of the problems discussed.

8.1 The problem of drag reduction, types of coherent vortex structures and combined methods of their control

The systematization of CVS in various problems of hydromechanics was first performed. General laws of CVS formation are shows. The problem of the interaction of various CVS in shear flows is developed. A study was made of the development of CVS in the boundary layer at various Reynolds numbers. Corresponding models of the boundary layer CVS during flow around a flat plate are constructed. Designed and systematized control methods for CVS. The interaction of various CVSs in shear flows and some control methods of CVS during flow around a plate, in particular, using special designs of elastic surfaces, using various ledges - generators of longitudinal CVS, using 21 options of three-dimensional deepening's, and others, were investigated experimentally. The features of CVS during interaction various deepening's, and also deepening's and ledges. Microbubble generators have been developed for combined drag reduction methods. Patented new design solutions for controlling the CVS and heat transfer. The combined methods of drag reduction are investigated:

- Elastic surfaces in combination with the formation of longitudinal vortex structures and two-dimensional disturbances in the boundary layer.
- The combination of elastic surfaces, injection into the boundary layer of polymer solutions and the formation of longitudinal disturbances.
- The combination of the shape of the nasal contours of a hard surface, in particular, various xiphoid tips, and injection of polymer solutions into the boundary layer.
- The same, but when forming around the model static electric fields.
- Injection into the boundary layer of a rigid model of polymer solutions from two slits with the simultaneous formation of two longitudinal vortex systems.
- The combination of elastic surfaces, injection into the boundary layer of polymer solutions, the formation of longitudinal vortex systems in the boundary layer and the use of nasal contours in the form of xiphoid tips.
- Injection into the boundary layer through three slits of polymer solutions, microbubbles and water flow in combination with three systems of longitudinal vortex structures.
- Various methods of injecting gas bubbles into the boundary layer using thin slots, conductive wires, strips or diaphragms placed along or across the flow, as well as a special microbubble generator.

8.2 Body form optimization

Shape of the body during its flow determines the distribution of pressure along the body and, as a consequence, its aerohydrodynamic characteristics. For example, the choice of the bow of the hull of a surface vessel affects its wave drag, and the shape of the wing profile determines the magnitude of its lift and drag. Analysis of the nasal contours of various species of aquatic organisms and birds allowed us to hypothesize that specific forms of the nasal part of the body play a significant role in their movement, which is stationary and unsteady. There are a number of different devices to reduce energy consumption. At the same time, the relationship between body shape and combined methods of drag reduction is essential. The results can be used on numerous types of vehicles moving in the water and air, as well as on two-environmed vehicles, for example, ekranoplans (wing-in-ground-effect vehicle—WIG).

8.3 Principle of additivity of aquatic and underwater bodies

Currently, an intensive search is underway to reduce the hydrodynamic drag of surface vehicles. One of the methods of significant drag reduction is the principle of dynamic maintenance. In this case, for example, the hull of a surface vessel is supported by two hydrofoils or two longitudinal cylinders with pointed end edges. At high speed, this leads to a significant decrease in wave drag and friction drag. A known project is when the main body at low speeds is supported by a displacement body, and with increasing speed it is supported by two longitudinal streamlined cylinders that are mounted on vertical profiles pivotally mounted on the main body. With increasing speed, the profiles rotate relative to the hull, increasing the distance between the cylinders, and the vessel rises and relies only on the cylinders due to the increase in lifting force. Projects of multihull vessels are also known.

8.4 Vortex chambers and control methods for coherent vortex structures in vortex chambers

Since 1998, V.V. Babenko together with V.N. Turik (National technical university of Ukraine "KPI", Kiev, Ukraine) has been conducting experimental investigations of the laws of flow in a vortex chamber (Part II Chapter 4.6). They developed an experimental setup and methodology for studying the features of the flow in a vortex chamber with different sizes of the dead-end section. A wide class of parameters of the inlet nozzle and other geometric parameters of the vortex chamber varied. Investigation of the development of CVS in various zones of the vortex chamber was carried out. A model of the vortex structures (Part II Chapter 4.6 Fig. 4.26) arising during the flow in the vortex chamber was constructed [128]. Experimental investigations of various control methods for the CVS of a vortex chamber have been performed. Based on the results of investigations of the vortex chamber, numerous technical solutions have been proposed:

- Purification of water from impurities, for example, in coal mines.
- Release of heavier gases from the air mixture.
- Significantly increase the efficiency of installations for the preparation of water or other mixtures with microbubbles.
- Improve the design and increase the efficiency of diesel and other engines.
- To improve the combustion chamber of aircraft turbines.
- Improve air conditioning and air purification systems and others.

A vortex chamber with an adjustable dead-end zone was designed and manufactured. For the first time, the development of vortex structures in a vortex chamber was experimentally investigated. The experiments were performed using visualization methods and a set of DISA equipment and a laser beam. Based on the results obtained, a model for the development of CVS in a vortex chamber was first constructed. Various methods for controlling the CVS in vortex chambers have been developed and experimentally investigated.

In the vortex chamber, experimental investigations of air flow around various types of deepening's were also performed. Two-dimensional and three-dimensional deepening's were investigated using smoke imaging and DISA equipment. For the first time, non-stationary patterns of the flow around recesses and the development of velocity profiles and spectra of the longitudinal velocity component in the vicinity and inside the deepening's were obtained. The results obtained allow us to propose the forms of deepening are for various technical applications:

- Control of vortex drag and friction drag reduction.
- Control of the formation of vortices in technological niches located on the surface of the apparatus in aviation and shipbuilding.
- Air conditioning systems.
- Flow control in vortex chambers.
- Control of longitudinal and transverse flow around the cylinders in various tasks of heat engineering and others.

The results obtained make it possible to effectively control the mixing processes of various types of liquids. In addition, a new property of the vortex chamber to perform a unique cleaning of the medium was discovered. Two technical solutions for cleaning the environment are proposed. In addition, a technical solution for the separation of air fractions was first proposed.

8.5 Environmental control concept of the coastal area

As a result of the modern development of civilization, there is an ever more intense pollution of the environment, including the aquatic environment, both in the seas and oceans, and in the coastal area. Control over the ecological state of the aquatic environment is carried out mainly from satellites or in the port area using surface ships. The proposed concept is based on many years of theoretical and experimental research of various problems of hydrodynamics and hydrobionics. To implement such control, it is planned to develop a complex or independent devices for various purposes, in particular, devices developed on the basis of the hydrobionic approach, including self-propelled devices and underwater gliders. In particular, it is envisaged to develop new designs and principles of operation. The developed program includes:

- Self-propelled buoys and bodies moving using the kinetic energy of the ocean.
- Apparatus for generating noise interference.
- Bionic devices having an optimal adaptive shape.
- Gliders using the potential energy of the ocean, including gliders−carriers and gliders−towing.
- Gliders: carriers designed on the basis of hydrobionic principles and having hydrobionic nontraditional propulsion devices.
- Various adaptive devices designed on the basis of the hydrobionic approach, having an optimal shape and a set of methods for reducing resistance, including the principles of control of Coherent Vortex Structures.
- New wave and vorticity detection systems.
- Devices of various functional and/or complex purposes, moving either in the planing mode or using the screen effect, having an adaptive shape and the ability to move under water at maximum speeds.

- New devices developed on the basis of the principle of movement of a body of variable mass and surface in an aqueous medium.

Some areas of technical applications are considered.

8.6 Reduction friction drag

Marine vehicles can be divided into three types:
- Devices moving under water.
- Devices moving on a water surface.
- Devices moving in two environments.

All devices have many specific subclasses, determined by purpose, size, scope, etc. Below are some examples of technical implementations for these three types of devices. Some of these examples are applied in practice. It is known that the drag of these types of devices has its own specific components. Common to all devices is friction drag.

Part II Chapter 3 in Figs. 3.33 and 3.40 shows model of the development of CVS at various stages of the transition, including a turbulent boundary layer. These results are fundamental. Each transition step shown on these models corresponds to a certain range of Reynolds numbers. Below are examples of the use of these results for the application of specific control methods for CVS. These models must be taken into account when conducting experimental and theoretical investigations in almost all problems of hydro-aerodynamics, thermo physics, etc.

Part II Chapter 5, Section 5.9 gives the results of an investigation of the physical picture of the development of CVS in the transition boundary layer during flow around an elastic surface. The properties of the boundary layer were discovered on rigid and elastic surfaces as a nonlinear asymmetric waveguide (Part II Chapter 5, Section 5.10). The waveguide in the boundary layer has completely different characteristics when flowing around elastic surfaces. These results are of practical importance for problems of acoustics, vibration absorption, and micro flows.

Part II Chapter 6, Section 6.1 gives a classification CVS control methods for boundary layer. These methods apply to all 12 types of CVS, which are discussed in [113].

Part II Chapter 5, Section 5.9 Fig. 5.61 provides a diagram of the distribution of hydrodynamic loads when flowing around a canonical body of revolution. A scheme of applying sectioning along a streamlined body is presented, according to which it is necessary to change the design and mechanical characteristics of elastic materials along the body. This scheme must be taken into account in all cases when various types of CVS control methods are used.

Part II Chapter 5, Section 5.1 gives designs for flexible coatings that have a wide range of uses. In this case, it is necessary to take into account the investigated mechanical characteristics of various elastic materials and the chemical formulation of the composition of elastomers.

Conclusion

Part II Sections 5.1, 5.7 present the results of measuring the dynamic characteristics of various elastomers. Other mechanical characteristics are measured. These results can be applied both for automatic control of vibration-absorbing characteristics of elastic materials, and for determining optimal vibration-absorbing characteristics of various types of devices.

The measured spectral characteristics of the boundary layer during flow around elastic plates and in studying the receptivity problem can be applied in solving problems of drag reduction, acoustics, and vibration absorption.

V.V. Babenko together with V.I. Korobov and V.V. Moroz, by order of Cortana corporation (president of K.J. Moore, USA), designs of a mixer and injector for preparation and injection of polymer solutions into the boundary layer were developed. Models of the mixer and injector are given in [83,88−90,92,93,105−107,119,120,133−135, etc.].

Chapter 7 Combined methods of drag reduction provides various combined drag reduction methods. Promising combined methods are:

- Injection of polymer solutions or surfactants through permeable elastic surfaces.
- Creating the optimal scale of microbubbles and their location in the boundary layer, where they are most effective.
- Use of waste heat to mix the polymer, control the mechanical characteristics of the elastic material, and to change the wall viscosity.
- Development of the concept of waveguides in the boundary layer, on the surface of an elastic material and in a liquid (including the injection of polymer solutions) to control CVS.
- Formation of large-scale vortex structures to reduce shape drag.

To reduce drag during injection of microbubbles into the boundary layer and to use combined methods to reduce drag, V.V. Babenko developed the principles and design of a compact and effective generator of microbubbles dissolved in a water flow [144]. In Fig. 8.1 shows one of the options for placing the microbubble generator.

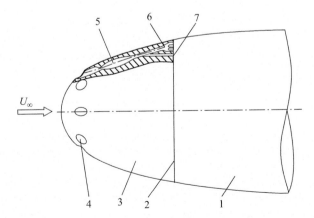

Figure 8.1 Scheme of the bow of the body of revolution or the underwater profile of the wing, in which the microbubble generators are located: (1) body of rotation, (2) slot, (3) bow, (4) nose of the hole, (5) generators of microbubbles, (6) channels with microbubbles, (7) channels with bubbles [144].

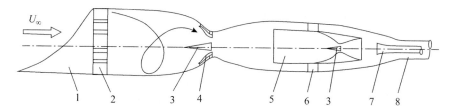

Figure 8.2 Longitudinal section of microbubble generators: (1) microbubble generator, (2) guide curved blades, (3) axisymmetric cavitator, (4) peripheral curved cavitators, (5) Laval nozzle, (6) streamlined plates, (7) tube with micro bubbles, (8) tube with bubbles [144].

Fig. 8.2 shows the cross section of the microbubble generator (1), in the bow of which there is guide curved blades (2) that twist the input stream (shown by a curved arrow).

The design of the microbubble generator (1) is made in the form of a Laval nozzle, in the narrow part of which an axisymmetric cavitator (3) and peripheral curvilinear cavitators (4) are installed in the center, installed along the streamlines of the swirling stream. Inside the housing (1), a streamlined Laval nozzle (5) is installed on the streamlined plates (6) located in accordance with the direction of the swirling flow. In the narrow part of the Laval nozzle (5), an axial cavitator (3) of a smaller size is installed. Behind the Laval nozzle (5), a tube is installed in which the axial part of the stream flowing from the nozzle enters. The peripheral part of the swirling stream flowing out of the nozzle (5) is mixed with the swirling flow around the nozzle (5) and moves in the pipe (8). The guide vanes (2) occupy only part of the cross section of the inlet pipe, which reduces hydraulic losses. When approaching the bottleneck of the Laval nozzle, the flow completely swirls over the entire cross section, and its speed increases significantly and has a maximum value in the narrowest cross section of the nozzle and at a certain distance behind it. The axisymmetric cavitator (3) and the peripheral curvilinear cavitators (4) (Fig. 8.2) are located in this region in such a way that their final sections are located at the minimum nozzle section. Curved cavitators (4) installed along the streamlines of the swirling flow allows you to maintain a swirl flow and reduce hydraulic loss. In addition, the cavitators (4) are mounted on plate curvilinear vertical supports at a distance from the inner surface of the nozzle, which exceeds the local thickness of the boundary layer. This allows you to place these cavitators in the main stream and ensure maximum efficiency of the cavitators.

Part II shows the advantage of combined CVS control methods and drag reduction methods. In 1974, the authorship certificate was obtained [25], according to which a combined method was proposed: the boundary layer was simultaneously affected by the elastic surface and plane waves formed by this surface or longitudinal vortex systems. In 1975, an author's certificate was obtained [27], according to which a new combined method was proposed. It differs from the previous one in that it adds the ability to inject polymer solutions into the boundary layer. Subsequently, other combined methods were developed. As a result of creative discussions with the president of Cortana corporation K.J. Moor, patents were obtained [83,88–90 ets.].

Conclusion

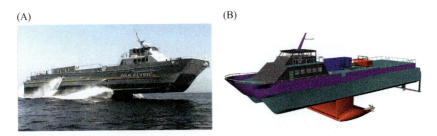

Figure 8.3 Photographs SEA FLYER (A); general scheme of SEA FLYER (B) [453].

Figure 8.4 Nasal cowl for the hydrofoil.

K.J. Moor did a lot of organizational work and recruited Research Faculty at the Applied Research Laboratory of Pennsylvania State University, USA. As a result, one of the combined drag reduction methods was tested in a full-scale experiment on SEA FLYER (Fig. 8.3). In [452], the results of field experimental investigations of the effectiveness of multilayer injection of various additives into the boundary layer of a SEA FLYER vessel are presented. On the hydrofoil in the region of the struts supporting the wing, three consecutive slots were installed in the longitudinal direction for the injection of water, polymer solutions, and microbubbles. In the design, along with others, the idea of forming a vertical density gradient in the boundary layer was realized. Field tests of the method were carried out in Hawaii on the Sea Flyer (Navatek) ship. The hydrofoil friction resistance decreased by 60%, the integral resistance of SEA FLYER decreased by 11%.

Figs. 2–6 Part II Chapter 6, Section 6.3 show the design of the vertical wing, in which the nose and tail fairings were removable. It was planned to carry out experiments with fairings of various shapes. The nose cone also had a removable xiphoid shape (Fig. 8.4).

Part II shows that it is preferable to form a system of longitudinal vortices in the boundary layer. In particular, in Part II Chapter 7, Section 7.7 it is shown that the xiphoid tip has advantages over standard cowls only when polymer solutions are injected into the boundary layer. If the nose of the hydrofoil is made in the form shown in Fig. 6.83, an improvement in the results obtained on SEA FLYER can be expected.

For the first time, the physical picture of the formation of flow during flow around a longitudinal cylinder was studied in the case when the temperature in its nose decreased significantly compared to the flow temperature. This led to a change in the density gradient across the boundary layer and a change in the pressure distribution along the longitudinal cylinder (V.I. Korobov, V.V. Babenko).

8.7 Devices moving on and under water

V.V. Babenko and V.G. Tarasenko developed a fin mover for canoe [38] (Fig. 8.5), introduced into production.

At low speeds under water, on the basis of the hydrobionic approach, various devices with elastic body coatings and nontraditional types of propulsion devices were developed (Fig. 8.6).

The designs of automatically controlled fin and piston (squid type) movers were developed, as well as the design of flexible stems for the operation of fin movers. A project of a two-seater underwater vehicle that implements many hydrobionic principles is proposed. In particular, the shape of the body and stabilizers were similar to those of fast-swimming animals. At the same time, the stabilizers had the ability to change the shape, sweep and streamlined surface area. The hydrobionic type propulsors were driven by batteries. It was also possible to use the muscles of the legs to drive.

The main goal is to reduce the resistance when moving the device. Therefore, with an increase in the speed of movement, an automatic decrease in the surface

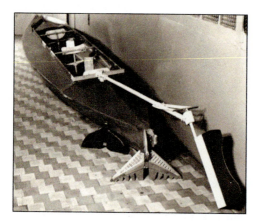

Figure 8.5 Photograph of canoe with various fin movers.

Conclusion

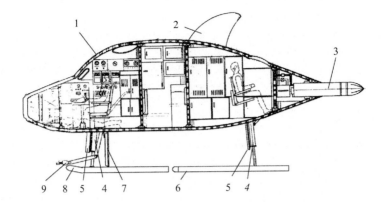

Figure 8.6 The project of a two-seater underwater vehicle with nontraditional types of movers: (1) body, (2) vertical stabilizer, (3) piston mover, (4) telescopic stand, (5) manhole cover, (6) rear ski, (7) hatch cover of a mechanical arm, (8) front ski, (9) mechanical arm.

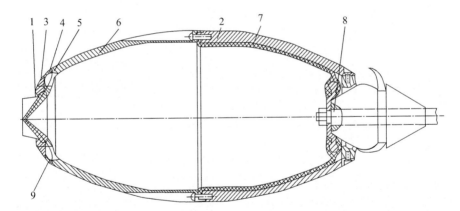

Figure 8.7 Piston mover with an elastic piston: (1) rigid outer nozzle, (2) front part of the body, (3) ring, (4) flat spring, (5) rotary nozzle leaves, (6) back part of the body, (7) elastic piston shirt, (8) piston, (9) rotary flaps [475].

area and a change in the angle of inclination of the leading edge of all stabilizers are provided. In addition, the telescopic racks (4) are retracted and rotated back, while the niches of the hatch where the racks (4) fit are simultaneously closed by covers (5), (7), and the skis (6), (8) are pressed tightly against the side bodies of the apparatus. To increase the speed of movement, in addition to the main fin movers, the piston mover (3) is included in the operation. Moreover, the jet exits the piston mover at the moment when the fin movers move away from the longitudinal axis of the apparatus.

The two-seater apparatus is equipped with two fin movers mounted on the sides, and a piston mover is installed between them (Figs. 8.6 and 8.7) [475].

Fig. 8.7 shows the arrangement of a piston propulsion device with an elastic piston. The assembly located on the right in Fig. 8.7 is of the same design as the piston movers shown in Part III. The difference is that the piston (8) in Fig. 8.7 is made in the form of a hard flat disk with rounded external contours on which the trailing edge of the elastic piston jacket is pressed. The front edge of the elastic piston jacket is fixed between the front (2) and rear (6) parts of the corps. The assembly located on the left in Fig. 8.7 has some structural elements, like the piston propeller shown in Part III Chapter 9 Section 9.2. The difference is that the outside of Fig. 8.7 has a cylindrical rigid nozzle (1) mounted on the ring (3). This ring is also fixed flat springs (4) and elastic reinforced sashes (5). In front of the corps (6) installed ring sashes (9), spring-loaded with springs (4).

At the Institute of Hydromechanics of the National Academy of Sciences of Ukraine, in the Department of Hydrobionics and Boundary Layer Control, investigations were carried out on various characteristics of a standard propeller mover, fin mover and the simultaneous operation of both types of movers. The necessary equipment was designed and manufactured in the Special Design Technological Bureau (SDTB) at the IGM NASU (Fig. 8.8). V.I. Korobov supervised the design work. M. V. Kanarsky supervised the manufacture of a catamaran. Experimental investigations were performed in Crimea (Ukraine). V.V. Babenko led the conduct of experiments from the IGM NASU (Fig. 8.8). The results of an experimental investigation of the wake behind two identical bodies are known, for example, the wake behind two identical transversely streamlined cylinders, the wake behind two transverse plates, etc.

In Fig. 8.8 shows an experimental setup that allowed us to investigate the interaction of a wake formed behind an oscillating plate and behind a standard screw. The two corps of the catamaran (1) were connected using a rigid frame (2), on which a strain gauge streamlined column (3) and a streamlined suspension with a standard screw (5) were installed along the longitudinal axis. A drive for oscillating a rectangular wing (4) was installed below on a strain gauge streamlined column (3), behind which a suspension was installed downstream with screw (5). Strain gauge

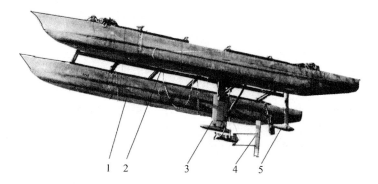

Figure 8.8 Photograph is an experimental setup for investigation the interaction of the fin mover and a standard propeller mover: (1) catamaran body, (2) rigid frame, (3) streamlined flow column, (4) oscillating wing, (5) suspension with a screw.

Conclusion

streamlined column (3) and streamlined suspension with standard screw (5) had the ability to change the relative position in the longitudinal and vertical directions.

The results of experimental investigations of the characteristics of an oscillating rectangular plate, performed on a catamaran, showed the possibility of using such mover on underwater vehicles. To test this hypothesis, under the guidance of V.G. Belinsky (Institute of Hydromechanics of NASU), a fin mover was designed and manufactured at SDTB of the IGM NASU, for which a copyright certificate was issued. At the same time, a linear electric motor was developed at the Institute of Electrodynamics of the National Academy of Sciences of Ukraine under the guidance of Professor A.G. Afonin, which was manufactured at SDTB ot the IE NASU. The fin mover and drive were installed on a semi-full-scale model of the underwater vehicle (Fig. 8.9).

On the model of the apparatus, a horizontal flat wing was installed in the rear part, inside of which there was an Afonins linear electric motor. From the ends of this wing extended along a pair of rods on which symmetrical rectangular wings with a NACA-0015 profile were fixed. The wing elongation was $\lambda = 5.0$, and the swing span was $2a = 3$. Thrust coefficient K_1 and the stop factor K_V of rectangular wings were also calculated. The electric motor moved one pair of rectangular wings in the transverse direction in one direction, and the other pair simultaneously in the opposite direction. At the same time, automatically during the movement of the rectangular wings according to a given program, the angle of attack changed during the movement. Having reached the extreme position, the rectangular wings began to move in the opposite direction. At this moment, the angle of attack shifted to the opposite position, and then during the movement the angle of attack changed according to a given law. Mooring tests were conducted for a model of an underwater vehicle with a fin mover.

Rudolf Bannash performed morphological and hydrobionic investigations of penguins. He designed and built autonomously floating models of the penguin and manta birostris.

Fig. 8.10 shows a general view (1) and front view (2) of a device designed for towing in an aquatic environment and developed at the Special Design Bureau of Physical and Mechanical Institute of NASU. The device is a rectangular box, inside of which there are six partitions freely installed in the guide grooves (the two extreme partitions in Fig. 8.10 (2) are not installed).

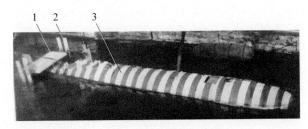

Figure 8.9 Fin mover mounted on an underwater vehicle: (1) linear electric motor, (2) fin mover, (3) model of the apparatus.

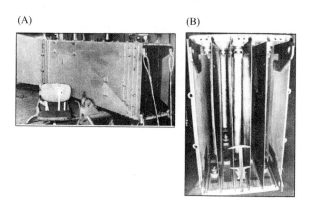

Figure 8.10 Device designed for towing in an aquatic environment.

These partitions form pair wise rectangular channels in which round cylinders are installed along the channels. The principle of the device is to form a vortex wake behind the cylinders, interacting with the side walls of the channels. To improve the operation of the device, A.I. Tsyganiuk and V.V. Babenko performed investigations to optimize the efficiency of hydrodynamic noise generators when using the Coand effect and the CVS arising from the flow around a cylinder located across the flow in the design of the generator. The results are introduced into production.

V.V. Babenko and S.V. Polishchuk developed various devices for underwater vehicles. Below is a list of these offers.

For large vessels moving on a water surface, wave drag and shape drag are of primary importance. Moreover, the shape of the hull is determined not only for reasons of drag reduction, but also the purpose of this class of ships. To reduce friction resistance, elastic coatings can be used. At the same time, the method of fixing the elastic coating to the hull of the vessel, and the chemical composition of the components of the mixture from which the coating is made, as well as the technology for making the coating is important. When applying antifouling paint to the body, it is also necessary to take into account the results of studies of the development of CVS at various stages of the transition and with a turbulent boundary layer. High-speed vessels have various versions of the principles of movement. There are multihull high-speed vessels, vehicles moving with dynamic principles of support and using an air cushion. Recently, much attention has been paid to gas saturation of the liquid under the hull. In Part II Chapter 7, Section 7.12, Fig. 7.53 shows photographs of devices for generating longitudinal layers of micro bubbles during flow around a rigid plate, and Fig. 7.56 shows the results of experiments generating longitudinal layers of micro bubbles during flow around an elastic plate.

V.V. Moroz (left) and V.V. Babenko (right) performed experimental investigations of the planing trimaran (Fig. 8.11) [114–116]. The model made it possible to regulate the relative position of the bodies: the location of the bodies along the length and the distance of the side bodies from the central body. In addition, the

Figure 8.11 Trimaran models developed by V.V. Moroz [116].

relative depth position of the trimaran shells is important. A five-body model was also developed. It is assumed that such an arrangement of the apparatus reduces wave impedance. A nasal wave and a wave from the trailing edge of the bottom leave the central body. These waves are damped by the side casings of the apparatus. To improve the damping of these waves, elastic damping surfaces having various mechanical characteristics are installed on the lateral inner walls of the side housings and on the bottom of the housing space.

To increase efficiency, the mechanical characteristics of elastic coatings should vary in length. In addition, to increase the efficiency, gas bubble generators are installed in the outer layer of elastic coatings, shown in Fig. 7.53 (Part II Chapter 7, Section 7.12). Coating is multilayer and also has the function of absorbing wave pressure and noise.

An elastic coating in the form of petals was also installed on the planing surface of the trimaran. In the area of the protruding edges of the petals, in comparison with the space between the petals, the properties of the elastic coating are different. Petals allow the generation of micro bubbles during their tear-off flow.

With the wrong choice of the mutual arrangement of the multihull apparatus due to discharge in the interspace, the liquid sticks to the surface of the interspace space. In this case, the internal channel is locked. Dramatically increases resistance. Therefore, it is very important to choose the ratio between the hulls in depth so that the trapped air between the hulls does not lead, due to the increased vacuum, to the liquid locking the space between the hulls.

The multihull M80 Stiletto has a classic planning body shape. On both sides of these cases semicircular body forms are installed. In the middle, two semicircular hulls are connected, forming a thin keel. The same thin keels end the side semicircular hulls. In the region of the two main supporting bodies, typical diverging waves form downstream. A wave is formed in the region of three thin keels, which smoothly twists along a circular surface. In the two extreme round channels, this swirling flow advances to the stern, where it flows onto the slightly swirling wave that has departed from the main supporting bodies and has interacted with it. In the middle part of the body, the same picture of the interaction of the swirling flow with a weakly swirling wave originating from the gliding surface from the opposite side occurs. Such interaction of large CVS corresponds to the laws of interaction of vortex systems according to the receptivity problem that we studied. Thus, behind the stern there follows a system of four swirling flows, in which semisubmerged screws are placed.

Chapter 4.6 (Part II) gives a picture of the formation of vortices in a vortex chamber. Based on the results of our experimental investigations, we can determine the patterns of vortex formation in the curved channel of the Stiletto apparatus and determine the patterns of interaction of the corresponding CVS. We have developed methods for controlling vortex structures. Thus, it is possible to optimize the Stiletto design and apply methods to reduce eddy resistance and friction resistance inside curved channels.

When conducting an experimental investigations of Görtler stability on curvilinear plates (Part II Chapter 4), it was found that the forming systems of longitudinal Görtler vortices remain stable on a horizontal surface located after the end of the curved section [34,81,82,620,628].

All fast-moving aquatic animals and birds in the bow of the body have one, and some birds, and two sections of longitudinal concave curvature. Fig. 8.12 shows a photograph of a petrel in which two sections of concave curvature are clearly visible.

Rudolph Bannash and D.W. Bechert investigated the effect of wing end feathers on the decrease in inductive resistance [158].

Due to Görtler's stability, and after the end of the curvature, a stable boundary layer is retained on the body. In aquatic animals and birds, the eyes are located in the area of the body, where there should not be disturbances in the boundary layer that can distort the perception of objects and the environment. In Refs. [140,151]

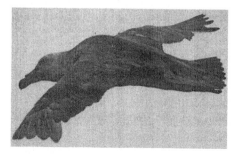

Figure 8.12 Photograph of the petrel.

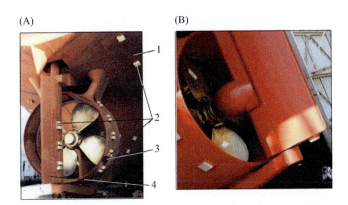

Figure 8.13 Photographs of the standard (A) and new (B) steering complex of the fishing trawler: (1) feed of the trawler, (2) tread protection elements, (3) ring nozzles of the screw, (4) steering wheel of the trawler.

gives a hypothetical view of the role of double curvature on the technical body of revolution.

V.V. Moroz carried out experimental studies of models of standard types of ship steering devices and the proposed steering device design, similar to the nose of a dolphin [147].

The physical justification for these experiments was the above results of experimental investigations of Görtler's stability. A patent has been obtained for the design of the steering device [99]. The steering gear is used on serial fishing trawlers (Fig. 8.13).

8.8 Technical proposals for two-medium devices

These devices include projects of modern wing-in-ground devices (WIG). This class of devices is being studied in the USA, Russia, China, Korea, and other countries. The main problems have not been resolved at the present time. These include the process of flight from a water surface, movement near the screen while maintaining the screen effect, and others. V.V. Moroz performed numerous experimental investigations of WIG models with various designs. Towing investigations were carried out in a high-speed hydrodynamic channel (Part II Chapter 6, Section 6.3 Fig. 6.7). A lot of experimental material has been accumulated to determine ways to solve WIG problems.

At the Institute of Hydromechanics of the National Academy of Sciences of Ukraine A.I. Tsyganyuk performed experimental investigations of the characteristics of a wing with double sweep in plan. Investigations were carried out far from of the water surface and near the screen. The effect of discrete slots located in different directions on the outer surface of the wing is investigated. A boundary layer was sucked out through the slits to study the effect on the vortex structures arising on the lateral edges of the wing. When moving near the screen, the circulation around the wing, the character of the flow around, and the structure of the vortices change. Experimentally, the most acceptable arrangement of discrete slots was found for the suction of the boundary layer in case of wing movement near the screen.

In Ref. [136], experimental investigations were performed of the effect of longitudinal vortex generators on the kinematic characteristics of the boundary layer during flow around a plate. Vortex generators developed by V.V. Babenko (Part II Chapter 4, Section 4.1 Fig. 4.2) and Abbas Fadil Machmud (Fig. 8.14) were used. Models of the development of CVS in the boundary layer (Part II Chapter 3, Section 3.6, Figs. 3.33, 3.40) were adjusted in accordance with the Reynolds numbers at each stage of the transition. The interaction of disturbances generated by the indicated vortex generators with disturbances existing in the boundary layer at each stage of the transition (receptivity [53,56]) is considered. It is shown that when the laws obtained in the study of receptivity are fulfilled, the optimal characteristics of the boundary layer are obtained.

In Ref. [137], the results of experimental investigations of the influence of vortex generators installed on the RSG-36 wing model in an open wind tunnel are presented (Fig. 8.15).

The influence of the design and various options for the placement of vortex generators, as well as their relative positions in the longitudinal and transverse directions on the aerodynamic characteristics of the wing model is investigated. The

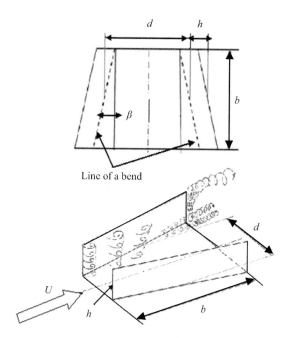

Figure 8.14 Scheme of a diffuser-type vortex generator [136].

Conclusion

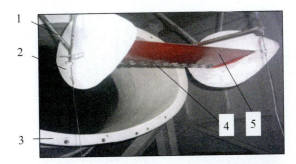

Figure 8.15 Photograph of the wing model in the working section: (1) strain gauge suspension, (2) end washers of the wing model, (3) cross section of the working section, (4) vortex generators, (5) wing model *RSG*-36 [137].

Figure 8.16 Photograph of the Aeroprakt A-20 model installed in the TAD-2 wind tunnel [1].

parameters of vortex generators are revealed at which the greatest influence on the aerodynamic characteristics of this wing model is found.

In Ref. [1], the results of experimental investigations of the influence of diffuser-type vortex generators installed on the AEROPRAKT A-20 airplane wing model are presented. The investigations were conducted in the wind tunnel TAD-2 of the National Aviation University, Kiev, Ukraine (Fig. 8.16). The effect of various options for the placement of vortex generators along the wing chord on the aerodynamic characteristics of the wing model is investigated. The parameters of vortex generators are revealed at which the greatest influence on the aerodynamic characteristics of the model of the wing of the aircraft is found.

With the help of riblets, it is possible to organize the corresponding longitudinal vortex systems when flowing around curved internal surfaces to reduce hydraulic losses. To test this hypothesis, the sizes of riblets and their direction on the curved

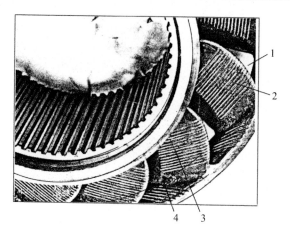

Figure 8.17 Photograph of the torque converter parts: (1) outer cylinder, (2) blades (profiles), (3) riblets on the blades, (4) riblets on the inner surface of the cylinder.

internal surfaces of the torque converter (hydro transformer) were calculated (Fig. 8.17). Riblets were applied to the finished parts of the torque converter assembly. Therefore, it was technologically possible to produce riblets only on a part of the streamlined curved surface, as can be clearly seen in the photograph of the torque converter.

In addition, in the corner junctions of the blades with the cylinder there are pairs of large vortex structures, which also could not be stabilized technologically. Despite the indicated trial attempt to influence the structure of the internal flow using riblets, it was still possible to increase the efficiency of the torque converter by 2%.

8.9 Vortex devices

Invention [145] refers to the separation of gas/liquid (2 phases), gas/liquid/solid particles (3 phases), as well as for the separation of purified gas or air into its components. The proposed invention is universal - it also allows you to effectively mix several types of gases. The invention can be used in various industries, for example, in energy, chemical, mining and petroleum industries, and to perform various ecology tasks.

The basis of the invention is the task of developing a universal design that improves the wet gas purification device, will be able to separate the purified gas and air into fractions, and can be used to mix gas flows by substantially changing the design of the vortex chamber taking into account the control of the revealed vortex structures in a classical vortex chamber [95–97,108,122,128,131,146,561–563,610]. Fig. 8.18 a shows the proposed device, which was constructed on the basis of quasistationary rotating pairs of vortex rings and vortex structures at the end of the vortex chamber that

Conclusion

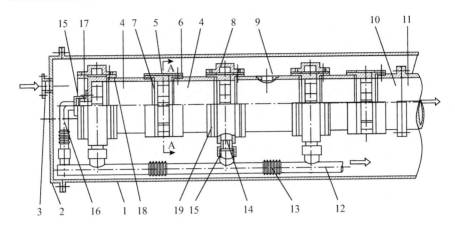

Figure 8.18 The design of the universal vortex chamber [145].

were first recorded during experimental investigations. Universal vortex chamber consists of a cylindrical body (1), cover (2), in which a flange (3) is installed for supplying moist gas. Universal vortex chamber is made of a set of identical structural elements of the pipe (4), at the end sections of which is an external thread; rings (5) and (6), on the horizontal external surface of which a through thread is made, and on the horizontal internal surface, an internal thread is made; from ring (8) for the selection of heavier components of gas and air; from thrust rings (18) for fixing the position of the rings (8); from the control rotary nozzle (9); from flanges (10) and (11) made on the pipe (4), at the end sections of which an external thread is made; from pipes (12) having threaded branches and connected between each other corrugated sections (13).

These branches, with the help of union nuts (14) are connected to bushings (15) which are screwed into rings (8). At the end of the device, ring (5) does not have a flowing central section, but a continuous surface to which the sleeve (17) is screwed axially connected by a threaded connection to the union nut (15) and to the corner pipe (16); the tightness of the connections of the device is provided by gaskets (7) and (19).

8.10 CVS and thermal conductivity control

To increase the saturation of the liquid with gas and the simultaneous formation of longitudinal CVSs, numerous experimental investigations of the flow around various deepening's were performed. Part II Chapter 7, Section 7.10 shows the formation of longitudinal CVS in a hemispherical deepening's.

One of the effective methods for generating CVS is the use of various types of deepening's on a streamlined surface. V.V. Babenko, V.P. Musienko, V.N. Turik, A.V. Voskoboinik and V.A. Voskoboinik performed experimental investigations of the flow of various types of deepening's in water and air flows [79,81,82,95,96,110,112,117,124,125,127,129,130,138,141, etc.]. Since 1989, V.V.

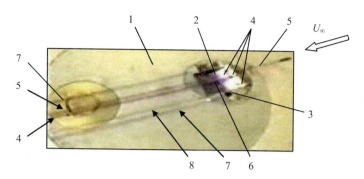

Figure 8.19 Visualization of the flow around a semi cylindrical deepening. Designations are given in the text.

Babenko and V.P. Musienko V.P. at the Institute of Hydromechanics of the National Academy of Sciences of Ukraine (Kiev) began to conduct experimental investigations of the flow of three-dimensional cavities in a water flow.

Fig. 8.19 shows a top view of the flow around the deepening (2) located on the streamlined surface (1).

The deepening (2) has geometric parameters: $L = 20$ mm, $H = 14$ mm, $h = 3$ mm, the cross section of the vertical grooves (3) has a square shape with a side of 3 mm. Visualization was performed in the water flow using tinted jets (4) [moving along the streamlined surface (1)] and jet (5) [moving over the streamlined surface (1) at a height of 5 mm]. The jets (4) flow out of the holes as seen from above at the flow around the deepening (2) located on the streamlined surface (1). The cross section of the vertical grooves (3) has a square shape with a side of 3 mm. Visualization was performed in the water flow using tinted jets (4) [moving along the streamlined surface (1)] and jet (5) [moving over the streamlined surface (1) at a height of 5 mm].

The jets (4) flow out of the holes in the bottom of the deepening and indicate that a transverse vortex (6) is formed at the bottom of the deepening (2), and longitudinal vortices (7), which are formed in the vertical grooves (3) and flow out of them at the level of the streamlined surface (1), follow from the depending (2) at the edges, and the longitudinal vortices (8) formed in the corners between the bottom and the side walls of the deepening.

Two-dimensional rectangular and half-cylindrical deepening's located across the flow, and various three-dimensional deepening's with passive and active methods for controlling their configuration (Part II Chapters 7, Section 7.10, Figs. 7.43, 7.44, 7.46, and 7.47) were investigated.

In 1991, for the first time, a methodology was developed for targeted experiments to obtain tornado-like vortices. When shooting on a video camera, the steady formation of a tornado of such vortices, forming behind the hole and moving downstream during flow around the plate, was recorded. Fig. 8.20 shows the frame of the video film, which shows the visualization of the formation of vortex structures. On the streamlined surface (1) there is a round hole (2) with the parameter $h = 6$ mm,

Conclusion

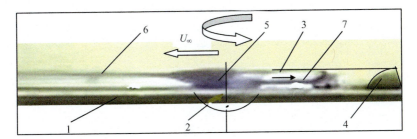

Figure 8.20 Video frame with visualization of the formation of vortex structures during flow around the device. Designations are explained in the text.

$d = 18$ mm, $h/d = 0.33$ and the transverse ledge (3) located in front of it, located at an angle of 45 degrees to the direction of the main flow. The height of the ledge was $h_1 = 5$ mm. Channels (4) were made along the main flow direction in the ledge with a step $z_1 = 0.46\,\hat{H}$, where $\hat{H}$ is the channel height, $\hat{H} = 0.02$ m, and the channel depth was $h_2 \approx 0.55 h_1$. The hole (2) is located behind the ledge (3) so that the recess is located along the right edge of the hole, while a constant tornado-like vortex (5) is formed in the hole, rotating counterclockwise. The upper surface of this vortex has a convex shape similar to a lens. From this surface, the vortex is carried out at the top in the form of a layer (6), which is folded into a longitudinal vortex. In the vicinity of the bottom of the groove, a stream (7) is carried out from the vortex, which moves in a transverse direction along the rear wall of the ledge, and then is folded into longitudinal structures directed along the main stream.

The complexes consist of a hole and a located front ledge. The complexes can be mounted on the heat exchange surface in the transverse direction parallel to the rows so that the holes are installed one after the other along their transverse axis of symmetry. In the longitudinal direction, the distance between adjacent rows of holes should be at least 2.5 times the diameter of the hole.

The complexes can also be mounted on the heat exchange surface in the transverse direction so that the ledge of the next complex is a continuation of the ledge of the previous complex. Various combinations of interaction as indicated holes, as well as combinations of holes with various ledges, were also investigated.

8.11 Devices for environmental monitoring of the coastal area

The contents of the developed concept of environmental control of the coastal area are given above. A list of devices developed on the basis of this concept is given. Most of these devices will be discussed in Part III. Based on the results of experimental study investigations presented in Parts I and II, the **"Principle of motion of a body of variable mass and surface in an aqueous medium"** was first developed.

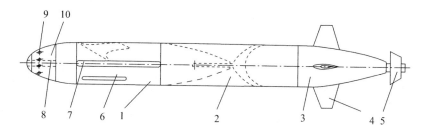

Figure 8.21 General view of a universal underwater vehicle with variable resistance [142].

The following is a variant of the apparatus developed on the basis of this principle [142]. The apparatus includes the designs of some of the apparatuses indicated above in Sections 8.12 − 8.18. The body of a universal underwater vehicle (Fig. 8.21) consists of a bow (1), a central (2) and tail (3) parts. The main electric batteries are located in the central part (2) of the device (not shown in Fig. 8.21). Corps of the apparatus is externally covered with a damping surface, equipped with a fin (indicated by a dashed line) and marching movers. In tail part (3) of the apparatus there is a standard system of hydrodynamic profiles (4) for stabilization and maneuvering of the apparatus, a standard engine with screw (5), as well as a compressed gas cylinder connected to a rubber container (not shown in Fig. 8.21).

The nose (1) is connected to the central part (2) of the device using controlled electric locks (not shown in Fig. 8.21). In the nose cylindrical part there are electric batteries and a container secured by controlled electric locks, with a system for recharging electric batteries from the energy of waves traveling on a water surface (not shown in Fig. 8.21), as well as a compressed gas container connected to a rubber cylinder (in Fig. 8.21), two containers (6) for accommodating extendable lateral fins, two containers (7) for accommodating extendable lateral complex wings, one extendable vertical fin (indicated by a dashed line). The front part (10) extends forward and contains a container (8) with a xiphoid tip extending forward, and eight holes (9) connected to the microbubble generators [144].

Universal underwater vehicle with variable resistance, depending on the given swimming modes, operates as follows. In the standard swimming mode (Fig. 8.21), when the power consumption of the main electric batteries ends, the device rise to the surface and the main batteries are charged.

In another swimming mode, using the controlled electric locks, the bow (1) and the central (2) parts of the body (Fig. 8.21) are disconnected, while the mover (5) is turned off, while the central (2) and tail (3) parts are separated from the body of the device and remain behind the bow (1) of the body of the device, which continues to move due to the fluctuation of the fin mover (5) (Fig. 8.22). Compressed gas is supplied to the rubber cylinder, which is located in the central Part II (Fig. 8.21), while the rubber cylinder expands, which leads to positive buoyancy of the disconnected parts (2), (3) of the apparatus body, which rise to the surface of the water.

At the same time, the fin mover (5) (Fig. 8.22) begins to oscillate, which is made in the shape and design of a swordfish made of elastic material and is installed in the

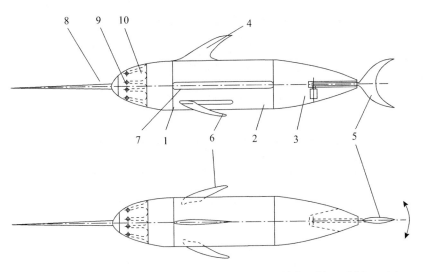

Figure 8.22 Top and side views of a universal underwater vehicle with variable resistance after separation of the tail and center parts of a standard vehicle [142].

tail part (3) of the disconnected apparatus, the fin mover (5) mounted on the tail stem, which is also made in shape and designs like swordfish [143].

The movement and amplitude of the tail stem are provided by an electric motor, and the stem is mounted on the axis of the electric motor according to [362]. The cylindrical surface (2) and the tail part (3) are covered with a damping surface according to the inventions [25,27,32,37]. At the same time, a vertical fin (4), which acts as a stabilizer, extends, and from the side containers - side fins (6), which act as a stabilizer and a rudder, and an xiphoid tip (8) extends from the central nose container, and the fins and tip are also made of elastic materials similar in shape and design like a swordfish. The design of the xiphoid tip (8) is made as in an automatic umbrella. The front nasal part (10) extends forward so that an annular slot is formed between it and the nasal part (1) along the end edge (10), and the shape and design of the under-slit chamber are performed as in [83]. Water through holes (9) enters the microbubble generators (shown in dashed lines in Fig. 8.22), and a mixture of water and microbubbles enters the near-wall zone of the boundary layer of the apparatus through the slot. It is known that a damping surface located on the cylindrical (2) and tail (3) parts of the apparatus, together with the water-microbubble mixture in the boundary layer and using the xiphoid tip (8) form a combined method of drag reduction, which can significantly reduce the hydrodynamic drag of the body of the moving apparatus.

Designed and manufactured small-sized remote-controlled autonomous apparatus for environmental monitoring [86].

In accordance with the programs and the results of experimental investigations, various design options have been developed. The list of patents of V.V. Babenko is given in Part III (73 patents). The following is a list of the systematization of some patents.

8.12 Adaptation of the form of the body

1. The device with nontraditional movers (1973) V.V. Babenko
2. Stem of fin propulsion (1975) V.V. Babenko
3. Body of an apparatus 29634 A (2000) S.V. Polishuk, V.V. Babenko
4. Body of an apparatus 46639 A (2002) S.V. Polishuk, V.V. Babenko Universal body of the device (2005) S.V. Polishuk
5. Underwater apparatus with fin mover 49237 A (2002) S.V. Polishuk, V.V. Babenko
6. Body of an apparatus 62707 A (2003) S.V. Polishuk, V.V. Babenko, V.V. Moros
7. Fin maneuverable complex 65249 A (2004) S.V. Polishuk, V.V. Babenko, V.V. Moros
8. Body of an apparatus 73850 A (2005) S.V. Polishuk, V.V. Babenko, V.V. Moros
9. Aircraft 13144 (2006) S.V. Polishuk, V.V. Babenko, V.V. Moros
10. Body of an apparatus (application № 2002021603, 22.07.02) S.V. Polishuk, V.V. Babenko.
11. Apparatus with fin mover 25646 (25.12.98) S.V. Polishuk, V.V. Babenko, V.I. Korobov
12. Apparatus with fin mover 25799 (25.12.98) S.V. Polishuk, V.V. Babenko Underwater apparatus with 41724 (17.09.01) S.V. Polishuk, V.V. Babenko, V.I. Majster fin mover

8.13 Self-propelled buoys

1. Mobile buoy 28271 A (16.10.2000) S.V. Polishuk, V.V. Babenko, V.I. Korobov
2. Mobile buoy 44208 A (15.01.2002) S.V. Polishuk, V.V. Babenko, V.I. Korobov with fin mover V.V. Moros, V.I. Majster

8.14 Underwater glider

1. Underwater glider 28282 A (2000) S.V. Polishuk, V.V. Babenko, V.I. Korobov
2. Body of an apparatus 29634 A (2000) S.V. Polishuk, V.V. Babenko, V.I. Korobov
3. Underwater glider 42202 A (2001) S.V. Polishuk, V.V. Babenko, V.I. Majster, V.I. Korobov, V.V. Moros
4. Body of an apparatus 46638 A (2002) S.V. Polishuk, V.V. Babenko
5. Body of an apparatus 46639 A (2002) S.V. Polishuk, V.V. Babenko
6. Universal apparatus body (2005) S.V. Polishuk
7. Underwater glider with an annular wing 99067 (2015) V.V. Babenko

8.15 Nontraditional movers

1. Fin mover 529104 (1976) V.V. Babenko
2. Flapping fin mover 484129 (1976) V.V. Babenko, Koslov L.F. V.P. Kajan
3. Fin mover for boats 796074 (1981) V.F. Tarasenko, V.V. Babenko, V.V. Gaponenko, N.V. Davidov
4. Fin mover 843397 (1981) S.V. Polishuk, L.F. Kozlov, V.E. Pyatetsky, V.M. Shakalo
5. Device mover 1092887 (1984) S.V. Polishuk

Conclusion

6. Fin maneuver complex 1221057 (1984) S.V. Polishuk
7. Piston mover 1221844 (1985) V.V. Ostistj, V.V. Babenko
8. Fin mover 1297377 (1987) V.V. Ostistj, V.V. Babenko, L.F. Koslov, S.D. Holjavchuk
9. Fin mover 1671515 (1991) V.I. Korobov, V.V. Babenko, V.G. Belinskj
10. Fin mover 1754578 (1992) S.V. Polishuk, V.V. Babenko
11. Fin mover 2013305 (1994) S.V. Polishuk, V.V. Babenko
12. Fin mover 2033938 (1995) S.V. Polishuk, V.V. Babenko
13. Fin mover of the apparatus 25355 (1998) S.V. Polishuk, V.V. Babenko, V.I. Korobov
14. Composite fin mover 25356 (1998) S.V. Polishuk, V.V. Babenko, V.I. Korobov
15. Fin mover 25621 (1998) S.V. Polishuk, V.V. Babenko, V.I. Korobov
16. Stem of fin propulsion 99030 (2015) V.V. Babenko

8.16 Changing the shape of the wing profile and its configuration

1. Fin mover 529104 (1976) V.V. Babenko
2. Flapping fin mover 484129 (1976) V.V. Babenko, L.F. Koslov, V.P. Kajan
3. Fin maneuver complex 1221057 (1984) S.V. Polishuk
4. Fin mover 1297377 (1987) V.V. Ostistj, V.V. Babenko, L.F. Koslov, S.D. Holjavchuk
5. Fin mover 2013305 (1994) S.V. Polishuk, V.V. Babenko

8.17 Change wing span

1. Fin mover 529104 (1976) V.V. Babenko
2. Fin mover 1297377 (1987) V.V. Ostistj, V.V. Babenko, L.F. Koslov, S.D. Holjavchuk
3. Fin mover 1689211 (1991) S.V. Polishuk
4. Fin mover of the apparatus 25355 (1998) S.V. Polishuk, V.V. Babenko, V.I. Korobov
5. Tandem with fin mover (application № 100229, 1990) S.V. Polishuk, V.V. Babenko

8.18 Changing the curvature of the wing profile

1. Device for automatically controlling (application 1973) V.V. Babenko, L.F. Koslov aerodynamic characteristics
2. Flapping fin mover 484129 (1976) V.V. Babenko, L.F. Koslov, V.P. Kajan
3. Fin mover 1754578 (1992) S.V. Polishuk, V.V. Babenko
4. Profile of a steering complex 52328 A (2002) V.V. Babenko, V.V. Babij, V.V. Moros of a ship

8.19 Changing wing profile chords

1. Apparatus with fin movers 25646 (1998) S.V. Polishuk, V.V. Babenko

8.20 Maneuverability control when moving the device

1. Fin maneuver complex 94117453 (1994) S.V. Polishuk
2. Tandem with fin mover 94117455 (1994) S.V. Polishuk, V.V. Babenko
3. Ships fin mover 1544638 (1990) S.V. Polishuk
4. Fin mover 1689211 (1991) S.V. Polishuk
5. Fin mover 1754578 (1992) S.V. Polishuk, V.V. Babenko
6. Fin mover 2013305 (1994) S.V. Polishuk, V.V. Babenko.
7. Fin mover of the apparatus 25355 A (1998) S.V. Polishuk, V.V. Babenko, V.I. Korobov
8. Composite fin mover 25356 A (1998) S.V. Polishuk, V.V. Babenko, V.I. Korobov
9. Apparatus with fin mover 25646 A (1998) S.V. Polishuk., V.V. Babenko
10. Underwater apparatus with 41616 (2001) S.V. Polishuk, V.V. Babenko, V.I. Korobov, fin mover V.V. Moros, I.I. Martinenko
11. Fin maneuverable complex 65249 (2004) S.V. Polishuk, V.V. Babenko, V.V. Moros

8.21 Vortex devices

1. Apparatus for damping the energy of two-phase 1325243 (1987) V.V. Ostistj, V.V. Babenko, flows and the separation L.F. Koslov, V.D. Ribko
2. Chamber for mixing 54767 U (2003) V.V. Babenko, V.N. Turik, V.A. Blohin, A.V. Voskoboynik, V.V. Krasovskj, V.G. Rozhavskj
3. Stand for research of compression 54768 (2003) V.V. Babenko, V.N. Turik, V.A. Blohin, processes of the twirled flows in A.V. Voskoboynik, V.V. Krasovskj, V.G. Rozhavskj engines of internal combustion
4. Free - piston engine 70779 (2004) V.N. Turik, V.V. Babenko, V.G. Rozhavskj
5. Chamber of mixture with 3443 (2004) V.N. Turik, V.V. Babenko, V.G. Rozhavskj additional nozzles
6. Method of control of mixing 557879 (2010) V.V. Babenko, V.N. Turik, D.E. Milukov. intensity of flows in the vortical chamber
7. Archimedes spiral for clearing gas and air 99029 (2015) V.V. Babenko
8. Universal vortex chamber 99280 (2015) V.V. Babenko, V.N. Turik
9. Intensity control method mixing flows in a 110914 (2016) V.V. Babenko, V.N. Turik vortex chamber V.A. Kochin, M.V. Kochin

8.22 Combined drag reduction methods

1. Damping coating 413286 (1974) V.V. Babenko, Koslov L.F, Pershin S.V.
2. Damping coating 483538 (1975) V.V. Babenko, Koslov L.F, V.I. Korobov
3. Regulating damping coating 597866 (1978) V.V. Babenko, Koslov L.F, V.I. Korobov
4. Damping coating для твердых тел 802672 (1981) V.V. Babenko, N.F. Yurchenko

8.23 CVS control methods

1. Damping coating 483538 (1975) V.V. Babenko, L.F. Koslov, V.I. Korobov
2. Regulating damping coating 597866 (1978) V.V. Babenko, L.F. Koslov, V.I. Korobov
3. Damping coating for solid bodies 802672 (1981) V.V. Babenko, N.F. Yurchenko
4. Method for determination 1183864(1985) V.V. Babenko, M.V. Kanarskj, of dynamic viscous-elastic characteristic N.F. Yurchenko
5. Coating that dampens and stabilizes 18126 (2006) V.I. Коробов
6. Fin mover 1754578 (1992) S.V. Polishuk, V.V. Babenko
7. Fin mover 2033938 (1995) S.V. Polishuk, V.V. Babenko
8. Fin mover 25621 (1998) S.V. Polishuk, V.V. Babenko
9. Apparatus with fin mover 25799 (1998) S.V. Polishuk, V.V. Babenko
10. Underwater apparatus with 41724 (2001) S.V. Polishuk, V.V. Babenko, V.I. Majster fin mover
11. Underwater apparatus 65250 (2004) S.A. Dovgy, V.V. Moros, V.V. Babenko, S.V. Polishuk
12. Chamber of mixture with 3443 (2004) V.N. Turik, V.V. Babenko, V.G. Rozhavskj additional nozzles
13. Method of control of mixing 557879 (2010) V.V. Babenko, V.N. Turik, D.E. Milukov intensity of flows in the vortical chamber
14. Intensity control method mixing flows in a 110914 (2016) V.V. Babenko, V.N. Turik, vortex chamber V.A. Kochin, M.V. Kochin

8.24 Heat and mass transfer

1. Device for improving heat transfer 100165 (2015) V.V. Babenko
2. Device for intensification of the heat and mass transfer 100216 (2015) V.V. Babenko
3. Device to enhance the mass transfer 100227 (2015) V.V. Babenko, V.N. Turik
4. Device for improving heat and mass transfer 101419 (2015) V.V. Babenko.

8.25 Ecology

1. Mobile buoy with fin mover 44208 (2002) S.V. Polishuk, V.V. Babenko, V.I. Korobov, V.V. Moros, V.I. Majster
2. Stem of fin propulsion 99030 (2015) V.V. Babenko
3. Wind turbine with vertical axis of rotation 107188 (2016) V.V. Babenko
4. Generator bubbles for underwater vehicles 99031 (2015) V.V. Babenko
5. Device for detecting waves and vortices 104525 (2015) V.V. Babenko
6. Underwater glider with an annular wing 99067 (2015) V.V. Babenko
7. Universal underwater device with variable resistance 98592 (2015) V.V. Babenko

Below is a table of contents for Part III.

Bibliography

References

[1] Abbas FM, Babenko VV, Ischenko SA. Influence of vortex generators on the aerodynamic characteristics of airplane modelInfluence of vortex generators on the aerodynamic characteristics of airplane models Aeropract A-20. National Academy of Sciences of Ukraine, Institute of Hydromechanics, Kyiv. Appl Fluid Mech 2012; 14(86):3−13 No. 1.

[2] Abramovich GN. Theory of turbulent jets. Moscow: Fizmatgiz; 1960.

[3] Aihara Y. Görtler vortices in the nonlinear region. In: Müller U, Roesner KG, Schmidt B, editors. Recent Developments in Theoretical and Experimental Fluid Mechanics. Berlin: Springer; 1979. p. 331−8.

[4] Aizik LB, Volodin AG. Stability of the boundary layer above the surface of a wave running along a plate. Appl Mech Theor Phys 1979;5:49−52.

[5] Aleev YuG, Leonenko IV. Hydrodynamic value of swordfish rostrum. National Academy of Sciences of Ukraine, Institute of Hydromechanics, Kyiv. Bionica 1974;8:21−3.

[6] Aleksin VA, Mikhail AA, Pilipenko VN. Effect of distributed injection of a polymer solution on the characteristics of a turbulent boundary layer. News of the Academy of Sciences of the USSR. Fluid and gas mechanics 1983; 5.

[7] Alimpiev AI, Mamonov VN, Mironov BP. Energy spectra of velocity pulsations in a turbulent boundary layer on a permeable plate. PMTF 1973;3:115−19.

[8] Amfilokhiev VB, Ivlev YuN. Speed pulsations in a laminar boundary layer of models with an elastic surface. National Academy of Sciences of Ukraine, Kyiv. Hydromechanics 1973;23:61−3.

[9] Amfilokhiev VB, Zolotov SS, Ivlev YuN. Estimation of the reduction of friction resistance of rotating bodies when using an elongated nasal tip. National Academy of Sciences of Ukraine, Institute of Hydromechanics, Kyiv. Bionica 1980;14:22−6.

[10] Antonia RA, Luxton RE. The responce of a turbulent boundary layer to a stein change in surface roughness. Part 2. Rough-smooth. J Fluid Mech 1972;53 No. 4.

[11] Ash RL. On the theory of compliant wall drag reduction in turbulent boundary layers. In: NASA CR-2387−National aeronautics and space administration. Washington, D. C.; April 1974.

[12] Avramenko AA, Khalatov AA. Taylor-Görtler instability in a laminar gradient boundary layer. Ind Heat Eng 1993;15(3):29−33.

[13] Avramenko AA, Kobzar SG. The effect of longitudinal pressure gradient on heat transfer in the turbulent boundary layer at a concave surface under the condition of centrifugal instability. Ind Heat Eng 2000;2(1-2):42−6 (Publisher: Begell House, USA).

[14] Azizov AN. Investigation of the influence of the temperature factor on the transition from laminar to turbulent flow in the boundary layer. Eng Phys J 1969;XVI(2):831−7.

[15] Babenko VV, Surkina RM. Some hydrodynamiic features of dolphin swimming. National Academy of Sciences of Ukraine, Institute of Hydromechanics, Kyiv. Bionica 1969;3:19−26.
[16] Babenko VV. Main characteristics of flexible coatings and similarity criteria. National Academy of Sciences of Ukraine, Institute of Hydromechanics, Kyiv. Bionica 1971;5:73−5.
[17] Babenko VV, Kozlov LF. Experimental investigations of hydrodynamic stability of a laminar boundary layer on an elastic-damping surface in a water flow. National Academy of Sciences of Ukraine, Institute of Hydromechanics, Kyiv. Bionica 1972;6:22−4.
[18] Babenko VV, Gnitetsky NA, Kozlov LF. Hydrodynamic stand of small turbulence, equipment and methods for conducting studies of the stability of a laminar boundary layer. National Academy of Sciences of Ukraine, Institute of Hydromechanics, Kyiv. Bionica 1972;6:84−90.
[19] Babenko VV, Kozlov LF. Experimental investigation of the hydrodynamic stability of a laminar boundary layer on a rigid plate in a water flow. National Academy of Sciences of Ukraine, Kyiv. Hydromechanics 1972;21:70−3.
[20] Babenko VV, Koslov LF. Experimental investigation of hydrodynamic stability on rigid and elastic damping surfaces. J Hydraul Res 1972;10(4):383−408.
[21] Babenko VV, Koslov LF. Experimental investigation of the hydrodynamic stability of laminar boundary layers at solid surfaces in water flow. Fluid Mech Sov Res 1973;2(6):163−6.
[22] Babenko VV, Kozlov LF. Experimental investigation of hydrodynamic stability on rigid and elastic-damping surfaces. News of the Academy of Sciences of the USSR. Fluid and gas mechanics 1973; 1: 122−7.
[23] Babenko VV, Kozlov LF. Experimental investigation of the behavior of the amplitude of a disturbing motion in a laminar boundary layer. National Academy of Sciences of Ukraine, Kyiv. Hydromechanics 1973;23:41−7.
[24] Babenko VV. Methodology for determination of mechanical properties and justification of the choice of the design of flexible coatings. National Academy of Sciences of Ukraine, Institute of Hydromechanics, Kyiv. Bionica 1973;7:71−9.
[25] Babenko VV, Kozlov LF, Pershin SV. Damping coating. USSR Author's Certificate 413286; 1974.
[26] Babenko VV. Experimental investigation of hydrodynamic stability during the flow around complex membrane surfaces. National Academy of Sciences of Ukraine, Institute of Hydromechanics, Kyiv. Bionica 1974;8:9−13.
[27] Babenko VV, Kozlov LF, Korobov VI. Damping coating. USSR Author's Certificate 483538; 1975.
[28] Babenko VV, Nikishova OD. Some hydrodynamic regularities of the structure of the skin of marine animals. National Academy of Sciences of Ukraine, Institute of Hydromechanics, Kyiv. Bionica 1976;10:27−33.
[29] Babenko VV. Experimental investigation of disturbing motion in a laminar boundary layer when flowing around damping surfaces. National Academy of Sciences of Ukraine, Institute of Hydromechanics, Kyiv. Bionica 1976;10:40−5.
[30] Babenko VV, Korobov VI. On the technique of injection of anomalous fluids into the boundary layer of bodies moving in water. In: Mechanics of anomalous systems. Proc II All-Union conference. Baku: Az. Institute of Petrochemistry; 1977. p. 32−3.
[31] Babenko VV. Experimental investigation of the beginning of turbulence. National Academy of Sciences of Ukraine, Institute of Hydromechanics, Kyiv. Bionica 1977;11:50−9.

[32] Babenko VV, Kozlov LF, Korobov VI. Adjustable damping coating. USSR Author's Certificate 597866; 1978.
[33] Babenko VV. Beginning of turbulence when flowing around an elastic plate. National Academy of Sciences of Ukraine, Institute of Hydromechanics, Kyiv. Bionica 1978;12:33−40.
[34] Babenko VV, Yurchenko NF. Experimental investigation of Görtler stability on rigid and elastic flat plates. National Academy of Sciences of Ukraine, Kyiv. Hydromechanics 1980;41:103−8.
[35] Babenko VV. To the interaction of flow with an elastic surface. In: Mechanics of turbulent flows. Moscow: Science; 1980. p. 292−301.
[36] Babenko VV, Voropaev GA, Yurchenko NF. On the problem of modeling the interaction of the outer covers of aquatic animals with the boundary layer. National Academy of Sciences of Ukraine, Kyiv. Hydromechanics 1980;41:73−81.
[37] Babenko VV, Yurchenko NF. Damping coating. USSR Author's Certificate. 802672; 1981.
[38] Babenko VV, Tarasenko VG, Gaponenko V, Davidov VV. Fin mover. Patent of Ukraine N796074, Bull N2, 15.01.81.
[39] Babenko VV, Ivanov VP, Blokhin VA, Kozlov LF. Laser Doppler velocity meter with electronic signal recording system. Instruments and equipment of experiment 1981;3:192−5.
[40] Babenko VV, Ivanov VP, Yurchenko NF. Measurement of the receptivity of the boundary layer to plane and three-dimensional perturbations with a laser anemometer. Autometry 1982;3:91−6.
[41] Babenko VV, Ivanov VP, Blokhin VA, Kozlov LF. Investigation of the stability of a laminar boundary layer with the help of LDMS. Autometry 1982;3:97−100.
[42] Babenko VV, Kozlov LF, Pershin SV, Sokolov VE, Tomilin AG. Self-tuning of cetacean skin damping with active swimming. National Academy of Sciences of Ukraine, Institute of Hydromechanics, Kyiv. Bionica 1982;16:3−10.
[43] Babenko VV, Koval AP. On hydrodynamic functions of the gill apparatus of swordfish. National Academy of Sciences of Ukraine, Institute of Hydromechanics, Kyiv. Bionica 1982;16:11−15.
[44] Babenko VV, Yurchenko NF. On the modeling of the hydrodynamic functions of the outer covers of aquatic animals. In: Hydrodynamic issues of bionics. Kiev: Naukova Dumka; 1983. p. 37−46.
[45] Babenko VV. On a mechanism for the flow of aquatic animals. National Academy of Sciences of Ukraine, Institute of Hydromechanics, Kyiv. Bionica 1983;17:39−45.
[46] Babenko VV, Kozlov LF, Ivanov VP. Experimental investigation of the interaction of the boundary layer with an injected semi bounded jet. National Academy of Sciences of Ukraine, Kyiv. Hydromechanics 1983;47:28−35.
[47] Babenko VV, Ivanov VP. Investigation of the structure of the boundary layer in the event of turbulence. National Academy of Sciences of Ukraine, Institute of Hydromechanics, Kyiv. Bionica 1984;18:28−35.
[48] Babenko VV. Investigation of the structure of the near-wall region of a turbulent boundary layer. In: Wall turbulent flows. Novosibirsk: Institute of Thermal Physics, Siberian Branch of the USSR Academy of Sciences; 1984. p. 5−12.
[49] Babenko VV, Kozlov LF, Yurchenko NF, Ivanov VP. The boundary layer under the influence of plane and three-dimensional disturbances. In: Wall turbulent flows. Novosibirsk: Institute of Thermal Physics, Siberian Branch of the USSR Academy of Sciences; 1984. p. 95−105.

[50] Babenko VV, Kozlov LF, Dovgj SA. et al. The influence of the outflow generated vortex structures on the boundary layer characteristics. In: Proc. second IUTAM symposium on laminar-turbulent transition. Novosibirsk/Berlin: Springer-Verlag; 1985. p. 509–13.

[51] Babenko VV, Kanarskiy MV. Method for determining the elastic-damping mechanical characteristics of materials. USSR Author's Certificate. 1183864; 1985.

[52] Babenko VV. The problem of turbulence development in a boundary layer. In: Proc. of the 17th sci. and method. Seminar on ship hydrodynamics, Bulgarian, Varna. 1988. p. 13-1–13-7.

[53] Babenko VV. The problem of boundary layer receptivity to various disturbances. National Academy of Sciences of Ukraine, Institute of Hydromechanics, Kyiv. Bionica, 22. 1988. p. 15–23.

[54] Babenko VV. Yurchenko NF. Development of organised vortex structures in boundary layers over rigid and elastic plates. In: Turbulence 89: organised structures and turbulence in fluid mechanics. Grenoble, France; 1989. p. 63.

[55] Babenko VV, Koval AP. On the hydrodynamic properties of the skin of aquatic animals. National Academy of Sciences of Ukraine, Institute of Hydromechanics, Kyiv. Bionica 1989;23:38–42.

[56] Babenko VV. Problem of interaction between different disturbances in a boundary layer. In: Colloquim on Görtler vortex flows "Euromech-261." Nantes, France; 1990. p. 70–4.

[57] Babenko VV. Methods of influence on coherent structures of a boundary layer. In: Res. workshop ordered and turbulent patterns in Taylor–Couette flow. Columbus, USA; 1991.

[58] Babenko VV. The boundary layer as a non-uniform asymmetrical waveguide. National Academy of Sciences of Ukraine, Institute of Hydromechanics, Kyiv. Bionica 1992;25:40–6.

[59] Babenko VV, Gnitetsky NA. The boundary layer on an adjustable elastic plate. National Academy of Sciences of Ukraine, Institute of Hydromechanics, Kyiv. Bionica 1993;26:21–7.

[60] Babenko VV, Kanarskiy MV, Korobov VI. The boundary layer on elastic plates. Kiev: Naukova Dumka; 1993. p. 263.

[61] Babenko VV, Musienko VP, Korobov VI, Ptukha YA. Experimental investigation of spherical groove influence on the intensification of heat and mass transfer in the boundary layer. In: Euromech Colloquium 327 "Effects of organized vortex motion on heat and mass transfer." Kiev, Ukraine; 1994. p. 23–4.

[62] Babenko VV. Hydrobionic Principles of drag reduction. In: Second int. conf. EAHE engineering aero-hydroelasticity. Pilsen, Czech Republic, V. II; 1994. p. 277–93.

[63] Babenko VV. Interaction of biological systems during the motion of marine animals. Review meeting on "Bio-locomotion and rotational flow over compliant surfaces." Baltimore, USA: Johns Hopkins Univ.; 1995.

[64] Babenko VV. Methods for control of coherent vortical structures in a boundary layer. Review meeting on "Bio-locomotion and rotational flow over compliant surfaces." Baltimore, USA: Johns Hopkins Univ.; 1995.

[65] Babenko VV. General principles of hydrobionics investigations and perspectives of research in hydrobionics. Review meeting on "Bio-locomotion and rotational flow over compliant surfaces." Baltimore, USA: Johns Hopkins Univ.; 1995.

[66] Babenko VV, Voropaev GA. Control of coherent structures of the boundary layer. National Academy of Sciences of Ukraine, Kyiv. Appl. Hydromechanics 1996;70:12–20.

[67] Babenko VV. Development of three-dimensional disturbances over concave elastic surface and with help of spherical grooves. In: Proc. 10th intern. Couette−Taylor workshop. Paris, France; 1997. p. 11−12.

[68] Babenko VV. Polymer's submission optimisation with the help of sword-shaped tips. In: Proc. 10th European drags reduction working meeting. Berlin: Technical Univ. Germany; 1997. p. 19−21.

[69] Babenko VV. Longitudinal vortex structure forming with the help of elastic wall. In: Proc. 10th European drags reduction working meeting. Berlin: Technical Univ. Germany; 1997. p. 336.

[70] Babenko VV. Experimental investigation of the boundary layer over rigid and elastic plates. In: Proc. Euromech third European Fluid Mechanics conference. Göttingen, Germany; 1997. p. 25.

[71] Babenko VV. Methods for control of coherent structures in a boundary layer of animals. In: Proc. workshop of the soc. for techn. biology and bionics. Jena, Germany; 1997. p. 166.

[72] Babenko VV. General principles of hydrobionics investigations and perspectives of research in hydrobionics. In: Proc. workshop of the soc. for techn. biology and bionics. Jena, Germany; 1997. p. 167.

[73] Babenko VV. Experimental investigation of the boundary layer over rigid and elastic plates. In: Pros. AGARD FDP workshop on "High Speed Body Motion in Water," Report 827. Kiev, Ukraine; 1998. p. 9-1−9-14.

[74] Babenko VV. Methods of influence on coherent vortices structures of a boundary layer. In: Proc. of the intern. symposium on seawater drag reduction. Newport, Rhode Island, USA; 1998. p. 113−20.

[75] Babenko VV., Latorre R. Role of bubble injection technique drags reduction. In: Proc. of the intern. symposium on seawater drag reduction. Newport, Rhode Island, USA; 1998. p. 319−25.

[76] Babenko VV, Yaremchuk AA. On biological foundations of dolphin's control of hydrodynamic resistance reduction. In: Proc. of the intern. symposium on seawater drag reduction. Newport, Rhode Island, USA; 1998. p. 451−2.

[77] Babenko VV. Hydrobionic principles of drag reduction. In: Proc. of the intern. symposium on seawater drag reduction. Newport, Rhode Island, USA; 1998. p. 453−5.

[78] Babenko VV, Kanarskiy MV. Turbulent boundary layer on an elastic surface. National Academy of Sciences of Ukraine, Institute of Hydromechanics, Kyiv. Bionica 1998; 27-28:3−24.

[79] Babenko VV, Musienko VP, Korobov VI, Pyadishus AA. The choice of geometrical parameters of the hole, generating disturbances in the boundary layer. National Academy of Sciences of Ukraine, Institute of Hydromechanics, Kyiv. Bionica 1998; 27-28:42−7.

[80] Babenko VV. Combined method of drag reduction. In: AGENDA. ONR workshop on gas based surface ship drag reduction. Newport, Rhode Island, USA; 1999.

[81] Babenko VV, Korobov VI, Musienko VP. Formation of vortex structure on curvilinear surfaces and semi-spherical cavities. In: 11-th Int. Couette−Taylor workshop. Bremen, Germany; 1999. p. 103.

[82] Babenko VV, Korobov VI, Musienko VP. Development of three-dimensional disturbances over concave elastic surface and with the help of spherical grooves. In: 11-th Int. Couette−Taylor workshop. Bremen, Germany; 1999. p. 107−8.

[83] Babenko VV. Method for reducing dissipation rate of fluid ejected into boundary layer. United States Patent 6,138,704; 2000.

[84] Babenko VV, Korobov VI, Moroz VV, Shkvar EA. Experimental investigation of a combined method of drag reduction. In: Proc. intern. conf. on ship and shipping research, NAV-2000. Venice, Italy; 2000.

[85] Babenko VV. Hydrobionic principles of resistance reduction. National Academy of Sciences of Ukraine, Institute of Hydromechanics, Kyiv. Appl Hydromech 2000; 2(74):3−17 No. 2.

[86] Babenko VV, Moros VV, Martynenko II. Experience of creation of underwater remotely operated vehicles in Ukraine and perspective of their application in geological oceanography. Miner Resour Ukr 2000;1:21−5.

[87] Babenko VV, Turick VN, Makarenko RO. Coherent structures and turbulence of limited flows in vortex chambers. In: 12-th Inter. Couette−Taylor workshop (sponsored by ONR and NSF) at Northwester Univ., Evanston (Chicago), IL, USA; 2001.

[88] Babenko VV. Method and apparatus for mixing high molecular weight materials with liquids. United States Patent 6,200,014; 2001.

[89] Babenko VV. Apparatus for reducing dissipation rate of fluid ejected into boundary layer. United States Patent 6,237,636; 2001.

[90] Babenko VV. Method for reducing dissipation rate of fluid ejected into boundary layer. United States Patent 6,305,399; 2001.

[91] Babenko VV. Control of the coherent vortical structures of a boundary layer. In: Aerodynamic drag reduction technologies. Proc. of the CEAS/DragNet European drag reduction conference, 19−21 June 2000, Potsdam, Germany. Springer-Verlag Berlin Heidelberg; 2001. p. 341−50.

[92] Babenko VV. Method and apparatus for reducing dissipation rate of fluid ejected into boundary layer. Assigment of Korean Patent Application 2001−708420 (PCT/US 99/ 31145); 2001.

[93] Babenko VV. Method and apparatus for reducing dissipation rate of fluid ejected into boundary layer. The Japanese Patent Application 2000−590768, based on international application PCT/US9931146; 2001.

[94] Babenko VV. External and internal flows on elastic surfaces. In: Proc.IUTAM symposium on flow in collapsible tubes and past other highly compliant boundaries. Coventry, England: University of Warwick; 2001.

[95] Babenko VV, Turik VN, Voskoboinik AV. Visualization of the flow structure in a vortex chamber. Bulletin of the National Technical University Kharkov Polytechnic, Institute Eng Mech Eng 2001;129(1):215−21.

[96] Babenko VV, Turik VN, Dembitsky VS. Stand for the study of the characteristics of turbulent swirling flows with their longitudinal compression. Bulletin of the National Technical University of Ukraine, Kyiv Polytechnic Institute Mech Eng 2001;40:422−5.

[97] Babenko VV, Turik VN, Voskoboinik AV. Investigation of coherent vortex structures in bounded swirling flows by the method of high-speed film recording. Bulletin of the National Technical University of Ukraine, Kyiv Polytechnic Institute Mech Eng 2001;40:426−32.

[98] Babenko VV, Sokolov VE, Kozlov LF, Pershin SV, Tomilin AG, Chernyshev OB. The skin quality of cetaceans actively regulate hydrodynamic resistance to swimming by controlling the local interaction of the skin with a stream. Diploma of Discavery No 265, 4. 11. 1982. Pub.15.5.1983. Bul №17. In: Institute of Hydromechanics. The book is dedicated to the 75th anniversary of the institute. Kiev: Institute of Hydromechanics NAS of Ukraine; 2002. p. 240.

[99] Babenko VV, Babiy VV, Moroz VV. Profile of the steering complex of the vessel. Declaration patent for the invention of Ukraine 52328; 2002.

[100] Babenko VV, Moroz VV, Shkvar EA. Experimental investigation of the combined drag reduction method. In: Intern. conference actual problems of continuous environment mechanisms. Series A: Natural sciences. Donetsk, Ukraine: Donetsk University Herald; 2002. p. 227–32.

[101] Babenko VV, Blokhin VA, Voskoboinik VA, Turik VN, Voskoboinik AV. Structure of a swirling flow entering a vortex chamber. Technological systems. J Sci Technol 2002;2(13):102–6.

[102] Babenko VV, Blokhin VA, Voskoboinik AV, Turik VN. Speed pulsations in a swirling jet of a vortex chamber. National Academy of Sciences of Ukraine, Institute of Hydromechanics, Kyiv. Acoustic Vesnik 2002;5(1):3–12.

[103] Babenko VV, Shkvar EA. Drag reduction with the aid of the combined method. In: 12-th European drag reduction meeting programme. Herning, Denmark; 2002.

[104] Babenko VV, Shkvar EA. Combined method of drag reduction.In: Intern. summer scientific school "High speed hydrodynamics." Cheboksary, Russia; 2002. p. 321–6.

[105] Babenko VV. Method for reducing dissipation rate of fluid ejected into boundary layer. United States Patent 6,349,734; 2002.

[106] Babenko VV. Method for reducing dissipation rate of fluid ejected into boundary layer. United States Patent 6,357,464; 2002.

[107] Babenko VV. Apparatus for reducing dissipation rate of fluid ejected into boundary layer. United States Patent 6,435, 214; 2002.

[108] Babenko VV, Turik VM, Blokhin VA Voskoboinik AV, Krasovsky VV, Rzhavsky VG. Mixing chamber. Declarative patent for the invention of Ukraine 54767; 2003.

[109] [61] Babenko VV, Voskoboynik VA, Voskoboinik AV, Turik VN. Field of velocity pulsations in the wall jet of a vortex chamber. Vib Eng Technol 2003;2(28):105–10.

[110] Babenko VV, Voskoboynik VA, Voskoboinik AV, Turik VN. The structure of the flow in a transversely streamlined semi-cylindrical recess on the plate. In: First international scientific conference problems, methods and ways of investigation of the World Ocean, Zaporozhye: STC PAS NAS Ukraine; 2003. p. 96–104.

[111] Babenko VV, Carpenter PW. Dolphin hydrodynamics. In: Carpenter PW, Pedley TJ, editors. IUTAM symposium on flow past highly compliant boundaries an din collapsible tubes. 26–30 Mach, 2001, University of Warwick Coventry England, Dordrecht/Boston/London: Kluwer Academic publishers; 2003. p. 293–323.

[112] Babenko VV, Turick VN, Voskoboinick VA, Voskoboinick AV. Vortices structures in cross-streamlined half-cylindrical cavity on a plate and interaction with boundary layer. In: The 5-th Euromech fluid mechanics conference, EFMC 2003. Toulouse, France; 2003. p. 414.

[113] Babenko VV, Turick VN. Coherent vortices structures control in flat and curvilinear parietal flows. In: The World Congress "Aviation in the XXI-st Century." Kyiv, Ukraine; 2003. p. 2.54–2.58.

[114] Babenko VV, Moroz AV Shkvar EA. Combined method of body stabilization in shear flow. In: Proc. of the fifth. intern. symp. on cavitation (Cav2003). Osaka, Japan; 2003. p. 321–6.

[115] Babenko VV. Hydrobionic principles of drag reduction. Int J Fluid Mech Res 2003; 30(2):125–46 (Translated from Applied hydromechanics 2000; 2(74), 2: 3-17).

[116] Babenko VV, Kuznetsov AlI, Kuznetsov AnI, Moroz VV. Technique of towing tests in the test basin with the help of two models of a gliding vessel. National Academy of Sciences of Ukraine, Institute of Hydromechanics, Kyiv. Appl Hydromech 2003; 5(77):5–11 No. 4.

[117] Babenko VV, Voskoboynik VA, Voskoboinik AV, Turik VT. Speed profiles in the boundary layer of a plate with grooves. National Academy of Sciences of Ukraine, Institute of Hydromechanics, Kyiv. Acoustic Bull 2004;7(3):14–27.

[118] Babenko VV, Koval AP. Structure and hydrodynamic properties of sword-fish. skin National Academy of Sciences of Ukraine, Institute of Hydromechanics, Kyiv. Appl Hydromech 2004;6(78):3–19 No. 3.

[119] Babenko VV. Method and apparatus for mixing high molecular weight materials with liquids. The European Patent Publication 1140341; 2004 (Application 99967725.5).

[120] Babenko VV. Method and apparatus for mixing high molecular weight materials with liquids. The International Patent Publication NWO0038827 (Application N PCT/US9931146); 2004.

[121] Babenko VV. Interaction quickly floating hydrobionts with flow. In: First international industrial conf. bionic 2004. Hannover, Germany; 2004. p. 153–9.

[122] Babenko VV, Turick VN, Rozhavskj VG. Chamber of mixture with additional nozzles. Declarative patent for useful model, Ukraine N3443 U, Bull.N11, 15.11.2004; 2004

[123] Babenko VV. Interaction of quickly swimming hydrobionts with flow. In: Proceedings of the second international symposium on seawater drag reduction (ISSDR 2005). Busan, Korea; 2005. p. 579–92.

[124] Babenko VV, Voskoboinick AV, Voskoboinick VA, Turick VN. Vortex formation in the half-cylindrical cavity on a plate. In: Fourth Intern. Conf. Problems of Industrial Heat Engineering, Kiyv, Ukraine; 2005. p. 41.

[125] Babenko VV, Turick VN Voskoboinick VA, Voskoboinick AV. Coherent vortical structures in half-cylindrical cavity on a plate. In: Proc. of PHYSMOD 2005. Univ. of Western Ontario, London, Ontario, Canada; 2005. p. 76–7.

[126] Babenko VV, Voskoboynik VA, Voskoboinik AV, Turik VN. Görtler vortices on the curvilinear surface of the vortex chamber. National Academy of Sciences of Ukraine, Institute of Hydromechanics, Kyiv. Appl Hydromech 2007;9(81):2–3 25-36.

[127] Babenko VV, Voskoboinick AV, Voskoboinick VA, Turick VN. Vortex formation in the hemispherical cavity on the surface of streamlined plate. In: Proc. of fifth Intern. Conf. Industrial Heat Engineering. Kyiv, Ukraine; 2007; p. 29–30.

[128] Babenko VV, Turik VN. Breadboard model of vortex flow structures in a vortex chamber. National Academy of Sciences of Ukraine, Institute of Hydromechanics, Kyiv. Appl Hydromech 2008;10(82):3–19 3.

[129] Babenko VV, Voskoboynik VA, Turik VN, Voskoboinik AV. Generation of vortices in local cavities. Bulletin of Donetsk National University, serA Nat Sci 2009;2:33–40.

[130] Babenko VV, Musienko VP, Turik VN, Milyukov DE. Visualization of the flow around hemispherical cavities. National Academy of Sciences of Ukraine, Institute of Hydromechanics, Kyiv. Appl Hydromech 2010;12(84):3–25 No. 2.

[131] Babenko V.V., Turik V.N., Milukov D.E. (2010) Method of control of mixing intensity of flows in the vortical chamber. Declarative patent for useful modelб for useful model, Ukraine N557879 U, M. Cl. B01F 5/00, F15D 1/00, Bull.N 24, 27.12.2010.

[132] Babenko V.V., H.H. Chun and I. Lee. Coherent vortical Structures and Methods of their Control for Drag Reduction of Bodies. In: Proc. of the 9-th International Conference on Hydrodynamics (ICHD-2010) Shanghai, China; 2010. p. 45–50.

[133] Babenko VV. Method and apparatus for mixing high molecular weight materials with liquids. The patent Sweden, based on European Patent publication 1140341, 2004 (Application 99967725.5); 2010.

[134] Babenko VV. Method and apparatus for mixing high molecular weight materials with liquids. The Patent Italy, Application 154250/ac, based on European Patent publication 1140341, 2004 (Application N 99967725.5); 2010.
[135] Babenko VV. Method and apparatus for mixing high molecular weight materials with liquids. The Patent Greck, based on European Patent publication 1140341, 2004 (Application 99967725.5); 2010.
[136] Babenko VV, Abbas FM, Gnitetsky NA. The interaction of the boundary layer with three-dimensional disturbances. National Academy of Sciences of Ukraine, Institute of Hydromechanics, Kyiv. Appl Hydromech 2011;13(85):3−22 No. 3.
[137] Babenko VV, Abbas FM, Ishchenko SA. Control of the boundary layer of the wing profile when generating three-dimensional disturbances. National Academy of Sciences of Ukraine, Institute of Hydromechanics, Kyiv. Appl Hydromech 2011; 13(85):55−67 No. 4.
[138] Babenko VV, Musienko VP, Turik VN. Kinematic characteristics of the flow around spherical cavities. National Academy of Sciences of Ukraine, Institute of Hydromechanics, Kyiv. Appl Hydromech 2012;14(86):3−21 3.
[139] Babenko VV, Chun HH, Lee I. Boundary layer flow over elastic surfaces. Compliant surfaces and combined methods for marine vessel drag reduction. —Amsterdam, Boston, Heidelberg, London and others: Elsevier publishers. Butterworth-Heinemann; 2012. p. 613.
[140] Babenko VV, Moroz VV. Experimental investigation of the influence of body shape on its integral characteristics. National Academy of Sciences of Ukraine, Institute of Hydromechanics, Kyiv. Appl Hydromech 2013;15(87):3−24 No. 3.
[141] Babenko VV, Musienko VP, Turik VN. Flow of spherical and oval cavities. National Academy of Sciences of Ukraine, Institute of Hydromechanics, Kyiv. Appl Hydromech 2014;16(88):3−25 No. 1.
[142] Babenko VV. Universal underwater device with variable resistance. Declarative patent for useful model, Ukraine UA N 98592, M. Cl. B63B 3/13, B63H 1/36, B63G 8/42, F15D 1/12, Bull.N 8, 27.04.2015.
[143] Babenko VV. Stem of fin propulsion. Declarative patent for useful model, Ukraine, RU N 99030, − publ. 12.05.2015, bull. N 9.
[144] Babenko VV. Generator bubbles for underwater vehicles. Declarative patent for useful model, Ukraine, UA N 99031, − publ. 12.05.2015, bull. N 9.
[145] Babenko VV, Turik VN. Universal vortex chamber. Declarative patent for useful model, Ukraine, UA N 99280, − publ. 25.05.2015, bull. N 10.
[146] Babenko VV, Turik VN, Kochin VA, Kochina MV. A control method for mixing flows in a vortex chamber. Declarative patent for useful model, Ukraine, UA. 110914, publ. 10.25.2016, bull. N 20.
[147] Babiy VV, Moros VV. On choice of the profile on a ship's steering complex. Int. J Fluid Mech Res 2002;29(5):534−43.
[148] Babkin VN, Belotserkovsky SM, Gulyaev VV, Dvorak AV. Jets and carrying surfaces: computer modeling. Moscow: Science; 1989. p. 208.
[149] Balaras E, Piomelli U, Wallace JM. Self-similar states in turbulent mixing layers. J Fluid Mech 2001;446:1−24.
[150] Bandyopadhyay PR. Review of mean flow in turbulent boundary layers disturbed to alter skin friction. Trans ASME J Fluid Eng 1986;108:127−40.
[151] Bandyopadhyay PR. Viscous drag reduction of a nose body. AIAA J 1989; 27(23):274−82.

[152] Bandyopahyay PR, Gad-el-Hak M. Reynolds number effects in wall-bounded turbulent flows. Appl Mech Rev 1994;47(28):139.
[153] Barbanel BA, Bogdevich VG, Maltsev LI, Malyuga AG. Some practical applications of the theory of boundary layer control. St. Petersburg Marine Engineering Bureau "Malachite"; 1994. p. 47.
[154] Barst F, Rastogi AK. Calculation of turbulent boundary layer flows with drag reducing polymer additions. Phys Fluids 1977;20(12).
[155] Basin AM, Korotkin AI, Kozlov LF. Control of the boundary layer of the vessel. Leningrad: Shipbuilding; 1968. p. 492.
[156] Batchov Z, Kriminal V. Questions of hydrodynamic stability. Moscow: World; 1971. p. 352.
[157] Bayley BJ. The dree-dimensional centrifugal-type instablities in inviscid two-dimensional flows. Phys Fluids 1987;31(1).
[158] Bechert DW, Bruse M, Hage W, Meyer R. Fluid mechanics of biological surfaces and their technological application. Naturwissenschaften 2000;87:157−71 − Springer − Verlag −.
[159] Belinsky VG, Ivanishin BP, Kochin VA, Moroz VV. On the drag coefficient of conveniently flowing bodies in acceleration and deceleration modes. National Academy of Sciences of Ukraine, Institute of Hydromechanics, Kyiv Bionica 1992;25:46−50.
[160] Belinsky VG, Zinchuk PI. Resistance of a disk and a sphere during accelerated motion from a state of rest. National Academy of Sciences of Ukraine, Institute of Hydromechanics, Kyiv Bionica 1998;27-28:88−99.
[161] Belinsky VG. Laboratory of hydrodynamics of hydrophysical systems (hydraulic stand) Intergraph; Kiev: Institute of Hydromechanics; 2002. p. 225−39.
[162] Belotserkovsky SM, Violinist BK. Aerodynamic derivatives of aircraft vehicle and wing at subsonic speeds. Moscow: Science; 1975. p. 437.
[163] Belotserkovsky SM, Kotovsky VM, et al. Mathematical modeling plane-parallel separation of the flow around the body. Moscow: Science; 1988. p. 232.
[164] Belousov AG. Unsteady motion of bodies in water. National Academy of Sciences of Ukraine, Institute of Hydromechanics, Kyiv Bionica 1998;27-28:100−3.
[165] Bendat J, Pirsol A. Application of correlation and spectral analysis. Moscow: World; 1983. p. 312.
[166] Benjamin TB. The three fold classification of unstable disturbances in flexible surfaces bounding inviscid flows. J Fluid Mech 1963;16:436−50.
[167] Betchov Z, Criminale V. Issues of hydrodynamic stability. Moscow: World; 1971. p. 352.
[168] Bippes H, Görtler H. Dreidimensionale Strömungen in der Grenzschicht an einer konkaven wand. Acta Mech 1972;14:251−67.
[169] Bippes H. Experimentelle untersuchungen des laminar turbulent umschlags an einer parallel angeströmten konkaven wand. Heidelb. Akad. Wiss., Math. Naturwissenschaften Kl., Sitzungsberg. Abhandlung. 1972; 3: 103−111.
[170] Blackwelder RF. Analogies between transitional and turbulent boundary layers. Phys Fluids 1983;26(10):2807−16.
[171] Blake WK, Anderson JM. The acoustics of flow over rough elastic surfaces. In: Conf. paper: flow induced noise and vibration issues and aspects. 2014. p. 1−23.
[172] Blick E.F. Turbulent boundary layer characteristics of compliant surfaces. In: 5th AIAA aerospace sciences meeting. New York; 1967. p. 67−128.
[173] Blick EF, Walters R. Turbulent boundary layer characteristics of compliant surfaces. J Aircr 1968;5(1):11−16.

[174] Blick EF, Walters R, Smith RL, Chu H. Compliant coating skin friction experiments. AIAA papers N69−165. In: 7th AIAA aerospace sciences meeting. New York; 1969. p. 9.

[175] Blick EF. Skin friction drag reduction by compliant coating. In: Proc. int. conf. drag reduct. Cambridge. Cranfield, s.a. F2 123-F23/36; 1974.

[176] Blick EF. Theories of compliant coating drag reductions. Phys Fluids 1977;20(10):132−6 pt. 2.

[177] Bödewadt UT. Die Drehströmung über festen Grund. Zeit Angew Math Mech 1940;20:241−53.

[178] Bogdevich VG, Malyuga AG. Distribution of surface friction in a turbulent boundary layer in water behind the place of gas injection. In: The study of control by the boundary layer. Novosibirsk: Institute of Engineering Thermal Physics (ITTF), Siberian Branch of the USSR Academy of Sciences; 1970.

[179] Bogdevich VG, Evseev AR, Maluga AG, Migirenko GS. Gas-Saturation effect on near-wall turbulence characteristics. In: Second int. conf. on drag reduction, Paper D2 (BHRA Fluid Eng., Cambridge, England), 1977. p. 25.

[180] Bogdevich VG, Kobets GF, Kozuk GS, et al. Some issues of the control of near-wall flows. PMTF 1980;5:99−109.

[181] Bogdevich VG, Malzev LI, Maluga AG. Optimization of the distributed gas injection into a turbulent boundary layer for the drag reduction. In: Proc. of the intern. symposium on seawater drag reduction. Newport, Rhode Island, USA; 1998. p. 327−30.

[182] Bottaro A., Zebib A. Görtler vortices promoted by wall roughness. Fluid Dynamics Res. 1997;19:343−362.

[183] Boyko AV, Grek AV, Dovgal AV, Kozlov VV. Beginning of turbulence in near-wall flows. Novosibirsk: Science; 1999. p. 328.

[184] Bradshaw P. Introduction to turbulence and its measurement. Moscow: World; 1974. p. 278.

[185] Brekhovskikh LM. Waves in complex media. Moscow: World; 1973. p. 540.

[186] Brennen C. Cavity surface wave patterns and general appearance. J Fluid Mech 1970;44(part 1):33−49.

[187] Britter RE, Hunt JGR, Mumford JC. The distortion of turbulence by a circular cylinder. J Fluid Mech 1979;92(pt.2):269−301.

[188] Brown SN. The physical model of boundary layer transition. In: Proc. of the ninth midwestern mechanics conf. Madison, WI: Univ. of Wisconsin−Madison, 1965. p. 117−29.

[189] Brown SN, Stewartson K. On the non-linear evolution of critical layers. Laminar-Turbulent Transit 1980;163−72.

[190] Brungard TA, Holmberg WJ, Fontane AA, Deutsch S. The scaling of the wall pressure fluctuations in polymer-modified turbulent boundary layer flow. J Acoust Soc Am 2000;108(1):71−5.

[191] Bugliarello G. La resistenza at moto accelerato di sfere in acgua. La ricerca Sci 1956;26(2):437−61.

[192] Bushnell DM, Hefner JN, Ash RL. Effect of compliant wall motion on turbulent boundary layers. Phys Fluids 1977;20(10):31−48 part 2.

[193] Bushnell DM. Compliant surfaces introduction. In: Viscous flow drag reduction symposium technical paper, Dallas, Texas, 1979. New York; 1980. p. 357−90.

[194] Bushnell DM. Turbulent drag reduction for external flow. AIAA-83−0227. In: AIAA 21st aerospace sciences meeting. Reno, Nevada; January 10−13, 1983. p. 1−20.

[195] Callenaere M, Franc JP, Mishel JM, Riondet M. The cavitation instability induced by the development of a re-entrant jet. J Fluid Mech 2001;444:223−56.

[196] Carpenter PW, Garrad AD. The hydrodynamic stability of flow over Kramer-type compliant surface. Part 1. Tollmien-Schlichting instabilities. J Fluid Mech 1985;155:465–510.
[197] Carpenter PW, Garrad AD. The hydrodynamic stability of flow over Kramer-type compliant surface. Part 2. Flow-induced surface instabilities. J Fluid Mech 1986;170:199–232.
[198] Carpenter PW. Biology-based drag reduction. In: Proceedings of the second international symposium on seawater drag reduction (ISSDR 2005). Busan, Korea; 2005. p. 45–52.
[199] Carr LV. Progress in the study and methods for calculating dynamic separation. Aerosp Appl 1988;12:59–73.
[200] Cebeci T, Smith AM. Analysis of turbulent boundary layers. New York; London: Acad. Press; 1974. p. 404.
[201] Chang PK. Separated flows: In 3V., V. 1. Moscow: Peace; 1972. p. 300.
[202] Chang PK. Control of flow separation. Washington, D.C: Hemisphere Publ. Corp.; 1976.
[203] Chang PK. Control separation flow. Moscow: World; 1979. p. 552.
[204] Chernyshov OB, Koval AP, Drobakha AA. Some features of the morphology of the gill apparatus of fish, associated with the speed of their swimming. National Academy of Sciences of Ukraine, Institute of Hydromechanics, Kyiv Bionica 1978;12:103–8.
[205] Chizhikov DN, Schastlivy VP. Tellur and his properties. Moscow: Science; 1966. p. 89.
[206] Ciolkosz LD, Spina EF. An experimental study of Görtlel vortices in comprecssible flow. AIAA Pap. 2006; N 4512: 1–21.
[207] Clark H, Deutsch S. Microbubble skin friction reduction on an axisymmetric body under the influence of applied axial pressure gradients. Phys Fluids 1991;A3:2948.
[208] Coles D. Coherent structures of turbulent boundary layers. Lechish Univ.; 1979. p. 462.
[209] Thill C, Toxopeous S, van Warlee F. Project energy-saving air-lubricated ships (PELS). In: Proc. of the 2-nd intern. symp.on seawater drag reduction (ISSDR 2005). Busan, Korea; 2005. p. 93–108.
[210] Davis MR. Design of flat plate leading edges to avoid flow separation. AIAA J 1980;18(5):598–600.
[211] Davies C, Carpenter PW, Ali R, Lockerby DA. Disturbance development in boundary layers over compliant surfaces. In Sixth IUTAM symposium on laminar-turbulent transition. vol. 78. 2006. p. 225–30. (Fluid Mechanics and its Applications).
[212] De Lange DF, Bruin GJ De, Wijngaarden L. On the mechanism of cloud cavitation experiment and modeling. In: H Kato, editor. Proc. Second Intern. Symp. on Cavitation. 1994. p. 45–9.
[213] Deutsch S, Castano J. Microbubble skin friction reduction on an axisymmetric body. Phys Fluids 1986;29:3590.
[214] Deutsch S, Pal S. Local shear stress measurements on an axisymmetric body in a microbubble modified flow field. Phys Fluids 1990;A 2:2140.
[215] Deutsch S., Money M. et al. Microbubble drag reduction in rough wall turbulent boundary layer. In: Proc. ASME Fluids Eng. 2003. p. 1–9.
[216] [411] Deutsch S, Fontaine AA, Money MJ, Petry HL. Drag reduction with combined micro-bubble and polymer injection. In: Proceedings of the second international symposium on seawater drag reduction (ISSDR 2005). Busan, Korea; 2005. p. 459–67.
[217] Dinkelacker A. Preliminare experiments on the influence of flexible walls on boundary layer turbulence. J Sound Vib 1966;4(2):187–214.

[218] Doherty BI. Investigation of drag reduction obtained through boundary layer injection of dilute solutions of poly ethylene oxide. U.S. Naval Academy Trilent Scholar Report; 1965.
[219] Dovgy SA, Kayan VP. To a method for determining thrust generated by an oscillating wing. National Academy of Sciences of Ukraine, Institute of Hydromechanics, Kyiv Bionica 1979;15:55−9.
[220] Dovgy SA, Kopeyka OV. The influence of a solid surface on the hydrodynamic characteristics of two oscillating wings with their nonlinear interaction. National Academy of Sciences of Ukraine, Institute of Hydromechanics, Kyiv Bionica 1991;24:28−33.
[221] Dovgy SA, Kopeyka OV. Investigation of hydrodynamic characteristics of two oscillating profiles in a biplane type system. National Academy of Sciences of Ukraine, Institute of Hydromechanics, Kyiv Bionica 1993;26:69−74.
[222] Dovgy SA, Lifanov IK. Methods for solving integral equations. Theory and applications. Kiev: Naukova Dumka; 2002. p. 343.
[223] Sanders JV. Drag coefficients of spheres in polyethylene oxide solutions. Int Shipbuld Prog 1967;14(152):140−9.
[224] Drazin PC, Reid WH. Hydrodynamic stability. USA: Cambridge; 1981. p. 317.
[225] Droblenkov V. Investigation of the influence of the external pressure gradient and changes in the physical constants of the fluid on the development of disturbances in the laminar boundary layer. Proceedings of the Leningrad Shipbuilding Institute (LCI); 1970. p. 34−41.
[226] Duong T-S, Gougat P, Lengeile G. Etude de la transition sur une plaque a paror sinusoidale en ecoulenment incompressible. C r Acad Sci 1970;271(3):179−82.
[227] Efimtsov BM, Kuznetsov VB. Pulsations of shear stress on the plate surface. Acoustics of turbulent flows. Moscow: Science; 1983. p. 46−53.
[228] Efimtsov BM. Criteria for the similarity of the spectra of near-wall pressure pulsations of a turbulent boundary layer. Acoustic J 1984;30(1):58−61.
[229] Kutateladze SS, Mironov BP, Nakoryakov VE, Khabakhpasheva EM. Experimental investigation of near-wall turbulent flows. Novosibirsk: Science; 1974. p. 166.
[230] Experimental study of the structure of the near-wall pulsation fields of the turbulent boundary layer. In: Proc. of TsAGI No. 579, Moscow; 1980. p. 80.
[231] Faddeev YuI, Zhurava VN. Equations of a laminar boundary layer on a deformed body of revolution. National Academy of Sciences of Ukraine, Institute of Hydromechanics, Kyiv Bionica 1969;3:40−6.
[232] Falco RE. Combined simultaneous flow visualization − hot wire anemometry for the study of turbulent flows. J Fluid Eng 1980;102:174−82.
[233] Fedyaevsky KK, Blumina LKh. Hydrodynamics separation flow around the bodies. Moscow: mechanical engineering; 1977. p. 120.
[234] Fibig M, Brockmeier U, Mitra NK, Guntermann T. Structure of velocity and temperature fields in laminar channel flows with longitudinal vortex generators. Numer Heat Transf Part A 1989;15:281−302.
[235] Fischer MC, Weinstein LM, Bushnell DM, Ash RL. Compliant wall-turbulent skin-friction research. AIAA Paper No. 75−833. 1975. p. 1−20 (AIAA 8th fluid plasma dynamics conference. Hartford, Connecticut/June 16−18, 1975).
[236] Fish FE, Legac P, Williams TM, Wei T. Measurement of hydrodynamic force generation by swimming dolphins using bubble DPIV. J Exp Biol 2014;217:252−60.
[237] Fisher MK. Investigations on change of friction in a turbulent boundary layer on pliable walls. The all-Union center of translations1980; №A-4985: 48.

[238] Fisher MK. Study on the measurement of friction in a turbulent boundary layer on pliable walls. In: Translation of VCP № A-4985; 1980.
[239] Floryan JM, Saric WS. Stability of Görtler vortices in boundary layers. AIAA J 1982;20(3):316–24.
[240] Floryan J, Saric WS. Wavelength selectioll alld growth of Görtler- vortices. AIAA J 1984;22:1529–38 NQ 11.
[241] Floryan JM. Görtler instability of boundary layers over concave and convex walls. Phys Fluids 1986;29(12).
[242] Floryan JM. Görtler instability of wall jets. AIAA J 1989;27(2):112–14.
[243] Floryan JM. On the Görtler instability of boundary layers. Prog Aerosp Sci 1991;28:235–71.
[244] Fontane AA, Deutsch S. The influence of the type of gas on the reduction of ckin friction drag by microbubble injection. Exp Fluids 1992;13(2–3):128–36.
[245] Fontane AA, Petrie HL, Brungard TA. Velocity profile statistics in a turbulent boundary layer with slot-injected polymer. J Fluid Mech 1992;238:435–66.
[246] Fontaine A.A., Deutsch S., Brungart T.A., Petrie H.L. & Fenstermacker M. Drag reduction by coupled systems: microbubble injection with homogeneous polymer and surfactant solutions. J. Exp Fluids 1999; 26: 397–403.
[247] Kozlov LF, Tsyganyuk AI, Babenko VV. etc Formation of turbulence in shear flows. Kiev: Naukova Dumka; 1985. p. 283.
[248] Fruman DH, Tulin MP. Diffusion of a tangential drag-reducing polymer injection on a flat plate of high Reynolds number. J Ship Res 1976;20(3):171–80.
[249] Gad-el-Hak M, Blackwelder RF, Riley JF. On the interaction of compliant coatings with boundary layer flows. J Fluid Mech 1984;140:257–80.
[250] Gad-el-Hak M, Chin-Ming H. The pitching delta wing. AIAA J 1985;23(11):1660–5.
[251] Gad-el-Hak M. Compliant coatings research: a guide to the experimentalist. J Fluid Struct 1987;1:55–70.
[252] Gad el Hak M. Compliant coatings for drag reduction. Prog Aerosp Sci 2002;38:77–99.
[253] Galway RD. An investigation into the possibility of laminar boundary layer stabilization using flexible surfaces. Phys Fluids 1977;20(10):31 Pt. II.
[254] Ganiev RF, Ukrainsky LE, Telalov AN. An experimental study of fluid flow in pipelines with pliable walls. Bionics 1980;14:46–50.
[255] Gaponov SA, Maslov AA. Development of disturbances in compressible flows. Novosibirsk: Science; 1980. p. 144.
[256] Garwood GC, Winkel ES, Vanapalli S, Elbing B, Walker DT, Ceccio SL, et al. Drag reduction by a homogenous polymer solution in large diameter, high shear pipe flow. In: Proceedings of the second international symposium on seawater drag reduction (ISSDR 2005). Busan, Korea; 2005. p. 509–15.
[257] Gasljevic K, Hall K, Chapman D, Matthus EF. Biotechnology-based drag reduction for naval application: production of biopolymers from marine microalgae and scale-up for type B reducers. In: Proceedings of the Second International Symposium on Seawater Drag Reduction (ISSDR 2005). Busan, Korea; 2005. p. 559–67.
[258] Gaster M. On transition to turbulence in boundary lyers. Transit turbulence 1981;95–112.
[259] Ginevsky AS, Fedyaevsky KK. Some regularities in case of unsteady translational motion in a viscous fluid. USSR Academy of Sciences. Ser. ONT 1959; 3: 207–9.
[260] Ginevsky AS, Vlasov EV, Kolesnikov AV. Aeroacoustic interactions. Moscow: Mechanical Engineering; 1978. p. 177.

[261] Glushko VN, Kayan VP, Kozlov LF. Hydrodynamic characteristics of a rectangular oscillating wing. National Academy of Sciences of Ukraine, Institute of Hydromechanics, Kyiv Bionica 1984;18:40−4.

[262] Glushko VN, Kayan VP. Experimental investigation of the hydrodynamics of a rigid oscillating wing. National Academy of Sciences of Ukraine, Institute of Hydromechanics, Kyiv Bionica 1994;25:71−5.

[263] Goldshtik MA. Vortex flows. Novosibirsk: Science, Siberian Branch; 1981. p. 366.

[264] Golovchenko VV, Gorelov DN. On arbitrary profile movement near the screen. Continuum Dynamics, 1975:22:99−106.

[265] Golubev VV. Works on the theory of the flapping wing. Works on aerodynamics. Moscow-Leningrad: State Technical Publishing House; 1957. p. 399−596.

[266] Gorelov DN. On the effectiveness of a flapping wing as a mover. National Academy of Sciences of Ukraine, Institute of Hydromechanics, Kyiv Bionica 1976;10:49−53.

[267] Gorelov DN. Experimental investigation of the flapping wing trust. National Academy of Sciences of Ukraine, Institute of Hydromechanics, Kyiv Bionica 1980;14:42−5.

[268] Gorelov DN. Propulsive characteristics of a flapping wing with a resiliently fixed aileron. National Academy of Sciences of Ukraine, Institute of Hydromechanics, Kyiv Bionica 1991;24:18−24.

[269] Gorlin SM, Slezinger NN. Aeromechanical measurements. Moscow: Science; 1964. p. 720.

[270] Görtler H. Über eine dreidimensionäle instabilität laminärer Grenzschichten an concave Wänden. Nach. Ges. Wiss. Göttingen Math.-Phys. Kl. 1940; Bd. 2, (1): 581−3.

[271] Görtler H. Hanssler. Einige neue experimentelle Beobachtungen über das Auftreten von Langswirbeln in Staupunct-strömungen. Schiffstechnik 1973;102H(20B):67−72.

[272] Grebeshov EP, Sagoyan OA. Hydrodynamic characteristics of an oscillating wing performing the functions of the carrier element and mover. Works of TsAGI 1966; No. 1725: 3−30.

[273] Greiner R. Hydrodynamics and energy underwater vehicles. Leningrad: Shipbuilding; 1978. p. 159.

[274] Greshilov EM, Evtushenko AV, Lyamshev LN, Shirokova NL. Some features of the influence of polymer additives on the near-wall turbulence. Eng Phys J (IFI) 1973; 25(6):999−1005.

[275] Gromeko IS. Collected works. Moscow: Science; 1952. p. 295.

[276] Gromov PS, Korotkin AI. Resistance in the boundary layer of a partially cavitating surface. National Academy of Sciences of Ukraine, Institute of Hydromechanics, Kyiv Bionica 1976;10:58−65.

[277] Grosskreutz R. Wechselwirkungen zwischen turbulenten Grenzschichten und weichen Wanden. Mitt. M. Plank Inst. Strömungsforschung und der Aerodyn. Verstichsangt 1971; № 53: 85−93.

[278] Guin MM, Kato H, Yamaguchi H, Maeda M, Miyanaga M. Skin friction reduction by microbubbles and its relation with near-wall bubble concentration in a channel. J Mar Sc Tech 1996;1:241.

[279] Guin M.M., Kato H, Takahashi Y. Experimentale vidence for alink between microbubble drag reduction phenomena and periodicall exicited wall-bounded turbulentflow. In: Proceedings of the second international symposium on seawater drag reduction (ISSDR 2005). Busan, Korea; 2005. p. 313−18.

[280] Gupta AK, Laufer J, Kaplan RE. Spatial structure in the viscous sublayer. J Fluid Mech 1971;50:493−512.

[281] Gupta A, Lilly D, Sayred N. Swirling streams. Moscow: World; 1987. p. 588.

[282] Gyorgyfalvy D. The possibilities of drag reduction by the use of flexible skin. AIAA 4th Aero. Sci. Meet. Los. Angeles, California. 1966. AIAA Paper N 66−430. p. 37.
[283] Halatov AA, Avramenko AA. Turbulent Taylor-Görtler instability on a concave surface with suction. National Academy of Sciences of Ukraine, Institute of Thermophysics, Kyiv Thermophys high Temp 1994;32(1):69−71.
[284] Halatov AA, Avramenko AA. Laminar instability of Taylor-Gertler near a concave surface. Eng Phys J (IJF) 1994;67(1−2):3−9.
[285] Halatov AA, Shevchuk IV, Avramenko AA, Kobzar SG, Zheleznaya TA. Thermogas dynamics of complex flows around curvilinear surfaces. Kiev: Institute of Thermal Physics, NAS of Ukraine; 1999. p. 299.
[286] Halatov AA, Avramenko AA, Shevchuk IV. Heat exchange and hydrodynamics in the fields of centrifugal mass forces. Twisted streams, V. 3. Kiev: Institute of Thermal Physics, NAS of Ukraine; 2000. p. 474.
[287] Halatov AA, Avramenko AA, Shevchuk IV. Heat transfer and hydrodynamics in the fields centrifugal mass forces. Engineering and technological equipment, Vol. 4. Kiev: Institute of Thermal Physics, NAS of Ukraine; 2000. p. 209.
[288] Hama ER, Nutant J. Detailed flow-field observation in the transitions process in a thick boundary layer. In: Proc. heat transfer and fluid mech. New York: Inst. Stanford Univ. Press; 1960. p. 77−93.
[289] Hashim A, Yaakob OB, Koh KK, Ismail N, Ahmed YM. Review of micro-bubble ship resistance reduction methods and the mechanisms that affect the skin friction on drag reduction from 1999 to 2015. J Teknol 2016;74(5).
[290] Hashim A, Yaakob OB, Koh KK, Ismail N, Ahmed YM. Influence of bubble size on micro-bubble drag reduction. Exp Fluids 2006;41(3):415−24.
[291] Hashimoto T, Yoshida M, Watanabe M, Kamijo K, Tsujimoto Y. Experimental study on rotating cavitation of rocket proppelant pump inducers. In: 32nd AIAA/ ASME/ SAE/ASEE Jont Propulsion Conf. 1996.
[292] Hassan YA, Gutierrez −Torres CC, Jimenez-Bernal JA. Temporal correlation modification by microbubbles injection in a boundary layer channel flow. Intern Commun Heat Mass Transf 2005;32(8):1009−15.
[293] Hatuntsev VN. On the question of the work of a propulsion system such as a flapping wing on regimes with a large relative gait. National Academy of Sciences of Ukraine, Institute of Hydromechanics, Kyiv Bionica 1984;18:57−62.
[294] Hatuntsev VN. Features of the creation of thrust on the model of a wing propulsor in an oblique flow. National Academy of Sciences of Ukraine, Institute of Hydromechanics, Kyiv Bionica 1985;19:47−50.
[295] Hatuntsev VN. Experimental investigation of non-stationary vortex structures. National Academy of Sciences of Ukraine, Institute of Hydromechanics, Kyiv Bionica 1992;25:54−8.
[296] Hatuntsev VN. On the method of processing and presenting the results of an experiment in the study of voluntary motions of bodies in a viscous incompressible fluid. National Academy of Sciences of Ukraine, Institute of Hydromechanics, Kyiv Bionica 1993;26:98−104.
[297] Hatuntsev VN, Babenko VV, Musienko VP, Mertvechenko OA. Investigation of nonstationary flow around bodies in the case of plane motion. National Academy of Sciences of Ukraine, Institute of Hydromechanics, Kyiv Bionica 1998;27−28:104−8.
[298] Hertel H. Structure, form and movement (biology and technology). NewYork: Reinhold Publishe Corporation; 1966. p. 251.
[299] Hinze IO. Turbulence. Its mechanism and theory. Moscow: Fizmatgiz; 1963. p. 681.

[300] Hintze JO. Turbulence. 2-nd ed. New York etc: McGraw — Hill; 1975. p. 790.
[301] Hizli H and Kurtulus DF. Numerical and experimental analysis of purely pitching and purely plunging airfoils in hover. In: 5-th Intern. micro air vehicle conference and flight competition (IMAV), Braunschweig, Germany; 2012.
[302] Howe MS. The influence of surface compliance on the production of sound by a turbulent boundary layer. J Vib Acoust Stress Reliab 1984;106(3):383—8.
[303] Howe MS. On the generation of sound by turbulent boundary layer flow over a rough wall. Proc. R Soc A: Math Phys Eng Sci 1984;247—63.
[304] Hoyt JW, Taylor JJ. Turbulence structure in a water jet discharging in air. Phys Fluids 1977;20(10):253—7 Pt. II.
[305] Hoyt JW. Hydrodynamic drag reduction due to fish slimes. In: Wu TY, Brokaw CJ, editors. Swimming and flying in nature, Vol. 2. New York: Plenum Press; 1975.
[306] Hoyt JW. Polymer solution effects on turbulent friction mechanisms. In: Proceedings of the second international symposium on seawater drag reduction (ISSDR 2005). Busan, Korea; 2005. p. 1—5.
[307] Yoon H-S, Park Y-H, Van S-H, Kim H-T, Kim W-J. Experimental investigation on the drag reduction for an axi-symmetric body by micro-bubble and polymer solution. In: Proc. of intern. workshop on frontier technology in ship and ocean engineering; 2003. p. 359—65.
[308] Ignatiev VN, Kuznetsov BG. Diffusion model of a boundary layer with a polymer. Numer Methods Contin Mech 1973;4(4).
[309] Levchenko VYa, editor. Instability of subsonic and supersonic flows (collection of works). Novosibirsk: Academy of Sciences of the USSR SB, ITAM; 1982. p. 132.
[310] Ionov AV. Temperature-frequency dependence of the loss coefficient of multilayer coatings for power equipment. New vibration-absorbing materials and coatings and their application in industry. Leningrad: Leningrad House of Scientific Technical Education; 1980. p. 44—8.
[311] Ioselevich VA, Pilipenko VN. On the resistance of a plate in a flow of a polymer solution of variable concentration. Proceedings of the Academy of Sciences of the USSR, Fluid and Gas Mechanics (MZHG) 1974; No. 1: 63—8.
[312] Ito A. Breakdown Structure of longitudinal vortices along a concave wail. J Jpn Soc Aero Space Sci 1988;36:272—9.
[313] Ivanov VP, Klochkov VP Kozlov LF. Measurement of the structure of velocity in liquid flows of large volumes using a laser Doppler velocity meter. News of the USSR Academy of Sciences. Fluid and gas mechanics 1977; 5: 170—3.
[314] Ivanov VP, Klochkov VP, Kozlov LF. Investigation of jet flow rotational ellipsoid using a laser Doppler anemometer. Eng Phys (IJF) 1978;34(1):99—103.
[315] Ivanov VP, Babenko VV, Blokhin VA, Kozlov LF, Korobov VI. Investigation of the velocity field in a hydrodynamic stand of small turbulence using a laser Doppler velocity meter. Eng Phys J (IFJ) 1979;37(5):818—24.
[316] Iversen HW, Balent RA. Correlating modulus for fluid resistance in accelerated motion. J Appl Phys 1951;22(3):324—8.
[317] Jacob B, Olivieri A, Miozzi M, Campana EF, Piva R. Drag reduction by microbubbles in a turbulent boundary layer. Phys Fluids 2010;22:115—24.
[318] Jones KD and Platzer MF. On the use of vortex flows for the propulsion of micro-air and sea vehicles. In: RTO AVT symposium on "advanced flow management: part A — vortex flows and high angle of attack for military vehicles." Loen, Norway. Published in RTO-MP-069(I). 2001. p. 40-1—40-13.

[319] Kachanov YuS, Kozlov VV, Levchenko VYa. Occurrence of turbulence in the boundary layer. Novosibirsk: Science; 1982. p. 149.
[320] Kaind R, Dj, Guden E, Dvorak FA. An experimental research of currents with a tangential blow in and comparison of experimental data with results of settlement methods. Rocket Eng Astronaut 1979;17:74−80.
[321] Kanarskiy MV, Teslo AP. On the turbulent flow past a plate with an elastic surface National Academy of Sciences of Ukraine, Kyiv Hydromechanics 1977;35:51−3.
[322] Kanarsky MV, Babenko VV, Kozlov LF. Experimental investigation of a turbulent boundary layer on an elastic surface. Stratified and turbulent flows. Kiev: Naukova Dumka; 1979. p. 59−67.
[323] Kanarskiy MV. Experimental study of the dynamic modulus of elasticity of an elastic plate. National Academy of Sciences of Ukraine, Institute of Hydromechanics, Kyiv Bionica 1981;15:98−101.
[324] Kanarskiy MV, Babenko VV, Voropaev GA. Measurement of the kinematic characteristics of a turbulent boundary layer on a plate and processing of the obtained information on a computer. National Academy of Sciences of Ukraine, Kyiv Hydromechanics 1982;45:30−6.
[325] Kaplan RE. The stability of laminar incompressible boundary layers in the presence of compliant boundaries. Rep. Aeroelastic and Structures Research Lab. TR 116-1 (Contract Nom-1841 (89)). Cambridge: Massachusetts Inst. of Technology; June 1964.
[326] Karavosov RK, Prozorov AG. Artificial laminarization of the boundary layer on a straight wing when irradiated with its sound. Works of TsAGI 1976; 1790.
[327] Kawamata S, Kato T, Matsumura Y, Sato T. Experimental research on the possibility of reducing the drag acting on a flexible plate. Theor Appl Mech 1973;21:507−21.
[328] Kawanami Y, Kato H, Yamaguchi H, Tagaya Y, Tanimura M. Mechanism and control of cloud cavitation. Trans ASME J Fluids Eng 1997;119:788−95.
[329] Kawanami Y, Kato H & Yamaguchi H. Three-dimensional characteristics of the cavitaties formed on a two-dimensional hydrofoil. In: Michel JM, Kato H, editors. Proc. third intern. symp. on cavitation. Vol. 1. 1998. p. 191−6.
[330] Kayan VP, Pyatetsky VE. Closed-type bio-hydrodynamic installation for studying the hydrodynamics of swimming of marine animals. National Academy of Sciences of Ukraine, Institute of Hydromechanics, Kyiv Bionica 1971;5:121−5.
[331] Kayan VP, Kozlov LF, Pyatetsky VE. Kinematic characteristics swimming some aquatic animals. News of the USSR Academy of Sciences. Fluid and gas mechanics 1978;5:3−9.
[332] Kayan VP. On the hydrodynamic characteristics of the fin mover dolphin. National Academy of Sciences of Ukraine, Institute of Hydromechanics, Kyiv Bionica 1979;13:9−15.
[333] Kayan VP. Experimental investigation of a hydrodynamic mover created by an oscillating wing. National Academy of Sciences of Ukraine, Institute of Hydromechanics, Kyiv Bionica 1983;17:45−9.
[334] Keith WL, Foley AV, Cipolla KM. Transmission on a turbulent boundary layer wall pressure field through an elastomeric coating. Ocean Eng 2012;47.
[335] Kelly HR, Rentz AW, Siekmann J. Experimental studies on the motion of a flexible hydrofoil. J Fluid Mech 1964;19(1):30−48.
[336] Kilian FP. Verfahren zur Wiederstands Verminderung. 1. Forteschriftsbereich. Versuchsanstalt für Wasserbau und Schiffbau, Bericht 384/67; 1967.

[337] Kim S, Tagori T. Drag measurements on flat plates with uniform injection of polymer solutions and their direct application to the wall. Proc. annual meeting, ASME; 1969.
[338] Kim HT, Kim OS, Kim WJ. Experimental results of friction-drag reduction by injection of PEO solution: report (1). In: Proc. of the annual spring meeting of society of naval architecture of Korea. Chochiwon, Korea; 2003.
[339] Kim OS, Kim HT, Kim WJ. Experimental results of friction-drag reduction by injection of microbubbles. In: Proc. of the annual spring meeting of society of naval architecture of Korea. Chochiwon, Korea; 2003.
[340] Kind RJ, Gouden K, Dvorak FA. Experimental investigation of flows with tangential injection and comparison of experimental data with results calculation methods. Rocket Technol Astronaut (AIAA) 1979;17:74−80.
[341] Klebanoff PS, Diehl ZW. Some features of artificially thickened fully developed turbulent boundary layers with zero pressure gradient. Report 1110 NACA T № 2475. 1951. p. 1−27.
[342] Klebanoff PS. Characteristics of turbulence in a boundary layer with zero pressure gradient. NACA Reprt N 1247. 1955. p. 1135−53.
[343] Klebanoff PS, Tidstrom KD, Sargent LM. The three-dimensional nature of boundary layer instability. J Fluid Mech 1962;12(pt. 1):1−34.
[344] Klein A. Development of turbulent current in a pipe: (Review). Theor. Bases Eng. Calculation 1981;103(2):180−8.
[345] Kline SJ, Reynolds U, Schraub FA, Rundstadler PW. The structure of turbulent boundary layers. J Fluid Mech 1967;30(4):741−73.
[346] Kline SJ, Reynolds U, Schraub FA, Rundstadler PW. Structure of turbulent boundary layers. Mechanics 1969;4:41−8.
[347] Knapp CF, Roach PI. A combined visual and hot-wire anemometer investigation of boundary-layer transition. AIAA J 1968;6(1):29−36.
[348] Knapp CF, Roach PI. Investigation of the transition of the boundary layer by the visual method and with the help of a thermo-anemometer. AIAA J 1968;6(11):32−42.
[349] Kobashi Y, Hayakawa M. The Transition mechanism of an oscillating boundary layer. Laminar-turbulent transition. 1980. p. 102−9.
[350] Kochin NE, Kibel IA, Roze NV. Theoretical fluid mechanics. Moscow−Leningrad: Gostekhizdat; 1948. p. 612.
[351] Kochin VA, Moroz VV. Automated data acquisition and processing system for a high-speed hydrodynamics of supercavitating bodies at an angle of attacks under conditions of considerable effect of fluid weightiness and closeness of free border 265 towing tank. J Mod Technol Autom 2009;3:15−24.
[352] Kochin V, Moroz V, Serebryakov V, Nechitailo N. Hydrodynamics of supercavitating bodies at an angle of attacks under conditions of considerable effect of fluid weightiness and closeness of free border. J Ship Ocean Eng 2015;5:255−65.
[353] Kodama У, Kakugawa A, Takahashi T, Nagaya S, Sugiyama K. Microbubbles: drag reduction mechanism and applicability to ships. In: 24-th Symp. on naval hydrodynamics. Fukuoka, Japan; 2002.
[354] Kohama Y. Three-dimensional boundary layer transition on a concave−convex curved wall. In: Liepman HW, Narasimha R, editors. Turbulence management and relaminarisation. Berlin, Heidelberg, New York, London, Paris, Tokyo: Springer−Verlag; 1988. p. 215−26.
[355] Korobov VI. Experimental investigation of the influence of compliant surfaces on the integral characteristics of the boundary layer. Eng Phys J (IFJ) 1979;37(3):518−23.

[356] Korobov VI. Methods of experimental investigations of hydrodynamic resistance of a plate with gradient-free flow. National Academy of Sciences of Ukraine, Institute of Hydromechanics, Kyiv Bionica 1980;14:50−3.
[357] Korobov VI. Experimental investigation of integral characteristics of the boundary layer on compliant plates. National Academy of Sciences of Ukraine, Institute of Hydromechanics, Kyiv Bionica 1980;14:53−7.
[358] Korobov VI, Babenko VV, Kozlov LF. Integral characteristics of the boundary layer on elastic plates. Eng Phys J (IFZH) 1981;41(2):351−2.
[359] Korobov VI, Babenko VV. On the method of measuring the integral characteristics of the boundary layer. National Academy of Sciences of Ukraine, Institute of Hydromechanics, Kyiv Hydromechanics 1983;48:57−62.
[360] Korobov VI, Babenko VV. On one mechanism of the interaction of an elastic wall with a flow. Eng Phys J (IFI) 1983;44(5):730−3.
[361] Korobov VI, Babenko VV, Kozlov LF. Interaction of a turbulent boundary layer with an elastic plate. Eng Phys J (IFJ) 1989;56(2):220−5.
[362] Korobov VI, Babenko VV, Belinskj VG. Fin mover. Patent of Ukraine N1671515 A. Bull N31, 23.08.91; 1991.
[363] Korobov VI. Combined effect of surface compliance and polymer additives on the boundary layer. National Academy of Sciences of Ukraine, Institute of Hydromechanics, Kyiv Bionica 1993;26:27−31.
[364] Korobov VI. Complex effect of surface compliance and high-molecular polymer additives on turbulent friction. National Academy of Sciences of Ukraine, Institute of Hydromechanics, Kyiv Appl Hydromech 2000;2(74):59−63.
[365] Korobov VI. Influence of longitudinal plate finning on hydrodynamic friction. National Academy of Sciences of Ukraine, Institute of Hydromechanics, Kyiv Appl Hydromech 2005;7(79):90−2 1.
[366] Korobov VI. Hydrodynamics of oscillating wing on the pitch angle. Proc Natl Aviat Univ 2017;2(71):70−5.
[367] Koval AP. Ampoules of sword-fish skin and their possible functional significance. National Academy of Sciences of Ukraine, Institute of Hydromechanics, Kyiv Bionica 1977;11:86−91.
[368] Koval AP. Crypto-like mucus-forming structure of the skin and gill covers of swordfish. National Academy of Sciences of Ukraine, Institute of Hydromechanics, Kyiv Bionica 1978;12:108−11.
[369] Koval AP. Histological structures of the skin of a sailboat and swordfish. National Academy of Sciences of Ukraine, Institute of Hydromechanics, Kyiv Bionica 1982;16:21−7.
[370] Koval AP. Roughness and some properties of the skin structure of swordfish. National Academy of Sciences of Ukraine, Institute of Hydromechanics, Kyiv Bionica 1987;21:7377.
[371] Koval AP, Butuzov SV. Some properties of the microrelief of the caudal fin of blue marlin and yellowfin tuna. National Academy of Sciences of Ukraine, Institute of Hydromechanics, Kyiv Bionica 1991;24:88−91.
[372] Koval AP, Koshovsky AA. Structure of the heat exchanger for tuna of various high-speed groups. National Academy of Sciences of Ukraine, Institute of Hydromechanics, Kyiv Bionica 1992;25:94−8.
[373] Kozlov LF, Pyatetsky VYe. Effect of copolymers and fish mucus on the hydrodynamic resistance of models and fish. In Mechanisms of movement and orientation of animals. Kiev: Naukova Dumka; 1968. p. 22−9.

[374] Kozlov L.F., Leonenko I.V. Influence of the xiphoid tip on reducing the resistance of the sphere. Reports of the Academy of Sciences of the USSR 1971; 6: 622−4.
[375] Kozlov LF, Leonenko IV. Research of the effect of the xiphoid tip on the resistance of the body of revolution. National Academy of Sciences of Ukraine, Institute of Hydromechanics, Kyiv Bionica 1973;7:8−14.
[376] Kozlov LF. Investigations of the laminar boundary layer and its transition to turbulence. Kiev: Naukova Dumka; 1974. p. 176.
[377] Kozlov LF, Oleinik AJ. Hydrodynamics of aquatic animals swimming in a combo way. Reports of the USSR Academy of Sciences 1976;11:1001−4.
[378] Kozlov LF, Babenko VV. Experimental studies of the boundary layer. Kiev: Naukova Dumka; 1978. p. 184.
[379] Kozlov LF. Hydrodynamics of aquatic animals with a lunate tail fin. National Academy of Sciences of Ukraine, Institute of Hydromechanics, Kyiv Bionica 1979;13:3−9.
[380] Kozlov LF. Theoretical studies of the boundary layer. Kiev: Naukova Dumka; 1982. p. 293.
[381] Kozlov LF. Theoretical biohydrodynamics. Kiev: Higher School; 1983. p. 240.
[382] Kramer MO. Boundary layer stabilization by distributed damping. J Aeron Sci 1957;24(6):459−60.
[383] Kramer MO. Boundary layer stabilization by distributed damping. J Am Soc Nav Eng 1960;72(1):25−33.
[384] Kramer MO. Improvements in method of making drag-reducing covering. Patent UK 855.224; 1960.
[385] Kramer MO. Improvements in means for reducing frictional drag in fluids. Patent UK 864.593; 1961.
[386] Kramer MO. Device for Stabilizing Laminar Boundary Layer Flow. Patent UK 881.570; 1961.
[387] Kramer MO. Boundary layer stabilization by distributed damping. Nav Eng J 1962;72 (1):25−73.
[388] Kramer MO. Means and method for stabilizing laminar boundary layer flow. Patent USA 3.161.385; 1964.
[389] Kramer MO. Hydrodynamics of the Dolphin. Advances in hydroscience, v. 2. New York, London: Academic Press; 1965.
[390] Kramer MO. Means and method for stabilizing laminar boundary layer flow. Patent USA 3.585.953; 1971.
[391] Krasilnikova TN. Calculation of the statistical characteristics of pressure pulsations on the surface using the second moments of the longitudinal velocity pulsations in the boundary layer. In Turbulent flows. Moscow: Science; 1970. p. 44−9.
[392] Kravets AS. Characteristics of aviation profiles. Moscow-Leningrad: Oborongiz; 1939. p. 264.
[393] Kreichnan RH. Pressure fluctuation in turbulent flow over a flat plate. JASA 1956;28 (N3):378−90.
[394] Kurava VM, Faddeev YuI. On the question of the influence of the nonstationarity of the translational motion of hydrobionts on friction resistance. National Academy of Sciences of Ukraine, Institute of Hydromechanics, Kyiv Bionica 1973;7:56−9.
[395] Kutateladze SS, Styrikovich MA. Hydrodynamics of gas-liquid systems. Moscow: Energy; 1976.
[396] Kutateladze SS, Volchkov EP, Terekhov VI. Aerodynamics and heat and mass transfer in limited vortex flows. Moscow: Energoatomizdat; 1987. p. 283.

[397] Landhal MT. On the stability of laminar incompressible boundary layer over a flexible surface. J Fluid Mech 1962;13(2):609–32.
[398] Lang B. Drag measurement on a plate with ejection of polymer solution. In: Proceedings of the 7-th Intern. Congress on Rheology. Groningen, Netherlands; 1976.
[399] Latto B, Osama KF, Riedy EL. Diffusion of polymer additives in a developing turbulent boundary layer. J Hydronaut 1976;10(4).
[400] Latorre R. Ship hull drag reduction using bottom air injection. Ocean Eng 1997; 24(2):161–75.
[401] Latorre R, Babenko VV. Role of bubble injection technique drags reduction. In: Proceedings of the second international symposium on seawater drag reduction (ISSDR 2005). Busan, Korea; 2005. p. 319–25.
[402] Latorre R, Miller A, Philips R. Ship hull drag reduction using bottom air injection. Ocean Eng 2003;30:161–76.
[403] Lavrentyev MA, Shabbat BV. Problems of hydrodynamics and their mathematical models. Moscow: Science; 1973. p. 215.
[404] Lee T, Fischer MC, Schwarz WH. The measurement of flow-induced surface displacement on a compliant surface by optical holographic interferometry. Exp Fluids 1993;14:159–68.
[405] Lee T, Fischer M, Schwarz WH. Investigation of the stable interaction of a passive compliant surface with a turbulent boundary layer. J Fluid Mech 1993;257:373–401.
[406] Leneweit G. Roesner KG, Koehler R. Experimental study of the surface instabilities of a thin liquid film on a rotating disk. In: Third European fluid mech. conf. Göttingen; 1997. p. 210.
[407] Leonenko IV. On the hydrodynamic properties of the swordfish rostrum. National Academy of Sciences of Ukraine, Institute of Hydromechanics, Kyiv Bionica 1974;8:21–3.
[408] Levchenko VYa, Volodin AG, Gaponov SA. Characteristics of the stability of boundary layers. Novosibirsk: Science; 1975. p. 313.
[409] Levin, VB. On the stabilizing effect of flow rotation on turbulence. Thermal physics of high temperatures 1964;2, No. 6:892–900.
[410] Li WP, Zhou H. The Delay of transition of a laminar boundary layer over compliant wall. Project supported by the National Natural Science Foundation of China; 1990. p. 1–9.
[411] Li WP, Zhou H. Static divergence instability of the turbulent boundary layer over compliant walls. Project supported by the National Natural Science Foundation of China; 1990.
[412] Liepman NW, Narasimka R, editors. Turbulence management and relaminarisation. Berlin, Heidelberg, New York, London: Springer-Verlag; 1988. p. 524.
[413] Lighthill J. Mathematical biofluid dynamiecs. Philadelphia, PA: SIAM; 1975. p. 281.
[414] Lin CC. On the stability of two-dimensionale parallel flows. Parts 1, 2, 3. Quart Appl Math 1945;3:117–42 218–234, 277–301.
[415] Lin CC. Theory of hydrodynamic stability. Moscow: publishing house of foreign lit; 1958. p. 194.
[416] Lissamen PVS, Harris GL. Turbulent superficial friction on deformable surfaces. AIAA J 1966;7(8):243–4.
[417] Lissamen PVS, Harris GL. Turbulent surface friction on deformable surfaces. Rocket Technol Astronaut (AJAA) 1969;7(8):243–4.
[418] Logvinovich GV. Hydrodynamics of a flow with free boundaries. Kiev: Naukova Dumka; 1969. p. 215.

[419] Logvinovich GV. Hydrodynamics of swimming fish National Academy of Sciences of Ukraine, Institute of Hydromechanics, Kyiv Bionica 1973;7:3−8.
[420] Loitsyansky LG. Mechanics of fluid and gas. Moscow: Science; 1987. p. 848.
[421] Loonay HW, Blick EF. Skin friction coefficient of compliant surfaces in turbulent flow. J Spacecr Rockets 1966;3(10):1562−4.
[422] Lopatyuk MM. The use of spline functions for solving problems of flow around airfoils and surfaces. In: Collection of theses of the scientific and technical conference of young scientists. Novosibirsk: Sibniat; 1979. p. 11−13.
[423] Love RH. The effect of ejected polymer solutions on the resistance and wake of a flate plate in a water flow. Hydromechanics Inc. Tech. Rep. 1965. p. 353−62.
[424] Lucey AD, Carpenter W. Boundary layer instability over compliant walls. Comparison between theory and experiment. Phys Fluids 1995;7(10):2355−63.
[425] Madavan NK, Deutsch S, Merkle CL. Reduction of turbulent skin friction by microbubbles. Phys Fluids 1984;27:356−63.
[426] Madavan NK, Deutsch S, Merkle CL. Measurements of local skin friction in a microbubble modified turbulent boundary layer. J Fluid Mech 1985;156(2):37−56.
[427] Madavan NK, Deutsch S, Merkle CL. Measurements of local skin friction in a micobubble-modified turbulent boundary layer. J Fluid Mech 1985;156:237−56.
[428] Makarenko RA, Turik VN. Kinematics of flow in a dead end part of a vortex chamber National Academy of Sciences of Ukraine, Institute of Hydromechanics, Kyiv Appl Fluid Mech 2001;3(75):46−51 No. 1.
[429] Makhshutov EG. Visualization of the flow around holographic interferometry methods in biohydrodynamic experiments. National Academy of Sciences of Ukraine, Institute of Hydromechanics, Kyiv Bionica 1988;2:59−67.
[430] Maksimov VP. Transformation of external disturbances into waves of the boundary layer. Numer Methods Contin Mech 1978;9(2):49−50.
[431] Malyuga A., Mikuta V., Nenashev A. Local drag reduction at flow of polymer solutions aerated by air bubbles. In: Proc. of the 18th sci. and method. Seminar on ship hydrodynamics. Varna, Bulgaria; 1989. p. 743-1−743-6.
[432] Malzev LI. Jet methods of gas injection into fluid boundary layer for drag reduction. Appl Sci Res 1995;54.
[433] Mangalam SM, Dagenhart JR, Hepner TE, Meyers JF. The Görtler instability on an airfoil. AIAA Pap, 491. 1985. p. 1−22.
[434] Marie JL. A simple analytical formulation for microbubble drag reduction. J Physico-Chem Hydrodyn 1987;13:213.
[435] Martem'yanov SA, Vorotyntsev MA, Grafov BM. Derivation of the equation of non-local transport of matter in a turbulent diffusion layer. Electrochemistry 1979; 15(6):913−17.
[436] Martynov AK. Applied aerodynamics. Moscow: Mechanical Engineering; 1972. p. 447.
[437] Culbreth M, Allaneau Y and Jameson A. High-fidelity optimization of flapping airfoils and wings. In: 29th AIAA applied aerodynamics conference. Honolulu, Hawaii; 2011.
[438] McCarthy JH. Flat plate frictional drag reduction with polymer injection. J Ship Res 1971;15(7).
[439] McCarthy JH. Some fundamental problems of ship resistance and flow; new methods to reduce frictional drag. In: Scientific and methodological seminar on ship hydrodynamics. Bulgarian ship hydrodynamics center; 1983. p. 37−9.
[440] McCormick ME, Bhattacharyya R. Drag reduction of a submersible hull by electrolysis. Nav Eng J 1993;85:11−16.

[441] McMichael JM, Klebanoff PS, Meas NE. Experimental investigation of drag on a compliant surface. In: Techn. pap. symposium. Viscous flow drag reduct. Dallas/New York; 1980. p. 410–38.
[442] Yu M. Numerical and experimental investigations on unsteady aerodynamics of flapping wings [Graduate theses and dissertations]. Iowa State University; 2012.
[443] Meng JCS. Wall layer microturbulence phenomenological model and a semi-Markov probability predictive model for active control of turbulent boundary layers. In: Panton RL, editor. Self-sustaining mechanism of wall turbulence. Southampton, UK and Boston, USA: Computational Mechanics Publications; 1997. p. 201–52.
[444] Meng JCS and Uhlman JS, Jr. Microbubble formation and splitting in a turbulent boundary layer for turbulence reduction. In.: Proc. of the intern. symposium on seawater drag reduction. Newport, Rhode Island, USA; 1998. p. 341–55.
[445] Meng JCS. Engineering insight of near-wall micro turbulence for drag reduction and derivation of a design map for seawater electromagnetic turbulence control. In.: Proc. of the intern. symp.on seawater drag reduction. Newport, Rhode Island, USA; 1998. p. 359–67.
[446] Merkle CL, Deutsch S, Pal S, Cimbala J and Seelig W. Microbubble drag reduction. In: Proc. sixteenth symposium on naval hydrodynamics. Berkeley, California, USA; 1986. p. 199.
[447] Merkle CL, Deutsch S. Microbubble drag reduction. In: Gad-el-Hak M, editor. Frontiers in experimental fluid mechanics, lecture notes in Eng., vol. 46. Berlin: Springer-Verlag; 1989. p. 291.
[448] Migirenko GS, Evseev AR. Turbulent boundary layer with gas saturation. Problems of thermal physics and physical hydrodynamics. Novosibirsk: Science; 1974.
[449] Mitsudharmadi H, Winoto SH, Shah DA. Development of most amplified wavelength Görtler vortices. Phys Fluids 2006;18(1):014101-1–014101-12.
[450] Moore KJ, Rajan T, Gorban VA, Babenko VV. Method and apparatus for increasing the effectiveness and efficiency of multiple boundary layer control techniques. Patent US 6,357,374; 2002.
[451] Moore KJ, Rajan TD, Gorban VA, Babenko VV. Method and apparatus for increasing the effectiveness and efficiency of multiple boundary layer control techniques. European Patent 1305205; 2003.
[452] Moore KJ. Engineering an efficient shipboard friction drag reduction system. In: Proceedings of the second international symposium on seawater drag reduction (ISSDR 2005). Busan, Korea; 2005. p. 345–58.
[453] Moore KJ, Moore CM, Stern MA, Deutch S. Design and test of a polymer drag reduction system on sea flyer. In: 26-th Symp. on naval hydrodynamics. Rome, Italy; 2006.
[454] Morkovin MV, Reshotko E. Dialogue on progress and issues in stability and transition research. Opeing invited lecture. Third IUTAM symp.on laminar turbulent transition. Toulouse, France; Sept. 1989. p. 1–24.
[455] Moroz V.V., Kochin V.A. Energy aspects of movement of a body in a mode of acceleration and deceleration. In: First intern. industrial conf. Bionic 2004, Hannover, Germany. Reihe 15, Umwelttechnik, Nr. 249. 2004; p. 201–7.
[456] Morrison JR, O'Brien MP, Johnson SW, Schaat SA. The forces exerted by surface wave on piles. J Pet Technol 1950;189:149–89.
[457] Mueller TI, et al. Smoke visualization of boundary-layer transition on a spinning axisymmetric body. AIAA J 1981;12:1607–8.
[458] Mukoseev BI. A flow of a viscous incompressible fluid around an oscillating surface. Fluid and gas mechanics (MZHG) 1971; No. 4: 60–5.

[459] Musienko VP. On the formation of Tollmien-Schlichting waves by a mechanical vibrator. National Academy of Sciences of Ukraine, Institute of Hydromechanics, Kyiv Bionica 1992;25:50−3.
[460] Musienko VP. Experimental study of localized flow around cavities. National Academy of Sciences of Ukraine, Institute of Hydromechanics, Kyiv Bionica 1993;26:31−4.
[461] Nachtigal W. Über Kinematik, Dynamik und Energetik des schwimmenes einheimischer Dytisciden. Z vergltichende Physiol 1960;43:48−118.
[462] Naotsugu I, Hirohisa M, Hisashi K, et al. The study on a propulsion system by fin stroke. Bull Mar Eng Soc Jap 1980;8(1):71−9.
[463] Newman BG. The deflection of plane jets by adjacent boundaries, Coanda effect. In: Lachmen GV, editor. Boundary layer and flow control. Pergamon Press; 1964. p. 232.
[464] Nickel K, Schönauer W. Eine einfache exeprimentelle methode zur Sichtbarmachung von Tollmien − Wellen und Görtler wirbein. ZAMM 1961;41:145−7.
[465] Nikishova OD, Babenko VV. Flowing around an elastic body by a fluid flow. National Academy of Sciences of Ukraine, Institute of Hydromechanics, Kyiv Bionica 1975;9:55−60.
[466] Nikishova OD. Stability of flow near a curvilinear moving surface to three-dimensional disturbances. National Academy of Sciences of Ukraine, Institute of Hydromechanics, Kyiv Appl Fluid Mech 2000;2(2):64−75.
[467] Nikitin IK. Turbulent bed flow and processes in the bottom region. Kiev: Publishing House of the Ukrainian Academy of Sciences; 1963. p. 142.
[468] Nikitin IK. Complicated turbulent flows and heat and mass transfer processes. Kiev: Naukova Dumka; 1980. p. 238.
[469] Nikulin VA. A model of near-wall turbulence in weak solutions of polymers. Physical fluid mechanics. Donetsk: Visscha School; 1977. p. 34−46.
[470] Nisewanger CR. Flow noise and drag measurement of vehicle with compliant coating. Notes Nav. Weps Rep. 8518; 1964.
[471] Nonweiler TRF. Qualitative solution of the stability equation for a boundary layer in contact with various forms of flexible surface. Aeronaut. Res. Council Current Pap. 1963; Ra M ARC Rep. N 22, 670. p. 75.
[472] Obremsky HJ, Fejer AA. Transition in oscillating boundary layer flows. J Fluid Mech 1967;29(1):147−61.
[473] Obremsky HJ, Morkovin MV. Using the model of quasistationary stability to calculate the periodic flow in the boundary layer. Rocket Technol Astronaut (AJAA) 1969; 7(7):102−6.
[474] Olivieri A, Jacob B, Cancello A, Van Oostyrum P, Campana EF and Piva R. The effect of microbubbles on a flat plate turbulent boundary layer. In: Proceedings of the second international symposium on seawater drag reduction (ISSDR 2005). Busan, Korea; 2005. p. 145−54.
[475] Ostistiy VV, Babenko VV. Piston mover. Patent of Ukraine N1221844, Bull N12, 1.12.85; 1985
[476] Ovcharov OP. On the hydrodynamic functions of the gill apparatus of a swordfish. National Academy of Sciences of Ukraine, Institute of Hydromechanics, Kyiv Bionica 1982;16:11−15.
[477] Ovchinnikov VV. Swordfish and sailerfish. Kaliningrad: AtlantNIRO; 1970. p. 269.
[478] Pal S, Merkle CL, Deutsch S. Bubble characteristics and trajectories in a microbubble boundary layer. Phys Fluids 1988;31:744.

[479] Pang M, Wei J. Experimental investigation on the turbulence channel flow laden with small bubbles by PIV. Chem Eng Sci 2013;94:302–15.
[480] Patrashev AN, Kivako LA, Gogiy SI. Applied fluid mechanics. Moscow: Voenizdat; 1970. p. 684.
[481] Pershin SV. Biohydrodynamic laws of swimming of aquatic animals as a principle of optimization in the nature of the motion of immersed bodies. Questions of bionics. Moscow: Science; 1967. p. 555–60.
[482] Pershin SV. Some results of hydrodynamic studies of a wave propulsor. National Academy of Sciences of Ukraine, Institute of Hydromechanics, Kyiv Bionica 1974;8:35–43.
[483] Pershin SV, Chernyshov OB, Kozlov LF, Koval AP, Tsayts VA. Regularities of the covers of high-speed fish. National Academy of Sciences of Ukraine, Institute of Hydromechanics, Kyiv Bionica 1976;10:3–21.
[484] Pershin SV. Biogidrodynamic phenomenon of swordfish as a limiting case of high-speed hydrobionts. National Academy of Sciences of Ukraine, Institute of Hydromechanics, Kyiv Bionica 1978;12:40–8.
[485] Pershin SV. Fundamentals of hydrobionics. Leningrad: Shipbuilding; 1988. p. 263.
[486] Petrie H, Fontaine AA, Money M, Deutsch S. Experimental study of slot injected polymer drag reduction. In: Proceedings of the second international symposium on seawater drag reduction (ISSDR 2005). Busan, Korea; 2005. p. 605–19.
[487] Petrichenko MR. The blocking effect of the rotational movement of gas on heat transfer in the compression (combustion) chamber. Engine building 1990; No. 4: 57–8.
[488] Petrova IM. Hydrobionics in shipbuilding. Leningrad: TSNIITEI; 1970. p. 270.
[489] Petrovsky VS. Hydrodynamic problems of turbulent noise. Leningrad: Shipbuilding; 1966. p. 314.
[490] Pfenninger W, Viken J, Vemuru CS, Volpe G. All laminar supercritical LFC airfoils with natural laminar flow in the region of the main wing structure. In: Liepmann NW, Narasimka R, editors. Turbulence management and relkaminarisation. Berlin, Heidelberg, New York, London: Springer-Verlag; 1988. p. 349–405.
[491] Philips RB, Castano JM and Stace J. Combined polymer and microbubble drag reduction. In: Proc. of the intern. symposium on seawater drag reduction. Newport, Rhode Island, USA; 1998. p. 335–40.
[492] Pierre G. Almost everything about waves. Moscow: World; 1976. p. 176.
[493] Pogrebnyak VG, Voloshin VS. Energy saving and Thoms effect. Kiev: Osvita of Ukraine; 2017. p. 440.
[494] Polishchuk SV, Babenko VV, Korobov VI. Underwater glider. Declaration patent for the invention of Ukraine 28282A; 2000.
[495] Polishchuk SV, Babenko VV, Meister VI. Underwater vehicle with a fin mover. Declaration patent for the invention of Ukraine 41724A; 2001.
[496] Polishchuk SV, Babenko VV. Apparatus hull. Declaration patent for the invention of Ukraine 46638A; 2002.
[497] Polyakov NF, Domaratsky AN, Skuratov AI. On the interaction of a sound field with an incompressible laminar boundary layer. News SB AS USSR, a series of technical sciences 1976; 13(3): 6–15.
[498] Porih M, Shu KS. Diffusion of drag reducing polymer in a turbulent boundary layer. J Hydronaut 1972;6(1):27–33.
[499] Poturaev VN, Dyrda VI, Krush II. Applied rubber mechanics. Kiev: Naukova Dumka; 1975. p. 215.

[500] Povh IL, Finoshin NV. Calculation of impedances in pipes of variable section. Theoretical and applied mechanics, 21. Kharkov: Higher school; 1990. p. 120−4.
[501] Prandtl L. Hydroaeromechanics. Moscow: Foreign Literature Publishing House; 1951. p. 576.
[502] Protasov VR, Staroselskaya AG. Atlas Hydrodynamic properties of fish. Moscow: Science; 1978. p. 103.
[503] Putiata VI. Approximate formulas for the distribution of pressure along a profile at subcritical velocities depending on its geometrical characteristics. In: Proc. of the Kuibyshev Aviation Institute 1961; 12: p. 223−31.
[504] Pyatetsky VE, Kayan VP, Kravchenko AM. Experimental devices, apparatus and methods for studying the hydrodynamics of swimming in aquatic animals. National Academy of Sciences of Ukraine, Institute of Hydromechanics, Kyiv Bionica 1973;7:91−102.
[505] Pyatetsky VE, Khatuntsev VN, Sizov II. Features of determining the efficiency of the mover model of a hydrobiont like an oscillating wing. National Academy of Sciences of Ukraine, Institute of Hydromechanics, Kyiv Bionica 1984;18:87−91.
[506] Pyatetsky VE, Khatuntsev VN. On the effectiveness of a cetacean propulsion complex. National Academy of Sciences of Ukraine, Institute of Hydromechanics, Kyiv Bionica 1985;19:43−7.
[507] Rao DM, Kariys TT. Boundary layer submerged vortex generators for separation control − an exploratory study. A collection of technical papers. Pt. 2. Vigyan Research Associattes, Inc. Published by the American Inst. of Aeronautics and Astronautics. AIAA/ASME/SIAM/ARS Inst. Nat. Fluid Dyn. Cogr.Cincinnati Ohio; 1988. p. 839−46.
[508] Rayner JMV. Dynamics of the vortex wakes of flying and swimming vertebrates. In: Paper for SEB symposium 49 biological fluid dynamics. Leeds; 1994. p. 1−26.
[509] Riley JJ, Gad-el-Hak M, Metcalfe RW. Compliant coating. Annu Rev Fluid Mech 1988;20:393−420.
[510] Ritter I, Porteous JS. Water tunnel measurement of skin friction on a compliant coating. Teddingston, England: Admiralty Research Lab ARL (G); 1965; 8.
[511] Romanenko YeV. Hydrodynamics of fish and dolphins. Moscow: KMK Publishing House; 2001. p. 411.
[512] Roshko A, Fishdon U. On the role of the transition in the near wake. Mechanics 1969; No. 6 50−8.
[513] Rotta IK. Turbulent boundary layer in incompressible fluid. Leningrad: Shipbuilding; 1967. p. 232.
[514] Rudenko AO. Approximate method for calculating hydrodynamic characteristics oscillating wing. National Academy of Sciences of Ukraine, Institute of Hydromechanics, Kyiv Bionica 1991;24:37−40.
[515] Saric WS. Görtler vortices. Annu Rev Fluid Mech 1994;26:379−409.
[516] Saric WS. Control of Görtler yortices. In: Pros. AGARD FDP workshop on "High Speed Body Motion in Water," Report 827. Kiev, Ukraine; 1998. p. 8-1−8-5.
[517] Schilz W. Untersuchungen über den Einfluss biegegeformiger Wandschwingungen auf die Entwicklung der Strömungsgrenzschicht. Acustika 1965;15(1):6−10.
[518] Schlichting H. Zur Entstehung der Turbulenz bi der Plattenströmumg. Nachr. Ges. Wiss. Göttingen Math. - Phys. Kl. 1933;1:- S. 115−81.
[519] Schlichting H. Ampltudenvertelung end energyebalanz der klener Störungen bei der Platterstromung. Nachr. Ges. Wss. Göttingen Math.-Phys. Kl. 1935;1:98−125.

[520] Schlichting G. Beginning of the turbulence. Moscow: Foreign literature; 1962. p. 204.
[521] Schlichting H, Gersten K. Boundary - layer theory. 8th Revised Enlarger Edition Berlin Heidelberg NewYork: Sringer-Verlacg; 2000. p. 801.
[522] Schmidt FS. Zubeschleunigter Bewegung des Sphärekörpers in aufsetzenden sich Mitten. Ann Phys LPZ 1920;61:633−64 Bd.
[523] Schubauer GB, Skramstad HK. Laminar boundary layer oscillation and stability of laminar flow. J Aeronaut Sci 1947;14(2):69−81.
[524] Panton RL, editor. Self-sustaining mechanism of wall turbulence. Southampton, UK and Boston, USA: Computational Mechanics Publications; 1997. p. 422.
[525] Serebryakov VV, Moros VV, Kochin VV, Dzielski JE. Experimental study on planning motion of a cylinder at angle of attack in the cavity formed behind an axisymmetric cavitator. J Ship Res 2019;6.
[526] Shebalov AN. Some questions of the influence of non-stationarity on the mechanisms of formation of resistance National Academy of Sciences of Ukraine, Institute of Hydromechanics, Kyiv Bionica 1969;3:61−5.
[527] Shen SF. Some considerations in the laminar stability of time dependent basic flow. J Aeronaut Sci 1961;15(5):397−404.
[528] Shlanchauskas AA, Vegite NI. A model of a turbulent boundary layer in the form of large-scale movement. In: Near-wall turbulent flow, vol. 2. Novosibirsk; 1975. p. 203−8.
[529] Shmakov YuI. Approximate formulas for the characteristics of the boundary layer on the profile. Abstract of the dissertation of the candidate of technical sciences. Kiev, Ukraine; 1963. p. 12.
[530] Shmakov YuI., Kalion VA. Model of nonstationar flow past the wing profile. In: Proc. of the conference on aerohydromechanics. Ivano-Frankivsk, Ukraine; 1967. p. 225−6.
[531] Scott JR. Physical testing of rubber and elastic. Moscow: Chemistry; 1968. p. 315.
[532] Sedov LI. Methods of similarity and dimension in mechanics. Moscow: Science; 1977. p. 438.
[533] Sedov LI, Ioselevich VA, Pilipenko VN. Features of the structure of near-wall turbulence and the mechanism for reducing friction with polymer additives. In: Turbulent biphasic flows and experimental technique (V All-Union Conference on Theoretical and Applied Aspects of Turbulent Flows. Part II). Tallinn; 1985.
[534] Smith AMO. On the growth of Taylor-Görtler vortices along highly concave walls. Quart Appl Math 1955;13(3):233−62.
[535] Smith RL, Blick E. Skin friction of compliant surfaces with foamed material substrate. J Hydronaut 1969;3(2):100−2.
[536] Smolyakov AV, Tkachenko VM. Measurement of turbulent pulsations. Leningrad: Energy; 1980. p. 164.
[537] Sodha MS, Gkhatak AK. Inhomogeneous optical waveguides. Moscow: Communication; 1980. p. 216.
[538] Somandepalli V, Hou YX and Mungal MG. Streamwise evolution of drag reduction in a boundary layer with polymer injection. In: Proceedings of the second international symposium on seawater drag reduction (ISSDR 2005). Busan, Korea; 2005. p. 569−78.
[539] Ceccio SL. Friction drag reduction of external flows with bubble and gas injection. Annu Rev Fluid Mech 2010;42:183−203.
[540] Stuart JT. Hydrodynamic stability. Appl Mech Rev 1965;18(7):223−31.

[541] Sukhovich EP. Turbulent mixing of bounded swirling jets. News of the Academy of Sciences of Latvian SSR. Ser. Physical and Technical Sciences 1982;no. 1:72–80.

[542] Swearingen JD, Blackrwelder RF. The growth and breakdown of streamwise vortices in the presence of a wall. J Fluid Mech 1987;182:255–90.

[543] Sydney R. Review of the results of the investigation of the influence of small protrusions on the flow in the boundary layer. Rocket Technol Astronaut (AJAA) 1973; 6(11):16–30.

[544] Tagori T, Ashiate I. Some experiments on friction reduction on flat plate by polymer solutions. In: Proceedings 12-th intern. Towing tank conference. Rome; 1969.

[545] Takahashi T, Kakugawa A, Kodama Y. Streamwise distribution of the skin friction reduction by microbubbles. J Soc Nav Archit Jpn 1997;182:1–8.

[546] Taneda S, Honji H. The skin friction drag on flat-plates coated with flexible material. Rep Res Inst Appl Mech 1967;15(49):121–37.

[547] Tani I. Production of longitudinal vortices in a boundary layer along a curved wall. J Geophys Res 1962;67:3075–80.

[548] Tani I. Boundary layer transition. Annu Rev Fluid Mech 1969;12(1):169–97.

[549] Taylor O. Statistical theory of turbulence effect of turbulence on boundary layer. Proc Roy Soc A 1936;156:307–17.

[550] Tetyanko VA Experimental investigation of the statistical characteristics of velocity pulsations during the transition of a laminar boundary layer to a turbulent one. In: Preprint no. 70, Novosibirsk: Siberian Branch of the USSR Academy of Sciences Institute of Theoretical Physics; 1981. p. 43.

[551] Thill Cornel, Toxopeous Serge and Frans van Warlee. Project energy-saving air-lubricated ships (PELS). In: Proceedings of the second international symposium on seawater drag reduction (ISSDR 2005). Busan, Korea; 2005. p. 93–108.

[552] Thuinston S, Jones RD. Experimental Model Steadies of Non-Newtonian Soluble Coatings for Drag Reduction. J Aircr 1965;2(2):122–6.

[553] Tokunaga K. Reduction of frictional resistance of a flat plate by microbubbles. Trans. West Japan Soc. of Nav. Arch. 1986; 73: 79.

[554] Tollmien W. Über die Entstehung der Turbulenz. Mitteilung. Nachr. Ges. Wiss. Göttingen Math. Phys. Kl. 1929: 21–44.

[555] Toms BA. Some observation on the flow of linear polymer solution through straight tubes at large Reynolds numbers. In: Proc. first inter. congress on rheology, vol. II. North-Holland, Amsterdam. 1948. p. 135–41.

[556] Tsujimoto Y. Rotating cavitation: known's and unknown's. In: Proc. ASME/JSME fluids eng. div. summer meeting; 1995.

[557] Tsyganyuk AI, Kozlov LF, Vovk VN, Maksimov SL. Technique for controlling the near-wall layer of a stream flowing around a solid body using the method of controlled jets and a device for implementing this technique. USSR Author's Certificate 1585569; 1990.

[558] Tuncer IH, Kaya M. Optimization of flapping airfoils for maximum thrustand propulsive efficiency. AIAA J 2005;43(11):2329–36.

[559] Frost U, Moulden T, editors. Turbulence. Principles and applications. Moscow: World; 1980. p. 527.

[560] Liepmann NW, Narasimka R, editors. Turbulence management and relaminarisation. Berlin, Heidelberg, New York, London: Springer-Verlag; 1988. p. 524.

[561] Turik VN, Makarenko RA. Generalization of the characteristics of tangential flows in the dead-end part of the vortex chamber Bulletin of the National Technical University of Ukraine KPI Mech Eng 2000;1(38):38–44.

[562] Turik VN, Makarenko RA. Aerodynamics of a vortex chamber with a uniform tangential air supply. Collection of works of the Kirovograd State Technical University 2000;7:38−43.

[563] Turik VM, Babenko VV, Rozhavsky VG. Mixing chamber with additional nozzles. Declaration patent for useful model of Ukraine 3443; 2004.

[564] Turik VN, Babenko VV, Voskoboynik VA, Voskoboynik AV. Vortex motion in a hemispherical hole on the surface of a streamlined plate. Bulletin of NTUU Kiev Polytechnic Institute, Mashinostroenie 2006;vol. 1, 48:79−85.

[565] Turik VN, Babenko VV, Voskoboinik VA, Voskoboynik AV. Vortex movement in a semi-cylindrical hole on the surface of a streamlined plate Bulletin of the National Technical University of Ukraine KPI Mech Eng 2006;48:79−85.

[566] Turik VN, Babenko VV, Voskoboinik VA, Voskoboinik AV. Speed in the boundary layer above the plate with semi-cylindrical deepening. Scientific news of the National Technical University of Ukraine Kyiv Polytechnic Institute 2008;4:46−54.

[567] Turik VN, Babenko VV, Voskoboinik VA, Voskoboinik AV. Frequency-wave characteristics of coherent vortex structures in three-dimensional deepening. Industrial hydraulics and pneumatics 2009;1(23):21−28.

[568] Turik VN, Babenko VV, Voskoboinik VA, Voskoboynik AV. Kinematic features of the boundary layer in the neighborhood of hemispherical deepenings on the plate. Bulletin of the National Technical University of Ukraine, Kiev Polytechnic Institute Mech Eng 2010;59:110−17.

[569] Turik VN, Babenko VV, Voskoboinik VA, Voskoboynik AV. Vortex movement in semi-cylindrical deepening on a plate. Industrial hydraulics and pneumatics 2011; 3 (31): 23−27.

[570] Turik VN, Babenko VV, Voskoboinik VA, Voskoboynik AV. Controlling the flow structure inside the semi-cylindrical deepening. In: Industrial hydraulics and pneumatics. Materials of the XV international scientific and technical conference association of industrial hydraulics and pneumatics. Melitopol, Ukraine; 2014. p. 26−7.

[571] Turik VN, Babenko VV, Voskoboynik VA, Voskoboynik AV. Formation of vorticity inside and near a transversely streamlined semi-cylindrical trench on a flat surface. Bulletin of the National Technical University of Ukraine KPI Mech Eng 2015; No. 74: 90−9.

[572] Golyamin IP. Ultrasound. Small encyclopedia. Moscow: Soviet Encyclopedia; 1979. p. 400.

[573] Ehrenstein Uwe, Rossi Maurice. Nonlinear Tollmien-Schlichting waves for a Blasius flow over compliant coatings. Phys Fluids 1996;8(4):1036−51.

[574] Van Driest ER, Brumer CB, Welis CS. Boundary layer transition on blunt bodies. Effect of roughness. AIAA J 1967;5(10):1913−15.

[575] Van Dyke M. Album of liquid and gas flows. Moscow: World; 1986. p. 181.

[576] Vanin YuP, Migirenko GS. Experimental studies of the distribution of polymer additives in the boundary layer behind the injection site. In: Studies on boundary layer management. Novosibirsk: Institute of Engineering Thermophysics (ITTF) Siberian Branch of the USSR Academy of Sciences; 1976.

[577] Vanin YuP, Khodaev AM. Investigation of the characteristics of near-wall turbulence in a stream with variable concentration of polymer additives. Influence of polymer additives and surface elasticity on wall turbulence. Novosibirsk: Institute of Engineering Thermophysics (ITTF) Siberian Branch of the USSR Academy of Sciences; 1978.

[578] Vasetskaya NT, Ioselevich VA, Pilipenko VN. Mechanical destruction of polymer molecules in a turbulent flow. Some issues in continuum mechanics. Moscow: Science; 1978. p. 55−69.
[579] Vasilyev AS, Illichev AF, Mikheev GN. Some aspects of the hydrodynamic resistance of fish and their models National Academy of Sciences of Ukraine, Institute of Hydromechanics, Kyiv Bionica 1985;19:50−6.
[580] Vdovin AV, Smolyakov AV. Diffusion of polymer solutions into turbulent boundary layer. Applied Mathematics and Technical Physics 1978;2.
[581] Vdovin AV, Smolyakov AV. Turbulent diffusion of polymers in the boundary layer. Applied Mathematics and Technical Physics 1981;4:98−104.
[582] Virk PS. An elastic sublayer model for drag reduction by dilute solutions of linear macromolecules. J Fluid Mech 1971;45:417.
[583] Virc PS, Chen RH. Type B drag reduction by aqueous and saline solution of two biopolymers at high Reynolds number. In: Proceedings of the second international symposium on seawater drag reduction (ISSDR 2005). Busan, Korea; 2005. p. 545−58.
[584] Vogel WM, Patterson AM. An experimental investigation of the effect of additives injected into the boundary layer of an underwater body. In: 5-th Symposium of naval hydrodynamic. Ship motions and drag reduction, Bergen. Washington; 1966. p. 51−83.
[585] Voitkunsky Ya I. Water resistance to vessel traffic. Leningrad: Shipbuilding; 1964. p. 412.
[586] Voropaev GA, Babenko VV. Absorption of pulsation energy by a damping coating. National Academy of Sciences of Ukraine, Institute of Hydromechanics, Kyiv Bionica 1975;9:60−9.
[587] Voropaev GA, Babenko VV. Turbulent boundary layer on an elastic surface. National Academy of Sciences of Ukraine, Institute of Hydromechanics, Kyiv Hydromechanics 1978;38:71−7.
[588] Voropaev GA, Kozlov LF, Leonenko IV. Potential flow around complex-shaped axisymmetric bodies. National Academy of Sciences of Ukraine, Institute of Hydromechanics, Kyiv Bionica 1982;16:41−4.
[589] Voropaev GA, Svirskaya EA. Absorption of pulsation energy by a damping coating. National Academy of Sciences of Ukraine, Institute of Hydromechanics, Kyiv Bionica 1982;16:47−53.
[590] Voropaev GA, Babenko VV. Turbulent gradient flows on a compliant surface. In: Proceedings of the all-union seminar on hydrodynamic stability and turbulence. Novosibirsk: SB Science; 1989. p. 186−90.
[591] Voropaev GA, Ptukha YA. Modelling of turbulent complex flows. Kiev: Naukova Dumka; 1991. p. 168.
[592] Voropaev GA & Rozumntuk NV. Turbulent boundary layer over a compliant surface. In: Pros. AGARD FDP workshop on "High Speed Body Motion in Water," Report 827. Kiev, Ukraine; 1998.
[593] Voropaev GA, Zagumenny YaV. Dynamic and kinematic characteristics of a viscoelastic layer of variable thickness under the action of a pulsed load. Acoustic gazette 2005;8(4):29−37.
[594] Vovk VN, Kalion VA, Tsyganyuk AI. Calculation of the boundary layer resistance in a polymer stream of constant concentration without pressure gradient. National Academy of Sciences of Ukraine, Institute of Hydromechanics, Kyiv Hydromechanics 1984;43:44−7.

[595] Vovk VN, Pyatetsky VE, Khatuntsev VN. On the theoretical substantiation of the experimental method for studying the periodic motions of bodies in a viscous incompressible fluid. National Academy of Sciences of Ukraine, Institute of Hydromechanics, Kyiv Bionica 1987;21:29−34.

[596] Vovk VN. The use of a system of standing vortices for turning the flow in hydraulic systems. National Academy of Sciences of Ukraine, Institute of Hydromechanics, Kyiv Bionica 1993;26:52−6.

[597] Voitkunsky YaI, Faddeev YuI. Some problems of technical hydrobionics. National Academy of Sciences of Ukraine, Institute of Hydromechanics, Kyiv Bionica 1976;10:21−6.

[598] Volshanik VV, Zuikov AL, Mordasov AP. Wrapped streams in hydraulic structures. Moscow: Energoatomizdat; 1990. p. 280.

[599] Walker D, Tiederman G. The concentration field in a turbulent channel flow with polymer injection at the wall. Exp Fluids 1989;8:86−94.

[600] Watanabe S, Tsuimoto Y, Frans JP & Mishel JM. Linear analyses of cavitation instabilities. In: Proc. intern. symp. on cavitation, vol. 1. 1998. p. 347−52.

[601] Watanabe S, Sato K, Tsuimoto Y, Kamijo K. Analyses of rotating cavitation in a finite pitch cascade using a closed cavity model and a singularity method. Trans ASME J Fluids Eng 1999;121:834−40.

[602] Wehrmann OH. Tollmien-Schlichting waves under the influence of a flexible wahl. Phys Fluids 1965;8(7):1389−90.

[603] Weinstein LM, Fisher MK. Experimental confirmation of the reduction of friction resistance on compliant walls. Rocket technique and cosmonautics 1975;13(7): 144−6.

[604] White D. Drag coefficients for spheres in high Reynolds numbers of dilute solutions of high polimer. Nature 1966;212(5059):277−83.

[605] Willmarth WW, Jang CS. Wall pressure fluctuation beneath turbulent boundary layers on a flat plate and a cylinder. J Fluid Mech 1970;41(1):47−80.

[606] Willmarth WW, Sharma KK. Study of turbulent structure with hot wires smaller than the viscous length. J Fluid Mech 1984;142:121−49.

[607] Wilson DDJ. An experimental investigation of the mean velocity, temperature and turbulence fields in plane and curved now-dimensional wall jets: Coanda effect [Ph.D. thesis]. Univ. of Minessota; 1970.

[608] Wilson DDJ, Goldstein RJ. Turbulent near-wall jet on a cylindrical surface. Theoretical bases of engineering calculations. Series D 1976;8(3):328−36.

[609] Winoto SH, Durao DFG, Grane BI. Measurement within Görtler vortices. Trans. ASME. J. Fluid Eng, 101. 1979. p. 517−20.

[610] Wormley. Analytical model of incompressible flow in short vortex chambers. Proc. of the American Society of Mechanical Engineers. Theoretical foundations of engineering calculations 1969;91(2):145−59.

[611] Wortmann PX. Eine Methode zur Beobachtung und Messung von Wasserströmung mit Tellur. Z für angew Phys 1953;5(6):201−15.

[612] Wortmann PX. Investigation of unstable oscillations of a boundary layer in a water channel by a tellurium method. Boundary problems and heat transfer issues. Moscow−Leningrad: Science; 1960. p. 385−94.

[613] Wortmann PX. Visualization of transition. J Fluid Mech 1969;38(3):473−80.

[614] Wu J. Suppreshed diffusion of drag reducing polymer in a turbulent boundary layer. J Hydronaut 1972;6(1):46−50.

[615] Wu J, Fruman DH, Tulin MP. Drag reducting by polymer diffusion at high Reynolds number. J Hydronaut 1973;12(3).
[616] Wu TY. Introduction to the scaling of aquatic animal locomotion. In: Pedley T, editor. Scale effects in animal locomotion. London: Academic Press; 1977. p. 203−32.
[617] Shen X, Ceccio SL, Perlin M. Effects of seawater and the resulting bubble size on micro-bubble drag reduction. In: Proceedings of the second international symposium on seawater drag reduction (ISSDR 2005). Busan, Korea; 2005. p. 9−15.
[618] Yoshida Y, Takahashi Y, Kato H, Masuko A, Watanabe O. Simple Lagrangian formulation of bubbly flow in a turbulent boundary layer (bubbly boundary layer flow). J Mar Sci Technol 1996;1(5):241−54.
[619] Kodama Y. Effect of microbubble distribution on skin friction reduction. In: Proc. of the intern. symposium on seawater drag reduction. Newport, Rhode Island, USA; 1998. p. 331−4.
[620] Yurchenko NF, Babenko VV, Kozlov LF. Experimental study of the Görtler instability in the boundary layer. Stratified and turbulent flows. Kiev: Sciences. Dumka; 1979. p. 50−9.
[621] Yurchenko NF, Babenko VV. On the stabilization of longitudinal vortices by the skin of dolphins. Biophysics 1980;XXV(N2):299−304.
[622] Yurchenko NF. On the method of experimental study of the system of longitudinal vortices in the boundary layer. Eng Phys J 1981;41(6):966−1002.
[623] Yurchenko NF, Babenko VV. On modeling hydrodynamic functions of the external covers of aquatic animals. Hydrodynamic questions of bionics. Kiev: Naukova Dumka; 1983. p. 37−46.
[624] Yurchenko NF, Babenko VV, Kozlov LF. The control of the three-dimensional disturbances development in the transitional boundary layer. In: Laminar-turbulent transition IUTAM symp. Novosibirsk, 1984. Springer-Verlag, Berlin, Heidelberg; 1985. p. 329−35.
[625] Yurchenko NF, Babenko VV, Kozlov LF. Experimental determination of the stability of longitudinal vortex disturbances. Eng Phys J 1986;50(2):201−7.
[626] Yurchenko NF. Experimental investigation of the system of longitudinal vortices in the boundary layer. Eng Phys J 1987;41(6):996−1002.
[627] Yurchenko NF, Babenko VV. Stability criteria for three-dimensional disturbances on a concave elastic surface. Eng Phys J 1987;52(5):781−8.
[628] Yurchenko NF. Fluid flow in a channel with a concave section. National Academy of Sciences of Ukraine, Institute of Hydromechanics, Kyiv Bionica 1987;21:59−63.
[629] Yurchenko NF, Babenko VV, Kozlov LF. Development of three-dimensional disturbances over compliant surface. In: Internat. conf. EAHE (engineering aero-hydroelasticity). Prague, Czeskoslovensko; 1989. p. 63−9.
[630] Yampolsky DD. A non-contact micrometer for measuring micro-displacements of an elastic surface. Proceedings of the Central Research Institute named after A. N. Krylov. 1965; 219: 96−103.
[631] Zaat JA. Numerische Beiträge zur Stabilitatstheorie der Grenzschichten. Grenzschichtforschung Symp. Freiburg; 1957. p. 157−63.
[632] Zolotov SS, Khodorkovsky YaS. Features of the resistance to friction of the body of the form of a sword-fish National Academy of Sciences of Ukraine, Institute of Hydromechanics, Kyiv Bionica 1973;7:14−18.

Index

Note: Page numbers followed by "*f*" and "*t*" refer to figures and tables, respectively.

A

Accelerated movement features in water of bodies of various shapes, 31–40
    dependence of acceleration
        of combined body of rotation, 38*f*
        of flat plate, 39*f*
    device of experimental models, 34*f*
    hydrodynamic pipe scheme, 32*f*
    scheme of combined model, 35*f*
    weight of investigated models, 36*t*
"Accelerating vortices", 69–70
Acceleration
    modes
        drag coefficient of fluid bodies in, 20–25
        energy aspects of movement of bodies in, 25–31
    of motion, 1
Acoustic methods, 361
Acoustic wave-guides, 163–164
Active impact of caudal fin, 41
Additivity of aquatic and underwater bodies, 579–580
Aeroprakt A-20 airplane wing model, 595, 595*f*
Air pressure receiver (APR), 403
Amplification coefficient, 170
Amplitude of angular oscillations, 43
Anisotropy coefficient, 328
Area of flow toward butt-end, 229

B

Band of stabilization, 132–134
Benny–Lin vortices, 136–137, 152, 218–220, 315–316, 379
    method of investigation of, 102–103
Bernoulli equation, 455–456
Bionic approach, 242

Blasius' theoretical velocity profile, 105–106
Bottlenose dolphins (*Tursiops truncatus*), 70
Boundary layer (BL), 41–42, 93, 189
    on analogs of skin covers of hydrobionts
    distribution of disturbing movements across thickness of laminar BL of elastic plates, 281–295
    elastomers' mechanical characteristics, determination of, 320–334
    experimental investigation of CVS in transition BL of elastic plates, 259–272
    Görtler neutral curve in flow around elastic curvilinear plates, 311–320
    LDVM of BL structure of elastic plates, 295–311
    neutral curves of linear stability of laminar BL of elastic plates, 272–281
    oscillations on surface of elastic plates, 334–344
    physical substantiation of interaction mechanism of flow, 345–350
    structures of elastic surfaces, 237–258
    bubbles interaction with, 546–549
    combined method of forming longitudinal vortex systems in, 554–557
    flow around controlled elastic plate, 350–357
        effect of heating elastic plate, 356*f*
        influence of duration of heating of elastic plate, 353*f*
        longitudinal averaged velocity, 352*f*
        longitudinal pulsation velocity, 355*f*
    as heterogeneous asymmetrical wave-guide, 158–171
    of hydrobionts, 93–94

Boundary layer (BL) (*Continued*)
  influence of small and large disturbance on characteristics of rigid plate BL, 171–185
  influence of streamlined surface on, 549–551
  interaction with injected semilimited jet, 506–514
    dependence of thickness and maximum speed of near-wall jets, 514*f*
    longitudinal averaged velocity profiles, 510*f*
    profiles longitudinal fluctuation velocity, 512*f*
    speed profiles in flooded jet, 508*f*
  LDVM measurements of structure of boundary layer of rigid plate, 150–158
  on smooth rigid curved plate
    experimental construction of Görtler neutral curve in, 209–213
    experimental equipment and methodology for Görtler stability investigation, 187–191
    experimental investigation of Görtler stability on rigid curvilinear plates, 200–206
    experimental investigations of Görtler's vortex systems, 191–200
    GV on concave surface behind vortex chamber nozzle, 232–236
    LDV measurements of BL longitudinal velocity profiles, 206–209
    suction method, 196
    vortex structures model in vortex chamber, 213–232
  on smooth rigid plate
    distribution of disturbing motion over thickness of laminar BL of rigid plate, 123–130
    experimental investigation of coherent vortex structures of rigid plate transition BL, 103–117
    experimental methods for investigation flow structures of rigid plate transition BL, 100–103
    hydrodynamic stand of low turbulence, 93–100, 94*f*
    laminar-turbulent transition of BL of rigid plate, 130–150

    neutral curves of linear stability of laminar BL of rigid plate, 117–123
Braking process, 24
Bubbles. *See also* Microbubbles (MB)
  injection from
    nozzles, 557–558
    upper screens, 557
  interaction with boundary layer, 546–549
  test with electrolysis of, 551–554
  microbubble generation scheme in boundary layer, 553*f*
  plates for various types of combined control of BL, 552*f*

C

Calibration signals, 33
Carbon black method, 227
Centimetric waves, 165–166
Channel, 76–77, 97
Character of curves, 367
Civilization, 581–582
Clusters, 439
Coanda effect, 506, 543–544, 590
Coefficient, 18
  of friction, 546
  full friction resistance, 469
  of hydrodynamic thrust, 57
  of nonstationary resistance, 12, 21, 23
  of viscosity, 15
Coherent vortex structures (CVS), 93, 146–148, 150, 191, 214, 359, 437, 577. *See also* Elastic surface structures
  control methods for CVS, 605
  of boundary layer, 359–361, 360*t*
  in vortex chambers, 580–581
  experimental investigation in transition BL of elastic plates, 259–272
    dependence of resistance coefficient, 260*f*
    development of disturbing movement, 264*f*
    disturbing motion during flow around viscoelastic surfaces, 267*f*
    disturbing motion in flow around viscoelastic surfaces, 268*f*
    graphic copies of photograph of disturbing motion, 266*f*
    profiles of average velocity, 265*f*

Index

velocity amplitudes, 263f
velocity profiles in flow around simple membrane surface, 270f
problem of drag reduction, types of CVS and combined methods, 578–579
of rigid plate transition BL, 103–117
and thermal conductivity control, 597–599
Combined drag reduction method, 437–440, 518, 604
  drag of the model with OT and at injection of polymer solutions into BL, 480–489
  cone-shaped fairing of tail of the model, 487f
  drag of the model with xiphoid tips and injection of polymer solutions into BL, 490–498
  experimental equipment and investigation methodology for, 472–480
  drag of the model without injection of polymer solutions, 478–480
  experimental equipment and measurement techniques, 473–477
  experimental investigations
    of bodies with xiphoid tips, 441–451
    of influences of bubbles injection on drag reduction, 545–560
    of interaction of BL with injected semilimited jet, 506–514
  flow around elastic surfaces, 438
  hydrodynamic peculiarities of skin structure and body of swordfish, 515–531
  combined drag reduction method, 518
  economical consumption of mucus of skin, 517–518
  features of fluid injection through gill slits, 515–516
  methods of stabilization of vortex disturbances, 518–531
  structure of disturbing movement in BL of swordfish skin, 516–517
  investigation of influences of microbubbles injection into BL on drag reduction, 531–545
  modeling of disturbances development in flow behind ledge, 560–575
  cavitation flow around body, 574f
  coherent vortex structure visualization of BL, 562f
  formation of coherent vortex structures, 563f
  mechanical characteristics of elastic plates, 568t
  neutral curves in coordinates of dimensionless frequency, 570f
  relapse of return jet, 564f
  shape of cavitation cavity, 575f
  visualization of cavitation flow, 561f
  physical mechanism of influence of xiphoid tip on drag reduction, 498–506
  problem of injection of polymer solutions into boundary layer, 459–472
  theoretical investigations of bodies with xiphoid tips, 451–459
Composite elastic coatings, 409–410
Control system, 72
Coordinate device, 232–233
"Crisis of resistance", 4–5, 28
Critical frequency, 159–160, 165
Curvilinear plates, 187

D

Deceleration modes
  drag coefficient of fluid bodies in, 20–25
  energy aspects of movement of bodies in, 25–31
Devices moving on and under water, 586–593
  canoe with various fin movers, 586f
  device designed for towing in aquatic environment, 590f
  fin mover mounted on underwater vehicle, 589f
  interaction of fin mover and standard propeller move, 588f
  petrel, 592f
  piston mover with elastic piston, 587f
  project of two-seater underwater vehicle, 587f
  trimaran models, 591f
Diffuser, 94–95
Diffusion process, 461–462, 546–548
Digital code, 72
Digital particle image velocimetry (DPIV) method, 70

Dimensionless acceleration, 8
Dimensionless coefficient of average thrust force, 49
DISA 55P11 hot-wire anemometer, 197–198
DISA thermoanemometric instrument, 101
Docking washers, 524
"Dolphin" model, 262
Drag. *See also* Reduction friction drag
  coefficient of fluid bodies, 20–25
  of longitudinally streamlined cylinders, 389–404
  of the model
    without injection of polymer solutions, 478–480
    with OT and at injection of polymer solutions into BL, 480–489
    with xiphoid tips and injection of polymer solutions into BL, 490–498
Drag reduction, 579–580, 583
  experimental investigation of influences of bubbles injection on, 545–560
  bubbles interaction with boundary layer, 546–549
  combined method of forming Lνs. in BL, 554–557
  influence of streamlined surface on boundary layer, 549–551
  injection of bubbles from nozzles, 557–558
  injection of bubbles from upper screens, 557
  microbubbles effect on integral characteristics of axisymmetric models, 558–560
  test with electrolysis of bubbles, 551–554
  investigation of influences of microbubbles injection into BL on, 531–545
  dependence of friction coefficient on wall, 532f
Dryden's measurements, 499
Dynamic methods, 361

E

Ecology, 605
Economical consumption of mucus of skin, 517–518
Ekranoplans, 579
Elastic coatings, 237, 346, 367–368, 433
Elastic longitudinally streamlined cylinders. *See also* Longitudinally streamlined cylinders
  friction drags of, 404–420
  geometric parameters of elastic cylinders, 406t
  long cylinders, 407–420
  short cylinders, 407–420
Elastic plates
  engineering method for selection of elastic plates to drag reduction, 428–435
  experimental investigations of friction drag on, 376–389
    dependence of friction drag coefficient, 376f
    design schemes elastic plates, 382f
    frequency dependences of energy absorption coefficient, 384f
    friction drag coefficients of plates in airflow, 389t
    friction drag coefficients on rigid and elastic plates, 381f
  influence on friction resistance, 367
  review of experimental investigations of integral characteristics of, 361–368
    friction drag coefficients on rigid and elastic plates, 364f
Elastic surface structures, 237–258. *See also* Flow around elastic surfaces
  composite elastic plates cross-sections, 246f
  device for determining dependence of elasticity, 241f
  dimensionless mechanical and kinematic parameters of elastic plates, 249t
  geometric parameters of elastic plates, 251t, 254t
  inner surface of inverted riblets, 256f
  measurement location and dimensionless parameters of elastic plates, 247t
  mechanical characteristics of elastic plates, 248t, 257t
  mechanical parameters of elastic plates, 258t
  membrane and viscoelastic surfaces
    without border frame, 240f
    design of, 239f

Index 645

panel structure for fixing elastic surface, 243*f*
physical substantiation of interaction mechanism of flow with, 345–350
　multivariate system with arbitrary, 349*f*
　standard distributions of loadings along wing, 347*f*
strain gage
　placement on hydrodynamic stand, 245*f*
　suspension in working section of hydrodynamic stand, 244*f*
types of elastic plates, 255*f*, 256*f*
Elasticity, 324
Elastomer(s), 413
　absorption coefficient, 415–416
　frequencies, 432
　mechanical characteristics of, 320–334, 333*t*
　　apparatus for determining dynamic characteristics, 321*f*
　　conditions for performing experiments, 327*t*
　　dynamic modulus of elasticity, 326*t*
　　experimental device, 323*f*
　　oscillogram of surface oscillations, 332*f*
　　oscilloscope on elastic plate, 329*f*, 331*f*, 332*f*
　　oscilloscope screen of oscillations, 324*f*
　　range of parameters of balls, 328*t*
　　surface oscillations of elastomers, 325*f*
Electric methods, 361
Electro-motive force (EMF), 33
Electromagnetic waves, 165
"End effect", 214
Energy aspects of movement of bodies, 25–31
Engineering method for selection of elastic plates to drag reduction, 428–435
　neutral and limiting neutral curves of hydrodynamic stability flat rigid, 430*f*
　power-density spectra of energy-carrying pressure pulsations, 429*f*
Environmental control concept of coastal area, 581–582
Environmental monitoring of coastal area, 599–601
　adaptation of form of body, 602
　change wing span, 603

changing curvature of wing profile, 603
changing shape of wing profile and configuration, 603
changing wing profile chords, 603
combined drag reduction methods, 604
CVS control methods, 605
ecology, 605
general view of universal underwater vehicle, 600*f*
heat and mass transfer, 605
maneuverability control, 604
nontraditional movers, 602–603
self-propelled buoys, 602
top and side views of universal underwater vehicle, 601*f*
underwater glider, 602
vortex devices, 604
Experimental complex for friction drag investigations, 368–376
Experimental studies of nonstationary motion, 19, 40

F

Faraokhi model, 518–519
Fast-floating aquatic organisms, 577
Fast-swimming hydrobionts, 368, 577
Faulkner resistance law, 16
Flow around elastic surfaces, 438
　influence of shape of nasal contours, 438–439
　injection of polymer solutions
　　from one nasal slit when flowing around a cylinder with elastic surface, 439–440
　　through three slits, 440
　　from two slots, 439
　when flowing around elastic surfaces, 438
　methods of injection into BL of microbubbles of gas, 440
　static electric field effect on BL characteristics, 439
Flow around three-dimensional deepening, 515–516
Flow behind ledge
　modeling of disturbances development in, 560–575
　　cavitation flow around body, 574*f*
　　coherent vortex structure visualization of BL, 562*f*

Flow behind ledge (*Continued*)
  formation of coherent vortex structures, 563*f*
  mechanical characteristics of elastic plates, 568*t*
  neutral curves in coordinates of dimensionless frequency, 570*f*
  relapse of return jet, 564*f*
  shape of cavitation cavity, 575*f*
  visualization of cavitation flow, 561*f*
"Flow crisis", 6–7
Foam elast, 257
"Folding" waves, 265
Fortran numerical program, 456–457
Friction coefficients, 486
Friction drag
  control methods for CVS of boundary layer, 359–361, 360*t*
  drag of longitudinally streamlined cylinders, 389–404
  of elastic longitudinally streamlined cylinders, 404–420
  engineering method for selection of elastic plates to drag reduction, 428–435
  experimental complex for investigations of, 368–376
  boundary layer, 369*f*
  high-speed towing tank, 375*f*
  model of body rotating with diameter, 374*f*, 375*f*
  vertical axisymmetric wing for towing plates, 370*f*
  wing design–strain gauge, 371*f*
  experimental investigations of friction drag on elastic plates, 376–389
  for injection of polymer solutions into BL, 491–498
  polymer additives influence on friction drag of elastic cylinders, 420–428
  review of experimental investigations of integral characteristics of elastic plates, 361–368
Full-scale microbubble tests, 539

G

Geometric functions, 458
Geometric parameters models, 391–392
Görtler neutral curve in flow around elastic curvilinear plates, 311–320
  change effect in z of properties of streamlined elastic plate FE-3, 312*f*
  characteristic thickness of boundary layer, 309*t*, 310*t*
  parameters of disturbed boundary layer in cylindrical channel, 317*t*
  reaction of boundary layer of elastic surface of FE-3, 315*f*
  three-dimensional disturbances stability on concave elastic surface of FE-3, 318*f*
  visualization of $U(z)$ profiles in flow around uniform elastic curved plate, 313*f*
Görtler number, 318
Görtler stability, 591–592
  experimental construction of Görtler neutral curve in, 209–213
  Görtler's stability diagram on a concave rigid surface, 212*f*
  parameters of Görtler's neutral curve, 213*t*
  experimental equipment and methodology for investigation, 187–191
  interaction of Görtler vortices with longitudinal vortices, 190*f*
  three-dimensional disturbation generators, 188*f*
  experimental investigations
  disturbing motion development, 204*f*
  of Görtler's vortex systems, 191–200
  on rigid curvilinear plates, 200–206
  $U(z)$ profiles visualization in flow around curvilinear plates, 203*f*
  Görtler's vortices on concave surface behind vortex chamber nozzle, 232–236
  LDV measurements of BL longitudinal velocity profiles, 206–209
  method of Görtler stability investigation, 102–103
  vortex structures model in vortex chamber, 213–232
Görtler vortices (GVs), 146–147, 190–191, 190*f*
  on concave surface behind vortex chamber nozzle, 232–236
  experimental investigations of Görtler's vortex systems, 191–200, 455–456
  formation of GV, 194*f*

Görtler's instability visualization, 192*f*
stability of longitudinal vortex systems, 198*f*
stream-lining of supercritical wing LFC NASA 998 A profile, 195*f*
visualization of transition on concave wall, 192*f*
Görtler's diagram, 211

H
"Hairpin-shaped vortex structures", 146
"Hairpin-shaped" vortices, 139–140, 156–158
Harmonic process, 159
Helmholtz theorem, 135–136
Hemisph-2 frame, 529
Heterogeneous asymmetrical wave-guide, BL as, 158–171
  acoustic wave-guides types for surface acoustic waves, 164*f*
  distribution of averaged and pulsation longitudinal velocity component, 179*f*, 181*f*
  disturbing motion
    caused by plate oscillation, 175*f*, 177*f*
    on external boundary, 183*f*
    generated in unperturbed flow, 173*f*
  homogeneous and heterogeneous types of wave-guides, 170*f*
  superficial waves, 162*f*
  transverse normal symmetric wave, 161*f*
  two wave-guides with one common site of length, 166*f*
  wave parameters dependence from wave-guide geometry, 167*f*
Heterogeneous wave-guide, 169
High-speed fish hypothesis, 504–505
Hot-wire anemometer, 193, 450
"Hump of resistance", 4
Hydrobionics, 66, 350, 438, 531, 577, 586
Hydrobiont(s), 20, 498–499, 577
  BL of, 93–94
  skin covers, 238
Hydrodynamic
  characteristics of oscillating wing, 44–48
  efficiency, 58–59
  instability, 100
  nonstationary influence on hydrodynamic drag formation mechanism, 12–19

peculiarities of skin structure and body of swordfish, 515–531
pipe scheme, 31–32, 32*f*
stability, 126–127, 567
  criterion, 532
  method for hydrodynamic stability study, 101–102
  stand of low turbulence, 93–100, 94*f*
    cross section of hydrodynamic stand, 96*f*
    device for transverse oscillations of different bodies, 97*f*
    microcoordinate, 99*f*
    supports and oscillator in working part of hydrodynamic stand, 98*f*
  thrust generated by oscillating wing, 52–56

I
Immersion, 77
Impurity diffusion in TBL, 461
Inertial resistance, 15
Inhomogeneous wave, 162–163
Instantaneous angle of attack, 45–46
Integral efficiency coefficients, 414–415
Interaction principle, 360
"Inverted riblets", 413

K
Kinematic methods, 361
Kinematic-dynamic principle, 345
Klein vortices, 150
Knife-pylon, 476–477
Kramer curves, 367
Kramer test, 261–262

L
Lamb's waves, 159–160
Laminar BL of elastic plates
  distribution of disturbing movements across thickness of, 281–295
  change in longitudinal pulsating speed, 294*f*
  change of plane wavelength and longitudinal vortices along plates, 290*f*
  dependence of wave number and phase velocity, 287*f*
  dependence of wavelength on Reynolds number, 292*f*

Laminar BL of elastic plates (*Continued*)
  dependences of amplitude of velocity, 283*f*
  dimensionless oscillation frequencies, 285*t*
  distribution of amplitudes of velocity, 282*f*, 288*f*
  distribution of average kinetic energy, 289*f*
  distribution of wave number and phase velocity, 284*f*
  neutral curves of linear stability of, 272–281
    growth rates in flow around rigid surface, 280*f*
    neutral curves in coordinates of dimensionless frequency, 274*f*
    neutral curves in flow past viscoelastic surfaces, 279*f*
Laminar BL of rigid plate
  distribution of disturbing motion over thickness of, 123–130
    coherent vortex structures development scheme, 149*f*
    coherent vortex structures of turbulent boundary layer, 146*f*
    conditions of experiments, 129*t*
    distribution of dimensionless velocity over thickness of BL, 128*f*
    distribution of the wave number, 124*f*
    influence of amplitude of vibrations of vibrator tape, 125*f*
    large coherent vortex structures of turbulent BL, 147*f*
    longitudinal coherent vortex structures, 148*f*
    phase velocity dependence on oscillation frequency, 130*f*
  neutral curves of linear stability of, 117–123
    dependence of growth factors on wave number, 123*f*
    dependence of wave number, 120*f*
    growth curves of disturbances in BL, 122*f*
    limit values of neutral oscillations, 121*t*
    neutral curves for dimensionless wave number, 119*f*
Laminar flow control (LFC), 193

Laminar flow regime, 428
Laminar-turbulent transition process, 295–311
  of rigid plate BL, 130–150
    change in plane wavelength and wavelength of longitudinal vortices, 137*f*
    disturbing motion in longitudinal flow past flat plate, 133*f*
    evolution of nonlinear disturbances, 136*f*
    flow modeling in viscous sublayer of turbulent BL, 145*f*
    model of coherent vortex structures, 138*f*
    smoke visualization of flow, 141*f*
    velocity profiles in front of slit, 144*f*
Large longitudinal vortices (LLV), 171
Laser anemometer optical system, 187
Laser Doppler anemometer (LDA), 507
Laser Doppler velocity (LDV), 206
  measurements of BL longitudinal velocity profiles, 206–209
    averaged and pulsating longitudinal velocities of BL, 207*f*
    dimensionless velocity profiles along concave surface, 208*f*
    profiles of concave surface areas, 206*f*
Laser Doppler velocity meter (LDVM), 101, 206, 295
  of BL structure of elastic plates in laminar-turbulent transition process, 295–311
    design of elastic plates, 298*t*
    longitudinal averaged and pulsation velocities at flow around plate, 300*f*, 302*f*, 307*f*
    longitudinal pulsating velocity with flow around plate, 308*f*
    mechanical characteristics of elastic plates, 303*t*
    profiles of averaged and pulsating longitudinal velocities, 296*f*, 306*f*
  measurements of structure of boundary layer of rigid plate, 150–158
    placement on trolley of equipment for BL investigation, 151*f*
    structure of longitudinal vortices, 153*f*
    value of root-mean-square values, 155*f*

Index 649

velocity profiles, 157f
working section of hydrodynamic test, 153f
Law of torsional oscillations, 48–49
Least squares method, 236
Lifting body (LB), 542
Limit neutral curves, 429–430, 569
Linear stability of laminar BL
    of elastic plates, 272–281
    of rigid plate, 117–123
Long sword-shaped tip (LXT), 473–474
Longitudinal oscillating vortex generators, 360
Longitudinal stationary vortex generators, 360
Longitudinal vortex systems (LVS), 189–191, 198f, 203, 554
    combined method of forming Lvs. in BL, 554–557
    visualization of velocity, 555f, 556f
    sinusoidal development of, 103
Longitudinally streamlined cylinders
    calculating the areas of washed surface, 397f
    dependencies of drag coefficients, 401f
    drag of, 389–404
    geometric parameters of models, 393t
    model parameters of, 395f
    relative embedding models, 393t
    with respect to hydrodynamic weights, 392f
    strain-gauge device, 390f
    variants of investigated models, 394f

M

Maneuverability control, 604
Marine vehicles, 582
Membrane, 239–240, 240f
Microbubbles (MB), 438, 542
    effect on integral characteristics of axisymmetric models, 558–560
    generators, 578–579, 584
    influences into BL on drag reduction, 531–545
        xiphoid nose of hydrofoil model, 544f
Microcoordinate, 98–100, 99f
Model of vortex structures of vortex chamber, 229–232, 230f
Modern two-component PIV system, 540–541

Multihull M80 Stiletto, 591–592
Multilayer composite materials oscillations, 322
Multilayered elastic-damping surfaces, 361

N

Nasal xiphoid-shaped fairings, 474
Navier-Stokes equation, 227
Neutral curves of linear stability
    of laminar BL of elastic plates, 272–281
    of linear stability of laminar BL of rigid plate, 117–123
Neutral oscillation, 124–125, 130
Newtonian fluid, 461
Nonharmonic wave, 168
Nonlinear effects, 159
Nonlinear Tollmien–Schlichting waves, 272
Nonsinusoidal oscillations, 135
Nonstationary flow, 203–204
    of viscous fluid, 44–45
Nonstationary motion
    accelerated movement features in water of bodies of various shapes, 25–31
    drag coefficient of fluid bodies in acceleration and deceleration modes, 20–25
    drag of disk and sphere at accelerated movement from state of rest, 1–12
    energy aspects of movement of bodies in modes of acceleration and deceleration, 25–31
    nonstationary influence on hydrodynamic drag formation mechanism, 12–19
Nonstationary movement, 1
Nonstationary resistance, 6
    coefficient, 1, 5, 7
Nonstationary vortex structures behind oscillating bodies, 65–75
Nontraditional movers, 602–603
NUWC closed-loop hydrodynamic research tunnel, 537

O

Ogival tip (OT), 474
    drag of the model with, 480–489
ONR SEA FLYER technology, 541–542
Optical system, 187
Orr-Sommerfeld equation, 166–168
Oscillating wing

Oscillating wing (*Continued*)
  approximate method for calculating of hydrodynamic characteristics of, 44–48
  experimental investigation
    of hydrodynamic thrust generated by, 52–56
    of oscillating wing thrust, 48–52
Oscillation frequency (ω), 16, 126
Oscillations on surface of elastic plates, 334–344
  amplitudes of oscillations of elastic plate surface, 336t
  comparison of
    oscillation spectra of hard surface and elastic plates, 340f
    spectra of pressure pulsations on rigid and elastic plates, 342f
  dependences of average dimensionless transverse interval, 344f
  device for measuring vibrations on elastic plate surface, 335f
  oscillation spectra of hard surface and elastic plates, 338f
  spectra of pressure pulsations on rigid and elastic plates, 341f
Oscillatory cycle, 28
Oscillogram, 16, 55, 331–332
Oscilloscope C1–18, 321

P

Passive strike of caudal fin, 41
Peripheral region of flow, 229
Photographing, 74
Polarization, 159
Polyethylene oxide (PEO), 536, 543
  PEO-309, 543
Polymer additives, 459
  influence on friction drag of elastic cylinders, 420–428
    coefficient of drag reduction, 427f
    coefficient of volumetric flow rate, 425f
    combined method of drag reduction, 426f
    drag coefficient of streamlined cylinder with a solid surface, 423f
    experimental setup, 422f
Polyoxyethylene (POE), 460, 482, 485
Polyurethane, 257

foam, 238, 240, 267–268
Power, 165
Pre-burst modulation mechanism, 431
Precrisis mode, 30
Pressure pulsation
  sensors, 336, 340
  spectra, 429
"Primary" near-wall jet, 222
Propulsion unit, 64
Protrusions, 560, 577
Pulse propagation, 320–334

Q

Quasi stationary hypothesis, 8–9
Quasi-solid flow, 229
Quasi-Taylor's vortex systems, 228–229
Quasistationary theory, 28, 46

R

Raleigh's waves, 161–163
Receptivity, 354
  of BL to external disturbances, 171
  method, 572–573
  problem, 569–571
Rectangular rigid oscillating wing, hydrodynamic characteristics of, 56–65
Reduction friction drag, 582–586. *See also* Friction drag
  longitudinal section of microbubble generators, 584f
  Nasal cowl for hydrofoil, 585f
  SEA FLYER, 585f
Resembling Görtler-Ludwig vortices, 218–220
Resistance
  coefficient, 13
  force, 1, 20
  hump, 7
Reverse design approach, 458
Reynolds numbers (Re), 1–3, 11–12, 16, 21–22, 41, 65–66, 104–105, 142, 209, 236, 246–250, 260, 272–275, 362–363, 372, 386–387, 441, 461–462, 474, 499–500, 541–542, 555–556
Reynolds stress, 238–239, 471–472
"Riblets", 549–550
Rigid curvilinear plates

Index

experimental construction of Görtler neutral curve in flow around, 209–213
Görtler stability, experimental investigation on, 200–206

S

"Secondary" parietal jet, 222
Self-propelled buoys, 602
Sensors, 322
Separation zones, 499–501, 504–505
Shear stress, 347
Short sword-shaped tip (SXT), 473–474
Similarity theory, 13
Simple Voigt-Kelvin model, 362
Single-layer elastic-damping surfaces, 361
Sinusoidal
    development of longitudinal vortices system, 103
    law, 16–17
    process, 131–132
Smoke imaging, 193, 580–581
Soot visualization method, 218–220
Spatial disturbance generator, 98–100
Special Design Technological Bureau (SDTB), 588
Speed of movement, 14
Stabilization zone, 265
"Standard" model, 479
Static electric field effect on BL characteristics, 439
Stationarity hypothesis, 43–44
Stationary tests, 6–7
Strain dynamometer, 374
Strain gage dynamometer, 444, 444f
Strain gauges, 73
Streamlined surface influence on boundary layer, 549–551
    flow structure at initial stages of nonlinear plane wave deformation, 550f
    influence of ribbed surface regular, 551f
Strouhal number (Sh), 42, 46, 66
Supercritical flow around, 30
Supercritical wing profile model (SC wing profile model), 193
Superficial waves, 161
"Swing-immersion" mode, 66, 77
Swordfish (*Xiphias gladius*), 443
    hydrodynamic peculiarities of skin structure and body of, 515–531

Symmetric strain-gauge beams, 372–373
Symmetric waves movement, 160–161
Synergistic drag reduction, 536
Synthetic glue, 490–493
Synthetic wallpaper glue, 486
Systematic model tests, 1

T

Tabulated functions, 46–47
Taylor vortex systems, 225
Taylor–Görtler-lice vortices, 457
Telescopic racks, 586–587
Tellurium
    clouds, 555–556
    jets, 263–264, 271–272
    method, 151–152, 499, 548–549
    tellurium-method, 200–201, 429–430
    wire, 98, 101, 106–107, 111–113, 201
Tensor dynamometers, 374–375
Thermal conductivity control, 597–599
Thermo-anemometer, 243–244
    sensors, 187
Thermoregulation, 350
Thrust force, 43
Tollmien-Schlichting (T-S)
    plane wave, 136–137, 379
    stability, 238–239
    wavelength, 120–121
    waves, 196, 211, 237, 294, 517
Towing tests, 16, 370
Transition boundary layer of rigid plate. *See also* Boundary layer (BL)
    experimental investigation of coherent vortex structures of plate, 103–117
    deformation of velocity profile, 109f
    development of disturbing motion, 108f
    distribution of amplitudes of velocity v' over thickness of BL, 116f
    laws of resistance of smooth longitudinally streamlined flat plate, 104f
    photographing disturbing motion in various places, 111f
    simultaneous photographing of disturbances motion, 110f
    simultaneous photographing of velocity profile, 106f
    successive phases of disturbing motion development, 112f

Transition boundary layer of rigid plate (*Continued*)
  velocity amplitude v' in different places along plate, 114*f*
  velocity amplitude v' as function of oscillation frequency, 112*f*
  velocity profile of laminar boundary layer on flat rigid plate, 105*f*
  velocity profiles $U(z)$ with positive pressure gradient, 114*f*
  velocity profiles with positive pressure gradient, 113*f*
  experimental methods for investigation flow structures of, 100–103
  method for study of hydrodynamic stability, 101–102
  method for studying sinusoidal development of longitudinal vortices system, 103
  method of investigation nonlinear stages of natural transition, 101
  method of investigation of Benny–Lin vortices and Görtler stability, 102–103
Trimaran models, 590–591, 591*f*
TsAGI towing tank, 42
Turbulent boundary layer (TBL), 138, 145–146, 455, 459, 470
Two-dimensional flow pattern, 73–74
Two-medium devices, technical proposals for, 593–596
  Aeroprakt A-20 model, 595*f*
  diffuser-type vortex generator, 594*f*
  steering complex of fishing trawler, 593*f*
  torque converter parts, 596*f*
  wing model in working section, 595*f*

U
Underwater glider, 602
Universal vortex chamber, 596–597, 597*f*

V
Van Drist method, 453
Velocity profiles, 465, 508–509, 511, 531–532, 550–551
Vibrations, 269–270
Viscous drag, 546
  coefficient, 21–22
Viscous fluid model, 46

Viscous incompressible fluid flow, 75
Viscous sublayer, 459, 461, 471–472, 543
Visualization, 4–5
  methodology, 73–74
  of vortex structures, 68*f*, 75–91
Vortex chamber, 577
  and control methods for CVS in vortex chambers, 580–581
  flow in area of nozzle, 216–220
  flow in inner area of vortex chamber, 215–216
  flow in peripheral area of vortex chamber, 220–225, 222*f*
    vortex structures formation in peripheral region of vortex chamber, 223*f*
  flow near butt, 225–229
    flow structure visualization at butt of vortex chamber, 227*f*
    flow structure visualization in area of butt of vortex chamber, 225*f*
    flow structure visualization near butt of vortex chamber, 226*f*
    rotational motion of fluid over fixed base, 228*f*
  model of vortex structures of vortex chamber, 229–232, 230*f*
    model of vortex structures at butt-end, 231*f*
    shape of quasi-solid cylinder, 230*f*
    vortex structures interaction in butt-end area, 232*f*
  visualization, 217*f*
    with short butt-end, 218*f*
    using soot method, 219*f*
  vortex structures model in, 213–232, 215*f*
Vortex devices, 596–597, 604
  design of universal vortex chamber, 597*f*
  video frame with visualization of vortex structures formation, 599*f*
  visualization of flow around a semi cylindrical deepening, 598*f*
Vortex disturbances
  cinema development flow patterns, 527*f*
  small wind tunnel, 525*f*
  stabilization of, 518–531
  visualization
    of flow pattern development, 522*f*
    of flow structure in cavities of various shapes, 530*f*

Index 653

of vortices in angular conjugations, 519f
water flow around cavity, 521f
Vortex formation, 41, 74–75
Vortex generators (VG), 188–189, 438
Vortex structures model in vortex chamber, 213–232

W

Water resistance, 14
"Water-air" boundary, 560–561
Wave
  dispersion, 159
  number, 191, 196–197, 246–250
  polarization, 159, 160f
  propagation, 320–334
  wave-guide, 159
    amplification coefficient, 169
Waving fin mover modeling
  approximate method for calculating of hydrodynamic characteristics of oscillating wing, 44–48
  experimental investigation
    of hydrodynamic thrust generated by oscillating wing, 52–56
    of nonstationary vortex structures behind oscillating bodies, 65–75
    of oscillating wing thrust, 48–52
    of waving wing, 41–44
  hydrodynamic characteristics of rectangular rigid oscillating wing, 56–65
  visualization of vortex structures in wake of oscillating profiles, 75–91
Whiskers, 215–217, 224–225
  vortex systems, 229
Wing profile
  changing curvature of, 603
  changing shape of wing profile and configuration, 603
  changing wing profile chords, 603
Wing-in-ground devices (WIG devices), 579, 593
Wortmann tellurium method, 187

X

Xiphoid tips, 438–439
  drag of the model with xiphoid tips and injection of polymer solutions into BL, 490–498

friction drag for injection of polymer solutions into BL, 491–498
experimental investigations of bodies with, 441–451
  change in momentum thickness along length of body, 451f
  dependence of hydrodynamic drag of sphere, 444f
  dependence of length of laminar section of boundary layer, 450f
  dependence of sphere drag coefficient on Reynolds number, 445f
  forms of models and bow parts, 447f
  kinematic scheme of strain gage dynamometer, 444f
  pressure distribution along swordfish model, 449f
  strain gage dynamometer with sphere, 443f
  swordfish model installed in a water tunnel, 448f
  theoretical drawing of swordfish and location of drainage holes, 448f
  types of swordfish research, 442t
influence on pressure distribution, 446f
installation for towing a model with, 445f
physical mechanism of influence of, 498–506
  body forms of swordfish and marlin, 502f
  coefficient of models drag, 503f
  flow around an infinite cylinder at different Reynolds numbers, 500f
  visualization by tellurium method, 499f
theoretical investigations of bodies with, 451–459
  dependence of coefficient of resistance of models, 454f
  dependence of viscosity coefficient of model, 451f
  distribution of excess pressure, 455f
  distribution of pressure coefficient along length of body, 452f
  numerical calculation of surface pressure distribution for cylinder, 456f, 457f
  pressure distribution along bodies, 453f

Printed in the United States
By Bookmasters